Arid Zone Geomorphology

Process, Form and Change in Drylands

SECOND EDITION

Edited by

David S. G. Thomas
University of Sheffield, UK

JOHN WILEY & SONS

Chichester • New York • Weinheim • Brisbane • Singapore • Toronto

Other Wiley Editorial Offices

John Wiley & Sons, Inc., 605 Third Avenue,
New York, NY 10158-0012, USA

Jacaranda Wiley Ltd, 33 Park Road, Milton,
Queensland 4064, Australia

John Wiley & Sons (Canada) Ltd, 22 Worcester Road,
Rexdale, Ontario M9W 1L1, Canada

John Wiley & Sons (Asia) Pte Ltd, 2 Clementi Loop #02-01,
Jin Xing Distripark, Singapore 129809

Library of Congress Cataloging-in-Publication Data

Arid zone geomorphology : process, form, and change in drylands /
    edited by David S.G. Thomas. — 2nd ed.
        p.   cm.
    Includes bibliographical references and index.
    ISBN 0-471-95557-4 (pbk.)
    1. Geomorphological—Arid regions.   I. Thomas, David S. G.
GB611.A75   1997
551.4'15—dc20   96-28792
                                                            CIP

British Library Cataloguing in Publication Data

A catalogue record for this book is available from the British Library

ISBN 0-471-97160-X
ISBN 0-471-97610-5

Typeset in 10/11pt Plantin from the author's disks by Mathematical Composition Setters Ltd, Salisbury
Printed and bound in Great Britain by Bookcraft (Bath) Ltd
This book is printed on acid-free paper responsibly manufactured from sustainable forestation,
for which at least two trees are planted for each one used for paper production.

For Mair

Contents

Preface to the first edition

Arid environments may not be the most hospitable places on Earth, but the 30% or more of the global land surface which they cover does support an ever-growing human population and has fascinated travellers, explorers and scientists for centuries. Early geomorphological studies were frequently carried out indirectly, sometimes even unwittingly, by those whose main purpose and motives lay elsewhere: inevitably, but with some notable exceptions, their accounts were descriptive and unscientific. Some would even argue that these traits persisted and dominated desert geomorphological studies well into the second half of this century. Recent years have, however, seen an enhanced rigour in the investigation and explanation of landforms and geomorphological processes in arid lands. New data have been gathered by techniques ranging in scale from the detailed monitoring of processes in the field to remote sensing from space; old theories have been questioned and new ones, based on evidence rather than surmise, have been proposed.

The idea for this volume grew out of these advances and the absence of a recent book which encapsules them (Cooke and Warren's *Geomorphology in deserts* is 15 years old and Mabbutt's *Desert landforms* is 11). There have been valuable volumes produced in recent years which deal with specific topics of interest to desert geomorphologists, but none (to my knowledge) which attempts a broader view of arid zone geomorphology. It is hoped that this book fills this gap.

The decision to invite others to contribute chapters was made easily. The geomorphology of arid environments is a huge topic, embracing much of the subject matter of geomorphology as a whole: desert landforms consist of much more than piles of unvegetated sand. Arid and semi-arid environments are very varied, too; involving the expertise of others has therefore inevitably broadened and deepened the basis of the text. While there are inevitably gaps, these have hopefully been kept to a minimum. Many people have provided the help and inspiration needed to turn *Arid zone geomorphology* from an idea to a book. Andrew Goudie introduced me to deserts, since which time many people and funding bodies have enabled me to visit them and to conduct research in them: I would particularly like to thank the Shaws in Botswana and Sleaze and Val for showing me Death Valley and other Californian hotspots. During the production of the book the contributors have efficiently met the tasks I have set them, including refereeing other people's chapters; Rod Brown provided additional help in this respect, too, while Chapter 13 also passed through refereeing within the US Geological Survey. The cartographers of many institutions, but especially Paul Coles of the Geography Department, University of Sheffield, produced the diagrams. At the publishers, Iain Stevenson and Sally Kilmister gave me valuable advice and logistical help. Steve Trudgill inspired me to put a book together in the first place.

Lastly, but most importantly, Liz Thomas not only suffered me during the book's gestation, but helped in a multitude of practical ways and provided a valuable, independent, geomorphological viewpoint. To all of the above, my parents, and any I have forgotten to mention, my sincere thanks.

David S.G. Thomas
Sheffield
August 1988

Preface to the second edition

It is almost eight years since the text of the first edition of *Arid zone geomorphology* was written, and seven years since the book was first published. Coincidently, on the day of publication, 7 December 1988, the 'Finger of God', pictured on the cover of the first edition, collapsed. So, along with a substantially changed content, this second edition has a new cover.

Since the first edition was produced, much has happened in terms of both geomorphic research in arid environments (or drylands, or deserts: such terms are commonly used interchangeably) and in general and non-scientific interest for such areas. Arid regions are areas of concern, because of population growth, the impacts of desertification and of natural phenomena, particularly droughts. The impending impacts of global warming on these areas and their peoples are also of growing concern. Scientists, including geomorphologists, are responding to the need to know more about the nature and operation of processes in drylands by conducting more research, both fundamental and applied.

This new edition of *Arid zone geomorphology* aims to reflect the changes and advances in geomorphological knowledge that have occurred, especially since the publication of the first edition. This has been done in two ways. First, the chapters from the first edition have been updated, in some cases radically. Second, the content of the book has been expanded, with the number of chapters all but doubled and arranged in a new framework of six sections. This has been done to fill gaps in coverage or expand areas of particular interest. In both cases, as with the first edition, experts have been invited to write the chapters of this text rather than one person attempting to summarise and review what amounts to a vast chunk of geomorphology. In the majority of cases the authors of chapters from the first edition have rewritten their own material. In some cases where circumstances have prevented this, new co-authors have conducted the task. For a few themes covered in the first edition, new authors have written material afresh.

In all, the production of this new edition has resulted in 34 researchers from over 25 academic or research institutions making contributions, all involved with research in the fields on which they write. It is this wealth of expertise and the wide-ranging and diverse experience of drylands that it represents that make this book. As editor I am indebted to the cooperation of the contributors for meeting deadlines and to those who have conducted last-minute tasks at my request. The willingness with which writing, updating, changing text, reviewing and other tasks have been taken up is enormously appreciated. The involvement of some new contributors to the second edition has come about through conducting fieldwork in deserts in Africa and Asia with them: Dave Nash and Jo Bullard, whose PhDs I had the privilege to supervise; Stephen Stokes, Giles Wiggs and Sarah O'Hara. For others, listening to their papers at conferences and meetings in Ahmedabad, India, and Hamilton, Canada, and even the UK, or casual conversations over coffee or on fieldtrips, led me to ask them to contribute: David Dunkerley, Gerald Nanson, Jacky Croke, Ed Derbyshire, Helen Rendell, Lillan Berger, Vatche Tchakerian (and, indirectly, Julie Laity) all became victims in this way.

The production of this book has been greatly helped by Kate Schofield, Sam Rewston and Sarah Harmston in the Geography Department office at Sheffield and Paul Coles and Graham Allsopp in Cartographic Services who have produced or updated many of the figures. Iain Stevenson and Katrina Sinclair at the London

office of John Wiley and Sons have eased production matters. To all those named above, the
undergraduates, postgraduates and academics
who used the first edition and passed on comments for possible future changes, and
especially my wife Lucy, who painstakingly
prepared the index, and our daughter Mair,
who has tolerated the production of this volume
since her birth, my sincere thanks.

David S.G. Thomas
Sheffield
January 1996

Acknowledgements

The following individuals, organisations and publishers are thanked for granting permission to reproduce figures and tables (in some instances material may have been redrawn or modified slightly). Alberta Forestry, Lands and Wildlife for Figure 13.13; the Australian Bureau of Meteorology for Figure 18.5; Quarternary Research Centre for Figure 26.8 (from *Quaternary Research*); The Society of Economic Paleontologists and Mineralogists for Figures 11.23 and 17.19b (from *Journal of Sedimentary Petrology*); the US Geological Survey Office of Scientific Publications for Figures 11.12 and 11.14 (from *Professional Papers*); Blackwell Scientific Publications Ltd for Figure 11.18 (from L.E. Frostick and I. Reid (eds), *Desert sediments ancient and modern*) and Figure 11.21 (from *Sedimentology*); Elsevier Science Publishers BV for Figure 26.1 (from *Sedimentary Geology*); and John Wiley & Sons Ltd for Figure 11.17 (from *Hydrological Processes*). NASA are thanked for many of the photographs in Chapter 29 and the US Space Shuttle astronauts for Figure 17.3. UNESCO for Figure 24.1; E. Schweizerbart'sche Verlagsbuchhandlung for Figure 24.6; Interciencia Journal of Science and Technology of the Americas for Figure 24.13; and Keith Mitchell, ANU, for Figure 25.1. A few authors or publishers may inadvertently not have been contacted to obtain permission to use copyright material, although every effort has been made to do so. The editor and publishers will be grateful for any information that will enable them to obtain permission for any material reproduced, where the source is unacknowledged.

List of contributors

Dr Robert J. Allison, Department of Geography, Science Laboratories, University of Durham, South Road, Durham DH1 3LE, UK.

Dr Andrew J. Baird, Department of Geography, University of Sheffield, Sheffield S10 2TN, UK.

Dr I.A. Lillan Berger, Geography Laboratory, University of Sussex, Brighton BN1 9RA, UK.

Dr Carol S. Breed, US Geological Survey, Branch of Astrogeology, 2255 North Gemini Drive, Flagstaff, AZ 86001, USA.

Katherine J. Brown, Department of Geography and Environmental Science, Clayton, Melbourne, Victoria 3168, Australia.

Dr Joanna E. Bullard, Department of Geography, University of Keele, Keele ST5 5BG, UK.

Dr Ian A. Campbell, Department of Earth and Atmospheric Sciences, 1–26 Earth Sciences Building, University of Alberta, Edmonton T6G 2E3, Canada.

Dr Jacky Croke, Cooperative Research Centre for Catchment Hydrology, CSIRO Division of Water Resources, GPO Box 1666, Canberra, ACT 2601, Australia.

Professor Edward Derbyshire, Department of Geography, Royal Holloway, University of London, Egham TW20 0EX, UK.

Dr David L. Dunkerley, Department of Geography and Environmental Science, Clayton, Melbourne, Victoria 3168, Australia.

Professor Lynne E. Frostick, School of Geography and Earth Resources, University of Hull, Hull HU6 7RX, UK.

Professor Andrew S. Goudie, School of Geography, University of Oxford, Mansfield Road, Oxford OX1 3TB, UK.

Dr Adrian M. Harvey, Department of Geography, University of Liverpool, Roxby Building, PO Box 147, Liverpool L69 3BX, UK.

Dr David Knighton, Sheffield Centre for International Drylands Research, Department of Geography, University of Sheffield, Sheffield S10 2TN, UK.

Dr Julie E. Laity, Department of Geography, California State University, Northridge, CA 91330, USA.

Dr John F. McCauley, US Geological Survey, Branch of Astrogeology, 2255 North Gemini Drive, Flagstaff, AZ 86001, USA.

Dr Nick J. Middleton, School of Geography, University of Oxford, Mansfield Road, Oxford OX1 3TB, UK.

Professor Gerald Nanson, Department of Geography, University of Wollongong, Wolongong, New South Wales 2500, Australia.

Dr David J. Nash, Earth and Environmental Science Research Unit, University of Brighton, Mithras House, Lewes Road, Brighton BN2 4AT, UK.

Dr Theodore M. Oberlander, formerly: Department of Geography, University of California Berkeley, Earth Sciences Building, Berkeley, CA 94720, USA.

Dr Sarah L. O'Hara, Sheffield Centre for International Drylands Research, Department of Geography, University of Sheffield, Sheffield S10 2TN, UK.

Professor Ian Reid, Department of Geography, Loughborough University, Loughborough, LE11 3TU, UK.

Professor Helen Rendell, Geography Laboratory, University of Sussex, Brighton BN1 9RA, UK.

Dr Paul A. Shaw, School of Geology and Environmental Science, University of Luton, Park Square, Luton LU1 3JU, UK.

Dr Bernard Smith, School of Geosciences, The Queen's University of Belfast, Belfast BT7 1NN, UK.

Dr Stephen Stokes, School of Geography, University of Oxford, Mansfield Road, Oxford OX1 3TB, UK.

Dr Vatche P. Tchakerian, Department of Geography, Texas A & M University, College Station, TX 77843, USA.

Professor David S.G. Thomas, Sheffield Centre for International Drylands Research, Department of Geography, University of Sheffield, Sheffield S10 2TN, UK.

Dr Patricia Warke, School of Geosciences, The Queen's University of Belfast, Belfast BT7 1NN, UK.

Dr Andrew Watson, 23 Prospect Street, Holliston, MA 01746, USA.

Dr Gordon L. Wells, Route 1, Box 93X, Mount Enterprise, TX 75681, USA.

Dr Marion I. Whitney, Emeritus, Central Michigan University, Mount Pleasant, MI 48859, USA.

Dr Giles F.S. Wiggs, Sheffield Centre for International Drylands Research, Department of Geography, University of Sheffield, Sheffield S10 2TN, UK.

Dr James R. Zimbelman, NASM MRC315, Smithsonian Institution, Washinton DC 20560, USA.

Section 1
Framework

1

Arid environments: their nature and extent

David S.G. Thomas

Geomorphology in arid environments

For much of history and for many human races, arid environments have been areas to avoid. Lack of surface water, limited foodstuffs and climatic extremes have generally made arid areas unfavourable places for habitation, except for resourceful hunter–gatherer and pastoral–nomadic peoples, or persecuted population groups. Even with twentieth-century technological advances which have made travel and existence in drylands possible for a greater range of people, arid environments still provide major limitations to the range and extent of human occupations and activities.

European interest in arid environments grew from the late eighteenth century onwards (Heathcote 1983), and much of the early scientific knowledge concerning such areas came from those whose aims were to convert indigenous peoples to Christianity, explore, exploit or colonise. It has been noted (see, for example, Cooke and Warren 1973; Cooke *et al.* 1993) that geomorphological research in arid areas has been dogged by excessive description, superficiality and secular national terminology. The first characteristic, that of description, has often been criticised, especially at times when quantification has been a central paradigm in

geomorphology. Yet description can be an important prerequisite of rigorous explanation, analysis and deeper investigation. In the case of early works, the descriptive component is hardly surprising. For European writers with temperate world origins, desert landscapes must have represented spectacular, bizarre and unusual contrasts to the plant- and soil-mantled landscapes of many of their homelands. Before early descriptive accounts are totally pilloried, it should also be remembered that geomorphological accounts in the works of early Europeans were usually but byproducts of the reasons for their being in deserts in the first place. This also helps to explain the second characteristic, the superficiality of early reports and studies.

The third characteristic attributed above to early works, secularity, arose because national groups tended, until relatively recently, to confine their interests to particular deserts. In Africa, Asia and Australia, early geomorphological investigations were heavily influenced by the distribution of the impacts of European colonialism. Thus Flammand's (1899) account from the Sahara, Passarge's (1904) two volumes on the Kalahari and numerous reports from Australia (e.g. Sturt 1833; Mitchell 1837; Spencer 1896) reflect broader colonial interests of their time. It has been noted, and now increasingly realised with importance as solutions are sought to dryland environmental problems (e.g. Mortimore 1989; Thomas and Middleton 1994), that the knowledge of the environment of indigenous peoples was and remains considerable. Yet this was usually either ignored or unrealised by early Europeans entering arid environments for the first time, who often saw deserts through eyes more accustomed to their starkly contrasting points of origin, leading to a preoccupation in some cases with the spectacular and unusual landforms they encountered in deserts.

There were, of course, exceptions to these characteristics, even in the nineteenth century. Perhaps most notable were the investigations in the southwestern United States, often with a geomorphological slant, of John Wesley Powell (1875; 1878) and Grove Karl Gilbert (1875; 1877; 1895), regarded by many as the father of modern geomorphology. Their activities were driven by a governmental quest to expand the frontiers of (European) utilisation of North America, and their works were essentially early forms of resource appraisal. Some of the early accounts of the geomorphology of the Australian deserts had a similar basis, for example Thomas Mitchell wrote:

After surmounting the barriers of parched deserts and hostile barbarians, I had at last at length the satisfaction of overlooking from a pyramid of granite a much better country.

(Mitchell 1837, p. 275)

We had at last discovered a country ready for the reception of civilised man.

(Mitchell 1939, p. 171)

The distinctiveness of the arid zone and the quest for explanation

Early accounts may however be of restricted geomorphological value in the sense that in many of them a focus on unusual and spectacular features may have been at the expense of representativeness. A further significant consequence of the nature of many early investigations has also been noted:

A prime feature of desert geomorphological research over the past century or so has been the rapidity with

which ideas have changed, and the dramatic way in which ideas have gone in and out of fashion. This reflects the fact that hypothesis formulation has often preceded detailed and reliable information on form and process, and the fact that different workers have written about different areas where the relative importance of different processes may vary substantially.

(Goudie 1985, p. 122)

Within these changing ideas has often been the view that arid environments are distinct, even unique, in terms of the operation of geomorphological processes and their resultant landscape outcomes. Early quests for synthesising explanations sought generalisations that were distinct from those developed for other environments. Davis (1905) produced his cycle of erosion for arid environments based on the belief that fluvial processes in drylands produced distinct outcomes at the landscape scale. This notion of distinctiveness was clearly also present in morphogentic or climatic geomorphology models of explanation (e.g. Birot 1960; Budel 1963; Tricart and Cailleux 1960). While the very terms 'drylands', 'arid zone' and so on clearly imply a climatic delimitation of the extent of these environments, it is debatable whether sweeping models of desert geomorphic explanation are justified. First, this is because drylands are clearly not internally homogeneous. They vary, for example, in tectonic settings (see Chapter 2), which are geomorphologically significant. Second, climate change, certainly in the Quaternary period, has affected arid environments as much as any others, so they may contain landform expressions inherited from past, different, climatic regimes. Third, as Parsons and Abrahams (1994, p. 10) succinctly note:

...the emphasis of geomorphology has shifted away from morphogenesis within specific areas towards the study of processes *per se*. This shift...in large measure undermines the distinctiveness of desert geomorphology.

None the less, the relative importance of individual processes and the magnitude and frequency of their operation may differ in arid environments compared with other areas. This, together with growing human populations in drylands and the common treatment of them as environmentally distinguishable, is reason enough to pursue arid zone geomorphology in its own right.

The last two decades or so have seen a new rigour enter geomorphological research in arid environments. New techniques have been employed, new methodologies pursued. Landform description for its own sake has largely been eschewed, though it does, of course, still have a valid role in geomorphological research, and has been replaced by studies of process and form, measurement, explanation and application.

Arid zones defined

Definitions and delimitations of arid environments and deserts abound, varying according to the purpose of the enquiry or the location of the area under consideration. Literary definitions, thoroughly reviewed by Heathcote (1983), commonly employ terms such as 'inhospitable', 'barren', 'useless', 'unvegetated' and 'devoid of water'. Scientific definitions have been based on a number of criteria, including erosion processes (Penck 1894), drainage patterns (de Martonne and Aufrère 1927), climatic criteria based on plant growth (Köppen 1931) and vegetation types (Shantz 1956). Whatever criteria are used, all schemes involve a consideration of moisture availability, at least indirectly, through the relationship between precipitation and evapotranspiration.

Meigs' (1953) classification of arid environments was produced on behalf of UNESCO, and has been widely utilised in geographical works. This classification has been extensively used, for example in the first edition of this book. As it was ultimately concerned with global food production, it is not surprising that arid areas too cold for plant growth (such as polar deserts) were excluded from the classification. Meigs based his scheme on Thornthwaite's (1948) indices of moisture availability (Im): $Im = (100S - 60D)/PET$, where PET is potential evapotranspiration, calculated from meteorological data, and S and D are, respectively, the moisture surplus and moisture deficit, aggregated on an annual basis from monthly data, and taking stored soil moisture into account. Meigs (1953) identified three types of arid environments, delimited by different Im index values: semi-arid, arid and hyper-arid. Grove (1977) subsequently attached mean annual precipitation values to the first two categories (200–500 mm and 25–200 mm, respectively), though these are only approximate. Hyper-arid areas that have no consecutive months without precipitation have been recorded (Meigs 1953).

The UN (1977) delimitation of drylands is used as the spatial framework in another recent desert geomorphology volume (Abrahams and Parsons 1994). UN (1977) also provided the climatic input to the UNESCO (1979) survey of arid lands utilised in Cooke *et al.*'s (1993) text. The UN (1977) approach defines aridity zones using a P/PET index. P is annual precipitation; PET is calculated using Penman's formula, which requires a large body of directly measured meteorological data for its calculation and which in practice is not consistently available at the global scale required for dryland delimitation (Hulme and Marsh 1990). A new assessment of the extent of drylands, based on an aridity index (AI), where $AI = P/PET$ and PET is calculated using the simpler Thornthwaite method, has been conducted by Hulme and Marsh (1990) on behalf of UNEP. This new assessment differs from earlier ones by using meteorological data from a fixed time period (a 'timebounded' study), rather than simply from mean data from the full length of records available, to calculate index values. This is significant given that climate variability can cause mean data to differ depending on the period data come from (Hulme 1992). The data used in this new scheme, adopted and utilised in dryland studies such as UNEP (1992), Thomas (1993) and Thomas and Middleton (1994), covers the period 1951–1980 and is based on data from over 2000 meteorological stations worldwide.

The UNEP (1992) classification of drylands also differs from previous estimates by including dry-subhumid areas. This was done because these areas experience many of the climatic characteristics of semi-arid areas (Thomas and Middleton 1994) and with UNEP (1992) concerned with desertification, embraces the original application of that term proposed by Aubreville (1949). The delimitation of the different types of dryland environments by AI values are dry-subhumid ($AI = 0.50-<0.65$), semi-arid ($AI = 0.20-<0.50$), arid ($AI = 0.05-<0.20$) and hyper-arid ($AI = <0.05$). According to this scheme, these four environments cover about 47% of the global land area (Figure 1.1). This is significantly more than the areas considered in other schemes (Table 1.1), with this difference principally due to the inclusion of dry-subhumid areas.

In this volume the global *arid zone* is considered to include all elements of this four-fold classification. There are several reasons for this:

1. The divisions between the elements of the classifications are somewhat arbitrary. For example, in UN (1977) the boundary between arid and hyper-arid areas was taken as $P/PET = 0.03$, while in UNEP (1992) it is 0.05.

2. In studies prior to UNEP (1992) the climatic data input to the calculation of indices was from non-timebounded and therefore temporally variable data sets.

3. In drylands, annual precipitation frequently varies substantially from year to year, so that in dry-subhumid, semi-arid, arid and hyper-arid areas, the only safe assumption is that any year could be extremely arid (Shantz 1956).

4. Distinct geomorphic thresholds in terms of processes and landforms have not been identified between the four elements of the scheme.

5. Climatic fluctuations and anthropogenic activities in the twentieth century have caused the

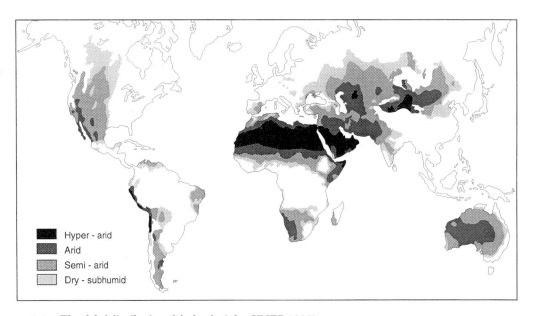

Figure 1.1   *The global distribution of drylands (after UNEP 1992)*

Table 1.1   *The extent of the global drylands (expressed as a percentage of the global land area)*

Classification	Dry-subhumid	Semi-arid	Arid	Hyper-arid	Total
Köppen (1931)	–	14.3	12.0	–	26.3
Thornthwaite (1948)	–	15.3	15.3	–	30.6
Meigs (1953)	–	15.8	16.2	4.3	36.3
Shantz (1956)	–	5.2	24.8	4.7	34.7
UN (1977)[*]	–	13.3	13.7	5.8	32.8
UNEP (1992)	9.9	17.7	12.1	7.5	47.2

[*] In Heathcote (1983).

expansion of arid surface conditions, especially a decrease in vegetation cover, into some semi-arid environments.

6. Semi-arid areas are often called 'deserts' by their inhabitants.

The distribution of arid zones

Table 1.2 shows the distribution of arid zones according to continent. While Africa and Asia each contain almost a third of the global arid zone, inspection of Figure 1.1 clearly shows that Australia is the most arid continent, with approximately 75% of the land area being arid or semi-arid.

Annual temperature regimes in arid areas vary considerably, affected by latitude, altitude and continentality. Cloudsley-Thompson (1969, in Heathcote 1983) noted that the only common element of temperatures between different arid areas is their range. Meigs (1952) divided arid lands into those that are hot all year round and those with mild, cool and cold winters (Tables 1.3 and 1.4). Variations in temperature affect the seasonal availability of moisture, by influencing evapotranspiration

Table 1.2 *Distribution of the arid zone by continent (expressed as a percentage of the total global arid zone)*

Africa	31.9
Asia	31.7
Australasia	10.8
North America	12.0
South America	8.8
Europe	4.9

Source: UNEP (1992).

Table 1.3 *Arid land climates*

	Percentage of arid lands	Mean temperature (°C)	
		Coldest month	Warmest month
Hot	43	10–30	>30
Mild winter	18	10–20	10–30
Cool winter	15	0–10	10–30
Cold winter	24	<0	10–30

Source: Meigs (1952).

Table 1.4 *Examples of arid zones with different climates (classification as in Table 1.3)*

Hot	Central Sahara; Arabia; Great Sandy Desert (Australia)
Mild	Southern Sahara; Kalahari; Mexican; Simpson (Australia)
Cool	Northern Sahara; Turkish Steppes; Atacama; Mojave
Cold	Canadian Prairies; Gobi; Turkmenistan; Chinese deserts

rates and affecting the form of precipitation in relatively high-latitude arid areas. For example, in the arid areas of Canada and central Asia, winter snowfall forms an important component of the annual precipitation budget.

Causes of aridity

Aridity is characterised by net surface-water deficits. It results from climatic, topographic and oceanographic factors which prevent moisture-bearing weather systems reaching an area of the land surface. Four main influences can be identified, which are not mutually exclusive.

ATMOSPHERIC STABILITY

Tropical and subtropical deserts cover about 20% of the global land area (Glennie 1987). These are concentrated in zones of descending, stable air: the tropical high-pressure belts. In these areas large arid zones are composed of central arid areas surrounded by relatively small, marginal, semi-arid and dry-subhumid belts. Precipitation is very unreliable and largely associated with the seasonal movements of the intertropical convergence zone.

CONTINENTALITY

Distance from the oceans prevents the penetration of rain-bearing winds into the centre of large continents, for example in central Asia. Precipitation and evapotranspiration are both usually lower than in arid areas owing their origins to atmospheric stability, while cold winters are common. In other continents, the failure of dominant easterly trade winds to penetrate to continental interiors (Thompson 1975), such as in southern Africa, contributes to the continentality effect. Relatively small arid areas are surrounded by an extensive zone of semi-aridity.

TOPOGRAPHY

Arid areas can occur in the rain shadow of mountain barriers. The Rockies contribute a rain-shadow effect in western North America, while in Australia the penetration of easterly trade winds to the interior is further inhibited by the north–south orientation of the Great Divide. Aridity primarily due to atmospheric stability or continentality can therefore also be enhanced by topographic effects.

COLD OCEAN CURRENTS

Cold ocean currents affect the western coastal margins of South America, southern Africa and Australia, giving rise to five west coast subtropical deserts (Meigs 1966; Lancaster 1989). These currents reinforce climatic conditions, causing low sea-surface evaporation, high atmospheric humidity, low precipitation (very low rainfall, with precipitation mainly in the form of fog and dew) and a low temperature range. Lack of rainfall in the Namib Desert, western southern Africa, is due both to the impact of the Benguela Current on local climates and the failure of easterly rain-bearing winds to penetrate across the continent (Schulze 1972).

The age of aridity on Earth

Desert dune and evaporite sediments preserved in the solid-rock record indicate that aridity has occurred on Earth since Precambrian times (Glennie 1987), with perhaps the earliest recorded aridity being represented by the *c.* 1800-million-year-old dune sediments in the Hornby Bay Group in the Canadian North-west Territories (Ross 1983). The changing configuration of land masses and oceans due to

the effects of tectonic plate movements and orogeny, and changes in global climate, have, however, caused the positions and extent of arid zones to change through geologic time.

Sedimentary evidence suggests that the Namib is probably the oldest and most persistent of current arid zones on Earth, dating back 80 million years to the Cretaceous (Ward *et al.* 1983), though such a great antiquity by no means meets with full agreement (see Tankard and Rogers 1978; Vogel *et al.* 1981). Notwithstanding the effects of subsequent climatic perturbations on the extent and intensity of aridity, many other deserts would seem to date from the Tertiary. Deep-sea core evidence indicates that aeolian material off West Africa dates from 38 million years ago in the Oligocene (Sarnthein 1978), while other deserts such as those of Australia (Bowler 1976) and the Atacama (Clark *et al.* 1967) date from the Miocene.

Arid zone geomorphology

The low moisture availability in arid areas has a profound effect on plant growth. As Bloom (1978, p. 314) has noted:

In the United States, the boundary between humid and semi-arid climates is approximated by the transition westward from medium-height grasses with a continuous turf or sod in the humid regions to short, shallow rooted bunch grasses on otherwise bare ground in semi-arid regions. In arid regions, even the bunch grasses disappear, and the vegetation is, at best, widely spaced shrubs and salt tolerant bushes.

The limited (or absent) vegetation cover is of considerable importance for the operation of geomorphological processes and the development of landforms (Thomas 1988b). The wind can take on the role of geomorphological agent to a degree which cannot occur in other terrestrial environments, except in some coastal locations or places where human activities have interfered with the plant cover. None the less, even limited vegetation can be a very important variable in the operation of arid land geomorphological processes.

The role of moisture is often underrated in the assessment of geomorphological activity in arid environments. Surface runoff, whether occurring ephemerally or episodically, is of considerable importance. Even in the driest areas, high-magnitude sheet floods can have significant geomorphic effects, though they have rarely been observed or recorded (but see McGee 1897 and Rahn 1967). More important is that even low-intensity rainfall events can generate runoff (Cooke *et al.* 1982; Goudie 1985) because of the nature of desert surface conditions. However, the 'spottiness' of desert rainfall events can result in considerable spatial and temporal inequalities in its hydrological and geomorphological effects. The interval between individual rainfall events may also have significant implications for both surface conditions and the operation of specific geomorphic processes. Drought events are a normal, not an unusual, component of dryland environments (Agnew and Anderson 1992; Thomas and Middleton 1994). The concept of interannual rainfall variability, which can exceed 50% in many arid areas, may in fact be a more useful descriptor of arid zone moisture availability than mean rainfall values (Thomas and Middleton 1994).

Many, perhaps all, landforming processes and their morphological expressions are not unique to arid environments. Some processes may operate more favourably or assume a greater relative importance than in other environments,

and some landforms may be better developed or better exposed, the latter at least in part due to the limited vegetation cover. On the other hand, for many features, arid conditions may set the possibilities for their development, but their ultimate formation is dependent on suitable materials, lithologies or topographic settings being available. Arid regions occupy a variety of structural and tectonic settings (Murphy 1968), discussed in Chapter 2, which range from tectonically active montane settings to stable continental cratons (Heathcote, 1983; Thomas 1988a). The climatic and structural variations present within and between arid regions, together with the impact of climatic changes during the Quaternary period, are therefore important variables contributing to the present appearance of arid zone landscapes.

Clements *et al.* (1957) and Mabbutt (1976) attempted to classify the landforms present in some arid areas (Table 1.5). Notwithstanding the limitations of such generalisations, the data in Table 1.5 serve to illustrate the importance of erosional and water-worked features in arid areas with high and frequent changes in relative relief, such as in much of the southwestern United States, and the greater (but not necessarily predominant) importance of the wind as a geomorphological agent in regions with extensive areas of more limited relative relief, such as in the Australian arid zone.

Table 1.5 *Arid zone landscapes in different regions (expressed as a percentage of area[*])*

	SW USA	Sahara	Libya	Arabia	Australia[†]
Mountains	38.1	43	39	47	16
Low angle bedrock surfaces	0.7	10	6	1	14
Alluvial fans	31.4	1	1	4	
River plains	1.2	1	3	1	13
Dry watercourses	3.6	1	1	1	
Badlands	2.6	2	8	1	–
Playas	1.1	1	1	1	1
Sand seas	0.6	28	22	26	38
Desert flats[‡]	20.5	10	18	16	18
Recent volcanic deposits	0.2	3	1	2	–

[*] Percentages given are only approximate, with the degree of accuracy differing between areas.

[†] From Mabbutt (1976). Categories used by Mabbutt do not necessarily coincide with those used in other areas: included for comparison only. The remaining data are from a study by Clements *et al.* (1957) for the US Army.

[‡] Undifferentiated: includes areas bordering playas.

Table 1.6 *Some geomorphological hazards in drylands*

Water hazards	Wind hazards	Materials hazards
Flooding following rainfall	Dust storms	Landslides in mountainous areas
Subsidence after water extraction	Dune encroachment	Desiccation contraction of sediments
Gully development	Dune reactivation	Salt weathering
Soil erosion	Soil erosion	

Arid zone geomorphology and people

The human population of the global arid zone increased by 63.5% between 1960 and 1974. By 1979, 15% (651 million) of the world's population lived in arid lands (Heathcote 1983). It is estimated that at least 900 million people live in the drylands as defined in UNEP (1992) (Rosanov 1990; UNEP 1991). In some continents, arid areas are central to human occupation. In Africa, for example, 49.5% of the total population live in arid areas (UNEP 1992).

Arid areas present a wide range of environmental hazards for their occupants, many of which are geomorphological (Table 1.6). This has prompted a strong and growing correlation (Goudie 1985) between arid zone geomorphology and applied research (see, for example, Cooke *et al.* 1982). To carry out applied geomorphological research requires a strong underpinning in the salient aspects of geomorphology and allied disciplines. It is to these, in the context of the arid zone, that this book is addressed.

Organisation of this book

There are many ways in which a book covering the geomorphology of arid environments could be organised. In this case, the main chapters have been divided among six sections for convenience, each devoted to a major theme in arid zone geomorphology.

Section 1 sets the physical framework for considering arid zone geomorphology, with the tectonic characteristics and settings of drylands examined in Chapter 2. Section 2 considers surface processes and characteristics. It is often noted that the relatively vegetation-free surfaces of drylands allow bedrock and material characteristics to exert a greater direct (but not necessarily overriding or all-determining) influence on geomorphic development than in other environments. The five chapters in this section each consider a major aspect of surface conditions and processes. The weathering of rocks is discussed in Chapter 3, while uncertainties in understanding dryland weathering are explored in Chapter 4. Chapter 5 considers soils, while chemical sediments (crusts and varnishes) are dealt with in Chapter 6. Arid zone surface and penesurface crusts are usually described and distinguished in terms of the dominant or characterising mineral constituent, and may occur within weathering residua and other deposits, soils, or bedrock. While rock varnish, also examined in Chapter 6, is by no means exclusively found in arid environments, the ample presence of bare rock surfaces in deserts provides circumstances favourable for the formation and study of surface patinas. Finally in Section 2, the ways in which vegetation can interact with and influence geomorphological processes are considered in Chapter 7.

Sections 3 and 4 examine the arid zone geomorphological process domains of water and wind, respectively. Section 3 commences with a chapter that considers a fundamental of geomorphology: slopes. Both of these sections contain a chapter which examines the ways in which sediment entrainment is realised: by runoff (Chapter 9) and by windflow (Chapter 16). Subsequent chapters in Section 3 each deal with a major water-shaped component of arid landscapes: desert drainage basins and networks, channels, alluvial fans, badlands and badland gullies, and pans, playas and salt lakes. The final chapter in this section considers how in drylands, so often devoid of surface water, groundwater may influence geomorphological processes. In the remainder of Section 4, aeolian sand deposits, desert dust and erosional forms are examined.

Section 5 contains six chapters that examine the geomorphology of the world's main arid environments. As well as explanations of the main geomorphological features, these chapters also examine the nature of geomorphological investigations in each of the areas considered.

Section 6 has four chapters which look both at changes in arid environments and beyond the present earthly arid zone. Chapter 26 looks at the geomorphic and sedimentary evidence for, and causes of, expansions and contractions of arid areas on Earth during the Quaternary period of geological time, while Chapter 27 considers how these changes may be dated. Human influences on arid geomorphology are explored in Chapter 28. Chapter 29 looks even further afield by considering the current state of knowledge of extraterrestrial arid zones. While the investigation of the surface morphology of planets other than our own is obviously a highly specialised area of study which is in its infancy, it is yielding information which helps clarify the nature of processes operating on Earth (Greeley 1985), notably in some subfields of arid geomorphology. Finally, an overview of some linking issues in arid zone geomorphology is provided in Chapter 30.

References

Abrahams, A.D. and Parsons, A.J. (eds) 1994. *Geomorphology of desert environments*. Chapman & Hall, London.

Agnew, C.T. and Anderson, E. 1992. *Water resources in the arid realm*. Routledge, London.

Aubreville, A. 1949. *Climats, Forêts et désertification de l'Afrique tropicale*. Soc d'editions géographiques maritimes et coloniales, Paris.

Birot, P. 1960. *Le cycle d'erosion sons les differents climats*. University of Brazil, Rio de Janeiro.

Bloom, A.L. 1978. *Geomorphology. A systematic analysis of late Cenzoic landforms*. Prentice-Hall, Englewood Cliffs, NJ.

Bowler, J.M. 1976. Aridity in Australia: age, origins and expression in aeolian landforms and sediments. *Earth Science Reviews*, 12: 279–310.

Büdel, J. 1963. Klima-genetische Geomorphologie. *Geographische Rundschau*, 15: 269–285.

Clark, A.H., Meyer, E.S., Mortimer, C., Sillitoe, R.H., Cooke, R.U. and Snelling, N.J. 1967. Implications of the isotope ages of ignimbrite flows, southern Atacama desert, Chile. *Nature*, 215: 723–724.

Clements, T., Merriam, R.H., Eymann, J.L., Stone, R.O. and Reade, H.L. 1957. *A study of desert surface conditions*. US Army Environmental Protection Research Division Technical Report EP-53, Natck, MA.

Cooke, R.U. and Warren, A. 1973 *Geomorphology in deserts*. Batsford, London.

Cooke, R.U., Brunsden, D., Doornkamp, J.C. and Jones, D.K.C. 1982. *Urban geomorphology in drylands*. Oxford University Press, Oxford.

Cooke, R.U., Warren, A. and Goudie, A.S. 1993. *Desert geomorphology*. UCL Press, London.

Davies, W.M. 1905. The geographical cycle in an arid climate. *Journal of Geology*, 38: 1–27, 136–138.

de Martonne, E. and Aufrère, L. 1927. Map of interior basin drainage. *Geographical Review*, 17: 414.

Flammand, G.B.M. 1899. La traversée de l'erg occidental (grands dunes du Sahara oranais). *Annales de Géographie*, 9: 231–241.

Gilbert, G.K. 1875. Report on the geology of portions of Nevada, Utah, California and Arizona. *Geographical and geological explorations and surveys west of the 100th meridian, Part 1*. Engineers Department, US Army: 21–187.

Gilbert, G.K. 1877. *Report on the geology of the Henry Mountains*. Government Printing Office, Washington, DC.

Gilbert, G.K. 1895. Lake basins created by wind erosion. *Journal of Geology*, 3: 47–49.

Glennie, K.W. 1987. Desert sedimentary environments, past and present — a summary. *Sedimentary Geology*, 50: 135–166.

Goudie, A.S. 1985. Themes in desert geomorphology. In A. Pitty (ed.), *Themes in geomorphology*. Croom Helm, London: 122–140.

Greeley, R. 1985. *Planetary landscapes*. Allen & Unwin, London.

Grove, A.T. 1977. The geography of semi-arid lands. *Philosophical Transactions of the Royal Society of London*, Series B, 278: 457–475.

Heathcote, R.L. 1983. *The arid lands: their use and abuse*. Longman, London.

Hulme, M. 1992. Rainfall changes in Africa: 1931–60 to 1961–90. *International Journal of Climatology*, 12: 685–699.

Hulme, M. and Marsh, R. 1990. *Global mean monthly humidity surfaces for 1930–59, 1960–69 and projected for 2030*. Report to UNEP/GEMS/GRID, Climate Research Unit, University of East Anglia, Norwich.

Köppen, W. 1931. *Die Klimate der Erde*. Berlin.

Lancaster, N. 1989. *The Namib sand sea: dune forms, processes and sediments*. Balkema, Rotterdam.

Mabbutt, J.A. 1976. Physiographic setting as an indication of inherent resistance to desertification. *WGDAL*, 1976: 189–197.

McGee, W.J. 1897. Sheetflood erosion. *Bulletin of the Geological Society of America*, 78: 93–98.

Meigs, P. 1952. Arid and semiarid climatic types of the world. In *Proceedings, Eighth General Assembly and Seventeenth International Congress, International Geographical Union*, Washington: 135–138.

Meigs, P. 1953. World distribution of arid and semi-arid homoclimates. In *Arid zone hydrology*. UNESCO Arid Zone Research Series, 1: 203–209.

Meigs, P. 1966. Geography of coastal deserts. *UNESCO Arid Zones Research*, 28: 1–40.

Mitchell, T.L. 1837. Account of the recent exploring expedition to the interior of Australia. *Journal of the Royal Geographical Society*, 7: 271–285.

Mitchell, T.L. 1839. *Three expeditions into the interior of Eastern Australia*. Boone, London.

Mortimore, M. 1989. *Adapting to drought: farmers, famine and desertification in West Africa*. Cambridge University Press, Cambridge.

Murphy, R.E. 1968. Landform regions of the world. *Annals of the Association of American Geographers*, 58: Map supplement 9.

Parsons, A.J. and Abrahams, A.D. 1994. Geomorphology of desert environments. In Abrahams, A.D. and Parsons, A.J. (eds), *Geomorphology of desert environments*. Chapman & Hall, London: 3–12.

Passarge, S. 1904. *Die Kalahari*. Dietrich Riemer, Berlin.

Penck, A. 1894. *Morphologie der Erdoberfläche*. Stuttgart.

Powell, J.W. 1875. *Exploration of the Colorado River of the West*. Washington, DC.

Powell, J.W. 1878. *Report on the lands of the arid region of the United States*. Harvard University Press, Cambridge, MA (1962 edn).

Rahn, P.H. 1967. Sheetfloods, streamfloods and the formation of pediments. *Annals of the Association of American Geographers*, 57: 593–604.

Rosanov, B.G. 1990. Global assessment of desertification: status and methodologies. In *Desertification revisited*. Proceedings of an ad hoc consultative meeting on the assessment of desertification. UNEP-DC/PAC, Nairobi: 45–122.

Ross, G.M. 1983. Bigbear erg: a Proterozoic inter-montane eolian sand sea in the Hornby Bay Group, Northwest Territories, Canada. In M.E. Brookfield and T.A. Ahlbrandt (eds), *Eolian sediments and processes*. Elsevier, Amsterdam: 483–519.

Sarnthein, M. 1978. Neogene sand layers off north-west Africa: composition and source environment. In Y. Lancelot and E. Seibold (eds), *Initial reports of the Deep Sea Drilling Project*, 41. US Government Printing Office, Washington, DC: 939–959.

Schulze, B.R. 1972. South Africa. In J.F. Griffiths (ed.), *World survey of climatology, volume 10, Climates of Africa*. Elsevier, Amsterdam.

Shantz, H.L. 1956. History and problems of arid lands development. In G.F. White (ed.), *The future of arid lands*. American Association for the Advancement of Science, Washington, DC: 3–25.

Spencer, B. 1896. *Report on the work of the Horn scientific expedition to Central Australia*. Dulan, London.

Sturt, C. 1833. *Two expeditions into the interior of Southern Australia during the years 1828, 1829, 1830 and 1831*. Smith, Elder, London.

Tankard, A.J. and Rogers, J. 1978. Late Cenozoic palaeoenvironments on the west coast of southern Africa. *Journal of Biogeography*, 5: 319–337.

Thomas, D.S.G. 1988a. The nature and depositional setting of arid to semi-arid Kalahari beds, central southern Africa. *Journal of Arid Environments*, 14: 17–26.

Thomas, D.S.G. 1988b. Arid and semi-arid areas. In H.A. Viles (ed.), *Biogeomorphology*. Blackwell, Oxford: 191–221.

Thomas, D.S.G. 1993. Sandstorm in a teacup? Understanding desertification in the 1990s. *Geographical Journal*, 159: 318–331.

Thomas, D.S.G. and Middleton, N.J. 1994. *Desertification: exploding the myth*. Wiley, Chichester.

Thompson, R.D. 1975. *The climate of the arid world*. University of Reading, Geographical Paper 35.

Thornthwaite, C.W. 1948. An approach towards a rational classification of climate. *Geographical Review*, 38: 55–94.

Tricart, J. and Cailleux, A. 1963. *Traite de geomorphology IV: le modelé des regions seches*. Le Cours de Sorbonne, Paris.

UN 1977. *Status of desertification in the hot arid regions, climatic aridity index map and experimental world scheme of aridity and drought probability, at a scale of 1:25,000,000. Explanatory note*. UN Conference on Desertification A/CONF. 74/31, New York.

UNEP 1991. *Status of desertification and implementation of the United Nations plan of action to combat desertification*. Report of the executive director to the governing council, third special session, Nairobi.

UNEP 1992. *World atlas of desertification*. Edward Arnold, Sevenoaks, UK.

UNESCO 1979. *Map of the world distribution of arid regions*. MAP Technical note 4, UNESCO, New York.

Vogel, J.C., Rogers, J. and Seely, M.K. 1981. Summary of SASQUA Congress, May 1981. *South African Journal of Science*, 77: 435–436.

Ward, J.D., Seely, M.K. and Lancaster, N. 1983. On the antiquity of the Namib. *South African Journal of Science*, 79: 175–183.

2
Tectonic frameworks

Helen Rendell

Introduction

The tectonic setting of dryland areas has received comparatively little attention in the literature on arid zone geomorphology. References to tectonics are often limited to discussions of alluvial fans or of pans and playas, where active tectonics may have played some part in the development of these landforms. However, tectonics in the broadest sense provide the backdrop, in terms of both absolute and relative relief, against which dryland processes operate. Tectonic controls operate on sediment sources and sediment sinks. Tectonic settings also influence the timescales over which relative stability or instability dominate particular areas, and thereby help to determine the continuity or discontinuity of sedimentary records.

The relationship between tectonics and landforms works both ways. Much of tectonic geomorphology is concerned with using evidence from landforms to infer rates of operation of tectonic processes (Ollier 1981; Mayer 1985; Morisawa and Hack 1985). Conversely, some geomorphological studies invoke tectonics in order to explain landform evolution (e.g. Currey 1994b).

The history of global climate change during the Quaternary is now well-documented in ocean sediment cores and ice cores, and, in common with other environments, dryland landscapes and sediments can be interpreted in the context of these changes. These landscapes and sediments can also be used to establish or refine local sequences of climate change (see Chapters 26 and 27). In addition, since landforms may be used to provide insights into recent crustal movements, unravelling climatic and tectonic influences remains a key problem (Frostick and Steel 1993b). Timescales are of critical importance in attempting to resolve the relative dominance of these influences.

A major problem is that tectonic influences at the Earth's surface have operated more or less continuously over the last 10^7 yr whereas, over the same timescale, climatic conditions have changed in a cyclic manner. The loci of tectonic activity are well specified at the global level, and are predominantly associated with plate margins. Tectonic activity may appear episodic in the short term (10^1–10^3 yr) (McCalpin 1993), but is effectively continuous in many areas when viewed over the longer term (10^4–10^7 yr). Rates of tectonic uplift are normally expressed in units of mm yr^{-1} or m 1000 yr^{-1}, even though vertical displacements during individual events may be of the order of 10^2–10^3 mm. Many contemporary

Arid Zone Geomorphology: Process, Form and Change in Drylands, 2nd edition. Edited by David S. G. Thomas.

dryland areas have only been arid or semi-arid since some point during the late Tertiary. The major exception may be the Namib Desert, which is thought to date from the Cretaceous (8×10^7 yr), but both the Atacama and Australian deserts are considered to date from some point during the Miocene ($22–5 \times 10^6$ yr) (see Chapter 26). The onset of aridity is recognised to have occurred at a global level *c.* 3×10^6 yr ago (Williams 1994). In addition, all desert areas have experienced climate change during the Late Quaternary, with some changes involving a move to humid or sub-humid conditions (Figure 26.13). Although Late Quaternary glacial/interglacial cycles have a frequency of 1.20×10^5 yr, some extremely rapid changes have been identified within these periods, at frequencies of $10^2–10^3$ yr (Taylor *et al.* 1993).

The emphasis in this chapter is on the tectonic setting of contemporary dryland areas, but given the issue of timescales mentioned above, problems arise from the fact that the various landforms considered may be out of phase with contemporary climatic conditions. This is a particular problem in the case of erosional landforms such as pediments, for example, which may have evolved over a much longer period of time than that over which arid or semi-arid conditions are thought to have persisted in a particular area (Dohrenwend 1994; see also Chapter 8).

Arid zones are in many ways ideal for the study of tectonics. Large structures, fold and fault zones are clearly visible on remotely sensed images of arid zones. Lack of or limited vegetation leads to enhanced visibility of surface expressions of tectonic activity like fault scarps and, in hyper-arid areas such as the Atacama or southern Negev, scarps cutting alluvial fans may remain undegraded for substantial periods of time. Visibility is particularly important given the transient nature of many small features which may result from seismic disturbance of the ground surface.

Although the scale at which tectonics operate may allow correlations at the macro- or mega-geomorphological scale of landform evolution, even at a mesoscale, tectonic influences may have considerable significance in arid zones. In an arid context, for example, fault-related springlines (see, for example, Chapter 15),

have a much greater significance than in environments in which water supply is far more plentiful. Also, salt diapirs provide one source of the salts which play a key role in many arid zone rock weathering processes.

This chapter will consider the tectonic setting of contemporary drylands, the controls which tectonics may have in terms of uplift and erosion and subsidence and sedimentation, the problems of developing a chronological framework for dryland landforms and sediments, the existing record of erosion and sedimentation and, finally, the interaction of active tectonics and contemporary processes.

Tectonic setting of drylands

Tectonic settings of contemporary dryland areas control the development of landforms and of sedimentary sequences by influencing factors such as sediment supply.

It is possible to identify five types of tectonic setting: cratons (shield and platform areas); active continental margins, associated with Cenozoic orogenic belts; older, Phanerozoic, orogenic belts; inter-orogenic basin and range and inter-cratonic rift zones; and passive continental margins. Examples of each of these are given in Table 2.1.

The currently accepted view of what 'tectonically stable' means has undergone a radical transformation since the late 1960s. The recognition that stability is relative rather than absolute has resulted, among other things, in the requirement for the assessment of seismic risk for large structures in the UK, as well as in California. However, it is possibile to identify particular tectonic settings on the basis of relative stability or instability and the term craton is still used to describe the 'central stable portion of a continent' (Ollier 1981, p. 75).

The Earth's crust comprises interlocking lithospheric plates which move relative to each other. New oceanic crust is continuously created at spreading centres (mid-ocean ridges and rifts), and this increase in crustal material is accommodated by 'loss' of crust by subduction, by folding and thrust faulting, or by relative movement along transcurrent faults. The most spectacular tectonic activity is associated with active plate margins, and with earthquakes and

Table 2.1 *Tectonic settings of arid zones*

Contemporary tectonic setting	Examples	Comments
1. Cratons	Kalahari Great Karoo Australian Desert Saudi Arabia	Relative stability since the late Tertiary
2. Active continental margins and Cenozoic orogenic belts	Atacama (Peru–Chile) Sahara Sinai–Negev Arabia–Zagros Thar (Great Indian Desert)	Compressional setting, thrust and transcurrent faulting
3. Older orogenic belts	Sahara China	Some reactivation of existing fault zones
4. Inter-orogenic, inter-cratonic	Sahara (Afar, Ethiopia) Mojave Great Basin (W USA) Sonora Desert Chihuahua Desert Monte Desert	Extensional tectonic setting, 'pull-apart' basins
5. Passive continental margins	Namib Desert Patagonian Desert	

Note: The physical extent of the Sahara is such that it features in several of the categories listed above.

volcanic activity. Passive (or trailing-edge) continental margins are not without interest, with many of them characterised by the presence of so-called 'great escarpments' (Ollier 1985a; 1985b). Not all crustal movements are of course compressional, and the development of major rift systems (grabens), and so-called 'basin and range' provinces are products of either mantle-generated or lithosphere-generated rifting (Frostick and Steel 1993b).

Uplift and erosion, subsidence and sedimentation

Tectonically driven changes can involve both uplift and subsidence. Whether these changes achieve topographic expression is a function of the degree to which uplift exceeds erosion and to which subsidence exceeds sedimentation. Since the exact meaning of the term uplift has been the subject of some debate, a definition is necessary. According to Brookfield (1993, p. 16):

uplift is positive vertical elevation with respect to the geoid (basically mean sea level). This normally refers

to net uplift, combining the effects of gross (or total) uplift minus the effects of erosion. Thus: net uplift = gross uplift – erosion.

Differences in rates of change are of critical importance. If erosion keeps pace with gross uplift there is of course no net uplift, but while incision by a major river may keep pace with uplift, as has been the case in the Himalayan region, areal erosion rates within the catchment may not do so, with the consequent development of high relative relief. In arid zones, the ability of erosion processes to respond to tectonic displacements is of course affected by the fact that many geomorphic and hydrological processes within these zones are ephemeral in nature.

In an attempt to address the fundamental issue of the comparative rates of uplift and erosion at the global level, Schumm (1963) compared data on maximum rates of uplift (*c.* 7 mm yr^{-1}) and maximum rates of erosion (*c.* 1 mm yr^{-1}) and concluded that uplift greatly exceeds maximum erosion. The problem with this approach, as noted by Vita-Finzi (1986), is that the erosional data are effectively averaged out areal data whereas the uplift rates may be unrepresentative of uplift at a regional level; that is, erosional data maxima are underestimates while uplift data are overestimates. One can equally well compare data for erosion rates of badland areas with current rates of uplift in tectonically active areas (Table 2.2). For this particular data set, the rates of uplift and erosion are comparable. However, it should be noted that all measurements of erosion are notoriously scale-dependent. It is well known that it is impossible to extrapolate the results from erosion plot studies to the field scale, and from the field scale to the catchment scale, with any confidence. One may question the value of areal erosion values calculated in $mm\,yr^{-1}$ where linear features, such as gullies, are concerned. Timescales are also important, and generally, the greater the length of record considered, the lower the estimate of the erosion rate, as is the case in the study by Wells and Gutierrez (1983).

The study of neotectonics has included the documentation of recent vertical crustal movements (RVCMs) which can be determined by instrumentation and measurement, and which are therefore limited to observations made during the last century (Fairbridge 1981). The order of magnitude of these very recent movements, and that of neotectonic measurements determined over longer timescales, are summarised in Table 2.3 for the various arid zone tectonic settings identified in Table 2.1. Again there appears to be a scale dependency, with the RVCMs exceeding the neotectonic estimates by up to an order of magnitude.

Depositional sequences may also reflect tectonic influences. In a recent review, Frostick and Steel (1993a, p. 2) identify six ways in which fluvial or lacustrine sedimentation may be affected by tectonic controls:

1. Changes in the overall accommodation space available for filling within the basin.

Table 2.2   *Comparison of uplift and erosion rates*

	References
Uplift rates	
$10\,mm\,yr^{-1}$ (Karakoram, N. Pakistan)	Brookfield (1993)
$1\,mm\,yr^{-1}$ (Gt Himalaya, India)	Cited in Brookfield (1993)
$2\,mm\,yr^{-1}$ (Makran, Iran)	Vita-Finzi (1986)
$2-10\,mm\,yr^{-1}$ (Zagros, Iran)	Vita-Finzi (1986)
$15\,mm\,yr^{-1}$ (Himalaya)	Brookfield (1993)
Erosion Rates	
$17.9\,mm\,yr^{-1}$ (S. Dakota Badlands)	Cited in Bryan and Yair (1982)
$4.0\,mm\,yr^{-1}$ (Alberta Badlands)	Bryan and Yair (1982)
$0.45\,mm\,yr^{-1}$ over $70\,000\,yr$ (Zin Badlands, Israel)	Yair *et al.* (1982)
$3\,mm\,yr^{-1}$ over $5600\,yr$ (Colorado, USA)	Wells and Gutierrez (1982)
$3-20\,mm\,yr^{-1}$ (Colorado, USA)	Wells and Gutierrez (1982)

Table 2.3   *Estimates of tectonic uplift rates as a function of tectonic setting*

Tectonic setting	Recent vertical crustal movements	Nontectonic warping
1. Cratons	Less than $1\,mm\,yr^{-1}$	
2. Active continental margins	Up to $20\,mm\,yr^{-1}$	Up to $10\,mm\,yr^{-1}$
3. Older orogenic belts	Up to $5\,mm\,yr^{-1}$	Up to $1\,mm\,yr^{-1}$
4. Inter-orogenic, inter-cratonic	Up to $10\,mm\,yr^{-1}$	Up to $5\,mm\,yr^{-1}$
5. Passive continental margins	Up to $10\,mm\,yr^{-1}$	Up to $1\,mm\,yr^{-1}$

Source: Fairbridge (1981).

2. Changes in the direction of tilt and the location and size of the depocentre.
3. Changes in the orientation and character of basin margins, and in overall basin size and shape (e.g. by backstepping of marginal faults).
4. Changes in the gradient both at the margins of the basin, through faulting, and within the basin (e.g. through tilting of the basin floor).
5. Deflection of the sedimentary systems where tectonically controlled morphological changes act as barriers to sediment transfer both into the basin from the hinterland and between various areas within the basin.
6. Changes in the rate of sediment supply due to uplift or erosion of the supplying hinterland.

One key characteristic of the supply of water and sediment to arid zone lake basins, and therefore of arid zone fluvial activity, is the coexistence of far-travelled perennial rivers, with catchments outside the desert, with ephemeral desert streams (Currey 1994a). The latter have a capacity for transporting sediment during flood events that far exceeds that of similar-size perennial channels. One explanation for this high rate of bedload transport is that armouring of the channel bed has no time to become established in ephemeral bedload streams (Laronne and Reid 1993). These differences between ephemeral and perennial streams have implications for both the rate and the nature of fluvial and lacustrine sedimentation in arid zone depocentres.

Lengths of record

From the point of view of process geomorphology, the different tectonic settings are associated with different potentials for sediment generation and sediment supply, and for the disruption of sedimentary sequences in particular climatic contexts. The record of change can be read in both erosional landforms and sequences of sedimentary deposits. The status of different variables changes as a function of the timescales considered. However, the timescales required for the development of erosional and depositional sequences in arid zones are still the subject of considerable speculation, since progress depends on the availability of both appropriate dating techniques and of potentially datable materials. Prior to the development of radiometric dating techniques, the evolutionary state of a particular landscape was assessed with reference to Davisian or Penckian models, and this analysis was used to provide the relative age of that landscape (Thornes and Brunsden 1977). The development of radiometric dating has shifted the burden of proof of antiquity from the landscape itself, but dating is by no means straightforward and depends not only on the presence of suitable materials but also on the existence of techniques with appropriate time ranges of application. The most well-known technique of ^{14}C dating, for example, can only be used for materials containing organic or inorganic carbon and within the age range $0-4.5 \times 10^4$ yr. But even this technique has calibration problems for material older than 1.2×10^4 yr (see Chapter 27).

In the arid zone context, the main challenges lie in the dating of exposure age of surfaces and of sedimentary deposits (fluvial, lacustrine, aeolian). Taking these in turn, the exposure age of surfaces can be approached via:

1. Dating of varnish coatings (AMS ^{14}C, or relative dating using cation-ratios) (Dorn 1994).
2. Dating of surfaces using cosmogenic isotopes.
3. Dating of palaeosurfaces — sealed by lava flows (dated using K/Ar or $^{40}Ar/^{39}Ar$) (Dohrenwend 1994).

The dating of sedimentary deposits (fluvial, lacustrine, aeolian) may involve the application of the following techniques:

1. ^{14}C dating of organic or inorganic carbon.
2. Uranium-series dating of inorganic carbonates.
3. ^{36}Cl dating of evaporites (within lacustrine sequences).
4. Dating of component sediment grains using luminescence techniques, provided the sediment grains were exposed to light, prior to deposition and burial, or have been heated to high temperature, e.g. in hearths or by proximity to lava flows.
5. Conventional K/Ar, $^{40}Ar/^{39}Ar$ or fission-track dating can be used in contexts in which volcanic ashes are intercalated with fluvial or lacustrine sequences.

Although a whole battery of dating techniques is now available, examples of application are relatively limited and thus, even though the preservation of organic carbon is notoriously poor in arid environments, much of the palaeoenvironmental reconstruction work already undertaken is based on ^{14}C, and ultimately limited by it. Access to potentially datable material may also be difficult and, depending on the depositional environment, it is often limited to natural cuttings and sections, sometimes revealed by tectonic disruption of sedimentary basins, and to boreholes. Seismic profiling may add considerably to the understanding of sedimentary basin architecture, but such remotely sensed data yield only relative chronological information.

Existing erosional and depositional records in arid environments

Given the problems with the application of radiometric dating to arid zone sequences and landforms, it is difficult to substantiate any assumptions about the lengths of record of erosion or sedimentation in different tectonic settings. Longer records are not necessarily associated with stable tectonic areas; instead tectonic activity may play a key role in ensuring continuity of sediment supply (erosion) and deposition. Data on existing lengths of record are summarised in Table 2.4.

Table 2.4   *Length of record of erosional landforms and depositional sequences in arid zones*

Erosional landform or depositional sequence	Length of record and comments	References
Pediments	Minimum age of relict surfaces $0.85-8.90 \times 10^6$ yr Max. incision (downwearing) rates $8-47$ m per 10^6 yr Max. slope retreat rates $37-365$ m per 10^6 yr	Data from W USA: Eastern Mojave Desert and Great Basin, constrained by dated lava flows. Dohrenwend (1994)
Desert pavements	Evolutionary setting for cumulic pedogenesis. Surface development sequence in Cima volcanic field. Time required for development: $0.2-0.7 \times 10^6$ yr	McFadden *et al.* (1986)
Alluvial fans and alluvial fan surfaces	Well-constrained record in W USA largely limited to range of AMS ^{14}C dating (4.5×10^4 yr) of varnish	Dorn (1994)
Playa lake sequences	Age constraint largely limited to range of AMS ^{14}C dating (4.5×10^4 yr) [36]Cl dating of evaporites yields a 2×10^6 yr sequence for the Owens River–Death Valley system	Jannick *et al.* (1991); Currey (1994a; 1994b)
Lacustrine sequences	Intercalated volcanic ashes dated by $^{40}Ar/^{39}Ar$, K/Ar, fission track, used to constrain lacustrine sequences in East African Rift (Lake Turkana) to beyond 3×10^6 yr	Frostick and Reid (1987)
Aeolian sequences–sand seas	Age estimates of sand seas are derived from estimates of sand fluxes and exceed 1×10^6 yr. Some but not all major sand seas are located in relatively stable (cratonic) areas of Sahara, S Africa, Australia, Rub' al Khali	Cooke *et al.* (1993)
Aeolian sequences–sand ramps	A series of sand ramps in the E Mojave yielded luminescence ages in the range $0-4.0 \times 10^4$ yr	Rendell (1995)

PEDIMENTS

At the global level, pediments tend to be concentrated on granitic terrains. The highest spatial concentration of pediments is reported from the southwest USA, from a basin and range area of inter-orogenic rifts (Cooke *et al.* 1993). The more stable cratonic areas of southern Africa and central Australia also feature pediments, but it is the association with active volcanism that has allowed the development rate of the pediments in the Mojave Desert to be established (Dohrenwend 1994).

At a more local level, a limited amount of information exists on the slope angles of pediments in Arizona and California in relation to proximity to faults. Data from Cooke *et al.* (1993, table 13.1c) indicate that fault-associated pediments have slopes of $2°55' \pm 1°12'$, whereas those not associated with faults had slopes of $2°10' \pm 48'$. Although difference in the means is statistically significant at the 0.01% level, the problem is that, in physical terms, mean slope angles are very similar and there is significant overlap of the standard deviations.

DRAINAGE PATTERNS AND FLUVIAL SYSTEMS

There is considerable evidence of the long-term impact of tectonic activity over time periods of 10^6–10^7 yr on patterns of drainage in contemporary drylands. Frostick and Steel (1993b) discuss the development of patterns of radial drainage in southern Africa in response to doming of the crust at a regional scale. The development of rift systems may also result in the reorganisation of drainage patterns, with some surprising results. In their model for the development of the East African Rift, Frostick and Reid (1987) note that the backtilting and uplift of footwall blocks essentially diverts drainage away from the main axial depocentre. A similar example is provided by the Red Sea–Gulf of Aden region, where rifting began in the Miocene, in which rivers approaching within 2 km of the west bank of the Red Sea are diverted into the Nile catchment (Frostick and Steel 1993b).

Tectonic disruption of drainage systems in the southwest USA is invoked by Zimbelman

et al. (1994) in their argument that the ancestral Mojave River originally continued to flow eastwards to join the Colorado River rather than terminating in the Soda Lake–Silver Lake playas (Brown *et al.* 1990).

PLAYAS

Playa lakes (see Chapter 14) develop in local topographic lows within arid zone drainage systems. Many lakes are set in basins of inland drainage, and it is generally recognised that tectonic activity is the fundamental cause of basin closure (Currey 1994a). The astonishing profusion of playa lakes in the southwest USA, for example, is associated with tectonically active inter-orogenic basin and range settings. Some drainage systems may be complex, with spillways coming into operation during lake high stands (e.g. the Owens River drainage system).

With the exception of the sequences associated with the East African Rift, which incorporate datable volcanic ashes and extend back to beyond 3×10^6 yr (Frostick and Reid 1987), and the 2×10^6 yr record of sedimentation in the Owens River–Death Valley system based on ^{36}Cl (Jannick *et al.* 1991), data on lacustrine/playa sequences tend to be limited to the age range of ^{14}C of 4.5×10^4 yr (Brown *et al.* 1990; Currey 1994b).

AEOLIAN SEQUENCES

At the moment, much of the record of Quaternary aeolian activity is effectively limited to the last 4.5×10^4 yr (Tchakerian 1994). The major exception is in North America for sand ramps in Nevada where the Bishop Ash (K/Ar date 7.4×10^5 yr) is preserved at the base of several of the ramps. Antiquity is often inferred on the basis of the volume of sand incorporated in sand seas, or because of the presence of distinct bounding surfaces between different generations of dunes (Lancaster 1992). Although it is recognised that some sand seas have developed, and been reworked, over timescales of the order of 10^6 yr (Cooke *et al.* 1994, p. 409), direct chronological information is often lacking. The influence of tectonic activity on aeolian activity is likely to be indirect, and a function of the timescale considered. Aeolian activity and sedimentation tends to be episodic in nature. In the short term, sediment availability may, for example, be a function of fluvial processes responsible for transporting material to a site for deflation. In the longer term, tectonics may be a major control on erosion, and therefore sediment supply. If only short-term records are available then it is likely that they will only be amenable to interpretation in terms of climate change rather than tectonism.

Selected examples of the geomorphological impact of active tectonics in arid environments

TECTONIC DISRUPTION OF FLUVIAL SYSTEMS

The 1980 El Asnam earthquake in Algeria provides one example of tectonic disruption of a fluvial system as a result of surface deformation. Active fold development resulted in the blocking of the Chelif River with the production of a lake some 5 km^2 in extent, and stratigraphical and archaeological evidence is used by King and Vita-Finzi (1981) to argue that local topography is the result of some 30 similar earthquakes over a period of 1.5×10^4 yr.

Large river systems tend to be prone to avulsion which may or may not be tectonically triggered. Avulsion is when the river relocates itself during a flood event by occupying the lowest point on the floodplain. The history of the Indus River system provides some evidence of depositional sequences responding to crustal deformation. McDougall (1989) hypothesises that a major easterly switch of the Indus course occurred at some point in the last $5 \times 10^5 - 1 \times 10^5$ yr owing to oblique compression in the active right lateral Kalabagh fault zone. Jorgensen *et al.* (1993, p. 310) discuss the interaction of climatic and tectonic influences on the course of the Lower Indus River. They note that:

all of the significant river pattern changes take place at tectonic boundaries... anomalously steep and gently sloping reaches of the river conform to suspected reaches of uplift and subsidence, and

avulsions may mark the points of relative upwarp in contrast to basin subsidence.

TECTONIC CONTROLS ON ALLUVIAL SEDIMENTATION

Although the potential role of tectonics in alluvial fan development clearly involves the topographic setting of the fan, which in many cases may be at a fault-bounded mountain front, the role of active tectonics in fan development is still debated, and climatic arguments tend to be dominant (Blair and McPherson 1994). In their examination of alluvial fan sequences in the Dead Sea rift in Israel, Frostick and Reid (1989) demonstrate that the discrimination between climatically induced and tectonically induced facies changes is extremely difficult. On the basis of field evidence, they argue (p. 537) that:

there is no reason to invoke tectonic destabilization of the river system as a cause of influx of coarse alluvium in the Dead Sea fan sequences... punctuation of the sedimentary sequence by coarse sediment was achieved by *climatic* [their emphasis] rather than tectonic factors.

However, Rockwell *et al.* (1985) argue that active folding and faulting of alluvial fans in the Ventura Valley, California, have produced complex fan morphologies that are quite distinct from those in less active areas along the same mountain front. The role of tectonics in alluvial fan development is discussed further in Chapter 12.

Conclusions

The chronological framework that is really needed to assess the erosional and depositional history of arid lands is still under construction. Such a framework is required before the roles of climate and tectonics can be properly disentangled. Although tectonic settings influence the topography of dryland areas, the critical role played by climate cannot be understated, since climate drives arid zone processes by dictating both the availability of water and the nature of aeolian activity.

References

Blair, T.C. and McPherson, J.G. 1994. Alluvial fan processes and forms. In A.D. Abrahams and A.J. Parsons (eds), *Geomorphology of desert environments*. Chapman & Hall, London: 354–402.

Brookfield, M.E. 1993. The interrelations of post-collision tectonism and sedimentation in Central Asia. In L.E Frostick and R.J. Steel (eds), *Tectonic controls and signatures in sedimentary successions*. Special Publication No. 20 of the International Association of Sedimentologists. Blackwell Scientific Publications, Oxford: 13–35.

Brown, W.J., Wells, S.G., Enzel, Y., Anderson, R.Y. and McFadden, L.D. 1990. The late Quaternary history of pluvial Lake Mojave–Silver Lake and Soda Lake Basins, California. In R.E. Reynolds, S.G. Wells and R.H.I. Brady (eds), *At the end of the Mojave: Quaternary studies in the Eastern Mojave Desert*. Special Publication of the San Bernardino County Museum Association, San Bernardino, CA: 55–72.

Bryan, R.B. and Yair, A. 1982. *Badland geomorphology and piping*. GeoBooks, Norwich.

Cooke, R.U., Warren, A. and Goudie, A.S. 1993. *Desert geomorphology*. UCL Press, London.

Currey, D.R. 1994a. Hemiarid lake basins: hydrographic patterns. In A.D. Abrahams and A.J. Parsons (eds), *Geomorphology of desert environments*. Chapman & Hall, London: 405–421.

Currey, D.R. 1994b. Hemiarid lake basins: geomorphic patterns. In A.D. Abrahams and A.J. Parsons (eds), *Geomorphology of desert environments*. Chapman & Hall, London: 422–444.

Dohrenwend, J.C. 1994. Pediments in arid environments. In A.D. Abrahams and A.J. Parsons (eds) *Geomorphology of desert environments*. Chapman & Hall, London: 321–353.

Dorn, R.I. 1994. The role of climatic change in alluvial fan development. In A.D. Abrahams and A.J. Parsons (eds) *Geomorphology of desert environments*. Chapman & Hall, London: 593–615.

Fairbridge, R.W. 1981. The concept of neotectonics: an introduction. *Zeitschrift für Geomorphologie*, Supplementband, 40: VII–XII.

Frostick, L.E. and Reid, I. 1987. Tectonic control of desert sediments in rift basins: ancient and modern. In L.E. Frostick and I. Reid (eds), *Desert sediments: ancient and modern*. Geological Society of London, Special Publication No. 35. Blackwell Scientific Publications, Oxford: 53–68.

Frostick, L.E. and Reid, I. 1989. Climatic versus tectonic controls of fan sequences: lessons from the Dead Sea, Israel. *Journal of the Geological Society, London*, 146: 527–538.

Frostick, L.E. and Steel, R.J. 1993a. Tectonics signatures in sedimentary basin fills: an overview. In L.E. Frostick and R.J. Steel (eds), *Tectonic controls and signatures in sedimentary successions*. Special Publication No. 20 of the International Association of Sedimentologists. Blackwell Scientific Publications, Oxford: 1–9.

Frostick, L.E. and Steel, R.J. 1993b. Sedimentation in divergent plate-margin basins. In L.E. Frostick and R.J. Steel (eds), *Tectonic controls and signatures in sedimentary successions*. Special Publication No. 20 of the International Association of Sedimentologists. Blackwell Scientific Publications, Oxford: 111–128.

Jannick, N.O., Phillips, F.M., Smith, G.I. and Elmore, D. 1991. A ^{36}Cl chronology of lacustrine sedimentation in the Pleistocene Owens River system. *Bulletin of the Geological Society of America*, 103: 1146–1159.

Jorgensen, D.W., Harvey, M.D., Schumm, S.A. and Flamm, L. 1993. Morphology and dynamics of the Indus River: implications for the Mohen Jo Daro site. In J.F. Shroder (ed.), *Himalaya to the sea: geology, geomorphology and the Quaternary*. Routledge, London: 288–326.

King, G.C.P. and Vita-Finzi, C. 1981. Active folding in the Algerian earthquake of 10 October 1980. *Nature*, 292: 22–26.

Lancaster, N. 1992. Relations between dune generations in the Gran Desierto of Mexico. *Sedimentology*, 39: 631–644.

Laronne, J.B. and Reid, I. 1993. Very high rates of bedload sediment transport by ephemeral desert rivers. *Nature*, 366: 148–150.

Mayer, L. 1985. Tectonic geomorphology of the Basin and Range–Colorado Plateau boundary in Arizona. In M. Morisawa and J.T. Hack (eds), *Tectonic geomorphology*. Binghampton Symposia in Geomorphology, International Series, No. 15. Allen & Unwin, London: 235–259.

McCalpin, J.P. 1993. Neotectonics of the northeastern basin and range margin, western USA. *Zeitschrift für Geomorphologie*, Supplementband, 94: 137–167.

McDougall, J.W. 1989. Tectonically-induced diversion of the Indus River west of the Salt Range, Pakistan. *Palaeogeography, Palaeoclimatology, Palaeoecology*, 71: 301–307.

McFadden, L.D., Wells, S.G. and Dohrenwend, J.C. 1986. Influence of Quaternary climatic changes on processes of soil development on desert loess deposits of Cima volcanic field. *Catena*, 13: 361–389.

Morisawa, M. and Hack, J.T. (eds) 1985. *Tectonic geomorphology*. Binghampton Symposia in Geomorphology, International Series, No. 15. Allen & Unwin, London.

Ollier, C.D. 1981. *Tectonics and landforms*. Longman, London.

Ollier, C.D. 1985a. Morphotectonics of continental margins with great escarpments. In M. Morisawa and J.T. Hack (eds), *Tectonic geomorphology*. Binghampton Symposia in Geomorphology, International Series, No. 15. Allen & Unwin, London: 3–25.

Ollier, C.D. 1985b. The great escarpment of Southern Africa. *Zeitschrift für Geomorphologie*, Supplementband, 54: 37–56.

Rendell, H.M. 1995. Luminescence dating of sand ramps in the Eastern Mojave Desert. *Geomorphology* (in press).

Rockwell, T.K., Keller, E.A. and Johnson, D.L. 1985. Tectonic geomorphology of alluvial fans and mountain fronts near Ventura, California. In M. Morisawa and J.T. Hack (eds), *Tectonic geomorphology*. Binghampton Symposia in Geomorphology, International Series, No. 15. Allen & Unwin, London: 183–207.

Schumm, S.A. 1963. *The disparity between present rates of denudation and orogeny*. US Geological Survey, Professional Paper, 454H.

Summerfield, M.A. 1985. Plate tectonics and landscape development on the African continent. In M. Morisawa and J.T. Hack (eds), *Tectonic geomorphology*. Binghampton Symposia in Geomorphology, International Series, No. 15. Allen & Unwin, London: 27–51.

Taylor, K.C., Lamorey, G.W., Doyle, G.A., Alley, R.B., Grootes, P.M., Mayewski, P.A., White, J.W.C. and Barlow, L.K. 1993. The 'flickering switch' of late Pleistocene climate change. *Nature*, 361: 432–436.

Tchakerian, V.P. 1994. Palaeoclimatic interpretations from desert dunes and sediments. In A.D. Abrahams and A.J. Parsons (eds), *Geomorphology of desert environments*. Chapman & Hall, London: 631–643.

Thornes, J.B. and Brunsden, D. 1977. *Geomorphology and time*. Methuen, London.

Vita-Finzi, C. 1986. *Recent earth movements*. Academic Press, London.

Wells, S.G. and Gutierrez, A.A. 1982. Quaternary evolution of badlands in the southeastern Colorado Plateau, USA. In R.B. Bryan and A. Yair (eds), *Badland geomorphology and piping*. GeoBooks, Norwich: 239–258.

Williams, M.A.J. 1994. Cenozoic climatic changes in deserts: a synthesis. In A.D. Abrahams and A.J. Parsons (eds), *Geomorphology of desert environments*. Chapman & Hall, London. 644–670.

Yair, A., Goldberg, P. and Brimer, B. 1982. Long-term denudation rates in the Zin-Havarim badlands, Northern Negev. In R.B. Bryan and A. Yair (eds), *Badland geomorphology and piping*. GeoBooks, Norwich: 279–291.

Zimbelman, J.R., Williams, S.H. and Tchakerian, V.P. 1994. Sand transport paths in the Mojave Desert: characterization and dating. In *Abstracts of a workshop on response of eolian processes to global change*. Desert Research Institute, Quaternary Sciences Center, Reno, Nevada. Occasional Paper No. 2, 127–131.

Section 2
Surface Processes and Characteristics

3
Weathering processes

Andrew S. Goudie

Introduction

In desert landscapes weathering appears to be pervasive, yet it is generally selective and frequently superficial. The selectivity (Cooke *et al.* 1993) results from extreme local variations of temperature and humidity and, more importantly, from the fact that as the proportion of exposed rock is higher than in other climatic zones, structural and lithological properties of rocks impose strong differentiation on the effectiveness of superficial weathering processes.

This selectivity has also been noted by Mabbutt (1977, p. 20):

The sharp outlines and rocky slopes of desert uplands are shared to some degree by mountains fashioned under other climates, for such forms, with their clear expression of underlying geology, occur wherever the removal of waste keeps pace with preparatory rock weathering. What distinguishes desert uplands is the wider range of rock types over which these conditions are fulfilled and the extreme degree of exploitation of lithologic contrasts. Rectilinear facets, angular breaks of profile and a relatively abrupt change of slope at the hill foot reflect this structural control, which may also be expressed in detail in bizarre weathering forms.

These bizarre weathering forms (Figure 3.1) often intrigued early climatic geomorphologists, who employed a range of local or national terms to categorise such features and associated processes. The German influence on terminology is still apparent, reflecting the early (and good) contributions of men like Walther (1893), who laboured in the short-lived German African colonies: *Kernsprung* (split rock); *Schattenverwitterung* (shadow weathering); *Salzprengung* (salt splitting), etc.

Desert rock outcrops and superficial debris are indeed commonly affected by flaking, spalling, splitting and granular disintegration, and bizarre forms with ethnic names abound (tafoni, gnammas, etc.). However, the role of weathering in moulding arid landscapes is greater than this implies, for it is weathering which prepares the bedrock for the subsequent operation of erosional processes (e.g. dust deflation: Chapter 18) and so provides the wherewithal for landform evolution. In this chapter we shall consider three main variables that may affect the rate of operation of weathering processes: insolation, moisture and salt, while the importance of superficial alteration processes is analysed in Chapter 6 in the context of crusts and varnishes.

Arid Zone Geomorphology: Process, Form and Change in Drylands, 2nd edition. Edited by David S. G. Thomas.
© 1997 John Wiley & Sons Ltd.

Figure 3.1 *The so-called Finger of God, Mukorob, in Namaland, Namibia, illustrates the selectivity of desert weathering processes operating on sedimentary rocks of varying character. It collapsed in 1988, shortly after this photo was taken*

Insolation

The most frequently proposed mechanism for rock breakdown in the early days of research in desert regions was that rocks expand and contract as diurnal temperatures rise and fall and that the contraction may be sufficient to set up stresses which exceed the tensile strength of the rock. Common terms for this process are 'insolation weathering' and 'thermoclasty'. Various arguments were brought forward in support of this theory.

First, rock surfaces in deserts, as already noted, are characterised by an unusual profusion of sharp, angular disintegration forms, including split pebbles and boulders (Bosworth 1922).

Second, there have been reports of rocks 'exploding' like pistol shots as they cool very rapidly after sundown (Walther 1893; George 1978).

The missionary, David Livingstone (1857, pp. 149–150), noted this while at the Bamangwato Hills in Bechuanaland:

When we were sitting in the evening, after a hot day, it was quite common to hear these masses of basalt split and fall among each other with the peculiar ringing sound which makes people believe that this rock contains much iron. Several large masses, in splitting thus by the cold acting suddenly on parts expanding by the heat of the day, have slipped down the sides of the hills.

Third, under conditions of clear skies, desert rock surfaces are subjected to extreme diurnal temperature ranges.

Figure 3.2 demonstrates selected ground surface temperature changes during a daily cycle while Table 3.1 lists maxima of ground surface temperatures that have been recorded and compares them with air temperatures at the same location. Ground surface temperatures in excess of 70°C are not uncommon, and diurnal ranges can exceed 50°C.

Fourth, there is a difference in temperature response between the outer and inner portions of a rock mass, with surface layers experiencing more severe temperature regimes than interior layers. This might promote exfoliation and spheroidal weathering.

Fifth, different mineral constituents of a rock will display unequal coefficients of expansion, different mineral types may expand unequally along different crystallographic 'axes', and minerals of different colour will warm and cool to differing degrees (Hockmann and Kessler 1950).

The classic portrayal of the insolation process is given, for example, by Ball (personal communication, in Hume 1925, p. 11):

When an exposed rock-surface becomes heated by the sun, its uppermost layers tend to expand, and if lateral expansion is forcibly prevented by the surrounding material, a horizontal compressive stress will be developed within the heated layers.

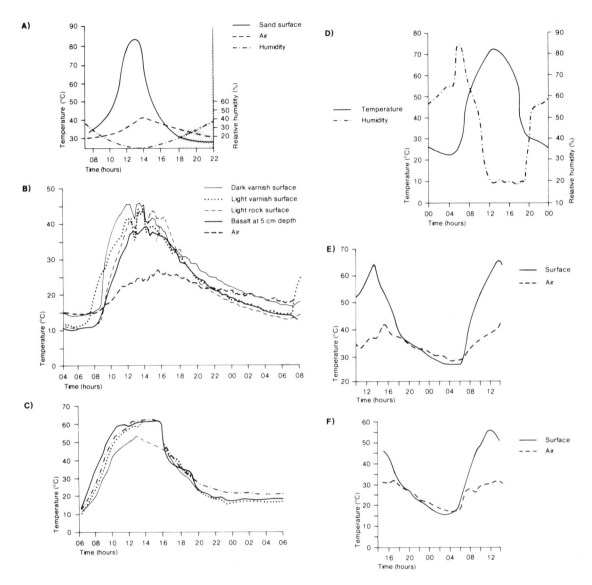

Figure 3.2 *Temperature and humidity cycles for desert stations (A) Wadi Halfa, Sudan, late summer (September 1962) (from Cloudsley-Thompson and Chadwick 1964). Air and sand surface temperatures on a dune. (B) Nagar, Karakoram Mountains, Pakistan, summer (11–12 August 1980) (from Whalley et al. 1984). Air and surface temperatures on different rock surface types. (C) Southern Namib, summer (19–20 December 1977) (from Robinson and Seely 1980). Surface temperatures on different parts of a sand dune. (D) Wadi Digla, Egypt, summer (9 August 1922). Ground surface temperature and relative humidity (from Williams 1923). (E) Tibesti, Sahara, summer (15–16 August 1961). Air and surface temperatures on a sand dune (from Peel 1974). (F) Pre-Saharan steppe, Tunisia, summer (14–15 August 1973). Air and ground surface temperatures (after Heatwole and Muir 1979).*

Conversely, when the surface is cooled by radiation at night, and the rock is prevented from contracting laterally, there will result a state of horizontal tension within the cooled layers. The diurnal changes of temperature must therefore give rise to alternations of horizontal compression and tension in the superficial layers of exposed rocks. If the stresses thus developed transcend the resisting power of the rocks, fracture must follow, and since most rocks are less resistant to tension than to compression, it is the

Table 3.1 *Maximum ground and air surface temperatures in deserts under clear skies*

Location	References	Month	Surface type	Air	Ground
Algeria	George (1986)	June	Rock	39	56
			Sand	41	64
Tibesti	Jäkel and Dronia (1976)	March	Rock (sandstone)	30	57
			Sand	28	59
Tibesti	Peel (1974)	August	Varnished sandstone	47	79.3
			Pale sandstone	47	78.8
			Basalt	47	78.5
Sudan	Cloudsley-Thompson and Chadwick (1964)	September	Sand	41	82
			Rock	40	72.5
Karakorams	Whalley *et al.* (1984)	July/August	Basalt	17.4	41
			Basalt	14.3	33.5
			Varnish	26.5	46.3
			Sandstone	41.0	54.0
Tunisia	Heatwole and Muir (1979)	June	Soil	35	52
		July	Soil	35	55
		August	Soil	36	54
Tucson (Arizona)	Sinclair (1922)	June/September	Adobe (silty mud)	49.7	71.5
Namib	Desert Ecological Research Unit (pers. comm.)	Mean	Soil	28.5	53.6
Egypt	Happold (1984)	August	Sand	31.1	51.1
Oman	Royal Geographical Society (1986)	February	Sand	26	42
Namib	Holm and Edney (1973)	January	Sand	33	59
California	Chappell and Bartholomew (1981)	June	Soil	52.5	72.5

tension resulting from cooling, rather than the compression due to heating which must be regarded as the most important geological consequence of diurnal change of temperature.

At an even earlier date, the German geomorphologist, Walther (1893, p. 169) had championed the combined role of insolation and wind in moulding desert landscapes:

In the desert it rains but seldom. The time in which water may there destroy rock and transport debris is at most 65 days in the year. It has been thought that during the remaining 300 days denundation in the desert is at a standstill; yet careful, unbiased study teaches that in these 300 dry days denudation is intense. A burning sun beats down on the rock surface, unprotected by any plant cover. In Texas daily variations in temperature of 40°C are not at all rare; and large and small stones are cracked by the heat...In Texas I saw granite blocks as high as houses divided by wide cracks, and Mr. Von Streeruwitz told me that he had seen and heard the cracking of such blocks.

The views of Ball and Walther were shared by various French (e.g. Velain 1910) and American scientists (e.g. Hobbs 1917).

More recently Smith (1994), in a useful review, has pointed to the fact that within rocks exposed to insolation there is a wide range of potential weaknesses that can be exploited during thermally induced surface expansion and contraction. He pointed out that small samples of recently quarried rock might not display all of these weaknesses in the way that larger, longer exposed *in situ* samples might. Microfractures and lines of potential weathering can be caused by stress relief within rocks (e.g. by dilation following pressure release).

CRITICISMS OF THE INSOLATION MECHANISM

In contrast to the views discussed above, doubts have often been expressed about the existence, speed and efficiency of this process. Thus various arguments have been brought forward to refute this theory (Schattner 1961).

First, moisture, in the form of rain, fog, dew, atmospheric humidity and groundwater seepage, is more prevalent than often assumed and therefore may play a potent role in weathering. Second, some laboratory experiments have failed to produce appreciable rock disintegration when heating and cooling cycles were applied in the absence of moisture, but they did produce breakdown when water was present (e.g. Blackwelder 1925; Griggs 1936). Third, supposedly characteristic weathering forms, such as exfoliation and spheroidal weathering, have been found under thick debris covers (e.g. of regolith) where temperature fluctuations are likely to be minimal, and under tropical wet conditions. Fourth, rock disintegration can be achieved by a multitude of processes which do not involve chemical alteration of rock minerals, including unloading (pressure release or, to use the terminology of Farmin (1937), 'hypogene exfoliation'), salt crystallisation and hydration, or moisture expansion (Blackwelder 1925). Fifth, the study of locations where weathering appears especially severe often reveals some local microclimatic association promoting moisture availability (Barton 1916). Sixth, exfoliation, a possible product of insolation, is much rarer in deserts than is often proposed (Blackwelder 1925).

Recent work tends to be rather equivocal or inconsistent about the power of insolation and the situation appears to be inconclusive. Ollier (1963), for example, made a strong case for boulder cleavage in central Australian conglomerate outcrops being the result of insolation, while Roth (1965), working in the Mojave Desert, thought otherwise. Roth found that water determinations on quartz monzonite boulders which he had blown up with explosives indicated an appreciable water content even in their interiors, while the monitoring of temperature profiles through the boulders suggested to him that insufficient stress would be created to cause rock failure (p. 467):

...the temperature gradients of the rock are fairly uniform, most of the rock heats or cools at approximately the same time. There is little lag in the temperature response of the rocks which means that the entire rock is probably able to expand and contract with little differential stress between various portions of it.

The role of laboratory simulations in attempting to solve the issue has also been far from clear. Thus Rice (1976) has criticised some early simulations for using a limited range of rock types, excessively small samples, and unnatural temperature cycles. There have also been some studies (e.g. Richter and Simmons 1974; Aires-Barros 1975) which have used photomicrographic techniques to identify that some microcracking can indeed occur, but the degree of cracking appears to be modest and other processes appear to be more immediately effective (e.g. salt, Goudie 1974; and frost weathering, e.g. Journaux and Coutard 1974). There is plainly a need for a more intensive use of non-destructive testing techniques to see whether insolation can cause significant changes in rock properties even before spalling or disintegration occurs.

Related to insolation is the role of fire. While this plainly requires some level of vegetation cover to be effective, if resinous shrubs are present and are subjected to natural or anthropogenic fires, then it may be an important process even in arid and semi-arid areas. Areas of *Spinifex* in Australia, when burnt, produce various forms of boulder weathering (see, for example, Dragovich 1993).

The role of moisture

Although deserts are by definition areas of moisture deficit (see Chapter 1), moisture is widely available in a variety of forms and so may play a role in rock decomposition.

First, and most importantly of all, desert rainfall occurs on a greater number of occasions and in a less intensive way than is often supposed. Table 3.2 presents data on the number of rainy days that occur in some major areas, together with the average amount of precipitation that occurs on each rainy day. Average daily falls are directly comparable to those in more humid areas, particularly in the mid-latitude deserts (e.g. former USSR, China and North Africa). Even in quite dry locations there may be an appreciable number of days in which

Table 3.2   *Rainfall per rainy day in arid areas*

Area	Number of stations	Range of mean annual rainfall (mm)	Range of number of rainy days receiving > 0.1 (mm) rainfall	Average rainfall per rainy day mm
USSR Karakum	12	92–273	42–125	2.56
CHINA Gobi	6	84–396	33–78	4.51
ARGENTINA Patagonia	11	51–542	6–155	5.41
N AFRICA N. Sahara	18	1–286	1–57	3.82
W AFRICA N. Sahara	20	17–689	2–67	9.75
KALAHARI	10	147–592	19–68	9.55
COMBINED	77	1–689	1–155	6.19

measurable quantities of rain are received (Table 3.3).

Second, coastal deserts may receive substantial amounts of moisture from fogs. This is, for example, the case in the Namib Desert where, at Gobabeb, the mean number of foggy days (based on 22 years of records) is over 37 per year and in some years the amount of precipitation deposited by fog is greater than that deposited by rain (Table 3.4). Gobabeb is over 70 km inland and even higher values are received at the coast.

A third significant source of moisture to promote chemical weathering is dew deposition. There are, unfortunately, few available data on

Table 3.3   *Number of days per annum experiencing measurable rainfall*

	Mean annual rainfall (mm)	No. of rainy days
Gafsa, Libya	160	29
Khartoum, Sudan	164	18
Keetmanshoop, Namibia	147	19
Turgaj, former USSR	177	79
Fergana, former USSR	169	52
La Serena, Chile	133	63
Cipoletti, Argentina	176	39

Table 3.4   *Sources of moisture at Gobabeb, Namibia (mm)*

Year	Rainfall	Fog
1963	12.8	na
1964	22.9	na
1965	18.4	na
1966	14.2	35.0
1967	26.6	30.4
1968	5.3	45.7
1969	29.8	na
1970	3.5	na
1971	20.6	28.4
1972	25.1	44.7
1973	7.9	29.0
1974	23.2	8.0
1975	7.0	26.7
1976	127.4	9.6
1977	14.3	25.8
1978	109.6	23.5
1979	31.0	33.5
1980	5.5	44.5
1981	4.6	48.6
1982	15.8	36.0
1983	11.6	41.8
1984	1.7	27.3
Mean	24.5	31.7

Source: Desert Ecological Research Unit, Gobabeb.

this phenomenon, even though desert travellers are often subjected to its unwelcome effects. There are, however, some good data from Israel where in the Avdat area of the Negev, dews occur in great profusion (Table 3.5), occurring on average on 195 nights in the year at the ground surface (Evenari *et al.* 1982). In that location the annual average production of dew water exceeds 30 mm, though the amount formed per dew night is small, amounting to no more than 0.35 mm even on the heaviest dewy night. Dew fall may create a host of microforms (Smith 1988).

Moisture also occurs in specific topographic situations as in proximity to rivers, springs, playas, sabkhas and cliff-base seeps. Another important feature of moisture in arid areas is that it often possesses an alkaline pH. This is especially true in proximity to salt lakes (Table 3.6). The significance of this is that at high pH levels, especially above 9, silica mobility in rocks may be greatly increased (Figure 3.3), due to the ionisation of H_4SiO_4. The dissolution of silica in alkali solutions probably involves the depolymerisation of silica through hydration and dissolution of $Si(OH)_4$ followed by addition of OH^- to form silicate ions. Silica mobility may also be increased in the presence of sodium chloride, and Young (1987) has argued that various cavernous weathering forms (tafoni) normally attributed to salt crystal growth or hydration could result from quartz grains being loosened by solution of silica.

It is now well proven that the application and removal of moisture can cause significant changes in rock properties. Winkler (1977) calculates that entrapped water in a rock (e.g. a

Table 3.6 *Reported pH values for selected lake waters in arid zones*

Lake	pH
Pyramid Lake, Nevada	9.4
Van Gulu, Turkey	9.9
Turkana, Kenya	9.5–9.7
Wadi Natrun, Egypt	10.9–11.0
Great Salt Lake, Utah	7.4–7.7
Deep Springs Lake, Oregon	7.4–9.6
Harney Lake, Oregon	9.8
Abert Lake, Oregon	9.7

Source: Goudie (1985, table 4).

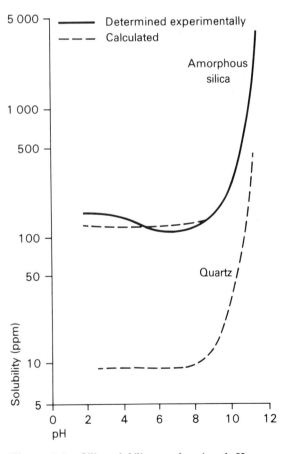

Figure 3.3 *Silica solubility as a function of pH*

Table 3.5 *Dewfall at Avdat (Negev Desert, as measured with a Duvdevani dew gauge*

Year	Height above ground (cm)	Number of dew nights	Dewfall (mm)
1963	0	176	32
	100	162	29
1964	0	216	37
	100	198	31
1965	0	198	35
	100	189	33
1966	0	190	26
	100	166	26
Mean	0	195	33
	100	172	28

Source: Evanari *et al.* (1982, table 5, p. 35).

granite) if heated from 10 to 50°C can develop about 250 atmospheres as tensile strength, a force strong enough to disrupt the soundest rock with time. Likewise, Hudec and Sitar (1975) indicate that moisture content of rock has an effect on its thermal expansivity and that a rock may expand or contract isothermally with changes in moisture content. They suggest that the saturated thermal contraction coefficient of sorption-sensitive rocks is approximately twice the dry coefficient, and in the case of carbonate rocks they found a large isothermal expansion upon wetting, which, on average, was equivalent to approximately 78°C dry thermal expansion. An earlier experimental confirmation of this conclusion is provided for granites by Hockmann and Kessler (1950), while the distintegration and slaking of shale by cyclic wetting and drying is well-attested in the field and the laboratory (Muggridge and Young 1983).

Changes in atmospheric humidity may be sufficient for such effects to occur, as made evident in the laboratory experiments undertaken on Caen Stone by Pissart and Lautridou (1984). Likewise Ravina and Zaslavsky (1974) theorise that when water condenses due to the increase of relative humidity at night, interaction of electrical double layers at the rock–water interfaces in the edges of cracks occurs. There, very high pressures are produced in the water because of very high electrical gradients. Two examples of substantial diurnal changes in humidity are shown in Figure 3.2a and 3.2d.

The role of processes such as solution in arid areas, though evident through the existence of features which include rillen on limestone and marble outcrops (e.g. Sweeting and Lancaster 1982) and shallow dayas, is by necessity limited under present climatic conditions. Under conditions of areic drainage and modest surface runoff, the amount of removal of solutes outside individual drainage systems is restricted, though local translocation (as made evident by the widespread occurrence of calcretes—see Chapter 6) is more important. On a global basis (Smith and Atkinson 1976) there is a very close relationship between limestone denudation and mean annual runoff. Thus the amount of chemical denudation in arid zone stations is low (generally <4 mm 10^{-3} yr^{-1}).

However, given the existence of past climatic changes, whether the wetter periods of the Pleistocene and the Holocene, or long-continued pre-Quaternary humid phases, there may be weathering features which are relicts of past moist conditions. Fossil soils and weathering crusts are of widespread occurrence, and analysis of cave sediments (e.g. Cooke and Verhagen 1977) has revealed alternations of lime solution and precipitation. The German climato-genetic geomorphologist Büdel (1982) was a fervent exponent of the view that many of the most important geomorphological features of the world's arid lands were formed under humid tropical regimes which, he believed, pertained extensively in pre-Pleistocene times.

Salt weathering

The ability of salt to cause rock breakdown has been known for many years (Goudie 1994) but it was not until the 1950s and 1960s that there was any substantial corpus of work focused on this mechanism. The work of various French geomorphologists greatly extended available knowledge (e.g. Tricart 1960), but the greatest stimulus to research was provided by two notable papers that explored the theoretical basis to salt weathering (Wellman and Wilson 1965; Evans 1970).

SALT CRYSTAL GROWTH

The most cited cause of mechanical weathering by salt is that of salt crystal growth from solutions in rock pores and cracks, for substantial pressures may be set up (Winkler and Singer 1972). Crystal growth occurs for a variety of reasons. Some salts (e.g. sodium sulphate, sodium carbonate, sodium nitrate and magnesium sulphate) rapidly decrease in solubility as temperatures fall. Thus nocturnal cooling could encourage salt crystallisation. Another cause of crystallisation is evaporational concentration of solutions, and when this occurs highly soluble salts will produce large volumes of crystals. The third cause of crystallisation is the 'common ion effect', whereby the mixing of two different salt solutions with the same major cation can cause salt precipitation.

The power of different crystallising salts will vary according to their solubility, the temperature dependence of that solubility, crystal habit (e.g. needle-shaped crystals like those of mirabilite may have a high disruptive capability), and their equilibrium relative humidities. Those with low values (e.g. calcium chloride) will be prone to dissolution rather than crystallisation in humid air.

A related process to salt crystallisation is *mass segregation* of salt masses. In some sediments and bedrock outcrops, such may be the accumulation of crystalline material that ground-heaving occurs, producing mounds and pseudo-anticlines, respectively (e.g. Watts 1977; Watson 1985), as in the calcretes of the Kalahari. This can create severe problems for road surfaces, runways, etc., as in the Algerian Sahara (Horta 1985).

HYDRATION

The second main category of salt weathering processes is that associated with hydration, for certain common salts hydrate and dehydrate relatively easily in response to changes in temperature and humidity (Mortensen 1933). As phase change occurs to the hydrated form, water is absorbed, thereby increasing the volumes of the salt which thus develops pressure against pore walls. The volume increases for sodium sulphate and sodium carbonate may be in excess of 300%. For some salts phase changes may occur at the sorts of temperatures encountered widely in nature. As Figure 3.4 demonstrates, for instance, in most months of the year in Khartoum, daily temperature regimes for the soil surface cross the 32.4°C threshold for sodium sulphate. Different salts generate different hydration pressures with the largest being achieved when anhydrite changes into gypsum (Winkler and Wilhelm 1970). Experimental hydration pressures generated by sodium sulphate have been found to be higher than the tensile strengths of many natural stones and ceramic materials (Knacke and von Erdberg 1975).

It is far from easy to separate out the relative power and importance of crystallisation and of hydration, though experimental work (Sperling and Cooke 1985) has succeeded in establishing that hydration alone can be effective.

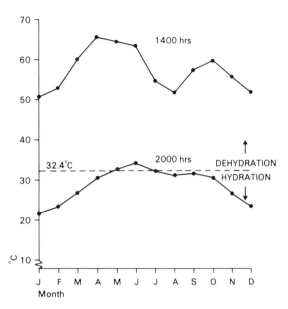

Figure 3.4 *Mean soil temperature means (at depth of 1 cm) for 1400 and 2000 hours at Khartoum (1958–62), after data in Oliver (1966, table 1, p. 49). The dashed line shows the transition temperature for the phase change from thenardite ($Na_2SO_4 \cdot 10H_2O$) to mirabilite (Na_2SO_4)*

THERMAL EXPANSION OF SALT

The third possible category of salt weathering process involves the differential thermal expansion of salts (Cooke and Smalley 1968), for salts entrapped in rock pores may have greater expansion coefficients than the rock minerals. For example, when temperatures increase from around 0°C to 60°C, halite (NaCl) expands by 0.5% whereas granite minerals expand by between 0.1 and 0.2%. Early experimental simulations (Goudie 1974) suggest that this process is less effective than salt crystallisation or hydration.

SALT SOURCES IN ARID ENVIRONMENTS

There are many different sources of salts in arid areas and because of low rates of leaching out of soil profiles and catchments under low rainfall regimes, such salts tend to accumulate in a variety of topographic situations. One obvious source of saline material is seawater, and lakes or land surfaces that have become isolated from, or uncovered by, the sea may retain some of the latter's chemical characteristics. This is 'relict sea water salt' (Johnson 1980). 'Cyclic salts' are those derived from inputs from the atmosphere, either as rainfall, aeolian dust, volcanic vapours, or submarine gas emissions. Rock weathering, especially where bedrock contains old evaporitic beds, may transfer 'connate salts' to rivers and playas. Salt may be redistributed from marine sabkhas and inland playa surfaces by wind action and may accumulate on hillslopes (Coque 1961) and stone pavements (Yaalon 1963).

There are many different processes (Table 3.7) which interact to create particular salt types in particular places, and the range of salts that appear to cause rock weathering is large. Some quite complex minerals may be involved (Goudie and Cooke 1984). In the USA, for example, playa and efflorescence minerals include natron, mirabilite, thenardite, epsomite, hexahydrite, gypsum and bloedite (Mustoe 1983; Last 1984; Keller *et al.* 1986; Timpson *et al.* 1986).

Although Matsukura and Yatsu (1985) indicate that the presence of salt may reduce the speed at which slaking of shales and tuffs occurs, there is, none the less, a large body of evidence that salt weathering processes are potent causes of rock disintegration.

Part of this evidence comes from laboratory simulations (e.g. Goudie *et al.* 1970; Kwaad 1970; Goudie 1974; 1985; Robinson and Williams 1982; Smith and McGreevy 1983; Fahey 1985). Other important indications of their power come from field observations

(Figure 3.5) and field monitoring (e.g. Goudie and Day 1981; Goudie and Watson 1984; Goudie 1985; Smith and McAlister 1986), and from studies of the decay of ancient structures (e.g. Hayden 1945; Goudie 1977; Bromblet and Bocquier 1985) and of modern engineering structures (Doornkamp *et al.* 1980; Nettleton *et al.* 1982; Bulley 1986).

Different desert climatic regimes will have differing levels of susceptibility to salt weathering. Temperature and humidity fluctuations need to be of such a nature as to enable cycles of hydration and crystallisation to occur. Experimental work (Goudie 1993) has suggested that high nocturnal humidity levels at times of low temperature, and low daytime humidity levels at times when temperatures exceed around 32°C, are ideal conditions for promoting salt attack.

Given the speed with which salt weathering occurs, it is evident that it may have a large range of geomorphological consequences in arid lands (Table 3.8). It is also apparent that salt weathering provides an arid zone process which is active even in hyper-arid situations. It prepares rock for subsequent erosional attack by wind and water and generates silt-sized debris (Goudie 1986). Arid zones are not, therefore, *contra* Walton (1969), zones of geomorphological inactivity in which most forms are the results of past pluvial events and influences. Indeed, salt weathering is also a major environmental hazard which is ignored to the peril of successful implementation of engineering schemes (Sayward 1984).

Finally, the weathering processes discussed in the chapter are not the only ones that occur in

Table 3.7   *Miscellaneous influences on salt chemistry*

Inputs	Changes within systems
Seawater	Cationic exchanges with clays
Atmosphere inputs:	Organic influences
dust, rainfall, volcanic emissions, submarine gas emissions	Temperature changes
Solutions derived from weathering of bedrock and ancient evaporites	Chemical reactions
	Tectonic changes
	Vertical leaching processes
	Deflational removal
	Evaporative concentration
	Capillary migration

Figure 3.5 *A field illustration of the power of salt weathering occurs in Death Valley, California, where fan boulders, which enter the saline zone on the edge of the playa, rapidly disintegrate*

Table 3.8 *Examples of the geomorphological effects of salt weathering*

Enhancement of deflation potential	Tricart (1953)
Pan formation and development of 'billard-table' rock flowers	Du Toit (1906); Goudie and Thomas (1985)
Tor formation	Watts (1981)
Rock meal formation	Wellman and Wilson (1965)
Loess formation and silt generation	Goudie *et al.* (1979)
Tafoni development	Bradley *et al.* (1978)
Cobble disintegration	Beaumont (1968)
Cliff sapping	Jutson (1918); Laity (1984)
Ground heaving and pseudo-anticline formation	Watts (1977); Horta (1985)
Pedestal rocks	Cooke (1981)

deserts. In cold deserts (Figure 3.6), especially those with winter rainfall regimes, frost activity may cause rock splitting and granular disintegration. Likewise, desert surfaces are not sterile, and micro-organisms and lower plants (e.g. lichens and micro-colonial fungi) may cause biological weathering to occur, mobilising elements and contributing to the formation of some desert varnishes (Dorn and Oberlander 1982). In some cases they may contribute to the destruction of rock coatings through the secretion of organic acids or by taking up elements from the substrate (Dragovich 1993b). Lichens may also play a physical role in rock disintegration. They contract and swell, creating stresses in and on rock surfaces.

Figure 3.6 *In high-altitude and higher latitude deserts, especially where there are winter rainfall regimes, much slope debris may be produced by frost attack. This is the probable mechanism for much of the scree that mantles the slopes of the Hunza Valley, near Pasu, in the Karakoram Mountains of Pakistan*

References

Aires-Barros, L. 1975. Dry and wet laboratory tests and thermal fatigue of rocks. *Engineering Geology*, 9: 249–265.

Barton, D.C. 1916. Notes on the disintegration of granite in Egypt. *Journal of Geology*, 24: 382–393.

Beaumont, P. 1968. Salt weathering on the margin of the Great Kavir, Iran. *Bulletin of the Geological Society of America*, 79: 1683–1684.

Blackwelder, E. 1925. Exfoliation as a phase of rock weathering. *Journal of Geology*, 33: 793–806.

Bosworth, T.O. 1922. *Geology of the Tertiary and Quaternary periods in the north-west part of Peru.* Macmillan, London: 434 pp.

Bradley, W.C., Hutton, J.T. and Twidale, C.A. 1978. Role of salts in development of granite tafoni, South Australia. *Journal of Geology*, 86: 647–654.

Bromblet, P. and Bocquier, G. 1985. Données petrologiques concernant l'alteration des grés des temples de karnak (Egypte). In *Proceedings Vth International Congress on Deterioration and Conservation of Stone*, Lousanne, 361–370.

Büdel, J. 1982. *Climatic geomorphology*. Princeton University Press, Princeton: 443 pp.

Bulley, B.G. 1986. The engineering geology of Swakopmund. *Communications, Geological Survey of South West Africa/Namibia*, 2: 7–12.

Chappell, M.A. and Bartholemew, G.A. 1981. Standard operative temperatures and thermal energetics of the antelope ground squirrel *Ammospermophilous leucurus*. *Physiological Zoology*, 54: 81–93.

Cloudsley-Thompson, J.L. and Chadwick, M.J. 1964. *Life in deserts*. Foulis, London: 218 pp.

Cooke, H.J. and Verhagen, B.Th. 1977. The dating of cave development — an example from Botswana. In *Proceedings of the 7th International Speleological Congress*, Sheffield: 122–124.

Cooke, R.U. 1981. Salt weathering in deserts. *Proceedings of the Geologists Association*, 92: 1–16.

Cooke, R.U. and Smalley, I.J. 1968. Salt weathering in deserts. *Nature*, 220: 1226–1227.

Cooke, R.U. and Warren, A. 1973. *Geomorphology in deserts*. Batsford, London: 374 pp.

Cooke, R.U., Warren, A. and Goudie, A.S. 1993. *Desert geomorphology*. UCL Press, London.

Coque, R. 1961. *La Tunisie présaharienne, étude géomorphologique*. Colin, Paris.

Doornkamp, J.C., Brunsden, D. and Jones, D.K.C. (eds) 1980. *Geology, geomorphology and pedology of Bahrain*. GeoBooks, Norwich: 443 pp.

Dorn, R.I. and Oberlander, T.M. 1982. Rock varnish. *Progress in Physical Geography*, 6: 317–367.

Dragovich, D. 1993a. Fire-accelerated boulder weathering in the Pilbara, Western Australia. *Zeitschrift für Geomorphologie*, NF, 37: 295–307.

Dragovich, D. 1993b. Distribution of chemical composition of microcolonial fungi and rock coatings from arid Australia. *Physical Geography*, 14: 323–341.

Du Toit, A.L. 1906. Geological survey of portions of the divisions of Vryburg and Mafeking. In *10th Annual Report Geological Commission of the Cape of Good Hope*: 205–258.

Evans, I.S. 1970. Salt crystallisation and rock weathering: a review. *Revue de Géomorphologie Dynamique*, 19: 153–177.

Evenari, M., Shanan, L. and Tadmor, N. 1982. *The Negev: the challenge of a desert*, 2nd edn. Harvard University Press, Cambridge, MA: 437 pp.

Fahey, B.D. 1985. Salt weathering as a mechanism

of rock breakup in cold climates: an experimental approach. *Zeitschrift für Geomorphologie*, 29: 99–111.

Farmin, R. 1937. Hypogene exfoliation in rock masses. *Journal of Geology*, 45: 625–635.

George, U. 1978. *In the deserts of this earth*. Hamish Hamilton, London.

George, W. 1986. The thermal niche: desert sand and desert rock. *Journal of Arid Environments*, 10: 213–224.

Goudie, A.S. 1974. Further experimental investigation of rock weathering by salt and other mechanical processes. *Zeitschrift für Geomorphologie*, NF, 21: 1–12.

Goudie, A.S. 1977. Sodium sulphate weathering and the disintegration of Mohenjo-Daro, Pakistan. *Earth Surface Processes*, 2: 75–86.

Goudie, A.S. 1985. *Salt weathering*. School of Geography, University of Oxford Research Paper, 33: 30 pp.

Goudie, A.S. 1986. Laboratory simulation of the 'wick effect' in salt weathering of rock. *Earth Surface Processes and Landforms*, 11: 275–285.

Goudie, A.S. 1993. Salt weathering simulation using a single-immersion technique. *Earth Surface Processes and Landforms*, 18: 369–376.

Goudie, A.S. 1994. Keynote lecture: salt attack on buildings and other structures in arid lands. In P.G. Fookes and R.H.C. Parry (eds), *Engineering characteristics of arid soils*. Balkema, Rotterdam: 15–28.

Goudie, A.S. and Cooke, R.U. 1984. Salt efflorescences and saline lakes: a distributional analysis. *Geoforum*, 15: 563–582.

Goudie, A.S. and Day, M.J. 1981. Disintegration of fan sediments in Death Valley, California, by salt weathering. *Physical Geography*, 1: 126–137.

Goudie, A.S. and Thomas, D.S.G. 1985. Pans in southern Africa with particular reference to South Africa and Zimbabwe. *Zeitschrift für Geomorphologie*, NF, 29: 1–19.

Goudie, A.S. and Watson, A. 1984. Rock block monitoring of rapid salt weathering in southern Tunisia. *Earth Surface Processes and Landforms*, 9: 95–99.

Goudie, A.S., Cooke, R.U. and Evans, I.S. 1970. Experimental investigation of rock weathering by salts. *Area*, 4: 42–48.

Goudie, A.S., Cooke, R.U. and Doornkamp, J.C. 1979. The formation of silt from dune sand by salt weathering processes in deserts. *Journal of Arid Environments*, 2: 105–112.

Griggs, D.T. 1936. The factor of fatigue in rock exfoliation. *Journal of Geology*, 74: 783–796.

Happold, D.C.D. 1984. Small mammals. In J.L. Cloudsley-Thompson (ed.), *Sahara desert*. Pergamon Press, Oxford: 251–275.

Hayden, J.D. 1945. Salt erosion. *American Antiquity*, 10: 373–378.

Heatwole, H. and Muir, M. 1979. Thermal microclimates in the pre-Saharan steppes of Tunisia. *Journal of Arid Environments*, 2: 119–136.

Hobbs, W.H. 1917. The erosional and degradational processes of deserts, with especial reference to the origin of desert depressions. *Annals of the Association of American Geographers*, 7: 25–60.

Hockmann, A. and Kessler, D.W. 1950. Thermal and moisture expansion studies of some domestic granites. *Journal of Research, US Bureau of Standards*, 44: 395–410.

Holm, E. and Edney, E.B. 1973. Daily activity of Namib desert arthropods in relation to climate. *Ecology*, 54: 45–56.

Horta, J.C. 1985. Salt heaving in the Sahara. *Géotechnique*, 35: 329–337.

Hudec, P.P. and Sitar, N. 1975. Effect of water sorption on carbonate rock expansivity. *Canadian Geotechnical Journal*, 12: 179–186.

Hume, W.F. 1925. *Geology of Egypt*, Vol. I. Cairo.

Jäkel, D. and Dronia, H. 1976. Ergebnisse von Boden und Gesteintemperaturmessungen in der Sahara. *Berliner Geographische Abhandlungen*, 24: 55–64.

Johnson, M. 1980. The origin of Australia's salt lakes. *Records of the New South Wales Geological Survey*, 19: 221–266.

Journaux, A. and Coutard, J.P. 1974. Expériences de thermoclastie sur des roches siliceuses. *Bulletin, Centre de Géomorphologie de Caen*, 18: 1–20.

Jutson, J.T. 1918. The influence of salts in rock weathering in sub-arid western Australia. *Proceedings of the Royal Society of Victoria*, 80: 165–172.

Keller, L.P., McCarthy, G.J. and Richardson, J.L. 1986. Mineralogy and stability of soil evaporites in North Dakota. *Journal of the Soil Science Society of America*, 50: 1069–1071.

Knacke, O. and von Erdberg, R. 1975. The crystallisation pressure of sodium sulphate decahydrate. *Berichte der Bunsen-Gesellschaft für Physikalische Chemie*, 79: 653–657.

Kwaad, F.J.P.M. 1970. Experiments on the granular disintegration of granite by salt action. *University Amsterdam Fys. Geogr. en Boden Kundig Laboratorium Publication*, 16: 67–80.

Laity, J.E. 1984. Diagenetic controls on groundwater sapping and valley formation, Colorado Plateau, revealed by optical and electron microscopy. *Physical Geography*, 4: 103–125.

Last, W.M. 1984. Sedimentology of playa lakes of the northern Great Plains. *Canadian Journal of Earth Science*, 21: 107–125.

Livingstone, D. 1857. *Missionary travels and researches in South Africa*. Murray, London: 687 pp.

Mabbutt, J.A. 1977. *Desert landforms*. MIT Press, Cambridge MA: 340 pp.

Matsukura, Y. and Yatsu, E. 1985. Influence of salt water on flaking rate of Tertiary shale and tuff. *Transactions of the Japanese Geomorphological Union*, 6: 163–167.

Mortensen, H. 1933. Die 'Salzsprengung' und ihre Bedeutung für die regional klimatische Gliederung der Wüsten. *Petermann's Geographische Mitteilungen*, 79: 130–135.

Muggridge, S.J. and Young, H.R. 1983. Disintegration of shale by cyclic wetting and drying and frost action. *Canadian Journal of Earth Sciences*, 20: 568–576.

Mustoe, G.E. 1983. Cavernous weathering in the Capital Reef Desert, Utah. *Earth Surface Processes and Landforms*, 8: 517–526.

Nettleton, W.D., Nelson, R.E., Brasher, B.R. and Derr, P.S. 1982. Gypsiferous soils in the western United States. *Soil Science Society of America Special Publication*, 10: 147–168.

Oliver, J. 1966. Soil temperatures in the arid tropics with reference to Khartoum. *Journal of Tropical Geography*, 23: 47–54.

Ollier, C.D. 1963. Insolation weathering examples from central Australia. *American Journal of Science*, 261: 376–381.

Peel, R.F. 1974. Insolation weathering: some measurements of diurnal temperature changes in exposed rocks in the Tibesti region, Central Sahara. *Zeitschrift für Geomorphologie*, Supplementband, 21: 19–28.

Pissart, A. and Lautridou, J.P. 1984. Variations de longuer de cyclindres de pierre de Caen (calcaire bathonien) sous l'effet de séchage et d'humification. *Zeitschrift für Geomorphologie*, Supplementband, 49: 111–116.

Ravina, I. and Zaslavsky, D. 1974. The electrical double layer as a possible factor in desert weathering. *Zeitschrift für Geomorphologie*, Supplementband, 21: 13–18.

Rice, A. 1976. Insolation warmed over. *Geology*, 4: 61–62.

Richter, D. and Simmons, G. 1974. Thermal expansion behaviour of igneous rocks. *International Journal of Rock Mechanics and Mining Sciences*, 11: 403–411.

Robinson, D.A. and Williams, R.B.G. 1982. Salt weathering of rock specimens of varying shape. *Area*, 14: 293–299.

Robinson, M.D. and Seely, M.K. 1980. Physical and biotic environments of the southern Namib dune system. *Journal of Arid Environments*, 3: 183–203.

Roth, E.S. 1965. Temperature and water content as factors in desert weathering. *Journal of Geology*, 73: 454–468.

Royal Geographical Society 1986. Oman Wahiba Sands Project, Rapid Assessment Document April 1986. Royal Geographical Society, London (unpublished).

Sayward, J.M. 1984. *Salt action on concrete*. US Army Corps of Engineers Cold Region Research and Engineering Laboratory Special Report, 84-25: 69 pp.

Schattner, I. 1961. Weathering phenomena in the crystalline rocks of the Sinai, in the light of current notions. *Bulletin of the Research Council of Israel*, 10G: 247–265.

Sinclair, J.G. 1922. Temperatures of the soil and air in a desert. *Monthly Weather Review*, 50: 142–144.

Smith, B.J. 1988. Weathering of superficial limestone debris in a hot desert environment. *Geomorphology*, 1: 355–367.

Smith, B.J. 1994. Weathering processes and forms. In A.D. Abrahams and A.J. Parsons (eds), *Geomorphology of desert environments*. London: Chapman & Hall: 39–63.

Smith, B.J. and McAlister, J.J. 1986. Observations on the occurrence and origins of salt weathering phenomena near lake Magadi, southern Kenya. *Zeitschift für Geomorphologie*, 30: 445–460.

Smith, B.J. and McGreevy, J.P. 1983. A simulation study of salt weathering in hot deserts. *Geografiska Annaler*, 65A: 127–133.

Smith, D.I. and Atkinson, T.C. 1976. Process, landforms and climate in limestone regions. In E. Derbyshire (ed.), *Geomorphology and Climate*. Wiley, Chichester: 367–409.

Sperling, C.H.B. and Cooke, R.U. 1985. Laboratory simulation of rock weathering by salt crystallisation and hydration processes in hot, arid environments. *Earth Surface Processes and Landforms*, 10: 541–555.

Sweeting, M.M. and Lancaster, N. 1982. Solutional and wind erosion forms on limestone in the Central Namib Desert. *Zeitschrift für Geomorphologie*, 26: 197–207.

Timpson, M.E., Richardson, J.L., Keller, L.P. and McCarthy, G.J. 1986. Evaporite mineralogy associated with saline seeps in southwestern North Dakota. *Journal of the Soil Science Society of America*, 50: 490–493.

Tricart, J. 1953. Influence des sols salées sur la déflation éolienne en basse Mauritanie et dans le Delta du Sénégal. *Revue de Géomorphologie Dynamique*, 4: 124–132.

Tricart, J. 1960. Expérience de désagrégation de roches granitiques par la cristallisation du sel marin. *Zeitschrift für Geomorphologie*, Supplementband, 1: 239–240.

Velain, Ch. 1910. L'érosion éolienne et ses effects dans les régions désertiques. *Revue de Géographie Annuelle*, 4: 359–421.

Walther, J. 1893. The North American deserts. *National Geographic Magazine*, 4: 163–208.

Walton, K. 1969. *The arid zones.* Hutchinson, London.

Watson, A. 1985. Structure, chemistry and origin of gypsum crusts in Southern Tunisia and the central Namib Desert. *Sedimentology*, 32: 855–875.

Watts, N.L. 1977. Pseudo-anticlines and other structures in some calcretes of Botswana and South Africa. *Earth Surface Processes*, 2: 63–74.

Watts, S.H. 1981. Near coastal and incipient weathering features in the Cape Herschel–Alexandra Fjord area, District of Franklin. *Geological Survey of Canada Paper*, 81-1A: 389–394.

Wellman, H.W. and Wilson, A.T. 1965. Salt weathering: a neglected geological erosive agent in coastal and arid environments. *Nature*, 205: 1097–1098.

Whalley, W.B., McGreevy, J.P. and Ferguson, R.I. 1984. Rock temperature observations and chemical weathering in the Hunza region, Karakoram: preliminary data. In K.J. Miller (ed.), *The International Karakoram Project*, Vol. 2. Cambridge University Press, Cambridge: 616–633.

Williams, C.B. 1923. A short bio-climatic study in the Egyptian desert. *Ministry of Agriculture, Egypt, Technical and Scientific Service Bulletin*, 81: 567–572.

Winkler, E.M. 1977. Insolation of rock and stone, a hot item. *Geology*, 5: 188–189.

Winkler, E.M. and Singer, P.C. 1972. Crystallisation pressure of salts in stone and concrete. *Bulletin of the Geological Society of America*, 83: 3509–3514.

Winkler, E.M. and Wilhelm, E.J. 1970. Saltburst by hydration pressures in architectural stone in urban atmosphere. *Bulletin of the Geological Society of America*, 81: 567–572.

Yaalon, D.H. 1963. On the origin and accumulation of salts in groundwater and in soils of Israel. *Bulletin of the Research Council of Israel*, 11G: 105–131.

Young, A. 1987. Salt as an agent in the development of cavernous weathering. *Geology*, 15: 962–966.

4

Controls and uncertainties in the weathering environment

Bernard Smith and Patricia Warke

Introduction

The study of weathering in hot deserts has many parallels to the classic domestic murder mystery. There is a limited range of characters, but from page one it is clear that there is a prime suspect with circumstantial evidence and eyewitness reports overwhelmingly pointing to him as the guilty party. This view is reinforced by the expert testimony of the forensic scientist who, often with the aid of complex laboratory tests, places the accused at the scene of the crime and demonstrates his capability of the act. The rest of the book then goes on to chronicle the doubts of the detective, who mistrusts the obvious and is convinced of the complexity of the crime. In this he is often supported by a sympathetic scientist who disputes the standard forensic tests. Eventually, through a combination of luck and dogged determination, they reveal that other members of the household had motive, means and opportunity. Through their endeavours it is finally discovered that the origins of the murder lie a long time in the past, often triggered by some seemingly unimportant event, and that the real felon had a number of collaborators who covered each others' tracks. Ultimately, original witnesses are shown to be unreliable, alibis are discredited and resulting confessions trap the conspirators.

In the case of weathering in hot deserts the obvious culprit is 'temperature', in the shape of high absolute temperatures and large diurnal temperature ranges supposedly characteristic of these areas. Circumstantial evidence of guilt is provided by selective observations of temperature extremes, the apparent paucity of moisture and the general absence of vegetation and soil. The accusatory finger is then pointed by eyewitness accounts of shattered rocks with no evidence of alteration or other weathering agents being present and tales of stones mysteriously cracking once the Sun goes down.

In this chapter we will demonstrate that the most obvious explanation is not necessarily the correct or complete answer, that in reality weathering in deserts is complex and that weathering environments are variable both geographically and over time. Because of this, decay and breakdown can result from a number of mechanisms acting singly, in conjunction or in sequence — combinations about which we still have much to learn. In studying these mechanisms, most geomorphological examinations have concentrated on environmental rather than lithological controls. This chapter will also, therefore, consider the role of the 'victim' (i.e. the stone itself) in contributing to its demise — in particular the influences of past

Arid Zone Geomorphology: Process, Form and Change in Drylands, 2nd edition. Edited by David S. G. Thomas.

environmental conditions on stone susceptibility to particular forms of decay. Through this it will be demonstrated that, as in many good mysteries, an inheritance is involved.

Background

In Chapter 3 the history of desert weathering, particularly the history of ideas on 'insolation weathering', was reviewed and the continuing influence of descriptions provided by early desert explorers was documented. The *a priori* arguments for the significance of high diurnal rock temperature ranges were also rehearsed, and evidence presented of the high absolute rock surface temperatures recorded by some workers in the field. The chapter then went on briefly to describe air difficulties in trying to reproduce insolation weathering under controlled laboratory conditions, as compared to the obvious efficacy of weathering simulation tests that combine diurnal temperature cycling with moisture and/or salt applications. It further argues that, in reality, conditions required for insolation weathering *sensu stricto* are unlikely to be encountered in hot deserts, because of the widespread availability of salts and, contrary to popular belief, the frequent wetting of stone surfaces by a combination of dew, fog and occasional rainfall. The presence of this moisture also provides the opportunity for chemical decay and dissolution in addition to the physical disruption of stone traditionally associated with these environments.

In describing all of these developments, the importance of laboratory simulations in demonstrating the operation of different weathering mechanisms is apparent. In harsh and unpopulated environments the collection of detailed environmental information is, understandably, difficult. Observations are frequently limited to a few days or the duration of a specific project. It is much easier and more efficient to attempt to reproduce hot desert conditions in the laboratory where environmental cycles can be precisely replicated and repeated, stone responses accurately monitored and, most importantly, the effects of controlling variables isolated. The difficulty with simulation experiments is that they necessarily simplify reality. Results are only as valid as the assumptions made about

weathering environments when establishing test conditions and, however long a test, it is brief compared with the age of most desert surfaces. Lastly, assessments of weathering-induced breakdown are very much at the mercy of the technique(s) used for damage measurement. In short, there is a risk that, without careful management, results of weathering simulations may be artefacts of experimental design rather than reflections of change under meaningful hot desert conditions. It is therefore important that we critically examine the experimental procedures on which our understanding of desert weathering processes has largely been based. This should also provide us with an insight into those aspects of natural hot desert environments that exert an influence over rates and patterns of weathering.

Importance of thermal regimes

Most published data on rock surface temperatures have been collected to demonstrate how high such temperatures can rise. Measurements are characteristically of flat, unshaded surfaces or small blocks that receive insolation from dawn to dusk, they are frequently made on cloudless days in high summer and many studies have used dark-coloured surfaces with a low albedo. Under such conditions it is not difficult to achieve surface temperatures in excess of 50°C and occasionally over 70°C (Chapter 3). Equally, however, it should be possible to obtain similar results on hot days in most climatic zones as the temperature reached is primarily a function of the duration of insolation related to angle of the Sun above the horizon and, most importantly, the thermal characteristics of the stone concerned. High rock temperatures are not, therefore, unique to hot deserts, although it could be argued that they are more frequent, widespread and often linked to a rapid fall in surface temperature once the Sun sets. Even this rapid drop in temperature is, however, more accentuated in many mountain environments where clear skies combine with low ambient air temperatures to produce particularly rapid temperature decreases in early evening (Whalley *et al.* 1984).

The occurrence of high absolute temperatures in deserts is, however, influenced by spatial and temporal variations in heating conditions. These include the effects of factors such as cloud cover, seasonality (most deserts experience a cool/cold season), aspect and angle in relation to position of the sun, shading related to local topography, maritime cooling and altitude. All of these influences conspire to produce potentially complex thermal regimes and highlight the potential problems of characterising any morphoclimatic zone using simple climatic criteria. This is further emphasised by the fact that much active weathering in deserts, especially that associated with salts, is found in, and is responsible for the creation of, shadow areas. This includes features such as tafoni and honeycombs characterised by granular disintegration, scaling and flaking, in which absolute temperatures and temperature ranges are lower than on nearby exposed surfaces (e.g. Dragovich 1967; 1981; Rögner 1987). Shadow weathering provides a graphic illustration that extreme temperatures are not required to produce rock breakdown and suggests that moisture absorption on surfaces protected from direct insolation may be just as important in facilitating processes such as salt weathering.

Despite these observations, many laboratory experiments designed to reproduce desert weathering have employed repetitive exposure to 'record breaking' diurnal temperature ranges and absolute temperatures in excess of 60–70°C. The dangers of attempting to simulate fatigue effects by repeated exposure to extreme conditions are three-fold. First, work on damage to building stone has demonstrated the efficacy of temperatures above 60–70°C in causing, for example, microfracturing at depth in samples (Marschner 1978; Yong and Wang 1980) that is not experienced at lower temperatures. High temperatures enhance salt crystallisation pressures, complete crystallisation of salts from solution and possibly salt crystal development (McGreevy and Smith 1982). It is possible, therefore, that patterns and intensities of weathering may be created in tests that are simply not achievable by the lower temperatures experienced in shaded areas. Second, even in the most barren and exposed environments it is unlikely that surfaces will be subjected day after day to extreme conditions. Third, fatigue failure results from repeated exposure to ordinary stresses and produces a structurally different pattern of failure to that produced by the dramatic exceedance of shear or tensile strength induced by extreme heating.

There would thus seem to be a reasonable argument for the use of moderated thermal regimes in simulation experiments, more characteristic of active weathering environments. These would also be more in keeping with maximum temperatures of 50–60°C now being reported from deserts by researchers using more reliable thermometers and thermistors (Smith 1994).

Those experiments that have used reduced diurnal regimes, for example, Smith and McGreevy (1988), have demonstrated the continued effectiveness of processes such as salt weathering. However, even this experiment, in common with most other studies, still employed a diurnal heating and cooling regime. This reliance on diurnal cycles is again largely a legacy of explorers' tales of 'roasting days' and 'freezing nights'. It also reflects early, manually recorded plots of air and rock temperatures sampled at time intervals which missed short-term fluctuations. As the technology of temperature measurement has improved it is now possible to record temperatures continuously without disturbing the rock surface using, for example, infra-red pyrometers. When this is done it is noticeable that daytime temperatures, in particular, are very prone to rapid short-term fluctuations of the order of 15–30 minutes duration. Rates of temperature change during these fluctuations can significantly exceed averaged rates of diurnal heating and cooling (Jenkins and Smith 1990).

Although short-term fluctuations — possibly related to variations in cloud cover or wind-speed — may not penetrate far into stone, they will affect the outermost few millimetres in which phenomena such as granular disintegration and flaking occur. Future experiments seeking to reproduce such patterns of breakdown could possibly, therefore, combine long- and short-term fluctuations. Such a combination is, however, difficult to produce because of the equipment used in most simulation experiments. Invariably samples are cycled in an oven or 'climatic cabinet' in which temperature change is achieved by heating or cooling the air in contact with the stone, which limits the rapidity with which samples can be cooled. Recent experiments by Warke (1994) and Warke and Smith (1994) have managed to effectively reproduce rapid temperature fluctuations using a directional heating source (infra-red lamps) and a rotating aluminium blade which can be adjusted to 'expose' and 'shade' samples for variable lengths of time. Although this system has demonstrated the effectiveness of short-term (15 minutes) fluctuations in facilitating salt-induced decay, it has not yet proved possible to combine these fluctuations with a diurnal cycle. Another disadvantage of

using oven-based systems to simulate real-world conditions is that they force different stone types through the same temperature regime — irrespective of sometimes very different thermal properties. This is of considerable significance, in that relative rates of decay under such test conditions have been used to rank stones on the basis of their susceptibility to salt weathering (e.g. Goudie *et al.* 1970) and they still form the basis of relative durability testing of building stones. It will be seen from Figures 4.1a and 4.1b, however, that stones which follow the same heating and cooling paths under 'oven' conditions perform very differently when

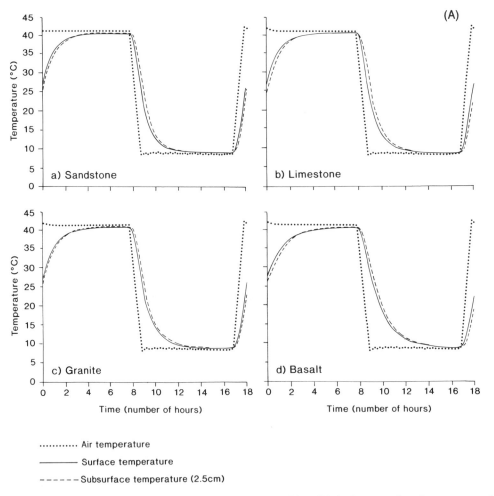

Figure 4.1a *Surface and subsurface temperature characteristics of four lithologies exposed to the same controlled temperature cycle in an environmental chamber. The temperature curves of all four samples closely correspond to air temperature within the 'oven'*

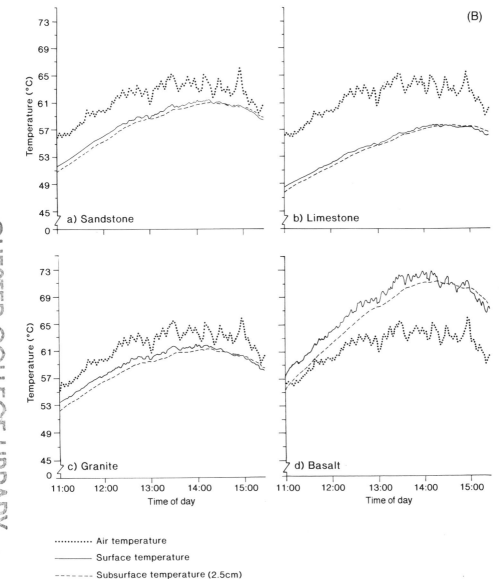

Figure 4.1b *When exposed to natural hot desert conditions in Death Valley, California (August), the same four lithologies shown in (a) display quite different surface/subsurface temperature response characteristics. The basalt sample, in particular, exhibits surface and subsurface temperatures well in excess of ambient air temperatures (5 cm above ground surface) — a feature attributed primarily to its low albedo*

exposed to insolation in a natural desert environment. These differences are, as pointed out by Kerr *et al.* (1984) and McGreevy (1985), due to differing thermal properties of albedo, thermal conductivity and heat storage capacity. Tests which do not take these into account are unlikely, therefore, to provide a sound basis for comparison between stone types. Because of this, radiation-based tests rather than, or in combination with climatic cabinets, may reproduce more accurately conditions experienced at the desert surface.

Moisture and salt: how much and how effective?

As noted previously, the realisation of the significance of dew and fog as moisture sources in deserts has been of considerable importance in advancing our understanding of desert weathering processes. At one level it has given credance to those who champion the role of biota — principally algae and lichen — in producing a range of surface microfeatures (e.g. Danin *et al.* 1982; Danin and Garty 1983). These can not only feed off the underlying geology but can also be sustained by a regular moisture supply — often rich in nutrients (Yaalon and Ganor 1968). This moisture supply also permits us to invoke the more frequent operation of mechanisms such as salt hydration/dehydration and crystallisation/dissolution, as well as allowing the possibility of hydrolysis and solution. But just how effective is direct deposition of moisture at wetting rock and activating weathering processes?

To begin with, nightly deposition is of the order of fractions of millimetres. Evenari *et al.* (1982), for example, measured an average of 33 mm of dew per annum at ground level at Avdat in the Negev Desert over a three-year period with an average of 195 dew nights per year (see Table 3.5). Such amounts (an average 0.17 mm per dew night) will not penetrate far into stone and will be rapidly lost from surfaces exposed to direct insolation. Even in sheltered locations, wind-driven evaporation is likely to rapidly remove such moisture. There is also the difficulty experienced by thin films of moisture in penetrating very dry porous materials. Theoretical analysis and work carried out on building stone has shown that it is particularly difficult for condensed moisture to penetrate pores that are open to the atmosphere via small capillaries (Camuffo 1983; 1984). Moisture fills these capillary openings, effectively trapping air within the stone and preventing further penetration. Large pores that are more open to the atmosphere fill more rapidly, but even these may eventually join at depth to other pores via narrow 'throats' at which point further moisture penetration is effectively halted. Similar constraints on moisture penetration also apply to dry surfaces that are subjected to sudden showers. Camuffo (1993) has pointed out that

the penetration of water running over the surface of stone is only possible when the complete surface (external and internal) is covered by a single or double layer of water molecules in the solid state. Thus (Camuffo 1991, p. 55) notes:

in the hot dry climate which is predominant during the summer in southern Europe, the external and internal surfaces of the monuments are completely dried out because of the warming associated with the strong solar radiation. This means that water, from a typical shower in the afternoon, cannot initially penetrate into the pores and capillaries of the stone, because these are not lined by a monolayer of water molecules in the solid state that would allow for the liquid water to run over it.

If rain persists, eventually the monolayer of water required to permit moisture penetration develops and rock begins to become wet from the surface inwards (Figure 4.2). However, under conditions of rapid heat loss at night associated with clear skies, an alternative mechanism of moisture absorption can operate. If surface temperature falls below ambient air temperature condensation will occur at relative

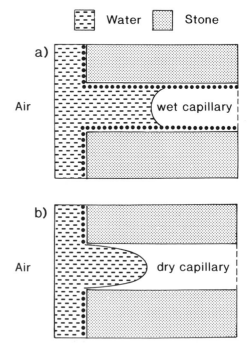

Figure 4.2 *Moisture penetration into wet (a) and dry (b) stone surface capillaries (after Camuffo 1991)*

humidities progressively below 100% — especially when hygroscopic salts are present (Arnold 1982). Moisture movement from atmosphere into the stone may, however, take place initially as vapour before moisture condenses. Condensation may first occur as an absorbed film of molecular water, still at relative humidities below 100%, which will continue to accumulate as humidity increases, before the film eventually collapses as bulk water into large pores (Camuffo 1983). An advantage of this mechanism is that it allows frequent deep penetration of moisture into rock without the necessity for dew point to be reached. It could also hydrate any salts present within the stone.

If dew and light rain are not absorbed they can still contribute to surface alteration of chemically complex stone and are likely to play a significant role in the formation of features such as rock varnish. This could involve the possible mobilisation of iron and manganese from surficial dusts (Smith and Whalley 1988) or indirectly through encouraging biological modification of dusts and substrate (see Chapter 6).

The final source of moisture is groundwater, which is increasingly viewed as important, and studies of urban areas within arid zones have demonstrated just how effective this can be in transporting salts into stone and concrete structures, ultimately leading to severe decay (e.g. Cooke *et al.* 1982). There are also numerous descriptions of salt deposition associated with the margins of saline and alkaline lakes (see Goudie and Cooke 1984, for a review) and the widespread occurrence of salts within desert soils suggests that, in lowland areas at least, upward movement of groundwater may be a viable means of introducing salts into pores and joint spaces.

In the light of these observations, it is clear that the method of moisture and salt presentation to samples must be an important consideration in any weathering simulation experiment. Traditionally, and again derived from building stone durability tests, samples are soaked in saturated salt solutions prior to each heating and cooling cycle. The aim in these circumstances is to disintegrate small test samples as quickly as possible. Little attention is paid to whether salt is available in such

quantities in nature, whether wetting is through inundation, whether significant weathering occurs between natural wetting cycles or whether the features to be simulated are larger than the samples. In attempts to more realistically replicate weathering conditions, mechanisms and features a number of modifications to simulation procedures have recently been employed. These include salt introduction by capillary rise (Goudie 1986), application of heat, moisture and salt through one exposed face (Smith and McGreevy 1983), wetting with 10% salt solutions (Smith *et al.* 1987) and one-off soaking with saturated salt solutions followed by wetting cycles using water only (Smith and McGreevy 1988; Goudie 1993).

Despite these improvements in experimental design, problems still remain. Although partial insulation and unidirectional environmental cycling of small samples can go some way towards replicating conditions experienced on large rock surfaces, many studies remain essentially examinations of debris weathering. Similarly, salts such as sodium and magnesium sulphates do occur in deserts and are very effective in simulations. It has not, however, been possible yet to reproduce the disruption frequently observed on natural outcrops in association with calcium sulphate and calcium carbonate crystallisation (McGreevy and Smith 1982). Problems also remain over the tendency for simulations to isolate individual mechanisms. Some work has been carried out on combined frost and salt attack (e.g. McGreevy 1982a; Jerwood *et al.* 1990a; 1990b), which may be applicable to high-altitude deserts, but we possess little knowledge of how, for example, salt weathering is enhanced by prior chemical alteration. Such combinations could be very important in initially non-porous stone, where processes such as salt weathering must rely on other agencies to create microfractures for them to exploit. Combined weathering has also taken on a greater significance in light of a growing acceptance that, in the presence of moisture from a variety of sources, chemical alteration does take place under hot desert conditions. This ranges from enhanced silica solution under saline and alkaline conditions common in deserts to hydrolysis and clay mineral formation, iron oxidation and solution — see Smith (1994) and Pope *et al.* (1995) for reviews.

One way to investigate the contribution of prior alteration to subsequent breakdown would be to use samples with a known weathering history in simulation experiments. This could extend beyond prior chemical alteration to samples with surface coatings or those that have previously been subject to a range of physical stresses. In carrying out such experiments it would be possible to examine the weathering 'memory' of stone and its role in determining rates and patterns of decay. This memory can derive from a wide range of sources which are examined in the next section.

Inheritance: the role of weathering history

Many studies of weathering in hot deserts have sought to explain weathering features in terms of a single, prevailing environmental condition. Similarly, because of difficulties in monitoring desert weathering processes there has been a reliance upon the inference of process from form alone. Growing awareness of the complexity of desert weathering environments has increasingly questioned this approach. In particular, it is clear that as rock weathers it is exposed to a range of weathering environments. The role of variability within the present-day climatic regime has already been discussed, but there is the additional component of possible exposure to completely different climatic regimes. Such a movement may be brought about by real, long-term climatic change or, in the case of debris, rock may be spatially relocated from high to low altitudes. As climate or the rock itself shifts it will carry with it morphological, structural and stress legacies derived from any earlier weathering environment or environments.

Exposure to various environmental conditions through either long-term climate change and/or spatial relocation gives rock a complex weathering history. Debris mantling the surface of contemporary hot deserts therefore carries within its fabric an inheritance of structural and mineralogical weaknesses incurred under former conditions but which influence response to present-day processes. Table 4.1 illustrates the variety of factors that can produce structural and/or chemical inheritance effects.

STRUCTURAL INHERITANCE

The presence of micro-scale structural discontinuities within rocks facilitates weathering and breakdown under subaerial conditions. Three types of feature were identified by Whalley *et al.* (1982):

1. Microfractures present in rock before it reaches the Earth's surface and formed by endogenetic processes.
2. Mineralogical concentrations which are particularly susceptible to subaerial processes and which form potential weathering lines and lines of fracture development.
3. Microfractures extended by subaerial processes.

While Whalley *et al.* (1982) note that 3 can only develop from 1 and 2, most arid weathering studies have tended to emphasise crack-propagating abilities of subaerial processes and neglected inherited endogenetic features which, following exposure, facilitate the ingress of moisture and salt. Therefore, assessment of the efficacy of a particular weathering mechanism may be misleading without due consideration of the geological and tectonic history of the rock concerned.

Endogenetic processes affect rock prior to exposure at the Earth's surface and arise from mainly tectonic processes which create weaknesses within the rock mass and constituent grains. Fractures form when local stress applied exceeds local strength, a condition which may be caused, for example, by non-hydrostatic stress, shock waves and differential expansion/contraction of mineral grains because of changes in temperature and pressure (Simmons and Cooper 1978). The structural weaknesses incurred are manifested as intra-, trans- and intergranular microfractures. Simmons and Richter (1976) identified eight types of microfracture attributable to particular endogenetic processes (see Table 4.2).

Ascent to the surface through gradual removal of overburden and resulting pressure release not only initiates fracturing but can lead to the opening of pre-existing fractures. The process of dilatation is closely associated with intrusive igneous rock such as granite emplaced under conditions of great heat and pressure but it can also affect other relatively homogeneous

Table 4.1 *Factors that contribute to the weathering of rock by creating an inheritance of structural and mineralogical weaknesses*

Causative factor	Inheritance effects
Pre-exposure	
Endogenetic processes	Development of structural weaknesses in rock-mass and individual grains
Dilatation	Erosion and removal of overburden causing stress release, expansion, opening of joint planes and intergranular microfractures
Post-exposure	
Insolation weathering	Differential thermal expansion of surface grains and layers in response to long- and short-term temperature fluctuations leading to microfracture development
Salt weathering	Deposition and accumulation of salt within rock fabric facilitating microfracture development through stresses related to crystallisation pressure, volume change associated with phase changes and thermal expansion/contraction
Frost weathering	Repeated freezing and thawing of moisture in pore spaces, joints and fractures leading to fracture propagation
Chemical weathering processes	Dissolution of rock fabric alters pore dimensions facilitating ingress of salt and moisture. Surface roughness may also be increased through etching and dissolution with clay mineral formation and eventually complete crystal lattice breakdown
Biological weathering	Acidic secretions cause dissolution and chelation of selected minerals while penetration and expansion of algal hyphae and lichen thalli result in granular release
Transportation	Impact damage sustained during gravity-related mass movements and fluvial transport causing micro- and macrofracture development
Surface alteration features	
Rock varnish	Changes albedo by darkening surfaces, increasing absorption of solar radiation and hence rock surface temperatures. Rock with low thermal conductivity will also experience increased surface/subsurface temperature gradients
Case-hardening	May form at the expense of a weakened substrate increasing susceptibility to rock breakdown if the case-hardened layer is breached. Also creates potential fracture plane between rock surface and substrate

lithologies such as sandstone and marble (Bradley 1963) where the orientation of individual grains is random creating conditions of unequal expansion.

While endogenetic processes can alter rock properties and indirectly increase susceptibility to subaerial processes by facilitating salt and moisture ingress, other structural anomalies may also influence rates and patterns of breakdown. Certain mineralogical concentrations, while not forming identifiable fractures, create potential weathering lines in both igneous (McGreevy 1982b) and sedimentary rock (McGreevy and Smith 1984), which remain

intact until exposure. In the latter example preferential weathering of fine-grained, clay-rich laminations in a Triassic sandstone occurred during salt weathering simulations suggesting that the presence of swelling clays enhanced salt weathering effects. It further demonstrates that, although inherited structural weaknesses are often overlooked, these pre-existing weaknesses are characteristically exploited, modified and extended by subaerial processes and act as major controls on rock response to contemporary processes.

Given that most rock debris which mantles present-day desert surfaces has been exposed for

Table 4.2   *Rock microfractures attributed to endogenetic processes (from Simmons and Richter 1976)*

(a) **dP/dT cracks** — mainly grain boundary cracks produced in rock during ascent to the Earth's surface
(b) **Stress-induced cracks** — caused by non-hydrostatic stress with crack development related to the principal directions of this stress
(c) **Radial and concentric cracks around enclosed grains** — caused by a mismatch between the volumetric properties of the host grain and the enclosed grain
(d) **Tube cracks** — three possible origins:
  (1) simple subsurface solution features/magmatic fluid solution
  (2) groundwater etching of dislocations
  (3) intermediate stage in healing of flat cracks
(e) **Thermal cycling cracks** — caused by repeated stressing arising from heating and cooling
(f) **Thermal gradient cracks** — caused by stresses initiated by high temperature gradients with rapid temperature change increasing fracture development
(g) **Shock-induced cracks** — caused by major shock events such as meteorite impact and blasting (nuclear and non-nuclear)
(h) **Cleavage cracks** — cracks formed parallel to cleavage planes in minerals

often thousands of years, the structural weaknesses currently exploited, for example, by salt weathering mechanisms may have been previously modified by processes associated with quite different environmental conditions. For example, in basin–range deserts, material released at high altitude under subalpine conditions by frost-related processes undergoes brief periods of high-energy fluvial transport interspersed with often prolonged periods of *in situ* weathering. Eventually this material reaches a low-altitude weathering environment characterised by high-temperature and salt-rich conditions which often lead to relatively rapid disintegration (see Figure 4.3). However, salt weathering processes and high-temperature conditions merely finish a process begun in a quite different weathering environment. Too often weathering processes immediately associated with observed weathering forms are judged to be solely responsible for their formation when in actual fact this 'guilt by association' obscures the role of previous events and conditions — the weathering history of the material.

CHEMICAL INHERITANCE

While weathering features relict from more humid conditions are widely recognised in many contemporary hot arid regions, the precise role of past chemical weathering in present-day rock breakdown is not often considered. However, environmental change whether caused by translocation and/or long-term climatic fluctuation can mean that chemical weathering may have played a significant role in the weathering history of material now primarily influenced by physical weathering processes.

Modifications arising from chemical weathering include mineral alteration, changes in pore dimensions and development of varnishes and case-hardened layers. Through alteration of primary minerals under previous humid conditions, lines of less resistant material can be created which may be exploited by, and enhance the action of, contemporary weathering processes (Gill 1981). Formation of clay minerals may also change rock properties such as porosity and permeability (McGreevy and Smith 1984), as will the removal of material through dissolution which can also reduce rock strength by weakening constituent mineral grains or the cement binding them. Young (1988) attributed etching of quartz grains at a site in northwest Australia to subaerial weathering under humid conditions in the Tertiary, a process that has rendered the rock friable and particularly susceptible to weathering processes associated with prevailing hot arid conditions.

Changes in pore dimensions can deleteriously affect rock durability. Honeyborne and Harris (1958) observed that increases in pore size caused by dissolution may alter subsequent rock response to freeze/thaw mechanisms by enhancing moisture penetration, while Dunn and Hudec (1966) suggested that intergranular stress development related to ordered water

Figure 4.3 *View westwards across Death Valley, California, from Dantes View in the Black Mountains to the Panamint Range (snow covered) with extensive alluvial fan development. The proximity of high mountains and deep valley troughs so characteristic of basin–range deserts creates a wide range of weathering environments from high-altitude, subalpine conditions to low-altitude, salt-rich playas*

within pore spaces would be less effective in rock with enlarged pore dimensions (>5 μm). Consequently, rock debris may only become susceptible to the action of a particular weathering process once dissolution has modified its superficial pore characteristics. As moisture supply is limited in hot deserts, windblown accumulations of salts and dust more readily gain access into the rock fabric when rainfall events do occur if pore spaces are large. Through facilitating moisture absorption and penetration, prior modification of pore characteristics may also enhance chemical alteration under hot desert conditions.

Chemical alteration can also produce other forms of surface change which may enhance the weathering susceptibility of rock. Rock varnish (see Chapter 6) and case-hardened layers, for example, are widespread in hot deserts and their development can greatly alter rock properties. The environmental conditions under which rock varnishes develop in deserts and other areas are not fully understood but it is accepted that they form and survive over thousands of years (Oberlander 1994). Present-day varnishes may, therefore, have formed under previous climatic conditions or, alternatively, they may have been initiated under more humid conditions but continue to develop at a reduced rate under prevailing aridity.

Regardless of rates and conditions of formation it is clear that rock varnish reduces rock surface albedo, increasing absorption of solar radiation and hence rock temperatures. A decrease in surface albedo can greatly increase rock surface temperatures (Warke *et al.* 1996) and therefore it is not unreasonable to assume that temperature-driven weathering mechanisms may become more effective. Case-hardening, like rock varnish, is not confined to hot deserts but does occur extensively in such regions. Hard surface crusts may initially fulfil a protective role as long as they remain intact, but once breached are often followed by rapid breakdown of a weakened substrate frequently leading to the formation of cavernous weathering features (Conca and Rossman 1982; 1985).

These few brief examples have sought to demonstrate that weathering features associated with contemporary hot desert environments are rarely if ever the result of the action of a single mechanism but instead reflect the cumulative and sequential effects of a range of processes. Unless freshly exposed at the Earth's surface or at the quarry face, all rock carries within its fabric an inheritance of structural and mineralogical weaknesses incurred during previous exposure and weathering activity. The complexity of this weathering history and rock fabric alteration from its pristine state varies greatly. Material with a complex history is invariably less durable than 'fresh' material of the same lithology when exposed to the same conditions. This may go some way to explaining weathering anomalies where, for example, lithologically similar clasts exposed to the same micro-environmental conditions respond quite differently, with one showing evidence of extensive breakdown while the other remains comparatively intact. In particular, a complex inheritance may lower thresholds of material strength with the result that weathering activity of a much lower magnitude is required to effect change than if the material had been fresh.

Conclusion

When the observations in this chapter are combined with those in Chapter 3, it is clear that the causes of weathering in hot deserts are complex, and quite obviously there is no single culprit. The temperature extremes found in these areas are important, but the thermal regimes experienced are far more varied than simple diurnal cycles. Moisture exerts a crucial influence, but this is predominantly indirect. It drives a variety of salt weathering mechanisms, chemically weathers rock to leave it more susceptible to physical disintegration, encourages biological activity and modifies thermal response.

Hot deserts, therefore, encompass a range of weathering environments, between which rock can be moved by a variety of erosion processes. Because of this translocation, rock may experience an apparent climatic change in addition to the real climatic change that many deserts have experienced. These changes mean that debris and bedrock invariably carry with them a weathering legacy inherited from exposure to one or more previous environments and weathering mechanisms. So that, although breakdown may be triggered by an extreme temperature or temperature change, it may represent the culmination of a long history of exposure that progressively reduced mass strength. In seeking to identify the contribution of any stress inheritance it may be possible to trace the specific environmental history of the stone in question. However, it is more feasible to investigate its structural, mineralogical and chemical characteristics. This materials based approach to understanding weathering has, as indicated in the introduction, generally been neglected by geomorphologists in favour of the

supremacy of climate over geology. If we are to understand more fully geological controls, there is a need for more precise and consistent methods of assessing surface and internal changes in stone properties. This could include:

- non-destructive testing for internal damage mainly using ultrasonic procedures;
- improved assessment of small-scale surface changes using, for example, electron microscopy, surface profilometers and optical reflectance;
- rock hardness measurement using, for example, the Schmidt Hammer; and
- the application of a rigorous terminology for defining damage (Smith 1994).

Only with this precision will we advance beyond the 'guilt by association', produced, for example, by simple measures of weight loss used in many simulation experiments.

Finally, even if it were possible to fully understand weathering mechanisms, to have any meaning these must be integrated with an explanation of the features they produce. This must occur at the scale of microfeatures such as granular disintegration, flaking and scaling, but also at the level of landforms produced by these phenomena and ultimately between weathering landforms and landscape. If this is to be achieved, the first stage must be to establish closer links between field and laboratory studies. One way of doing this (Smith 1994) may be to make greater use of deserts as natural laboratories, through the increased use of exposure trials with stones of known stress histories but exposed in different microenvironments and with differing pre-treatments.

By making the analogy with the classic murder mystery we have sought to demonstrate the underlying complexity of rock weathering in hot desert environments. Processes traditionally identified as prime suspects are now assuming less central roles while others, not usually associated with hot desert conditions, may in fact have made quite significant contributions to the final outcome. Hopefully, through a combination of improved investigative techniques and a greater understanding of the role of past and present environmental conditions, 'red herrings' can be identified and miscarriages of justice avoided.

References

Arnold, A. 1982. Rising damp and saline minerals. In K.L. Gauri and J.A. Gwinn (eds), *Proceedings of the 4th International Congress on Deterioration and Preservation of Stone Objects*. University of Louisville, Louisville: 11–28.

Bradley, W.C. 1963. Large-scale exfoliation in massive sandstones of the Colorado Plateau. *Bulletin of the Geological Society of America*, 74: 519–528.

Camuffo, D. 1983. Indoor dynamic climatology: investigations on the interactions between walls and indoor environment. *Atmospheric Environment*, 17: 1803–1809.

Camuffo, D. 1984. Condensation–evaporation cycles in pore and capillary systems according to the Kelvin model. *Water Air and Soil Pollution*, 21: 151–159.

Camuffo, D. 1991. Physical weathering of monuments. In F. Zezza (ed.), *Weathering and air pollution. Course notes for first course, Venice/Milan, 2–9 September 1991*. Community of Mediterranean Universities, School of Monument Conservation, Bari: 51–66.

Camuffo, D. 1993. Pores, capillaries and moisture movement in the stone. In F. Zezza (ed.), *Stone material in monuments: diagnosis and conservation. Course notes for second course, Heraklion, 24–30 May 1993*. Community of Mediterranean Universities, School of Monument Conservation, Bari: 27–42.

Conca, J.L. and Rossman, G.R. 1982. Case hardening of sandstone. *Geology*, 10: 520–523.

Conca, J.L. and Rossman, G.R. 1985. Core softening of cavernously weathered tonalite. *Journal of Geology*, 93: 59–73.

Cooke, R.U., Brunsden, D., Doornkamp, J.C. and Jones, D.K.C. 1982. *Urban geomorphology in drylands*. Oxford University Press, Oxford: 324 pp.

Danin, A. and Garty, J. 1983. Distribution of cyanobacteria and lichens on hillsides of the Negev Highlands and their impact on biogenic weathering. *Zeitschrift für Geomorphologie*, 27 (4): 423–444.

Danin, A.R., Gerson, R., Marton, K. and Garty, J. 1982. Patterns of limestone and dolomite weathering by lichens and blue-green algae and their palaeoclimatic significance. *Palaeogeography, Palaeoclimatology, Palaeoecology*, 37: 221–233.

Dragovich, D. 1967. Flaking, a weathering process operating on cavernous rock surfaces. *Bulletin of the Geological Society of America*, 78: 801–804.

Dragovich, D. 1981. Cavern microclimates in relation to preservation of rock art. *Studies in Conservation*, 26: 143–149.

Dunn, J.R. and Hudec, P.P. 1966. Water, clay and rock soundness. *Ohio Journal of Science*, 66: 153–168.

Evenari, M., Shanan, L. and Tadmor, N. 1982. *The Negev: the challenge of a desert*, 2nd edn. Harvard University Press, Cambridge, MA: 437 pp.

Gill, E.D. 1981. Rapid honeycomb weathering (tafoni formation) in greywacke, S.E. Australia. *Earth Surface Processes and Landforms*, 6: 81–83.

Goudie, A.S. 1986. Laboratory simulation of the 'wick effect' in salt weathering of rock. *Earth Surface Processes and Landforms*, 11: 275–285.

Goudie, A.S. 1993. Salt weathering simulation using a single immersion technique. *Earth Surface Processes and Landforms*, 18: 369–376.

Goudie, A.S. and Cooke, R.U. 1984. Salt efflorescences and saline lakes: a distributional analysis. *Geoforum*, 15: 563–582.

Goudie, A.S., Cooke, R.U. and Evans, I. 1970. Experimental investigation of rock weathering by salts. *Area*, 4: 42–48.

Honeybourne, D.B. and Harris, P.B. 1958. The structure of porous building stone and its relation to weathering behaviour. *The Colston Papers*, 10: 343–365.

Jenkins, K.A. and Smith, B.J. 1990. Daytime rock surface temperature variability and its implications for mechanical rock weathering: Tenerife, Canary Islands. *Catena*, 17: 449–459.

Jerwood, L.C., Robinson, D.A. and Williams, R.B.G. 1990a. Experimental frost and salt weathering of chalk, I. *Earth Surface Processes and Landforms*, 15: 611–624.

Jerwood, L.C., Robinson, D.A. and Williams, R.B.G. 1990b. Experimental frost and salt weathering of chalk, II. *Earth Surface Processes and Landforms*, 15: 699–708.

Kerr, A., Smith, B.J., Whalley, W.B. and McGrevy, J.P. 1984. Rock temperature measurements from southeast Morocco and their significance for experimental rock weathering studies. *Geology*, 12: 306–309.

Marschner, H. 1978. Application of salt crystallization test to impregnated stones. In *Deterioration and protection of stone monuments*. UNESCO/RILEM Conference, Paris: Section 3.4.

McGreevy, J.P. 1982a. Frost and salt weathering: further experimental results. *Earth Surface Processes and Landforms*, 7: 475–488.

McGreevy, J.P. 1982b. Hydrothermal alteration and earth surface rock weathering: a basalt example. *Earth Surface Processes and Landforms*, 7: 189–195.

McGreevy, J.P. 1985. Thermal rock properties as controls on rock surface temperature maxima and possible implications for rock weathering. *Earth Surface Processes and Landforms*, 10: 125–136.

McGreevy, J.P. and Smith, B.J. 1982. Salt weathering in hot deserts: observations on the design of simulation experiments. *Geografiska Annaler*, 64A: 161–170.

McGreevy, J.P. and Smith, B.J. 1984. The possible role of clay minerals in salt weathering. *Catena*, 11: 169–175.

Oberlander, T.M. 1994. Rock varnish in deserts. In A.D. Abrahams and A.J. Parsons (eds), *Geomorphology of desert environments*. Chapman & Hall, London: 106–119.

Pope, G.A., Dorn, R.I. and Dixon, J.C. 1995. A new conceptual model for understanding geographical variations in weathering. *Annals of the Association of American Geographers*, 85: 38–64.

Rögner, K. 1987. Temperature measurements of rock surfaces in hot deserts (Negev, Israel). In V. Gardiner (ed.), *International geomorphology*. Wiley, Chichester: 1271–1287.

Simmons, G. and Cooper, H.W. 1978. Thermal cycling cracks in three igneous rocks. *International Journal of Rock Mechanics, Mineral Science and Geomechanics Abstracts*, 15: 145–148.

Simmons, G. and Richter, D. 1976. Microcracks in rocks. In R.J. Stevens (ed.), *The physics and chemistry of minerals and rocks*. Wiley, London: 105–137.

Smith, B.J. 1994. Weathering processes and forms. In A.D. Abrahams and A.J. Parsons (eds), *Geomorphology of desert environments*. Chapman & Hall, London: 39–63.

Smith, B.J. and McGreevy, J.P. 1983. A simulation study of salt weathering in hot deserts. *Geografiska Annaler*, 65A: 127–133.

Smith, B.J. and McGreevy, J.P. 1988. Contour scaling of a sandstone by salt weathering under simulated hot desert conditions. *Earth Surface Processes and Landforms*, 13: 697–706.

Smith, B.J. and Whalley, W.B. 1988. A note on the characteristics and possible origins of desert varnishes from southeast Morocco. *Earth Surface Processes and Landforms*, 13: 251–258.

Smith, B.J., McGreevy, J.P. and Whalley, W.B. 1987. The production of silt-size quartz by experimental salt weathering of a sandstone. *Journal of Arid Environments*, 12: 199–214.

Warke, P.A. 1994. Inheritance effects in the weathering of debris under hot arid conditions. Unpublished PhD thesis, Queen's University of Belfast: 423 pp.

Warke, P.A. and Smith, B.J. 1994. Short-term rock temperature fluctuations under simulated hot desert conditions: some preliminary data. In D.A. Robinson and R.B.G. Williams (eds), *Rock Weathering and Landform Evolution*. Wiley, Chichester: 57–70.

Warke, P.A., Smith, B.J. and Magee, R.W. 1996. Thermal response characteristics of stone: implications for weathering of soiled surfaces in urban environments. *Earth Surface Processes and Landforms* (in press).

Whalley, W.B., Douglas, G.R. and McGreevy, J.P. 1982. Crack propagation and associated weathering in igneous rocks. *Zeitschrift für Geomorphologie*, 26: 33–54.

Whalley, W.B., McGreevy, J.P. and Ferguson, R.I. 1984. Rock temperature observations and chemical weathering in the Hunza region, Karakoram: preliminary data. In K.J. Miller (ed.), *Proceedings of the International Karakoram Project*. Cambridge University Press, Cambridge: 616–633.

Yaalon, D.H. and Ganor, E. 1968. Chemical composition of dew and dry fallout in Jerusalem. *Nature*, 217: 1139–1140.

Yong, C. and Wang, C.Y. 1980. Thermally induced acoustic emission in Westerly granite. *Geophysical Research Letters*, 7: 1089–1092.

Young, R.W. 1988. Quartz etching and sandstone karst: examples from the East Kimberleys, northwestern Australia. *Zeitschrift für Geomorphologie*, NF, 32 (4): 409–423.

5

Desert soils

David. L. Dunkerley and Katherine J. Brown

Introduction

The surfaces of deserts are only patchily covered with soil, while in wetter areas deeply decomposed bedrock that has undergone pedogenesis mantles the landscape. Indeed, the desert surface over vast areas is formed of outcropping bedrock or aeolian or fluvial sediments so little modified that they hardly amount to soils by any common definition (e.g. the *entisol* order of the US system of taxonomy). In semi-arid areas, the effects of pedogenesis are more apparent, and a richer array of soil types has been described for these environments.

In the overview presented in this chapter, we will not concentrate on the nutrient status or cropping potential of desert soils. Rather, we will review those aspects which relate to the geomorphic evolution of the materials and surfaces upon which desert soil development takes place, and on the particular properties of desert soils that make their hydrologic response and erosional behaviour distinctive. In interesting ways, the development of desert soils is intimately connected with landscape development and with the history of environmental change that has been considerable in the arid and semi-arid regions, especially through the Quaternary period (see Chapter 26). We begin with a short examination of the classification and naming of desert soils.

Taxonomy of desert soils

Desert soils are not widely documented. Those classification schemes for soils developed primarily to support food and fibre production in wetter areas do not in general lend themselves well to the classification of desert soils. Desert soils may show very little horizon differentiation, a key taxonomic tool in humid areas, but often show features such as accumulations of soluble salts, that are rare elsewhere. Furthermore, many of the most important features from the hydrologic and geomorphic perspectives, such as desert stone mantles, surface sealing, microbiotic crusts, and vesicularity in near-surface layers, are developed with little dependence on pedologic characteristics such as the nutrient status, mineralogy or horizonation of the deeper regolith. Thus, though of extreme importance in landscape behaviour, these features too are of relatively little use taxonomically.

Table 5.1 lists the major soil groups of arid regions recognised in the US classification system. A fuller examination of the properties

Arid Zone Geomorphology: Process, Form and Change in Drylands, 2nd edition. Edited by David S. G. Thomas.

Table 5.1   *The major soil orders of the arid and semi-arid regions (modified after Dregne 1976)*

Soil order	Primary characteristics	Total global land area occupied (km^2)	Percentage of global land area occupied
Aridisols	Plant growth restricted by dryness and/or salinity all year	16.6×10^6	11.3
Alfisols	Moderate base saturation; an argillic horizon; some plant-available water seasonally	3.1×10^6	2.1
Entisols	Almost no horizon differentiation; little-altered sedimentary materials	19.2×10^6	13.1
Mollisols	Thick, dark, base- and organically-rich epipedon	5.5×10^6	3.7
Vertisols	Deep, cracking clay soils, with shrink–swell features common	1.9×10^6	1.3

of these soil groups (see, for example, Nettleton and Peterson 1983; US Department of Agriculture 1988; Watson 1992) will confirm that many geomorphically significant features that are mentioned below are not included in the classification rules, and we thus side-step soil taxonomy, turning to the important issues of soil function in the landscape.

Some distinctive aspects of desert soil development

Desert soils develop in a wide range of parent materials, including the extensive fluviatile deposits of the piedmont slopes flanking upland areas, as well as fluvial and aeolian materials of diverse origins and ages. Averaged over regional scales, erosion rates in many desert areas are low, and leaching also occurs at low rates because of aridity. On balance many desert soils are consequently typified by *accumulations* of materials such as salts and windblown dusts entrained from playa lakes and other erodible surfaces upwind, since these materials are laid down at rates exceeding local erosion rates. Particular surfaces within deserts, of course, lose dusts to the global atmospheric transport system, and dust fallout and washout in rain are known from widespread and remote locations, and contribute to sedimentation over large areas of the oceans (see Chapter 18). This material includes a substantial fraction of desert quartz, foreign to the deep ocean. From dust accumula-

tion in north Pacific deep ocean core V21-146, located downwind of the Chinese deserts, Rea et al. (1994) have shown that dust accumulated there three to four times faster in glacial times of the late Quaternary than in interglacial times (see Figure 5.1). Further, as Rea *et al.* (1994) note, the increasing dust accumulation rates up the core evident in the figure are suggestive of increasing aridity in the late Quaternary. In the case of the desert of western New South Wales, Australia, it has been estimated that at the end of the last glacial arid phase, regional landscapes may have been blanketed by significant depths of windblown materials from the Australian interior, with A-horizons more recently emplaced by slopewash processes (see Chartres 1982; 1983). For the Negev Desert, Goossens (1995) reports long-term dust accumulation rates of $15–30 \, t^{-1} km^2 a^{-1}$. Tiller *et al.* (1987) cite aeolian dust accumulation rates from sites in Australia of up to $32 \, t^{-11} km^2 a^{-1}$ (equivalent to more than $15 \, mm \, ka^{-1}$). Rates of loess accumulation from sites in China also cited by these authors are up to $70 \, mm \, ka^{-1}$. Windblown materials can be incorporated into desert soils, providing allogenic minerals, salts and much more abundant clay than could be produced *in situ* by ordinary weathering processes. Thus, argillic horizons in desert soils are often attributable to dust incorporation. In contrast, many humid zone soils are dominated by autogenic clays weathered *in situ* from the parent materials, and any small accessions of dust become insignificant.

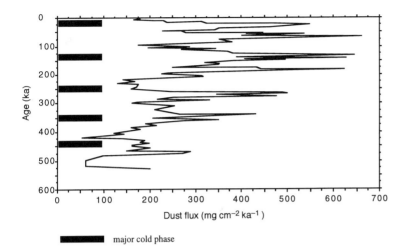

Figure 5.1 *Accumulation of dust recorded in deep-ocean core V21-146, north Pacific Ocean (after Rea et al. 1994). A trend of increasing dust accumulation towards the present is evident, together with an association of rapid dust buildup and Quaternary cold phases*

Because of this slow landscape change, desert soil-forming processes involve events spanning long time periods. Materials delivered in solution in rainwater also accumulate over these long periods. Rainwater often carries significant loads of carbonate compounds, and these may progressively accumulate in soils where infiltration takes place. For example, Reheis *et al.* (1992) report calcium carbonate accumulation rates of $3-5 \, g^{-1} m^2 a^{-1}$ for an arid site in Nevada. Subsoils, often alkaline because of accumulated carbonates, may slowly develop carbonate enrichment to the point where fully indurated *petrocalcic* horizons evolve (see Chapter 6). The very long timescales over which this occurs have been studied by various methods (e.g. Liu *et al.* 1994). Continuing accretion in carbonate horizons can result in their upper surface progressively approaching the ground surface, over substantial fractions of Quaternary time, and it may be possible to use the depth and thickness of carbonate layers to estimate the age of desert surfaces within which they occur (e.g. Marion 1989).

Given that many events in the formation of desert soils, such as dust accumulation during Quaternary arid periods, happen so slowly, it follows that climatic and environmental changes necessarily have impacts on these soils. The slow pace of developmental change then permits signatures of past events to endure in these soils for considerable periods. In more humid areas, rapid weathering can much more speedily erase signs of past events. Wetting and drying

cycles from Quaternary times of higher rainfall or more positive water balance have been inferred in various places (e.g. Dunkerley and Brown 1995) from the properties of the present-day soils. Processes affecting the soils have included enhanced shrink–swell cycles in clay-rich subsoils with the consequent development of stone sorting, gilgai microrelief (hummocky surface topography of various kinds), and other features.

The history of most desert soils generally remains unknown, and is likely to be complex. However, the present-day morphology of such soils has been studied in some detail. We will turn now to consider some of the key morphologic features found in many desert soils that are important in setting their place in the hydrologic and geomorphic functioning of the landscape. The features present often constitute a mixture of ages, some being rather young and some quite old. Desert soils are truly features that are *polygenetic*, forming a palimpsest of past events.

Stone-mantled surfaces and desert pavements

Many desert soils carry a surface veneer of stones, often coated with a desert varnish containing at least some allogenic components (see Figure 5.2). The stone veneer often overlies a stone-free subsoil. Dregne (1976, p. 42) argued that this feature was usually the result of the

Figure 5.2   *Desert pavement in northern South Australia. The gibbers which make up the pavement are rounded quartzite pebbles, strongly coated with desert varnish, and generally securely embedded within the finer matrix. Lens cap diameter 54 mm*

removal of fines by wind or water, leaving the gravel as a *lag* deposit. However, it appears that stones may also be concentrated at the surface by other means. A mechanism now widely envisaged is that windblown dusts, settling onto a desert surface, are washed down between surface stones, perhaps passing into the regolith along desiccation cracks. Weathered rock debris is thus kept exposed and continually rides to the top of the accumulating soil materials, being itself too large to be washed into soil crevices. Thus, far from signifying wind erosion, desert stone mantles may sometimes reflect quite the opposite, a local *accumulation* of significant amounts of material; the soils may indeed deepen with time. The mechanism just outlined could only operate, of course, if erosional stripping by water and other agencies were slower than the accumulation of aeolian materials. Thus, it has been argued (e.g. Cooke 1970) that pavement development may relate to the de-activation of a desert surface. This could happen, for example, when part of a piedmont becomes isolated by the incision of a local drainage, so that it no longer carried flood-waters. Indeed, it has been speculated (e.g.

McFadden *et al.* 1986) that the progressive enrichment of clay and carbonates in the subsoil slowly restricts the depth of water penetration and increases the fraction shed as surface runoff. This surface runoff may be the very agency that initiates land-surface dissection, isolating and de-activating the soil formation process. Water shed from the surface would then remove fresh accretions of dust, leaving the soil and stone veneer as relicts, with only slow processes such as disintegration of the surface stones, and accumulation of desert varnish on them, remaining active. This inevitable linking of soil and landscape evolution is an aspect of desert soil development that is at variance with the conventional view of soil development in humid areas, where it is envisaged more as a process taking place on and in a passive, slowly decomposing parent material.

Microphytic crusts

Arid and semi-arid zones have long been regarded as regions of low vegetative cover; however, if one considers the non-vascular component of the cover this is generally not correct. Many soil surfaces in these regions are covered in assemblages of mosses, liverworts, algae, lichens, fungi, bacteria and cyanobacteria — collectively known as microphytes (West 1990). Microphytic crusts can be observed growing in many different environments, but are particularly common in arid and semi-arid regions (see Figure 5.3). Microphytes are common on stone-free surfaces with low vascular plant cover, and in areas which are not greatly affected by blowing sand or other disturbances (West 1990). In extremely dry environments such as Atacama (Rauh 1985) and Namib (Walter 1985), microphytic crusts may be the only vegetation present.

Owing to the blanketing effect of the microphytic crusts one would assume that they have a stabilising effect on the soil surface, but this is a contentious issue and one which may depend on the soil, microclimate, and location in question. It would seem logical to assume that some microphytic crusts would be able to intercept and absorb some rainfall from light showers, however their role in modifying infiltration is less clear. Some researchers have found that infiltration increases when microphytic crusts are present (Fletcher and Martin 1948; Gifford 1972; Malcolm 1972; Blackburn 1975; Hacker 1986), others have found that infiltration decreases (Jackson 1958; Bolyshev 1962; Rogers 1977; Danin 1978; Stanley 1983), while others still have found no relationship between infiltration and the presence of microphytic crusts (Booth 1941; Faust 1971; Walker 1979; Penning de Vries and Djiteye 1982). In a study of a lichen-dominated surface in a chenopod shrubland in South Australia, Graetz and Tongway (1986) found that the infiltration rate increased by a factor of three when the crust was removed from the soil. More investigation of the processes involved is clearly needed.

Moisture storage in soils beneath microphytic crusts is believed to be increased (e.g. Schwabe 1960; Rogers 1977; Galun *et al.* 1982); however, how this relates to soil evaporation rates is unclear. Brotherson and Rushforth (1983) suggested that crusts decreased the amount of evaporation by acting as a surface seal, while Harper and Marble (1988) found that crusts increased the amount of evaporation. Their study in Utah determined that water loss in the upper 7.5 cm of soil was greater from crusted sites than bare sites owing to the dark

Figure 5.3 *Microbiotic soil crusting, Utah, USA. Here the regolith surface has a very rough microrelief, with amplitude of more than 5 cm, produced by dense colonies of non-vascular plants. The efficient water-trapping properties of this surface are readily damaged if the surface is trampled by stock or by people. Lens cap diameter 54 mm*

nature of the crust and therefore its higher absorption of solar radiation and resulting warmer temperature.

Not too surprisingly, the effect of biological crusts on erosion rates is also ambiguous. Various studies have associated the presence of a microphytic crust with a decrease in erosion rates; a result of a rough surface increasing the flow-path length of the overland flow and hence decreasing its erosivity (West 1990). On the other hand, Stanley (1983) argues that there is increased surface flow from crust-covered upper slopes which causes greater erosion on lower slopes. This study took place on a calcareous soil in the semi-arid rangelands of South Australia and different behaviour might be found elsewhere. Cooke and Warren (1973) concluded that microphytic crusts were generally associated with increased erosion rates.

Clearly, there is little agreement on the role of microphytic crusts in various hydrological parameters or in affecting erosion. This illustrates the danger of considering 'drylands' as a single unit and also considering the problem to be a simple two-parameter system (e.g. microphytic crust and erosion). The differing results obtained from the various studies considered in this section indicate that arid and semi-arid environments are complex. It is more likely that there are numerous factors which are important in determining the relationship between microphytic crusts and erosion rates. Such factors might include the physico-chemical properties of the soil, the moisture status of the soil, the geological history of the site in question (e.g. one would expect extremely old soils in the Australian landscape to behave quite differently to the younger post-glacial soils in the semi-arid regions of Canada), and the microclimate of the location in question. The spatial scale over which the processes are considered is, as is so often the case, itself a factor that has an influence on the conclusions that are drawn. Much work remains to be done in integrating local- or plot-scale process information with the more spatially extensive hillslope and landscape scales.

Inorganic (rain-beat) crusts

In addition to the biological or organic crusts described above, another class of soil crust

develops when physico-chemical changes occur at the soil–air interface during and after rainfall. These inorganic crusts are often referred to as 'soil seals' or 'soil crusts', depending on whether the soil is in a wet or dry state, and occur in numerous environments. For example, they develop on agricultural soils when these are left without a protective crop cover, and are common in arid and semi-arid regions in part because a protective plant canopy is so rarely present to intercept raindrop impact energy.

Various soil properties undoubtedly have a role to play in the formation of soil crusts. These include water content (Luk 1985), with dry aggregates being more resistant to destruction; clay content (Frenkel et al. 1978; Benhur et al. 1985), as increasing clay content increases the cohesive forces within aggregates; and the exchangeable sodium percentage (ESP), as soils with high ESP are more dispersive (Painuli and Abrol 1986). In addition, rainfall characteristics may play an important role in the formation of surface seals and crusts. It appears that the kinetic energy of a rain event is more important than the cumulative rainfall (Eigel and Moore 1983), while decreased surface porosity has been associated with increased raindrop diameter (Epstein and Grant 1973). It is worth remembering that even the raindrops that strike a desert surface are not quite like those that reach the surface in more humid areas. Often, for example, evaporation below the cloud base removes many of the smaller drops, so that the average size of drop striking the surface is increased. This effect is compounded if rain comes from intense convective storms over the desert, since then only very large drops, capable of falling through the strong updrafts, will reach the surface.

Early work on the nature of soil crusts by McIntyre (1958a; 1958b) suggested that the impact of rain falling upon the soil was the main factor in the formation of surface crusts. He argued that aggregates at the soil surface broke down during rainfall and released fine particles such as clays in the breakdown products. Some of these fine particles were washed into the soil profile to create an impermeable layer, blocking further water infiltration. McIntyre (1958b) identified two crust layers within the soil profile: a thin, compacted layer of 0.1 mm at the soil surface called the 'skin' and a less dense layer beneath it of 1.2 to 2.5 mm called the 'washed in' layer.

Much of the subsequent literature on surface crusts has been centred on explaining the presence of these layers. Investigations on the bulk density of the soil (e.g. Tackett and Pearson 1965; Eigel and Moore 1983) support the idea of the crust resulting from raindrop impact with highest bulk density measurements at the surface. Also supporting the raindrop impact argument are data from scanning electron microscope (SEM) images (Epstein and Grant 1973) which were examined for variation in the number of voids with depth. Table 5.2 shows that in the region of highest raindrop impact, i.e. at the surface, soil voids are absent, but they become more common with depth.

Debate has occurred with regard to the origin of the clay 'skin' identified by McIntyre (1958a; 1958b) and later by Chen *et al.* (1980). McIntyre (1958a; 1958b) explains the clay seal in terms of compaction of fines during rainfall. Mualem *et al.* (1990) claim that a thin clay skin could not possibly withstand the forces of raindrop impact and thus it must form after the rainfall event, during the drying process when disaggregated particles fall out of suspension. Different settling velocities explain the deposition of fine clays over coarser particles. Further SEM work undertaken on a structural crust by Pagliai *et al.* (1983) identified three layers within the surface crust: a clay layer of some micrometres thick; an intermediate silty layer of 0.15 mm thick; and a lower layer containing silt and clay which was 0.45 mm thick. The results of Pagliai *et al.* (1983) tend to support those of Mualem, and certainly indicate that these crusts should be regarded as complex features.

Table 5.2 *Presence of soil voids with depth beneath surface (Epstein and Grant 1973)*

Depth beneath surface (mm)	% volume of pores
upper 0.1	0
0.1–0.5	1
1–2	10–13
2–6	20–25

Vesicularity in near-surface horizons

In the soils of many dryland regions spherical voids of up to a few millimetres diameter are abundant in the uppermost parts of the soil profile (see Figure 5.4). These voids are called vesicles and were first observed by Lapham (1932) who speculated that they were formed when air was trapped beneath a soil seal during rainfall. Leshin (1937) showed that vesicles could be reproduced in the laboratory by pouring a soil paste into a beaker and allowing it to dry, while Springer (1958) found that pouring water into beakers filled with soil produced the same effect. He claimed that the vesicles resulted from air which was trapped ahead of the infiltrating water, and that vesicles were transitory and unstable features which were rearranged with each wetting and drying phase. Recent evidence that vesicles can contain clay and silt coatings (Sullivan and Koppi 1991) or precipitated calcium carbonate (Evenari *et al.* 1974) suggests that this may not be the case.

It appears that in order for vesicles to form, there must be a surface seal of some sort. This seal may take the form of a surface gravel layer or a hard surface crust such as the inorganic crusts described above (Evenari *et al.* 1974). The role of this seal is to behave as a barrier to escaping air as the matrix is wetted, and hence positive air pressure deforms the regolith into spherical vesicles.

Many dryland soils have low organic matter content and a high proportion of 2:1 clay minerals which are conditions which favour the formation of a surface seal (Van der Watt and Valentin 1992). Vesicle layers have been reported to depths ranging from 1 mm (Ringrose-Voase 1989) to 30 mm (Evenari *et al.* 1974; Chartres *et al.* 1985) and have been noted in a number of environments including the United States (Miller 1971), Israel (Evenari *et al.* 1974), Argentina (Figueira and Stoops 1983) and Australia (Chartres *et al.* 1985).

Vesicular structures in soils diminish the amount of soil matrix available to conduct infiltrating water, and reduce the surface tension generated at the wetting front. They are also inherently associated with restricted air escape and with greater tortuosity of flow paths through the soil. In concert, these effects act to limit infiltration rates and to reduce hydraulic

Figure 5.4 *Vesicular regolith near Broken Hill, NSW, Australia. The vesicular horizon was exposed by carefully removing a desert pavement of small stones, an undisturbed part of which is visible on the left of the photograph. The vesicles average about 1.5 mm in diameter, while the largest vesicles visible are about 2.5 mm in diameter. Lens cap diameter 54 mm*

conductivities of affected soils. The association of vesicular layers in desert soils with diminished infiltration and greater runoff generation has been demonstrated in a number of experimental studies (see, for example, Pierson *et al.* 1994).

Hydrologic and erosional response

The role of desert soils in supporting ecosystem functioning is significantly affected by many of the soil surface properties so far discussed. Because the distribution of soil materials, water and plant nutrients is such a vital aspect of landscape behaviour, we will consider briefly several of the key processes involved before turning to some very interesting consequences of their operation within the desert landscape.

The desert stone pavements described earlier act in various ways. The environment below stones, especially translucent ones, is a niche exploited by cryptogamous plants which enjoy some shelter and enhanced moisture availability

there. More importantly, the entry of water and dissolved materials delivered in rain into the soil is modified by the presence of a stone pavement. Laboratory and field studies have shown that stones may protect the soil surface from raindrop impact, leaving narrow annuli of unsealed soil near the sheltered margins of stones. These can act as sites for water entry and air escape. If the stones are quite immobile, and are really embedded within the soil surface, then this effect does not occur. Soil seals run right up to the stone margins, and so no avenue for water entry remains. Thus, the soils beneath very stable pavement surfaces are even more arid than the local rainfall would suggest. Being so well sealed and impervious, they act as runoff source areas, so that both water and dissolved nutrients are transferred downslope. The soils in runoff source areas like these remain extremely dry and are very inhospitable sites for plant growth. The absence of organic matter and canopy protection reinforce the effect.

Where downslope stone movement takes place episodically, or where the pavement is disrupted by ground ice or shrink–swell behaviour of the soil, the stones may rest more loosely upon the soil surface. Consequently, sheltered zones may form and infiltration may be promoted.

Surface stones also influence the stability of the soil surface by affecting the roughness that it exhibits during downslope waterflow. Small stones are rapidly submerged and roughness diminishes as flow becomes deeper. In the case of large stones, which may protrude through overland flow, increasing depth is merely associated with an increasing extent of submerged obstacles, and flow resistance increases with flow depth. Thus the velocities reached by surface runoff, and the resulting intensity of surface shear and entrainment of soil particles, are greatly influenced by the size, position and sorting of the surface stone veneer. Many intriguing investigations of the details of these effects exist (e.g. Poesen *et al.* 1994) but much remains to be worked out.

The widespread occurrence of overland flow is itself a characteristic distinctive of desert soils. Interestingly, though, forms of soil erosion typical of wetter areas, such as rill erosion, are far less common in desert soils. In this context, the lack of a protective canopy of vascular plants is again a determining factor. It has been proposed that the innumerable droplet impacts occurring across the soil surface permeate overland flow with small-scale turbulence that inhibits the flow concentration that leads to rill development. The numbers involved are always surprising to contemplate. For example, 10 mm of rain delivered in drops of 2.5 mm diameter involves over 12×10^6 impacts on each square metre of the soil surface. Clearly, splash and rainflow transportation, which we cannot discuss here, have the potential to be highly significant sediment dislodgement and transport agencies where the soil surface is exposed to this attack.

Spatial heterogeneity

A very distinctive characteristic of many desert soils is that their properties, at least those of their near-surface horizons, vary in dramatic ways over quite small distances (see Figures 5.5 and 5.6). The kinds of changes involved mean

Figure 5.5 *Mosaic landscape in mixed grassland and shrubland, southeast of Broken Hill, NSW, Australia. The contour-parallel bands of plants, which efficiently trap runoff shed from the intervening bare regolith surfaces, are here spaced about 40–50 m apart along slopes of less than 0.5°*

Figure 5.6   *Mosaic landscape of Figure 5.5 seen from the ground, with the view directly across the striking parallel bands of tussock grasses. The view is downslope, and the surface spread of stones that characterises the bare zones can be seen clearly within the downslope part of the first two unvegetated bands*

that, while taxonomically the soils might be similar, in terms of their function in the landscape they most certainly are not. Very often the spatial variability of desert soils can be regarded as involving repeated patterns of two distinct regolith forms, forming what is termed a *binary mosaic* or *two-phase mosaic*. One component of the mosaic consists of soils supporting relatively dense vascular plant cover (grasses, shrubs, small trees) while the other is relatively devoid of plants. For example, Tongway and Ludwig (1990) have described binary mosaics in semi-arid mulga woodland (dominant tree species *Acacia aneura*) where groves of trees alternate downslope with treeless intergroves. In association with this pattern are changes in the abundance of soil fauna, notably termites, which preferentially inhabit the wooded sub-areas (see, for example, Whitford *et al.* 1992). Thus, the mosaic actually involves a spatial patterning in biological activity, nutrient cycling, organic matter decomposition, and related aspects of soil development.

Spatial mosaics of soil like this have been reported widely from arid and semi-arid areas.

Their functioning is important to the continued existence of the plant communities that grow in these spatial mosaics. The key to this lies in the hydrologic response of the soil surfaces. The intergroves have bare, sealed surfaces often underlain by vesicular horizons. They are quite impermeable and absorb very little of the rain that falls on them. Rather, the rainwater is efficiently shed as runoff, and this trickles downslope to infiltrate in the more protected, more organically rich, and more porous soil of the vegetated phase of the mosaic. Little or no water passes through a typical grove onto the next intergrove downslope. Thus, with a 50/50 mosaic pattern, the groves receive water equivalent to almost twice the climatological rainfall, while the intergroves are much more arid than the rainfall would suggest. It has been argued recently that binary mosaics may be an ancient feature of the desert landscape, developing from changes triggered by the transition from last-glacial to Holocene climatic environments (Dunkerley and Brown 1995). For the genetic taxonomy of soils, this behaviour is problematic, because it results in soils having quite different leaching characteristics, salinity and moisture contents within metres of each other, and developed under a uniform external climatic environment. For example, Montaña *et al.* (1988), in a study of patchy vegetation in the Mexican Chihuahuan Desert, found mean organic matter levels of 2.6% within vegetation patches but only 0.8% in the intervening bare spaces. These differences occurred over spatial scales of 10–100 m. In Western Australia, Mabbutt and Fanning (1987) reported that, beneath mulga bands 10–20 m in width, a siliceous hardpan was typically located more deeply within the soil beneath groves than beneath intergroves. In fact, they described the depression of the hardpan as forming a 'trench' beneath the groves, which might act as a trap for water infiltrating there. Apart from this, the main differences in the grove and intergrove soils were restricted to the upper few centimetres.

The hydrologic efficiency of the runoff/runon process in two-phase mosaics was demonstrated by Tongway and Ludwig (1990) using artificial rainfall. In their mulga intergroves, runoff began after only seven minutes of rain at 29 mm h^{-1}, equivalent to 3.4 mm rain depth. In contrast, mulga grove soils showed no runoff from this rain application. Some of the other clear differences between the mulga grove soils and those of the intergrove runoff areas are summarised in Table 5.3. It is evident that the contrasts are especially marked in the top few centimetres of the soil.

Other forms of spatial heterogeneity have been reported from desert sites. For example, Rostagno *et al.* (1991) describe a Patagonian landscape of shrub mounds and mound interspaces. Beneath shrub mounds, total nitrogen, organic carbon, electrical conductivity, exchangeable sodium percentage, exchangeable cation levels and infiltration rates were significantly higher than in interspace soils. Rostagno *et al.* infer that erosion dominates in interspace areas while accumulation promotes soil evolution within the shrub mounds. Similar mounded topography, with mounds 3–5 m long and 1–3 m wide, and a relief of about 6.4 cm, was reported from arid acacia shrubland in Western Australia by Mott and McComb (1974). Once again the mound soils had higher organic carbon and total nitrogen levels. Moisture characteristics for the two soils were very similar, but a hardpan was located typically 10 cm deeper beneath mounds, so that soils there were overall about 16 cm deeper than those of the mound interspace. The greater soil depth thus provided a greater reservoir of soil moisture to support the mound vegetation.

Conclusions

Desert soils remain incompletely known. In many ways, they differ from the soils of wetter areas both in origin and development, and in their present-day role in landscape hydrology and geomorphology. Given the many possible forms of anthropogenic climate change presently under way, and the mounting pressure from a burgeoning world population, it is important that more attention be paid to the understanding and management of dryland soils. Some have already envisaged elevated levels of soil loss under enhanced-Greenhouse

Table 5.3 *Summary of grove and intergrove soil properties, semi-arid mulga woodland, NSW, Australia (after Tongway and Ludwig 1990)*

Depth (cm)	Electrical conductivity (μS cm^{-1})		Exchangeable calcium (meq 100 g^{-1})		Cation exchange capacity (meq 100 g^{-1})		Organic carbon (%)	
	Intergrove	Grove	Intergrove	Grove	Intergrove	Grove	Intergrove	Grove
0–1	22.0	57.9	2.69	4.78	7.38	10.50	0.71	1.97
1–3	16.5	36.3	2.46	3.99	6.63	9.37	0.43	0.95
3–5	22.1	29.4	3.68	5.05	8.40	9.21	0.40	0.71
25–50	44.2	17.9	5.05	3.56	9.21	7.87	0.23	0.30

environmental scenarios (see, for example, Dregne 1990).

References

Benhur, M., Shainberg, I., Bakker, D. and Keren, R. 1985. Effect of soil texture and CaCO₃ content on water infiltration in crusted soil as related to water salinity. *Irrigation Science*, 6: 281–294.

Blackburn, W.H. 1975. Factors influencing infiltration rates and sediment production of semi-arid rangelands in Nevada. *Water Resources Research*, 11: 929–937.

Bolyshev, N.N. 1962. Role of algae in soil formation. *Soviet Soil Science*, 1964: 630–635.

Booth, W.E. 1941. Algae as pioneers in plant succession and their importance in erosion control. *Ecology*, 22: 38–46.

Brotherson, J.D. and Rushforth, S.R. 1983. Influence of cryptogamic crusts on moisture relationships of a soil in Navajo National Monument, Arizona. *Great Basin Naturalist*, 43: 73–79.

Chartres, C.J. 1982. The pedogenesis of desert loam soils in the Barrier Range, western New South Wales. I. Soil parent materials. *Australian Journal of Soil Research*, 20: 269–281.

Chartres, C.J. 1983. The pedogenesis of desert loam soils in the Barrier Range, western New South Wales. II. Weathering and soil formation. *Australian Journal of Soil Research*, 21: 1–13.

Chartres, C.J., Greene, R.S., Ford, G.W. and Rengasamy, P. 1985. The effects of gypsum on macroporosity and crusting of two duplex soils. *Australian Journal of Soil Research*, 23: 467–479.

Chen, Y., Tarchitzky, J., Brouwer, J., Morin, J. and Banin, A. 1980. Scanning electron microscope observations on soil crusts and their formation. *Soil Science*, 130: 49–55.

Cooke, R. 1970. Stone pavements in deserts. *Annals of the Association of American Geographers*, 60: 560–577.

Cooke, R.U. and Warren, A. 1973. *Geomorphology of deserts*. University of California Press, Berkeley, California: 353 pp.

Danin, A. 1978. Plant species diversity and plant succession in a sandy area in the Northern Negev. *Flora*, 167: 409–422.

Dregne, H.E. 1976. *Soils of arid regions*. Elsevier, Amsterdam: 237 pp.

Dregne, H.E. 1990. Impact of climate warming on arid region soils. In H.W. Scharpenseel, M. Schommaker and A. Ayoub (eds), *Soils on a warmer earth*. Elsevier, Amsterdam: 177–184.

Dunkerley, D.L. and Brown, K.J. 1995. Runoff and runon areas in patterned chenopod shrubland, arid

western New South Wales, Australia: characteristics and origin. *Journal of Arid Environments*, 30: 41–55.

Eigel, J.D. and Moore, I.D. 1983. *Effect of rainfall energy on infiltration into bare soil*. Proceedings of the National Conference on Advances in Infiltration, 12–13 December, 1983, Chicago, Illinois. American Society of Agricultural Engineers.

Epstein, E. and Grant, W.J. 1973. Soil crust formation as affected by raindrop impact. In A. Hadas *et al.* (eds), *Physical aspects of soil water and salts in ecosystems*. Springer-Verlag, New York: 195–201.

Evenari, M., Yaalon, D.H. and Gutterman, Y. 1974. Note on soil with vesicular structure in deserts. *Zeitschrift für Geomorphologie*, NF, 18: 163–172.

Faust, W.F. 1971. Blue-green algal effects on some hydrologic processes at the soil surface. In *Hydrology and water resources in Arizona and the Southwest*. Proceedings 1971 Meeting Arizona Section, American Water Association and the Hydrology Section, Arizona Academy of Sciences, Tempe, Arizona: 99–105.

Figueira, H. and Stoops, G. 1983. Application of micromorphometric techniques to the experimental study of vesicular layer formation. *Pédologie*, 23: 77–89.

Fletcher, J.E. and Martin, W.P. 1948. Some effects of algae and moulds in the raincrust of desert soils. *Ecology*, 29: 95–100.

Frenkel, H., Goertzen, J.O. and Rhoades, J.D. 1978. Effects of clay type and content, exchangeable sodium percentage and electrolyte concentration on clay suspension and soil hydraulic conductivity. *Journal of the Soil Science Society of America*, 42: 32–39.

Galun, M., Burbrick, P. and Garty, J. 1982. Structural and metabolic diversity of two desert lichen populations. *Journal of the Hattori Botanical Laboratory*, 53: 321–324.

Gifford, G.F. 1972. Infiltration rate and sediment production on a plowed big sagebrush site. *Journal of Range Management*, 25: 53–55.

Goossens, D. 1995. Field experiments of aeolian dust accumulation on rock fragment substrata. *Sedimentology*, 42: 391–402.

Graetz, R.D. and Tongway, D.J. 1986. Influence of grazing management on vegetation, soil structure, nutrient distribution and the infiltration of applied rainfall in a semi-arid chenopod shrubland. *Australian Journal of Ecology*, 11: 347–360.

Hacker, R.B. 1986. Effects of grazing on chemical and physical properties of an earthy sand in the Western Australian mulga zone. *Australian Rangelands Journal*, 8: 11–17.

Harper, K.T. and Marble, J.R. 1988. A role for nonvascular plants in management of arid and semiarid rangelands. In P.T. Tueller (ed.),

Application of Plant Sciences to Rangeland Management and Inventory. Martinus Nijhoff, Amsterdam: 135–169.

Jackson, E.A. 1958. Soils and hydrology at Yudnapinna Station, South Australia. *CSIRO Soils and Land Use Series Bulletin*, 24: 1–66.

Lapham, M.H. 1932. Genesis and morphology of desert soils. *Bulletin of the American Soil Survey Association*, 13: 34–52.

Leshin, S.A. 1937. A study of the vesicular structure of certain desert soils. MS thesis, University of California, Berkeley. Cited in M.E. Springer, 1958. Desert pavement and vesicular layer of some soils of the desert of the Lahontan Basin, Nevada. *Proceedings of the Soil Science Society of America*, 22: 63–66.

Liu, B., Phillips, F.M., Elmore, D. and Sharma, P. 1994. Depth dependence of soil carbonate accumulation based on cosmogenic ^{36}Cl dating. *Geology*, 22: 1071–1074.

Luk, S.H. 1985. Effect of antecedent soil moisture content on rainwash erosion. *Catena*, 12: 129–139.

Mabbutt, J.A. and Fanning, P.C. 1987. Vegetation banding in arid Western Australia. *Journal of Arid Environments*, 12: 41–59.

Malcolm, C.V. 1972. *Establishing shrubs in saline environments.* Technical Bulletin No. 21. Department of Agriculture, Western Australia: 141 pp.

Marion, G.M. 1989. Correlation between long-term pedogenic $CaCO_3$ formation rate and modern precipitation in deserts of the American southwest. *Quaternary Research*, 32: 291–295.

McFadden, L.D., Wells, S.G. and Dohrenwend, J.C. 1986. Influences of Quaternary climatic changes on processes of soil development on desert loess deposits of the Cima volcanic field, California. *Catena*, 13: 361–389.

McIntyre, D.S. 1958a. Permeability measurements of soil crusts formed by raindrop impact. *Soil Science*, 85: 185–189.

McIntyre, D.S. 1958b. Soil splash and the formation of surface crusts by raindrop impact. *Soil Science*, 85: 261–266.

Miller, D. E. 1971. Formation of vesicular structure in soil. *Proceedings of the Soil Science Society of America*, 35: 635–637.

Montaña, C., Ezcurra, E., Carrillo, A. and Delhoume, J.P. 1988. The decomposition of litter in grasslands of northern Mexico: a comparison between arid and non-arid environments. *Journal of Arid Environments*, 14: 55–60.

Mott, J.J. and McComb, A.J. 1974. Patterns in annual vegetation and soil microrelief in an arid region of Western Australia. *Journal of Ecology*, 62: 115–126.

Mualem, Y., Assouline, S. and Rohdenburg, H.

1990. Rainfall induced soil seal. (A) A critical review of observations and models. *Catena*, 17: 185–203.

Nettleton, W.D. and Peterson, F.F. 1983. Aridisols. In L.P. Wilding, N.E. Smeck and G.F. Hall (eds), *Pedogenesis and soil taxonomy. II. The soil orders.* Elsevier, Amsterdam: 165–215.

Pagliai, M., Bisdom, E.B.A. and Ledin, S. 1983. Changes in surface structure (crusting) after application of sewage sludge and pig slurry to cultivated agricultural soils in Northern Italy. *Geoderma*, 30: 34–53.

Painuli, D.K. and Abrol, I.P. 1986. Effect of exchangeable sodium percent on surface sealing. *Agricultural and Water Management*, 11: 247–256.

Penning de Vries, F.W.T. and Djiteye, M.A. (eds) 1982. *La productive des Pâturages Saheliens.* PUDOC Centre for Agricultural Publishing and Documentation, Wageningen: 525 pp.

Pierson, F.B., Blackburn, W.H., Van Vactor, S.S. and Wood, J.C. 1994. Partitioning small scale spatial variability of runoff and erosion on sagebrush rangeland. *Water Resources Bulletin*, 30: 1081–1089.

Poesen, J.W., Torri, D. and Bunte, K. 1994. Effects of rock fragments on soil erosion by water at different spatial scales: a review. *Catena*, 23: 141–166.

Rauh, W. 1985. The Peruvian–Chilean deserts. In M. Evenari, I. Noy-Meir and D.W. Goodall (eds), *Hot deserts and arid shrublands*, Vol. 12A: *Ecosystems of the world.* Elsevier, Amsterdam: 239–268.

Rea, D.K., Hovan, S.A. and Janecek, T.R. 1994. Late Quaternary flux of eolian dust to the pelagic ocean. In *Material fluxes on the surface of the earth.* Studies in Geophysics Series, National Research Council, USA: 116–124.

Reheis, M.C., Sowers, J.M., Taylor, E.M., McFadden, L.D. and Harden, J.W. 1992. Morphology and genesis of carbonate soils on the Kyle Canyon fan, Nevada, U.S.A. *Geoderma*, 52: 303–342.

Ringrose-Voase, A.J., Rhoades, D.W. and Hall, G.F. 1989. Reclamation of a scalded, red duplex soil by waterponding. *Australian Journal of Soil Research*, 27: 779–795.

Rogers, R.W. 1977. Lichens of hot arid and semiarid lands. In M.R.D. Seward (ed.), *Lichen ecology.* Academic Press, London: 211–252.

Rostagno, C.M., del Valle, H.F. and Videla, L. 1991. The influence of shrubs on some chemical and physical properties of an aridic soil in northeastern Patagonia, Argentina. *Journal of Arid Environments*, 20: 179–188.

Schwabe, G.H. 1960. Zur autotropen Vegetation in ariden Boden, Blaualgen und Lebensraum IV. *Osterreichische Botanische Zeitschrift*, 107: 281–309.

Springer, M.E. 1958. Desert pavement and vesicular layer of some soils in the desert of the Lahontan Basin, Nevada. *Proceedings of the Soil Science Society of America*, 22: 63–66.

Stanley, R.J. 1983. Soils and vegetation: an assessment of current status. In J. Messer and G. Mosley (eds), *What future for Australia's arid land?* Australian Conservation Foundation, Canberra: 8–18.

Sullivan, L.A. and Koppi, A.J. 1991. Morphology and genesis of silt and clay coatings in the vesicular layer of a desert loam soil. *Australian Journal of Soil Research*, 29: 579–586.

Tackett, J.L. and Pearson, R.W. 1965. Some characteristics of soil crusts formed by simulated rainfall. *Soil Science*, 99: 407–413.

Tiller, K.G., Smith, L.H. and Merry, R.H. 1987. Accession of atmospheric dust east of Adelaide, South Australia, and the implications for pedogenesis. *Australian Journal of Soil Research*, 25: 43–54.

Tongway, D.J. and Ludwig, J.A. 1990. Vegetation and soil patterning in semi-arid mulga lands of eastern Australia. *Australian Journal of Ecology*, 15: 23–34.

US Department of Agriculture, Soil Conservation Service 1988. *Soil taxonomy. A basic system of soil classification for making and interpreting soil surveys.* Krieger Publishing, Malabar, Florida, 754 pp.

Van der Watt, H.v.H. and Valentin, C. 1992. Soil crusting: the African view. In M.E. Sumner and B.A. Stewart (eds), *Soil crusting: chemical and physical processes*. Lewis Publishers, Boca Raton, FL: 301–338.

Walker, B.H. 1979. Game ranching in Africa. In B.H. Walker (ed.), *Management of semi-arid ecosystems*. Elsevier, Amsterdam: 55–81.

Walter, H. 1985. The Namib Desert. In M. Evenari, I. Noy-Meir and D.W. Goodall (eds), *Hot deserts and arid shrublands. Vol. 12B: Ecosystems of the world*. Elsevier, Amsterdam: 245–282.

Watson, A. 1992. Desert soils. In I.P. Martini and W. Chesworth (eds), *Weathering, soils and paleosols*. Developments in Earth Surface Processes 2. Elsevier, Amsterdam: 225–260.

West, N.E. 1990. Structure and function of microphytic soil crusts in wildland ecosystems of arid to semi-arid regions. *Advances in Ecological Research*, 20: 179–223.

Whitford, W.G., Ludwig, J.A. and Noble, J.C. 1992. The importance of subterranean termites in semi-arid ecosystems in south-eastern Australia. *Journal of Arid Environments*, 22: 87–91.

6

Desert crusts and varnishes

Andrew Watson and David J. Nash

Introduction

The landscapes of the Earth's desert regions are often perceived as seas of mobile sand dunes or vast expanses of featureless gravel plains where the action of water and development of soils are negligible. Even in the most arid deserts, however, the occurrence of crusts (often termed *duricrusts* if they are indurated) at or near the land surface attests to mobilisation and precipitation of minerals in the presence of water. Sometimes, these accumulations are relict lacustrine deposits or weathering crusts which formed when the climate was more humid. More often, however, desert crusts are the products of specific hydrological or pedological processes which are characteristic of arid and semi-arid environments. As such, the different types of crust can provide useful environmental indicators in areas where detailed hydrological or climatic data are often sparse.

Some desert crusts are ephemeral features — for example, salcretes composed of sodium chloride are prone to rapid dissolution by rainwater. Nevertheless, these and crusts such as gypcretes may have a significant geomorphic effect when they encase and temporarily immobilise sand dunes (Watson 1985b; Chen *et al.* 1991). Other crusts, such as silcretes, are

persistent surficial phenomena (Milnes 1986) — some in the Namib Desert may have remained at the land surface for more than 50 million years (Beetz 1926, pp. 19-27; Kaiser 1926, p. 308). Calcrete crusts tens of metres thick are not uncommon (Figure 6.1) and such features can mantle extensive areas of desert terrain, protecting the underlying materials from subaerial weathering and erosion by wind and water. By retarding the denudation of desert landscapes, the crusts often preserve relict landforms and can play an important role in the accumulation of thick sedimentary sequences in arid environments (Nash *et al.* 1994a; Twidale and Campbell 1995). In doing so, desert crusts influence the geomorphology and sedimentology of these environments to a far greater extent than is indicated in much of the geological literature (Pain and Ollier 1995). However, a growing interest in the study of palaeoenvironments of the Earth's arid and semi-arid regions has spurred an increased awareness of the significance of desert crusts and rock varnish.

The occurrence of desert crusts in the geological record or as relict surficial deposits can provide valuable information on palaeoclimates. Yet the processes involved in crust formation — and, indeed, the precise

Arid Zone Geomorphology: Process, Form and Change in Drylands, 2nd edition. Edited by David S. G. Thomas.

Figure 6.1   *(A) An 8-m-thick calcrete exposed in the flanks of the Auob valley, southwest Kalahari, near Kalkheuval Farm in southeastern Namibia. Upper sections of the calcrete are partially silicified creating an indurated caprock. (B) A view across the Auob valley at Kalkheuval Farm showing a flat plateau landscape created by incision through the silicified calcrete. The 8-m-section shown in (A) is at the centre of the photograph, located above a further 40 m of calcrete which is partially mantled by sil-calcrete debris*

environmental conditions under which they develop — are not always well understood. Moreover, some desert crusts closely resemble features which form under markedly different environmental conditions — for example, some types of silcrete and calcrete are the products of weathering in the humid tropics. Hence, it is essential that the morphological and chemical characteristics of desert crusts and the processes involved in their genesis are fully appreciated before they are employed as evidence in palaeoenvironmental reconstructions.

Types of desert crust

There are four main types of crust which are commonly found in many of the Earth's warm deserts: calcrete, silcrete, gypcrete (gypsum crusts), and salcrete (halite crust). The terms 'calcrete' and 'silcrete' were coined by Lamplugh (1902; 1907) to describe indurated masses cemented together by either calcium carbonate (calcite) or silica. However, the precise definitions have been revised often as our understanding of the origins and characteristics of these materials has grown. A number of minerals occur as hardened crusts in deserts but they are less widespread and will not be discussed in detail here; they include ferricretes, dolocretes, and a wide variety of evaporite crusts composed of minerals such as the sodium sulphates thenardite (Na_2SO_4) and mirabilite ($Na_2SO_4 \cdot 10H_2O$), glauberite ($Na_2Ca(SO_4)_2$) and epsomite ($MgSO_4 \cdot 7H_2O$). Even more uncommon evaporites, such as nitratine ($NaNO_3$), natron ($Na_2CO_3 \cdot 10H_2O$) and trona ($NaH(CO_3) \cdot 2H_2O$), form salt crusts in desert basins in volcanic regions such as the East African Rift Valley (Eugster 1986; Renaut *et al.* 1986) and parts of Chile (Stoerz and Ericksen 1974; Chong Diaz 1984).

Throughout this chapter the different types of desert crust are, for convenience, treated separately. It should, however, be noted that forms such as silcrete, calcrete and gypcrete are only the end members of a spectrum of duricrust types. Intermediate hybrid forms can occur when, for example, diagenetic alteration of a pre-existing crust results from the movement of porewater through the crust. Thus it is possible to find not only silcretes and calcretes but also

siliceous calcretes (or sil-calcretes) and calcareous silcretes (or cal-silcretes) dependent upon the relative proportions of silica and calcium carbonate within the crust.

CALCRETE

Calcrete (or caliche) has been defined by Wright and Tucker (1991, p. 1) as:

a near surface, terrestrial, accumulation of predominantly calcium carbonate, which occurs in a variety of forms from powdery to nodular to highly indurated. It results from the cementation and displacive introduction of calcium carbonate into soil profiles, bedrock and sediments, in areas where vadose and shallow phreatic groundwater become saturated with respect to calcium carbonate.

This definition, based upon earlier versions by Watts (1980) and Goudie (1973; 1983a), is probably the most useful general description arising from a quagmire of terminological debate.

Goudie (1973, p. 18; 1983a) presented the mean composition of 300 calcretes; they comprised about 80% calcium carbonate, 12% silica, 3% MgO, 2% Fe_2O_3, and 2% Al_2O_3. Calcretes are, however, very variable chemically because of their diverse origins and different stages of development and this gross chemical composition masks a complex mineralogy. Watts (1980) showed that individual profiles through Kalahari calcretes exhibit significant variations in the percentages of high-magnesium calcite and dolomite. These variations in turn appear to be related to the occurrence of authigenic silica and silicates, notably the clay minerals palygorskite and sepiolite.

Calcretes are generally white, cream or grey in colour, though pink mottling and banding are common. It has been suggested that many forms of calcrete fall within an evolutionary continuum, their structural characteristics being dependent on their maturity and stratigraphic position within a profile (Netterberg 1969; 1980; Reeves 1970; Bachman and Machette 1977; Goudie 1983a). Calcified soils and powder calcretes develop into nodular calcretes as calcium carbonate concretions grow within the host material (Wieder and Yaalon 1974; Magaritz *et al.* 1981). These may eventually coalesce to form honeycomb calcrete with the voids being filled with residual, unconsolidated host material. As the surface horizons become plugged, a mature hardpan calcrete evolves (Gile *et al.* 1966; Yaalon and Singer 1974). Laminar calcretes consisting of finely banded horizons often cap hardpans (James 1972; Arakel 1982; Warren 1983) but they may also encase nodules and boulders (Goudie 1983a). Though this is an idealised evolutionary model, mature calcrete profiles often comprise a laminar zone (sometimes beneath a loose soil) upon a massive hardpan overlying a zone of nodules in a matrix of host material (Arakel 1982; Goudie 1983a). Such well-developed profiles are usually between 1.0 and 5.0 m thick; laminar zones are rarely more than 0.25 m thick (Goudie 1983a) and hardpans are commonly 0.3–0.5 m thick (Goudie 1984). Boulder calcrete may be the final stage in the evolution because it seems to represent the onset of solutional degradation of a hardpan (Netterberg 1969; 1978). However, many calcretes are undoubtedly polygenetic; often several generations of fragmented boulder calcrete are reincorporated within hardpans. Moreover, the superimposition of numerous calcrete horizons can create complex composite sequences (Watts 1980; Goudie 1983a). As noted above, admixing of calcareous and siliceous cements creates sil-calcretes (Wright 1978; Smith and Whalley 1982; Arakel *et al.* 1989) (Figure 6.6) as a result of mineral replacement or emplacement during diagenesis.

While much work has focused on pedogenic calcretes, the formation of some crusts is related to littoral processes or precipitation from phreatic water (Mann and Horwitz 1979; Arakel and McConchie 1982; Semeniuk and Searle 1985; Arakel 1986; 1991; Jacobson *et al.* 1988; Nash *et al.* 1994a — see Chapter 15). These are not pedogenic calcretes, at least in terms of Watt's (1980) definition, but neither are they beach rocks or tufas which Goudie (1973) excluded from his definition of calcrete. It is essential to acknowledge that the environmental conditions under which these various forms of calcrete develop are very different.

SILCRETE

Siliceous duricrusts are widespread in some low-latitude and other environments and are termed silcrete when they contain at least 85% silica (Summerfield 1983a) and exhibit a porphyroclastic texture in thin-section. This distinguishes them from orthoquartzites where the texture is more even-grained (Hutton *et al.* 1978). Some particularly pure silcretes may contain in excess of 95% silica. Silcrete has been described as 'very brittle, intensely indurated rock composed mainly of quartz clasts cemented by a matrix which may be a well-crystallised quartz, or amorphous (opaline) silica' (Langford-Smith 1978, p. 3). It is a product of the cementation or replacement of surficial materials such as rocks, sediments, saprolite or soils by various forms of secondary silica, including opal, cryptocrystalline quartz or well-crystallised quartz (Milnes and Thiry 1992). This silicification is not associated with high-temperature volcanism or metamorphism but is a low-temperature physico-chemical process (Summerfield 1979; 1981; 1983a; 1983b; Milnes and Thiry 1992).

Silcrete varies in appearance from opaline or cryptocrystalline material — the 'Albertina'-type silcrete of Frankel (1952) and Smale (1973) — to conglomeratic material, called 'Puddingstone' in Britain (Summerfield 1983a). Most commonly, silcrete consists of silica-cemented quartz grains (Figure 6.2), a variety which is sometimes referred to as 'terrazzo silcrete' (Smale 1973; Soegaard and Eriksson 1985), or a quartzitic material with a subconchoidal fracture through the grains and matrix cement (Summerfield 1983a). Silcrete may also occur as lenses within calcretes or in the form of a late-stage patina on rock outcrops (Hutton *et al.* 1972; McFarlane *et al.* 1992). The colour is also variable: grey, brown and green silcretes are common. A variety of terms have been used to describe silcrete profiles, including massive, columnar, bulbous, nodular, glaebular and mammilated, reflecting the numerous surface morphologies of many exposures. In general, well-developed silcrete horizons are between 0.5 and 3 m thick, though multiple-horizon crusts may exceed 4 or 5 m (Mabbutt 1977, p. 141), and thicknesses in excess of 7 m have been documented for single-

Figure 6.2   *Silcrete exposed within the floor of the Boteti River, Botswana, at Samedupe Drift. The silcrete consists of quartz-rich Kalahari Sand cemented by various forms of secondary silica. Note the surface texture of the outcrop which consists of a partially dissected and pitted weathering rind overlying unweathered material*

horizon silcretes in the Kalahari (Nash *et al.* 1994b).

There is considerable debate concerning the origin and mode of development of silcrete. Some evidence suggests that many thick silcretes are associated with deep weathering profiles which perhaps developed under humid conditions (Summerfield 1983d). Siliceous crusts may develop in highly alkaline evaporitic basins (Ambrose and Flint 1981; Renaut *et al.* 1986) and in association with micro-organisms (Shaw *et al.* 1991), but whether the extensive crusts of inland Australia and southern Africa, for example, originated in this way is vehemently disputed. Though many have supported the contention of early workers such as Woolnough (1930) and Frankel and Kent (1938) that silcretes form contemporaneously within weathering profiles, others have argued in favour of hydrological (Stephens 1971; Smale 1973; Senior 1978) or pedogenic (Auzel and Cailleux 1949; Hutton *et al.* 1978; Callen 1983) accretion. What is apparent is that silcretes may form by a variety of different mechanisms and that no one model is likely to be universally applicable.

GYPSUM CRUSTS

Gypsum crusts have received far less attention than silcretes or calcretes. In part, this reflects a more restricted geographical distribution which may itself result from the greater susceptibility of gypsum to dissolution by meteoric water. This limits the crusts to some of the Earth's most arid regions. In addition, the less widespread occurrence of sources of gypsum ($CaSO_4 \cdot 2H_2O$) compared with calcium carbonate or silica sources hinders crust formation. Gypsum crusts have been defined as 'accumulations at or within 10 m of the land surface from 0.10 m to 5.0 m thick containing more than 15% by weight gypsum... and at least 5.0% by weight more gypsum than the underlying bedrock' (Watson 1985a, p. 855).

On the basis of stratigraphic and structural criteria, three main forms of gypsum crust have been identified: horizontally bedded crusts; subsurface crusts composed either of large, lenticular crystals (between 1 mm and 0.50 m in diameter) — known as desert rose crusts — or of mesocrystalline material (crystal diameters from 50 μm to 1.0 mm); and surface crusts, composed mainly of alabastrine gypsum (crystallites less than 50 μm in diameter), occurring as columnar crusts, powdery deposits or superficial cobbles (Watson 1979; 1983a; 1985a). Each of these forms exhibit characteristic micromorphological fabrics and textures, as well as being chemically distinct. The bedded crusts generally contain between 50 and 80% gypsum, the desert rose crusts between 50 and 70%, and the mesocrystalline subsurface form and the surface forms contain up to about 90% gypsum. The other main constituents are quartz grains and variable amounts of calcium carbonate. Desert rose crusts can reach thicknesses of up to 5 m (Knetsch 1937; Kulke 1974). They range in colour from white or grey to green or red, depending on the coloration of the host material. The columnar surface crusts are usually 1–2 m thick and white or grey in colour. The roughly hexagonal columns, 0.25–0.75 m in diameter, extend through the full thickness of the crust (Figure 6.3).

As in the case of nodular, honeycomb and hardpan calcretes, the mesocrystalline, columnar and cobble forms of gypsum crust are probably genetically related. The morphological and chemical differences result from diagenesis and, possibly, also degradation of pedogenic crusts. However, just as some calcretes are non-pedogenic, the bedded and desert rose forms of gypsum crust are also genetically distinct. This has profound implications on the significance of

these desert crusts as palaeoenvironmental indicators (Watson 1988).

HALITE CRUSTS

Halite (sodium chloride) deposits in deserts and in the geological record have almost always been interpreted as lagoonal or lacustrine evaporites. Halite occurs most commonly in sabkhas (desert basins or littoral flats where the water table lies just below the land surface) or in the basins of ephemeral lakes which are subject to periodic evaporation to dryness (Lowenstein and Hardie 1985) (Figure 6.4). Here, hard crusts of almost pure halite form horizontal beds on the surface; these are often white, though the presence of micro-organisms such as the flagellate *Dunaliella salina* can result in a strikingly pink coloration. Periodic influxes of comparatively fresh rainwater or runoff often result in the complex dissolution and subsequent reprecipitation of the deposits, as in Lake Eyre in South Australia (Dulhunty 1975; 1982). In some lagoonal environments, where halite dissolution is not so pronounced, halite and gypsum–halite deposits up to 10 m thick have been reported (Morris and Dickey 1957).

Powdery halite efflorescences (Beckmann *et al.* 1972; Schwenk 1977; Eswaran *et al.* 1980) and encrustations (Pye 1980) are common, though usually ephemeral, desert phenomena. Despite their impersistence, their influence on

Figure 6.3   *Columnar surface gypsum on Neocomian clays south of Chott el Fadjedj, southern Tunisia. Columnar structures are probably formed through repeated volumetric fluctuations resulting from either gypsum dehydration (Tucker 1978) or periodic desiccation of an illuvial gypsum horizon (Watson 1985a)*

Figure 6.4 *Polygons on the surface of a 1.5-m-thick halite crust, Umm as Samim, Sultanate of Oman*

rock weathering processes (Mortensen 1927; 1933) and on the stabilisation of unconsolidated materials can be significant (Nickling and Ecclestone 1981; Nickling 1984) (Figure 6.5).

There are few examples of halite crusts which have accreted beneath the land surface either close to the water table or in the soil zone in the same way as many calcretes and gypsum crusts. Subsurface, phreatic crusts have been reported from coastal sabkhas along the Arabian Gulf (Shearman 1963; Patterson and Kinsman 1978) and in lagoonal environments in Western Australia (Arakel 1980). Pedogenic halite crusts which have a strong structural resemblance to columnar gypsum crusts have been reported from the Namib Desert (Kaiser and Neumaier 1932; Watson 1983b; 1985a) and the Chilean Tacna Desert (Mortensen 1927). They are restricted to the most arid regions of the Earth.

OTHER DESERT CRUSTS

A number of other types of surface crust are found in some deserts; usually they occur as very localised features rather than having the widespread distribution of many silcretes,

calcretes and gypsum crusts. Dolocretes (magnesium carbonate crusts), for example, often resemble phreatic calcretes, though dolomite predominates over calcite as the principal mineral or cement (Sochova *et al.* 1975; Hutton and Dixon 1981; Briot 1983; Abdel Jaoued 1987; El Sayed *et al.* 1991; Khalaf and Abdal 1993). Ferricretes (iron-oxide-cemented crusts) are found in some deserts, notably in Australia, but these features are usually the products of weathering and pedogenesis under humid, tropical conditions (McFarlane 1983) and are therefore relict. They attest to major climatic changes, but it is often extremely difficult to ascertain the age of these ancient accumulations.

Most types of evaporite crust (excluding the gypsum and halite crusts described above) are confined to saline sabkhas and ephemeral lakes in arid environments (Stoerz and Ericksen 1974; Eugster and Kelts 1983; Goudie and Cooke 1984). However, in some extremely arid regions, crusts composed of highly soluble salts can accumulate as pedogenic horizons. Nitrate deposits in the Atacama Desert of Chile are associated with gypsum and halite crusts (Ericksen 1981; Goudie 1983b) similar to the pedogenic crusts of the central Namib Desert.

Figure 6.5   *Thin salcrete beds in a barchan dune in eastern Saudi Arabia. The sand is cemented by halite deflated from surrounding sabkhas. As the dune migrates, the lee face accretes and the salcrete is buried; the cemented horizons are exposed again as the windward face of the advancing dune ablates*

Distribution of desert crusts

None of the types of desert crust which have been described are found exclusively in warm desert environments, as some varieties can persist when the climate becomes wetter. Similar crusts may also form in humid climates or in regions where, despite low temperatures, high rates of evaporation promote aridity. As a result, desert crusts often have distributional characteristics which defy classification according to climatic parameters, making a detailed description of the geographical distribution of desert crusts beyond the scope of this brief review. For more information the reader is referred to the following work: Summerfield (1983a) and Milnes and Thiry (1992) on silcrete; Goudie (1973; 1983a), Reeves (1976) and Wright and Tucker (1991) on calcrete (caliche); Watson (1983a; 1985a) and Eswaran and Zi-Tong (1991) on gypsum crusts; and Watson (1983b), Goudie and Cooke (1984) and Lowenstein and Hardie (1985) on halite and other evaporites.

CALCRETE

Calcretes are the most widespread variety of desert crust, with petrocalcic soil horizons covering approximately 20 million km² or some 13% of the Earth's land surface (Yaalon 1988). Given the vast number of studies of global calcretes it might be expected that the environmental boundaries delimiting calcrete formation would be well-understood. Many are suggested to have formed where mean annual rainfall is between 400 and 600 mm (Goudie 1983a), but phreatic calcretes may accrete under much more arid conditions and rhizoconcretionary calcretes in more humid environments (Mann and Horwitz 1979; Semeniuk and Searle 1985). Annual precipitation alone, however, does not provide a precise distributional parameter for calcretes or any other type of desert crust. The seasonality of rainfall and average temperature during the wetter months determine the broad pattern of evapotranspiration. This, more than any other factor, probably limits the formation and preservation of most desert crusts. Aristarain (1971, p. 284) defined calcrete (caliche) as pedogenic calcite accumulations formed 'in climates in which moisture is deficient during all seasons'. In southwest Australia, calcretes are best developed where mean annual rainfall reaches 800 mm but mean annual evaporation is 1900 mm (Semeniuk and Searle 1985). Notwithstanding these limiting factors, some case-hardened limestones, morphologically identical to calcretes, are found in the humid tropics: in Guatemala (Blount and Moore 1969); on the Yucatan Peninsula (Quinones and Allende 1974); and in Puerto Rico (Ireland 1979). Exposed reef carbonates may also be similar (Multer and Hoffmeister 1968; Walls *et al.* 1975; Kahle 1977). It has also been demonstrated experimentally that the freezing of calcium-carbonate-saturated water leads to calcite precipitation (Ek and Pissart 1965; Adolphe 1972). In France some relict calcretes may have formed during cold phases of the Pleistocene (Vogt 1977) and might be similar to calcrete currently forming in Greenland (Swett 1974).

Most pedogenic calcretes mantle undulating or gently sloping terrain, with desert fans (Rutte 1958; 1960; Lattman 1973; Blümel 1976) and pediments (Boulaine 1961; Hüser 1976; Van Arsdale 1982) often calcretised. Extensive plateau calcretes are widespread in North America (Bretz and Horberg 1949; Brown 1956; Reeves 1970) and throughout North Africa (Flandrin *et al.* 1948; Moseley 1965) and the Middle East (Chapman 1974). In southern Africa exposures of calcrete occur throughout the Kalahari with greatest thicknesses in the southern Kalahari and in association with palaeolake basins and palaeodrainage features (Mabbutt 1955; 1957; Blümel and Eitel 1994; Nash *et al.* 1994a). There are comparatively few descriptions from Asia, though calcretes can be common (Goudie 1983a). In Australia, many calcretes have been interpreted as phreatic crusts (Mabbutt 1977, p. 137; Mann and Horwitz 1979; Arakel 1986; 1991; Jacobson *et al.* 1988) but pedogenic crusts are also extensive (Warren 1983; Semeniuk and Searle 1985; Semeniuk 1986).

The development of both phreatic and pedogenic (vadose) calcretes often bears little relationship to the materials upon which they accrete. Hence, typical profiles or characteristic geological associations are rare (Figure 6.6). Calcretes can occur on most sediments and rock types, both fresh or weathered, ranging from granites and schists (Scholz 1972) and volcanics (Hay and Reeder 1978) to gypsum bedrock (Lattman and Lauffenburger 1974), alluvium (Harden and Biggar 1983), dune sand (Warren 1983) and loess (Reeves 1970). It has been suggested that pedogenic calcretes form preferentially on basic rocks (Lattman 1973; Wells and Schultz 1979) but this would presuppose that the major source of carbonate was from bedrock as opposed to surface inputs. Arakel (1986) also felt that bedrock geology can be important in determining the distribution of vadose and phreatic calcretes in central

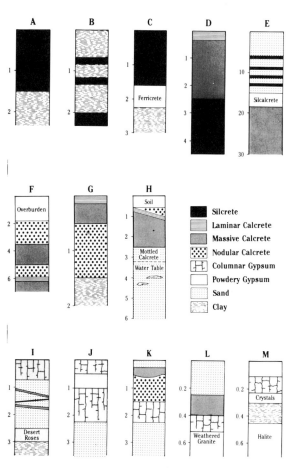

Figure 6.6 *Profiles of some desert crusts. All depths are in metres. (A) – (E) after Summerfield (1983a), (F) and (G) after Watts (1980), (H) after Arakel (1986), (I), (J), (L) and (M) after Watson (1985a) and (K) after Page (1972). (A) Massive silcrete on weathered shale, Cape Province, South Africa. (B) Massive silcrete interlayered with ferruginous clay near Albertinia, Cape Province, South Africa. (C) Massive silcrete over ferricrete on weathered tillite near Grahamstown, Cape Province, South Africa. (D) Calcrete on massive silcrete near Albertinia, Cape Province, South Africa. (E) Silcrete interbedded with Kalahari Sand on silcalcretised kimberlite, Botswana. (F) Complex calcrete, Botswana. (G) Typical laminar–hardpan–nodular calcrete profile on kimberlite, Botswana. (H) Typical phreatic calcrete profile, Australia. (I) Desert rose gypsum crust beneath calcretised and gypsum-encrusted dune sand, Tunisia. (J) Columnar surface gypsum crust above mesocrystalline subsurface crust, Tunisia. (K) Gypsum crust beneath calcrete, Tunisia. (L) Gypsum crust beneath calcrete, central Namib Desert. (M) Halite crust beneath gypsum crust, central Namib Desert*

Australia. There are numerous examples of calcretes formed by *in situ* alteration of limestone (e.g. Blank and Tynes 1965; El-Aref *et al.* 1985) or chalk (Yaalon and Singer 1974). Clearly, an immediate supply of calcium carbonate is conducive to calcrete formation but hydrological or atmospheric inputs from distant sources can be predominant.

SILCRETE

Silcretes are found in many parts of the world which did not undergo late Tertiary and Quaternary glaciation, including Australia, southern Africa, parts of the Sahara, the Arabian Gulf and, to a lesser extent areas of Brazil and western Europe (Milnes and Thiry 1992; Nash *et al.* 1994b). Despite the fact that silcretes have been identified on all continents except Antarctica, they are less globally significant than calcrete. Siliceous duricrusts are exposed in a variety of geomorphological settings, of which the most important are in association with remnants of palaeosurfaces or with drainage features.

Over large parts of southern, central and western Australia, silcretes encrust extensive areas of almost flat erosion surfaces and top low buttes and mesas where these surfaces have been incised (Pain and Ollier 1995; Twidale and Campbell 1995). However, silcretes also occur beneath basaltic flows (Brown 1926; Young and McDougall 1982) and within weathering profiles, so there is no typical stratigraphic or topographic form (Figure 6.6). Summarised studies of Australian silcretes are provided by Langford-Smith (1978), Callen (1983), Milnes and Twidale (1983), Twidale and Milnes (1983), Wells and Callen (1986), Thiry and Milnes (1991), and Milnes and Thiry (1992), among others.

In southern Africa, silcretes occur along the southern coast of the Cape Province of South Africa, in parts of the Namib Desert and over a large area of the centre of the subcontinent broadly coincident with the extent of the terrestrial Kalahari Group sediments. Silcretes occur in a variety of geomorphological settings including exposures associated with perennial, ephemeral and fossil drainage systems, escarpment caprocks and pan margins. There appears

to be a distinct genetic difference between silcretes from the Cape coastal areas (Frankel and Kent 1938; Frankel 1952; Mountain 1952; Smale 1973; 1978; Summerfield 1981; 1983a) and those from the Namib (Storz 1926; pp. 254–282; Knetsch 1937) and the Kalahari (Lamplugh 1907; MacGregor 1931; Maufe 1939; Dixey 1942; Summerfield 1982; 1983a; 1983b; 1983c; 1983d; Shaw and de Vries 1988; Nash *et al.* 1994a; 1994b). Those from the southern coast have been suggested to have formed in association with weathering profiles while those from the Namib and greater Kalahari may be lacustrine, phreatic or pedogenic (Figure 6.6). The origins of silcretes in these locations are still, however, the subject of debate.

Elsewhere in Africa, silcretes have been reported from Tanzania (Bassett 1954), Sudan (Kikoine and Radier 1949; 1950; Millot *et al.* 1959) and the Sahara (Auzel and Cailleux 1949; Elouard and Millot 1959; Belair *et al.* 1962; Jux 1983) where they are also interpreted as lacustrine or phreatic deposits or desert soils. There are few reports of silcrete from Asia, although limited exposures do occur in the Middle East (Khalaf 1988). Though there is some evidence of silicification of Quaternary sediments and soils (Flach *et al.* 1969), most European and North American silcretes are probably of Tertiary age (e.g. Summerfield and Whalley 1980; Thiry 1981; Thiry *et al.* 1988). Their interpretation as the products of either humid weathering or lacustrine evaporation and pedogenic processes in deserts has major palaeoenvironmental implications (Summerfield 1979; 1983c; Twidale and Hutton 1986; Milnes and Thiry 1992; Nash *et al.* 1994b).

There is a general lack of agreement over the distributional controls of silcrete formation. For example, in Australia, many present-day silcretes occur in areas which are less humid than where laterites are found but not as arid as those where calcretes predominate (Young 1978; Mann and Horwitz 1979; Milnes and Thiry 1992) (Figure 6.7). This has prompted the suggestion that the silcretes originated under semi-arid or markedly seasonal, wet and dry climates. However, even those that accept this view have differing opinions as to whether the crusts are the result of weathering (e.g. Brückner 1966; McGowran *et al.* 1977),

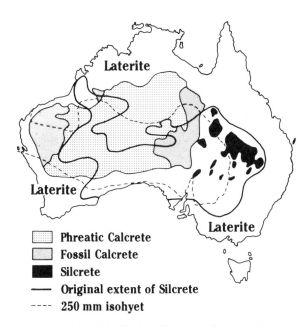

Figure 6.7 *Distribution of laterite, silcrete and phreatic calcrete in Australia in relation to mean annual rainfall (after Stephens 1971; Mann and Horwitz 1979)*

hydrological processes (Smale 1973; 1978; Watts 1975) or pedogenesis (Hutton *et al.* 1978; Callen 1983). It has also been argued that silcretes are relict horizons from deep chemical weathering profiles which formed under humid or seasonally wet, tropical conditions (Woolnough 1930; S.H. Watts 1975; 1977; 1978a; 1978b; Butt 1983; 1985) and were preserved under subsequent arid conditions (Twidale 1983). It has been suggested that silica may have been mobilised and precipitated in environments as diverse as riverine swamps and well-drained upland soils (Alley 1973; Langford-Smith and Watts 1978; Van De Graaff 1983; Wopfner 1983). Though the presence of silcrete in lateritic weathering profiles (Wright 1963; Alley 1977) may provide evidence of tropical silcrete formation, such occurrences do not prove simultaneous development (Smale 1973). It is unlikely that present climatic zonation reflects that at the time when many of the crusts formed; nor are the various types necessarily coeval even when they occur in the same profile (Langford-Smith and Dury 1965; Mabbutt 1965). It has also been argued that the extent of silcrete development varies according to the type of bedrock

(Williamson 1957; Mabbutt 1965). In such cases climatic controls may be relatively unimportant.

GYPSUM CRUSTS

Gypsum crusts are found in all the Earth's warm deserts, though their extent varies. Factors such as climate, bedrock geology, topography and hydrology are critical to their formation and preservation. Most gypsum crusts are found in areas where mean annual rainfall is less than about 250 mm (Watson 1985a). In North Africa, for example, there appears to be a transition from calcretes to gypsum crusts as mean annual rainfall drops below this amount (Pervinquière 1903, pp. 234–239). Rarely the upper rainfall limit reaches 300 mm; this is the case in Iraq (Tucker 1978) and Rajasthan (Srivastava 1969). Here, high mean monthly temperatures, especially during the wetter months, promote high rates of evaporation and it is likely that a net monthly soil moisture deficit is maintained throughout the year. As with calcrete distribution, there also appears to be an annual rainfall threshold below which the crusts are less common. In the central Namib Desert, there is a gradual transition from gypsum as rainfall drops below about 25 mm (Watson 1983b; 1985a).

The main areas of widespread gypsum crusts are in North Africa, particularly central Algeria (Durand 1949; 1963; Kulke 1974; Horta 1980) and Tunisia (Coque 1955a; 1962; le Houérou 1960; Vieillefon 1976; Watson 1979; 1985a; 1988); the central Namib Desert (Martin 1963; Scholz 1972; Watson 1981; 1985a; 1988); parts of Australia (Jack 1921; Warren 1982; Jacobson *et al.* 1988; Magee 1991; Milnes *et al.* 1991); and central Asia (Akhvlediani 1962; Tolchel'nikov 1962; Kondorskaya 1967; Evstifeev 1980; Eswaran and Zi-Tong 1991). They are also found throughout the Middle East from Egypt (Blanckenhorn 1921; Ali and West 1983), Israel (Dan *et al.* 1972; 1982; Amit and Gerson 1986) and Syria (Eswaran and Zi-Tong 1991) to Iraq (Tucker 1978), Iran (Gabriel 1964) and Kuwait (El Sayed 1993). Occurrences in the southwestern USA (Hunt *et al.* 1966; Reheis 1987; Harden *et al.* 1991) and

South America (Risacher 1978) seem to be more sporadic. They have been reported from Antarctica, where strong winds promote high rates of evaporation resulting in evaporite crystallisation (Gibson 1962; Lyon 1978).

As with many silcretes and calcretes, gypsum crusts often form an erosion-resistant horizon at the land surface. Perhaps because of their greater susceptibility to dissolution, however, gypsum crusts rarely create the mesa-and-butte landscapes associated with more durable crusts. Nevertheless, in southern Tunisia gypsum crusts mantle several generations of pediment slopes (glacis), apparently protecting relict surfaces from erosion (Coque 1955b; 1962). In many regions, gypsum crusts are found in and around large hydrological basins such as the chotts of Tunisia and Algeria, the salinas of South America, and salt lakes of Australia. While some crusts are lacustrine evaporites (Warren 1982) and others are phreatic precipitates (Kulke 1974; Risacher 1978), many occur on hill crests and on steep slopes beyond the phreatic zone. These are pedogenic crusts which blanket the landscape. Often the likely source of the gypsum is aeolian sand and dust deflated from desiccated evaporitic basins (Coque 1955a; 1955b; 1962). Elsewhere, sulphate-rich aerosols may be the main source, especially in the coastal Namib (Martin 1963; Carlisle *et al.* 1978) and Atacama (Ericksen 1981) Deserts. Such crusts may develop directly on unweathered bedrock such as granite, basalt, marble, limestone and clay (Watson 1985a; 1988), or on unconsolidated sediments such as colluvium, alluvium and dune sand (Watson 1979). As with silcretes and calcretes, there is no typical profile (Figure 6.6).

HALITE AND OTHER EVAPORITE CRUSTS

Halite and evaporite crusts are persistent features only in areas of extreme aridity. Thin, saline encrustations and efflorescences are usually dissolved by even small amounts of rainfall, but reform during subsequent dry, evaporitic conditions. These deposits are common in parts of Antarctica (Smith 1965; Torii *et al.* 1966). In lagoonal and lacustrine environments, influxes of water of low salinity can cause salt dissolution. However, this increases the brine's salinity, preventing further dissolution of the underlying evaporites and allowing thick salt sequences to accumulate (Morris and Dickey 1957; Busson and Perthuisot 1977; Perthuisot 1980). The distribution of these saline basins is, therefore, as much dependent on local hydrology as on climate. In general, they occur in areas where mean annual rainfall is less than about 200 mm (Stankevich *et al.* 1983; Goudie and Cooke 1984). In contrast, pedogenic halite crusts in the Namib and Atacama Deserts occur only where mean annual rainfall is less than about 25 mm.

As with gypsum crusts, those halite crusts associated with lacustrine and phreatic evaporative processes occur in topographic basins and littoral settings. Some salcretes, however, are probably the result of cementation by airborne salts and aerosols deposited on the land surface. Such crusts may consolidate the sand on dune slipfaces inclined at about 32° (Figure 6.5). Though these evaporite crusts do not have a lasting effect on the landscape owing to their susceptibility to dissolution, the salts are import in desert weathering (Cooke 1981; Sperling and Cooke 1985) and pedogenesis (Yaalon 1964b; Orlova 1980). They can also be of economic significance; Chilean nitrate deposits and many petroleum reserves are associated with evaporite sequences.

Since the sources of salts forming pedogenic halite crusts and certain salcretes are often atmospheric, the underlying materials can be very variable. Though pedogenic halite crusts are found in association with gypsum crusts (Figure 6.6), usually there are no distinct stratigraphic associations. Lagoonal, lacustrine and sabkha halite crusts may be interbedded with gypsum (Phleger 1969) or anhydrite. Other characteristic facies, as well as structural and textural features, are well-documented (Shearman 1966; Arakel 1980; Warren and Kendall 1985; Schreiber 1986).

Micromorphology, chemistry and origins of desert crusts

In order for any desert crust to form there needs to be a source of cementing material and mechanisms to mobilise, transport and ultimately precipitate the cementing agent. Conditions both within the source area and the site of precipitation therefore have an important influence upon the development of any crust (Ollier 1991).

The preceding sections have shown that there are several structural forms of most of the main types of desert crust. However, it is often difficult to identify the mode of origin of a particular crust by simply studying it in profile. Few profile-scale morphological characteristics can be ascribed to discrete genetic processes and, as such, simple morphogenetic models for crust formation are inappropriate. It is therefore important that the micromorphology and geochemistry of crusts also be taken into account. In some cases micromorphological fabrics and crystalline textures can be attributed to genetic or diagenetic processes which are specific to different chemical environments. In effect, the micromorphology and chemistry of many crusts provide the best available information on their origins.

CALCRETE

The carbonates which ultimately form a calcrete can come from a variety of sources (Goudie 1983a). These include carbonates introduced into the profile both from above and below, such as surface inputs of dust, carbonate-rich plant materials and shells, and subsurface contributions from ground water and weathered bedrock. A variety of mechanisms to transfer the source carbonate to the location of calcrete formation have been described, including vertical transfers within the soil profile driven by capillary rise mechanisms and water-table fluctuations and lateral tranfers associated with movement of carbonate in solution within lacustrine or channel-margin settings (see Chapter 15).

Calcium carbonate dissolved in water may be precipitated by a variety of processes including evaporation, an increase in pH to above pH 9.0, a decrease in the partial pressure of CO_2 (Schlesinger 1985), a loss of CO_2 as temperature increases (Barnes 1965), freezing (Ek and Pissart 1965; Adolphe 1972), the common ion effect (Wigley 1973), evapotranspiration, and biological activity (Goudie

1983a). The calcium carbonate in many calcretes is mainly micrite — aggregates of calcite crystallites less than about 100 μm in diameter. The presence of micrite can indicate specific conditions under precipitation; for example, rapid evaporation. However, it is not always syngenetic with calcretisation; sparry calcite may alter to micrite as a result of dissolution and reprecipitation (Kahle 1977).

Other common calcite crystal textures include flower spar and both random and tangential fibres (James 1972) which may crystallise from water supersaturated with calcium carbonate (Knox 1977). Rhombic calcite crystals found in calcrete pisoliths may accrete by displacement of host material (Folk 1971; Chafetz and Butler 1980). Other nodular structures are usually micritic, as are the laminae which characterise many calcretes.

The role of organisms in calcrete formation is being increasingly recognised and it has been suggested that some calcretes may be formed predominantly by the action of microorganisms, especially algae (Kahle 1977), fungi (Knox 1977; Verrecchia 1990; Verrecchia *et al.* 1990; 1993) and bacteria (Lattman and Lauffenburger 1974; Folk 1993). This has led Wright (1990) to propose a micromorphological classification of calcretes, with end members comprising Alpha calcretes consisting of skeletal grains within a micritic or microsparitic groundmass and Beta calcretes which are dominated by biogenic features. Laminae, in particular, have been recognised to be of organic origin (Klappa 1979a; 1979b; Wright 1989; Verrecchia *et al.* 1991) and one crystal texture consisting of millimetre-long calcite prisms has been attributed to a hypothetical colonial bacterium, *Microcodium* (Esteban 1974; Klappa 1978; Laurain and Meyer 1979). Undoubtedly, the formation of some calcretes is related to calcite precipitation around plant roots (Klappa 1979a; 1979b; Warren 1983; Semeniuk and Searle 1985) but it is uncertain whether this is a widespread phenomenon.

Brecciation structures which occur in many calcretes have been attributed to root growth (Klappa 1980), but such features, along with fibrous textures, certain nodules and floating grain fabrics, have been cited as evidence of displacive crystallisation (Watts 1978) (Figure 6.8) although this has been disputed (Klappa

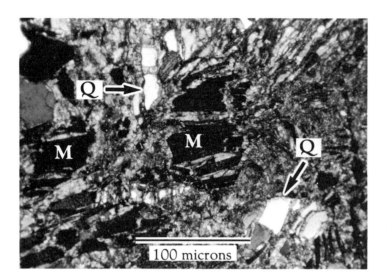

Figure 6.8 *Lath-like mica fragments (M) and quartz grains (Q) brecciated by micrite in a calcrete from the central Namib Desert. Exploded textures of book-leaf micas and fragmented quartz clasts may be indicative of displacive calcite crystallisation. Cross-polarised light; not orientated*

1979c; 1980; Braithwaite 1983). Since the calcium carbonate content of many calcretes often exceeds the host material's volumetric porosity (Gardner 1972), there has been considerable debate as to whether displacive or replacive processes predominate during subsurface accretion. Evidence of displacement, such as brecciated cobbles and bedrock (Young 1964; Blount and Moore 1969; Asserto and Kendall 1977), may not result primarily from calcite crystallisation. Large-scale features such as pseudo-anticlines which form domes and ridges at the land surface (Price 1925; Jennings and Sweeting 1961) have been attributed to volumetric expansion owing to displacive accretion of calcrete (N.L. Watts 1977; 1978; 1980). However, some are formed by chemical rather than physical processes (Blank and Tynes 1965). In order to explain the high purity of some crusts, chemical replacement rather than physical displacement of host material has been advocated. Hubert (1978), for example, held that palaeosol calcretes in Connecticut show signs of 50–95% replacement; McFarlane (1975) suggested that a fabric of floating garnet grains in some Kenyan calcretes is the only residuum of the now replaced host material. Many calcretes exhibit signs of replacement of clay (Hay and Reeder 1978) and feldspars and quartz (Burgess 1961; Chapman 1974), but the latter may be diagenetic rather than syngenetic features. It is likely that displacive, replacive and passive, void-filling processes can all take place (Yaalon and Singer 1974; Watts 1980; Nash *et al.* 1994a), but their relative importance varies according to host lithology and environmental conditions.

Though considerable debate has focused on the origin of arid zone calcretes, there has been increasing agreement that most are either pedogenic or phreatic. A lacustrine origin (Durand 1963) is unlikely because evaporitic limestones are structurally distinct (Eugster and Kelts 1983), although calcretes are commonly associated with pans and playas in the USA, Australia and southern Africa (see Chapters 14 and 15; Figures 15.11 and 15.12). A pedogenic, as opposed to a phreatic, origin should be detectable through the presence of calcified filaments, needle-fibre calcites and fungal microfossils thoughout a calcrete (Vaniman *et al.* 1994). In many areas, both genetic types occur, with phreatic forms often developing pedogenic features as hydrological conditions change (Mann and Horwitz 1979; Arakel 1982; 1986). Clear chemical distinctions are often masked by diagenesis, although Manze and Brunnacker (1977) showed that $^{12}C/^{13}C$ and $^{16}O/^{18}O$ ratios can distinguish the two types. At present, there is little information on stable isotopes from phreatic calcretes (Manze and Brunnaker 1977; Jacobson *et al.* 1988), but with further research it may be possible to recognise phreatic and pedogenic origins. Aristarain (1970) pointed out that since the calcium carbonate in pedogenic crusts is introduced from above, vertical variations in ionic concentrations may provide evidence of the genetic processes involved. In such cases, surface deposits of carbonate may be fluvial or colluvial sediments, volcanic ash, aeolian sand and dust, aerosols or organic detritus (Goudie 1983a). Phreatic calcretes develop by crystallisation of calcium carbonate from ground water near the water table (Figure 15.13). A variety of mechanisms can be involved, but evaporation and loss of CO_2 are predominant. A common characteristic of these crusts is the enrichment of certain ions such as strontium (Kulke 1974) or, more unusually, the concentration of minerals such as carnotite $(K_2(UO_2)_2(VO_4)_2 \cdot 3H_2O)$ (Barbier 1978; Carlisle *et al.* 1978; Arakel and McConchie 1982; Briot 1983).

Diagenesis markedly alters the micromorphological and chemical characteristics of calcretes. A detailed account of the diagenetic features of arid zone calcretes is beyond the scope of this review. However, the formation of minerals

such as quartz (Reeves 1970), chert (Walls *et al.* 1975) and opal (Brown 1956; Sowers 1983; Nash *et al.* 1994a) has important implications for silcrete and sil-calcrete genesis. Also, many dolocretes are diagenetic. The complex chemistry of the interactions between high- and low-magnesium calcite, silica and dolomite during diagenesis may produce authigenic clay minerals (Figure 6.9). In some calcretes these may be detrital (Beattie and Haldane 1958; Shadfan and Dixon 1984) or related to waterlogging (Hodge *et al.* 1984), but there is strong evidence that under alkaline conditions sepiolite and palygorskite (attapulgite) are authigenic (Hassouba and Shaw 1980; Watts 1980; Mackenzie *et al.* 1984; Singer 1984; Gauthier-Lafaye *et al.* 1993). The diagenetic processes may be extremely varied and can be determined by both the mineralogy of the host materials (Hay and Reeder 1978; Hay and Wiggins 1980; Hutton and Dixon 1981) and the introduction of foreign ions (El-Aref *et al.* 1985).

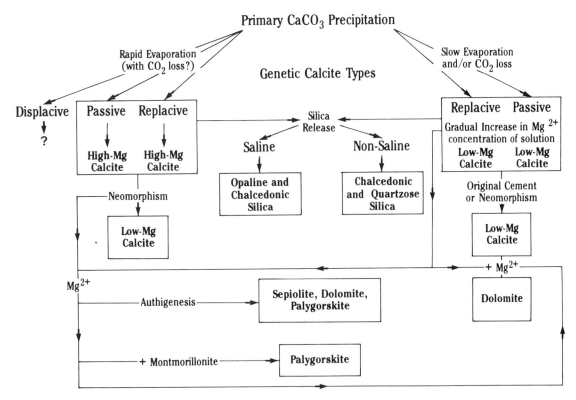

Figure 6.9 *Schematic summary of the main pedogenic and diagenetic processes in calcretes from the Kalahari (after Watts 1980)*

SILCRETE

As in the case of calcretes, the silica required for silcrete formation can come from a variety of sources. Of these, chemical weathering of silicate minerals is probably the most important, although replacement of quartz by carbonates may provide a major source in certain environments. Additionally, surface solution of quartz dust may be a silica source in deserts (Summerfield 1983a) together with biological inputs from certain silica-rich plants and micro-organisms such as diatoms. Again, the cementing agent may be transferred from its source by a variety of mechanisms of which the most important involve lateral or vertical movements of groundwater, porewater and surface water.

In arid environments, silica dissolved in water probably originates as quartz or silicates, especially clay minerals. However, quartz is only weakly soluble in neutral water — about 10 p.p.m. at 25°C (Siever 1972). It is only above pH 9.0 that the solubility of quartz increases significantly (Krauskopf 1956); that of amorphous silica reaches 800 p.p.m. at pH 10.6 (Alexander *et al.* 1954). Such highly alkaline conditions are not unusual in saline lakes. Following the dissolution of silicates at pH 9.5–10.5, desiccation can lower the pH to about 7.0, precipitating chert (Peterson and van der Borch 1965; Wheeler and Textoris 1978) or hydrous silicates such as magadiite ($NaSi_7O_{13}(OH_3) \cdot 3H_2O$) which converts to chert or opaline silica gel when leached by dilute water (Baker 1986; Renaut *et al.* 1986). There is evidence that silica gels may be the primary precipitates forming some silcretes, but it does not necessarily follow that such silcretes are of lacustrine origin because gels also occur in terrestrial settings (Kaiser 1926, p. 296; Storz 1926). They may precipitate from any silica-saturated alkaline solution (Elouard and Millot 1959; Millot *et al.* 1959). In contrast, quartz nucleates from dilute solutions, but only in the absence of inhibiting minerals (such as chlorite, illite and haematite) or ions (such as iron and magnesium) (Exon *et al.* 1971; Heald and Larese 1974; Ollier 1978). This may explain the restriction of many silcretes to ancient, highly chemically weathered landscapes where these inhibiting minerals are

absent. However, such broad relationships may be oversimplistic; in practice, the degree of silica saturation, rate of precipitation and presence of other ions can all contribute to the crystallographic form of the precipitate (Summerfield 1983a) and the presence of early silica precipitates will influence subsequent crystallisation (Landmesser 1995).

In addition to being a result of a drop in pH or evaporation, silica precipitation can occur as a result of cooling, organic processes (Shaw *et al.* 1991; McCarthy and Ellery 1995), absorption by solids, and reaction with cations (Summerfield 1983a). While the last mechanism may form clay minerals, it has been suggested that the mixing of silica-rich groundwater with downward-percolating water rich in sodium chloride may precipitate silica (Frankel and Kent 1938; Frankel 1952; Smale 1973). Accretion owing to groundwater evaporation has also been advocated (Lamplugh 1907; Waugh 1970) sometimes involving lateral flow (Mountain 1952; Bassett 1954). Others have argued that arid zone silcretes are pedogenic (Auzel and Cailleux 1949). In whatever case, silica precipitation would eventually infill all the voids in the host material, thereby sealing the horizon and preventing further accretion. In order to maintain the porosity, the host material must be displaced (Butt 1985) or chemically replaced (Soegaard and Eriksson 1985). It is likely that lacustrine, phreatic and pedogenic processes in arid environments can all form siliceous crusts under favourable chemical conditions (Williamson 1957; Callen 1983). Moreover, subsequent diagenesis may obscure the micromorphological and geochemical characteristics of silcretes originating in these different edaphic regimes. Hence, not only may silcretes form in either humid or arid settings, but the diversity of genetic and diagenetic processes complicates attempts to ascertain their provenance.

One of the main debates within the silcrete literature, as noted above, concerns the determination of the environment under which siliceous duricrusts form. Two approaches were adopted during the 1980s and early 1990s. One centred around the need to identify characteristics which would distinguish silcretes which had developed within weathered profiles from those which had formed in the absence of such a profile (e.g. Summerfield 1983c; Twidale and Hutton 1986; Nash *et al.* 1994b). Parallel to these studies, highly detailed petrographic and geochemical analyses of silcrete profiles in France and Australia were being undertaken by Medard Thiry and others in order to ascertain how the complex textures and fabrics found in many silcretes were formed (Milnes and Thiry 1992) (Figures 15.14 and 15.15). These latter studies have enabled the identification of distinctly different silcrete types dependent upon whether the silcrete formed via pedogenic processes or by groundwater silicification (see Chapter 15 for a full discussion). All of these studies have used microscale analyses in order to progress our understanding of silcrete formation.

Four main forms of silcrete can be identified at a microscale and may be classified using micromorphological criteria as having either grain-supported fabrics (quartzitic and conglomeratic silcretes) or fabrics with more than about 40% silica cement, and either quartzose, chalcedonic or opaline cement (porcellanite and terrazzo silcretes) (Figure 6.10). If the matrix

cement exceeds about 40% of the volume, clastic inclusions are dispersed, creating a fabric of floating grains. Porcellanite and silcretes with floating grain fabrics probably form mainly by pedogenesis within deeply weathered profiles (S.H. Watts 1975; 1977; 1978a; Summerfield 1979; 1983a), the fabrics having developed by silica replacement of the former matrix or possibly by displacive crystallisation (Butt 1985). Grain-supported fabrics are most common in arid zone silcretes (with matrix and floating fabrics also occurring) but have not been reported for silcretes developed within weathering profiles (Figure 6.10). In many cases, grain-supported silcretes, such as those occurring within the Kalahari, appear to have developed by silica infilling pore spaces within sandy host materials (Nash *et al.* 1994a).

It has been suggested that it is possible to distinguish two main environments of silcrete formation on the basis of the type of material cementing the crust. A low pH environment is indicated by either crystalline quartz growth or cryptocrystalline quartz aggregation, the latter cement developing within a soil profile and the former below the zone of pedogenesis. In contrast, alkaline oxidising conditions are indicated by a crystobalitic–tridymitic silica cement (Wopfner 1978; 1983) (Figure 6.10). However, the apparent association of quartzose and anhydrous cryptocrystalline silica cements with kaolinisation in weathering profiles is dubious, and the relationship between opaline cements and evaporite minerals is not universal (Folk and Pittman 1971; Folk 1975). Nevertheless, optically continuous overgrowths of chalcedony in host quartz grains may be absent from weathering-profile silcretes, so these crystalline textures may be genetically diagnostic (Summerfield 1979; 1983a; 1983b).

Other micromorphological features which may distinguish the arid zone and pedogenic silcretes associated with weathering profiles are colloform structures (cusp-like laminations and drapes within porespaces) and authigenic glaebules (globular structures) which occur within pedogenic silcretes associated with weathering-profiles (Frankel 1952; Williamson 1957; Summerfield 1979; 1983a; 1983b; Milnes and Thiry 1992). These features are usually enriched in anatase (TiO_2) (Smale 1978; Watts 1978a) and often zircon ($ZrSiO_4$), providing a

Fabrics	Textures
A. Grain supported (quartzitic silcrete)	i. optically continuous overgrowths ii. chalcedonic overgrowths iii. microquartz, cryptocrystalline or opaline matrix
B. Floating (> 5.0% skeletal grains) (terrazzo silcrete)	i. massive ii. glaebular (pisolitic or pseudo-pebbly)
C. Matrix (< 5.0% skeletal grains) (porcellanite/albertinia)	i. massive ii. glaebular
D. Conglomeratic (pebbles > 4.0 mm)	

Matrix Cements	Textures	Macrostructure
A. Crystalline quartz	i. irregular crystalline ii. optically continuous overgrowths	blocky or pillowy
B. Cryptocrystalline quartz	i. massive ii. glaebular (pisolitic) iii. laminated	botryoidal, columnar, platy or pillowy
C. Opaline (amorphous)		brecciated, conglomeratic or replacive and void-filling

Figure 6.10 *Silcrete classifications; top after Summerfield (1983a), bottom after Wopfner (1978). Silcretes developed in association with weathering profiles are limited to types (B) and (C) above and (A) and (B) below. Types (A) and (D) above and (C) below typically occur in arid environments and may be associated with highly alkaline conditions*

strong geochemical parameter for differentiating the two types of silcrete. Silcretes developed in weathered bedrock contain 1–3% or more anatase; silcretes which are not formed in weathering profiles contain less than 0.2% anatase (Hutton *et al.* 1972; Summerfield 1979; 1983a; 1983b). However, this distinction may be overly simple if, as in the case of some Kalahari silcretes, silcrete formation occurred in association with a deep-weathering environment but within TiO_2-poor sediments (Nash *et al.* 1994b).

These geochemical characteristics may result from concentration of relatively immobile minerals and trace elements, such as yttrium, within silcrete accretion zones in weathering profiles (Hutton *et al.* 1972; 1978; Twidale and Hutton 1986). However, they can be mobilised when pH falls below 4.0 (Summerfield and Whalley 1980; Butt 1983; 1985) or in the presence of micro-organisms (McFarlane *et al.* 1992), so these silcretes could be replacive of pre-existing material and are not necessarily residual accumulations. The many fascinating problems pertaining to the formation of weathering profile silcretes cannot be elaborated here; for a full discussion the reader is referred to Summerfield (1983a; 1983d) and Milnes and Thiry (1992). Suffice it to point out that arid and semi-arid conditions seem most conducive to the accretion of those silcretes which are not formed in weathering profiles.

GYPSUM CRUSTS

The solubility of gypsum in pure water is about 2.0 g $CaSO_4 L^{-1}$ — it varies very little with temperature (Hill 1937) but in the presence of sodium chloride gypsum solubility is raised to 8.0 g $CaSO_4 L^{-1}$ at 35°C when the solvent contains 100 g $NaCl L^{-1}$ (Zen 1965). Moreover, sodium ions appear to influence the crystalline habit of gypsum precipitates (Masson 1955; Edinger 1973). In effect, despite the greater solubility of alkaline environments, the micromorphology of gypsum crusts is often more genetically diagnostic than that of many calcretes and silcretes.

As in the case of lacustrine limestones, horizontally bedded gypsum crusts are structurally distinct from the other forms described above.

These crusts are characterised by size-grading of crystals in strata up to about 0.10 m thick; within individual beds there is a gradation from lenticular crystals about 0.50 mm in diameter at the base to alabastrine crystallites (less than 50 μm in diameter) at the top. Desert rose crusts are distinct not only because of the large, lenticular crystals but also because these crystals include clastic host material. The poikilitic inclusions are evident even in relict crusts which have experienced pronounced diagenesis (Watson 1988). Mesocrystalline subsurface crusts are pedogenic accretions which contain fibrous gypsum, a rare but highly significant crystalline texture. As with pedogenic calcretes, it is possible that this habit attests to displacive crystallisation within host material. The various forms of surface crust — columnar, powdery and cobble — are characterised by alabastrine textures. These are largely diagenetic, resulting from dissolution of lenticular crystals and reprecipitation of alabastrine material — a process akin to sparmicritisation in calcretes (Kahle 1977). Most surface crusts are exhumed mesocrystalline crusts (Watson 1985a; 1988), the cobble form representing a stage in the degradation of columnar crusts to powdery residua as a result of dissolution and leaching. Exhumation of the illuvial mesocrystalline crusts is often brought about by aeolian erosion of the unconsolidated overburden.

The lacustrine, phreatic and pedogenic forms of gypsum crusts are chemically distinct. In addition to significant differences in gypsum content and the clastic component, sodium levels are highest in the lacustrine evaporites followed by the phreatic crusts, and are lowest in the pedogenic forms. Magnesium levels show a similar trend. The evolution from the illuvial mesocrystalline form through columnar surface crusts to the cobble form results in a decrease in the concentration of sodium and strontium. This is probably related to the release of ions coprecipitated with the lenticular gypsum (Kushnir 1980; 1981) when it is dissolved and reprecipitated as alabastrine material. The lack of crystal growth following nucleation limits further coprecipitation.

The structural, micromorphological and chemical characteristics of desert gypsum crusts provide clear evidence of the different processes involved in their accretion. The bedded, lacustrine evaporites probably precipitated in shallow-water environments. The size-graded strata indicate that the water body evaporated to dryness because late in the evaporative cycle there was minimal ionic migration to the growth faces on alabastrine crystallites. The phreatic, desert rose crusts probably accrete as gypsum precipitates from evaporating groundwater (Castens-Seidell and Hardie 1984) where the water table is within 1 or 2 m of the land surface. Precipitation as a result of dilution by less saline meteoric water (Pouget 1968) is unlikely because it does not allow for the prolonged hydrochemical stability which is required for the

growth of large crystals. The pedogenic crusts are of illuvial origin. Gypsum deposited at the surface (either as sand, dust or aerosols) is dissolved by rainwater and leached into the soil zone where it precipitates during subsequent desiccation. Though some crusts exhibit replacive features, evidence of displacive crystallisation is more common (Figure 6.11). Displacive accretion is achieved mainly by uplift of the overlying host material — a process similar to that envisaged in the formation of some desert pavements (McFadden *et al.* 1987). The evolution of the surface forms is contingent upon the exhumation of illuvial crusts.

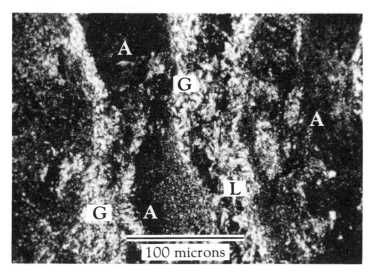

Figure 6.11 *(A) Large quartz grain (Q) in a groundmass of mainly alabastrine gypsum in a columnar surface crust from Tunisia. Though the clast shows signs of dissolution, this may have occurred in an alkaline sabkha environment prior to the grain's deflation rather than being syngenetic with gypsum crust formation. Cross-polarised light. (B) Bands of alabastrine gypsum (A), lozenge-shaped gypsum crystals (L) and granular textures (G) in the wall of a column of a surface gypsum crust from Tunisia. Such banding may result from displacive gypsum crystallisation in the fissures between columns. Cross-polarised light*

Little work has been undertaken on mineral authigenesis in gypsum crusts. Authigenic palygorskite may be more prevalent in some pedogenic gypsum crusts than in calcretes (Yaalon and Wieder 1976; Reheis 1987). This may be related to the development of two-tiered calcrete–gypsum crusts (Figure 6.6) as a result of differential leaching of soluble minerals (Drever and Smith 1978; Watson 1985a). This is described below.

HALITE CRUSTS

The formation of some terrestrial halite crusts is a pedogenic process. In the Namib Desert (Watson 1983b) and the Atacama Desert (Ericksen 1981), halite crusts are illuvial accretions formed by the same processes as pedogenic calcretes and gypsum crusts. Salt deposited at the surface in aeolian sand and dust (Orlova 1980), or in fog moisture in the Namibian and Chilean deserts, is leached into the solid zone where it recrystallises when the soil moisture evaporates. Provided that the soil's moisture storage capacity is not exceeded, the salts accumulate as an illuvial horizon. Highly soluble salts are less likely to persist because even infrequent saturation of the soil zone will cause flushing. The markedly different solubilities and resulting vertical mobilities (Yaalon 1964a) of the common crust-forming minerals can result in the formation of two-tiered crusts (Figure 6.6) if more than one salt is present and rainfall is sufficient to mobilise the less soluble salt but insufficient to flush the more soluble. The less soluble salt accumulates in an illuvial horizon above the more soluble. Calcretes above gypsum crusts are found in North Africa (Page 1972; Horta 1980) and gypsum crusts above halite crusts in the central Namib Desert.

Little information is available on the micromorphological characteristics of pedogenic halite crusts. However, there is evidence that crystal habits differ depending on the depositional environments. Surface efflorescences of halite can be made up of fibrous crystals while cubic or trigonal pyramids crystallise in porous media (Eswaran *et al.* 1980). Granular crystal textures predominate in phreatic crusts (Arakel 1980), and halite precipitated in shallow-water bodies by evaporation usually has a cubic

crystal habit — hopper crystals with negative-crystal brine inclusions are also common (Arakel 1980; Lowenstein and Hardie 1985). Cloudy crystals — some containing poikilitic inclusions — are probably primary precipitates, while clear crystals are syndepositional void-fill. Saccharoidal textures form during rapid, sporadic halite precipitation. They occur as bridge cements which consolidate the host grains in thin salcretes.

The mineralogical and chemical characteristics of halite and other evaporite crusts are also closely related to the depositional environments. Pedogenic halite crust may contain significant quantities of accessory minerals — particularly sulphates and, less commonly, nitrates. In lacustrine and sabkha settings, evaporite minerals associated with primary halite and gypsum deposits such as celestite ($SrSO_4$) (Evans and Shearman 1964; Magee 1991) and polyhalite ($K_2Ca_2Mg(SO_4)\cdot 2H_2O$) (Holster 1966; Perthuisot 1980) are formed by diagenesis.

Desert varnish

Unlike the various types of crust which occur at or just beneath the land surface, often mantling dryland landscapes, desert varnish forms only a thin coating on exposed pebbles and rock surfaces. Desert varnish is a variety of rock varnish which is a much more widespread phenomenon occurring in tropical (von Humboldt 1819, Vol. 2, pp. 299–304; Francis 1921), Antarctic (Glasby *et al.* 1981; Dorn *et al.* 1992) and high-altitude (Klute and Krasser 1940; Whalley 1984; Whalley *et al.* 1984) environments. It has been suggested that true desert varnish is restricted to areas with a mean annual rainfall of less than 130 mm (Engel and Sharp 1958), but it is also argued that in many areas the coatings formed under wetter conditions and are now relict (Hunt 1961; Hunt and Mabey 1966). Despite their widespread occurrence, it is only possible briefly to describe the nature of desert varnishes within this chapter — readers are referred elsewhere to recent summaries of this rapidly developing field (e.g. Dorn 1991; Oberlander 1994).

Desert varnish is a surficial stain or encrustation from 2.0–5.0 μm to 0.5–1.0 mm thick

(Krumbein and Jens 1981; Dorn and Oberlander 1982). Usually orange, grey, brown or black in colour, the coatings often have a lustrous appearance and have been referred to as 'desert lacquer' or 'patina' (Figure 6.12). Varnish is distinct from desert glaze (Fisk 1971), which is a siliceous precipitate, and also differs from various types of weathering rind which exhibit a transition zone several millimetres thick from chemically altered material at the surface to sound rock beneath. The coat of varnish is usually composed of about 70% clay minerals and the remainder of oxides and hydroxides of iron and manganese admixed with detrital silica and calcium carbonate (Dorn and Oberlander 1982). In general, the blacker varnishes contain more manganese (20% MnO_2 by weight), while iron predominates in orange varnish (10% FeO_2 by weight and less than 3% MnO_2) (Potter and Rossman 1977; Perry and Adams 1978; Elvidge and Moore 1979). However, some rock varnish contains little or no manganese and only minor amounts of iron (less than 5.0% FeO_2 by weight) (Glasby *et al.* 1981) and some may be composed mainly of organic matter (Krumbein and Jens 1981). In addition to Mn, Fe, Si, Al and Ca, other major elements found in rock varnish include K, Ba, Na, Sr, Cu and Ti, while minor elements such as V, P, Co, Cd, Ni, Pb, Zn, La, Y, Zr and B may be locally significant (Lakin *et al.* 1963; Potter 1979).

The mineralogy of desert varnish has proven difficult to decipher because much of the material is X-ray amorphous. The clay matrix is commonly a mixed-layer illite–montmorillonite, with birnessite and haematite as the main manganese and iron minerals (Potter and Rossman 1979). Trace element concentrations show considerable enrichment — sometimes several hundred times — relative to the levels in the host rock. This has been cited as evidence that the varnishes are the products of intense surficial weathering but it may also indicate that the source of the materials is external.

In warm deserts, high temperatures may be important in promoting thicker and blacker (manganese-rich) varnish. The type of rock which the varnish coats may also be a significant influence, although the fact that many recent investigations have identified the presence of elements within varnish which cannot have

Figure 6.12 *Desert varnish developed upon ophiolite-rich desert pavement near Barzaman, Sultanate of Oman. The upturned clast immediately left of the penknife (length 9 cm) allows the difference in colour between the blackened varnish and the underlying vesicular material to be seen*

been derived from weathering of the substrate material suggests that this may not always be the case (Oberlander 1994). While a strong lithological control has frequently been advocated (e.g. Engel and Sharp 1958; Marshall 1962), varnish has been described coating a wide variety of rock types ranging from glassy quartz and acid igneous rocks to basic rocks, and from siliceous sandstones to limestones and calcretes (Perry and Adams 1978; Dorn and Oberlander 1982). However, rocks which are susceptible to rapid weathering or dissolution do not exhibit well-developed varnishes (Whalley 1983). Though desert varnish is exclusively a subaerial phenomenon, usually coating only rock surfaces, occasionally varnish may spread from pebbles to adjacent unconsolidated material (Laudermilk 1931; Scheffer *et al.* 1963).

It has been suggested that desert varnish may be of great geomorphic importance in protecting rocks from weathering (Merrill 1898). However, Engel and Sharp (1958) found that it is readily destroyed by mechanical flaking and dissolution under wet conditions. Resistance to removal is related to varnish chemistry, clay mineral content and the roughness of the underlying substrate, with Si-rich varnish on originally rough substrate having the highest resistance and some clay- and Fe-rich varnishes on smooth rock being sufficiently soft to rub off with a fingertip (Oberlander 1994). Dorn and Oberlander (1982) reported hardnesses approaching 6.5 Moh (close to the hardness of quartz, which is 7.0 Moh) and Allen (1978) held that in the Sonoran Desert it is destroyed only by sandblasting.

Desert varnish is frequently made up of alternating layers of light and dark material ranging in thickness from the nanometre scale up to 20 μm (Perry and Adams 1978; Dorn 1984; Dragovich 1988; Krinsley *et al.* 1995). The dark laminae contain about 20% MnO_2 while the lighter ones have less than 3.0%. Their composition appears to be intimately linked to environmental conditions because manganese fixation is dependent on pH; alkaline conditions restrict fixation and orange-coloured varnish develops (Dorn and Oberlander 1982). Other micromorphological features of some varnishes include dendritic growth patterns (Billy and Cailleux 1968) and botryoidal textures (Dorn and Oberlander

1982). All are related to mechanisms of varnish accretion.

There is much debate over the origin of desert varnish (see, for example, Smith and Whalley 1988; 1989; Dorn 1989). Theories of desert varnish development can be divided into two broad groups dependent upon whether the coating formed from an internal or external source (Drake *et al.* 1993). Much of the early work on the origin of desert varnish suggested that the coatings developed as moisture was drawn out of rocks, precipitating the minerals that were dissolved within (Merrill 1898; Linck 1901; Blake 1905). Polishing of the varnish was felt to be a result of aeolian abrasion (Klute and Krasser 1940). However, this process of sweating would form a leached and weathered zone beneath the varnish; occasionally this is present (Allen 1978; Glasby *et al.* 1981) but the coating should then perhaps be classified as a weathering rind. Nevertheless, even when the rock cortex is unaltered, varnish has been attributed to surficial chemical weathering and the accumulation of an iron- and manganese-enriched residue (Engel and Sharp 1958; Marshall 1962; Whalley 1983; 1984; Smith and Whalley 1988). Others have advocated organic weathering by lichen (Laudermilk 1931) or micro-organisms such as fungi, algae and bacteria (Scheffer *et al.* 1963; Krumbein 1969; 1971; Krumbein and Jens 1981). Micro-organic weathering may account for manganese fixation at a pH below 9.0 — abiotic oxidation occurs only above pH 9.0 but most varnishes have a neutral pH (Dorn and Oberlander 1982). However, a biotic manganese fixation may be accomplished at a lower pH in the presence of bentonite or illite (Reddy and Perkins 1976).

In contrast with models of autistic formation of varnish, there are those which propose an external source of manganese, iron and clay minerals. Such sources may explain the presence of trace elements which are virtually absent from the host rock (Lakin *et al.* 1963; Knauss and Ku 1980). Again, either physico-chemical or micro-organic processes are involved in the fixation of deposits of airborne dust or aerosols — White (1924) suggested that pollen might be a source. Aeolian dust deposits may form varnish when the clay component effects manganese fixation (Potter and Rossman 1977). Alternatively, the manganese may also be allochthonous (Allen 1978; Perry and Adams 1978; Elvidge and Moore 1979). Micro-organic rather than physico-chemical fixation of introduced manganese has been widely regarded as the most likely process of varnish formation, with studies in North Africa (Drake *et al.* 1993), North America (Dorn and Oberlander 1981a; 1981b; 1982; Taylor-George *et al.* 1983; Palmer *et al.* 1985; Nagy *et al.* 1991), South America (Jones 1991), Australia (Staley *et al.* 1983; Dorn and Dragovich 1990) and the eastern Mediterranean (Krumbein and Jens 1981; Hungate *et al.* 1987) proposing a biological origin. However, there is still considerable debate over the specific types of micro-organism involved in varnish formation. While Krumbein and Jens (1981) suggested that lichens oxidise and concentrate Mn, Taylor-George *et al.* (1983) and Drake *et al.* (1993) found little evidence for this. The role of bacteria and fungi are less controversial (Grote and Krumbein 1992) with, for example, Dorn and Dragovich (1990) successfully culturing nine Mn-oxidising bacteria. Algae and fungi symbiotic with cyanophytes have also been implicated (Bauman 1976; Borns *et al.* 1980; Krumbein and Jens 1981). Yet, one of the main clay minerals in desert varnish, montmorillonite, inhibits fungal respiration (Stortsky and Rem 1967); this may suggest that oxidisation and fixation of manganese is accomplished mainly by bacteria which experience respiratory stimulation in the presence of montmorillonite (Dorn and Oberlander 1981a; 1982). Indeed, dendritic patterns which occur in some varnishes are produced during iron and manganese oxidation by *Bacillus cereus* (Billy and Cailleux 1968).

A physico-chemical original for some varnishes cannot be discounted, especially those which are poor in manganese (Dorn and Oberlander 1981b; Whalley 1983; 1984). Dorn and Oberlander (1982) suggested that the light and dark laminations in varnishes may have arisen due to oscillations in environmental conditions between mesic and hyper-arid. Alkaline conditions dominate during hyper-arid periods with associated aeolian deposition, minimal Mn-fixing by biota but continued physico-chemical fixation of Fe to form orange varnish. Bacterial manganese fixation occurred under more neutral conditions resulting in the formation of

alternating laminae. Shiny, black varnish and botryoidal textures may indicate deposition of material poorer in clay then forming lamellate coatings (Dorn and Oberlander 1982).

Palaeoenvironmental implications

Desert crusts are of value as palaeoenvironmental indicators only if their modes of origin, ages and the environmental conditions at the time of their formation are correctly interpreted. Ideally, modern environments where crusts are forming contemporaneously should be used as analogues; this is feasible for calcretes, gypsum and halite crusts, and desert varnish, but not for silcretes where contemporary formation is only taking place through fixation of silica by bacteria in highly alkaline environments (Shaw *et al.* 1991).

The possibility of determining the ages of calcretes by radiocarbon dating (Williams and Polach 1971; Magaritz *et al.* 1981) has led to their widespread use in palaeoenvironmental studies. However, great care must be exercised to ensure that only one generation of carbonate is represented in the dated sample and the appropriate correction factors are applied to take account of natural variations in stable isotope levels (Salomons and Mook 1976; Salomons *et al.* 1978). The same is true of $^{230}Th/^{234}U$ dating (Ku *et al.* 1979; Schlesinger 1985). Calcretes have been identified in rocks from every continent and dating from the Proterozoic (Sochova *et al.* 1975), Lower (McPherson 1979) and Upper Palaeozoic (Burgess 1961; Nagtegaal 1969; Adams and Cossey 1981), Mesozoic (Bernouilli and Wagner 1971; Hubert 1977; 1978; Laurain and Meyer 1979) and throughout the Cenozoic to the present (Semeniuk 1986). However, some have been interpreted as altered reef carbonates and others as cold-climate calcretes, so palaeoenvironmental inferences must be circumspect.

Compared with calcretes, siliceous duricrusts are extremely difficult to date. Radtke and Brückner (1991) used the electron spin resonance technique to date Australian silcretes but their use of crushed samples for analysis renders their derived dates almost meaningless since it is unlikely that silcrete formation was monophase or occurred over a short timescale. Using stratigraphic criteria, silcretes developed within saprolitic material in Australia have been dated to the early Jurassic (Twidale 1983) and mid-Tertiary or earlier (Exon *et al.* 1971; Wopfner 1974; S.H. Watts 1977). Phreatic silcretes attributed to either humid (Alley 1973; Young 1978) or arid (Stephens 1971) conditions and lacustrine silcretes (Ambrose and Flint 1981) have mostly been ascribed a mid-Tertiary age. Some Australian silcretes may date from the Pleistocene (Callen 1983) and Holocene (Soegaard and Eriksson 1985). In Europe, Cenozoic sarsens of southern Britain and meulières of the Paris Basin have been variously interpreted as lacustrine (Cayeux 1929, pp. 612–661) or either pedogenic or phreatic accretions (Alimen and Deicha 1958; Bellair *et al.* 1962; Summerfield 1979; Summerfield and Whalley 1980) of mid or late Tertiary age. However, since the micromorphological and chemical criteria which distinguish weathering-profile silcretes from those formed in other settings are disputed (Summerfield 1983c; 1986; Twidale and Hutton 1986; Nash *et al.* 1994b), stratigraphic and lithological information from rocks associated with ancient silcretes may provide the best guide to their origins (Waugh 1970; Isaac 1981; Ross and Chiarenzelli 1985).

Gypsum and halite crusts can be valuable palaeoclimatic indicators because they attest to extreme aridity. However, their susceptibility to dissolution when climate becomes wetter limits their preservation to regions where aridity has persisted. This can in itself provide valuable palaeoclimatic information on the ages of deserts if the crusts can be dated as in Sinai (Dan *et al.* 1982), Wyoming (Reheis 1987) and Namibia (Watson 1988). Since the phreatic and pedogenic forms accrete within host material, the age of incorporated artefacts and radiometrically datable substances may not be representative of the crusts' ages (Watson 1988). Furthermore, gypsum and halite crusts form in several different ways — in shallow-water bodies, in the phreatic zone, and by pedogenesis — so, environmental interpretations must be based on careful consideration of the crusts' origins. In addition to the morphological and chemical criteria that have been described, isotopic studies of gypsum crusts have provided valuable insights into their origins

(Carlisle *et al.* 1978; Sofer 1978; Vieillefon 1980). Most reported occurrences of these crusts in the stratigraphic record have been interpreted as lacustrine, lagoonal or sabkha evaporites (Schreiber 1986). In part, this reflects a lack of clear criteria for identifying pedogenic evaporite facies but, as more research is undertaken, the likelihood of their discovery in the rock record increases. The study of stable isotopes and major element compositions of fluid inclusions preserved within halite crystals offers perhaps the greatest potential for detailed palaeoenvironmental analysis. Fluid inclusions can yield homogenisation temperatures for the time at which halite precipitation took place and offers the possibility of obtaining actual, as opposed to proxy, palaeotemperatures (Roberts and Spencer 1995; Yang *et al.* 1995).

Desert varnish has also been used widely in a palaeoenvironmental context. Because the texture and chemistry of varnish may change depending on the alkalinity and clay content of aeolian deposits, the microlaminations within the coatings might be sensitive indicators of past environmental conditions (Dorn 1984; 1986). Moreover, desert varnish tends to darken with age so it can provide a useful tool for relative dating of rockslides, rock exposures and, in some cases, petroglyphs and artefacts. This property has been used in conjunction with remote sensing techniques to provide estimates of the relative age and stability of land surfaces (e.g. Rivard 1989; White 1990; 1993). It has been suggested that desert varnish may coat rock surfaces within a few decades (Engel and Sharp 1958; Dorn and Meek 1995), but it is more likely to take hundreds or possibly thousands of years (Elvidge 1982; Dorn and DeNiro 1985). Hence, some varnishes may provide an extremely long palaeoenvironmental record. Most research has been undertaken in the American southwest, so more work is needed to ascertain whether desert varnish from other regions has a similar potential. Furthermore, even in the American southwest, it remains uncertain whether the palaeoenvironmental oscillations suggested by variations within any one varnish sample have any regional significance.

There is a good potential for dating varnish (see Chapter 27 and Oberlander 1994) using the $^{230}Th/^{234}U$ technique (corroborated with

^{231}Pa/^{235}U) (Knauss and Ku 1980), as well as cation-ratio (K$^+$ + Ca^{2+}/Ti^{4+}) and radiocarbon methods (Dorn *et al.* 1986; 1989). Stable carbon isotope ratios may also provide detailed palaeoenvironmental data (Dorn and DeNiro 1985). However, all of these techniques, and particularly the use of cation-ratio methods, have been the subject of recent debate concerning their accuracy, comparability and reliability (Bierman and Gillespie 1991; 1992a; 1992b; 1994; 1995; Cahill 1992; Dorn 1992; 1995; Harrington and Whitney 1995).

Summary

The main types of desert crust — silcrete, calcrete, gypsum and halite — are often the products of mineral precipitation from either percolating or evaporating water. They occur mainly, but not exclusively, in arid regions where a consistent predominance of evaporation over precipitation enables soluble materials to accumulate in lake basins, in the phreatic zone, and in the soil. The accretion of some silcretes and calcretes may be related to processes other than evaporation; for example, chemical replacement. However, lacustrine, phreatic and pedogenic origins are most common. To some extent, the intensity of aridity determines the type of crusts which develop, with silcretes and calcretes in semi-arid to arid regions, gypsum crusts in more arid areas, and halite and other evaporite crusts only in hyper-arid deserts. There is, however, another important criterion — the availability of minerals making up the crusts. Hence, desert silcretes may form only in alkaline environments or where other edaphic conditions promote silica mobility and precipitation. Gypsum crusts may develop only where groundwater contains calcium sulphate dissolved from bedrock, or where aeolian or atmospheric deposition of sulphates occurs. Similarly, desert varnish may be a product of micro-organic alteration of atmospheric deposits, though it is likely that some varnishes are arid zone weathering phenomena.

Desert crusts have an important geomorphic influence in many of the Earth's arid regions. They consolidate and armour the land surface, protecting loose sediments and eluvia from erosion and limiting the effect of weathering processes on bedrock. The preservation of relict landscapes and fossil landforms in many deserts is a result of surficial encrustation by mineral precipitates. Certain types of crust, particularly silcretes and calcretes, may persist even when the climate becomes considerably wetter. Therefore, ancient crusts can be extremely useful palaeoenvironmental indicators provided care is taken to distinguish them from similar crusts which originate in markedly different environments. The morphological and chemical criteria which differentiate the various genetic forms for each of the crust types can be employed to refine palaeoenvironmental interpretations still further. Although diagenetic processes often mask these characteristics, by bearing in mind the potential pitfalls, desert crusts and desert varnish can provide a wealth of geomorphological and palaeoenvironmental information.

References

Abdel Jaoued, S. 1987. Calcretes and dolocretes of Upper Palaeocene–Miocene (Bou-Loufa Formation) in southern Tunisia. *Bulletin de la Société Géologique de France*, 3: 777–781.

Adams, A.E. and Cossey, P.J. 1981. Calcrete development at the junction between the Martin Limestone and the Red Hill Oolite (Lower Carboniferous), South Cumbria. *Proceedings of the Yorkshire Geological Society*, 43: 411–431.

Adolphe, J.-P. 1972. Obtention d'encroûtements carbonatés par gel expérimental. *Comptes rendus hebdomadaires des Séances de l'Académie des Sciences*, 274D: 1139–1142.

Akhvlediani, G.K. 1962. Classification of gypsum bearing soils in the Trans-Caucasus. *Soviet Soil Science*, 1962: 532–534.

Alexander, G.B., Heston, W.M. and Iler, R.K. 1954. The solubility of amorphous silica in water. *Journal of Physical Chemistry*, 58: 453–455.

Ali. Y.A. and West, I. 1983. Relationships of modern gypsum nodules in sabkhas of loess to composition of brines and sediments in northern Egypt. *Journal of Sedimentary Petrology*, 53: 1151–1168.

Alimen, H. and Deicha, G. 1958. Observations pétrographiques sur les meulières pliocenes. *Bulletin de la Société Géologique de France*, Series 6, 8: 77–90.

Allen, C.C. 1978. Desert varnish of the Sonoran Desert — optical and electron probe analysis. *Journal of Geology*, 86: 743–752.

Alley, N.F. 1973. Landscape development of the mid-north of South Australia. *Transactions of the Royal Society of South Australia*, 97: 1–17.

Alley, N.F. 1977. Age and origin of laterite and silcrete duricrusts and their relationship to episodic tectonism in the mid-north of South Australia. *Journal of the Geological Society of Australia*, 24: 107–116.

Ambrose, G.J. and Flint, R.B. 1981. A regressive Miocene lake system and silicified strandlines in northern South Australia: implications for regional stratigraphy and silcrete genesis. *Journal of the Geological Society of Australia*, 28: 81–94.

Amit, R. and Gerson, R. 1986. The evolution of Holocene reg (gravelly) soils in deserts — an example from the Dead Sea region. *Catena*, 13: 59–79.

Arakel, A.V. 1980. Genesis and diagenesis of Holocene evaporitic sediments in Hutt and Leeman Lagoons, Western Australia. *Journal of Sedimentary Petrology*, 50: 1305–1326.

Arakel, A.V. 1982. Genesis of calcrete in Quaternary soil profiles, Hutt and Leeman Lagoons, Western Australia. *Journal of Sedimentary Petrology*, 52: 109–125.

Arakel, A.V. 1986. Evolution of calcrete in palaeodrainages of the lake Napperby area, central Australia. *Palaeogeography, Palaeoclimatology, Palaeoecology*, 54: 283–303.

Arakel, A.V. 1991. Evolution of Quaternary duricrusts in Karinga Creek drainage system, central Australian groundwater discharge zone. *Australian Journal of Earth Sciences*, 38: 333–347.

Arakel, A.V. and McConchie, D. 1982. Classification and genesis of calcrete and gypsum lithofacies in palaeodrainage systems of inland Australia and their relationship to carnotite mineralization. *Journal of Sedimentary Petrology*, 52: 1149–1170.

Arakel, A.V., Jacobson, G., Salehi, M. and Hill, C.M. 1989. Silicification of calcrete in palaeodrainage basins of the Australian arid zone. *Australian Journal of Earth Sciences*, 36: 73–89.

Aristarain, L.F. 1970. Chemical analysis of caliche profiles from High Plains, New Mexico. *Journal of Geology*, 78: 201–212.

Aristarain, L.F. 1971. On the definition of caliche deposits. *Zeitschrift für Geomorphologie*, NF, 15: 274–289.

Asserto, R.L.A.M. and Kendall. C.G.St.C. 1977. Nature, origin and classification of peritidal tepee structures and related brecias. *Sedimentology*, 24: 153–210.

Auzel, M. and Cailleux, A. 1949. Silicifications nord-saharieenes. *Bulletin de la Société Géologique de France*, Series 5, 19: 553–559.

Bachman, G.O. and Machette, M.N. 1977. *Calcic soils and calcretes in the southwestern United States.* US Geological Survey Open File Report, 77–794.

Baker, B.H. 1986. Tectonics and volcanism of the southern Kenya Rift Valley and its influence on rift sedimentation. In L.E. Frostick, R.W. Renaut, I. Reid and J.J. Tericelin (eds), *Sedimentation in the African rifts*. Blackwell Scientific, Oxford: 45–57.

Barbier, J. 1978. A propos de calcrète, d'érosion et de la répartition des gîtes d'uranium intragrantiques Français. *Bulletin de BGRM*, Series 2, 2: 31–38.

Barnes, I. 1965. Geochemistry of Birch Creek, Inyo County, California. *Geochimica et Cosmochimica Acta*, 29: 85–112.

Bassett, H. 1954. Silicification of rocks by surface waters. *American Journal of Science*, 252: 733–735.

Bauman, A.J. 1976. Desert varnish and marine ferromanganese oxide nodules: congeneric phenomena. *Nature*, 259: 387–388.

Beattie, J.A. and Haldane, A.D. 1958. The occurrence of palygorksite and barytes in certain parna soils of the Murrumbridge Region, New South Wales. *Australian Journal of Science*, 20: 274–275.

Beckmann, H., Scharpenseel, H.W. and Stephan, S. 1972. Profilstudien an tunesischen Böden. 1: Beschreibung und Analyse. *Fortschritte in der geologie von Rheinland und Westfalen*, 21: 65–82.

Beetz, W. 1926. Die Tertiärablegerungen der Küstennamib. In E. Kaiser (ed.), *Die Diamentenwuste Südwest-Afrikas*, Vol. 2. Deitrich Riemer, Berlin: 1–54.

Bellair, P., Pomerol, C. and Prost, A. 1962. Les hamadas tertiares du bassin parisien. *Comptes rendus hebdomadaires des Séances de l'Académie des Sciences*, 255: 2479–2480.

Bernoulli, D. and Wagner, C.W. 1971. Subaerial diagenesis and fossil caliche deposits in the Calcare Massiccio Formation (Lower Jurassic, Central Apennines, Italy). *Neues Jahrbuch für Paläeontologie, Abhandlungen*, 138: 135–149.

Bierman, P.R. and Gillespie, A.R. 1991. Accuracy of rock varnish chemical analyses — implications for cation-ratio dating. *Geology*, 19: 196–199.

Bierman, P.R. and Gillespie, A.R. 1992a. Accuracy of rock varnish chemical analyses — implications for cation-ratio dating — reply. *Geology*, 20: 470.

Bierman, P.R. and Gillespie, A.R. 1992b. Accuracy of rock varnish chemical analyses — implications for cation-ratio dating — reply. *Geology*, 20: 471–472.

Bierman, P.R. and Gillespie, A.R. 1994. Evidence suggesting that methods of rock-varnish cation-ratio dating are neither comparable nor consistently reliable. *Quaternary Research*, 41: 82–90.

Bierman, P.R. and Gillespie, A.R. 1995. Evidence suggesting that methods of rock-varnish cation-ratio dating are neither comparable nor consistently reliable — reply. *Quaternary Research*, 43: 274–276.

Billy, C. and Cailleux, A. 1968. Depots dendritiques d'oxydes de fer et manganese par action bacterienne. *Comptes rendus hebdomadaires des Séances de l'Académie des Sciences*, 266D: 1643–1645.

Blake, W.P. 1905. Superficial blackening and discoloration of rocks especially in desert regions. *Transaction of the American Institute of Mining Engineers*, 35: 371–375.

Blanckenhorn, M. 1921. *Ägypten. Handbuch der regionalen geologie*, Band 7. Carl Winter, Heidelberg.

Blank, H.R. and Tynes, E.W. 1965. Formation of caliche *in situ*. *Bulletin of the Geological Society of America*, 76: 1387–1391.

Blount, D.N. and Moore, C. 1969. Depositional and non-depositional carbonate breccias, Chiantlas Quadrangle, Guatemala. *Bulletin of the Geological Society of America*, 80: 429–441.

Blümel, W.D. 1976. Kalkkrustvorkommen in Südwestafrika. Untersuchungsmethoden und ihre Aussage. *Mitteilungen der Basler Afrika Bibliographien*, 15: 17–50.

Blümel, W.D. and Eitel, B. 1994. Tertiäre Deckschichten und Kalkkrusten in Namibia: Entstehung und geomorphologische Bedeutung. *Zeitschrift für Geomorphologie*, 38: 385–403.

Borns, D.J., Adams, J.B., Curtiss, B., Farr, T., Palmer, F., Staley, J. and Taylor-George, S. 1980. The role of micro-organisms in the formation of desert varnish and other rock coatings: SEM study. *Geological Society of America, Abstracts with Programs*, 12: 390.

Boulaine, J. 1961. Observations sur les carapaces du piémont saharien. *Travaux de l'institut de Recherches Sahariennes*, 20: 79–89.

Braithwaite, C.J.R. 1983. Calcrete and other soils in Quaternary limestones: structures, processees and applications. *Journal of the Geological Society of London*, 140: 351–363.

Bretz, J.H. and Horberg, L. 1949. Caliche in southeastern New Mexico. *Journal of Geology*, 57: 491–511.

Briot, P. 1983. L'environnement hydrogéochimique du calcrète uranifère de Yeelirrie (Australie Occidentale). *Mineralium Deposita*, 18: 191–206.

Brown, C.N. 1956. The origin of caliche in the north-eastern llana Estacado, Texas. *Journal of Geology*, 64: 1–15.

Brown, I.A. 1926. Some Tertiary formations on the south coast of New South Wales. With special reference to the age and origin of the so-called 'silica' rocks. *Journal and Proceedings of the Royal Society of New South Wales*, 59: 387–399.

Brückner, W.D. 1966. Origin of silcretes in Central Australia. *Nature*, 209: 496–497.

Burgess, I.C. 1961. Fossil soils in the Upper Old Red Sandstone of South Ayrshire. *Transactions of the Geological Society of Glasgow*, 24: 138–153.

Busson, G. and Perthuisot, J.P. 1977. Intérêt de la Sabkha el Melah (Sud tunisien) opour l'interprétation des séries évaporitiques anciennes. *Sedimentary Geology*, 19: 139–164.

Butt, C.R.M. 1983. Aluminosilicate cementation of saprolites, grits and silcretes in Western Australia. *Journal of the Geological Society of Australia*, 30: 179–186.

Butt, C.R.M. 1985. Granite weathering and silcrete formation on the Yilgarn Block, Western Australia. *Australian Journal of Earth Sciences*, 32: 415–432.

Cahill, T.A. 1992. Accuracy of rock varnish chemical analyses — implications for cation-ratio dating — comment. *Geology*, 20: 469.

Callen, R.A. 1983. Late Tertiary 'grey billy' and the age and origins of surficial silicifications (silcrete) in South Australia. *Journal of the Geological Society of Australia*, 30: 393–410.

Carlisle, D., Merifield, P.M., Orme, A.R. and Kolker, O. 1978. *The distributuon of calcretes and gypcretes in southwestern United States and their uranium favorability. Based on a study of deposits in Western Australia and South West Africa (Namibia)*. University of California, Los Angeles, Open File Report 76-002-E.

Castens-Seidell, B. and Hardie, L.A. 1984. Anatomy of a modern sabkha in a rift valley setting, northwest Gulf of California, Baja California, Mexico. *Bulletin of the American Association of Petroleum Geologists*, 68: 460.

Cayeux, L. 1929. *Les roches sédimentaires de France. Roches siliceuses*. Imprimerie Nationale, Paris.

Chafetz, H.S. and Butler, J.C. 1980. Petrology of recent caliche pisolites, spherulites, and speleothem deposits from central Texas. *Sedimentology*, 27: 497–518.

Chapman, R.W. 1974. Calcareous duricrust in Al-Hasa, Saudi Arabia. *Bulletin of the Geological Society of America*, 85: 119–130.

Chen, X.Y., Bowler, J.M. and Magee, J.W. 1991. Aeolian landscapes in central Australia — gypsiferous and quartz dune environments from Lake Amadeus. *Sedimentology*, 38: 519–538.

Chong Diaz, G. 1984. Die Salare in Nordchile — Geologie, Struktur und Geochemie. *Geotektonische Forschungen*, 67: 1–146.

Cooke, R.U. 1981. Salt weathering in deserts. *Proceedings of the Geologists Association*, 92: 1–16.

Coque, R. 1955a. Les croûtes gypseuses du Sud tunisien. *Bulletin de la Société des Sciences naturelles de Tunisie*, 8: 217–236.

Coque, R. 1955b. Morphologie et croûte dans le Sud tunisien. *Annales de Geographie*, 64: 359–370.

Coque, R. 1962. *La Tunisie présaharienne*. Armand Colin, Paris.

Dan, J., Yaalon, D.H., Koymudjisky, H. and Raz, Z. 1972. The soil association map of Israel

(1:1 000 000). *Israel Journal of Earth Sciences*, 21: 29–49.

Dan, J., Yaalon, D.H., Moshe, R. and Nissim, S. 1982. Evolution of reg soils in southern Israel and Sinai. *Geoderma*, 28: 173–202.

Dixey, F. 1942. The age of silicified surface deposits in Northern Rhodesia, Angola and the Belgian Congo. *Transactions of the Geological Society of South Africa*, 44: 39–49.

Dorn, R.I. 1984. Cause and implications of rock varnish microchemical laminations. *Nature*, 310: 767–770.

Dorn, R.I. 1986. Rock varnish as an indicator of aeolian environmental change. In W.G. Nickling (ed.), *Aeolian geomorphology*. Allen & Unwin, Winchester, MA: 291–307.

Dorn, R.I. 1989. A comment on 'A note on the characteristics and possible origins of desert varnish from southeast Morocco' by Drs Smith and Whalley. *Earth Surface Processes and Landforms*, 14: 167–170.

Dorn, R.I. 1991. Rock varnish. *American Scientist*, 79: 542–553.

Dorn, R.I. 1992. Accuracy of rock varnish chemical analyses — implications for cation-ratio dating — comment. *Geology*, 20: 470.

Dorn, R.I. 1995. Evidence suggesting that methods of rock-varnish cation-ratio dating are neither comparable nor consistently reliable — comment. *Quaternary Research*, 43: 272–273.

Dorn, R.I. and DeNiro, M.J. 1985. Stable carbon isotope ratios of rock varnish organic matter: a new paleoenvironmental indicator. *Science*, 227: 1472–1474.

Dorn, R.I. and Dragovich, D. 1990. Interpretation of rock varnish in Australia: case studies from the arid zone. *Australian Geographer*, 21: 18–31.

Dorn, R.I. and Meek, N. 1995. Rapid formation of rock varnish and other rock coatings on slag deposits near Fontana, California. *Earth Surface Processes and Landforms*, 20: 547–560.

Dorn, R.I. and Oberlander, T.M. 1981a. Microbial origin of desert varnish. *Science*, 213: 1245–1247.

Dorn, R.I. and Oberlander, T.M. 1981b. Rock varnish origin, characteristics and usage. *Zeitschrift für Geomorphologie*, NF, 25: 420–436.

Dorn, R.I. and Oberlander, T.M. 1982. Rock varnish. *Progress in Physical Geography*, 6: 317–367.

Dorn, R.I., Bamforth, D.B., Chaill, T.A., Dohrenwend, J.C., Turrin, B.D., Donahue, D.J., Jull, A.J.T., Long, A., Macko, M.E., Weil, E.B., Whitney, D.H. and Zabel, T.H. 1986. Cation-ratio and accelerator radiocarbon dating of rock varnish on Mojave artifacts and landforms. *Science*, 231: 830–833.

Dorn, R.I., Jull, A.J.T., Donahue, D.J., Linick,

T.W. and Toolin, L.J. 1989. Accelerator mass spectrometry radiocarbon dating of rock varnish. *Bulletin of the Geological Society of America*, 101: 1363–1372.

Dorn, R.I., Krinsley, D.H., Liu, T.Z., Anderson, S., Clark, J., Cahill, T.A. and Gill, T.E. 1992. Manganese-rich rock varnish does occur in Antarctica. *Chemical Geology*, 99: 289–298.

Dragovich, D. 1988. A preliminary electron microprobe study of microchemical laminations in desert varnish in western New South Wales. *Earth Surface Processes and Landforms*, 13: 259–270.

Drake, N.A., Heydeman, M.T. and White, K.H. 1993. Distribution and formation of rock varnish in southern Tunisia. *Earth Surface Processes and Landforms*, 18: 31–41.

Drever, J.I. and Smith, C.L. 1978. Cyclic wetting and drying of the soil zone as an influence on the chemistry of groundwater in arid terrains. *American Journal of Science*, 278: 1448–1454.

Dulhunty, J.A. 1975. Salt crust distribution and lake bed conditions in southern areas of Lake Eyre North. *Transactions of the Royal Society of South Australia*, 98: 125–133.

Dulhunty, J.A. 1982. Holocene sedimentary environments in Lake Eyre, South Australia. *Journal of the Geological Society of Australia*, 29: 437–442.

Durand, J.H. 1949. Formation de la croûte gypseuse du Souf (Sahara). *Comptes rendus sommaire des Séances de la Société Géologique de France*, Series 5, 19: 303–305.

Durand, J.H. 1963. Les croûtes calcaires et gypseuses en Algérie: formation et âge. *Bulletin de la Société Géologique de France*, Series 7, 5: 959–968.

Edinger, S.R. 1973. The growth of gypsum. An investigation of the factors which affect the size and growth rates of the habit faces of gypsum. *Journal of Crystal Growth*, 18: 217–224.

Ek, C. and Pissart, A. 1965. Dépôt de carbonate de calcium par congélation et teneur en bicarbonate des eaux résiduelles. *Comptes rendus hebdomadaires des Séances de l'Académie des Sciences*, 260: 929–932.

El-Aref, M., Abdel Wahab, S. and Ahmed, S. 1985. Surficial calcareous crust of caliche type along the Red Sea coast, Egypt. *Geologische Rundschau*, 74: 155–163.

El Sayed, M.I. 1993. Gypcrete of Kuwait – field investigation, petrography and genesis. *Journal of Arid Environments*, 25: 199–209.

El Sayed, M.I., Fairchild, I.J. and Spiro, B. 1991. Kuwaiti dolocrete — petrology, geochemistry and groundwater origin. *Sedimentary Geology*, 73: 59–75.

Elouard, P. and Millot, G. 1959. Observations sur les silifications du Lutétien en Mauritanie et dans la vallée du Sénégal. *Bulletin du Service de la Carte Géologique d'Alsace et de Lorraine*, 12; 15–19.

Elvidge, C.D. 1982. Re-examination of the rate of desert varnish formation reported south of Barstow, California. *Earth Surface Processes and Landforms*, 7: 345–348.

Elvidge, C.D. and Moore, C.B. 1979. A model for desert varnish formation. *Geological Society of America, Abstracts with Programs*, 11: 271.

Engel, C.G. and Sharp, R. 1958. Chemical data on desert varnish. *Bulletin of the Geological Society of America*, 69: 487–518.

Ericksen, G.E. 1981. *Geology and origin of the Chilean nitrate deposits*. United States Geological Survey Professional Paper, 1188: 1–37.

Esteban, M. 1974. Caliche textures and 'Microcodium'. *Bollettina della Società Geologie Italiana*, Supplement, 92: 105–125.

Eswaran, H. and Zi-Tong, G. 1991. Properties, genesis, classification and distribution of soils with gypsum. In W.D. Nettleton (ed.), *Occurrence, characteristics and genesis of carbonate, gypsum and silica accumulation in soils*. Soil Science Society of America, Special Publication, 26: 89–119.

Eswaran, H., Stoops, G. and Abtahi, A. 1980. SEM morphologies of halite. *Journal of Microscopy*, 120: 343–352.

Eugster, H.P. 1986. Lake Magadi, Kenya: a model for rift valley hydrochemistry and sedimentation? In L.E. Frostrick, R.W. Renaut, I. Reid and J.J. Tiercelin (eds), *Sedimentation in the African rifts*. Blackwell Scientific, Oxford: 177–189.

Eugster, H.P. and Kelts, K. 1983. Lacustrine chemical sediments. In A.S. Goudie and K. Pye (eds), *Chemical sediments and geomorphology*. Academic Press, London: 321–368.

Evans, G. and Shearman, D.J. 1964. Recent celestite from the sediments of the Trucial Coast of the Persian Gulf. *Nature*, 202: 385–386.

Evstifeev, Y.G. 1980. Extra-arid Gobi soils. *Problems of Desert Development*, 1980 (2): 17–26.

Exon, N.F., Langford-Smith, T. and McDougall, I. 1971. The age and geomorphic correlations of deep-weathering profiles, silcrete, and basalt in the Roma–Amby region, Queensland. *Journal of the Geological Society of Australia*, 17: 21–30.

Fisk, E.P. 1971. Desert glaze. *Journal of Sedimentary Petrology*, 41: 1136–1137.

Flach, K.W., Nettleton, W.D., Gile, L.H. and Cady, J.G. 1969. Pedocementation: induration by silica, carbonates, and sesquioxides in the Quaternary. *Soil Science*, 107: 442–453.

Flandrin, J., Gautier, M. and Laffitte, R. 1948. Sur la formation de la croûte calcaire superficielle en Algérie. *Comptes rendus hebdomadaires des Séances de l'Académie des Sciences*, 226: 416–418.

Folk, R.L. 1971. Caliche nodule composed of calcite

rhombs. In O.P. Bricker (ed.), *Carbonate cements*. Johns Hopkins University Press, Baltimore: 167–168.

Folk, R.L. 1975. Third-party reply to Hatfield: discussion of Jacka, A.D., 1974, 'Fossils by length — slow chalcedony and associated dolimitization'. *Journal of Sedimentary Petrology*, 45: 952.

Folk, R.L. 1993. SEM imaging of bacteria and nannobacteria in carbonate sediments and rocks. *Journal of Sedimentary Petrology*, 63: 990–999.

Folk, R.L. and Pittman, J.S. 1971. Length-slow chalcedony: a new testament for vanished evaporites. *Journal of Sedimentary Petrology*, 41: 1045–1058.

Francis, W.D. 1921. The origin of black coatings of iron and manganese oxides on rocks. *Proceedings of the Royal Society of Queensland*, 32: 110–116.

Frankel, J.J. 1952. Silcrete near Albertina, Cape Province. *South African Journal of Science*, 49: 173–182.

Frankel, J.J. and Kent, L.E. 1938. Grahamstown surface quartzites (silcretes). *Transactions of the Geological Society of South Africa*, 15: 1–42.

Gabriel, A. 1964. Zum Problem des Formenschatzes in extrem-ariden Raumen. *Mitteilungen der Österreichischen Geographischen Gesellschaft*, 106: 3–15.

Gardner, L.R. 1972. Origin of the Mormon Mesa caliche, Clark county, Nevada. *Bulletin of the Geological Society of America*, 83: 143–155.

Gauthier-Lafaye, F., Taieb, R., Paquet, H., Chahi, A., Prudencio, I. and Sequeira Braga, M.-A. 1993. Composition isotopique de l'oxygène de palygorskitesassociées à des calcrètes: conditions de formation. *Comptes rendus hebdomadaires des Séances de l'Académie Sciences Paris*, Series 2, 316: 1239–1245.

Gibson, G.W. 1962. Geological investigations of southern Victoria Land, Antarctica. Part 8: Evaporite salts in the Victoria Valley region. *New Zealand Journal of Geology and Geophysics*, 5: 361–374.

Gile, L.H., Peterson, F.F. and Grossman, R.B. 1966. Morphological and genetic sequences of carbonate accumulation in desert soils. *Soil Science*, 101: 347–360.

Glasby, G.P., McPherson, J.G., Kohn, B.P., Johnson, J.H., Keys, J.R., Freeman, A.G. and Tricker, M.J. 1981. Desert varnish in Southern Victoria Land, Antarctica. *New Zealand Journal of Geology and Geophysics*, 24: 389–397.

Goudie, A.S. 1973. *Duricrusts in tropical and subtropical landscapes*. Clarendon Press, Oxford.

Goudie, A.S. 1983a. Calcrete. In A.S. Goudie and K. Pye (eds), *Chemical sediments and geomorphology*. Academic Press, London: 93–131.

Goudie, A.S. 1983b. Surface efflorescences and

nitrate beds. In A.S. Goudie and K. Pye (eds), *Chemical sediments and geomorphology*. Academic Press, London: 187–195.

Goudie, A.S. 1984. Duricrusts and landforms. In K.S. Richards, R.R. Arnett and S. Ellis (eds), *Geomorphology and soils*. Allen & Unwin, London: 37–57.

Goudie, A.S. and Cooke, R.U. 1984. Salt efflorescences and saline lakes: a distributional analysis. *Geoforum*, 15: 563–585.

Grote, G. and Krumbein, W.E. 1992. Microbial precipitation of manganese by bacteria and fungi from desert rock and rock varnish. *Geomicrobiology Journal*, 10: 49–57.

Harden, D.R. and Biggar, N.E. 1983. Pedogenic calcrete developed on early Pleistocene erosional surfaces, southeastern Utah. *Geological Society of America, Abstracts with Programs*, 15: 328.

Harden, J.W., Taylor, E.M., Rehels, M.C. and McFadden, L.D. 1991. Calcic, gypsic and siliceous soil chronosequences in arid and semi-arid environments. In W.D. Nettleton (ed.), *Occurrence, characteristics and genesis of carbonate, gypsum and silica accumulation in soils*. Soil Science Society of America, Special Publication, 26: 1–16.

Harrington, C.D. and Whitney, J.W. 1995. Evidence suggesting that methods of rock-varnish cation-ratio dating are neither comparable nor consistently reliable — comment. *Quaternary Research*, 43: 268–271.

Hassouba, H. and Shaw, H.F. 1980. The occurrence of palygorskite in Quaternary sediments of the coastal plain of south-west Egypt. *Clay Minerals*, 15: 77–83.

Hay, R.L. and Reeder, R.J. 1978. Calcretes of Olduvai Gorge and Ndolanya Beds of northern Tanzania. *Sedimentology*, 25: 649–673.

Hay, R.L. and Wiggins, B. 1980. Pellets, ooids, sepiolite and silica in three calcretes of the south-western United States. *Sedimentology*, 27: 559–576.

Heald, M.T. and Larese, R.E. 1974. Influence of coatings on quartz cementation. *Journal of Sedimentary Petrology*, 44: 1269–1274.

Hill, A.E. 1937. Transition temperature of gypsum to anhydrite. *Journal of the American Chemical Society*, 59: 2242–2244.

Hodge, T., Turchenek, L.W. and Oades, J.M. 1984. Occurrence of palygorskite in ground-water rendzinas (Petrocalcic calciaquolls) in south-east South Australia. In A. Singer and E. Galan (eds), *Palygorskite–Sepiolite: occurrences, genesis and uses*. Developments in Sedimentology, 37. Elsevier, Amsterdam: 199–210.

Holster, W.T. 1966. Diagnetic polyhalite in recent salt from Baja California. *American Mineralogist*, 51: 99–109.

Horta, J.C. de O.S. 1980. Calcrete, gypcrete and soil classification in Algeria. *Engineering Geology*, 15: 15–52.

Hubert, J.F. 1977. Paleosol caliche in the New Haven Arkose, Connecticut: record of semiaridity in Late Triassic–Early Jurassic time. *Geology*, 5: 302–304.

Hubert, J.F. 1978. Paleosol caliches in the New Haven Arkose, Newark Group, Connecticut. *Palaeogeography, Palaeoclimatology, Palaeoecology*, 24: 151–168.

Hungate, B., Danin, A., Pellerin, N.B., Stemmler, J., Kjellander, P., Adams, J.B. and Staley, J.T. 1987. Characterisation of manganese-oxidising (MnII–MnIV) bacteria from Negev Desert rock varnish — implications in desert varnish formation. *Canadian Journal of Microbiology*, 33: 939–943.

Hunt, C.B. 1961. *Stratigraphy of desert varnish*. US Geological Survey Professional Paper 424B: 194–195.

Hunt, C.B. and Mabey, D.R. 1966. *Stratigraphy and structure, Death Valley, California*. US Geological Survey Professional Paper 494A.

Hunt, C.B., Robinson, T.W., Bowles, W.B. and Washburn, A.L. 1966. *Hydrological basin, Death Valley, California*. US Geological Survey Professional Paper 494B.

Hüser, K. 1976. Kalkrusten in Namib-Randbereich des mittleren Südwestafrika. *Mitteilungen der Basler Afrika Bibliographien*, 15: 51–81.

Hutton, J.T. and Dixon, J.C. 1981. The chemistry and mineralogy of some South Australian calcretes and associated soft carbonates and their dolomitization. *Journal of the Geological Society of Australia*, 28: 71–79.

Hutton, J.T., Twidale, C.R., Milnes, A.R. and Rosser, H. 1972. Composition and genesis of silcretes and silcrete skins from the Beda valley, southern Arcoona plateau, South Australia. *Journal of the Geological Society of Australia*, 19: 31–39.

Hutton, J.T., Twidale, C.R. and Milnes, A.R. 1978. Characteristics and origin of some Australian silcretes. In T. Langford-Smith (ed.), *Silcrete in Australia*. Department of Geography, Unviersity of New England, Armidale, NSW: 19–39.

Ireland, P. 1979. Geomorphological variations of 'case-hardening' in Puerto Rico. *Zeitschrift für Geomorphologie, Supplementband*, 32: 9–20.

Isaac, K.P. 1981. Tertiary weathering profiles in the plateau deposits of East Devon. *Proceedings of the Geologists Association*, 92: 159–168.

Jack, R.L. 1921. The salt and gypsum resources of South Australia. *Bulletin of the Geological Survey of South Australia*, 8.

Jacobson, G., Arakel, A.V. and Chen, Y.J. 1988.

The central Australian groundwater discharge zone — evolution of associated calcrete and gypcrete deposits. *Australian Journal of Earth Sciences*, 35: 549–565.

James, N.P. 1972. Holocene and Pleistocene calcareous crust (caliche) profiles: criteria for subaerial exposure. *Journal of Sedimentary Petrology*, 42: 817–836.

Jennings, J.N. and Sweeting, M.M. 1961. Caliche pseudo-anticlines in the Fitzroy Basin, Western Australia. *American Journal of Science*, 259: 635–639.

Jones, C.E. 1991. Characteristics and origin of rock varnish from the hyperarid coastal deserts of northern Peru. *Quaternary Research*, 35: 116–129.

Jux, U. 1983. Zusammensetzung und Ursprung von Wüstengläsern aus der Grossen Sandsee Ägyptens. *Zeitschrift der Deutschen Geologischen Gesellschaft*, 143: 521–553.

Kahle, C.F. 1977. Origin of subaerial Holocene calcareous crust: role of algae, fungi and sparmicritisation. *Sedimentology*, 24: 413–435.

Kaiser, E. 1926. *Die Diamentenwüste Südwest-Afrikas, Vol. 2.* Dietrich Reimer, Berlin.

Kaiser, E. and Neumaier, F. 1932. Sand-Steinsalz-Kristellskellete aus der Namib Südwestafrikas. *Zentralblatt für Mineralogie, Geologie und Paläontologie*, 6A: 177–188.

Khalaf, F.I. 1988. Petrography and diagenesis of silcrete from Kuwait, Arabian Gulf. *Journal of Sedimentary Petrology*, 58: 1014–1022.

Khalaf, F.I. and Abdal, M.S. 1993. Dedolomitization of dolomite deposits in Kuwait, Arabian Gulf. *Geologische Rundschau*, 82: 741–749.

Kikoine, J. and Radier, H. 1949. Quartzites d'alteration au Soudan oriental. *Compte rendu sommaire des Séances de la Société Géologique de France*, Series 5, 19: 339–341.

Kikoine, J. and Radier, H. 1950. Silification au Soudan oriental; de calcaire a silex du Tertaire Continental (T.C.). *Compte rendu sommaire des Séances de la Société Géologique de France*, Series 5, 20: 168–170.

Klappa, C.F. 1978. Biolithogenesis of microcodium. *Sedimentology*, 25: 489–522.

Klappa, C.F. 1979a. Lichen stromatolites: criterion for subaerial exposure and a mechanism for the formation of laminar calcretes (caliche). *Journal of Sedimentary Petrology*, 49: 387–400.

Klappa, C.F. 1979b. Calcified filaments in Quaternary calcretes: organo-mineral interactions in the subaerial vadose environment. *Journal of Sedimentary Petrology*, 49: 955–968.

Klappa, C.F. 1979c. Comment on 'Displacive calcite: evidence from recent and ancient calcretes'. *Geology*, 7: 420–421.

Klappa, C.F. 1980. Brecciation textures and tepee

structures in Quaternary calcrete (caliche) profiles from eastern Spain: the plant factor in their formation. *Geological Journal*, 15: 81–89.

Klute, F. and Krasser, L.M. 1940. Über Wüstenlackbildung im Hochgebirge. *Petermann's Geographische Mitteilungen*, 86: 21–22.

Knauss, K.G. and Ku, T.-L. 1980. Desert varnish: potential for age dating via uranium-series isotopes. *Journal of Geology*, 88: 95–100.

Knetsch, G. 1937. Beiträge zur Kenntnis von Krustenbildung. *Zeitschrift der Deutschen Geologischen Gesellschaft*, 89: 177–192.

Knox, G.J. 1977. Caliche profile formation, Saldanha Bay (South Africa). *Sedimentology*, 24: 657–674.

Kondorskaya, N.I. 1967. Areas of present-day salt accumulations. *Soviet Soil Science*, 1967: 462–473.

Krauskopf, K.B. 1966. Dissolution and precipitation of silica at low temperatures. *Geochimica et Cosmochimica Acta*, 10: 1–26.

Krinsley, D., Dorn, R.I. and Tovey, N.K. 1995. Nanometre-scale layering in rock-varnish — implications for genesis and paleoenvironmental interpretation. *Journal of Geology*, 103: 106–113.

Krumbein, W.E. 1969. Uber die Einfluss der Mikroflora auf die exogene Dynamik (Verwitterung und Krustenbildung). *Geologische Rundschau*, 58: 333–363.

Krumbein, W.E. 1971. Biologische Entstehung von Wüstenlack. *Umschau in Wissenschaft und Technik*, 71: 240–241.

Krumbein, W.E. and Jens, K. 1981. Biogenic rock varnishes of the Negev Desert (Israel): an ecological study of iron and manganese transformation by cyanobacteria and fungi. *Oecologia*, 50: 25–38.

Ku, T.-L., Bull, W.B., Freeman, S.T. and Knauss, K.G. 1979. Th^{230}–U^{234} dating of pedogenic carbonates in gravelly desert soils of Vidal Valley, southeastern California. *Bulletin of the Geological Society of America*, 90: 1063–1073.

Kulke, H. 1974. Zur Geologie und Mineralogie der Kalk- under Gipskrusten Algeriens. *Geologische Rundschau*, 63: 970–998.

Kushnir, J. 1980. The coprecipitation of strontium, magnesium, sodium, potassium and chloride ions with gypsum. An experimental study. *Geochimica et Cosmochimica Acta*, 44: 1471–1482.

Kushnir, J. 1981. Formation and early diagenesis of varved evaporite sediments in a coastal hypersaline pool. *Journal of Sedimentary Petrology*, 51: 1193–1203.

Lakin, H.W., Hunt, C.B., Davidson, D.F. and Oda, U. 1963. *Variations in minor-element content of desert varnish.* US Geological Survey Professional Paper 475B: 28–31.

Lamplugh, G.W. 1902. Calcrete. *Geological Magazine*, 9: 575.

Lamplugh, G.W. 1907. The geology of the Zambezi Basin around the Batoka Gorge (Rhodesia). *Quarterly Journal of the Geological Society of London,* 63: 162–216.

Landmesser, M. 1995. Silica transport and accumulation at low temperatures. *Chemie der Erde — Geochemistry,* 55: 149–176.

Langford-Smith, T. 1978. A select review of silcrete research in Australia. In T. Langford-Smith (ed.), *Silcrete in Australia.* Department of Geography, University of New England, Armidale, NSW: 1–11.

Langford-Smith, T. and Dury, G.W.H. 1965. Distribution, character, and attitude of the duricrust in the northwest of New South Wales and adjacent areas of Queensland. *American Journal of Science,* 263: 170–190.

Langford-Smith, T. and Watts, S.H. 1978. The significance of coexisting siliceous and ferruginous weathering products at select Australian localities. In T. Langford-Smith (ed.), *Silcrete in Australia.* Department of Geography, University of New England, Armidale, NSW: 143–165.

Lattman, L.H. 1973. Calcium carbonate cementation of alluvial fans in southern Nevada. *Bulletin of the Geological Society of America,* 84: 3013–3028.

Lattman, L.H. and Lauffenburger, S.K. 1974. Proposed role of gypsum in the formation of caliche. *Zeitschrift für Geomorphologie, Supplementband,* 20: 140–149.

Laudermilk, J.D. 1931. On the origin of desert varnish. *American Journal of Science,* Series 5, 21: 51–66.

Laurain, M. and Meyer, R. 1979. Paléoaltération et paléosol: l'encroûtement calcaire (calcrète) au sommet de la craie sous les sédiments éocène de la Montagne de Reims. *Comptes rendus hebdomadaires des Séances de l'Académie des Sciences,* 289D: 1211–1214.

le Houérou, H.N. 1960. Contribution à l'étude des sols du Sud tunesien. *Annales agronomiques,* 11: 241–308.

Linck, G. 1901. Über die dunkeln Rinden der Gesteine der Wüste. *Jenaische Zeitschrift für Naturwissenschaften,* 35: 329–336.

Lowenstein, T.K. and Hardie, L.A. 1985. Criteria for the recognition of salt-pan evaporites. *Sedimentology,* 32: 627–644.

Lyon, G.L. 1978. The stable isotope geochemistry of gypsum, Miers Valley, Antarctica. *Bulletin of the New Zealand Department of Scientific and Industrial Research,* 220: 97–103.

Mabbutt, J.A. 1955. Erosion surfaces in little Namaqua-land and the ages of surface deposits in the south-western Kalahari. *Transactions of the Geological Society of South Africa,* 58: 13–29.

Mabbutt, J.A. 1957. Physiographic evidence for the age of the Kalahari sands of the Kalahari. In J.D. Clark (ed.), *3rd Pan-African Congress on Pre-History, Livingstone, 1955.* Chatto & Windus, London: 123–126.

Mabbutt, J.A. 1965. The weathered land surface in Central Australia. *Zeitschrift für Geomorphologie,* NF, 9: 82–114.

Mabbutt, J.A. 1977. *Desert landforms.* MIT Press, Cambridge, MA.

MacGregor, A.M. 1931. Geological notes on a circuit of the great Makarikari salt pan, Bechuanaland Protectorate. *Transactions of the Geological Society of South Africa,* 33: 89–102.

Mackenzie, R.C., Wilson, M.J. and Mashhady, A.S. 1984. Origin of palygorskite in some soils of the Arabian peninsula. In A. Singer and E. Galan (eds), *Palygorskite–Sepiolite: occurrences, genesis and uses.* Developments in Sedimentology, 37. Elsevier, Amsterdam: 177–186.

McCarthy, T.S. and Ellery, W.N. 1995. Sedimentation on the distal reaches of the Okavango Fan, Botswana, and its bearing on calcrete and silcrete (ganister) formation. *Journal of Sedimentary Research,* A65: 77–90.

McFadden, L.D., Wells, S.G. and Jercinovich, M.J. 1987. Influences of aeolian and pedogenic processes on the origin and evolution of desert pavements. *Geology,* 15: 504–508.

McFarlane, M.J. 1975. A calcrete from the Namanga-Bissel area of Kenya. *Kenyan Geographer,* 1: 31–43.

McFarlane, M.J. 1983. Laterites. In A.S. Goudie and K. Pye (eds), *Chemical sediments and geomorphology.* Academic Press, London: 7–58.

McFarlane, M.J., Borger, H. and Twidale, C.R. 1992. Towards an understanding of the origin of titanium skins on silcrete in the Beda Valley, South Australia. In J.-M. Schmitt and Q. Gall (eds), *Mineralogical and geochemical records of paleoweathering.* Ecole des mines de Paris, Mém. Sciences de la terre, 18: 39–46.

McGowran, B., Rutland, R.W.R. and Twidale, C.R. 1977. Discussion: age and origin of laterite and silcrete duricrusts and their relationship to episodic tectonism in the mid-north of South Australia. *Journal of the Geological Society of Australia,* 24: 421–422.

McPherson, J.G. 1979. Calcrete (caliche) palaeosols in fluvial redbeds of the Aztec Siltstone (Upper Devonian), southern Victoria Land, Antarctica. *Sedimentary Geology,* 22: 267–285.

Magaritz, M., Kaufman, A. and Yaalon, D.H. 1981. Calcium carbonate nodules in soils: $^{18}O/^{16}O$ and $^{13}C/^{12}C$ ratios and ^{14}C contents. *Geoderma,* 25: 157–172.

Magee, J.W. 1991. Late Quaternary lacustrine, groundwater, aeolian and pedogenic gypsum in the

Prungle Lakes, southeastern Australia. *Palaeogeography, Palaoclimatology, Palaeoecology*, 84: 3–42.

Mann, A.W. and Horwitz, R.C. 1979. Groundwater calcrete deposits in Australia: some observations from Western Australia. *Journal of the Geological Society of Australia*, 26: 293–303.

Manze, U. and Brunnacker, K. 1977. Über das Verhalten des Sauerstoff- under Kohlenstroof-Isotope in Kalkkursten und Kalktuffen des mediterranen Raumes und der Sahara. *Zeitschrift für Geomorphologie*, NF, 21: 343–353.

Marshall, R.R. 1962. Natural radioactivity and the origin of desert varnish. *Transactions of the American Geophysical Union*, 43: 446–447.

Martin, H. 1963. A suggested theory for the origin and a brief description of some gypsum deposits of South-West Africa. *Transactions of the Geological Society of South Africa*, 66: 345–351.

Masson. P.H. 1955. An occurrence of gypsum in south-west Texas. *Journal of Sedimentary Petrology*, 25: 72–77.

Maufe, H.B. 1939. New sections in the Kalahari Beds at the Victoria Falls, Rhodesia. *Transactions of the Geological Society of South Africa*, 41: 211–224.

Merrill, G.P. 1898. Desert varnish. *Bulletin of the United States Geological Survey*, 150: 389–391.

Millot, G., Radier, H., Muller-Feuga, R., Defossez, M. and Wey, R. 1959. Sur la géochemie de la silice et les silifications sahariennes. *Bulletin du Service de la Carte Géologique d'Alsace et de Lorraine*, 12 (2): 3–14.

Milnes, A.R. 1986. Armoured landscapes. *Geology Today*, 2: 73–74.

Milnes, A.R. and Thiry, M. 1992. Silcretes. In I.P. Martini and W. Chesworth (eds), *Weathering, soils and paleosols*. Developments in Earth Surface Processes 2. Elsevier, Amsterdam: 349–377.

Milnes, A.R. and Twidale, C.R. 1983. An overview of silicification in Cenozoic landscapes of arid central and southern Australia. *Australian Journal of Soil Research*, 21: 387–410.

Milnes, A.R., Thiry, M. and Wright, M.J. 1991. Silica accumulations in saprolite and soils in South Australia. In W.D. Nettleton (ed.), *Occurrence, characteristics and genesis of carbonate, gypsum and silica accumulation in soils*. Soil Science Society of America, Special Publication, 26: 121–149.

Morris, R.C. and Dickey, P.A. 1957. Modern evaporite deposition in Peru. *Bulletin of the American Association of Petroleum Geologists*, 41: 2467–2474.

Mortensen, H. 1927. Der Formenschatz der norchilenischen Wüste, ein Beitrag zum Gesetz der Wüstenbildung. *Abhandlungen der Gesellschaft der Wissenschaften zu Göttingen*, NF, 12.

Mortensen, H. 1933. Die 'Salzsprengung' und ihre

Bedeutung für die regionalklimatische Gliederung der Wüsten. *Petermann's Geographische Mitteilungen*, 79: 130–135.

Moseley, F. 1965. Plateau calcrete, calcreted gravels, cemented dunes and related deposits of the Maallegh-Bomba region of Libya. *Zeitschrift für Geomorphologie*, NF, 9: 167–185.

Mountain, E.D. 1952. The origin of silcrete. *South African Journal of Science*, 48: 201–204.

Multer, H.G. and Hoffmeister, J.E. 1968. Subaerial laminated crusts of the Florida Keys. *Bulletin of the Geological Society of America*, 79: 183–92.

Nagtegaal, P.J.C. 1969. Microtextures in recent and ancient caliche. *Leidse Geologische Mededelingen*, 42: 131–141.

Nagy, B., Nagy, L.A., Rigali, M.J., Jones, W.D., Krinsley, D.H. and Sinclair, N.A. 1991. Rock varnish in the Sonoran Desert — microbiologically mediated accumulation of manganiferous sediments. *Sedimentology*, 79: 542–553.

Nash, D.J., Shaw, P.A. and Thomas, D.S.G. 1994a. Duricrust development and valley evolution — process–landform links in the Kalahari. *Earth Surface Processes and Landforms*, 19: 299–317.

Nash, D.J., Thomas, D.S.G. and Shaw, P.A. 1994b. Siliceous duricrusts as palaeoclimatic indicators: evidence from the Kalahari Desert of Botswana. *Palaeogeography, Palaeoclimatology, Palaeoecology*, 112: 279–295.

Netterberg, F. 1969. Ages of calcrete in southern Africa. *South African Archaeological Bulletin*, 24: 88–92.

Netterberg, F. 1978. Dating and correlation of calcretes and other pedocretes. *Transactions of the Geological Society of South Africa*, 81: 379–391.

Netterberg, F. 1980. Geology of southern African calcretes. I. Terminology, description, macrofeatures and classification. *Transactions of the Geological Society of South Africa*, 83: 255–283.

Nickling, W.G. 1984. The stabilizing effect of bonding agents on the entrainment of sediment by wind. *Sedimentology*, 31: 111–117.

Nickling, W.G. and Ecclestone, M. 1981. The effects of soluble salts on the threshold shear velocity of fine sand. *Sedimentology*, 28: 505–510.

Oberlander, T.M. 1994. Rock varnish in deserts. In A.D. Abrahams and A.J. Parsons (eds), *Geomorphology of desert environments*. Chapman & Hall, London: 107–119.

Ollier, C.D. 1978. Silcrete and weathering. In T. Langford-Smith (ed.), *Silcrete in Australia*. Department of Geography, University of New England, Armidale, NSW: 13–17.

Ollier, C.D. 1991. Aspects of silcrete formation in Australia. *Zeitschrift für Geomorphologie*, NF, 35: 151–163.

Orlova, M.A. 1980. Role of the eolian factor in the

salt regime of desert solonchaks. *Problems of Desert Development* 1980 (3): 73–77.

Pain, C.F. and Ollier, C.D. 1995. Inversion of relief — a component of landscape evolution. *Geomorphology* 12: 151–165.

Page, W.D. 1972. The geological setting of the archaeological site at Oued el Akarit and the palaeoclimatic significance of gypsum soils, Southern Tunisia. Unpublished PhD thesis, Department of Geological Sciences, University of Colorado.

Palmer, F.E., Staley, J.T., Murray, R.G.E. and Counsell, T. 1985. Identification of manganese oxidising bacteria from desert varnish. *Geomicrobiology Journal*, 4: 343–360.

Patterson, R.J. and Kinsman, D.J.J. 1978. Marine and continental groundwater sources in a Persian Gulf coastal sabkha. *American Association of Petroleum Geologists, Studies in Geology* 4: 381–397.

Perry, R.S. and Adams, J.B. 1978. Desert varnish: evidence for cyclic deposition of manganese. *Nature*, 276: 489–491.

Perthuisot, J.P. 1980. Sebkha el Melah near Zarzis, a recent paralic salt basin (Tunisia). In Chambre Syndicale de la Recherche et de la Production du Pétrole et du Gaz Natural, *Evaporite deposits: illustration and interpretation of some environmental sequences*. Editions Technip, Paris: 11–17.

Pervinquière, L. 1903. *Étude Géologique de la Tunisie central*. De Rudeval, Paris.

Peterson, M.N.A. and van der Borch, C.C. 1965. Chert: modern inorganic deposition in a carbonate-precipitating locality. *Science*, 149: 1501–1503.

Phleger, F.B. 1969. A modern evaporite deposit in Mexico. *Bulletin of the American Association of Petroleum Geologists*, 53: 824–830.

Potter, R.M. 1979. The tetravalent manganese oxides: clarification of their structural variations and characterisation of their occurrence in the terrestrial weathering environment as desert varnish and other manganese oxide concentrations. PhD dissertation, California Institute of Technology.

Potter, R.M. and Rossman, G.R. 1977. Desert varnish: the importance of clay minerals. *Science*, 196: 1446–1448.

Potter, R.M. and Rossman, G.R. 1979. The manganese and iron oxide mineralogy of desert varnish. *Chemical Geology*, 25: 79–94.

Pouget, M. 1968. Contribution à l'étude des croûtes et encroûtements gypseux de nappe dans le sud tunisien. *Cahiers ORSTOM*, Série Pédologie, 6: 309–365.

Price, W.A. 1925. Caliche and pseudo-anticlines. *Bulletin of the American Association of Petroleum Geologists*, 9: 1009–1017.

Pye, K. 1980. Beach salcrete and eolian sand transport: evidence from North Queensland. *Journal of Sedimentary Petrology*, 50: 257–261.

Quinones, H. and Allende, R. 1974. Formation of the lithified carapace of calcareous nature which covers most of the Yucatan peninsula and its relationto the soils and geomorphology of the region. *Tropical Agriculture*, 51: 94–101.

Radtke, U. and Brückner, H. 1991. Investigation on age and genesis of silcretes in Queensland (Australia) — preliminary results. *Earth Surface Processes and Landforms*, 16: 547–554.

Reddy, M.R. and Perkins, H.F. 1976. Fixation of manganese by clay minerals. *Soil Science*, 121: 21–24.

Reeves, C.C. 1970. Origin, classification, and geological history of caliche on the southern High Plains, Texas and eastern New Mexico. *Journal of Geology*, 78: 352–362.

Reeves, C.C. 1976. *Caliche: origin, classification, morphology and uses*. Estacado Books, Lubbock, TX.

Reheis, M.C. 1987. *Gypsic soils on the Kane Alluvial Fans. Big Horn County, Wyoming*. Bulletin of the United States Geological Survey, 1590C.

Renaut, R.W., Tiercelin, J.J. and Owen, R.B. 1986. Mineral pricipatation and diagenesis in the sediments of the Lake Bogoria basin, Kenya Rift Valley. In L.E. Frostick, R.W. Renaut, I. Reid, and J.J. Tiercelin (eds), *Sedimentation in the African rifts*. Blackwell Scientific, Oxford: 159–175.

Risacher, F. 1978. Genèse d'une croûte de gypse dans un basin de l'Altiplano bolivien. *Cahiers ORSTOM*, série Géologie, 10: 91–100.

Rivard, M. 1989. Field results and laboratory measurements for coated rock surfaces for Meatiq Dome, Egypt. *Geological Society of America, Abstracts with Programs*, 21: 110.

Roberts, S.M. and Spencer, R.J. 1995. Paleotemperatures preserved in fluid inclusions in halite. *Geochimica et Cosmochimica Acta*, 59: 3929–3942.

Ross, G.M. and Chiarenzelli, J.R. 1985. Paleoclimatic significance of widespread Proterozoic silcretes in the Bear and Churchill Provinces of the northwestern Canadian Shield. *Journal of Sedimentary Petrology*, 55: 196–204.

Rutte, E. 1958. Kalkkrusten in Spanien. *Neues Jahrbuch für Paläontologie, Abhandlungen*, 106: 52–138.

Rutte, E. 1960. Kalkrusten im östlichen Mittelmeergebiet. *Zeitschrift der Deutschen Geologischen Gesellschaft*, 112: 81–90.

Salomons, W. and Mook, W.G. 1976. Isotope geochemistry of carbonate dissolution and reprecipitation in soils. *Soil Science*, 122: 15–24.

Salomons, W., Goudie, A. and Mook, W.G. 1978. Isotopic compostion of calcrete deposits from

Europe, Africa and India. *Earth Surface Processes*, 3: 43–57.

Scheffer, F., Meyer, B. and Kalk, E. 1963. Biologische Ursachen der Wüstenlackbildung. *Zeitschrift für Geomorphologie*, NF, 7: 112–119.

Schlesinger, W.H. 1985. The formation of caliche in soils of the Mojave Desert, California. *Geochimica et Cosmochimica Acta*, 49: 57–66.

Scholz, H. 1972. The soils of the central Namib Desert with special consideration of the soils in the vicinity of Gobabeb. *Madoqua*, 2: 33–51.

Schreiber, B.C. 1986. Arid shorelines and evaporites. In H.G. Reading (ed.), *Sedimentary environments and facies*. Blackwell Scientific, Oxford: 189–228.

Schwenk, S. 1977. Krusten und Verkrustung in Südtunisien. *Stuttgarter geographische Studien*, 37L: 83–103.

Semeniuk, V. 1986. Holocene climate history of coastal southwestern Australia using calcrete as an indicator. *Palaeogeography, Palaeoclimatology, Palaeoecology*, 53: 289–308.

Semeniuk, V. and Searle, D.J. 1985. Distribution of calcrete in Holocene coastal sands in relationship to climate, southwestern Australia. *Journal of Sedimentary Petrology*, 55: 86–95.

Senior, B.R. 1978. Silcrete and chemically weathered sediments in southwest Queensland. In T. Langford-Smith (ed.), *Silcrete in Australia*. Department of Geography, University of New England, Armidale, NSW: 41–50.

Shadfan. H. and Dixon, J.B. 1984. Occurrence of palygorskite in the soils and rocks of the Jordan Valley. In A. Singer and E. Galan (eds), *Palygorskite–sepiolite: occurrences, genesis and uses*. Developments in Sedimentology, 37. Elsevier, Amsterdam: 187–198.

Shaw, P.A. and de Vries, J.J. 1988. Duricrust, groundwater and valley development in the Kalahari of south-east Botswana. *Journal of Arid Environments*, 14: 245–254.

Shaw, P.A., Cooke, H.J. and Perry, C.C. 1991. Microbialitic silcretes in highly alkaline environments: some observations from Sua Pan, Botswana. *South African Journal of Geology*, 93: 803–808.

Shearman, D.J. 1963. Recent anhydrite, gypsum, dolomite, and halite from the coastal flats of the Arabian shore of the Persian Gulf. *Proceedings of the Geological Society of London*, 1607: 63–64.

Shearman, D.J. 1966. Origins of marine evaporites by diagenesis. *Transactions of the Institution of Mining and Metallurgy*, 75B: 208–215.

Siever, R. 1972. Silicon: solubilities of compounds which control concentrations in natural waters. In K.H. Wedepohl (ed.), *Handbook of geochemistry*. Springer-Verlag, Berlin: 14-H-1–14-H-7.

Singer, A. 1984. Pedogenic palygorskite in the arid environment. In A. Singer and E. Galan (eds), *Palygorskite–sepiolite: occurrences, genesis and uses*. Developments in Sedimentology, 37. Elsevier, Amsterdam: 169–175.

Smale, D. 1973. Silcretes and associated silica diagenesis in southern Africa and Australia. *Journal of Sedimentary Petrology*, 43: 1077–1089.

Smale, D. 1978. Silcretes and associated silica diagenesis in southern Africa and Australia. In T. Langford-Smith (ed.), *Silcrete in Australia*. Department of Geography, University of New England, Armidale, NSW: 261–279.

Smith, B.J. and Whalley, W.B. 1982. Observations on the composition and mineralogy of an Algerian duricrust complex. *Geoderma*, 28: 285–311.

Smith, B.J. and Whalley, W.B. 1988. A note on the characteristics and possible origins of desert varnish from southeast Morocco. *Earth Surface Processes and Landforms*, 13: 251–258.

Smith, B.J. and Whalley, W.B. 1989. A note on the characteristics and possible origins of desert varnish from southeast Morocco. A reply to comments by R.I. Dorn. *Earth Surface Processes and Landforms*, 14: 171–172.

Smith, G.J. 1965. Evaporite salts from the Dry Valleys of Victoria Land, Antarctica. *New Zealand Journal of Geology and Geophysics*, 8: 381–382.

Sochova, A.V., Savel'yev, A.A. and Shuleshko, I.K. 1975. Caliche in the Middle Proterozoic sequence of central Karelia. *Doklady Akademia Nauk SSSR, Earth Science Section*, 223: 173–176.

Soegaard, K. and Eriksson, K.A. 1985. Holocene silcretes from NW Australia: implications for textural modification of preexisting sediments. *Geological Society of America, Abstracts with Programs*, 17: 722.

Sofer, Z. 1978. Isotopic composition of hydration water in gypsum. *Geochimica et Cosmochimica Acta*, 42: 1141–1149.

Sowers, J.M. 1983. Pedogenic calcretes of Kyle Canyon, Nevada — morphology and development. *Geological Society of America, Abstracts with Programs*, 15.

Sperling, C.H.B. and Cooke, R.U. 1985. Laboratory simulation of rock weathering by salt crystallization and hydration processes in hot, arid environments. *Earth Surface Processes and Landforms*, 10: 541–555.

Srivastava, K.K. 1969. Gypsum. *Bulletin of the Geological Survey of India*, 31A.

Staley, J.T., Jackson, M.J., Palmer, F.E., Adams, J.B., Borns, D.J., Curtiss, B. and Taylor-George, S. 1983. Desert varnish coatings and microcolonial fungi on rocks of the Gibson and Great Victorian Deserts, Australia. *Bureau of Mineral Resources Journal of Australian Geology and Geophysics*, 8: 83–87.

Stankevich, E.F., Imameev, A.N. and Garanin, I.V. 1983. Formation of salt lake deposits in the arid regions of Middle Asia and Kazakhstan. *Problems of Desert Development*, 1983 (5): 19–24.

Stephens, C.G. 1971. Laterite and silcrete in Australia: a study of the genetic relationships of laterite and silcrete and their companion materials, and their collective significance in the formation of the weathered mantle, soils, relief and drainage of the Australian continent. *Geoderma*, 5: 5–52.

Stoerz, G.E. and Ericksen, G.E. 1974. *Geology of salts in northern Chile*. US Geological Survey Professional Paper 811.

Stortsky, G. and Rem, L.T. 1967. Influence of clay minerals on microorganisms.IV: Montmorillonite and kaolinite on fungi. *Canadian Journal of Microbiology*, 13: 1535–1550.

Storz, M. 1926. Zur Petrogenesis der sekunderären Kieselgesteine in der südlichen Namib. In E. Kaiser (ed.), *Die Diamentenwüste Südwest-Afrikas*, Vol. 2. Dietrich Reimer, Berlin: 254–282.

Summerfield, M.A. 1979. Origin and palaeoenvironmental interpretation of sarsens. *Nature*, 281: 137–139.

Summerfield, M.A. 1981. *The nature and occurrence of silcrete, Southern Cape Province, South Africa*. Oxford University School of Geography, Research Papers, 28.

Summerfield, M.A. 1982. Distribution, nature and genesis of silcrete in arid and semi-arid southern Africa. *Catena*, Supplement, 1: 37–65.

Summerfield, M.A. 1983a. Silcrete. In A.S. Goudie and K. Pye (eds), *Chemical sediments and geomorphology*. Academic Press, London: 59–91.

Summerfield. M.A. 1983b. Petrology and diagenesis of silcrete from the Kalahari Basin and Cape coastal zone, southern Africa. *Journal of Sedimentary Petrology*, 53: 895–909.

Summerfield. M.A. 1983c. Silcrete as a palaeoclimatic indicator: evidence from southern Africa. *Palaeogeography, Palaeoclimatology, Palaeoecology*, 41: 65–79.

Summerfield, M.A., 1983d. Geochemistry of weathering profile silcretes, southern Cape Province, South Africa. In R.C.L. Wilson (ed.), *Residual deposits: surface related weathering processes and materials*. Geological Society of London, Special Publication, 11. Blackwell Scientific, Oxford: 167–178.

Summerfield. M.A. 1986. Reply to discussion — silcrete as a palaeoenvironmental indicator: evidence from southern Africa. *Palaeogeography, Palaeoclimatology, Palaeoecology*, 52: 356–360.

Summerfield, M.A. and Whalley, W.B. 1980, Petrographic investigations of sarsens (Cainozoic silcretes) from southern England. *Geologie en Mijnbouw*, 59: 145–153.

Swett, K. 1974. Calcrete crusts in an Arctic permafrost environment. *American Journal of Science*, 274: 1059–1063.

Taylor-George, S., Palmer, F., Staley, J.T., Borns, D.J., Curtiss, B. and Adams, J.B. 1983. Fungi and bacteria involved in desert varnish formation. *Microbial Ecology*, 9: 227–245.

Thiry, M. 1981. Sédimentation continentale et altérations associées: calcitisations, ferruginisations et silicifications. *Les Argiles Plastiques de Sparnacien du Bassin de Paris. Sciences Géologiques, Memoire.* 64, Strasbourg.

Thiry, M. and Milnes, A.R. 1991. Pedogenic and groundwater silcretes at Stuart Creek opal field, South Australia. *Journal of Sedimentary Petrology*, 61: 111–127.

Thiry, M., Ayrault, M.B. and Grisoni, J. 1988. Ground-water silicification and leaching in sands: example of the Fontainebleau Sand (Oligocene) in the Paris Basin. *Bulletin of the Geological Society of America*, 100: 1283–1290.

Tolchel'nikov, Y.S. 1962. Calcium sulphate and carbonate neoformations in sandy desert soils. *Soviet Soil Science*, 1962: 643–650.

Torii, T., Muratas, S., Yoshida, Y., Ossaka, J. and Yamagata, N. 1966. Report of the Japanese summer parties in Dry Valleys, Victoria Land, 1963–65. 1: On the evaporites found in Miers Valley, Victoria Lane, Antarctica. *Antarctic Record (Japan)*, 27: 1–12.

Tucker, M.E. 1978. Gypsum crusts (gypcrete) and patterned ground from northern Iraq. *Zeitschrift für Geomorphologie*, NF, 22: 89–100.

Twidale, C.R. 1983. Australian laterites and silcretes: ages and significance. *Revue de geologie dynamique et de geographie physique*, 24: 35–45.

Twidale, C.R. and Campbell, E.M. 1995. Pre-Quaternary landforms in the low-latitude context — the example of Australia. *Geomorphology* 12: 17–35.

Twidale, C.R. and Hutton, J.T. 1986. Silcrete as a climatic indicator: discussion. *Palaeogeography, Palaeoclimatology, Palaeoecology*, 52: 351–356.

Twidale, C.R. and Milnes, A.R. 1983. Aspects of the distribution and disintegration of siliceous duricrusts in arid Australia. *Geologie en Mijnbouw*, 62: 373–382.

Van Arsdale, R. 1982. Influence of calcrete on the geometry of arroyas near Buckeye, Arizona. *Bulletin of the Geological Society of America*, 93: 20–26.

Van De Graaff, W.J.E. 1983. Silcrete in Western Australia: geomorphological settings, textures, structures, and their genetic implications. In R.C.L. Wilson (ed.), *Residual deposits: surface related weathering processes and materials*. Geological Society of London, Special Publication, 11. Blackwell Scientific, Oxford: 159–166.

Vaniman, D.T., Chipera, S.J. and Bish, D.L. 1994. Pedogenesis of siliceous calcretes at Yucca Mountain, Nevada. *Geoderma*, 63: 1–17.

Verrecchia, E.P. 1990. Litho-diagenetic implications of the calcium oxalate–carbonate biogeochemical cycle in semiarid calcretes, Nazareth, Israel. *Geomicrobiology Journal*, 8: 87–99.

Verrecchia, E.P., Dumont, J.-L. and Rolko, K.E. 1990. Do fungi building limestones exist in semi-arid regions? *Naturwissenschaften*, 77: 584–586.

Verrecchia, E.P., Ribier, J., Patillon, M. and Rolko, K.E. 1991. Stromatolitic origin for desert laminar crusts. *Naturwissenschaften*, 77: 505–507.

Verrecchia, E.P., Dumont, J.-L. and Verrecchia, K.E. 1993. Role of calcium oxalate biomineralisation by fungi in the formation of calcretes: a case study from Nazareth, Israel. *Journal of Sedimentary Petrology*, 63: 1000–1006.

Vieillefon, J. 1976. *Inventaire critique des sols gypseux en Tunisie. Étude préliminaire.* République Tunisienne, Ministère de l'Agriculture, Direction des Ressources en Eau et en Sol — Division des Sols. ORSTOM, Mission Tunisie.

Vieillefon, J. 1980. *Approche de l'intensité de la dynamique actuelle des accumulations gypseuses dans les sols an moyen dosage de la teneur en tritium de l'eau de constitution du gypse.* République Tunisienne, Ministère de l'Agriculture, Direction des Ressources de Eau et an Sol — Division des Sols. ORSTOM, Mission Tunisie E-S 181.

Vogt, T. 1977. Croûtes calcaires quaternaires de période froide en France méditerranéenne. *Zeitschrift für Geomorphologie*, NF, 21: 26–36.

von Humboldt, A. 1819. *Voyage aux Régions equinoxiales du nouveau Continent.* N. Maze, Paris.

Walls, R.S., Harris, W.B. and Nunan, W.E. 1975. Calcareous crust (caliche) profiles and early subaerial exposure of Carboniferous carbonates, northeastern Kentucky, *Sedimentology*, 22: 417–440.

Warren, J.K. 1982. The hydrological setting, occurrence and significance of gypsum in late Quaternary salt lakes in South Australia. *Sedimentology*, 29: 609–637.

Warren, J.K. 1983. Pedogenic calcrete as it occurs in Quaternary calcareous dunes in coastal South Australia. *Journal of Sedimentary Petrology*, 53: 787–796.

Warren, J.K. and Kendall, C.G.St.C. 1985. Comparison of sequences formed in marine sabkhas (subaerial) and salina (subaqueous) settings — modern and ancient. *Bulletin of the American Association of Petroleum Geologists*, 69: 1013–1023.

Watson, A. 1979. Gypsum crusts in deserts. *Journal of Arid Environments*, 2: 3–20.

Watson, A. 1981. Vegetation polygons in the central Namib Desert near Gobabeb. *Madoqua*, 2: 315–325.

Watson, A. 1983a. Gypsum crusts. In A.S. Goudie and K. Pye (eds), *Chemical sediments and geomorphology.* Academic Press, London: 133–161.

Watson, A. 1983b. Evaporite sedimentation in non marine environments. In A.S. Goudie and K. Pye (eds), *Chemical sediments and geomorphology.* Academic Press, London: 163–185.

Watson, A. 1985a. Structure, chemistry and origins of gypsum crusts in southern Tunisia and the central Namib Desert. *Sedimentology*, 32: 855–875.

Watson, A. 1985b. The control of wind blown sand and moving dunes: a review of methods of sand control, with observations from Saudi Arabia. *Quarterly Journal of Engineering Geology*, 18: 237–252.

Watson, A. 1988. Desert gypsum crusts as palaeoenvironmental indicators: a micropetrographic study of crusts from southern Tunisia and the central Namib Desert. *Journal of Arid Environments*, 15: 19–42.

Watts, N.L. 1977. Pseudo-anticlines and other structures in some calcretes of Botswana and South Africa. *Earth Surface Processes*, 2: 63–74.

Watts, N.L. 1978. Displacive calcite: evidence from recent and ancient calcretes. *Geology*, 6: 699–703.

Watts, N.L. 1980. Quaternary pedogenic calcretes from the Kalahari (southern Africa): mineralogy, genesis and diagenesis. *Sedimentology*, 27: 661–686.

Watts, S.H. 1975. An unusual occurrence of silcrete in adamellite — implications for duricrust genesis. *Search*, 6: 434–435.

Watts, S.H. 1977. Major element geochemistry of silcrete from a portion of inland Australia. *Geochemica et Cosmochimica Acta*, 41: 1164–1167.

Watts, S.H. 1978a. A petrographic study of silcrete from inland Australia. *Journal of Sedimentary Petrology*, 48: 987–994.

Watts, S.H. 1978b. The nature and occurrence of silcrete in the Tibooburra area of northwestern New South Wales. In T. Langford-Smith (ed.), *Silcrete in Australia.* Department of Geography, University of New England, Armidale, NSW: 167–185.

Waugh, B. 1970. Petrology, provenance and silica diagnesis of the Penrith Sandstone (lower Permian) of northwest England. *Journal of Sedimentary Petrology*, 40: 1226–1240.

Wells, R.T. and Callen, R.A. (eds) 1986. *The Lake Eyre Basin — Cenozoic sediments, fossil vertebrates and plants, landforms, silcretes and climatic implications.* Australasian Sedimentologists Group, Field Guide Series No. 4.

Wells, S.G. and Schultz, J.D. 1979. Some factors influencing Quaternary calcrete formation and distribution in alluvial fill of arid basins. *Geological Society of America, Abstracts with Programs*, 11: 538.

Whalley, W.B. 1983. Desert varnish. In A.S. Goudie and K. Pye (eds), *Chemical sediments and geomorphology*. Academic Press, London: 197–226.

Whalley, W.B. 1984. High altitude rock processes. In K.J. Miller (ed.), *The International Karakoram Project*, Vol. 1. Cambridge University Press, Cambridge: 365–373.

Whalley, W.B., McGreevy, J.P. and Ferguson, R.I. 1984. Rock temperature observations and chemical weathering in the Hunza region, Karakoram: preliminary data. In K.J. Miller (ed.), *The International Karakoram Project*, Vol. 2. Cambridge University Press, Cambridge: 616–633.

Wheeler, W.H. and Textoris, D.A. 1978. Triassic limestone and chert of playa origin in North Carolina. *Journal of Sedimentary Petrology*, 48: 765–776.

White, C.H. 1924. Desert varnish. *American Journal of Science*, Series 5, 7: 413–420.

White, K.H. 1990. *Spectral reflectance characteristics of rock varnish in arid areas*. Oxford University School of Geography Research Papers 46.

White, K.H. 1993. Image-processing of thematic mapper data for discriminating piedmont surficial materials in the Tunisian southern Atlas. *International Journal of Remote Sensing*, 14: 961–977.

Wieder, M. and Yaalon, D.H. 1974. Effect of matrix composition on carbonate nodule crystallisation. *Geoderma*, 11: 95–121.

Wigley, T.M.L. 1973. Chemical evolution of the system calcite–gypsum–water. *Canadian Journal of Earth Sciences*, 10: 306–315.

Williams, G.E. and Polach, H.A. 1971. Radiocarbon dating of arid-zone calcareous palaeosols. *Bulletin of the Geological Society of America*, 83: 3069–3086.

Williamson, W.O. 1957. Silicified sedimentary rocks in Australia. *American Journal of Science*, 255: 23–42.

Woolnough, W.G. 1930. The influence of climate and topography in the formation and distribution of products of weathering. *Geological Magazine*, 67: 123–132.

Wopfner, H. 1974. Post-Eocene history and stratigraphy of northeastern Australia. *Transactions of the Royal Society of South Australia*, 98: 1–12.

Wopfner, H. 1978. Silcretes of northern South Australia and adjacent regions. In T. Langford-Smith (ed.), *Silcrete in Australia*. Department of Geography, University of New England, Armidale, NSW: 93–141.

Wopfner, H. 1983. Environment of silcrete formation: a comparison of examples from Australia and the Cologne Embayment, West Germany. In R.C.L. Wilson (ed.), *Residual deposits: surface related weathering processes and materials*. Geological Society of London Special Publication 11. Blackwell Scientific, London: 151–157.

Wright, E.P. 1978. Geological studies in the northern Kalahari. *Geographical Journal*, 144: 235–249.

Wright, R.L. 1963. Deep weathering and erosion surfaces in the Daly River basin, Northern Territory. *Journal of the Geological Society of Australia*, 10: 151–163.

Wright, V.P. 1989. Terrestrial stromatolites and laminar calcretes: a review. *Sedimentary Geology*, 65: 1–13.

Wright, V.P. 1990. A micromorphological classification of fossil and recent calcic and petrocalcic microstructures. In L.A. Douglas (ed.), *Soil micromorphology: a basic and applied science*. Developments in Soil Science 19. Elsevier, Amsterdam: 401–407.

Wright, V.P. and Tucker, M.E. 1991. Calcretes: an introduction. In V.P. Wright and M.E. Tucker (eds), *Calcretes*. International Association of Sedimentologists Reprint Series, Vol. 2. Blackwell Scientific, Oxford: 1–22.

Yaalon, D.H. 1964a. Downward movement and distribution of ions in soil profiles with limited wetting. In E.G. Hallsworth and D.V. Crawford (eds), *Experimental pedology*. Butterworth, London: 159–164.

Yaalon, D.H. 1964b. Airborne salts as an active agent in pedogenetic processes. *Transactions of the 8th Congress of Soil Science*, 5: 997–1000.

Yaalon, D.H. 1988. Calcic horizon and calcrete in Aridic soils and paleosols: progress in the last twenty two years. *Soil Science Society of America Agronomy Abstracts*. Cited in V.P. Wright and M.E. Tucker (1991).

Yaalon, D.H. and Singer, S. 1974. Vertical variation in strength and porosity of calcrete (nari) on chalk, Shefala, Israel, and interpretation of its origin. *Journal of Sedimentary Petrology*, 44: 1016–1023.

Yaalon, D.H. and Wieder, M. 1976. Pedogenic palygorskite in some arid brown (Calciorthid) soils in Israel. *Clay Minerals*, 11: 73–80.

Yang, W.B., Spencer, R.J., Krouse, H.R., Lowenstein, T.K. and Cases, E. 1995. Stable isotopes of lake and fluid inclusion brines, Dabusin Lake, Qaidam Basin, western China — hydrology and paleoclimatology in arid environments. *Palaeogeography, Palaeoclimatology, Palaeoecology*, 117: 279–290.

Young, R.G. 1964. Fracturing of sandstone cobbles in caliche cemented terrace gravels. *Journal of Sedimentary Petrology*, 34: 886–889.

Young, R.W. 1978. Silcrete in a humid landscape: the Shoalhaven Valley and adjacent coastal plans of southern New South Wales. In T. Langford-Smith (ed.), *Silcrete in Australia*. Department of Geography, University of New England, Armidale, NSW: 195–207.

Young, R.W. and McDougall, I. 1982. Basalts and silcrete on the coast near Ulladulla, southern New South Wales. *Journal of the Geological Society of Australia*, 29: 425–430.

Zen, E.-A. 1965. Solubility measurements in the system $CaSO_4$–$NaCl$-H_2O at 35°, 50° and 70° and one atmosphere pressure. *Journal of Petrology*, 6: 124–164.

7

Vegetation and dryland geomorphology

Joanna E. Bullard

Introduction

Geomorphic processes in humid environments are greatly affected by the presence of vegetation. For example, plant cover affects runoff generation by intercepting rainfall, protects soil from erosion by wind and water and plays a part in determining rates of weathering. In dryland areas, however, it is the absence of a vegetation cover that is often of greater importance in determining the unique characteristics of the landscape. Soils are less protected and far more susceptible to aeolian and hydrological processes resulting in the increased availability of sediment to form features such as sand dunes or to supply alluvial fans. Although the general lack of vegetation gives many dryland environments their distinctive character, few arid areas are completely devoid of vegetation of all types. The range of vegetation types present in drylands extends from microscopic algae to annual or perennial grasses, shrubs and trees, all of which alter ground surface characteristics and consequently the operation of geomorphic processes. The distribution of all these vegetation types is often discontinuous, both spatially and temporally, but where and when it does occur it can assume greater relative significance in terms of geomorphology than would a continuous and persistent plant cover.

Vegetation in desert landscapes has been documented and studied extensively by ecologists and botanists but it is only comparatively recently that suggestions have been made as to the links between vegetation and geomorphology. During the past 25 years or so, the number of detailed studies considering the geomorphological role of vegetation at both the plant and landform scales, whether explicitly or implicitly, has risen rapidly to the extent that vegetation is now accepted as playing an important role in all aspects of dryland geomorphology (e.g. Viles 1988; Thornes 1990a; Millington and Pye 1994).

In this chapter the emphasis is on the nature, characteristics and distribution of vegetation in arid environments. The role of vegetation in modifying aeolian and hydrological processes is considered with an emphasis on its implications for landform development. Details of the interaction between vegetation and physical processes, and examples of specific vegetation — landform links, are also considered elsewhere (Chapters 5, 6, 9, 10, 16 and 17).

The characteristics of vegetation in arid environments

The environmental conditions associated with most arid environments — low or sporadic

Arid Zone Geomorphology: Process, Form and Change in Drylands, 2nd edition. Edited by David S. G. Thomas.

moisture availability and high or extremely variable temperatures — are not ideal for vegetation growth. These extreme conditions result in desert vegetation communities being very distinctive and often highly adapted to persist in such hostile environments. Specific adaptations contribute to the global diversity of dryland vegetation and these, combined with other factors such as tectonic, climatic and land-use histories, mean that not only are there considerable distinctions between the vegetation communities of different deserts but also within individual deserts (Table 7.1).

There are several classifications of desert vegetation types. These have generally been made from a botanical or ecological viewpoint and are frequently based upon the relationship between plants and the availability of moisture. They vary from that suggested by Goodall and Perry (1979), which describes common

Table 7.1 *Summary of selected environmental characteristics of arid areas and the dominant vegetation types (compiled from Evenari* et al. *1985; 1986)*

Area	Latitudinal range	Altitudinal range (m)	Range of mean annual rainfall (mm)	Dominant vegetation types
North America				
Sonoran	25–33°N	0–1200	100–300	Thorny succulent savanna, tall perennial shrubs (>5 m), both winter and summer annuals
Chihuahuan	23–32°N	700–1600	75–400	Cacti are abundant, perennial shrubs dominate, summer annuals
South America				
Peruvian–Chilean Coastal Desert	7°N–35°S	0–900	<50	Succulent desert dwarf shrubs, lomas (fog-dependent vegetation)
Australian				
Australian Desert	21–33°S	0–700	250–380	Low shrubland especially *Acacia* and eucalyptus, perennial *spinifex* grassland
Asia				
Thar	24–31°N	0–800	150–500	Thorny savanna, psammophytic vegetation on dunes
Gobi	41–47°N	500–1800	80–400	Chenopod–*Tamarix* desert shrubland
Arabian Peninsula	15–31°N	0–1200	Generally <200	Desert dwarf shrubs, psammophytic vegetation on sands, annuals prolific after rains
Africa				
Sahara	22–32°N	0–800	0–400	Open woodland in wetter areas, evergreen shrubs/small trees, perennial grasslands
Sahel	13–22°N	0–1300	100–200 to 400–500	Open savanna–steppe, woody vegetation dominated by thorny shrubs and low trees
Namib	18–30°S	0–1100	Generally <100	Succulent desert dwarf-shrubs, lichens and algae important near coast, perennial grasses inland
Kalahari	20–28°S	500–1500	<200 to >600	Mixed savanna grassland, low perennial trees and shrubs

assemblages associated with decreasing amounts of moisture input, to those suggested by Shantz (1956) or Levitt (1972), which are based upon how plants have adapted to cope with moisture shortages or drought. These classifications can be useful as the relationships between vegetation and water availability are geomorphologically important, for example controlling plant spacing, ground cover and vegetation persistence (Thomas 1988a). However, it is not only higher-order plants that affect geomorphological processes. Vegetation communities in arid areas may consist of ephemeral annuals, perennials, shrubs and trees, as well as incorporating algal crusts, lichens and plant litter. Goodall and Perry's (1979) classification (Table 7.2) gives some indication of the relative amounts of shrubs and grasses in 'steppe' and 'true' desert. This classification excludes the lower plants and also what is known as 'accidental' vegetation, the term given to ephemeral annuals that grow very rapidly, usually after a single rainfall event. The types of vegetation and the relative quantities of each type in a desert are controlled by a range of factors, but the most important in terms of the initial plant successions are moisture availability and the stability of the medium in which they are growing.

THE ROLE OF MOISTURE

The four main moisture sources in arid areas are rainfall, coastal fog, dew and groundwater, and their relative significance varies from desert to desert. Rainfall is usually the primary determinant of gross vegetation production in areas where the mean annual precipitation is less than 600 mm, although this relationship can be complicated by the input of moisture from other sources and losses from soil and plants. Where rainfall is in excess of this, other factors such as nutrient availability assume a greater importance.

Significant positive relationships between precipitation and above-ground primary production have been established for east and southern Africa (Desmukh 1984), north Africa, the Sahelo-Sudanian zone (Le Houérou and Hoste 1977) and parts of North America (e.g. Sims and Singh 1978) (Figure 7.1). Calculations such as these have tended to concentrate on grassland production and do not include shrub or tree biomass which may be less dependent upon rainfall as a moisture source. For example, an analysis of data from east and southern Africa predicts that for every 100 mm precipitation received above 20 mm, a maximum of 800 kg ha^{-1} of grassland biomass can be attained (Desmukh 1984). Most relationships between precipitation and primary productivity suggest that a certain amount of rainfall, usually 20–25 mm, is necessary to trigger the growth of plants (Went 1959; Seely 1978; Desmukh 1984), although not every 25 mm rainstorm event will trigger germination. The growth of most species of annuals is a response to a combination of factors such as the quantity of rainfall and the frequency of rainfall events as well as access to sufficient nutrients such as nitrogen or phosphorus (Gutiérrez and Whitford 1987). 'Accidental' vegetation grows very rapidly, usually following a one-off rainfall event (e.g. Kassas 1966). This vegetation is short-lived and characteristic of areas where precipitation is both low and irregular to the extent that no permanent vegetation can exist.

Table 7.2   *Classification of different assemblages of dryland vegetation (after Goodall and Perry 1979)*

Term	Vegetation description
Steppe	10–30% dwarf suffrutescent shrubs, with perennial grasses and cacti
True desert	Perennial vegetation with less than 10% dwarf shrubs or a few tall shrubs accompanied by a flush of herbaceous annuals in the rainy season; cover never exceeds 50% even in the most favourable years
Contracted desert	Vegetation only occurs in wadis where the groundwater is accessible to deeper-rooted plants including palms

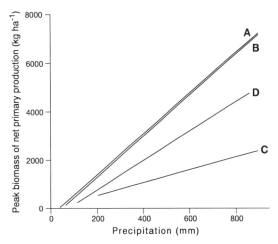

Figure 7.1 *Suggested relationships between precipitation and above-ground peak biomass (A, B, C) or primary production (D). (A) East and southern Africa (Desmukh 1984); (B) Mediterranean North Africa (Le Houérou and Hoste 1977); (C) the Sahelo-Sudanian zone (Le Houérou and Hoste 1977); and (D) grasslands of the USA (Sims and Singh 1978) (diagram adapted from Desmukh 1984)*

Studies from southwest Egypt suggest that accidental vegetation is the only type of precipitation-dependent vegetation that can exist in areas with less than an annual rainfall of 10 mm (Bornkamm 1987).

In arid areas, however, the irregular spatial and temporal distributions of rainfall are more significant than the actual amount available in determining vegetation patterns and their spatial and temporal persistence. Rainfall is usually very variable and arid regions are often characterised by strong seasonality. For example, in the southwest Kalahari, southern Africa, over 80% of the 200–300 mm yr^{-1} rainfall occurs between October and the end of March (Thomas and Shaw 1991a).

More reliable sources of moisture in some arid areas are fog, dew and groundwater. In coastal deserts, such as the Namib and the Atacama, condensation from fog can provide 50–300 mm yr^{-1} (Tivy 1993), exceeding rainfall amounts at the coast. In the Namib the impact of fog extends over 100 km inland, though decreasing with distance from the coast in terms of both the number of fog days (Oliver 1995) and consequent fog precipitation amounts (Lancaster *et al.* 1984).

Humidity is often very low in arid environments; however, the increase in relative humidity that accompanies a fall in night-time temperatures can cause atmospheric vapour to condense as dew. A 20-year study of the Negev Desert recorded an average of 195 dew-nights per year resulting in a mean annual dew deposit of 33 mm (Noy-Meir 1973). The presence of vegetation itself can affect the spatial distribution of dewfall. Anderson (1988) examined the relationship between dewfall and *Prosopis* vegetation in the Wahiba Sands, Oman, and found that in areas of bare ground the dewfall was up to seven times that recorded within the *Prosopis* clumps. This was attributed to slightly warmer air temperatures within the canopy and a possible reduction in ambient humidity caused by absorption of water through the plant leaves. However, within dense stands of *Prosopis* dewfall was more common than over comparatively clear dune or plain areas and this may indicate a relationship between dewfall and the transpiration of water by the trees (Anderson 1988).

Subterranean dew may also be an important moisture source for plants. Exposed surface sands can heat up very rapidly in response to high daytime temperatures and cool rapidly at night, whereas sand as little as 30 cm below the surface experiences little change in temperature (of the order of 1–2°C diurnally) due to the low thermal diffusivity of the soil (Tsoar 1990). The resulting temperature differences (up to 20°C over 30 cm depth) can lead to the establishment of a steep night-time temperature gradient between the surface and subsurface layers (Tsoar 1990). This gradient can instigate an upward flow of subsoil vapour that distills on the surface causing subterranean dew (Salisbury 1952; Monteith 1957; Migahid 1961). This moisture source was dismissed by Bagnold (1941), Noy-Meir (1973) and Willis *et al.* (1959) as being of little geomorphological or biological importance, but it has been suggested that 15–25 mm of moisture could be precipitated (Thornes 1994), which is a significant amount in very arid areas.

Plants, primarily trees and shrubs, that depend upon groundwater are known as phreatophytes. Phreatophytes are deep-rooted to reach the water table and, although water consumption varies considerably from species to species, in general they use more water than other vegetation in the same environment. Estimates of water 'loss' through phreatophytes vary widely; for example, estimates of annual evapotranspiration for Tamarisk in the United States range from 34 to 280 cm^3 (Graf 1985). At 100% coverage annual water consumption by cottonwood has been estimated at 1.8 m^3 yr^{-1} and by mesquite at 1 m^3 yr^{-1} (Bouwer 1975). Loss of water by evapotranspiration decreases as the depth of the water table increases (van Hylckama 1963). The use of groundwater means that the vegetation is not rainfall-dependent and so can persist through periods of prolonged drought as long as the water table remains within reach of the plant roots. In some parts of Arizona water pumping has lowered the water table beyond the reach of the dominant saltcedars and these plants are dying from lack of moisture; however, elsewhere near the San Carlos Reservoir, Arizona, the saltcedars appear to have adapted to a non-phreatophytic existence (Leppanen 1970).

SURFACE STABILITY AND VEGETATION

Loose sand deposits, characteristic of some arid areas, are highly susceptible to erosion by both wind and water and so the surfaces of the deposits may be highly mobile. This mobile layer is generally restricted to the top few centimetres of the surface, especially where aeolian processes dominate, but the resulting instability can make it very difficult for vegetation to establish. There is a very fine balance between where soil erosion or mobility is too great to allow vegetation to become established and where the vegetation cover is sufficient to stabilise the surface and keep erosion in check (see Thornes 1985; 1990b).

In humid environments, coarse, loose sand is a poorer soil for plant growth than finer-textured soils such as loams and clays, but in arid conditions sand has a number of advantages over finer soils. The major moisture losses from soils are through runoff, vertical infiltration beyond the reach of plant roots and evaporation from the soil surface (Noy-Meir 1973). Runoff is usually negligible on loose, coarse sands leaving deep percolation and evaporation as the major methods of moisture loss. Sand has a low moisture tension and high hydraulic conductivity which means that under heavy precipitation water can easily be lost by rapid downward movement beyond the reach of most plant roots (Tsoar 1990). In New Mexico the rate of hydraulic conductivity on sands is up to 20 m per day (Hennessy *et al.* 1985) and in Israel it has been estimated at 13 m per day (Goldschmidt and Jacobes 1956). However, under the light precipitation characteristic of many deserts, the water serves to keep sand moist below the layer of sand which is desiccated by evaporation, estimated at 30–60 cm below the surface (Prill 1968; Noy-Meir 1973; Tsoar 1990). This results in a dry, loose surface layer underlain by more moist conditions. Noy-Meir (1973) estimated that the threshold amount of precipitation which governs whether the properties of sand give it an advantage or a disadvantage over finer soils lies between 300 and 500 mm yr^{-1}. Where the sand depth is less than 60 cm, the deep moisture zone will be poorly developed and moisture will be lost by evaporation. This results in less vegetation or smaller, stunted vegetation growing on sands less than 60 cm deep (Chadwick and Dalke 1965; Yeaton 1988).

The existence of damp sand at depth creates a favourable rooting environment for established plants with deep roots, but the dry surface sand is still highly susceptible to erosion and so the establishment of vegetation from seed is difficult. A temporary reduction in surface mobility will occur after precipitation events which raise the cohesivity of sand particles making them more resistant to wind erosion (Bisal and Hsieh 1966). However, large amounts of moisture are required to reduce significantly the erosion potential and the effects may last as little as 24 hours (Hidore and Albokhair 1982; Sarre 1987). Surface mobility will also be reduced when wind velocities are very low. Even if they can establish themselves, in order to survive on a moving substrate the plants must be adapted to cope with sandblasting and burial by sand (Rempel 1936; Bendali *et al.* 1990). In a study of sand dunes in the Namib Sand Sea, Yeaton (1988; 1990) found that the distribution of species of plants on the dunes reflected the amount of surface sand movement taking place. Near the crests of the dunes, where the sands are highly

...ile, species establishing vegetatively from rnizomes (e.g. *Stipagrostis sabulicola*) were dominant, and those species that establish predominantly from seed (e.g. *Stipagrostis ciliata, Cladoraphia spinosa*) were found on the more stable lower slopes of the dunes.

Once a vegetation cover has been established on a coarse sand surface it can modify the surface to make conditions more favourable for plant growth. A sequence of events leading to the stabilisation of sand surfaces by organic agents has been proposed (e.g. Tsoar and Møller 1986; Danin *et al.* 1989). First, an initial pioneer vegetation cover must be established; this is usually by vegetative propagation or plants taking advantage of periods of rainfall or low wind speeds when the surface is less mobile. The presence of vegetation increases the roughness of the surface so reducing particle saltation, and will encourage the accumulation of fine particles such as silts and clays in the dune surface sediments (Danin and Yaalon 1982; Tsoar and Møller 1986; see Chapter 16). These finer particles can promote the growth of bio-, or mycophytic, crusts (algal, lichen or bacterial) that bind the surface particles and so reduce their erodibility (Danin 1978; Van den Ancker *et al.* 1985; Danin *et al.* 1989; see Chapter 5). These crusts can also increase the fertility of the soil by providing nutrients that further promote vegetation growth (Fletcher and Martin 1948; Fuller *et al.* 1960) and increase the water retention properties of the soil (Tsoar and Møller, 1986). As the surface becomes more stable with a more cohesive soil, greater densities of higher vegetation can be supported resulting in a plant succession (Danin 1978; Danin and Yaalon 1982; Danin *et al.* 1989) (Figure 7.2).

Yeaton (1988) found an inverse correlation between the amount of sand movement and plant density (Figure 7.3) for dunes in the Namib Sand Sea. Although more vegetation might be expected to survive on less mobile soils, equally, according to the model in Figure 7.2, the greater the amount of vegetation, the less dynamic a sand surface is likely to be. Several species of plants are adapted to withstand inundation by sand. These plants trap sand at their bases by acting as an obstacle in the path of saltating grains. As the plants are buried they grow new roots from their stem that

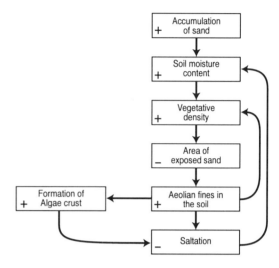

Figure 7.2 *A feedback mechanism for the natural stabilisation of desert sand dunes (after Tsoar and Møller 1986)*

lie under the new higher sand surface, and the growth in the plant roots and shoots trap more sand causing the formation of a sand mound. The older, original plant mass at the base of the mound may eventually die off but still acts to bind the sand. Some species are dependent upon the continued accumulation of sand and may die if sediment supplies decrease. *Rhanterium suaveolens* survives low levels of sand accumulation by producing new aerial structures and roots near the surface of the soil and may cause the development of 'micronebkhas', but cannot tolerate high levels of sand accumulation. In contrast, *Aristida pungens* thrives in areas of rapid sand accumulation and can create hillocks up to 2 m high (Bendali *et al.* 1990).

There are similar relationships between vegetation and other types of substrate. In arid badland areas the unconsolidated nature of the substrate and high rates of soil erosion are not conducive to vegetation establishment and the lack of vegetation means the soils are susceptible to erosion. However, lower-order vegetation may be able to colonise the unstable substrate and from that the vegetation assemblage can develop sufficiently that the slope may be stabilised. Alexander *et al.* (1994) describe a possible mechanism of slope stabilisation by vegetation from Tabernas, southern Spain. Initially the slopes are characterised by a few

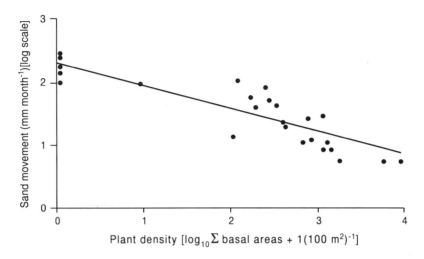

Figure 7.3   *The relationship between monthly rates of sand movement (mm) and plant density (basal cover per 100 m²) in the Namib Desert (after Yeaton 1988)*

shrubs and by lichens which increase runoff but decrease sediment concentration (Alexander and Calvo 1990). Where lichen covers are dense and beneath clumps of shrubs the slopes are protected from erosion, but elsewhere the increased runoff will lead to soil loss. In the protected areas grass and shrub cover will increase so reducing runoff and erosion rates by interception and increased soil consolidation. As vegetation cover increases, root densities and greater litter concentration will raise the infiltration capacity of the soil, further reducing runoff potential (Faulkner 1990).

The distribution of vegetation

SPATIAL PATTERNS

Although some deserts are characterised by a relatively continuous natural or semi-natural vegetation cover, the distribution of vegetation is more often sporadic both spatially and temporally. Commonly, vegetation is sparse but spatially aggregated, forming a mosaic of relatively well-vegetated areas separated by patches of virtually bare ground (see also Chapter 5). The spacing of vegetation, whether individual plants, aggregated groups of plants or patches of mycrophytic crust, is very important in terms of both modifying geomorphological processes

and determining the characteristics of vegetation-dependent landforms.

At a localised scale where moisture availability is relatively constant, such as in wadis and oases, there is a tendency towards permanent but spatially restricted patches of vegetation with the least drought-tolerant plants in the centre of the vegetation patch (Gimingham 1955). Over larger areas with more variable water supply, the vegetation cover is often more extensive but less dense. If populations have expanded to a level where environmental resources have to be competed for, plants of individual species tend to be regularly spaced. This has been attributed to intraspecific competition, for example for water (Woodell *et al.* 1969; Bhadresa and Moore 1982). Larger plants may be further away from their nearest neighbours than smaller plants to enable the former to develop more extensive root patterns (Nobel 1981). As well as taking place intraspecifically, resource partitioning also occurs between species; for example, although the shallow-rooted *Stipagrostis sabulicola* will not tolerate being near other plants of the same species, different species such as the deep-rooted *Stipagrostis namaquensis* or *Cladoraphia spinosa* can survive nearby (Yeaton 1990). This is thought to be because the different species are obtaining water from different sources so are not in direct competition.

The spatial distribution of vegetation can also be influenced by the substrate on which it grows. In gravel deserts the maturity of the gravel armour may be such that the gravel is so closely packed as to be impenetrable to plant roots and so predominantly sterile (Kassas and Imam 1959). More loosely packed gravels or those which are overlain by patches of aeolian sands are more likely to support higher plants. The thickness of sand cover can be crucial to the density and variety of vegetation that can survive. Sands more than 60 cm deep are likely to contain moisture at depth and so are favoured sites for plant growth, but shallower sands can provide a substrate for shallow-rooted ephemerals. This contributes to the development of vegetation patterns on landforms. For example, sand dune vegetation is often banded with low plant densities in the shallow sands of the interdunes, high plant densities supported on the deeper, relatively stable lower and mid-dune slopes, and sparse or highly adapted vegetation on the mobile dune crests. There is some evidence to suggest that although these vegetation bands may be related to topography, it is the soil texture rather than the topographic location itself that is of greater importance in determining the vegetation assemblages (Buckley 1981).

On very shallow slopes, with an overall gradient of less than 0.5%, vegetation may be highly organised into linear bands contouring the slope. These, usually dense, bands of vegetation (groves) alternate with bands of little or no vegetation (intergroves or interbands). The resultant patterning is known as 'brousse tigrée' or 'steppe tigrée' (Le Houérou 1981) and has been recognised in north Africa (MacFadyen 1950; Greenwood 1957; Worrall 1960; Boaler and Hodge 1962; White 1970; Wickens and Collier 1971), west Asia (White 1969) and Australia (Mabbutt and Fanning 1987; Tongway and Ludwig 1990). The vegetated groves are often tree- or shrub-dominated with a cover of 30–40%, whereas the wider intergroves generally have less than 10% vegetation cover which is made up of grasses and ephemeral plants. Groves vary in width from 10 to 50 m and intergroves are commonly two to five times as wide as the vegetated bands (Tongway and Ludwig 1990). The vegetation banding is associated with patterns of soil surface

characteristics, soil depth, runoff, infiltration and sedimentation (see Chapter 5).

In a study of vegetation banding in the Wiluna–Meekatharra area of Western Australia, Mabbutt and Fanning (1987) found that the groves are associated with narrow trenches in the silica-cemented hardpan underlying the area which act as a trap for water running off the intergroves. This, and the increased depth of soil overlying the trenches, makes conditions more favourable for vegetation growth. The shallow slopes on which these patterns are found are too gentle for the full development of drainage channels so sheetflow dominates (Mabbutt and Fanning 1987). Runoff is greatest from the intergrove areas where vegetation is sparse and the soil surfaces usually compact and sealed or armoured. Runoff is captured in the groves where infiltration is up to twice as high as in the intergroves, and microtopography, dense vegetation and surface litter all encourage surface ponding (Mabbutt and Fanning 1987). Any soil material eroded from the intergroves will be deposited in the groves and may contribute to the development of a 'stepped' slope topography, although the exact relationship between the topography and the position of the vegetation band can vary (Tongway and Ludwig 1990).

TEMPORAL PATTERNS

The density and distribution of vegetation can vary at a range of temporal scales from seasonal to decadal and longer. Rainfall in many desert regions is highly seasonal, restricting active vegetation growth to a few months of the year. Ephemerals and annuals are very sensitive to moisture inputs and will develop after rainfall within two to three days for ephemerals and a few days for annuals (Heathcote 1983). In the semi-arid Mtera region of Tanzania the growth of annuals swells the vegetation cover from a dry season low of 3% to over 80% in the wet season (Thomas 1988a). The time lag between the rainfall event and the subsequent plant growth will result in any runoff generated being very erosionally effective.

Ephemeral and perennial herbaceous species are highly dependent upon moisture availability, most often rainfall. The actual density and composition of the vegetation is dependent not only on the amount of rainfall but also on its timing in the year because of its effect on germination (Bowers 1987). Fluctuations in rainfall lead to rapid flushes of vegetation when water availability is high and to a reduction in ground cover where rainfall is absent or restricted.

The increases and decreases in ground cover associated with annual variations in rainfall do not only last for one year but can also affect vegetation patterns in subsequent years. For example a 25% decrease in annual precipitation in Wyoming between 1984 and 1986 resulted in a 73% reduction in perennial grass cover and a 23% decrease in root biomass in an ephemeral channel (Smith *et al.* 1993). Similar decreases in vegetation cover associated with periods of reduced rainfall have also been recorded for the Kalahari (Choudhury and Tucker 1987). Out of the growing season, dead plants and plant litter can still play a significant geomorphological role in stabilising surfaces (Wasson 1976; Gibbens *et al.* 1983; Brazel and Nickling 1987; Wiggs *et al.* 1994).

Shrubs and trees tend to be less dependent upon rainfall and so are less dynamic in terms of variations in ground cover. Even during droughts spanning a number of years, in the short to medium term shrub and tree cover remains constant, whereas herbaceous species are far more dynamic and tend to fluctuate in cover under the influence of seasonal variations (Westoby 1980; Bertiller 1984; Van Rooyen *et al.* 1984; Gutiérrez *et al.* 1993; Luk *et al.* 1993).

As in other environments, fire has a profound effect on vegetation cover. Burning vegetation removes ground cover and litter very rapidly. Lightning-ignited fires are most likely when vegetation cover is dense, for example after exceptionally heavy rains (Trollope 1980), and the extent and severity of fires is thought to be greatest following periods of above-average rainfall (van der Walt and Le Riche 1984). Vegetation is, however, more vulnerable to fire when it is dry and so may also be at risk towards the end of the dry season (Manry and Knight 1986). The recovery of vegetation will depend upon the time of year and the severity of the fire. Burning removes vegetation and litter leaving roots and short stems that offer little protection and leave the ground surface highly susceptible to erosion by water or wind

(Conacher 1971; Wiggs *et al.* 1994). Fires can also result in a complex mosaic of differential runoff and infiltration patterns, the precise implications of which are dependent upon the intensity of the fire (Lavee *et al.* 1995).

Vegetation and geomorphic processes

There are no simple relationships between the presence of vegetation and the modification or operation of geomorphological processes. Vegetation modifies surface flow regimes and can either increase or decrease flow velocities and flow volume. Sediment transport, infiltration and runoff may be enhanced or diminished and, while parts of the ground surface are protected, other parts are subject to heightened erosion. Vegetation characteristics (e.g. spacing, height) and external influences (e.g. substrate) combine to control the precise relationship between the vegetation cover and its effect on hydrological (discussed in Chapter 9), aeolian (Chapter 16) and soil processes.

AEOLIAN PROCESSES

The presence of vegetation has three main direct effects on aeolian processes. First, it protects the surface immediately beneath the plant from erosion; second, it increases the surface roughness so increasing friction, reducing wind velocity and hence erosive power; and third, it can act as an obstacle or trap for migrating particles (Figure 7.4) (Wolfe and Nickling 1993). Vegetation can also encourage the deposition of fines in the soil which alter the particle size distribution and can encourage crust development.

Higher vegetation modifies the interface between the atmospheric boundary layer and the ground surface by increasing surface roughness, decreasing the erosivity of the wind, reducing its near-surface velocity and encouraging sediment deposition (see Chapter 16). This decrease in erosivity is not ubiquitous especially when vegetation cover is sparse, as zones of accelerated and decelerated windflow may be created leading to differential erosion and deposition across the surface (Figures 7.5 and 7.6).

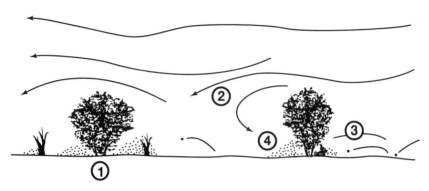

Figure 7.4 *Summary of the major effects of vegetation on sand movement. 1 = deposition in windward and lee zones; 2 = reattaching flow, 3 = upwind scour; 4 = lee deposition due to reserve flow*

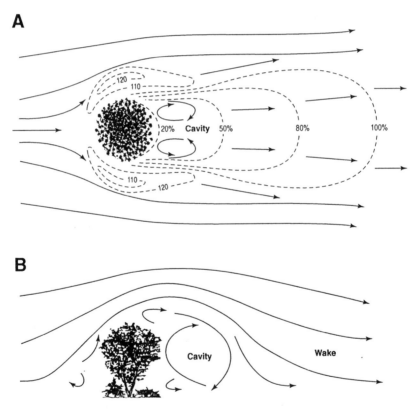

Figure 7.5 *Percentage wind speed and flow lines around a single bush. (A) Plan view, (B) section view (after Ash and Wasson 1983)*

Vegetation characteristics such as lateral cover, overall shape, porosity and flexibility are important to the modification of aeolian processes; however, the two most influential attributes are spacing and height (Bressolier and Thomas 1977; Lee and Soliman 1977; van de Ven *et al.* 1989; Lee 1991; Wolfe and Nickling 1993). Essentially, each individual plant acts as a roughness element. In the case of a single plant windflow may be accelerated around the sides of the plant, but a zone of low velocity, the wake, will develop in the lee of the plant and deposition may occur there (Ash and Wasson 1983) (Figure 7.7). Where plants are

Figure 7.6   *Ripples reflect changes in wind speed around individual grass clumps. Wavelength of ripples in the centre of the photograph is approximately 7 cm. The wind direction is towards the top left of the picture*

Figure 7.7   *Sand deposits in the lee of grass clumps on the crest of a sand dune*

more closely spaced, the wakes of individual plants can interact with one another.

Morris (1955) identified three types of airflow pattern that may occur depending upon the spacing of the roughness elements. These are isolated-roughness flow, wake-interference flow and skimming flow, and represent decreasing element spacing (Figure 7.8). The wind can act directly on the surface as a mechanism for erosion in isolated-roughness and wake-interference flow regimes. When skimming flow dominates, there may still be a considerable amount of exposed surface, but this is protected from direct erosion by the surrounding roughness elements (Wooding *et al.* 1973). Lee and Soliman (1977) used wind-tunnel experiments to investigate roughness element densities and suggest that a ground cover of as little as 40% could result in skimming flow over a surface (Table 7.3). This model is useful given that plants in many arid areas have a regular or random spacing (Woodell *et al.* 1969; Anderson 1971). Wolfe and Nickling (1993) suggest that the sparsely

vegetated areas of arid regions should be characterised by wake-interference or isolated-roughness flow, although skimming flow has been recorded (e.g. Lee 1991). Plant shape, porosity and flexibility also affect the flow regimes and the density of vegetation needed to fully protect the surface (Buckley 1987). Plant canopies can also be penetrated by strong gusts of wind (Allen 1968).

The wide range of factors influencing the way in which vegetation and wind interact means that estimates of the density of vegetation cover required to prevent wind erosion are varied. The most commonly cited thresholds are that aeolian erosion increases rapidly once vegetation cover is reduced below 15–20% (Marshall 1970; Wasson and Nanninga 1986; Wiggs *et al.* 1994), but that sand movement can still take place with densities of over 30% (Ash and Wasson 1983). Grasses and shrubs are the most effective at reducing wind erosion because most sand transport takes place near ground level. Heathcote (1983) estimated that 90% of sand is carried in the 50 cm immediately above the

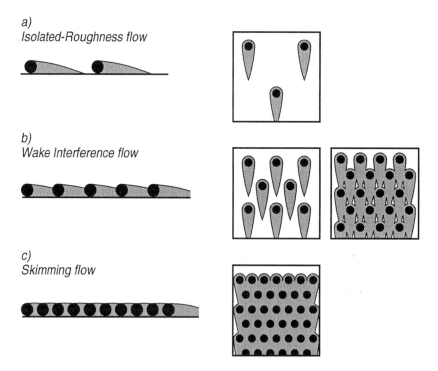

a)
Isolated-Roughness flow

b)
Wake Interference flow

c)
Skimming flow

Figure 7.8   *Flow regimes associated with different roughness element spacings. The lightly shaded area indicates zones of reduced wind speed (after Wolfe and Nickling 1993)*

Table 7.3   *Roughness element descriptions associated with different flow regimes (after Lee and Soliman 1977)*

Flow regimes	Roughness element description		
	Spacing to height ratio (Sp/h)	% cover	Roughness concentration (Lc)
Isolated-roughness flow	>3.5	<16	<0.082
Wake-interference flow	3.5–2.25	16–40	0.083–0.198
Skimming flow	<2.25	>40	>0.198

surface and Wiggs (1992) measured up to 35% of sand transport taking place in the lower 2.5 cm (see Chapter 16).

Algae and cyanobacteria can decrease the erodibility of a sandy surface by binding sand particles. This affects the upper few millimetres of the surface and may prevent ripple formation (Yair 1990). Van den Ancker *et al.* (1985) conducted experiments on the effect of algal crust development on rates of deflation and found that, in an airstream of $9 \, \mathrm{m \, s^{-1}}$, $148 \, \mathrm{mg \, cm^{-2} \, min^{-1}}$ of surface sand was removed where no crust was present whereas $<1 \, \mathrm{mg \, cm^{-2} \, min^{-1}}$ was removed where the crust was well-developed. Biocrusts reduce the amount of wind erosion that can take place but can be damaged by sandblasting in strong winds (Pluis and van Boxel 1993).

HYDROLOGICAL PROCESSES

The presence of vegetation affects not only the amount of water available for geomorphological work, but also the nature and characteristics of surface flow where it does occur. The precise role of vegetation depends upon its type and distribution and there can be a marked difference between the effects of higher-order plants (grasses, shrubs, trees) and lower-order plants (algae, fungi, lichen, etc.).

In arid areas infiltration capacity is generally greater where higher plants are present than on bare surfaces. Studies from New Mexico have concluded that total ground cover is the single most important variable influencing infiltration and sediment production (Wood *et al.* 1987; Wilcox *et al.* 1988). Soils underlying plants are often sandy, contain a high percentage of organic matter and are disturbed by animal activity, all of which lower the soil bulk density and so increase infiltration potential (Lyford and Qashu 1969; Abrahams and Parsons 1991) (Figure 7.9).

Interception of raindrops by the vegetation canopy also minimises the likelihood of surface sealing which would reduce infiltration potential (Wilcox *et al.* 1988; Abrahams and Parsons 1991). Lyford and Qashu (1969) found that the rate of infiltration under creosote bushes (*Larrea tridentata*) was three times greater than that between plants, at $0.644 \, \mathrm{cm \, min^{-1}}$ underneath the plant and $0.260 \, \mathrm{cm \, min^{-1}}$ in the interplant areas. In banded vegetation the infiltration rate

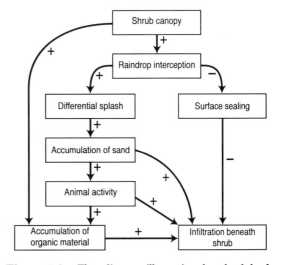

Figure 7.9   *Flow diagram illustrating the role of shrub vegetation and litterfall on infiltration (after Abrahams and Parsons 1991)*

in groves may be twice that of the intergroves (Mabbutt and Fanning 1987). Plant litter cover also increases infiltration (Tromble *et al.* 1974).

The effect of lower plants, such as cryptogams (mosses, lichens, algae, cyanobacteria) on infiltration rates varies. Cryptogams can reduce infiltration by binding the soil surface and blocking soil pores (Fogg 1952; Graetz and Tongway 1986; Greene *et al.* 1990) or they can enhance it by increasing surface roughness and the likelihood of surface ponding (Fletcher and Martin 1948; Loope and Gifford 1972; Blackburn 1975). There is some evidence to suggest that the precise effects of cryptogams on infiltration may be related to the physical properties of the soil (Eldridge and Rothon 1992), which may in turn be governed by the characteristics of the overall vegetation assemblage. If the surface is characterised by a mixture of grasses and cryptogams, then the soil is likely to be dominated by macropores (>0.75 mm diameter) caused by the roots of the higher plants (Greene and Tongway 1989) and infiltration will be high. In the absence of higher plants, matrix pores will dominate. These are susceptible to blocking by secretions from cyanobacteria (Fogg 1952) or fungal hyphae (Greene and Tongway 1989; Chartres 1992) which cause a reduction in infiltration. However, infiltration rates in soils dominated by matrix pores can be enhanced as cryptogams on the surface help to maintain the structural stability of the soil and protect the pore openings allowing water to enter (Eldridge 1993a). Cryptogam secretions can also bind soil particles into water-stable aggregates (Durrell and Shields 1961; Bond and Harris 1964), making them less susceptible to sealing by raindrop impact (Eldridge 1993b; Eldridge and Greene 1993).

Plant adaptations such as small leaf areas mean that interception of rainfall is generally low in desert areas. Any rainfall or moisture that is intercepted is usually rapidly redistributed via throughfall or stemflow to other parts of the plant such as the base of the stem where the water infiltrates and is taken up by the plant roots (Návar and Bryan, 1990). Stemflow improves the plant microenvironment and enables a closer plant spacing.

On semi-arid hillslopes almost all runoff occurs in the form of overland flow, generated when rainfall intensity exceeds the surface infiltration rate. Runoff is most likely to take place where the soil surface has been sealed, either by rainsplash or cryptogams, to form a crust. Experiments in the western Negev, Israel, suggest that on a sealed surface runoff takes place for any storm exceeding 2–3 mm, but that when the topsoil crust is removed, no runoff will take place (Yair 1990). The prevention of soil crust formation by higher vegetation reduces the likelihood of runoff generation. Both sheetflow and channel flow can take place in arid areas but it is common for surface flow to infiltrate before it reaches a channel (Yair and Lavee 1985; Lavee *et al.* 1995). Higher plants act as roughness elements and provide resistance to sheetflow which is then diverted around the obstacles. This dissipates the sheetflow into rectilinear flow and then channel flow (Emmett 1970; Abrahams *et al.* 1989; 1990; Thornes *et al.* 1990; Baird *et al.* 1992). Between obstacles, the volume and velocity of the water is increased (Abrahams *et al.* 1994) and turbulence may make the flow more erosive (Abrahams *et al.* 1992).

Phreatophytic vegetation places a heavy demand on groundwater and this can affect channel processes. These plants dominate on floodplains and riparian zones and obstruct surface and channel flow (Graf 1980). The increase in hydraulic roughness caused by vegetation blocking ephemeral channels means that when channel flow does occur, floods and turbulence are more likely and increased sedimentation will take place (Hadley 1961; Graf 1981).

SURFACE MATERIALS AND SOIL PROCESSES

Vegetation in desert environments contributes widely to the development of soils and surface materials. Plant litter does not only lower the soil bulk density and increase infiltration rates, but is also a main contributor of nutrients and minerals to the soil. The favourable, sheltered, damp and protected environment directly beneath plants means that litter that remains on the surface below the plant or is trapped by other plants will decompose more rapidly than that on bare ground (Binet 1981). This environment also attracts animals that help with litter decomposition. Litter cover that remains on the surface helps to protect sands from aeolian erosion (Wasson 1976; Gibbens *et al.* 1983; Wiggs *et al.* 1994) and if buried, for example by aeolian deposits, will decompose more rapidly and can stabilise the soil (Le Houérou 1981; Moorhead and Reynolds 1989).

Below the surface the growth of plant roots causes bioturbation which destroys primary sediment structures and causes mixing of sediments (Bigarella 1975; Cook and Stubbendieck 1986; Tengberg 1994). Roots also give rise to a range of organo-sedimentary structures known as rhizoliths (Klappa 1980). These form when material around roots becomes cemented, for example by calcium carbonate, gypsum, silica or iron oxides (Pye 1983). Voids left following the decay of roots are known as root moulds or may be filled with sediment or authigenic precipitates to become root casts (Klappa 1980).

The formation of rhizoliths is also associated with the development of calcretes, as are other biogenic processes. Higher plants and microorganisms play a major role in the formation of calcretes (Wright and Tucker 1991), by providing a source of calcium (Goudie 1973), possible triggers for calcium precipitation (e.g. Krumbein and Giele 1979) and deposition structures within the calcrete surfaces (e.g. Phillips *et al.* 1987; Verrecchia *et al.* 1993) (see Chapter 6). Plant uptake of water from soil can also affect the rate and depth of calcium carbonate deposition (Schlesinger *et al.* 1987). Many plants contain other compounds that are important for the development of surface materials, for example some grasses concentrate silica which may be an important source for the development of silcretes (Summerfield 1982). A biogeochemical method of rock varnish formation has also been advocated (e.g. Dorn and Oberlander 1982; see Chapter 6).

The link between vegetation and geomorphology

The effect of vegetation on the land surface has a number of implications for landscape development at a variety of scales. Some examples of landforms that have been linked to vegetation either directly or indirectly are given in Table 7.4.

It has been widely observed that individual shrubs and trees or clumps of grasses are located on raised mounds and these can form in a variety of ways. Mounds form as vegetation traps aeolian sands both by acting as an obstacle to saltating particles and by creating a zone of low wind velocity on the lee side of the plant. This can lead to the formation of small dunes known as 'lee' or 'shadow' dunes, the dimensions and shape of which are dictated by the plant size, shape and wind regime (Bagnold 1941; Hesp 1981). Nebkha dunes also develop around vegetation and bear some similarities to lee dunes but rather than the vegetation simply being an obstacle it becomes an integral part of the dune, growing as the sand mound grows and resulting in dunes up to 20 m long and 5 m high (Gibbens *et al.* 1983; Nickling and Wolfe 1994; Tengberg 1994; see Chapter 17) (Figure 7.10). The plants around which nebkhas form are usually phreatophytic (e.g. *Ziziphus lotus, Tamarisk, Prosopis*). These make large demands on water supply which often results in the areas between nebkhas being very sparsely vegetated (Gibbens *et al.* 1983).

Not all plant mounds are formed by trapping of aeolian sand. Bush and tree mounds of up to 20 cm in height may form by a local raising of the surface under the canopy caused by high root densities and termite activity (Biot 1990). In other cases the mounds may be formed not by a raising of the area under the plant but by erosion of the interplant areas. Swarms of mounds with a density of 450 ha^{-1} and an average height of 41 cm have been reported from northeastern Patagonia and are thought to represent a relict land surface (Rostagno and del Valle 1988). Sand mounds can also form under shrubs through a combination of sand accumulation beneath the shrub canopy by rainsplash erosion and selective erosion of intershrub areas by overland flow (Parsons *et al.* 1992). The size of the resulting mound is proportional to that of the plant and accumulation may account for up to 20 cm of the overall mound height (Parsons *et al.* 1992). Plant cover on hillslopes interacts with runoff and sedimentation processes to form microtopographic mounds (Sanchez and Puigdefabregas 1994) or, in the case of banded vegetation, a stepped slope profile (Tongway and Ludwig 1990).

Plant mounds, however they are formed, are common in all desert environments where

Table 7.4   *Examples of reported evidence of interactions of vegetation and geomorphology*

Landform	Selected references
Microtopographic mounds	Biot (1990)
	Parsons *et al.* (1992)
	Sanchez and Puigdefabregas (1994)
Nebkhas	Tenberg (1994)
	Wolfe and Nickling (1993)
Parabolic dunes	Verstappen (1968; 1970)
	Anton and Vincent (1986)
	Eriksson *et al.* (1989)
Blowouts	Watts (1937)
	Lancaster (1986)
	Warren (1988)
Linear dunes	Tsoar and Møller (1986)
	Wopfner and Twidale (1988)
Sand sheets	Kocurek and Nielson (1986)
Ephemeral channels	Graf (1981; 1985)
Desert varnish	Dorn and Oberlander (1981; 1982)
	Krumbein and Jens (1981)
Calcretes	Wright and Tucker (1991)
	Rossinsky and Wanless (1992)
Surface stabilisation	Tsoar and Møller (1986)
	Danin *et al.* (1989)
	Alexander *et al.* (1994)
Sand dune activity	Thomas and Shaw (1991b)
	Livingstone and Thomas (1993)

higher vegetation is present and can be superimposed on other landforms, such as alluvial fans (Blair and McPherson 1994) or large sand dunes. Plant mounds can also act as nuclei for the formation of larger dunes.

The geomorphological role of vegetation is not restricted to very small topographic features. Higher vegetation has been cited as a control on sand sheet development (Kocurek and Nielson 1986) and plays a fundamental role in the development of vegetated linear ridges and parabolic dunes (see Chapter 17).

The development of surface materials, such as calcretes, in which vegetation and especially microflora play a major part, also contribute to landscape development. For example, calcretes can act as a caprock or limit to erosion, can cause relief inversion and may control slope development (Goudie 1985; see Chapter 6).

Floodplain characteristics and channel morphology are also affected by vegetation. Vegetation in channel beds can cause sediment deposition which leads to the formation of sand bars and can cause the channel to be blocked completely. A study of the Gila River, Arizona, showed very close links between temporal trends in channel sinuosity and temporal trends in tamarisk growth (Graf 1981). This is attributed to the increased sediment deposition caused by higher vegetation cover which leads to increased channel sinuosity.

Vegetation has also been used as an indicator of environmental change, for example the issue of sand dune activity has often been linked to the presence or absence of vegetation (e.g. Flint

Figure 7.10 *Nebkha dunes formed on the floor of a pan. Note the lack of interdune vegetation and the accumulation of fine sands around the shrubs*

and Bond 1968; Lancaster 1981; see Chapters 17 and 26).

Conclusion

The rapid growth in the number of geomorphological studies considering the roles of vegetation confirms that its presence is increasingly being recognised as fundamental rather than incidental to the development of arid landscapes. Vegetation does not only affect the development of landforms in its immediate vicinity but can also indirectly influence landscape development by regulating the amount of water or sediment available. This recognition can offer insights not only into the direct effects of vegetation on geomorphology but also into the geomorphic consequences of vegetation removal and vegetation change (e.g. Burkham 1972; Tsoar and Møller 1986; Chapter 28).

References

Abrahams, A.D. and Parsons, A.J. 1991. Relation between infiltration and stone cover on a semiarid hillslope, southern Arizona. *Journal of Hydrology*, 122: 501–503.

Abrahams, A.D., Parsons, A.J. and Luk, S.-H. 1989. Distribution of depth of overland flow on desert hillslopes and its implications for modelling soil erosion. *Journal of Hydrology*, 106: 177–184.

Abrahams, A.D., Parsons, A.J. and Luk, S.-H. 1990. Field experiments on the resistance to overland flow on desert hillslopes, southern Arizona. In *Erosion, transport and deposition processes*. International Association of Hydrological Sciences, Publication, 189: 1–18.

Abrahams, A.D., Parsons, A.J. and Hirsch, P.J. 1992. Field and laboratory studies of resistance to interrill overland flow on semi-arid hillslopes, southern Arizona. In A.J. Parsons and A.D. Abrahams (eds), *Overland flow: hydraulics and erosion mechanics*. UCL Press, London: 1–23.

Abrahams, A.D., Parsons, A.J. and Wainwright, J. 1994. Resistance to overland flow on semiarid grassland and shrubland hillslopes, Walnut Gulch, southern Arizona. *Journal of Hydrology*, 156: 431–446.

Alexander, R.W. and Calvo, A. 1990. The influence of lichens on slope processes in some Spanish badlands. In J.B. Thornes (ed.), *Vegetation and erosion*. Wiley, Chichester: 386–398.

Alexander, R.W., Harvey, A.M., Calvo, A., James, P.A. and Cerda, A. 1994. Natural stabilisation mechanisms on badland slopes: Tabernas, Almeria, Spain. In A.C. Millington and K. Pye (eds), *Environmental change in drylands: biogeographical and geomorphological perspectives*. Wiley, Chichester: 85–111.

Allen, L.H. Jr 1968. Turbulence and wind speed spectra within a Japanese larch plantation. *Journal of Applied Meteorology*, 7: 73–78.

Anderson, D.J. 1971. Pattern in desert perennials. *Journal of Ecology*, 59: 555–560.

Anderson, E. 1988. Preliminary dew measurement in the eastern Prosopis zone of the Wahiba Sands. *Journal of Oman Studies, Special Report 3: The scientific results of the Royal Geographical Society's Oman Wahiba Sands Project 1985–1987*: 201–212.

Anton, D. and Vincent, P. 1986. Parabolic dunes of the Jafurah Desert, Eastern Province, Saudi Arabia. *Journal of Arid Environments*, 11: 187–198.

Ash, J.E. and Wasson, R.J. 1983. Vegetation and sand mobility in the Australian desert dunefield. *Zeitschrift für Geomorphologie*, Supplementband, 45: 7–25.

Bagnold, R.A. 1941. *The physics of blown sand and desert dunes*. Methuen, London.

Baird, A.J., Thornes, J.B. and Watts, G.P. 1992. Extending overland-flow models to problems of slope evolution and the representation of complex slope-surface topographies. In A.J. Parsons and A.D. Abrahams (eds), *Overland flow: hydraulics and erosion mechanics*. UCL Press, London: 199–224.

Bendali, F., Floret, C., Le Floc'h, E. and Pontanier, R. 1990. The dynamics of vegetation and sand mobility in arid regions. *Journal of Arid Environments*, 18: 21–32.

Bertiller, M.B. 1984. Specific primary productivity in arid ecosystems: a case study in Patagonia, Argentina. *Acta Ecologia*, 5: 365–381.

Bhadresa, R. and Moore, P.D. 1982. Desert shrubs: the implications of population and pattern studies for conservation and management. In B. Spooner and H.S. Mann (eds), *Desertification and development: dryland ecology in social perspective*. Academic Press, London: 269–276.

Bigarella, J.J. 1975. Structures developed by dissipation of dune and beach ridge deposits. *Catena*, 2: 107–152.

Binet, P. 1981. Short-term dynamics of minerals in arid ecosystems. In D.W. Goodall and R.A. Perry (eds), *Arid land ecosystems, structure, functioning and management*, 2. Cambridge University Press, Cambridge: 325–356.

Biot, Y. 1990. The use of tree mounds as benchmarks of previous land surfaces in a semi-arid tree savanna, Botswana. In J.B. Thornes (ed.), *Vegetation and erosion*. Wiley, Chichester: 437–450.

Bisal, F. and Hsieh, J. 1966. Influence of moisture on erodibility of soil by wind. *Soil Science*, 102: 143–146.

Blackburn, W.H. 1975. Factors influencing infiltration rate and sediment production of semiarid rangelands in Nevada. *Water Resources Research*, 11: 929–937.

Blair, T.C. and McPherson, J.G. 1994. Alluvial fan processes and forms. In A.D. Abrahams and A.J. Parsons (eds), *Geomorphology of desert environments*. Chapman & Hall, London: 354–402.

Boaler, S.B. and Hodge, C.A.H. 1962. Vegetation stripes in Somaliland. *Journal of Ecology*, 50: 465–474.

Bond, R.D. and Harris, J.R. 1964. The influence of the microflora on the physical properties of soils, I: effects associated with filamentous algae and fungi. *Australian Journal of Soil Research*, 2: 111–122.

Bornkamm, R. 1987. Growth of accidental vegetation on desert soils in SW Egypt. *Catena*, 14: 267–274.

Bouwer, H. 1975. Predicting reduction in water losses from open channels by phreatophyte control. *Water Resources Research*, 11: 96–101.

Bowers, M.A. 1987. Precipitation and the relative abundances of desert winter annuals: a 6-year study in the northern Mojave Desert. *Journal of Arid Environments*, 12: 141–149.

Brazel, A.J. and Nickling, W.G. 1987. Dust storms and their relation to moisture in the Sonoran–Mojave desert region of the southwestern United States. *Journal of Environmental Management*, 24: 279–291.

Bressolier, C. and Thomas, Y.F. 1977. Studies of wind and plant interactions on the French Atlantic coastal dunes. *Journal of Sedimentary Petrology*, 47: 331–338.

Buckley, R. 1981. Soils and vegetation of central Australian sand ridges III. Sand ridge vegetation of the Simpson Desert. *Australian Journal of Ecology*, 6: 405–422.

Buckley, R. 1987. The effect of sparse vegetation on the transport of dune sand by wind. *Nature*, 325: 426–428.

Burkham, D.E. 1972. *Channel changes of the Gile River in Safford Valley, Arizona*. US Geological Survey Professional Paper 655-J.

Chadwick, H.W. and Dalke, P.D. 1965. Plant succession on dune sands in Freemont County, Idaho. *Ecology*, 46: 765–780.

Chartres, C.J. 1992. Soil crusting in Australia. In M.E. Sumner and B.A. Stewart (eds), *Soil crusting: chemical and physical processes*. Lewis Publishers, Ann Arbor: 339–365.

Choudhury, B.J. and Tucker, C.J. 1987. Satellite observed seasonal and inter-annual variation of vegetation over the Kalahari, the Great Victoria Desert and the Great Sandy Desert: 1979–1984. *Remote Sensing of the Environment*, 23: 233–241.

Conacher, A.J. 1971. The significance of vegetation, fire and man in the stabilisation of sand dunes near the Warburton Ranges, central Australia. *Earth Science Journal*, 5: 92–94.

Cook, C.W. and Stubbendieck, J. 1986. *Range research: basic problems and techniques*. Society for Range Management, Denver, Colorado.

Danin, A. 1978. Plant species diversity and plant succession in a sandy area in the northern Negev. *Flora*, 167: 409–422.

Danin, A. and Yaalon, D.H. 1982. Silt plus clay sedimentation and decalcification during plant succession in sands of the Mediterranean coastal plain of Israel. *Israel Journal of Earth Sciences*, 31: 101–109.

Danin, A., Bar-Or, Y., Dor, I. and Yisraeli, T. 1989. The role of cyanobacteria in stabilization of sand dunes in southern Israel. *Ecologia Mediterranea*, 15: 55–64.

Desmukh, I.K. 1984. A common relationship between precipitation and grassland peak biomass for East and southern Africa. *African Journal of Ecology*, 22: 181–186.

Dorn, R.I. and Oberlander, T.M. 1981. Microbial origin of desert varnish. *Science*, 227: 1472–1474.

Dorn, R.I. and Oberlander, T.M. 1982. Rock varnish. *Progress in Physical Geography*, 6: 317–367.

Darrell, L.W. and Shields, L.M. 1961. Characteristics of soil algae relating to crust formation. *Transactions of the American Microbiological Society*, 80: 73–79.

Eldridge, D.J. 1993a. Cryptogam cover and soil surface condition: effects on hydrology on a semi-arid woodland soil. *Arid Soil Research and Rehabilitation*, 7: 203–217.

Eldridge, D.J. 1993b. Cryptogams, vascular plants and soil hydrological relations: some preliminary results from the semi-arid woodlands of eastern Australia. *Great Basin Naturalist*, 53: 48–58.

Eldridge, D.J. and Greene, R.S.B. 1993. Assessment of sediment yield by splash erosion on a semi-arid soil with varying cryptogam cover. *Journal of Arid Environments*, 26: 221–232.

Eldridge, D.J. and Rothon, J. 1992. Runoff and sediment production from a semi-arid woodland in eastern Australia, I: the effect of pasture type. *The Rangeland Journal*, 14: 26–39.

Emmett, W.W. 1970. The hydraulics of overland flow on hillslopes. *US Geological Survey Professional Paper*, 662–A.

Eriksson, P.G., Nixon, N., Snyman, C.P. and Bothma, J. duP. 1989. Ellipsoidal parabolic dune patches in the southern Kalahari Desert. *Journal of Arid Environments*, 16: 111–124.

Evenari, M., Noy-Meir, I. and Goodall, D.W. 1985. *Ecosystems of the world, Vol. 12A: Hot deserts and arid shrublands*. Elsevier, Amsterdam.

Evenari, M., Noy-Meir, I. and Goodall, D.W. 1986. *Ecosystems of the world, Vol. 12B: Hot deserts and arid shrublands*. Elsevier, Amsterdam.

Faulkner, H. 1990. Vegetation cover density variations and infiltration patterns on piped alkali sodic soils: implications for the modelling of overland flow in semi-arid areas. In J.B. Thornes (ed.), *Vegetation and erosion*. Wiley, Chichester: 317–346.

Fletcher, J.E. and Martin, N.P. 1948. Some effects of algae and moulds in the rain crust of desert soils. *Ecology*, 29: 95–100.

Flint, R.F. and Bond, G. 1968. Pleistocene sand ridges and pans in western Rhodesia. *Bulletin of the Geological Society of America*, 79: 299–314.

Fogg, G.E. 1952. The production of extra-cellular nitrogenous substances by blue green algae. *Proceedings of the Royal Society*, 139: 372–377.

Fuller, W.H., Cameron, R.E. and Raioa, N. Jr 1960. Fixation of nitrogen in desert soils by algae. *Transactions of the 7th International Congress of Soil Science*, Madison, 2: 617–624.

Gibbens, R.P., Tromble, J.M., Hennessy, J.T. and Cardenas, M. 1983. Soil movement in mesquite dunelands and former grasslands of southern New Mexico from 1933 to 1980. *Journal of Range Management*, 36: 145–148.

Gimingham, C.H. 1955. A note on water table, sand-movement and plant distribution in a North African oasis. *Journal of Ecology*, 43: 22–25.

Goldschmidt, M.J. and Jacobes, M. 1956. *Underground water in the Haifa-Acco sand dunes and its replenishment*. Hydrological Paper 2, State of Israel Hydrological Service, Jerusalem.

Goodall, D.W. and Perry, R.W. (eds) 1979. *Arid-land ecosystems: structures, functioning and management*. Cambridge University Press, Cambridge.

Goudie, A.S. 1973. *Duricrusts in tropical and subtropical landscapes*. Clarendon Press, Oxford.

Goudie, A.S. 1985. Duricrusts and landforms. In K.S. Richards, R.R. Arnett and S. Ellins (eds), *Geomorphology and Soils*. Allen & Unwin, London: 37–55.

Graetz, R.D. and Tongway, D.J. 1986. Influence of grazing management on vegetation, soil structure and nutrient distribution and the infiltration of applied rainfall in a semi-arid chenopod shrubland. *Australian Journal of Ecology*, 11: 347–360.

Graf, W.L. 1980. Riparian management: a flood control perspective. *Journal of Soil and Water Conservation*, 35: 158–161.

Graf, W.L. 1981. Channel instability in a braided, sand bed river. *Water Resources Research*, 17: 1087–1094.

Graf, W.L. 1985. *The Colorado River: instability and basin management*. Association of American Geographers, Washington, DC.

Greene, R.S.B. and Tongway, D.J. 1989. The significance of (surface) physical and chemical properties in determining soil surface condition of retreads in rangelands. *Australian Journal of Soil Research*, 27: 213–225.

Greene, R.S.B., Chartres, C.J. and Hodgkinson, K.H. 1990. The effect of fire on the soil in a degraded semi-arid woodland, I: physical and micromorphological properties. *Australian Journal of Soil Research*, 28: 755–777.

Greenwood, J.E.G.W. 1957. The development of vegetation patterns in Smallholding Protectorate. *Geographical Journal*, 123: 465–473.

Gutiérrez, J.R. and Whitford, W.G. 1987. Responses of Chihuahuan Desert herbaceous annuals to rainfall augmentation. *Journal of Arid Environments*, 12: 127–139.

Gutiérrez, J.R., Meserve, P.L., Jaksic, F.M., Contreras, L.C., Herrera, S. and Vásquez, H. 1993. Structure and dynamics of vegetation in a Chilean arid thorn scrub community. *Acta Œcologica*, 14: 271–285.

Hadley, R.F. 1961. Influence of riparian vegetation on channel shape, north-eastern Arizona. *US Geological Survey Professional Paper*, 424-C: 30–31.

Heathcote, R.L. 1983. *The arid lands: their use and abuse*. Longman, London.

Hennessy, J.T., Gibbens, R.P., Tromble, J.M. and Cardenas, M. 1985. Mesquite (*Prospis–Glandulosa*–Torr) dunes and interdune in southern New Mexico — a study of soil properties and soil water relations. *Journal of Arid Environments*, 9: 27–38.

Hesp, P.A. 1981. The formation of shadow dunes. *Journal of Sedimentary Petrology*, 51: 101–112.

Hidore, J.J. and Albokhair, Y. 1982. Sand encroachment in Al-Hasa Oasis, Saudi Arabia. *Geographical Review*, 72: 350–356.

Kassas, M. 1966. Plant life in deserts. In E.S. Hills (ed.), *Arid lands*. Methuen, London, and UNESCO, Paris: 145–180.

Kassas, M. and Imam, M. 1959. Habitat and plant communities in the Egyptian desert. *Journal of Ecology*, 47: 289–310.

Klappa, C.F. 1980. Rhizoliths in terrestrial carbonates: classification, recognition, genesis and significance. *Sedimentology*, 27: 613–629.

Kocurek, G. and Nielson, J. 1986. Conditions favourable for the formation of warm-climate aeolian sand sheets. *Sedimentology*, 33: 795–816.

Krumbein, W.E. and Giele, C. 1979. Calcification in a coccoid–cyanobacterium associated with the formation of desert stromatolites. *Sedimentology*, 26: 593–604.

Krumbein, W.E. and Jens, K. 1981. Biogenic rock varnishes of the Negev Desert (Israel): an ecological study of iron and manganese transformation by cyano-bacteria and fungi. *Oecologia*, 50: 25–38.

Lancaster, J., Lancaster, N. and Seely, M.K. 1984. Climate of the central Namib desert. *Madoqua*, 14: 5–61.

Lancaster, N. 1981. Palaeoenvironmental implications of fixed dune systems in southern Africa. *Palaeogeography, Palaeoclimatology, Palaeoecology*, 33: 395–400.

Lancaster, N. 1986. Dynamics of deflation hollows in the Elands Bay area, Cape Province, South Africa. *Catena*, 13: 139–153.

Lavee, H., Kutiel, P., Segev, M. and Benyamini, Y. 1995. Effect of surface roughness on runoff and erosion in a Mediterranean ecosystem: the role of fire. *Geomorphology*, 11: 227–234.

Le Houérou, H.N. 1981. Long-term dynamics in arid-land vegetation and ecosystems of North America. In D.W. Goodall and R.A. Perry (eds), *Arid land ecosystems, structure, functioning and management*, 2. Cambridge University Press, Cambridge: 357–384.

Le Houérou, H.N. and Hoste, H. 1977. Rangeland production and annual rainfall relations in the Mediterranean Basin and in the African sahelo-Sudanian zone. *Journal of Range Management*, 30: 181–189.

Lee, B.E. and Soliman, B.F. 1977. An investigation of the forces on three dimensional bluff bodies in rough wall turbulent boundary layers. *Transactions of the ASME, Journal of Fluids Engineering*, 503–510.

Lee, J.A. 1991. The role of desert shrub size and spacing on wind profile parameters. *Physical Geography*, 12: 72–89.

Leppanen, O.E. 1970. Evapotranspiration: its measurement and variability. In R.C. Culler *et al.* (eds), *Objectives, methods and environment — Gila River Phreatophyte Project, Graham County, Arizona*. US Geological Survey, Professional Paper, 655-A: A3–A5.

Lerner, D.N., Issar, A.S. and Simmers, I. 1990. *Groundwater recharge*. Verlag Heinz Heise, Hannover.

Levitt, J. 1972. *Responses of plants to environmental stress*. Academic Press, New York.

Livingstone, I. and Thomas, D.S.G. 1993. Modes of linear dune activity and their palaeoenvironmental significance: the southern African example. In K. Pye (ed.), *The dynamics and environmental context of aeolian sedimentary systems*. Geological Society Special Publication: 91–102.

Loope, W.L. and Gifford, G.F. 1972. Influence of a soil microfloral crust on select properties of soils under pinyon-juniper in southeastern Utah.

Journal of Soil and Water Conservation, 27: 164–167.

Luk, S.-H., Abrahams, A.D. and Parsons, A.J. 1993. Sediment sources and sediment transport by rill flow and interrill flow on a semi-arid piedmont slope, southern Arizona. *Catena*, 20: 93–111.

Lyford, F.P and Qashu, H.K. 1969. Infiltration rates as affected by desert vegetation. *Water Resources Research*, 5: 1373–1376.

Mabbutt, J.A. and Fanning, P.C. 1987. Vegetation banding in arid Western Australia. *Journal of Arid Environments*, 12: 41–59.

MacFadyen, W. 1950. Vegetation patterns in the semi-desert plains of British Somaliland. *Geographical Journal*, 115: 199–211.

Manry, D.E. and Knight, R.S. 1986. Lightning density and burning frequency in South African vegetation. *Vegetatio*, 66: 67–76.

Marshall, J.K. 1970. Assessing the protective role of shrub-dominated rangeland vegetation against soil erosion by wind. In *Proceedings of the XI International Grassland Congress*. University of Queensland Press, Brisbane: 19–23.

Migahid, A.A. 1961. The drought resistance of Egyptian desert plants. *Arid Zone Research*, 16: 213–233.

Millington, A.C. and Pye, K. (eds) 1994. *Environmental changes in drylands: biogeographical and geomorphological perspectives*. Wiley, Chichester.

Monteith, J.L. 1957. Dew. *Quarterly Journal of the Royal Meteorological Society*, 83: 322–341.

Moorhead, D.L. and Reynolds, J.F. 1989. Mechanisms of surface litter mass loss in the northern Chihuahuan desert: a reinterpretation. *Journal of Arid Environments*, 16: 157–163.

Morris, H.M. 1955. Flow in rough conduits. *Transactions of the American Society of Agricultural Engineers*, 120: 373–398.

Návar, J. and Bryan, R. 1990. Interception loss and rainfall redistribution by three semi-arid growing shrubs in northeastern Mexico. *Journal of Hydrology*, 115: 51–63.

Nickling, W.G. and Wolfe, S.A. 1994. The morphology and origin of nabkhas, Region of Mopti, Mali, West Africa. *Journal of Arid Environments*, 28: 13–30.

Nieman, W.A., Heyns, C. and Seely, M.K. 1978. A note on precipitation at Swakopmund. *Madoqua*, 11: 69–73.

Nobel, P.S. 1981. Spacing and transpiration of various sized clumps of a desert grass, *Hilaria rigida*. *Journal of Ecology*, 69: 735–742.

Noy-Meir, I. 1973. Desert ecosystems: environment and producers. *Annual Review Ecology and Systematics*, 4: 25–52.

Olivier, J. 1995. Spatial distribution of fog in the Namib. *Journal of Arid Environments*, 29: 129–138.

Parsons, A.J., Abrahams, A.D. and Simanton, J.R. 1992. Microtopography and soil-surface materials on semi-arid piedmont hillslopes, southern Arizona. *Journal of Arid Environments*, 22: 107–115.

Phillips, S.E., Milnes, A.R. and Foster, R.C. 1987. Calcified filaments: an example of biological influences in the formation of calcrete in south Australia. *Australian Journal of Soil Research*, 25: 405–428.

Pluis, J.L.A. and van Boxel, J.H. 1993. Wind velocity and algal crusts in dune blowouts. *Catena*, 20: 581–594.

Prill, R.C. 1968. Movement of moisture in the unsaturated zone in a dune area, southwestern Kansas. *US Geological Survey, Professional Paper*, 600-D: D1–D9.

Pye, K. 1983. Early post-depositional modification of aeolian dune sands. In M.E. Brookfield and T.S. Ahlbrandt (eds), *Eolian sediments and processes*. Elsevier, Amsterdam: 97–221.

Rempel, P.J. 1936. The crescentic dunes of the Salton Sea and their relation to the vegetation. *Ecology*, 17: 347–358.

Rossinsky, V. Jr and Wanless, H.R. 1992. Topographic and vegetative controls on calcrete formation, Turks and Caicos Islands, British West Indies. *Journal of Sedimentary Petrology*, 62: 84–98.

Rostagno, C. M. and del Valle, H. F. 1988. Mounds associated with shrubs in aridic soils of northeastern Patagonia: characteristics and probable genesis. *Catena*, 15: 347–359.

Salisbury, E.J. 1952. *Downs and dunes*. Bell, London.

Sanchez, G. and Puigdefabregas, J. 1994. Interactions of plant growth and sediment movement on slopes in a semi-arid environment. *Geomorphology*, 9: 243–260.

Sarre, R.D. 1978. Aeolian sand transport. *Progress in Physical Geography*, 11: 157–182.

Schlesinger, W.H., Fonteyn, P.J. and Marion, G.H. 1987. Soil moisture content and plant transpiration in the Chihuahuan Desert of New Mexico. *Journal of Arid Environments*, 12: 119–126.

Seely, M.K. 1978. Grassland productivity: the desert end of the curve. *South African Journal of Science*, 74: 295–297.

Shantz, H.L. 1956. History and problems of arid lands development. In G.F. White (ed.), *The future of arid lands*. American Association for the Advancement of Science Publication 43, Washington, DC.

Sims, P.L. and Singh, J.S. 1978. The structure and function of ten western North American grasslands. III. Net primary production and efficiencies of energy capture and water use. *Journal of Ecology*, 66: 573–597.

Smith, W.A., Dodd, J.L., Skinner, Q.D. and Rodgers, J.D. 1993. Dynamics of vegetation along and adjacent to an ephemeral channel. *Journal of Range Management*, 46: 56–64.

Summerfield, M.A. 1982. Distribution, nature and probable genesis of silcrete in arid and semi-arid southern Africa. *Catena*, Supplement, 1: 37–65.

Tengberg, A. 1994. Nebkhas — their spatial distribution, morphometry, composition and age — in the Sidi Bouzid area, central Tunisia. *Zeitschrift für Geomorphologie*, 38: 311–325.

Thomas, D.S.G. 1988a. The biogeomorphology of arid and semi-arid environments. In H.A. Viles (ed.), *Biogeomorphology*. Blackwell, London: 193–221.

Thomas, D.S.G. 1988b. The geomorphological role of vegetation in the dune systems of the Kalahari. In G.F. Dardis and B.P. Moon (eds), *Geomorphological studies in Southern Africa*. Balkema, Rotterdam: 145–158.

Thomas, D.S.G. and Shaw, P.A. 1991a. *The Kalahari environment*. Cambridge University Press, Cambridge.

Thomas, D.S.G. and Shaw, P.A. 1991b. 'Relict' desert dune systems: interpretations and problems. *Journal of Arid Environments*, 20: 1–14.

Thornes, J.B. 1985. The ecology of erosion. *Geography*, 70: 222–236.

Thornes, J.B. (ed.) 1990a. *Vegetation and erosion*. Wiley, Chichester.

Thornes, J.B. 1990b. The interaction of erosional and vegetational dynamics in land degradation: spatial outcomes. In J.B. Thornes (ed.), *Vegetation and erosion*. Wiley, Chichester: 41–53.

Thornes, J.B. 1994. Catchment and channel hydrology. In A.D. Abrahams and A.J. Parsons (eds), *Geomorphology of desert environments*. Chapman & Hall, London: 257–287.

Thornes, J.B., Francis, C.F., Lopez-Bermudez, F. and Romero-Diaz, A. 1990. Reticular overland flow with coarse particles and vegetation roughness under Mediterranean conditions. In J.L. Rubio and J. Rickson (eds), *Strategies to control desertification in Mediterranean Europe*. European Community, Brussels: 228–243.

Tivy, J. 1993. *Biogeography: a study of plants in the ecosphere*, 3rd edn. Longman, Essex.

Tongway, D.J. and Ludwig, J.A. 1990. Vegetation and soil patterning in semi-arid mulga lands of Eastern Australia. *Australian Journal of Ecology*, 15: 23–34.

Trollope, W.S.W. 1980. Controlling bush encroachment with fire in the savanna area of South Africa. *Proceedings of the Grassland Society of South Africa*, 15: 173–177.

Tromble, J.M., Renard, K.G. and Thatcher, A.P. 1974. Infiltration for three rangeland soil–vegetation complexes. *Journal of Range Management*, 27: 318–321.

Tsoar, H. 1990. The ecological background, deterioration and reclamation of desert dune sand. *Agriculture, Ecosystems and Environment*, 33: 147–170.

Tsoar, H. and Møller, J.T. 1986. The role of vegetation in the formation of linear dunes. In W.G. Nickling (ed.), *Aeolian geomorphology*. Binghampton Symposia in Geomorphology, International Series, No. 17. Allen & Unwin, Boston: 75–95.

van de Ven, T.A.M., Fryrear, D.W. and Spaan, W.P. 1989. Vegetation characteristics and soil loss by wind. *Journal of Soil and Water Conservation*, 44: 347–349.

Van den Ancker, J.A.M., Jungerius, P.D. and Mur, L.R. 1985. The role of algae in the stabilization of coastal dune blowouts. *Earth Surface Processes and Landforms*, 10: 189–192.

van der Walt, P.T. and Le Riche, E.A.N. 1984. The influence of veld fire on an *Acacia erioloba* community in the Kalahari Gemsbok National Park. *Koedoe*, Supplement, 27: 103–106.

van Hylckama, T.E.A. 1963. Growth, development and water use by saltcedar (*Tamarix pentandra*) under different conditions of weather and access to water. *International Association of Scientific Hydrology*, 62: 75–86.

Van Rooyen, N., Van Rensburg, D.J., Theron, G.K. and Bothma, J.duP. 1984. A preliminary report on the dynamics of the vegetation of the Kalahari Gemsbok National Park. *Koedoe*, Supplement, 27: 83–102.

Verstappen, H.T. 1968. On the origin of longitudinal (seif) dunes. *Zeitschrift für Geomorphologie*, 12: 200–220.

Verstappen, H.T. 1970. Aeolian geomorphology of the Thar desert and palaeo-climates. *Zeitschrift für Geomorphologie*, Supplementband, 10: 104–120.

Verrecchia, E.P., Dumont, J.-L. and Verrecchia, K.E. 1993. Role of calcium oxalcite biomineralization by fungi in the formation of calcretes: a case study from Nazareth, Israel. *Journal of Sedimentary Petrology*, 63: 1000–1006.

Viles, H.A. (ed.) 1988. *Biogeomorphology*. Blackwell Scientific Publications, London.

Warren, A. 1988. A note on sand movement and vegetation in the Wahiba Sands. *Journal of Oman Studies, Special Report 3: The scientific results of the Royal Geographical Society's Oman Wahiba Sands Project 1985–1987*: 251–255.

Wasson, R.J. 1976. Holocene aeolian bedforms in the Belarabon area, SW of Cobar, NSW. *Journal and Proceedings of the Royal Society of New South Wales*, 109: 91–101.

Wasson, R.J. and Nanninga, P.M. 1986. Estimating wind transport of sand on vegetation surfaces. *Earth Surface Processes and Landforms*, 11: 505–514.

Watts, A.S. 1937. On the origin and development of blowouts. *Journal of Ecology*, 25: 91–112.

Went, F.K. 1959. Ecology of desert plants. II: The effect of rain and temperature on germination and growth. *Ecology*, 30: 1–13.

Westoby, M. 1980. Elements of a theory of vegetation dynamics in arid rangelands. *Israel Journal of Botany*, 28: 160–194.

White, L.P. 1969. Vegetation arcs in Jordan. *Journal of Ecology*, 57: 461–464.

White, L.P. 1970. Brousse tigrée patterns in southern Niger. *Journal of Ecology*, 58: 544–553.

Wickens, G.E. and Collier, F.W. 1971. Some vegetation patterns in the Republic of the Sudan. *Geoderma*, 6: 43–59.

Wiggs, G.F.S. 1992. Sand dune dynamics: field experimentation, mathematical modelling and wind tunnel testing. Unpublished PhD thesis, University of London.

Wiggs, G.F.S., Livingstone, I., Thomas, D.S.G. and Bullard, J.E. 1994. Effect of vegetation removal on airflow patterns and dune dynamics in the southwest Kalahari desert. *Land Degradation and Rehabilitation*, 5: 13–24.

Wilcox, B.P., Wood, M.K. and Tromble, J.M. 1988. Factors influencing infiltrability of semiarid mountain slopes. *Journal of Range Management*, 41: 197–206.

Willis, A.J., Folkes, B.F., Hope-Simpson, J.F. and Yemn, E.W. 1959. Brauton Burrows: the dune system and its vegetation. *Journal of Ecology*, 47: 249–288.

Wolfe, S.A. and Nickling, W.G. 1993. The protective role of sparse vegetation in wind erosion. *Progress in Physical Geography*, 17: 50–68.

Wood, J.C., Wood, M.K. and Tromble, J.M. 1987. Important factors influencing water infiltration and sediment production on arid lands in New Mexico. *Journal of Arid Environments*, 12: 111–118.

Woodell, S.R.J., Mooney, H.A. and Hill, A.J. 1969. The behaviour of *Larrea divericata* (Creosote bush) in response to rainfall in California. *Journal of Ecology*, 57: 37–44.

Wooding, R.A., Bradley, E.F. and Marshall, J.K. 1973. Drag due to regular arrays of roughness elements of varying geometry. *Boundary-Layer Meteorology*, 5: 285–308.

Wopfner, H. and Twidale, C.R. 1988. Formation and age of desert dunes in the Lake Eyre depocenters in central Australia. *Geologische Rundschau*, 77: 815–834.

Worrall, G.A. 1960. Tree patterns in the Sudan. *Journal of Soil Science*, 11: 63–67.

Wright, V.P. and Tucker, M.E. 1991. Calcretes: an introduction. In V.P. Wright and M.E. Tucker (eds), *Calcretes*. Blackwell, Oxford: 1–22.

Yair, A. 1990. Runoff generation in a sandy area — the Nizzana sands, western Negev, Israel. *Earth Surface Processes and Landforms*, 15: 597–609.

Yair, A. and Lavee, H. 1985. Runoff generation in arid and semi-arid zones. In M.G. Anderson and T.P. Burt (eds), *Hydrological forecasting*. Wiley, Chichester: 183–220.

Yeaton, R.I. 1988. Structure and function of the Namib dune grasslands: characteristics of the environmental gradients and species distributions. *Journal of Ecology*, 76: 744–758.

Yeaton, R.I. 1990. The structure and function of the Namib dune grasslands: species interactions. *Journal of Arid Environments*, 18: 343–349.

Section 3
The Work of Water

8

Slope and pediment systems

Theodore M. Oberlander

Arid zone erosional landscapes

In the arid lands, theoretical models of slope development collide with the bedrock exposures of weathering-limited landscapes. Whereas specific bedrock properties can be all but ignored in process/response studies of slopes smoothed by nearly isotropic colluvial mantles, slope analyses and modelling in arid regions must consider a host of petrologic variables that can differ radically from point to point along a single slope profile, and from one profile to the next. In a sense, each rock slope is a law unto itself. Critical determinants of rock strength, process rates and slope forms begin with the nature and geometry of discontinuities of all types and extend to petrology, cementation, presence or absence of pre-weathering under prior climates, and susceptibility to frost action, salt crystal growth, saline corrosion and seepage effects. With bedrock properties exerting more direct influence than in vegetated and soil-covered landscapes, the slope forms of arid regions are extremely diverse, dissimilar process systems being relevant to dissimilar lithologies and structural architectures.

While it is possible to monitor processes and responses of relatively sensitive arid landform systems — fluvial channels, aeolian dunes and badland slopes — the rock slopes that dominate much desert scenery are the product of events so sporadic in space, time and intensity that they are rarely witnessed or measured. Further complicating traditional process/form analysis is the indisputable fact that existing slope forms in arid regions bear a heavy imprint of process systems that are close to moribund after prior phases of more vigorous or more frequent activity. Wherever the record has been analysed, it seems that desert climates of the past 7000 years have been more xeric than preceding climates over a period of several hundred thousand years. A realistic analysis of arid slope systems thus cannot avoid being an exploration of inherited features and structural geomorphology, with much of our understanding of arid slope processes consisting of inferences based on various aspects of morphology and correlated sediments.

THE PEDIMENT PROBLEM

Basic to discussions of arid slope processes and resulting landforms is the *pediment*, a gently inclined slope of transportation and/or erosion that truncates rock and connects eroding slopes or scarps to areas of sediment deposition at

Arid Zone Geomorphology: Process, Form and Change in Drylands, 2nd edition. Edited by David S. G. Thomas.

lower levels. The term 'pediment', originating with McGee (1897), has been applied to two geometrically similar but genetically different geomorphic elements, causing endless confusion regarding pediment-forming processes, and the general question of slope development in arid regions (see, for example, Tator 1992).

Gilbert (1877) described and fully explained the less problematical form: an erosional surface bevelling the substrate of a caprock escarpment or truncating erodible materials in contact with higher-standing, more resistant formations (Figure 8.1A and 8.1B; see also Royse and Barsch 1971; Twidale 1978b). Such pediments are known to French researchers as *glacis d'erosion* (see, for example, Birot 1968, fig. 16; Tricart 1972, p. 199) to distinguish them from true pediments that transgress onto resistant rock. Pediments truncating weak materials in contact with more resistant lithologies have hydraulic profiles, are clearly created by stream planation in the weak materials, and are veneered by alluvial gravels derived from the higher-standing resistant formations. Pediments of this type are commonly 'staged' at multiple levels as a consequence of climatic oscillation or tectonic instability, and are the unconfined equivalent of strath terraces in valleys. While staged surfaces bevelling weak materials pose questions of the causes of periodicity in landscape development, the processes creating them have not been a subject of controversy.

The form that has stimulated a century of argument is the *rock pediment* composed of the same material as the diverse relief rising about it — most commonly a plutonic rock: granite, granodiorite or quartz monzonite (Figure 8.1C). Most rock pediments have nearly straight profiles to the point of alluvial onlap, where gradients diminish perceptibly. They lack transported veneers of other lithologies, often carry a lag of angular vein fragments, and are at least lightly dissected. They show no evidence of planation by fluvial processes, and are seldom terraced. Some observers have suggested that accordant summits of conical inselbergs rising from rock pediments indicate older erosion surfaces. Conflicting views concerning the origin of rock pediments in deserts (and disregarding their presence in more humid regions) probably comprise the largest corpus in the literature on arid landforms.

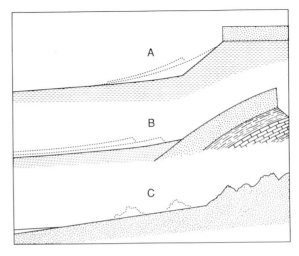

Figure 8.1 *Contrasting types of pediments. (A) Glacis pediment at the base of a caprock escarpment. The pediment cuts across erodible strata supporting the resistant caprock, and expands (to the right) as the escarpment retreats. Dotted lines indicate possible staged pediment surfaces and alternative condition of smooth transition from pediment to caprock as in areas of badland dissection. Pediment may be veneered with exotic gravels transported from resistant caprock. (B) Glacis pediment cut across erodible material in vertical or oblique contact with more resistant material. As erosion continues, the slope break continues to locate on the contact between erodible and resistant materials (shifting to the left here). Pediment may be veneered with exotic gravels transported from the resistant rock. Dotted lines show possible staged pediment surfaces. (C) Rock pediment in which there is no change in lithology from the upland to the peripheral erosion surface. Associated with crystalline rocks. Pediment lacks gravel veneer, cuts both sound and decomposed rock, and descends under onlapping alluvium (left). Dotted lines indicate possibility of isolated erosional residuals within pediment area. Pediment expands in upslope direction (to the right) as hillslopes erode back at constant angle*

Shale badlands display 'micropediments', initially created by slope retreat in uniform materials and subsequently smoothed by hydraulic action. Here hillslope erosion by combined creep and surface wash results in parallel rectilinear slope recession measurable over the span of a few years (Schumm 1956). The mechanism preventing slope foot anchoring and hillslope decline in badlands was identified by Schumm (1962) as a decrease in frictional retardation of flows moving from slopes to the flatter surface below them, so that transport velocities are maintained across the gradient change at the slope foot. The same principle explains the abrupt slope break at the apices of pediments in weak lithologies below caprock escarpments.

As Chapter 13 concerns badland processes, we shall be most concerned with pediments of the first two types, most particularly those cut in crystalline rock, which have been variously interpreted. However, in the absence of intermediating vegetation and soil, desert-slope phenomena are unusually diverse, and slope problems go well beyond those of pediment formation.

MEASURED SLOPE MODIFICATION

Precise information on process rates on slopes in arid settings comes principally from slope monitoring in badlands (Chapter 13) and the degradation of scarps in unconsolidated materials. Wallace (1977) pioneered the dating of ground-rupturing earthquakes in Nevada from the profiles of eroding fault sharps in alluvium. This work has been continued in various environments by many researchers (e.g. Hauks *et al.* 1984; Mayer 1984). In their analysis of alluvial terrace scarps in semi-arid eastern Idaho, Pierce and Colman (1986) demonstrate that microclimate and initial scarp height significantly influence slopewash, causing local rates of slope denudation to vary through two orders of magnitude. Apart from complications introduced by differences in texture and permeability (Dodge and Grose 1980), the range of temperature and moisture conditions encountered in drylands brings varying process combinations into play, some dominated by slope gradient and effectively modelled by the diffusion equation (soil creep related to freeze and thaw, wetting and drying, thermal effects, and biotic disturbance) and others more dependent on slope length as well as gradient (cohesion-limited creep and soil wash, with or without gullying).

For reasons indicated at the outset, studies of active process systems on rocky desert slopes have been rare. Those carried out in Australia and the southwestern United States seem to demonstrate the dominance of creep transport of debris on frost-prone slopes, with rates of

displacement of loose clasts proportionate to slope gradient (see, for example, Schumm 1967; Williams 1974). Abrahams *et al.* (1984) found that clast movement in a granitic area of less severe winters is correlated with both slope gradient and slope length, implying some transport by creep in a setting that is dominated by hydraulic action. Using artificially generated runoff, Abrahams *et al.* (1986) have begun to analyse the competence of Hortonian overland flow on realistic desert slopes. Their results show that the relationship between frictional resistance and erosivity of runoff can vary in character as a result of the interplay of micro-topography and flow depth. Thus, the erosional competence of overland flow will be altered in accordance with the degree of submergence of the smallest obstacles and the concentration of flow into discrete paths. Here again, the extreme variability of surface types in soil-free desert environments promotes wide divergence in the effect of the dominant process, making generalisation hazardous. This point, and the difficulties it raises, are considered further in Chapter 9.

GENERAL GEOLOGIC SETTINGS

In arid erosional landscapes the most obtrusive morphological and process contrasts are those differentiating terrains of layered and massive rocks — the 'platform' and 'shield' deserts of Mabbutt (1977). Layered rocks produce tabular landscapes in areas of both sedimentary and extrusive igneous rock. Form variations in layered rocks relate to lithology, jointing, deformational features, and the specific characteristic of exposed stratigraphic contacts. Massive crystalline rock, including both plutonic and metamorphic type, create extensive boulder slopes, tors, domed inselbergs and erosional pediments, with variety introduced by variations in petrography and jointing. Except where laterites and silcretes cap remnants of inherited deeply weathered mantles, the morphogenetic processes relevant to crystalline rock terrains are quite different from those applicable to arid sedimentary terrains. Recently formed fault blocks of complex internal structure often juxtapose layered and massive rocks and display slope forms quite unlike those where the same rocks have experienced long-continued denudation.

Slope elements in arid cuestaform landscapes

The deserts of north Africa, eastern Arabia and the Colorado Plateau (southwestern United States) consist of lightly deformed blankets of sedimentary rocks covering far more ancient crystalline rocks of continental cratons. Mabbutt (1977) terms such regions 'platform deserts'. Their scenery consists of rock plains produced by erosional stripping to resistant strata that commonly terminate in retreating cuestas, often with detached mesas and buttes. The cuesta is regarded as the diagnostic element in this landform assemblage.

COMPOSITE SCARPS

The cuesta of platform deserts are most commonly 'composite scarps' (Schumm and Chorley 1966), capped by resistant cliff- or ledge-making strata, underlain by more erodible beds that form a fringe of badland dissection or a debris-veneered ramp extending downward from the caprock. Erosion of the substrate leads to caprock failure and collapse. The substrate may express itself as a zone of intricate dissection with talus at gully heads, as a talus-mantled slope, as a staircase of talus-strewn ledges, or as a smooth 'slick rock' ramp, usually free of talus (Figure 8.2). This diversity of form seems to have prevented acceptance of a generic term for this fundamental landscape element. Several terms have been utilised: 'subtalus slope' (Koons 1955), 'badlands' (Schumm and Chorley 1966), 'debris-covered slope' (Cooke and Warren 1973), 'footslope' (Oberlander 1977), 'rampart' (Howard 1987). Since none of these terms specifies the important distinction between the substrate and the caprock, or is universally applicable, the term 'substrate ramp' will be employed.

The break in slope from the cliff to the substrate ramp almost always lies well below the stratigraphic contact, within the erodible substrate itself. This phenomenon was clearly explained by Koons (1955), who attributed it to downward erosion of the portion of the substrate that projects beyond the caprock, producing and unbuttressed vertical face in mechanically weak material (Figure 8.3). Failure of this face under the caprock load eventually causes collapse of the caprock, producing a composite talus. Caprock collapse temporarily returns the cliff foot to the stratigraphic contact, but renewed substrate erosion creates a new substrate face and dissects the previously formed talus, leaving remnant 'talus flatirons'. As substrate erosion continues, the caprock rim begins to be rounded, while the cliff face becomes darkened by rock varnish, and perhaps indented by alveolar weathering. The resulting morphology is wholly or partially removed by the next caprock collapse.

In locations as widely separated as the central Sahara and the Colorado Plateau there have been past periods of massive caprock slides, subsidence, or rotational slumps related to deep-seated failures in shale or mark substrates. This mode of scarp retreat effects permeable sandstones overlying great thicknesses of incompetent sediments, but does not appear to be active in existing arid environments. While it seems to provide evidence of past pluvial conditions in regions that are arid today, the processes involved will not be examined in detail here.

CAPROCK SUBSTRATES

The active element in the cuesta landscape, the caprock substrate, can range from marine shales that are fissile and impermeable, to massive, porous aeolian sandstones. The only requirement is that the material is removed more rapidly than the load resting on it, which collapses from lack of support. Accordingly, massive sandstone that forms bold cliffs where underlain by shale submissively forms a ramp where it underlies basalt, the latter assuming the cliff-forming role. This may be seen in Utah, in the vicinities of Moab and St George. The fact that the massive Navajo and Entrada sandstones of the Colorado Plateau alternatively form cliffs above shale ramps or ramps below basalt cliffs demonstrates that it is not the intrinsic properties of strata, but their properities relative to adjacent strata, that determines their effects on the landscape. In fact, Schumm

Figure 8.2   *Variations in caprock — substrate relationships and scarp form*

Figure 8.2A   *Massive sandstone caprock over thin-bedded sandstone and shale substrate. Active caprock retreat evident in fresh detachments and talus below caprock. Near Moab, Utah*

Figure 8.2B   *Thin caprock of conglomerate over highly erodible substrate. Active caprock retreat but insufficient talus to retard dissection of substrate. Near Hurricane, Utah*

Figure 8.2C   *Badland dissection of impermeable, high erodible shale substrate below thin sandstone caprock. East of San Rafael Swell, Utah*

Figure 8.2D   *Very thin caprocks with downward cliff extension into erodible substrates, producing substrate faces that are the dominant relief elements. Note substrate bedding in conical remnants at base of scarp. San Rafael Swell, Utah*

Figure 8.2E   *Sandstone caprock dominating scarp, with shale substrate largely obscured at base of scarp. Spring Mountains, southern Nevada*

Figure 8.2F   *Simple scarp in aeolian sandstone. Erodible unit is 1–2-cm-thich shale interbed at slope break. Although the sandstone units above and below the slope break have identical physical characteristics, the sandstone below the shale interbed mimics a more erodible substrate to form a slick-rock slope. The cliff above is relatively inactive. Arches National Monument, Utah. Compare with Figure 8.4a*

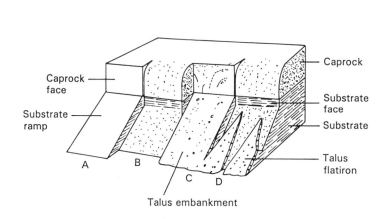

Caprock face

Substrate ramp

A   B

C   D

Talus embankment

Caprock

Substrate face

Substrate

Talus flatiron

Figure 8.3 *Model of compound scarp development, in left to right sequence. (A) Generalised form of compound scarp composed of caprock face and substrate ramp. (B) Erosion of caprock brow and substrate ramp, with downward cliff extension into substrate. (C) failure of substrate face and collapse of unsupported caprock, producing talus embankment on substrate ramp. (D) Erosion of caprock brow, talus embankment, and substrate ramp; erosion of talus embankment produces talus flatirons; downward cliff extension into substrate approaches threshold for next collapse*

and Chorley (1966) demonstrated that the cliff-forming sandstones of the Colorado Plateau are in general mechanically weak and easily disintegrated.

Erosion by combined creep and rainwash in shale substrates can range from 0.15 mm to 1.5 m ka^{-1}, as determined by stake exposure (Schumm 1956; 1962; 1964). Reservoir infilling in areas of sedimentary rock on the Colorado Plateau suggests sediment-generating events at intervals of approximately two years, and an extrapolated denudation rate of about 1.4 km ka^{-1} (Dendy and Champion 1978; Hereford 1987). Erosion rates on massive sandstone substrates have not been measured, but the weight-loss experiments of Schumm and Chorley suggest that in semi-desert regions of seasonal freeze and thaw activity, porous sandstone can be disintegrated by granulation and splitting two to three years after initial exposure. Substrate faces ('secondary walls' in Oberlander 1977) extend downward below caprock faces in all substrates from shale to massive sandstone, attaining lengths inversely proportional to the caprock load. Even in massive sandstone, substrate faces often have propagated downward 10 m or more indicating very effective removal on the ramp with no retreat of the cliff itself.

A variety of influences can differentiate the substrate from the caprock: tensile or compressive strengths; fracture density; the fissile nature of shales; slaking in some shales, marls and sandstones; shrink/swell effects promoting creep in shales; and impermeability that accelerates surface-wash processes. Clearly, a critical value of any one influence may negate a combination of contrary effects from other factors. Thus a study of the Organ Rock formation in Utah's Canyonlands found 'erodible' substrates to be *more* resistant than the caprock in all respects other than joint spacing (Nicholas and Dixon 1986). The present author has stressed that, in the explanation of scarp form, causes and effects can be enormously disparate in scale. Within the Entrada Sandstone in eastern Utah, a single 1–2-cm-thick shale interbed forms the active substrate initiating caprock collapse that spawns vertical faces up to 100 m high (Oberlander 1977). This phenomenon is discussed on page 143.

The processes that cause removal of the most common caprock substrates are those of slope denudation in badland topography in general, discussed separately in Chapter 13. Substrates not producing some form of badland topography are rhythmically bedded sediments of contrasting character that form micro-caprocks and substrate ramps (limestone and marl or shale; sandstones, siltstones and shales), and massive sandstones, dealt with on pages 143–145.

CAPROCK RESPONSE

Cuesta-forming caprocks in platform deserts vary in thickness from 1 to 100 or more metres, and may account for less than 10 to nearly 100% of the relief of the cuesta scarp. There has been little analysis of the mechanics of caprock collapse. Koons (1955) implied that large-scale collapses in massive sandstone followed the upward propagation of tensional and shear fractures oblique to load-bearing vertical faces in mechanically weak substrates. One such feature was described by Koons, and similar features on a small scale have been observed by the present author. Koons's point, also made by Schumm and Chorley (1966), was that cliff failure in massive caprocks involved infrequent large-scale slab collapses rather than frequent falls of smaller blocks.

Observation of scarp edges reveals that even in a single formation, the cliff edge may be controlled by clearly expressed joint sets, or may entirely disregard them. Nicholas and Dixon's (1986) study in Utah's Canyonlands indicates that caprock failure is a passive response to substrate erosion controlled by joint density. In the Organ Rock Formation, caprock promontories are points of reduced substrate jointing, while embayments are areas of two-fold increase in substrate joint density. The caprock is itself resistant only by virtue of sparse jointing.

Regional joint sets that influence topography (see, for example, Hodgson 1961) are distinct from those joints that are themselves a response to topography. In massive sandstones, offloading and dilation produce conspicuous fracture planes parallel to cuesta faces and canyon walls (Bradley 1963). These clearly facilitate slab collapse, frequently producing giant alcoves overhung by rock arches. While some alcoves are spring sites, the majority show no evidence of water concentration. The stress distributions creating such fracture-controlled forms have been analysed by Robinson (1970) and Rogers (1979).

Uniaxial stress tests of cross-bedded Navajo Sandstone have indicated compressive strengths in the range $400-1000 \, \mathrm{kg \, cm^{-2}}$ and tensile strengths of $<5-30 \, \mathrm{kg \, cm^{-2}}$ (Robinson 1970). Average uniaxial stress figures for sandstone are $700-750 \, \mathrm{kg \, cm^{-2}}$ in compression and $20 \, \mathrm{g \, cm^{-2}}$ in tension (Jaeger 1956; Birch *et al.* 1961). These tests ignore the actual stress distribution in large, partially supported rock masses, as well as primary sedimentary structures and variations in porosity and cementation that influence rock strength. Consequently, the tensile strength of large *in situ* masses is considerably overstated by uniaxial test results. According to Robinson (1970), a joint-bounded suspended column of sandstone having a tensile strength of $20 \, \mathrm{kg \, cm^{-2}}$ should detach when it is freed to a height of 70 cm. In fact, breakdown occurs in much smaller masses, even when no joints are visible, owing to local stress patterns and fabric imperfections. The production of arches itself implies an absence of joint influences other than concentric offloading fractures.

The crystallisation of common salts, such as halite, anhydrite and gypsum, develops pressures of 2000 to nearly 4000 atmospheres (Winkler and Singer 1972), such that any granular fissile, or microfractured rock might be expected to fail rapidly when subjected to periodic wetting by saline moisture, as at seepage faces, playa margins, or coasts. This could be a factor in the retreat of particular scarps, though it is more likely to produce scarp microrelief. Cooke (1981) found that paint applied to a weathered basal niche in limestone in Bahrain was removed by rock scaling within four months. However A.R.N. Young's (1987) SEM study of cavernous weathering has indicated that solution etching of quartz grain contacts and the silica of clay cements is vastly accelerated by sodium chloride in solution, while the mechanical effects of salt crystallisation and hydration are less effective than anticipated. This may account for the association of tafoni with siliceous rocks: sandstone, granite and rhyolite.

A climatic interpretation of variations in scarp morphology in massive sandstone in the southwestern United States was proposed by Ahnert (1960), who suggested that rounded caprock brows reflect surface-wash processes in the present arid climate, with angular caprock brows indicating collapse resulting from accentuated springline sapping during the Late Pleistocene pluvial epoch. However, all degrees of varnishing, rounding and alveolar pitting may be seen in a single caprock unit, as Ahnert himself noted, making it appear more likely that a constant tendency toward caprock rounding is periodically interrupted by collapse events resulting from substrate failure (Oberlander 1977). In thin-bedded caprocks, principally limestones, the constant tendency is increasing roughness induced by bedding-plane- and joint-influenced weathering and hydraulic erosion. Collapse events 'freshen' the scarp by removing such irregularities.

GROUNDWATER SAPPING

The importance of groundwater sapping in cuestaform deserts continues to be a subject of discussion (see Howard 1987). Investigation of groundwater sapping as an important factor in arid settings has been stimulated by the discovery of fretted terrain on Mars that seems most easily explained by outflow of subsurface water presently (or formerly) stored as permafrost (see, for example, Higgins 1982; Mars Channel Working Group 1983; Kochel *et al.* 1985).

In the Gilf Kebir of southwestern Egypt, Maxwell (1983) described steep-walled valleys with abrupt heads cut below flat surfaces unmarked by water action. He likened them to the Martian features attributed to sapping. Laity and Malin (1985) reiterated Ahnert's (1960) observation on up-dip versus down-dip (sapping-related) contrasts in scarp form and adduced further details suggesting that groundwater sapping is of crucial importance in the Colorado Plateau region. Laity's (1983) studies of the Navajo Sandstone indicate that groundwater seepage causes diagenetic changes that increasingly concentrate flow, cause salt efflorescence on seepage faces, and abet grain release and sloughing, leading to alcove, development. According to Laity and Malin, sapping processes create distinctive canyons with flat, seepage-controlled profiles and near-constant widths from source to outlet. Other erosional characteristics regarded by Laity and Malin as seepage-induced include up-dip growth axes of canyons and constant tributary angles related to jointing. Those characteristics would be anticipated even in the absence of groundwater sapping processes, but clearly would be accentuated by it. Groundwater sapping is explored further in Chapter 15.

SIMPLE SCARPS IN MASSIVE SANDSTONES

Schumm and Chorley (1966) recognised two categories of 'simple' scarp that do not involve a caprock over a less resistant substrate. One of these is the zone of badland dissection that rises to a drainage divide, or separated higher and lower surfaces composed of a single highly erodible lithology — normally shale. The other simple scarp is composed of resistant cliff-making rock, usually massive sandstone, that has no substrate exposed at the scarp foot. Simple scarps in massive sandstone (see, for example, Figure 8.4) have been investigated in the Colorado Plateau region by the present author (Oberlander 1977).

Figure 8.5 summarises the evolution of simple scarps in the massive Entrada, Navajo and Cedar Mesa sandstones of Utah and Arizona. The simple scar is initiated up-dip as a compound scarp including a weaker substrate that disappears below grade in the down-dip direction, exposing a second massive formation in front of the scarp. The latter forms a local base-level that halts surface lowering, preventing further exposure of the erodible substrate as the scarp continues to erode. Subsequent scarp form changes are controlled by fabric variation in the higher massive unit. These include localised thin shale interbeds, primary bedding structures, and secondary joint sets and pressure-release fractures. The latter encourage the maintenance of rounded forms and abet collapse when undercutting occurs. The shale interbeds, which are relicts of short-lived interdune playas, determine slope breaks and form transitions. Fragmentation and removal of the fragile 1–3-cm-thick shale interbeds (by slaking, ice and salt crystallisation, and seepage flushing) leaves small voids into which slabs of the porous, low-strength caprock subside. The stopping process feeds upwards, slab by slab, generating vertical cliffs above shale interbeds while rounded wash slopes form below them. There is no physical or chemical difference between the cliff- and slope-forming sandstones. The degree of cliff development appears to reflect the distance of retreat from initial exposure of the relevant shaly parting. High cliffs suggest lengthy retreat from the initial shale exposure, whereas low cliffs interrupting slick-rock slopes seem to imply more recent exposure or slower removal of the shale. The latter may be influenced by groundwater seepage paths, possibly during pluvial epochs.

In the absence of petrovariance between cliff faces and individual wash slopes below cliff

Figure 8.4A   *Simple scarp in aeolian sandstone, showing variations in activity above an effective parting (shale interbed). A cycle of collapse, rock varnish deposition, surficial weathering and rock varnish removal in evident on free faces above effective partings. Slick-rock slopes below effective partings are free of varnish, reflecting relative rapid erosion by surface wash. Arches National Monument, Utah*

Figure 8.4B   *Complex scarp, with older talus flatirons in front of scarp, younger talus flatirons at midddle elevations to left, undissected embankment of fine colluvium at middle elevations to right, coarse recent talus in centre. San Rafael Swell, Utah*

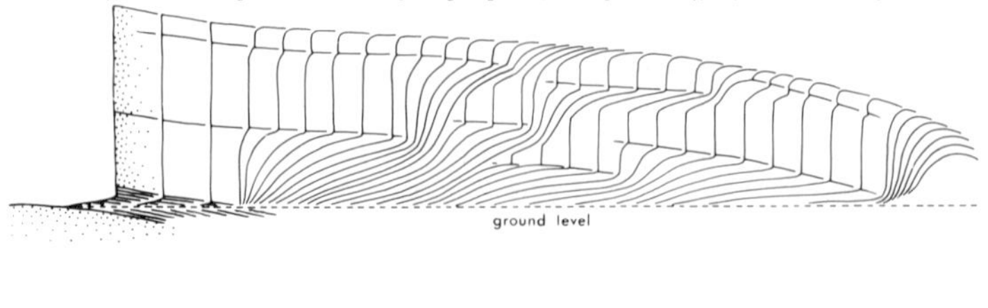

Figure 8.5   *Evolution of a simple scarp in massive sandstone. At A, scarp is compound, with erodible substrate exposed at base of sandstone free face. Sandstone scarp retreats by collapse resulting from substrate erosion, At B, substrate is not exposed; internal structures in massive sandstone assume control of scarp morphology. Dominant control is by discontinuous partings containing shale (1–2 cm thick), with secondary control by joints, offloading fractures, and aeolian cross-bedding. Letters indicate form changes induced by exposure or termination of shaly partings. Termination of parting results in stagnation of free face, brink rounding, and decrease in slope angle. Base-level control by second resistant unit below substrate of compound scarp may retard progressive exposure of substrate, forcing the evolution portrayed*

ground level

faces, the wash slopes of simple sandstone scarps do not develop in accordance with the prevailing concepts of weathering-limited slopes. As the active face retreats by collapse, the wash slope below it lengthens and flattens. No characteristic (equilibrium) slope angle is apparent below cliffs retreating at various levels. Wash slopes may be smooth 'slick-rock' slopes or rougher forms with etched-out aeolian cross-bedding, tafoni, and scattered pothole-like pits up to 6 m in depth. On slick-rock form assemblages in weakly cemented aeolian sandstones, grain detachment by decementation, salt wedging and frost action appear to act uniformly over the entire surface, with subsequent removal by wash being completely efficient. Continuous loss of material from slick-rock slopes is demonstrated by a near-total absence of rock varnish. The cliff, on the other hand, evinces more episodic acitivity, being streaked or thoroughly darkened by rock varnish deposition and removal by spalling and alveolar weathering.

TALUS PROCESSES

The talus (or scree) resulting from caprock failure is a conspicuous element of many arid landscapes. Whereas alpine talus has attracted much study, desert talus has received minimal attention. Talus accumulation is a function of the mechanical strength of falling rock as well as the rate of detachment from scarp faces. Aeolian sandstones often fragment to constituent grains on impact, and intact fallen blocks appear to disintegrate quickly, so that talus embankments may be sparse below active sandstone cliffs (see, for example, Schumm and Chorley 1966). By contrast, conglomerates and thin-bedded limestones in the same region have greater intact strength and generate conspicuous talus. This factor complicates the evaluation of relative rates of scarp retreat based on talus volumes (see page 146).

Yair and Lavee (1974) noted the basal concavity in many talus profiles in the eastern Sinai and attributed it to redeposition by debris flows originating on the talus itself. Using artificial rainfall on natural talus slopes, they were able to show how talus can produce runoff. In talus gullies the accumulation of large blocks, which shed rainfall, embedded in finer material with a low void ratio, encourages runoff with relatively shallow (5–10 cm) depths of wetting. Interfluvial areas, where fines are not present at the surface, easily absorb all precipitation. How the gullies critical to runoff generation are themselves initiated is not explained. However, the tracks left by sliding or rolling blocks could trigger the process of gully formation — trapping subsequent blocks and fines and encouraging more erosion in a self-enhancing manner. Yair and Lavee concluded that desert talus processes probably do not differ in any significant way from those of more humid areas.

As already noted, talus mantles erode to produce talus flatirons (see Figure 8.4b). These are often so voluminous as to suggest a past period of complete substrate blanketing by continuous talus embankments. Within the Grand Canyon of the Colorado River, the substrates below cliffs of Coconino Sandstone and Redwall Limestone appear to be undergoing exhumation from a formerly continuous colluvial veneer. Gerson (1982) and Gerson and Grossman (1987) have proposed that such tendencies are a significant indication of climatic change (see page 147). Conversely, enormous debris slopes in the hyper-arid northern Atacama Desert evince neither recent accretion nor surface erosion. They appear to have stagnated and been smoothed, possibly by long-continued salt weathering and aeolian redistribution (Figure 8.8).

RATES OF SCARP RETREAT

Rates of scarp retreat from structural highs and fluvial canyons in arid regions have been estimated from various types of evidence (see Table 8.1). Yair and Gerson (1974) considered escarpments composed of clastic rocks capped by datable reef limestones and granitic rocks retreating from fault lines in the hyper-arid southern Sinai, where the precipitation is approximately 25 mm yr^{-1}. They derived a maximum possible rate of scarp retreat of 1.2–2.0 mm yr^{-1} over the past 250 000 years in sedimentary rocks, or a more likely 0.1 to 0.4 mm yr^{-1} with earlier scarp initiation. Yair and Gerson estimated the rate of slope retreat in granite in the Sinai to be 0.1–2.0 mm yr^{-1}

Table 8.1   *Rates of scarp retreat*

Scarp retreat rate	Lithology	Evidence	Location
0.1–2.0	Reef Limestone	Age of limestone; date of fault offset	Sinai, Israel (Yair and Gerson 1974)
6.7	Moenkopi Formation conglomerate over Chinle shale	K/Ar dates on volcanic veneers over retreating scarp	Northern Arizona (Lucchitta 1975)
0.5–6.7	Mesa Verde Sandstone	Geometry of valleys beheaded by scarp retreat	Northern Arizona (Schmidt 1980; 1988)
0.4	Redwall Limestone	Packrat midden distance from cave threshold	Northern Arizona (Cole and Mayer 1982)
0.16–0.17 (post-early Eocene)	Kaibab Limestone	Age-diagnostic fossils and inferred Eocene scarp positions based on location of strike valleys	North West Arizona (Young 1985)

based on dating of fault events. This rate is an order of magnitude higher than that derived from the present author's studies of granitic inselbergs in the Mojave Desert (see pages 152–153).

Lucchita (1975) computed the rate of back-wearing of the Chocolate Cliffs of the 'Grand Staircase' north of Arizona's Grand Canyon by age determinations on lavas, calculating that the Chocolate Cliffs have retreated at a rate of 6.7 mm yr^{-1} over the last million years. Schmidt (1980; 1988) derived similar rates elsewhere on the Colorado Plateau through study of the widths of beheaded valleys on cuesta back-slopes. Schmidt's results, which are based on a number of arguable assumptions, range from 0.5 to 6.7 mm yr^{-1} since Middle Tertiary time. On a shorter timescale, Cole and Mayer (1982) computed the rate of retreat of the Redwall Limestone in the Grand Canyon at about 0.4 mm yr^{-1} on the basis of the proximity of ^{14}C dated *Neotoma* middens to the mouths of caves. This is in fair agreement with other approaches; however, some of the assumptions of this study have been disputed (Haman 1983; Hose 1983).

In an analysis of the western Grand Canyon region of Arizona, Young (1985) derived widely divergent rates of retreat of escarpments of Palaeozoic sedimentary rock in pre- and post-early Eocene time. Young's conclusions are based upon the date of Laramide deformation, the age of Tertiary erosion surfaces as established by fossil gastropods, and the past positions of scarps inferred from the location of ancient strike valleys. The computed post-early Eocene rate of scarp retreat is 0.16 mm yr^{-1}, an order of magnitude less than the pre-Laramide rate. Young attributed this discrepancy to the effects of post-Laramide stream incision and scarp dissection, which succeeded a long phase of simple scarp retreat in the low-relief pre-Laramide landscape. Young's suggestions that regional drainage incision greatly diminishes rates of scarp backwearing requires further scrutiny, especially as his argument appears to disregard the possible geomorphic effects of major climatic change from pre- to post-Eocene time. It is during the Eocene that Axelrod (1979) sees the global deserts beginning to emerge, to become accentuated after the Late Miocene, according to the botanical evidence, which seems very clear in the case of North America.

Ahnert (1960) suggested that talus is under-developed in the Colorado Plateau, indicating that the region's escarpments might presently be in a state of relative paralysis after periods of activity during Pleistocene pluvials. Schumm and Chorley (1966) responded with findings on the current high frequency of rock falls and the tendency of aeolian sandstones to revert to sand on impact or after a brief period of weathering. On the other hand, repeat photography in the Colorado Plateau region shows surprisingly little alteration of bedrock features, or even talus, over a 100-year period (Baars and Molenaar 1971), despite significant change in fluvial systems (Graf 1987). In fact, the near-constant form of retreating scarps implies that rates of cliff retreat are controlled by (and must be equivalent to) rates of substrate erosion, in many instances tantamount to measured rates of slope denudation in badlands; that is, $1.5-15$ mm yr^{-1}. The question of periodicity in scarp processes, raised by Ahnert (1960), is enmeshed with that of the effect of climatic oscillations, which is considered in the following subsection.

INFLUENCE OF OSCILLATING CLIMATES

Rotating slump blocks below Black Mesa and Vermillion and Echo Cliffs of Arizona subsided blocks below the scarp of New Mexico's Chuska Mountains, and both types of detachment in Nubian sandstones in the central Sahara, all strongly eroded and without recent counterparts, led Reiche (1937), Strahler (1940), Watson and Wright (1963), and Hagedorn *et al.* (1978) to conclude that substrate erosion (or failure) and caprock collapse were much more active in a past period of greater precipitation, presumably in the Late Pleistocene. Ahnert (1960) echoed this conclusion on the basis of scarp morphology. In a study of limestone scarp forms in Morocco, Algeria and Tunisia, Smith (1978) concluded that previously active cliff recession and seepage-related cavernous weathering are now stagnant, with inactive cliff faces presently being pitted by tafoni. The present author, while arguing against climatic change as an influence on the form of sandstone cliffs in Utah, has hypothesised that certain long, severely weathered slick-rock ramps in sandstone below high-level seepage lines were detachment planes, possibly related to past periods of greater moisture availability (Oberlander 1977).

An interesting argument for climatically controlled periodicity in cliff retreat has been made by Gerson (1982) and Gerson and Grossman (1987), based on the interpretation of talus flatirons in the Eilat Mountains of the eastern Sinai. Two types of 'talus' are recognised in the Eilat Mountains — coarse rubble resulting from cliff collapse in the present arid environment and more expansive aprons of debris-flow deposits formed in pluvial phases when cliffs are regarded as stable. Although the latter interpretation contradicts most opinion on scarp activity, caprock and substrate variations may account for the discrepancy. The North American and Saharan scarps apparently active during pluvials have massive permeable caprocks, could be vulnerable to groundwater sapping, and would not easily generate debris flows. The scarps investigated by Gerson and Grossman consist of thin-bedded carbonates capable of debris flow generation. According to their analysis, continuous veneers of pluvial debris flow 'talus' are eroded into flatirons during interpluvial phases when caprock collapse also accelerates due to undermining by intensified substrate gullying. Rock varnish microchemical analysis indicating three Mn to Fe cycles, interpreted as three pluvial/interpluvial transitions (see Dorn 1984), appears to support this interpretation. Isotopic analysis of rock varnish on alluvial fans in California's Death Valley by Dorn *et al.* (1987) likewise suggests fanhead aggradation during pluvial phases and incision during interpluvials, contrary to intuition and laboratory simulations (see, for example, Hooke 1967). Dorn's conclusions based on work in Death Valley are somewhat in conflict with findings in a nearby area by Wells *et al.* (1987), who stress the importance of pluvial–interpluvial transitions in sediment generation.

In the eastern Sinai, scarp processes seem less sensitive to minor climatic oscillations than fluvial systems, with two to three generations of talus flatirons linking into $5-12$ alluvial terraces. Thus the 'noise' affecting fluvial systems appears to make them a less useful indicator of major climatic periodicity than the more complacent slope systems. Gerson (1982) admits the possibility that cliff retreat occurs and talus and debris flow deposits accumulate and are eroded under both Pleistocene and Holocene climates, with episodic rainfall events in dry phases replicating more general tendencies of wet phases.

Pediment–inselberg landscapes

Among the most venerable of geomorphic problems in arid regions is the origin of ramp-like erosional surfaces known as 'pediments' that radiate from the bases of much steeper hillslopes composed of the same rock. This landform association is particularly striking in arid terrains of granite or gneiss that weathers into domes, spires, fins, boulders and sand (grus), without intermediate cobble- and pebble-sized debris. Indeed, bimodality in granitic weathering products were adduced by Bryan (1923) and reaffirmed by Kirkby and Kirkby (1974) and Kesel (1977), as the controlling factor in arid granitic landscapes, with hillslopes approximating the angle of repose of the usual boulder cover (upwards of 20°) and the lower (2–7°) ramp being a slope of transport for sand.

The bedrock erosional ramp, which truncates veins and metamorphic septa in granitic terrains, is normally masked by sandy alluvium where peripheral to a sizeable highland, but often exposes bedrock 3 or more kilometres beyond the bases of bouldery knolls of rubble-covered hills (Figure 8.6A). Until recently, these planar surfaces were assumed to be actively expanding in deserts. The processes creating such surfaces have long remained a matter of speculation and controversy.

While most observers have accepted the headward growth of erosional pediments at the expense of associated uplands in granitic terrains, Twidale (1978b) has adamantly opposed this view. Based on Australian evidence, he has insisted that all major relief variations in granite are structurally controlled, emerging as a consequence of irregular landscape lowering controlled by petrography and joint density. Twidale and Bourne (1975; 1978) have proposed that sparsely jointed, domed inselbergs in the Australian landscape have changed little over immense spans of time. Based on slope indentations indicating 'episodic exposure' of Australian bornhardts and on the presence of Tertiary silcretes on Cretaceous sediments below (and peripheral to) bornhardt-like features, they argue for an early Tertiary or Mesozoic origin for Australian bornhardts (Twidale and Bourne 1975; 1978; Twidale 1978a). While this interpretation of monolithic Australian inselbergs is not disputed, it should be pointed out that structure, climatic history and related geomorphic process system in other pediment/inselberg areas differ significantly from the Australian examples. Granitic inselbergs in the southwestern United States and northern Mexico are commonly a confusion of joint-derived boulders and pinnacles, often adjacent to (and accordant with) smooth domes that are themselves divided by visible joints. Bornhardt-like domes of unjointed granite similar to the Australian examples, Stone Mountain (Georgia), and the sugarloafs of Rio de Janeiro and the Serra do Orgãos (Brazil), are not conspicuous elements of the arid landscapes of western North America. Although North American granitic inselbergs can indeed be shown to be ancient, they are as closely jointed as surrounding pediments, which are often well-exposed and composed of very sound rock (Figure 8.6B). The age of these features does not preclude slope retreat as a formative process at an early stage in their evolution (see, for example, Oberlander 1972; 1974).

LITHOLOGY AND THE PIEDMONT ANGLE

Pediments are most conspicuous in granitic and gneissic rocks as a consequence of the characteristically abrupt 'piedmont angle' formed at the junction of granitic pediments and the hillslopes rising above them. The distinctiveness of the piedmont angle in crystalline rocks may have various origins. Australian researchers have attributed it to slope-foot notching by subsoil weathering (Mabbutt 1966; Twidale). While this may not seem feasible in drier environments where no weathered mantle is preserved, it is clear that many present areas of extreme aridity received significantly more moisture during Pleistocene pluvial phases, and for a much longer period prior to global climatic differentiation in late Tertiary time, and may have experienced phases of subsoil weathering. However, the weathering of well-jointed or granular rocks can proceed adequately in arid climates (see Chapters 3 and 4), which obviates the necessity for subsoil weathering as a factor in piedmont angle formation in lithologies with these characteristics.

Extensive pediments have developed in lithologies other than granite and gneiss, but attract less notice owing to the absence of well-defined piedmont angles. In the northern Great Basin region of the western United States, a semi-desert environment, broadly

Figure 8.6A   *Granitic pediment and upland, with bedrock exposed in an erosional trench. Alluvial fan superimposed on bedrock pediment at left. Coxcomb Mountains, Mojave Desert, California*

Figure 8.6B   *Minor granitic inselberg and adjacent pediment. Granite of upper pediment is sound (elastic), with joints that continue into the inselberg. Lucerne valley, Mojave Desert, California*

Figure 8.6C   *Form asymmetry related to lithology. To left, quartz monzonite with abrupt piedmont angle between pediment and irregular boulder slope. To right, colluvial mantle derived from metavolcanic rocks creates smooth transition from hillslope to pediment. Apple Valley, Mojave Desert, California*

concave piedmont slopes often merge smoothly into uplands. These slopes are occasionally broken and offset by recent extensional faulting. Many of the fault scarps expose bedrock (usually rhyolite) at the surface two or more kilometres beyond mountain fronts. These slopes clearly are pediments, though the absence of well-defined piedmont angles impedes their recognition as such. Occasionally an individual desert inselberg is composed of granite in contact with a lithology yielding a continuum of clast sizes, including a high proportion of fine, water-retaining material. A profile across such a feature is decidedly asym-

metrical, with a piedmont angle on the granite side and a smoothly concave transition from hillslope to pediment on the non-granitic side (Figure 8.6C). In the Mojave Desert the latter is evidence of a colluvial mantle usually bearing a strongly oxidized argillic soil. Such mantles appear to be undergoing stripping at present. The resulting colluvial flatirons are similar to those in the Sinai described by Gerson, and likewise appear to reflect process changes from Late Pleistocene to Holocene time (Figure 8.7A). Wells *et al.* (1987) have proposed that water-absorbing colluvial mantles in the eastern Mojave Desert exert a critical influence on

Figure 8.7A *Colluvial flatirons, foreshortened in upslope view from granitic pediment at centre of Figure 8.6c. Corestones derived from decomposed granite of pediment. Apple Valley, Mojave Desert, California*

Figure 8.7B *Sound granite at exposed weathering front, with pre-weathered granite at higher level. Undecayed granite is coloured only by surficial rock varnish and sheds coherent scales at depth of moisture penetration in sound rock (2–4 mm). Pre-weathered granite is oxidised (chroma > 2), sheds granules, and yields under impact. Lucerne Valley, Mojave Desert, California*

Figure 8.7C *Artificial excavation exposing decomposed granite in colluvium-covered hill. Colluvium is 1–2 m in depth (figure in centre for scale). Granite in exposed face crumbles under finger pressure. No corestones evident in exposure. El Mirage, Mojave Desert, California*

runoff processes and dissection in the upper piedmont zone.

On non-granite rocks the absence of well-defined piedmont angles may reflect a continuum of clast sizes in transit from hillslope to pediment as opposed to the common bimodal size distribution of granitic debris (see, for example, Kesek 1977). Kirkby and Kirkby (1974) found that in southern Arizona the 'slope break' (equivalent to the piedmont angle) is about 10 times wider in schist and basalt (250–270 m) than in granite (20–80 m).

In contrast to other lithologies, granitic bedrock in arid and semi-arid regions shows a tendency to be cleanly stripped of alluvium and regolith. When this is the case, the granite piedmont angle is abrupt — not a zone in any sense (Oberlander 1972). Selby's (1982a) study of inselbergs in the Namib Desert, an area of long sustained aridity, indicates abrupt piedmont angles in schist, marble, and vein and dyke complexes, as well as in granite.

QUANTITATIVE ANALYSES OF ROCK PEDIMENTS AND ASSOCIATED HILLSLOPES

A number of researchers have attempted quantitative process/form analyses of rock pediments and associated hillslopes (see, for example, Mammerickx 1964; Cooke 1970; Cooke and Reeves 1972; Kesel 1977). Such studies have dealt with classic pediment locales generally with rock hillslopes and conspicuous piedmont angles. Morphometric analyses have been hindered by lack of consensus as to the exact definition of the term 'pediment'. To some, the term signifies the narrow band of exposed rock that may or may not be present below the piedmont angle; others include the alluvial cover over suballuvial bedrock ramps, or restrict the term to the suballuvial bedrock surface alone. Cooke (1970) offered a set of measurable parameters to be utilised in future analyses of pediments and associated landforms. The comprehensive study by Mammerickx (1964) demonstrates especially well that the relationship between catchment areas, debris sizes, pediment lengths and pediment gradients are far from systematic or straightforward.

Irrespective of details, pediments cut in rock identical to that of adjacent hillslopes seem to be the product of backwearing of residual relief by hillslope processes. Even Twidale (1978a) concedes slope retreat of tens of metres in exposed granite and in the conglomerate of Ayers Rock. Hypotheses involving spur trimming by flood flows (for example, Johnson 1932; Howard 1942; Rahn 1966) fail to explain the sharp piedmont angles at the bases of isolated skyline inselbergs or the many pediment-encircled granitic massifs that lack valleys altogether. Early proposals that erosive sheetfloods could form pediments are defeated by the fact that sheetfloods require planar surfaces and are a consequence rather than a cause of planation.

Quantitative analyses of inselberg slopes seem to have been more conclusive than morphometric studies of pediments. On the basis of gradient and debris-size statistics from boulderclad Arizona hillslopes, Melton (1965) refuted Bryan's (1923) suggestion that the angles of granitic hillslopes in southern Arizona were closely determined by joint spacing and boulder dimensions. However, a study by Parsons and Abrahams (1987) relating particle size, joint spacing and slope gradient in the Mojave Desert concludes that there is, in fact, a continuum between slopes mantled by small (<30 cm) clasts, where slope angle is correlated with particle size, and slopes composed of larger (>500 cm) boulders and rooted outcrops, where slope angles and clast sizes are not correlated. The first type of slope, mantled by small, mobile debris sizes is regarded as adjusted to the current process system. Poor correlation between slope angles and clast sizes is interpreted as an indicator of inherited forms that are not adjusted to current processes. The slopes investigated by Melton are of the latter type.

In a study of conical inselbergs in the Namib Desert, Selby (1982b) found that the gradients of rock slopes that lack a debris cover correlated well with rock-mass strength ratings based on eight different measurable parameters. The slopes studied were composed of schist, gneiss, granite and marble, with angles of 5–90° and strength ratings between 55 and 90 on a scale of 100. Strength parameters included various parting characteristics, water content, weathering state, and intact strength measured by Schmidt hammer. Results coincided remarkably well with similar analyses in New Zealand and Antarctica. The study implies that rock slopes should retreat at constant equilibrium angles related to physical strength regardless of climate or climatic changes. The question of changing rates of retreat of such equilibrium slopes is not raised by Selby.

THE CASE FOR INHERITANCE OF DESERT PEDIMENTS

In the 1950s, French and German investigators established that pediments in the granitic

highlands of the central Sahara were extremely ancient, having been covered by 'tropical loam soils', fragments of which remain preserved under Tertiary volcanic rocks (Büdel 1955; Kubiena 1955; Dresch 1959). They proposed that these pediments are relict from periods of more humid morphogenesis in early Tertiary or even Late Mesozic time. Similarly, Mabbutt (1965) presented detailed evidence for the relict nature of planation surfaces in Australia, with arid processes inscribing a different form assemblage of much more recent origin. The inselberg-studded crystalline shield of western Arabia likewise emerged from a lengthy humid phase ending in the early Pleistocene (Al-Sayari and Zötl 1978; Jado and Zötl 1984). Birot and Dresch (1966) seem to have been the first to propose that pediments of the southwestern United States are relict forms. Busche (1972) concurred, based on comparative studies with Saharan (Tibesti) pediments, which were regarded as relict from a more humid pre-mid-Pleistocene morphogenetic phase.

Oberlander (1972; 1974) noted that the veneer of case-hardening and patination by rock varnish implies that inselberg slopes of well-jointed *in situ* granite in California's Mojave Desert are essentially paralyed at present, although surrounded by pediments suggesting kilometres of parallel rectilinear slope retreat. Nearby slopes covered by mobile boulder mantles are composed of granitic rock decomposed to depths of 10 m or more (far in excess of moisture penetration under current conditions). As in North Africa, remnants of Miocene and Pliocene basalt flows are present on the upper, presumably most recently formed, pediment surfaces. Detailed work on the Cima volcanic field of the eastern Mojave Desert has verified the antiquity of pediments buried by eruptive events in this area (Dohrenwend *et al.* 1984). Beneath the lava remnants are saprolites, some strongly oxidised, argillic, and indicative of non-desert conditions either during or subsequent to the period of active slope retreat and pediment expansion. Complete stripping of this saprolite has occurred on the higher portions of pediments surrounding small inselbergs, where the sediment supply was most limited (Figure 8.6B). the extremely sound granite exposed in such locations is interpreted as the basal surface of chemical weathering in the prior landscape (Figure 8.7B). Moss (1977) subsequently noted the thorough decay of the granite beneath the boulder cover on inselbergs in southern Arizona, and reiterated the earlier suggestion of Mensching (1973) that pediments in Arizona were developed by arid-phase denudation of an inherited deeply weathered mantle.

Regolith stripping in the Mojave region may have been initiated by reduction of vegetation in Late Miocene and Pliocene time, when uplift of the Sierra Nevada, Transverse and Peninsular Ranges of California began to cast a rain shadow eastward. However, arid zones began to be accentuated globally at this time. Palynological and stratigraphic evidence from the Mojave Desert indicate woodland to semi-desert conditions with non-desiccating freshwater lakes prior to approximately 6 million years BP (Dibblee 1967; Axelrod 1983). *Neotoma* middens demonstrate *Juniperus* woodlands and *Yucca* chaparral in lowlands during the Late Pleistocene (Van Devender and Spaulding 1979; Wells and Woodcock, 1985). Analysis of a 930-metre core from Searles Lake that penetrates to bedrock and spans 3.2 million years seems to indicate only one prior period as arid as the present in this region — an interval between 0.57 and 0.31 million years BP within which several perennial lake phases are also recorded (Smith *et al.* 1983) It is within this largely semi-arid context that we must consider the classic pediment landscapes of the Mojave Desert.

According to the present author's interpretation, pedimentation in the Mojave Desert occurred by parallel rectilinear slope retreat in a well-vegetated, saprolite-mantled landscape. Hillslope boulder litters in the present arid landscape appear to be relict corestones and rooted outcrops shaped by subsurface chemical weathering during a period (or periods) of enhanced moisture availability as compared with the present. Current piedmont angles are attributed to accentuated physical and chemical weathering at the slope foot during and subsequent to regolith stripping triggered by loss of vegetative protection. Although the evidence of pre-Holocene humidity would seem to suggest Pleistocene slope activity, radiometric ages of basalts on upper pediment surfaces indicate less than 100 m of slope retreat over the past 5 million years.

ENVIRONMENTS OF ACTIVE PEDIMENTATION

Pediment expansion seems active at present in regolith-mantled granitic landscapes in non-desert areas of the southwestern United States. Figure 8.8A illustrates an example of a chaparral landscape in central Arizona that appears to replicate Miocene climate, vegetation, soils and landforms in the Mojave Desert region, at the time of major pediment expansion in the latter (Oberlander 1974; Axelrod 1979; 1983).

Figure 8.8 *Active pediments and hillslopes in Arizona*

Figure 8.8A *Pediment apparently expanding by hillslope erosion under chaparral cover; Sierra Prieta, west of Prescott, Arizona. Sclerophyll shrub vegetation consists of scrub oak (*Quercus tubinella*) 1–2 m high, mountain mahogany (*Cerococarpus*), and* Acacia greggii, *with extensive steppe grasses. Annual precipitation varies from 400 to 600 mm*

Figure 8.8B *Gully headcut exposing mantle of oxidised saprolite (2.5 YR 6/6) protected by chaparral scrub on Sierra Prieta pediment. Pick at top for scale; long dimension of clast in centre about 25 cm*

Present eroding granitic hillslopes in southern Arizona and the Mojave Desert (Figures 8.7C and 8.8D) seem intermediate between the paralysed inherited slope systems in petrographically dissimilar granite in the Mojave region (Figures 8.6A and 8.6B) and actively retreating slopes and expanding pediments in the chaparral region of central Arizona.

Particularly in the chaparral landscape of central Arizona, and locally in the Sonoran and Mojave deserts, a mantle of thoroughly decomposed granite or granitic saprolite, frequently including corestones (Figures 8.8B and 8.8C), is presently being eroded on both hillslopes and

planar peripheral surfaces. In the chaparral setting, which receives 400–600 mm yr^{-1} of precipitation, hillslopes merge with peripheral pediments by broad concavities, with no discrete piedmont angle present (Figure 8.8A). Bedrock protrudes through the saprolite in scattered locations on hillslopes and exposed corestones are scattered thinly over all surfaces. The pediment surface and the steeper forms rising above it are multi-convex, being dissected by networks of lightly incised drainage lines that only rarely expose sound bedrock. All surfaces carry a lag of coarse granules and vein fragments. Except for conspicuous oxidation, which extends into the

Figure 8.8C *Unoxidised corestones in oxidised grus matrix, exposed in roadcut, Sierra Prieta pediment. Diameter of largest corestone approximately 1 m*

Figure 8.8D *Boulder slope, West Sacaton Buttes, south of Phoenix, Arizona. Two dwarfed but living saguaro cacti (*Cereus giganteus*) in foreground overturned and dragged downslope by the creeping boulder mantle while maintaining attenuated root connections upslope (beyond photo margin). Surface lowering evident at base of standing saguaros*

zone of corestones, no soil profile development is evident on the saprolite of interfluves, hillslopes, or valley slopes. Above the zone of alluvial onlap, 2 km from the mountain front, all surfaces appear to be eroding, causing backwearing of hillslopes and lowering of pediment surfaces and hillcrests by slope conjunction. A time-independent transport-limited geomorphic system presently seems to be expanding pediments at the expense of uplands in the chaparral region of central Arizona.

In the Sonoran and Mojave deserts, with higher potential evapotranspiration and one-fourth of the precipitation of the chaparral region, saprolite is usually replaced by grus, and a piedmont angle is evident, separating hillslope boulder mantles from boulder-free pediments. In the Sacaton region of the Sonoran Desert, south of Phoenix, Arizona, there is direct evidence of landscape denudation, but the marked difference between hillslopes and peripheral surfaces there, as in the Mojave, greatly retards pediment expansion. In the Sacaton region, corestones perched on 15–20-cm-high pedestals of decomposed granite, along with exposure of the roots of saguaro cacti (*Cereus giganteus*) and palo verde (*Cercidium microphyllum*) to as much as 20 cm below the original ground line, indicate localised active surface lowering along the upper margins of pediments under current conditions. Evidence of active downslope creep of the boulder covers on hillslopes is seen in the form of stunted saguaro cacti that have been overturned gradually and dragged downslope while maintaining ever-lengthening root connections 2 or more metres upslope (Figure 8.8D). Exposure of subsurface portions of undisturbed sagauros on some boulder slopes suggests active lowering of such slopes by hydraulic erosion of grus and thoroughly decomposed granite between case-hardened, varnished boulders and also beneath the boulder mantle. Thus the corestone lag on certain granitic slopes seems to be settling vertically at a rate equal to downward erosion on the pediment, stranding a fringe of dark, case-hardened boulders on barely coherent pedestals at the retreating piedmont concavity.

Both of the areas described are grazed by cattle, possibly accelerating erosion rates, and perhaps expanding the woody component in the present chaparral region at the expense of grasses, as has been the tendency elsewhere in western North America. However, soils in the chaparral region preserve no relict features to substantiate this. In any case, the pediments evident here appear to have formed in the presence of a thick, erodible, corestone-laden regolith in a semi-arid climate.

European climatic geomorphologists have largely agreed on the conditions most favourable to pediment expansion. Birot (1968) stated that pedimentation of crystalline rock was most rapid in seasonally wet, low-latitude thorn forests. Mensching (1978) has described 'active' pedimentation in a savanna landscape in the Sudan. On the other hand, Höverman (1972) asserted that pedimentation is currently most active in cold-winter deserts in which upland slopes are affected by cryergic processes — a view seized on by Büdel (1982), who saw deep weathering rather than slope retreat and pedimentation as the norm in the seasonally wet tropics. Thus Büdel, dismissing the possibility of pedimentation in a Sahara that had emerged from tropical savanna conditions, asserted that pediments did not exist there, and explained away exposed granitic surfaces peripheral to hillslopes in the central Sahara as 'shield inselbergs'. It seems more relevant that Rohdenburg (1982) and Embrechts and de Dapper (1987) have recently described ongoing pedimentation in seasonally wet tropical locations in Brazil and the Cameroon.

REQUIREMENTS FOR PEDIMENTATION

The problem of pediment formation has become the problem of maintaining parallel rectilinear retreat of permeable slopes in a saprolite-mantled landscape. In this respect, the detailed analyses of processes at piedmont angles in the Sudan by Ruxton (1958) and Ruxton and Berry (1957) continue to be of paramount importance, although seemingly ignored by most English-speaking pediment researchers. The predilection for parallel rectilinear slope retreat on saprolite-mantled crystalline rock can be explained in terms of negation of the creep process by deep permeability in grus, lateral eluviation of fines by slope throughflow, and accelerated corrosion by subsurface moisture moving down steep hillslope gradients, and being especially destructive at the slope foot. The critical processes relevant to slope retreat and pediment expansion thus may be subsurface in nature rather than surficial as widely supposed. They demand a weathered mantle, but do not require 'deep weathering' or a tropical savanna climate — only weathering into slopes at a rate the compensates slope erosion. The studies of throughflow processes and effects by Ruxton and Ruxton and Berry seem definitive and should lend themselves to verification elsewhere, as in the central Arizona example of active pediment expansion.

Tectogenic slopes

In tectonically active desert regions such as the Iranian Plateau, Xinjiang, the Chilean Atacama and the southwestern United States, crustal rupture has started the erosional clock ticking much more recently than in the more stable shield and platform deserts. The geologic settings for early slope development as a result of tectonism include fault scarps produced by compression, tension and shear; fault block backslopes generated by rotation; and giant fault mullions resulting from crustal extension along low-angle detachments. Erosional forms and rates in these situations reflect petrovariance and process systems that may have fluctuated greatly in intensity over a relatively brief span of time (see, for example, Wells *et al.* 1987). It is arresting to consider that such giant features as the profound rifts of Owens Valley and Death Valley in California, with the host of prominent erosional and depostional features they display over elevation ranges of more than 3000 m, are barely 3 million years old, and were covered by *Juniperus* woodland and *Yucca* chaparral some 13 000 years ago, with the ubiquitous *Larrea* shrubs of today being totally unrepresented in Late Pleistocene *Neotoma* middens (Wells and Woodcock 1985).

Most tectogenic landforms are well-known and constitute textbook illustrations of structural effects on landscapes lacking the softening effect of a soil–vegetation cuticle. Only the extensional fault mullions known as 'turtlebacks', which are being identified with increasing frequency in the western United States, are somewhat unfamiliar (see Wright *et al.*

1974). Erosion rates in lithologically complex tectogenic terrains have been estimated from alluvial fan areas in eastern California (Hooke and Rohrer 1977). While variations in fan slope related to debris size would seem to make fan volume preferable to fan area as an index of catchment erodibility, the principal conclusion of the study by Hooke and Rohrer anticipates the later findings of Nicholas and Dixon (1986) in reference to caprock scarps. Both analyses indicate that in arid landscapes erodibility is a function of fracture density rather than of mineralogy or intact strength of the material.

Hyper-aridity and erosional paralysis

Some geographic locations are arid enough to appear completely paralysed in the morphogenetic sense (Figure 8.9). In the eastern Sahara a sand sheet has entombed a prior fluvial erosional topography (Haynes 1982; McCauley *et al.* 1982), while aeolian abrasion has erased much of the fluvial relief in the Eocene limestones between the Nile Valley and Kharga. In the Atacama Desert of northern Chile, a strongly developed inherited relief in diverse lithologies seems to have persisted without change for a vast period of time.

In accord with the paucity of measured rainfall in the Atacama (see Ericksen 1981), many localities are conspicuously free of signs of recent fluvial action. Some segments of the deep *quebradas* between Iquique and Arica lack any trace of a hydraulic channel. Their 600-m-high walls are smooth, unrilled, and diversified by unusual mass-wasting phenomena (Figure 8.9B). So are large expanses of the great coastal escarpment, as well as extensive upland tracts.

The long unrilled slopes seen in the Atacama, in some places truncating rock, in others composed entirely of detritus, are puzzling, the more so as other slopes in their environs may be gullied to varying degrees. According to Abele (1983), the distinctive smooth valley walls of the deep quebradas of northern Chile are found in the rainless zone of coastal fog (*camanchaca*), up to an elevation of about 1100 m, and are the product of exaggerated salt weathering resulting from almost daily wetting and drying cycles. Above the fog limit near 1100 m, and to the north and south where rainfall occurs annually,

Figure 8.9   *Stagnant erosional forms in hyper-arid setings*

Figure 8.9A   *Sand-smothered cuesta scarp at Dakhla, Western Desert of Egypt*

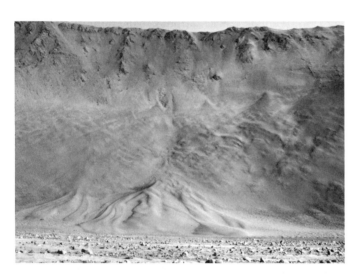

Figure 8.9B   *Qubrada wall, approximately 600 m high, with degraded debris-flow deposits and characteristic mass-wasting phenomena. Cuya, northern Atacama Desert, Chile*

normal fluvial erosion resumes. While Abele's explanation may be valid for particular lithologies, especially ash-flow tuffs, there are many exceptions to his generalisations. In the driest portion of the coastal zone (south of Arica), and again south of Tocopilla, cliffs are tafoni-riddled, ravined, and familiar in form; smooth and dissected slopes alternate vertically and horizontally in the Cordillera de la Costa; and smooth slopes are common far east of, as well as above, the fog zone. The smooth slopes have a variety of surface characteristics: open (non-interlocking) desert pavement; fluffy soft gilgae soils that are thoroughly impregnated with salts; massive crusts of salt, gypsum and nitrates; and enormous talus embankments with disproportionately small source outcrops. Beaty (1983) has noted that surficial patterning of granitic regolith on smooth slopes, which he attributed to salt-related processes and called 'tiger striping' — not an entirely appropriate term, as the patterns are chevron-like, resembling the markings of *Conidae* shells — especially the Textile Cone.

In an attempt to explain the region's unique nitrate deposits, Ericksen (1981) has suggested that the Atacama has been hyper-arid for perhaps 10–15 million years, possibly with brief humid episodes (see also Chapter 24). The region's unique salines are seen as being derived from the atmosphere as dry fallout and fog-condensate over this lengthy span of time, their source being the Pacific Ocean to the west and Andean volcanoes to the east. The present author's observations indicate periodicity in salt and loess deposition and striping, with large-scale solution and translocation of salts evident both in salars and on hillslopes (Oberlander 1988). Smooth slopes seem to be a result of lateral equiplanation, abetted by salt weathering during upward growth of massive talus embankments, with the various stages of the process visible in canyon walls between Arica and Iquique, and in the coastal escarpment between Arica and Tocopilla. The penultimate shreds of the original craggy scarps in many places survive at the brows of otherwise smooth slopes, which locally rise nearly 1000 m from the sea of from valley floors. Andean tephra is interbedded with colluvium on these slopes, offering the possibility of age and rate determinations, which are currently under investigation.

Salt plays a different role in granitic uplands in the northern Atacama, with clasts of all sizes from 3-m blocks to finger-scale prisms appearing to have been thrust upward through a fine-textured regolith. Accordingly, the crests of many granitic hills bristle with large and small 'sarsen stones'. Unusual concentrations of salts in a permeable regolith seem to produce fragmentation, sorting, pulling and heaving effects similar to those seen in permafrost areas as well as distinctive forms of slope activity. Conspicuous by their absence in the northern Atacama are the granitic boulder slopes and bornhardt-like forms that are widespread in the Namib Desert, another area of long-sustained aridity (see Selby 1977).

Conclusion

While exact knowledge of processes affecting rock slopes in deserts remains scanty, quantitative investigations and stratigraphic analyses are beginning to narrow the range of surmises concerning landscape history and rates of change in Late Pleistocene and Holocene time as well as the degree to which desert landforms have evolved from earlier non-desert configurations. Still, much is based on inference, and

owing to the difficulty of process measurements and the problem of establishing the periodicity of geomorphically significant events in arid environments, this problem will continue. While such basic matters as the optimum conditions for scarp retreat (pluvial or interpluvial?), and the very possibility of slope backwearing in granitic rocks continue to be debated, we cannot fail to appreciate how far we are from a satisfactory of desert slope processes.

References

Abele, G. 1983. Flachenhafte Hanggesaltung und Hangzerschneidung im chilenisch-peruanischen Trockengebiet. *Zeitschrift für Geomorphologie*, Supplementband, 48: 197–201.

Abrahams, A.D., Parsons, A.J., Cooke, R.U. and Reeves, R.W. 1984. Stone movement on hillslopes in the Mojave Desert: a 16-year record. *Earth Surface Processes and Landforms*, 9: 365–370.

Abrahams, A.D., Parsons, A.J. and Luk, S.H. 1986. Resistance to overland flow on desert hillslopes. *Journal of Hydrology*, 88: 343–363.

Ahnert, F. 1960. The influence of Pleistocene climates upon the morphology of cuesta scarps on the Colorado Plateau. *Annals of the Association of American Geographers*, 50: 139–156.

Al-Sayari, S.S. and Zötl, J.G. (eds) 1978. *Quaternary period in Saudi Arabia*, Vol. 1. Springer-Verlag, New York.

Axelrod, D.I. 1979. Desert vegetation: its age and origin. In J.R. Goodin and D.R. Northington (eds), *Arid land plant resources*, Texas Technical University, Lubbock: 1–72.

Axelrod, D.I. 1983. Paleobotanical history of the western deserts. In S.G. Wells and D.R. Harrigan (eds), *Origin and evolution of deserts*. University of New Mexico Press, Albuquerque: 113–129.

Baars, D.L. and Molenaar, C.M. 1971. *Geology of Canyonlands and Cateract Canyon*. Four Corners Geological Society, Sixth Field Conference, Cataract Canyon River Expedition.

Beaty, C.B. 1983. Tiger-striping: a curious form of surficial patterning in the Atacama Desert, northern Chile. *Geological Society of America Abstract with Programs*. Geological Society of America. Rocky Mountain Section 36th Annual Meeting and Cordilleran Section 79th Annual Meeting, 15: 387.

Birch, F., Schairer, J.E. and Spicer, H.C. (eds) 1961. *Handbook of physical constants*. Geological Society of America, Special Paper, 36.

Birot, P. 1968. *The cycle of erosion in different climates*. Translated by C.I. Jackson and K.M. Clayton, from *Le Cycle d'erosion sous les differents climats.* University of California Press, Berkeley, California.

Birot, P. and Dresch, J. 1966. Pediments et glasis dans l'ouest des Etats-Unis. *Annales de Géographie,* 411: 513–552.

Bradley, W.A. 1963. Large-scale exfoliation in massive sandstones of the Colorado Plateau. *Bulletin of the Geological Society of America,* 75: 519–528.

Bryan, K. 1923. Erosion and sedimentation in the Papago Country, Arizona. *Bulletin of the United States Geological Survey,* 730: 19–90.

Budel, J. 1955. Reliefgenerationen und plio-pleistocaner Klimawandel im Hoggar-Gebirge. *Erdkunde,* 9: 100–115.

Budel, J. 1982. *Climatic geomorphology.* Translated by L. Fischer and D. Busche, from *Klima-Geomorphologie.* Princeton University Press, Princeton, NJ.

Busche, D. 1972. Untersuchungen zur Pedimententwicklung im Tibesfi-Gebirge (Républic du Tschad). *Zeitschrift für Geomorphologie,* Supplementband, 15: 21–38.

Cole, K.L. and Mayer, L. 1982. Use of packrat middens to determine rates of cliff retreat in the Eastern Grand Canyon, Arizona. *Geology,* 10: 597–599.

Cooke, R.U. 1970. Morphometric analysis of pediments and associated landforms in the Western Mojave Desert. *American Journal of Science,* 269: 26–38.

Cooke, R.U. 1981. Salt weathering in deserts. *Proceedings of the Geologists Association,* 92: 1–16.

Cooke, R.U. and Reeves, R.W. 1972. Relations between debris size and slope of mountain fronts and pediments in the Mojave Desert, California. *Zeitschrift für Geomorphologie,* 16: 76–82.

Cooke, R.U. and Warren, A. 1973. *Geomorphology in deserts.* University of California Press, Berkeley and Los Angeles.

Dendy, F.E. and Champion, W.A. 1978. *Sediment deposition in U.S. reservoirs.* United States Department of Agriculture, Miscellaneous Publications 1362.

Dibblee, T. 1967. *Areal geology of the Western Mojave Desert, California.* US Geological Survey Professional Paper 522: 153 pp.

Dodge, R.L. and Grose, L.T. 1980. Tectonic and geomorphic evolution of the Black Rock Fault, northwestern Nevada. In P.C. Andrise (ed.), *Earthquake hazards along the Wasatch-Sierra Nevada frontal fault zones.* US Geological Survey Open File Report 80–801: 494–508.

Dohrenwend, J.C., McFadden, L.D., Turrin, B.D. and Wells, S.G. 1984. K–Ar dating of the Cima volcanic field, eastern Mojave Desert, California: late volcanic history and landscape evolution. *Geology,* 12: 163–167.

Dorn, R.I. 1984. Speculations on the cause and implications of rock varnish microchemical laminations. *Nature,* 310: 767–70.

Dorn, R.I., DeNiro, M.J. and Ajie, H.O. 1987. Isotopic evidence for climatic influence on alluvial-fan development in Death Valley, California. *Geology,* 15: 108–110.

Dresch, J. 1959. Notes sur la géomorphologie de l'Air. *Association Géographes Français Bulletin,* 280: 2–20.

Embrechts, J. and de Dapper, M. 1987. Morphology and genesis of hillslope pediments in the Febe area (South Cameroon). *Catena,* 14: 31–43.

Ericksen, G.E. 1981. *Geology and origin of the Chilean nitrate deposits.* US Geological Survey Professional Paper 1188.

Gerson, R. 1982. Talus relicts in deserts: a key to major climatic fluctuations. *Israel Journal of Earth Science,* 31: 123–132.

Gerson, R. and Grossman, S. 1987. Geomorphic activity on escarpments and associated fluvial systems in hot deserts as an indicator of environmental regimes and cyclic climatic changes. In M.R. Rampino, J.E. Sanders and W.S. Newman (eds), *Climate: history, periodicity, predictability.* Van Nostrand Reinhold, Stroudsburg, PA: 300–322.

Gilbert, G.K. 1877. *Report on the geology of the Henry Mountains: United States Geological Survey of the Rocky Mountain region.* Department of the Interior, Washington, DC.

Graf, W.L. 1987. Late Holocene sediment storage in canyons of the Colorado Plateau. *Bulletin of the Geological Society of America,* 99: 261–271.

Hagedorn, H., Busche, D., Grunert, J., Schäffer, K., Schulz, E. and Skowronek, A. 1978. Bericht über geowissen schaftliche Untersuchungen am Westrand des Murzuk-Beckens (zentrale Sahara). *Zeitschrift für Geomorphologie,* Supplementband, 30: 20–38.

Haman, J.F. 1983. Comments and replies on Cole, K.L. and Mayer, L. 'Use of packrat middens to determine rates of cliff retreat in the eastern Grand Canyon, Arizona'. *Geology,* 11: 315.

Hauks, R.C., Bucknam, R.C., Lajoie, K.R. and Wallace, R.E. 1984. Modification of wave-cut and faulting-controlled landforms. *Journal of Geophysics Research,* 89: 5771–5790.

Haynes, C.V. 1982. Great Sand Sea and Selima Sand Sheet, Eastern Sahara: geomorphology of desertification. *Science,* 217: 629–633.

Hereford, R. 1987. Sediment-yield history of a small basin in Southern Utah, 1937–1976: implications for land management and geomorphology. *Geology,* 15: 954–957.

Higgins, C.G. 1982. Drainage systems developed by sapping on Earth and Mars. *Geology,* 10: 147–152.

Hodgson, R.A. 1961. Regional study of jointing in Comb Ridge-Navajo Mountain area, Arizona and Utah. *Bulletin of the American Association of Petroleum Geologists*, 45: 1–38.

Hooke, R.LeB. 1967. Processes on arid-region alluvial fans. *Journal of Geology*, 75: 438–460.

Hooke, R.LeB. and Rohrer, W.L. 1977. Relative erodibility of source-area rock types, as determined from second-order variations in alluvial-fan size. *Bulletin of the Geological Society of America*, 88: 1177–1182.

Hose, L.D. 1983. Comments and replies on Cole, K.L. and Mayer, L. 'Use of packrat middens to determine rates of cliff retreat in the eastern Grand Canyon, Arizona'. *Geology*, 11: 314.

Höverman, J. 1972. Die periglaziale Region des Tibesti und ihr Verhältnis zu angrenzenden Formungsregionen. *Göttinger Geographische Abhandlungen*, 60: 261–283.

Howard, A.D. 1942. Pediment passes and the pediment problem. *Journal of Geomorphology*, 5: 3–31, 95–136.

Howard, A.D. 1987. Introduction to cuesta landforms and sapping processes on the Colorado Plateau. In A.D. Howard, C. Kochel and H.E. Holt (eds), *Sapping features of the Colorado Plateau: a comparative planetary geology field guide*. National Aeronautics and Space Administration Special Publication SP-49.

Jado, A.R. and Zötl, J.G. (eds), 1984. *Quaternary Period in Saudi Arabia*, Vol. 2. Springer-Verlag, New York.

Jaeger, J.C. 1956. *Elasticity, fracture, and flow with engineering and geological applications*. Methuen, London.

Johnson, D.W. 1932. Rock planes in arid regions. *Geographical Review*, 22: 656–665.

Kesel, R. 1977. Some aspects of the geomorphology of inselbergs in central Arizona, U.S.A. *Zeitschrift für Geomorphologie*, 21: 119–146.

Kirkby, A.T.V. and Kirkby, M.J. 1974. Surface wash at the semiarid break in slope. *Zeitschrift für Geomorphologie*, Supplementband, 21: 151–171.

Kochel, R.C., Howard, A.D. and McLane, C. 1985. Channel networks developed by groundwater sapping in fine-grained sediments: analogues to some Martian valleys. In M.J. Woldenberg (ed.), *Models in geomorphology*. Allen & Unwin, Boston: 313–341.

Koons, D. 1955. Cliff retreat in the southwestern United States. *American Journal of Science*, 253: 44–52.

Kubiena, W. 1955. Über die Braunlehmrelikte des Atakor (Hoggar-Gebirge, Zentrale Sahara). *Erdkunde*, 9: 115–132.

Laity, J.E. 1983. Diagenetic controls on groundwater sapping and valley formation, Colorado Plateau, as revealed by optical and electron microscopy. *Physical Geography*, 4: 103–125.

Laity, J.E. and Malin, M.C. 1985. Sapping processes and the development of theatre-headed valley networks on the Colorado Plateau. *Bulletin of the Geological Society of America*, 96: 203–217.

Lucchitta, I. 1975. Application of ERTS images and image processing to regional geologic problems and geologic mapping in Northern Arizona: Part IVB, The Shivwits Plateau. *National Aeronautical and Space Administration Technical Report*, 32–1597: 41–72.

Mabbutt, J.A. 1965. The weathered land surface in central Australia. *Zeitschrift für Geomorphologie*, 9: 82–114.

Mabbutt, J.A. 1966. Mante-controlled planation of pediments. *American Journal of Science*, 264: 79–91.

Mabbutt, J.A. 1977. *Desert landforms*. MIT Press, Cambridge, MA.

Mammerickx, J. 1964. Quantitative observations on pediments in the Mojave and Sonoran deserts. *American Journal of Science*, 262: 417–435.

Mars Channel Working Group 1983. Channels and valleys on Mars. *Bulletin of the Geological Society of America*, 94: 1035–1054.

Maxwell, T.A. 1983. Erosional patterns of the Gilf Kebir Plateau and implications for the origin of Martian canyonlands. In F. El Baz and T.A. Maxwell (eds), *Desert landforms of southwest Egypt: a basis for comparison with Mars*. National Aeronautical and Space Administration Contractor Report: 281–300.

Mayer, L. 1984. Dating Quaternary fault scarps formed in alluvium using morphologic parameters. *Quaternary Research*, 22: 300–313.

McCauley, J.F., Schaber, G.G., Breed, C.S., Grolier, M.J., Haynes, C.V., Issawi, B., Elachi, C. and Blom, R. 1982. Subsurface valleys and geo-archaeology of the eastern Sahara revealed by Shuttle radar. *Science*, 218: 1004–1019.

McGee, W.J. 1897. Sheetflood erosion. *Bulletin of the Geological Society of America*, 8: 87–112.

Melton, M.A. 1965; Debris-covered hillslopes of the southern Arizona desert – consideration of their stability and sediment contributions. *Journal of Geology*, 73: 715–729.

Mensching, H. 1973. Pediment und Glacis, ihre Morphogenese und Einordnung in das System der klimatischen Geomorphologie auf Grund von Beobachtung im Trockengebiet Nordamerikas (USA und Nordmexiko). *Zeitschrift für Geomorphologie*, Supplementband, 17: 133–155.

Mensching, H. 1978. Inselberge, Pedimente and Rumpfflächen im Sudan (Republik); Ein Beitrag zur morphogenetischen Sequenz in den ariden Subtropen und Tropen Afrikas. *Zeitschrift für Geomorphologie*, Supplementband, 30: 1–19.

Moss, J.R. 1977. The formation of pediments: scarp backwearing or surface downwasting. In D.O. Doehring (ed.), *Geomorphology in arid regions*. 8th Annual Geomorphological Symposium. Binghampton, NY: 51–78.

Nicholas, R.M. and Dixon, J.C. 1986. Sandstone scarp form and retreat in the land of standing rocks, Canyonlands National Park, Utah. *Zeitschrift für Geomorphologie*, 30: 167–187.

Oberlander, T.M. 1972. Morphogenesis of granitic boulder slopes in the Mojave Desert, California. *Journal of Geology*, 80: 1–20.

Oberlander, T.M. 1974. Landscape inheritance and the pediment problem in the Mojave Desert of southern California. *American Journal of Science*, 274: 849–875.

Oberlander, T.M. 1977. Origin of segmented cliffs in massive sandstones of southeastern Utah. In D.O. Doehring (ed.), *Geomorphology in arid regions*. 8th Annual Geomorphological Symposium. Binghampton, NY: 79–114.

Oberlander, T.M. 1988. Salt crust preservation of pre-Late Pleistocene slopes in the Atacama Desert, and relevance to the age of surface artifacts. *Association of American Geographers, Program Abstracts*. Phoenix, AZ: 143.

Parsons, A.J. and Abrahams, A.D. 1987. Gradient–particle size relations on quartz monzonite debris slopes in the Mojave Desert. *Journal of Geology*, 95: 423–442.

Pierce, K.L. and Colman, S.M. 1986. Effect of height and orientation on geomorphic degradation rates and processes: late glacial terrace scarps in central Idaho. *Bulletin of the Geological Society of America*, 97: 869–885.

Rahn, P.H. 1966. Inselbergs and knickpoints in southwestern Arizona. *Zeitschrift für Geomorphologie*, 10: 217–225.

Reiche, P. 1937. The Toreva-block—a distinctive landslide type. *Journal of Geology*, 45: 538–548.

Robinson, E.S. 1970. Mechanical disintegration of the Navajo Sandstone in Zion Canyon, Utah. *Bulletin of the Geological Society of America*, 81: 2799–2805.

Rogers, J.D. 1979. *The development of natural rock arches at Arches National Park, Utah*. Second Conference on Scientific Research in the National Parks, San Francisco.

Rohdenburg, H. 1982. Geomorphologisch-bodenstratigraphe Verleich zwischen dem Nordöstbrasilianschen Trockengebiet und immerfeucht-tropischen Gebeiten Südbrasiliens mit Ausführungen zum Problemkreis der Pediplain-Pediment-Terrassentreppan. *Catena*, Supplement, 2: 74–122.

Royse, C.F. Jr and Barsch, D. 1971. Terraces and pediment-terraces in the Southwest: an interpretation. *Bulletin of the Geological Society of America*, 82: 3177–3182.

Ruxton, B.P. 1958. Weathering and subsurface erosion in granite at the piedmont angle, Balos, Sudan. *Geological Magazine*, 95: 353–377.

Ruxton, B.P. and Berry, L. 1957. Weathering profiles and geomorphic position on granite in two tropical regions. *Bulletin of the Geological Society of America*, 68: 1263–1292.

Schmidt, K.-H. 1980. Eine neue Methode sur Ermittlung von Stufenrückwanderungsraten-dargestellt am Beispiel der Black Mesa Schichtstufe, Colorado Plateau, USA. *Zeitschrift für Geomorphologie*, 24: 180–191.

Schmidt, K.-H. 1988. The significance of scarp retreat for Cenozoic landform evolution on the Colorado Plateau. *Earth Surface Processes and Landforms*.

Schumm, S.A. 1956. The role of creep and rainwash on the retreat of badland slopes. *American Journal of Science*, 254: 693–706.

Schumm, S.A. 1962. Erosion on miniature pediments in Badlands National Monument, South Dakota. *Bulletin of the Geological Society of America*, 73: 719–724.

Schumm, S.A. 1964. Seasonal variations of erosion rates and processes on hillslopes in western Colorado. *Zeitschrift für Geomorphologie*, Supplementband, 5: 215–238.

Schumm, S.A. 1967. Rates of surficial rock creep on hillslopes in western Colorado. *Science*, 155: 560–561.

Schumm, S.A. and Chorley, R.J. 1966. Talus weathering and scarp recession in the Colorado Plateaus. *Zeitschrift für Geomorphologie*, Supplementband, 10: 11–35.

Selby, M.J. 1977. Bornhardts of the Namib Desert. *Zeitschrift für Geomorphologie*, 21: 1–13.

Selby, M.J. 1982a. Controls on the stability and inclinations of hillslopes formed on hard rock. *Earth Surface Processes and Landforms*, 7: 449–467.

Selby, M.J. 1982b. Rock mass strength and the form of some inselbergs in the central Namib Desert. *Earth Surface Processes and Landforms*, 7: 489–497.

Smith, B.J. 1978. The origin and geomorphic implications of cliff foot recesses and tafoni on limestone hamadas in the northwest Sahara. *Zeitschrift für Geomorphologie*, Supplementband, 22: 21–43.

Smith, G.I., Barczek, V.J., Moulton, G.F. and Liddicoat, J.C. 1983. *Core KM-3, a surface-to-bedrock record of Late Cenozoic sedimentation in Searles Valley, California*. US Geological Survey Professional Paper 256.

Strahler, A.N. 1940. Landslides of the Vermillion and Echo Cliffs, northern Arizona. *Journal of Geomorphology*, 3: 285–300.

Tator, B.A. 1952. Pediment characteristics and terminology. *Annals of the Association of American Geographers*, 42: 295–317.

Tricart, J. 1972. *Landforms of the humid tropics, forests and savannas*. Longman, London.

Twidale, C.R. 1967. Origin of the piedmont angle as evidenced in South Australia. *Journal of Geology*, 75: 393–411.

Twidale C.R. 1978a. On the origin of Ayers Rock, central Australia. *Zeitschrift für Geomorphologie*, Supplementband, 31: 177–206.

Twidale, C.R. 1978b. On the origin of pediments in different structural settings. *American Journal of Science*, 278: 1138–1176.

Twidale, C.R. and Bourne, J.A. 1975. Episodic exposure of inselbergs. *Bulletin of the Geological Society of America*, 86: 1473–1481.

Twidale, C.R. and Bourne, J.A. 1978. Bornhardts. *Zeitschrift für Geomorphologie*, Supplementband, 31: 111–137.

Van Devender, T.R. and Spaulding, W.G. 1979. Development of vegetation and climate in the southwestern United States. *Science*, 204: 701–710.

Wallace, R.E. 1977. Profiles and ages of young fault scarps, north-central Nevada. *Bulletin of the Geological Society of America*, 88: 1267–1281.

Watson, R.A. and Wright, H.E. 1963. Landslides on the east flank of the Chuska Mountains northwestern New Mexico. *American Journal of Science*, 261: 525–548.

Wells, P.V. and Woodcock, D. 1985. Full-glacial vegetation of Death Valley, California: juniper woodland opening to *Yucca* semi-desert. *Madrono*, 32: 11–23.

Wells, S.G., McFadden, L.D. and Dohrenwend, J.C. 1987. Influence of Late Quaternary climatic changes on geomorphic and pedogenic processes on a desert piedmont, eastern Mojave Desert, California. *Quaternary Research*, 27: 130–146.

Williams, M.A.J. 1974. Surface rock creep on sandstone slopes in the Northern Territory of Australia. *The Australian Geographer*, 12: 419–424.

Winkler, E.M. and Singer, P.C. 1972. Crystallization pressure of salts in stone and concrete. *Bulletin of the Geological Society of America*, 83: 3509–3513.

Wright, L.A., Otton, J.K. and Troxel, B.W. 1974. Turtleback surfaces of Death Valley viewed as phenomena of extensional tectonics. *Geology*, 2: 53–54.

Yair, A. and Gerson, R. 1974. Mode and rate of escarpment retreat in an extremely arid environment (Sharm el Sheikh, southern Sinai Peninsula). *Zeitschrift f̄ Geomorphologie*, Supplementband, 21: 202–215.

Yair, A. and Lavee, H. 1974. Areal contribution to runoff on scree slopes in an extreme arid environment. *Zeitschrift für Geomorphologie*, Supplementband, 21: 106–121.

Young, A.R.N. 1987. Salt as an agent in the development of cavernous weathering. *Geology*, 15: 962–966.

Young, R.A. 1985. Geomorphic evolution of the Colorado Plateau margin in west-central Arizona: a tectonic model to distinguish between the causes of rapid, symmetrical scarp retreat and scarp dissection. In M. Morisawa and J.T. Hack (eds), *Tectonic Geomorphology*. Allen & Unwin, Boston: 261–278.

9

Overland flow generation and sediment mobilisation by water

Andrew J. Baird

Introduction

Most runoff in arid and semi-arid areas occurs as overland flow, although natural soil pipes appear to be important in conducting water in some environments (Bryan and Yair 1982; Watts 1989). Overland flow can be generated in two basic ways. In the first, often called Hortonian overland flow, the rate of rain arriving at the soil surface exceeds the rate of infiltration. Water is stored in small surface depressions until they are overtopped when the water starts to flow downhill (Horton 1945). In the second, water is prevented from entering the soil surface by the rise of the water table in topographic lows and areas adjacent to stream channels. Effluent groundwater and rainwater then both contribute to overland flow after the surface detention store has been exceeded. Both types of overland flow generation occur in a range of environments, although in arid and semi-arid areas the former is the more important. Despite the low frequency of rainfall events and low annual amounts of rain in arid and semi-arid environments, overland flow is an important geomorphic agent in these areas. Rainfall intensities in the arid zone can be high, resulting in rapid delivery of water to the soil surface and surface ponding as the infiltration capacity is exceeded (Thornes 1994). More often, rainfall remains at the soil surface because of the low infiltration capacity of the soil (Yair 1992). Overland flows entrain surface sediment and carry it downslope.

The hydraulics of overland flow remain to be fully elucidated and a range of flow types have been observed on arid zone slopes. Despite the early work of Horton (1945) and Emmett (1970), which showed that sheetflow, *sensu stricto*, is rarely if ever present on natural hillslopes, many of the first hydraulic models assumed that overland flow could be acceptably described using a mean flow depth (Woolhiser and Liggett 1967; Kibler and Woolhiser 1970; Woolhiser 1975). Later models considered the importance of flow concentrations in erosional features such as rills and gullies (for example, Foster and Meyer 1975). The rill–interrill dichotomy is reflected in some relatively recent models such as that of Kirkby (1990b). However, recent models have attempted to characterise more realistic flows in which running water appears as numerous small threads of different depths and velocities controlled by the microtopography of the slope surface. This last type of flow has been described as anastomotic or reticular (see, for example, Abrahams *et al.* 1989; Thornes *et al.* 1990;

Arid Zone Geomorphology: Process, Form and Change in Drylands, 2nd edition. Edited by David S. G. Thomas.
© 1997 John Wiley & Sons Ltd.

Baird *et al.* 1992; Kirkby *et al.* 1996). Since the rate of erosion is related to flow velocity, an understanding of the microhydraulics of flow over arid zone slopes is very important. Raindrop impact also affects the hydraulic behaviour and erosional capacity of flowing water and, in its own right, can effect erosion by dislodging surface material (Mosley 1973; Mutchler and Young 1975; Brandt 1990; Parsons *et al.* 1994a; Wainwright *et al.* 1995). Advances in our understanding of all of these processes have been made since the first edition of this book appeared (see Scoging 1989). This chapter concentrates on a number of these advances. In the first section consideration is given to factors affecting infiltration in semi-arid soils. In the second section the hydraulics of overland flow are considered. Particular attention is given to the role of microtopography in controlling the properties of flow. The mobilisation and transport of sediment by running water are briefly considered in the final section where rainsplash erosion is also described.

Overland flow generation

INFILTRATION

Infiltration is defined as the process of water entry into the soil. Most overland flow in the arid zone is generated when the rainfall intensity exceeds the rate of infiltration. If the soil is not limiting, the rate of infiltration will be equal to the rainfall rate. If the soil is limiting and surface ponding occurs, the infiltration rate will depend on the physical properties of the soil material (which are partly affected by raindrop impact), and the antecedent moisture content of the soil (Hillel 1980; Marshall and Holmes 1989). The most obvious soil physical controls on infiltration are soil texture and soil structure. *Ceteris paribus*, the rate of infiltration can be expected to be lowest in a soil with a fine texture and a poorly developed structure. Infiltration is also controlled by the moisture conditions within the soil because these control the suction forces that operate to draw moisture through the soil profile (Hillel 1980; Marshall and Holmes 1989). Recently attention has focused on the role of rock fragments at and near the soil surface. If the density of rock fragments at the surface is sufficiently high then infiltration is reduced as water runs off their impermeable surfaces. A lower density of rock fragments can protect the finer surface material from raindrop impact while still allowing passage of rainwater to this finer soil so promoting infiltration. Where rock fragments are scattered or absent from the soil surface the soil is prone to a process called sealing or crusting. The mechanisms of sealing are not fully understood but appear to be related to compaction of the soil by raindrop impact, chemical dispersion of peds at the soil surface, and the blocking of larger soil pores by fine material mobilised by rainsplash. Sealing can also occur when fine material is deposited by flowing water and blocks surface pores. Microphytic soil crusts are also likely to have an impact on infiltration (see Chapter 5). Another control on infiltration, and one that has received little attention, is air encapsulation. During infiltration air can become trapped in the soil pores and acts to block the passage of water through these pores. The result is a lower hydraulic conductivity and a reduced rate of infiltration. The effects of rock fragments, surface sealing and air encapsulation on infiltration are not fully understood and are rarely taken into account in models of infiltration. Indeed, infiltration is often described using relatively simple analytical equations.

MODELLING INFILTRATION

The rate of infiltration under ponded conditions is usually initially high and then decreases in an approximate exponential fashion with time. The high rate in early times is associated with a high hydraulic gradient caused by near-saturated conditions giving low suctions at the soil surface and relatively high suctions immediately below this layer. As the soil wets up at depth, the hydraulic gradient is reduced (Hillel 1980). Many workers have attempted to describe the infiltration process mathematically either from empirical observations or from basic soil physics principles, and most have found that infiltration under ponded conditions can be expressed either as a power function of time or as a function of total amount of water infiltrated. These equations are called, respectively, time-based and storage-based infiltration equations. An example of a time-based equation is Philip's physically based equation (Philip 1957) in which infiltration rate is a function of the square root of time:

$$i = i_c + \frac{s}{2t^{1/2}} \qquad (9.1\text{A})$$

where i is the infiltration rate $(L\,T^{-1})$, i_c the asymptotic steady infiltration rate $(L\,T^{-1})$, t is time and s is a system defining constant. An empirical time-based equation originally attributed to Kostiakov and now to M.R. Lewis (Swartzendruber 1993) commonly used in infiltration studies is:

$$i = at^{-b} \qquad (9.2)$$

where a and b are system defining constants. Both of these equations can be rewritten to solve for the time taken for ponding and therefore the time taken to overland flow t_o (assuming minimal surface detention and $P > i_c$ where P is the rainfall rate) by replacing i with a constant value of P and rewriting the equation for t_o. For example, equation 9.1 then becomes:

$$t_o = \left(\frac{s}{2(P - i_c)} \right)^2 \qquad (9.1\text{B})$$

Equations 9.1 to 9.2 are simple to use and have been used successfully by some authors (for example, Cundy and Tento 1985) in models of semi-arid and arid overland flow generation. However, time-based equations can prove unreliable. Scoging *et al.* (1992) found that a time-based version of the well-known Green and Ampt (1911) equation solved for t_o could not be satisfactorily used in a combined infiltration–overland flow model. The equation was calibrated, using small $(1\,\text{m} \times 1\,\text{m})$ plots, for two types of desert surface — areas with small amounts of desert pavement and areas with high percentages of pavement cover — at Walnut Gulch, Arizona, under a constant artificial rainfall rate. The predictions of the overland flow model were compared with runoff data from a large experimental plot $(18\,\text{m} \times 35\,\text{m})$ subject to the same constant rainfall intensity as used for the calibration of the infiltration equation. Scoging *et al.* (1992) found that model predictions of infiltration on

the large plot did not adequately predict time to ponding for those areas with small amounts of desert pavement. This problem in prediction was due to overland flow being generated more rapidly in those areas with higher percentages of pavement cover which caused water to flow onto areas where ponding had not occurred. This water (called runon, R_o) would then infiltrate and effectively reduce the time to ponding in the areas with reduced pavement cover. In terms of equation 9.1B the effect of R_o is to increase the value of P. One way around this problem is to express the maximum possible rate of infiltration in terms of soil moisture storage. Kirkby (1990a) effectively does this by using another version of the Green and Ampt (1911) equation given by:

$$i = i_c + \frac{a}{S} \qquad (9.3)$$

where a is a constant for any soil type $(L^2\,T^{-1})$ and S is a soil moisture storage term (L). The rate of delivery of water to the soil surface $(P + R_o)$ can then be compared with i. Where $P + R_o > i$ ponding will occur. S, of course, has to be updated within the model. This is done by using equation 9.3 within a time loop. For each iteration or time step (Δt) the algorithm or procedure would be as follows:

1. Calculate i (maximum or potential infiltration rate)
2. Is $P + R_o > i$?
   If YES
       Runoff volume $= (P + R_o - i)\Delta t$
       Actual infiltration $(AI) = i\,\Delta t$
   If NO
       Actual infiltration $(AI) = (P + R_o)\,\Delta t$
3. Update S
   $$S = S + AI$$
4. Update model's internal clock
   $$t = t + \Delta t$$
5. Simulation over?
   If YES then STOP.
   If NO then go to line 1.

Despite the obvious attraction of this storage-based approach, Kirkby (1990a) found that equation 9.3 did not provide particularly accurate predictions of infiltration on desert surfaces at Sede Boqer in the northern Negev of

Israel (Yair and Lavee 1985; Yair 1992) (see Figure 9.1).

In practice, equations 9.1, 9.2 and 9.3 require extensive calibration for different rainfall intensities, soil types and antecedent moisture conditions and often do not provide reliable estimates of infiltration into all but the most simple, homogenous soils. Although easy to use, their very simplicity means that they do not always describe the complex flow that can occur in arid zone soils. To account for the effects of vertical soil variability and antecedent moisture content on infiltration, some workers have used more complex models which describe

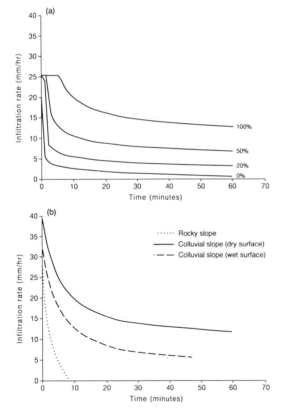

Figure 9.1 *(a) Simulated infiltration rates (using equation 9.3) for a rainfall of 26 mm h^{-1} into surfaces of different regolith cover. (b) Measured infiltration rates for the same rainfall intensity for different surfaces at Sede Boqer in the northern Negev of Israel (from Yair and Shachak 1987, cited in Kirkby 1990a). It can be seen that the simulations give initial infiltration rates that are too low, while the simulated rate after about 10 minutes is too high (after Kirkby 1990a)*

the drainage of water through the soil profile during and after rainfall, and the upward movement and subsequent loss of water from the profile caused by evaporation at the soil surface and transpiration losses from plants (Kirkby *et al.* 1996). These models are usually based on the unsaturated form of Darcy's empirical law for porous media flow, which, when combined with the continuity equation for vertical one-dimensional water movement in the soil profile, gives the well-known Richards equation (for a derivation see Dingman 1984, p. 314):

$$\frac{\partial \theta}{\partial t} = \frac{\partial}{\partial z}\left(K(\theta) + K(\theta)\,\frac{\partial \varphi(\theta)}{\partial z}\right) \quad (9.4)$$

where θ is moisture volume per unit volume of soil (L^3 L^{-3}, 1), t is time, z is vertical distance, K is hydraulic conductivity (L T^{-1}), and ψ is soil suction (L). In layered soils, $K(\theta)$ has to be defined for each layer. Because of its non-linearity, equation 9.4 is usually solved numerically.

Most soil infiltration models are one-dimensional. This is a reasonable approximation to the infiltration process where horizontal variability of the soil is significantly less than variability with depth. On many arid zone hillslope surfaces, however, this is not the case. It was noted above that changes in pavement cover can result in changes in infiltration capacity. Horizontal water movement from soils under fine-grained surface material and soils under pavement can be expected. In addition, there will often be significant horizontal movement of soil water from and immediately below erosional channels such as rills and, where surface sealing is effective, between vegetated and bare areas. Three-dimensional models require very detailed information on the spatial pattern of surface soil properties and are expensive in terms of their computational effort. As a way around this problem, Kirkby *et al.* (1996) divided infiltration into Mediterranean hillslopes into five representative domains. These were infiltration into: soil under trees, soil under shrubs, soil under perennial grasses, bare areas (where annual but not perennial grasses may be present) with no surface ponding, and bare areas under ponded water. Each domain is modelled using a vertical infiltration model. To allow for horizontal transfer of soil water there is a horizontal diffusion between domains based on the average distance between domains, which is computed as the area of the domains divided by their shared perimeter. The last two categories are highly dynamic in areal extent and can change during a rainfall event.

In equation 9.4 it is assumed that hydraulic conductivity is a function of soil moisture content alone and that this relationship is fixed for any soil type. In practice this is only approximately true and, as noted above, for many semi-arid and arid soils the hydraulic properties of the soil can change during the course of a rainfall event due to processes such as surface sealing and air encapsulation. Some infiltration models already take these factors into account (see Romkens *et al.* 1990). New models of infiltration into semi-arid soils will probably use mathematical descriptions of these dynamic processes in combination with the Richards equation. New models will also need to consider the effects of rock fragments on and within the soil.

SURFACE SEALING

Surface sealing causes a reduction in the infiltration capacity of the soil. Most seals form during and immediately after rainfall events. Bare soils, and therefore many arid zone soils, are particularly prone to the development of these inorganic or *rainbeat* crusts (see Chapter 5). Sealing is caused by the physical and chemical breakdown of soil aggregates by rainsplash and slaking, the washing of fines into larger pores (filtration), particle segregation, and compaction (Mualem *et al.* 1990; Romkens *et al.* 1990) (Figure 9.2). Sealing can also occur when fine particles in suspension in overland flow settle out on the soil surface to leave a thin relatively impermeable film. Mualem *et al.* (1990) term the former structural seals and the latter depositional seals. Other types of seals or crusts include those formed by precipitation of solutes from evaporating water at the soil surface and crusts formed by microscopic cryptogamous plants such as algae and mosses (Verrecchia *et al.* 1995). The term seal is usually used to describe the seal in its wet condition during a storm event. The dried seal is called a crust.

Descriptions of seal morphology vary. As discussed in Chapter 5, Mualem *et al.* (1990) have criticised simple descriptions of seal morphology, especially the presence of a thin clay skin. Seals are probably morphologically complex and are likely to include several of the elements of Figure 9.2.

Evidence for the effect of sealing on infiltration during rainfall has been presented by many authors. In laboratory simulations Poesen (1986) found that percolation of water from the base of soil trays subjected to artificial rainfall declined exponentially through time and attributed the decline entirely to surface seal development. It is important to remember that infiltration into soils in which sealing does not occur also often shows an exponential type decay over time (see equations 9.1 to 9.3). Such decay is caused by the soil wetting up and the reduction in the hydraulic gradient near the soil surface. Poesen (1986) effectively removed such effects by conducting his experiments on near-saturated soils and was, therefore, able to describe sealing as a dynamic process using a parameter called the sealing index, defined as the rate at which percolation from the base of the soil tray decreases over time. Poesen's (1986) study is important for two reasons. First, it demonstrates the effect of surface slope on sealing. Second, it shows the effects of rock fragments on seal formation. Poesen (1986) found that for the same soil type, sealing intensity decreased with increases in slope angle. He attributed this mainly to greater erosion on the steeper slope which prevented an effective seal developing (see Figure 9.3). The effect of rock fragments on seal formation depended on whether they were partially embedded in or merely placed on the soil surface, and also on soil type. On a silt soil, Poesen (1986) found that there was no significant difference in sealing intensity between experiments where artificial 'rock fragments' (glass marbles) were placed on the soil and where the 'fragments' were partially embedded in the soil. For a sandy soil, which was much more prone to sealing, there was a significant difference, with sealing intensity greater when rock fragments were embedded in the soil. Rock fragments on the soil surface have an 'overhang' which appears to protect some of the soil from raindrop impact and sealing. Embedded rock fragments do not afford such protection. Rock fragments have many other effects on infiltration and water

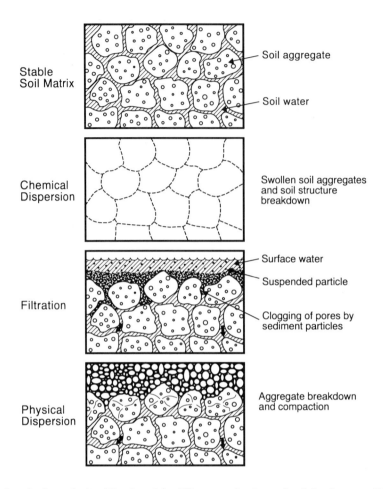

Stable
Soil Matrix

— Soil aggregate

— Soil water

Chemical
Dispersion

Swollen soil aggregates
and soil structure
breakdown

Filtration

— Surface water

— Suspended particle

Clogging of pores by
sediment particles

Physical
Dispersion

Aggregate breakdown
and compaction

Figure 9.2   *A schematic simplification of the different mechanisms of seal development (from Romkens* et al. *1990)*

movement on and in the soil. Some of these effects have been summarised by Poesen (1992) and some are detailed below.

ROCK FRAGMENTS

Many arid zone soils are characterised by rock fragments in their surface layers. For example, Poesen (1990 cited in Poesen 1992; see also Poesen and Lavee 1994) has estimated that such soils occupy 60% of the land area of the Mediterranean (see Figure 9.4). Reasons put forward for the preponderance of stones in surface horizons of arid soils include selective removal of fines by wind action and running water, and preferential upward movement of coarse particles through the soil (see Chapters 5 and 19). Recently, Wainwright *et al.* (1995) have speculated that such surfaces arise from removal of fines by rainsplash erosion in combination with overland flow. They note that, where shrub vegetation is present, desert pavement is often found in intershrub areas in association with mounds of fine-grained sediment around the bases of shrubs. Their model of rainsplash and overland flow processes was found successfully to predict observed mounds of fine material at Walnut Gulch in Arizona, but less successfully the development of pavement in intershrub areas. However, there is much debate over the development of stone pavements in drylands, with other explanations referred to in Chapters 5 and 19.

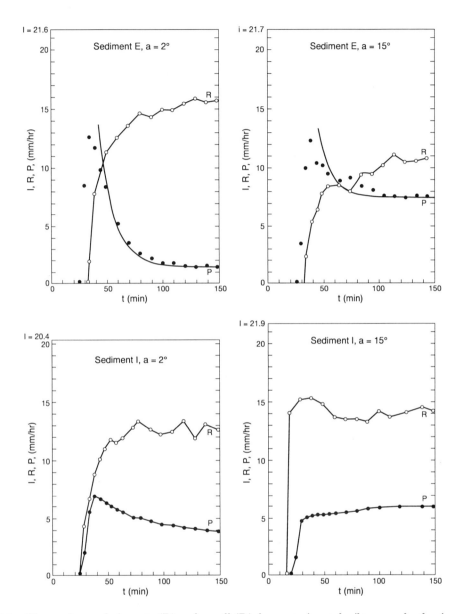

Figure 9.3 *Changes in percolation rate (P) and runoff (R) from experimental soil trays under the given rainfall intensities (I) (from Poesen 1986). Sediments E and I were a fine sand and a silt, respectively. It can be seen that surface sealing in the fine sand causes a dramatic reduction in the rate of percolation. This effect is less apparent in the silt. Increases in slope angle (a) reduce the effects of sealing, apparently because the surface is more readily eroded thus preventing seal formation. See text for more details.*

Coarse material at the soil surface can affect runoff generation in a number of ways. If the rock fragments are impermeable, the most obvious effect that might be expected is that infiltration will be reduced as rainwater runs off rock surfaces and is unable to rapidly penetrate the reduced area of finer permeable material between the rock fragments. Lavee and Poesen (1991) have shown that, although this is generally the case, the effects of stones on overland flow generation are more subtle. Using an experimental design similar to that of Poesen

Figure 9.4   *Rock fragments (pavement) and clump grasses on a semi-arid surface near Belmonte, Cuenca Province, Spain*

(1986) (see above), in which they placed artificial stones in the form of polystyrene blocks onto a loamy sand, they investigated the effect of stone size, stone spacing and stone position relative to the soil surface (that is either on the surface or partially embedded) on overland flow generation. They found that:

1. As stated above, a cover of stones tended to promote overland flow. However, small fragments, especially when the overall cover of stones was small, or when the stones were resting on the soil surface, produced less overland flow than bare soil.
2. The amount of overland flow was directly related to stone size but inversely related to the distance between stones, the latter, of course, decreasing with an increase in stone cover.
3. The amount of overland flow was always greater when stones were partially embedded in the soil surface rather than on the soil surface.

Lavee and Poesen (1991) conducted their experiments using air-dried soil samples and did not consider the effects of stones on the distribution of moisture within the upper soil horizons. The moisture content of surface soils will, of course, affect infiltration rates (see equations 9.1 to 9.3). The effects of stones on hydrological processes in and on the soil are illustrated in Figure 9.5 taken from Poesen and Lavee (1994). Whether soils below a stone cover are generally wetter or drier than soils without a stone cover is unclear. As shown in Figure 9.5, a layer of rock fragments will tend to act as a mulch, restricting evaporation losses from the soil. This will help keep a soil moist and encourage plant growth. By intercepting rainfall the plants will help protect the soil from sealing so that infiltration is increased and less overland flow is generated. Conversely, infiltration rates into moist soils are less than into dry soils so it could be the case that overland flow generation is increased. By reducing the amount of water that can infiltrate into a soil, stones may cause the soil to be generally drier than in their absence. These effects of stones on soil moisture conditions will also depend on time since rainfall and rates of potential evaporation. For example, given sufficient time a bare soil and a soil with a stone 'mulch' will probably dry out to similar moisture contents (Corey and Kemper 1968 cited in Poesen and Lavee 1994). The foregoing discussion illustrates some of the complex effects that rock fragments can have

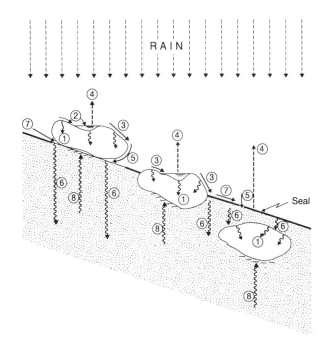

Figure 9.5 *Different hydrological processes occurring on and in soils with rock fragments at and near the soil surface. 1 = water absorption, 2 = interception and depression storage, 3 = rockflow, 4 = evaporation, 5 = infiltration, 6 = percolation, 7 = overland flow, 8 = capillary rise (from Poesen and Lavee 1994)*

on soil hydrology and overland flow generation. Rock fragments can also affect overland flow hydraulics. These effects are considered later in the chapter.

AIR ENTRAPMENT

During infiltration, soil air can become trapped in soil pores. Bond and Collis–George (1981) have used the term *air encapsulation* to describe this process. It appears that air remains in soils during infiltration because some pores fill before others causing air to become trapped in the pore space. This air will tend to block the pores to the passage of water and reduce the hydraulic conductivity and therefore the rate of infiltration. Since the early work of Christiansen (1944), many workers have investigated the effects of air encapsulation on the hydraulic conductivity of soils. In a study using field and laboratory infiltration experiments on sandy and gravelly loam soils, Constantz *et al.* (1988) found that air encapsulation reduced hydraulic conductivities to between 10 and 20% of the value of the saturated hydraulic conductivity. The volume of air encapsulated in these soils ranged between 4 and 19% of total pore

volume. The authors recommended that infiltration models which explicitly use values of hydraulic conductivity may be improved by using values of the order of 15 to 20% of the soil's saturated hydraulic conductivity. No work seems to have been done on air encapsulation in soils in the arid zone and it is difficult to speculate on the importance of the process during infiltration and runoff generation in these environments. However, in view of the apparent importance of the process in the experiments of Constantz *et al.* (1988) and other researchers such research could prove very useful.

VEGETATION EFFECTS

It seems that few attempts have been made to assess the importance of vegetation in affecting infiltration and surface runoff on arid hillslopes. Recently, Dunne *et al.* (1991) have concluded that vegetation may affect infiltration in two main ways. First, the vegetation may intercept rainfall energy and reduce the potential for surface sealing. Second, large pores formed from root holes and the spaces between large soil aggregates stabilised by relatively high levels of organic matter allow the escape of trapped air during infiltration and increase the bulk hydraulic conductivity of the soil. Infiltration can be expected to be higher under plants and this may explain why plants are often situated atop mounds of finer material. Erosive overland flow is more likely to be generated between plants leaving the plants 'stranded' on less eroded surfaces. A detailed account on vegetation and dryland geomorphology can be found in Chapter 7.

Overland flow hydraulics and erosion

If surface ponding occurs there will come a point when the surface detention store is exceeded and overland flow takes place. Overland flow is defined as a 'visible flow of water over the ground surface' (Kirkby 1985). A number of forces act on water flowing over hillslopes, the most important of which are the gravitational force which promotes flow, and forces which transmit the frictional effects of the slope surface into the flow. Of the latter, viscous forces are important in laminar flow while turbulent forces become more important at higher flow velocities. The relative magnitude of viscous and turbulent forces in a channel can be expressed using the Reynolds number (*Re*) given by:

$$Re \equiv \frac{vR}{v} \qquad (9.5)$$

where v is the velocity ($L\,T^{-1}$), R is the hydraulic radius (a measure of flow 'efficiency' given by ratio of the wetted cross-sectional area of the flow to the wetted perimeter — where the channel is very wide R is approximated by flow depth h) (L), and v is the kinematic viscosity ($L^2\,T^{-1}$). In classical hydraulics it is usually found that in flows where $Re < 500$ viscous forces dominate and laminar flow prevails, whereas for $Re > 2000$ fully turbulent flow occurs. For $500 < Re < 2000$ the flow is said to be transitional, exhibiting properties from both laminar flow and turbulent flow. In overland flow the picture is less clear because raindrop impact and a complex microtopography can induce turbulence even at relatively low Reynolds numbers. It is generally accepted that many overland flows are turbulent even when

Re values are well below 2000. Thus many authors use turbulent flow equations to describe overland flow. Two commonly used equations are the steady-state Chezy and Manning formulas,[1] both originally developed for river channels, and given, respectively, as:

$$v = C(RS)^{1/2} \qquad (9.6A)$$

and

$$v = \left(\frac{1}{n}\right) R^{2/3} S^{1/2} \qquad (9.6B)$$

where v is the flow velocity ($L\,T^{-1}$), R is the hydraulic radius, C is the Chezy roughness coefficient which accounts for frictional resistance to flow and is a measure of the conductance (the inverse of the resistance) of the surface or channel ($L^{1/2}\,T^{-1}$), S is the gradient of the hillslope, and n is the Manning roughness coefficient ($T\,L^{-1/3}$).

A similar equation that can be used to describe laminar and transitional as well as turbulent flows, and with the added advantage that the friction factor is dimensionless, is the Darcy–Weisbach equation given by:

$$v = \left(\frac{8gRS}{f}\right)^{1/2} \qquad (9.7)$$

where g is acceleration due to gravity ($L\,T^{-2}$) and f is the Darcy–Weisbach friction factor. Abrahams *et al.* (1986b) note that on many desert surfaces, flow even over small areas may be wholly laminar, wholly turbulent, wholly transitional or consist of patches of any of these three flow states. They term such a mixture of flow states *composite flow*. Where turbulent flow alone cannot be assumed, and a composite flow situation is expected, equation 9.7 will be more appropriate.

In models of overland flow, one of equations 9.6 to 9.7 or an equivalent formulation are often combined with the unsteady continuity equation (see Chow, 1959 pp. 525–526) to simulate changes in velocity and flow depth over time. Because this approach ignores the effect of changes in momentum it is called the kinematic approximation or kinematic wave

---

[1] The Chezy formula has a physical basis while that of Manning is purely empirical (see Chow 1959, pp. 89–127).

approach (Ponce *et al.* 1978; Ponce 1991). In essence this approach implies that unsteady overland flow can be considered as a succession of steady uniform flows, with the water surface slope remaining constant at all times. The success of the kinematic wave approach in simulating overland flow on a range of hillslope surfaces including those in the arid zone has been demonstrated and discussed, *inter alias*, by Henderson (1966), Woolhiser and Liggett (1967), Kibler and Woolhiser (1970), Foster and Meyer (1972), Woolhiser (1975), Ponce *et al.* (1978), Wu *et al.* (1978; 1982), Croley (1982), Rose *et al.* (1983a; 1983b; 1983c), Miller (1984), Cundy and Tento (1985), Yair and Lavee (1985), Scoging (1992) and Scoging *et al.* (1992).

A problem in the application of equations 9.6 to 9.7 is that the friction factor is not a constant and changes with discharge or the depth of flow. On smooth surfaces these changes can be readily defined and characterised (see Chow 1959, pp. 8–12 and Streeter and Wylie 1983, pp. 232–239). It is common to express the friction factor as a function of dimensionless discharge given by the Reynolds number (*Re*). The plot of the Darcy–Weisbach friction factor against *Re* on a log–log graph is called a Stanton diagram after Stanton and Pannell (1914 cited in Chow 1959). For flow over smooth surfaces and rougher surfaces in which all of the roughness elements are submerged the $f-Re$ relation shows two distinct linear relationships with f decreasing with *Re*. For laminar flow $f = 1000 \, Re^{-1.0}$ while for turbulent flow $f = 1000 \, Re^{-0.2}$ (Abrahams *et al.* 1989). For transitional flows the situation is less clear and in some experiments a positive relationship between f and *Re* has been reported (see Abrahams *et al.* 1989). In a series of papers starting in 1986, Athol Abrahams, Tony Parsons and their collaborators have shown that the $f-Re$ relation for desert surfaces is often more complex than the simple forms given above (Abrahams *et al.* 1986a; 1986b 1990; 1992; Parsons *et al.* 1990; Abrahams and Parsons 1994). In runoff plot experiments conducted at Walnut Gulch, Tombstone, Arizona, on gravelly fine loam soils, Abrahams *et al.* (1990) found two common forms of the $f-Re$ relationship. In some cases it was found that a simple linear relationship was present. In

others it was found that f tended to increase only to decrease later — a relationship termed convex upward by the authors. Stanton diagrams of these relationships are illustrated in Figure 9.6. Abrahams *et al.* (1990) note that the $f-Re$ relationships shown in Figure 9.6 bear no resemblance to the conventional $f-Re$ relation for shallow overland flow over a plane bed and suggest that the forms of $f-Re$ measured at Walnut Gulch are unlikely to be explained in terms of changes in the state of flow. For the convex upwards forms of $f-Re$ they attribute the initial rise in roughness to the successive inundation of roughness elements which causes the wetted upstream projected area of the elements to increase rapidly and so increase f. A point is reached when most of the elements are largely or wholly submerged, and any further increase in discharge is accompanied by a progressively slower rate of increase in wetted upstream projected area of the elements (Figure 9.7). The greater depth of flow outweighs the effect of the increase in wetted upstream projected area of the elements and f begins to decrease. Abrahams *et al.* (1986b) also note that f tends to decrease downslope for any given flow condition due to the tendency for flow to concentrate in microtopographic hollows. In other words, flows tend to become more efficient downslope.

Such a model of flow concentration has been proposed by Thornes *et al.* (1990) and Baird *et al.* (1992). In their RETIC model, Baird *et al.* (1992) propose that flow arriving at a slope cross-section will tend to occupy the deeper parts of the profile. They further suggest that friction is best described as a constant as long as the hydraulic radius of the various flow elements at each cross-section is properly estimated rather than being replaced by an average depth of flow. They present a diagram showing how flow efficiency can decrease and then increase again with successive inundation of a surface in a similar way to the upward convex Stanton diagrams of Abrahams *et al.* (1990). However, a limitation of their approach is that it ignores the plan configuration of microtopographic features which can also cause increases in roughness due to form resistance and wave resistance. Form resistance is exerted by stones that impede the flow and give rise to spatially varying downstream flow cross-sections. Wave

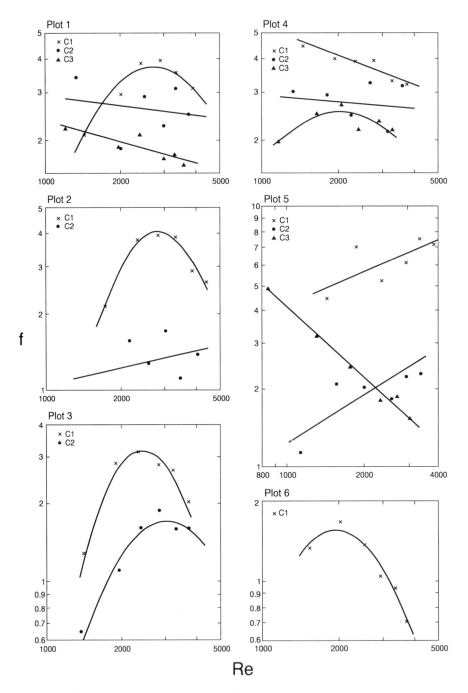

Figure 9.6   *Graphs of Darcy–Weisbach friction factor (f) against the Reynolds number (Re) for desert surfaces at Walnut Gulch, Tombstone, Arizona (from Abrahams et al. 1990). C refers to cross-sections within the experimental plots. See text for further details*

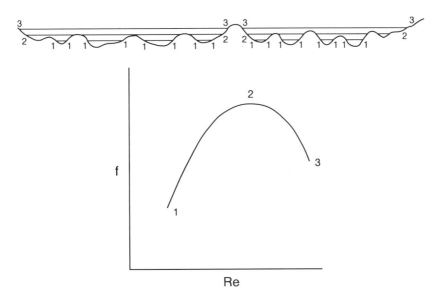

Figure 9.7 *Hypothetical ground surface on intermediate-sized roughness elements of varying height, showing how the progressive inundation of such a surface gives rise to a convex-upward relationship between the Darcy–Weisbach friction factor and the Reynolds number (see also Figure 9.6) (from Abrahams* et al. *1990)*

resistance occurs where stones disturb the free water surface and energy is dissipated in maintaining the uneven water surface. Abrahams and Parsons (1994) note that form resistance appears to be a dominant component of f in many overland flow situations. However, despite the relationship between f and Re, Abrahams *et al.* (1994b) have recently suggested that resistance to overland flow on many desert surfaces can effectively be treated as independent of flow state. In other words it can be assumed to be a constant for any particular surface. They present multiple regression models for calculating f and show that only a small percentage of the variance in f is explained by Re while much larger percentages are explained by the percentage cover of gravel on poorly vegetated pavement surfaces areas and by basal plant stem and litter cover in grassland areas.

SEDIMENT MOBILISATION

Virtually all overland flow erosion/deposition models describe surface lowering and aggradation using various forms of the sediment continuity equation which, for a laterally uniform hillslope, is given by (after Bennett 1974 and Croley 1982)

$$\frac{\partial (hc)}{\partial t} + \frac{\partial (hu_{\mathrm{p}})}{\partial x} + (1 - \lambda)\frac{\partial y}{\partial t} = 0 \quad (9.8)$$

where

$$(1 - \lambda)\frac{\partial y}{\partial t} = D - E$$

and where h is flow depth (L), c is the sediment concentration (L^3 L^{-3}), t is time, u_{p} is the average velocity of the sediment (L T^{-1}), x is distance positive in the direction of flow (L), λ is the porosity of the sediment (L^3 L^{-3}), D is the rate of sediment deposition (L T^{-1}), and E is the rate of sediment entrainment into the flow (L T^{-1}) (see Figure 9.8). In equation 9.8 a constant density of the sediment particles is assumed. It should be noted that entrainment is the rate of sediment temporarily leaving the ground surface and deposition is the actual rate of sediment reaching the ground surface (not the net amount which is the difference between the two). Thus if deposition (D) exceeds entrainment (E) (that is, $\partial y/\partial t$ is positive) *net* deposition takes place; if entrainment exceeds

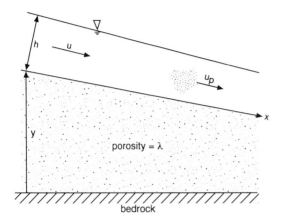

Figure 9.8 *Definition of flow transport system as considered in equation 9.8 (from Bennett 1974)*

deposition ($\partial y/\partial t$ is negative), then erosion takes place.

Equation 9.8 assumes that sediment is carried in suspension. Where the main mode of transport is at or near to the ground surface by rolling and saltation it is less appropriate to think of sediment concentration. Equation 9.8 might then be expressed as:

$$\frac{\partial Q_s}{\partial x} + (1 - \lambda)\frac{\partial y}{\partial t} = 0 \qquad (9.9)$$

where Q_s is the rate of sediment transport per unit width ($L^2\,T^{-1}$) and deposition (D) can be considered to have occurred when the sediment is temporarily at rest. Equation 9.9 might be more suitable as a description of sediment mobilisation where the size of the sediment grains is large and the supply of fine material is limited such as on desert pavement. For the purposes of the discussion that follows only equation 9.8 is considered.

In theory, providing E and D in equation 9.8 can be calculated for each flow condition, the amount of surface erosion or aggradation can be calculated. Deposition is fairly readily calculated as the balance between the downward movement of sediment, given by the product of sediment concentration and fall velocity, and the upward movement of sediment within the flow due to turbulent eddies. In contrast, it has proved very difficult accurately to calculate E on real hillslopes. This is because, as noted

above, flows on real hillslopes are complex and often consist of many threads of water flowing at different velocities and both entrainment and deposition are affected by the flow type and velocity. The complex form of flow on hillslopes also means that flow cannot be considered as having a constant cross-slope depth as assumed in equation 9.8 and many overland flow models. Figure 9.9 illustrates a typical rilled slope surface. During overland flow this surface will be occupied by deeper, faster-moving threads of flow in the microtopographic hollows and shallower, more slowly moving flows on the microtopographic highs. The problems of modelling erosion on real hillslopes as opposed to idealised laboratory surfaces are briefly discussed below.

A full account of the principles of entrainment would require a much more detailed treatment than is possible here. Useful discussions of entrainment can be found in Thornes (1979), Dingman (1984) and Allen (1985), and only a brief outline is given below. It is intuitive to expect entrainment to increase with velocity and to be affected by the density, size and cohesiveness of the surface sediments, and this is indeed the case. Entrainment can be considered as a function of the balance of forces acting on a sediment particle. Where the resisting forces are greater than the detaching forces, entrainment does not occur, while where the latter become greater, mobilisation occurs. Those forces promoting movement are fluid drag forces and lift forces due to Bernoulli lift and turbulence. Drag forces act parallel to the mean direction of flow, while lift forces act normally to the mean flow direction. Forces resisting movement are the submerged weight of the sediment and the shear strength of the sediment. Many workers express entrainment as a function of the shear stress acting on the slope surface which is given by:

$$\tau = \gamma h S \qquad (9.10)$$

where γ is the unit weight of water ($M\,L^{-2}\,T^{-2}$), h is flow depth (L) and S is the gradient of the hillslope. The relationship between velocity and shear stress are obvious in that both are dependent on h and S. Implicit in the force balance approach is that a critical shear stress is required to mobilise sediment so that detachment is proportional to the

Figure 9.9 *Rills on a small landslide, near Alora, Malaga Province, Spain. The central rill is approximately 15 cm wide and 20 cm deep at its base*

difference between actual shear stress and the critical shear stress. Croley (1982) has argued that a critical shear stress is not required for entrainment of fine-grained non-cohesive sediments and suggests that, even under small velocities, entrainment occurs. Most workers, however, assume that a threshold for detachment is required.

A problem in many studies is that a mean value of flow depth is used in calculations of shear stress. Where a variation of flow depths occurs, flow detachment may be underestimated if the mean value is used. Using a simple entrainment equation,

$$E = a(\tau - \tau_c)_b \qquad (9.11)$$

where τ_c is the critical shear stress and a and b are system defining parameters, together with detailed measurements of the distribution of flow depths on experimental hillslope plots at Walnut Gulch in Arizona, Abrahams *et al.* (1989) showed that the ratio of entrainment calculated using the distribution of depths (E') and entrainment calculated using a mean flow depth (E_m) varied between 1 and 6 depending on the value of the exponent b. Even where $E'/E_m = 1$, use of a mean flow depth will not describe the spatial distribution of entrainment. It will therefore fail to predict the concentration of erosion in the flow-wetted area, the microtopographic hollows, which will deepen further and become more efficient conduits for overland flow. Models that use a value of mean depth of flow are, therefore, likely to be in error when used for prediction of amounts of erosion due to overland flow even though they may be acceptable as purely hydrologic models.

Sediment mobilisation by raindrop impact

A great deal of the sediment transported by overland flow has been detached by the action of raindrops striking the ground surface. Sediment is also transported directly by raindrop action. Abrahams *et al.* (1994a) distinguish between two types of transport: rainsplash and rain dislodgement. In the former, dislodged particles are transported by splash droplets, while in the latter they are not. Statham (1977) suggests that raindrop impact moves sediment by a combination of rebound, pushing and undermining, while Torri (1987) contends that the detaching forces from an impacting raindrop are essentially the same as those occurring in overland flow. He notes that when a drop hits the ground surface it initially resists a change in shape after which radial flow away from the point of impact begins. This radial flow is characterised by very high velocities which may be up to an order of magnitude greater than the fall velocity of the drop. These high velocities give rise to high shear stresses which cause detachment so that the amount of detachment can be determined by the force balance approach mentioned earlier. Whatever the exact mechanisms, many empirical studies have shown that raindrop detachment can be related to the kinetic energy of the raindrop E_k:

$$E_k = \tfrac{1}{2} m v^2 \qquad (9.12)$$

where m is raindrop mass (M) and v velocity (LT^{-1}) (see for example, Quansah 1981; Poesen 1985). As noted by Hillel (1980), the total action of raindrops during a storm might be expected to be a function of the sum of their kinetic energies:

$$E_{k,total} = \tfrac{1}{2} \sum m_i v_i^2 \qquad (9.13)$$

where m_i and v_i are, respectively, the masses and velocities of successive drop size groups. On level surfaces there is no net movement of rain-splashed material whereas on a slope the splashes move the sediment downhill. Part of this net downslope movement is caused by the fact that, for the same trajectories, soil splashed in the downslope direction travels further in the air than that splashed uphill as shown in

Figure 9.10. Thornes (1979) notes that on a 10% slope the downhill movement is about three times that upslope.

Raindrop action is affected by a number of factors in addition to slope. For example, vegetation can intercept rainfall and thus reduce the number of drops reaching the ground surface. Rainfall that runs off vegetation, intercepted throughfall, will have a quite different drop size distribution and erosivity from that of uninterrupted rainfall. Erosivity decreases with vegetation cover but is also affected by the height of the vegetation so that the higher the vegetation the greater the velocity of the drop up to a limiting value, the fall velocity. This effect of vegetation has been modelled by Brandt (1990). Despite the low density of plants, the effect of vegetation on rainsplash in the arid zone has been observed by a number of workers. Wainwright *et al.* (1995), for example, note that desert pavement does not appear to form under shrubs and attribute this to the protective effect of vegetation against rain detachment combined with higher rates of infiltration under plants so that overland flow and sediment transport are reduced (see also Carson and Kirkby, 1972, p. 189 and Williams 1969, cited in Carson and Kirkby 1972).

The texture of the surface sediment also has an effect on rates of detachment and transport by raindrops. Coarser particles are more difficult to move because of their larger mass. Wainwright *et al.* (1995) note that both theoretical and empirical results from a number of studies suggest that detachment of particles with diameters greater than 12 to 20 mm is effectively zero. Finer particles subject to

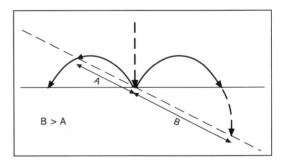

Figure 9.10 *Sketch showing how the same trajectories from an impacting raindrop will transport more sediment downslope than upslope*

cohesive forces can also effectively resist impact. Wainwright *et al.* (1995) use a model of differential splash detachment and transport to show how desert pavement can form under the action of raindrops.

Finally, the effectiveness of detachment and transport under raindrop impact depends on the depth of ponded or flowing water at the soil surface. Water at the surface tends to dissipate the impact energy of the raindrop and Mutchler and Young (1975) have suggested that a depth of water at the ground surface equal to three drop diameters protects the soil from raindrop impact. However, the impact of raindrops can enhance the transporting capacity of flowing water by inducing turbulence so that their erosive potential is greatest where a thin film of water is present. This point has been recently made by Parsons *et al.* (1994a) who found from rainfall simulation plot experiments that erosion was not closely related to raindrop detachment but depended critically on sediment supply and the presence of overland flow competent enough to carry sediment.

Conclusion

From the foregoing discussion it is clear that overland flow generation and sediment mobilisation in the arid zone are often complex and cannot be readily described using simple models. Part of this complexity of response arises from the unusual nature of arid soils, many of which have high concentrations of rock fragments near and at the ground surface. These fragments have a profound effect on infiltration and, through their control on surface microtopography, overland flow. Arid surfaces also tend to display extreme spatial variability over short distances (often of the order of centimetres) so that models that assume, for example, that overland flow depths and surface shear stresses can be described using a single mean value are likely to be in error when used to calculate erosion rates. Future attempts to improve understanding of overland flow generation and sediment mobilisation in the arid zone will need to take account of such variability. In short, improved understanding of overland flow generation and sediment mobilisation by water will require

much more fieldwork such as that carried out at Walnut Gulch in Arizona, but also more imaginative attempts at modelling these processes, perhaps by describing overland flow and sediment movement as a series of small-scale and partly independent fluxes which combine to produce an overall discharge of water and sediment from the hillslope.

References

Abrahams, A.D. and Parsons, A.J. 1994. Hydraulics of interrill overland flow on stone-covered desert surfaces. *Catena*, 23: 111–140.

Abrahams, A.D., Parsons, A.J. and Luk, S.-H. 1986a. Field measurements of the velocity of overland flow using dye tracing. *Earth Surface Processes and Landforms*, 11: 653–667.

Abrahams, A.D., Parsons, A.J. and Luk, S.-H. 1986b. Resistance to overland flow on desert hillslopes. *Journal of Hydrology*, 88: 343–363.

Abrahams, A.D., Parsons, A.J. and Luk, S.-H. 1989. Distribution of depth of overland flow on desert hillslopes and its implications for modeling soil erosion. *Journal of Hydrology*, 106: 177–184.

Abrahams, A.D., Parsons, A.J. and Luk, S.-H. 1990. Field experiments on the resistance to overland flow on desert hillslopes. In D.E. Walling, A. Yair and S. Berkowicz (eds), *Erosion, transport and depositional processes*. International Association of Hydrological Sciences Publication 189, Wallingford, UK.

Abrahams, A.D., Parsons, A.J. and Hirsch, P.J. 1992. Field and laboratory studies of resistance to interrill overland flow on semi-arid hillslopes, southern Arizona. In A.J. Parsons and A.D. Abrahams (eds), *Overland flow: hydraulics and erosion mechanics*. UCL Press, London.

Abrahams, A.D., Howard, A.D. and Parsons, A.J. 1994a. Rock-mantled slopes. In A.D. Abrahams and A.J. Parsons (eds), *Geomorphology of desert environments*. Chapman & Hall, London.

Abrahams, A.D., Parsons, A.J. and Wainwright, J. 1994b. Resistance to overland flow on semiarid grassland and shrubland hillslopes, Walnut Gulch, southern Arizona. *Journal of Hydrology*, 156: 431–446.

Allen, J.R.L. 1985. *Principles of physical sedimentology*. Allen & Unwin, London.

Baird, A.J., Thornes, J.B. and Watts, G.P. 1992. Extending overland flow models to problems of slope evolution and the representation of complex slope-surface topographies. In A.J. Parsons and A.D. Abrahams (eds), *Overland flow: hydraulics and erosion mechanics*. UCL Press, London.

Bennett, J.P. 1974. Concepts of mathematical modeling of sediment yield. *Water Resources Research*, 10: 485–492.

Bond, W.J. and Collis-George, N. 1981. Ponded infiltration into simple soil systems. 1. The saturation and transmission zones in the moisture content profile. *Soil Science*, 131: 202–209.

Brandt, C.J. 1990. Simulation of the size distribution and erosivity of raindrops and throughfall drops. *Earth Surface Processes and Landforms*, 15: 687–698.

Bryan, R.B. and Yair, A. (eds) 1982. *Badland geomorphology and piping*. GeoBooks, Norwich.

Carson, M.A. and Kirkby, M.J. 1972. *Hillslope form and process*. Cambridge University Press, Cambridge.

Chow, V.T. 1959. *Open-channel hydraulics*. McGraw-Hill, London.

Christiansen, J.E. 1944. Effect of trapped air upon the permeability of soils. *Soil Science*, 58: 355–365.

Constantz, J., Herkelrath, W.N. and Murphy, F. 1988. Air encapsulation during infiltration. *Journal of the Soil Science Society of America*, 52: 10–16.

Corey, A.T. and Kemper, W.D. 1968. *Conservation of soil water by gravel mulches*, Hydrology Paper 30, Colorado State University, Fort Collins.

Croley, T.E. 1982. Unsteady overland sedimentation. *Journal of Hydrology*, 56: 325–346.

Cundy, T.W. and Tento, S.W. 1985. Solution to the kinematic wave approach to overland flow routing with rainfall excess given by Philip's equation. *Water Resources Research*, 21: 1132–1140.

Dingman, S.L. 1984. *Fluvial hydrology*. Freeman, New York.

Dunne, T., Zhang, W. and Aubry, B.F. 1991. Effects of rainfall, vegetation, and microtopography on infiltration and runoff. *Water Resources Research*, 27: 2271–2285.

Emmett, W.M. 1970. *The hydraulics of overland flow*. US Geological Survey Professional Paper 662-A.

Foster, G.R. and Meyer, L.D. 1972. A closed-form soil erosion equation for upland areas. In H.W. Shen (ed.), *Sedimentation*. Colorado State University, Fort Collins.

Foster, G.R. and Meyer, L.D. 1975. Mathematical simulation of upland erosion by fundamental erosion mechanics. In *Present and prospective technology for predicting sediment yield and sources*. Agricultural Research Service Paper S-40, United States Department of Agriculture: 190–207.

Green, W.H. and Ampt, G.A. 1911. Studies on soil physics. 1. Flow of air and water through soils. *Journal of Agricultural Science*, 4: 1–24.

Henderson, F.M. 1966. *Open channel flow*. Macmillan, New York.

Hillel, D. 1980. *Fundamentals of soil physics*. Academic Press, New York.

Horton, R.E. 1945. Erosional development of streams and their drainage basins: hydrophysical approach to quantitative morphology. *Bulletin of the Geological Society of America*, 56: 275–370.

Kibler, D.F. and Woolhiser, D.A. 1970. *The kinematic cascade as a hydrologic model*. Colorado State University Hydrology Paper 39, Colorado State University, Fort Collins.

Kirkby, M.J. 1985. Overland flow. In A. Goudie (ed.), *The encyclopaedic dictionary of physical geography*. Blackwell, Oxford.

Kirkby, M. 1990a. A simulation model for desert runoff and erosion. In D.E. Walling, A. Yair and S. Berkowicz (eds), *Erosion, transport and deposition process*. International Association of Hydrological Sciences Publication 189, Wallingford, UK.

Kirkby, M.J. 1990b. A one-dimensional model for rill–interrill interactions. In R.B. Bryan (ed.), *Soil erosion: experiments and models*. Catena Supplement 17, Catena Verlag, Cremlingen.

Kirkby, M., Baird, A.J., Diamond, S.M., Lockwood, J.G., MacMahon, M.D., Mitchell, P.L., Shao, J., Sheehy, J.E., Thornes, J.B. and Woodward, F.I. 1996. The MEDALUS slope catena model: a physically-based process model for hydrology, ecology and land degradation interactions. In C.J. Brandt and J.B. Thornes (eds), *Mediterranean desertification and land use*. Wiley, Chichester.

Lavee, H. and Poesen, J.W.A. 1991. Overland flow generation and continuity on stone-covered soil surfaces. *Hydrological processes*, 5: 345–360.

Marshall, T.J. and Holmes, J.W. 1989. *Soil physics*. 2nd edn. Cambridge University Press, Cambridge.

Miller, J.E. 1984. *Basic concepts of kinematic wave models*. US Geological Survey Professional Paper 1302.

Mosley, M.P. 1973. Rainsplash and the convexity of badland divides. *Zeitschrift für Geomorphologie*, Supplementband, 18: 10–25.

Mualem, Y., Assouline, S. and Rohdenburg, H. 1990. Rainfall induced soil seal. A: A critical review of observations and models. *Catena*, 17: 185–203.

Mutchler, C.K. and Young, R.A. 1975. Soil detachment by raindrops. *United States Department of Agriculture Report ARS-S-40*, 113–117.

Parsons, A.J., Abrahams, A.D. and Luk, S.-H. 1990. Hydraulics of interrill overland flow on a semi-arid hillslope, southern Arizona. *Journal of Hydrology*, 117: 255–273.

Parsons, A.J., Abrahams, A.D. and Wainwright, J. 1994a. Rainsplash and erosion rates in an interrill area on semi-arid grassland, Southern Arizona. *Catena*, 22: 215–226.

Parsons, A.J., Abrahams, A.D. and Wainwright, J. 1994b. On determining resistance to interrill overland flow. *Water Resources Research*, 30: 3515–3521.

Philip, J.R. 1957. The theory of infiltration. 4. Sorptivity and algebraic infiltration equations. *Soil Science*, 84: 257–264.

Poesen, J. 1985. An improved splash transport model. *Zeitschrift für Geomorphologie*, 29: 193–211.

Poesen, J. 1986. Surface sealing as influenced by slope angle and position of simulated stones in the top layer of loose sediments. *Earth Surface Processes and Landforms*, 11: 1–10.

Poesen, J. 1990. Erosion process research in relation to soil erodibility and some implications for improving soil quality. In J. Albaladejo, M.A. Stocking and E. Diaz (eds), *Soil degradation and regeneration in Mediterranean environmental conditions*. Consejo Superior de Investigaciones Cientificas, Madrid.

Poesen, J. 1992. Mechanisms of overland-flow generation and sediment production on loamy and sandy soils with and without rock fragments. In A.J. Parsons and A.D. Abrahams (eds), *Overland flow: hydraulics and erosion mechanics*. UCL Press, London.

Poesen, J. and Lavee, H. 1994. Rock fragments in top soils: significance and processes. *Catena*, 23: 1–28.

Ponce, V.M. 1991. The kinematic wave controversy. *Journal of Hydraulic Engineering, American Society of Civil Engineers*, 117: 511–525.

Ponce, V.M., Li, R.M. and Simons, D.B. 1978. Applicability of kinematic and diffusion models. *Journal of the Hydraulics Division, ASCE*, 104: 353–360.

Quansah, C. 1981. The effect of soil type, slope, rain intensity and their interactions on splash detachment and transport. *Journal of Soil Science*, 32: 215–224.

Romkens, M.J.M., Prasad, S.N. and Whisler, F.D. 1990. Surface sealing and infiltration. In M.G. Anderson and T.P. Burt (eds), *Process studies in hillslope hydrology*. Wiley, Chichester.

Rose, C.W., Parlange, J.-Y., Sander, G.L., Campbell, S.Y. and Barry, D.A. 1983a. Kinematic flow approximation to runoff on a plane: an approximate analytic solution. *Journal of Hydrology*, 62: 363–369.

Rose, C.W., Williams, R., Sander, G.L. and Barry, D.A. 1983b. A mathematical model of soil erosion and deposition processes. 1. Theory for a plane land element. *Journal of the Soil Science Society of America*, 47: 991–995.

Rose, C.W., Williams, R., Sander, G.L. and Barry, D.A. 1983c. A mathematical model of soil erosion and deposition processes. 2. Application to data from an arid-zone catchment. *Journal of the Soil Science Society of America*, 47: 996–1000.

Scoging, H. 1989. Runoff generation and sediment mobilisation by water. In D.S.G. Thomas (ed.), *Arid zone geomorphology*. Belhaven Press, London.

Scoging, H. 1992. Modelling overland-flow hydrology for dynamic hydraulics. In A.J. Parsons and A.D. Abrahams (eds), *Overland flow: hydraulics and erosion mechanics*. UCL Press, London.

Scoging, H., Parsons, A.J. and Abrahams, A.D. 1992. Application of a dynamic overland-flow model to a semi-arid hillslope, Walnut Gulch, Arizona. In A.J. Parsons, and A.D. Abrahams (eds), *Overland flow: hydraulics and erosion mechanics*. UCL Press, London.

Stanton, T.E. and Pannell, J.R. 1914. Similarity of motion in relation to surface friction of fluids. *Philosophical Transactions of the Royal Society of London*, 214A: 199–224.

Statham, I. 1977. *Earth surface sediment transport*. Clarendon Press, Oxford.

Streeter, V.L. and Wylie, E.B. 1983. *Fluid mechanics*. First SI Metric edition, international student edition. McGraw-Hill, London.

Swartzendruber, D. 1993. Revised attribution of the power form infiltration equation. *Water Resources Research*, 29: 2455–2456.

Thornes, J.B. 1979. Fluvial processes. In C. Embleton and J.B. Thornes (eds), *Process in geomorphology*. Edward Arnold, London.

Thornes, J.B. 1994. Catchment and channel hydrology. In A.D. Abrahams and A.J. Parsons (eds), *Geomorphology of desert environments*. Chapman & Hall, London.

Thornes, J.B., Francis, C.F., Lopez-Bermudez, F. and Romero-Diaz, A. 1990. Reticular overland flow with coarse particles and vegetation roughness under Mediterranean conditions. In J.L. Rubio and J. Rickson (eds), *Strategies to control desertification in Mediterranean Europe*. Commission of the European Community, Brussels.

Torri, D. 1987. A theoretical study of soil detachability. In F. Ahnert (ed.), *Geomorphological models: theoretical and empirical aspects*. Catena Supplement 10, Catena Verlag, Cremlingen.

Verrecchia, E., Yair, A., Kidron, G.J. and Verrecchia, K. 1995. Physical properties of the psammophile cryptogamic crust and their consequences to the water regime of sandy soils, northwestern Negev Desert, Israel. *Journal of Arid Environments*, 29: 427–437.

Wainwright, J., Parsons, A.J. and Abrahams, A.D. 1995. A simulation study of the role of raindrop erosion in the formation of desert pavements. *Earth Surface Processes and Landforms*, 20: 277–291.

Watts, G.P. 1989. Modelling the sub-surface hydrology of semi-arid agricultural terraces. Unpublished PhD thesis, University of Bristol.

Williams, M.A.J. 1969. Prediction of rainsplash erosion in the seasonally wet tropics. *Nature*, 222: 763–765.

Woolhiser, D.A. 1975. Simulation of unsteady flows. In K. Mahmood and V. Yevyevich (eds), *Unsteady flow in open channels*. Water Resources Publications, Fort Collins.

Woolhiser, D.A. and Liggett, J.A. 1967. Unsteady, one-dimensional flow over a plane — the rising hydrograph. *Water Resources Research*, 3: 753–771.

Wu, Y.H., Yevjevich, V. and Woolhiser, D.A. 1978. *Effects of surface roughness and its spatial distribution on runoff hydrographs*. Hydrology Paper 96, Colorado State University, Fort Collins.

Wu, Y.H., Woolhiser, D.A. and Yevjevich, V. 1982. Effects of spatial variability of hydraulic resistance on runoff hydrographs. *Journal of Hydrology*, 56: 231–248.

Yair, A. 1992. The control of headwater area on channel runoff in a small arid watershed. In A.J. Parsons and A.D. Abrahams (eds), *Overland flow: hydraulics and erosion mechanics*. UCL Press, London.

Yair, A. and Lavee, H. 1985. Runoff generation in arid and semi-arid zones. In M.G. Anderson and T.P. Burt (eds), *Hydrological forecasting*. Wiley, Chichester.

Yair, A. and Shachak, M. 1987. Studies in watershed ecology of an arid area. In L. Berkofsky and M.G. Wurtele (eds), *Progress in desert research*. Rowman and Littlefield, Totowa, NJ.

10

Distinctiveness, diversity and uniqueness in arid zone river systems

David Knighton and Gerald Nanson

Rivers and their associated channels exist to convey the water and sediment delivered to them from the drainage basin. However, the supply, transmission and outflow of that water and sediment vary over a range of scales, between individual rivers depending on their particular physiographic characteristics and between groups of rivers defined according to various criteria. One of those criteria is climate.

In reality there is a continuum of conditions from hyper-arid to very humid, but it has become convenient to regard arid and humid rivers as distinct groups. The extent of that distinctiveness in terms of basin hydrology and channel morphology is the initial concern here. Most hydrological research has been carried out in humid regions but it has been argued that the results obtained therefrom should not be extrapolated to the arid zone (McMahon 1979; Finlayson and McMahon 1988). However, the hyper-arid to semi-arid zones cover well over 30% of the Earth's surface and their fluvial environments could show as much internal diversity as external differentiation. Certainly hydrological conditions are far from uniform. Of particular interest is the position of Australia's arid zone rivers, and especially Cooper Creek. Cooper Creek is one of the main contributors to the Lake Eyre Basin which, at

1.3×10^6 km^2, is the largest internally draining catchment area in the world. Thus the key themes of distinctiveness, diversity and uniqueness are explored in this chapter, while Chapter 11 goes on to examine the behaviour of dryland fluvial systems and the nature of the resultant channel deposition.

Distinctiveness: arid–humid contrasts

Based on the UNESCO (1979) classification, nearly half of the world's countries face problems of aridity. There is, therefore, a need for improved understanding of hydrological processes in arid areas and for the development of appropriate modelling strategies. The lack of good quality data is a major stumbling-block to that end, related partly to the problems of establishing the requisite monitoring networks over large areas where rainfall and runoff are highly variable.

Despite problems of data acquisition, the distinctive features of arid zone hydrology are reasonably well established (Figure 10.1). Most of the world's drylands are located beneath high-pressure cells which inhibit the penetration of rain-bearing low-pressure systems. Annual precipitation amounts are not only low but also

Arid Zone Geomorphology: Process, Form and Change in Drylands, 2nd edition. Edited by David S. G. Thomas.
© 1997 John Wiley & Sons Ltd.

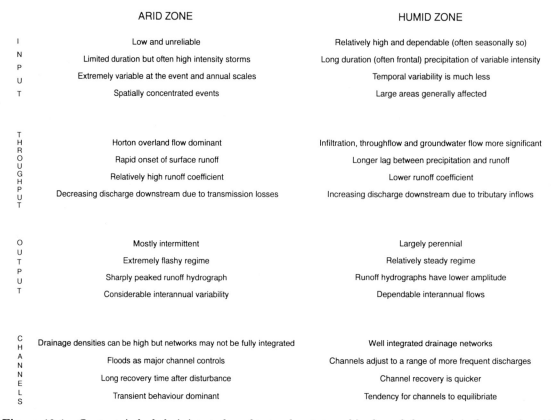

ARID ZONE | HUMID ZONE

INPUT

ARID ZONE	HUMID ZONE
Low and unreliable	Relatively high and dependable (often seasonally so)
Limited duration but often high intensity storms	Long duration (often frontal) precipitation of variable intensity
Extremely variable at the event and annual scales	Temporal variability is much less
Spatially concentrated events	Large areas generally affected

THROUGHPUT

ARID ZONE	HUMID ZONE
Horton overland flow dominant	Infiltration, throughflow and groundwater flow more significant
Rapid onset of surface runoff	Longer lag between precipitation and runoff
Relatively high runoff coefficient	Lower runoff coefficient
Decreasing discharge downstream due to transmission losses	Increasing discharge downstream due to tributary inflows

OUTPUT

ARID ZONE	HUMID ZONE
Mostly intermittent	Largely perennial
Extremely flashy regime	Relatively steady regime
Sharply peaked runoff hydrograph	Runoff hydrographs have lower amplitude
Considerable interannual variability	Dependable interannual flows

CHANNELS

ARID ZONE	HUMID ZONE
Drainage densities can be high but networks may not be fully integrated	Well integrated drainage networks
Floods as major channel controls	Channels adjust to a range of more frequent discharges
Long recovery time after disturbance	Channel recovery is quicker
Transient behaviour dominant	Tendency for channels to equilibriate

Figure 10.1 *Contrasts in hydrologic input, throughput and output, and in channel characteristics between the arid zone and the humid zone*

extremely variable, becoming more so as the degree of aridity increases (Bell 1979; see Chapter 1). On the northwest coast of Australia with an average annual rainfall of 250 mm, the annual totals in four consecutive years were 570, 70, 680 and 55 mm (Pilgrim *et al.* 1988). Long dry spells are not uncommon, which act to reduce the vegetation cover and promote the erosional effectiveness of rainfall when it does occur, often at high intensity. Coupled with greater temporal variability is greater spatial variability. Discrete convective storms, which are an important source of precipitation in many arid environments and contribute to the spottiness of dryland rainfall referred to in Chapter 1, generally affect an area of < 100 km², with the result that only parts of a drainage system may be activated during an individual event. In large catchments, therefore, different parts may contribute runoff from one event to another, increasing the

unpredictability of hydrologic output (see Chapter 11).

Runoff generation involves a different mix of processes from humid areas with familiar processes such as throughflow and groundwater flow making minimal contributions. Hortonian overland flow predominates (see Chapter 9) which, combined with a sparse plant cover and a relative absence of surface organic matter, leads to the rapid onset of runoff from small rainfall depths. This point is illustrated with data from New South Wales where runoff initiation requires ≥16 mm of rain in the arid west but ≥35 mm in the humid east (Cordery *et al.* 1983). The soil surface is much more likely to receive the direct impact of rainfall in arid environments so that soil type and soil surface properties play a more important role in runoff production.

The disposition of stones, vegetation and other forms of roughness can strongly influence overland flow characteristics (see Chapters 5 and 9; Dunne *et al.* 1991). Because of lower infiltration capacities and generally higher rainfall intensities, runoff coefficients (which express the proportion of rainfall converted to runoff) tend to be higher in the arid zone but, except in headwater areas, rainfall receipt and downstream runoff correlate poorly on the whole, due largely to transmission losses. Dryland channels characteristically experience considerable water loss through evaporation and seepage into the channel boundary (Table 10.1). Not only does this pose a serious modelling problem but, in the absence of appreciable tributary inflow, it produces a significant downstream decrease in flood peaks and total runoff volumes, which has major implications for sediment transport and channel morphology. The issue of transmission losses is explored further in Chapter 11.

Streamflow hydrographs tend to rise more steeply, have a sharper peak and recede more quickly in arid zone than humid zone rivers, as befits a regime in which flood events of irregular frequency and short duration dominate.

Regional flood frequency curves are usually very steep (Figure 10.2), underlining the large increase in relative flood magnitude characteristic of such rivers. The 12 largest floods ever recorded in the United States have all occurred in semi-arid or arid areas (Costa 1987). However, areas of low precipitation have, by their very nature, low runoff amounts and as that runoff decreases so it becomes more variable. In Australia and North America mean annual runoff is about twice as variable in the arid zones as in the continental areas as a whole (McMahon 1979), although the association between climatic aridity and runoff variability is not as straightforward as once thought (McMahon *et al.* 1992). Furthermore, the first-order serial correlation between annual flows is very low, suggesting that carry-over storage from one year to the next is insignificant in arid basins. Consequently, the potential for flow regulation is much reduced. The diversion of water for irrigation and other supply purposes on top of the natural depletion of discharge through seepage and evaporation can have deleterious effects on the downstream river, as

Table 10.1   *Transmission losses downstream*

References	River	Transmission loss		Comments
		Percentage decrease	Rate	
		Runoff volume / Peak discharge		
Renard and Keppel (1966)	Walnut Gulch, Arizona ($A = 150$ km^2)	57 / 63	0.97×10^3 m^3 km^{-1} / 0.49 m^3 s^{-1} km^{-1}	Single flood event; $L = 54$ km
Walters (1990)	Queen Creek, Arizona	38	62.4×10^3 m^3 km^{-1}	Average of 10 events; $L = 32$ km
	Habawnah Saudi Arabia	50	39.9×10^3 m^3 km^{-1}	Average of 4 events; $L = 20$ km
	Yiba Saudi Arabia	63	19.2×10^3 m^3 km^{-1}	Average of 6 events; $L = 33$ km
Dunkerley (1992)	Arid New South Wales: Fowlers Creek ($A = 434$ km^2) Sandy Creek ($A = 60$ km^2)		17 m^3 s^{-1} km^{-1} 19.8 m^3 s^{-1} km^{-1}	
Sharma *et al* (1994)	Luni basin, NW India ($A = 34\,866$ km^2)	8–56 / 41–89	38×10^3 m^3 km^{-1} (rocky terrain; $L = 73$ km) *to* 6500×10^3 m^3 km^{-1} (deep alluvium; $L = 34$ km)	Range in 15 reaches

Symbols: A = drainage area; L = distance between upstream and downstream gauges.

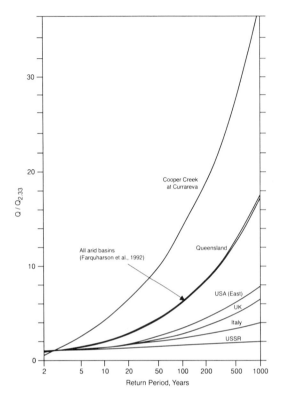

Figure 10.2 *Regional flood frequency curves, showing relative flood magnitude plotted against return period (based partly on Farquharson et al. 1992 and Lewin 1989)*

the situation in the Aral Sea Basin illustrates only too well (see page 645).

Hydrologically, therefore, arid zone rivers are often in a delicate state of balance which can easily be upset. Geomorphologically, also, arid and particularly semi-arid systems are very sensitive to disturbance if the steepness of sediment yield curves in the relevant precipitation range is regarded as a reliable indicator (Figure 10.3). Certainly a rapid initial rise and high peak values of suspended sediment concentration are distinctive features of arid rivers, largely because of the rapid mobilisation of readily available material at the onset of high flows. There is also evidence to suggest that ephemeral rivers are many times more efficient as transporters of coarse bedload material than their perennial counterparts (Laronne and Reid 1993; Figure 10.4). The large channel widths commonly associated with arid zone rivers, at

least in small- to moderate-sized basins (Wolman and Gerson 1978; Figure 10.5), may well be related to that greater transporting ability in conjunction with the highly skewed flow regime.

High-magnitude, low-frequency floods appear to control channel development in dryland rivers (Baker 1977; Graf 1983; 1988), which contrasts with the dominant influence of intermediate events in humid areas (Wolman and Gerson 1978). Along the Gila River in Arizona a series of large floods in the period 1905–17, the largest of which probably had a recurrence interval of 300 years, caused a 600% increase in channel width as most of the floodplain vegetation was destroyed (Burkham 1972; Figure 10.6). Recovery occurred during the ensuing 50 years as vegetation re-established itself, but in areas even drier than semi-arid Arizona recovery from flood damage may be much longer delayed in the absence of sediment-trapping vegetation or restorative flows. Consequently, the effects of high-magnitude events tend to be preserved for much longer periods in the arid zone. The sporadic action of runoff and the long recovery times characteristic of dry environments inhibit the complete adjustment of channel form to process so that a state of equilibrium may rarely, if ever, exist in dryland rivers. Transient behaviour is more likely to be the norm. In addition, the wide difference between high and low flows prevents the development of a close relationship between channel form and a particular discharge such as 'bankfull' (Graf 1988). Thus, concepts such as 'equilibrium' and 'dominant discharge' seem less relevant in the arid zone context. Despite being subject to the same basic controls, the arid zone fluvial environment is by dint of its greater variability of rainfall and flow, different mix of hydrological processes, decreasing discharge downstream and contrasting adjustment characteristics, quite distinct from its humid zone counterpart.

Diversity within the arid zone

While the arid–humid distinction is clear in all its variety, it would be wrong to infer that the fluvial environment behaves similarly in all dryland areas. Indeed Pilgrim *et al.* (1988)

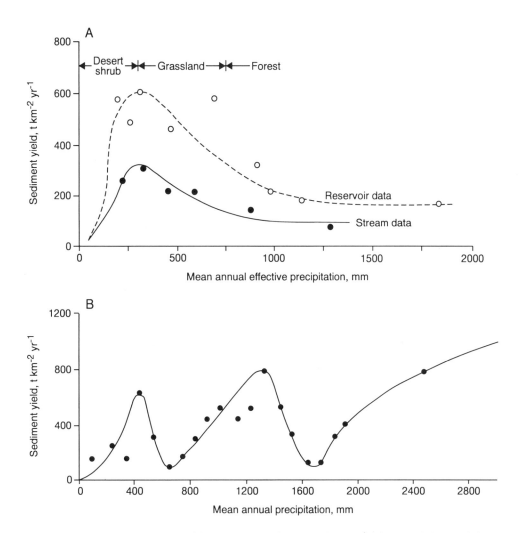

Figure 10.3   *Relationships of sediment yield to mean annual precipitation: (A) after Langbein and Schumm (1958); (B) after Walling and Kleo (1979)*

suggest that there is greater diversity in hydrological characteristics within arid regions than within humid ones. That diversity is visualised in terms of a continuous scale of flow occupancy (Figure 10.7), with the driest deserts where streamflow is virtually unknown at one end and perennial, mostly through-flowing rivers at the other. Whether the attendant characteristics are better represented in terms of a continuum or in terms of discrete classes is a matter of debate. Cooke *et al.* (1993) favour the latter in distinguishing between endogenic and allogenic drainage, being derived, respectively, from sources within and outside the arid

area. Although mutually exclusive, such a classification does not adequately reflect the range of conditions existing within the arid zone.

Local convective thunderstorms are commonly regarded as a significant contributor to precipitation in drylands but, given their variability and limited individual extent, they may not be the major source except in drier, more continental areas. The western and poleward margins of such areas are occasionally penetrated by frontal systems and the eastern and equatorward ones by tropical storms, both of which can make substantial contributions to

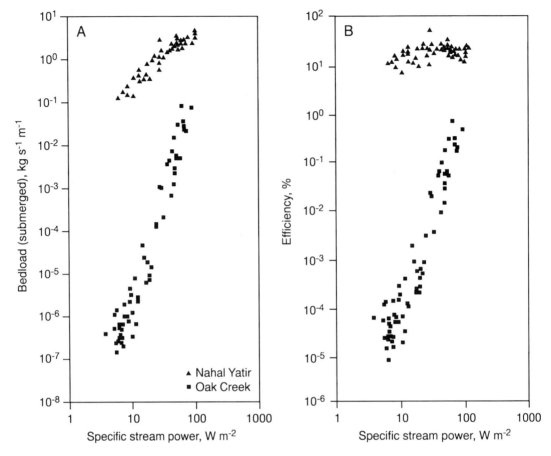

Figure 10.4   *Contrasts in the rate (A) and efficiency (B) of bedload transport between an arid zone (Nahal Yatir, Negev Desert, Israel) and a humid zone (Oak Creek, Oregon, USA) stream (after Laronne and Reid 1993)*

total precipitation amounts and lead to the generation of rare but important flood events. Of all the rainfalls yielding ≥5 mm in arid western New South Wales, local convective storms constitute less than 10% with a variety of general storms making up the remainder (Cordery *et al.* 1983). Even the arid centre of Australia can be reached by tropical monsoonal weather entering from the north to produce devastating floods (Williams 1970). In the higher parts of the arid western USA, a relatively high proportion of the annual precipitation falls as snow. Thus the mix of *endogenic sources* can be quite varied in composition, scale and seasonality, the last of which becomes more pronounced toward the margins. The perennial end of the scale also includes a range of supply conditions, from

internal springs to external rains and snowfall, but the former are relatively rare and generally give rise to rather small streams. Major through-flowing rivers are largely fed from mountainous headwaters with a water surplus which often occurs in the form of snow. In the case of the Euphrates, almost 90% of the total runoff is generated within the mountains of eastern Turkey with practically nothing produced in arid Iraq (Beaumont 1989). Supplied from water surpluses in the East African Plateau and Ethiopian Highlands, the Nile flows for much of its lower course through a region where mean annual rainfall is less than 50 mm. As the source becomes larger in size, from small convective storm centres at one end of the scale to extensive headwaters at the other, there is greater reliability of supply.

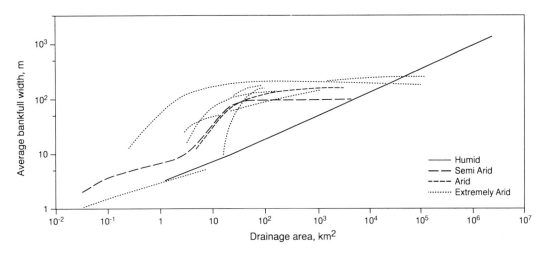

Figure 10.5 *Relationship of bankfull channel width to drainage area for different climatic environments (after Wolman and Gerson 1978)*

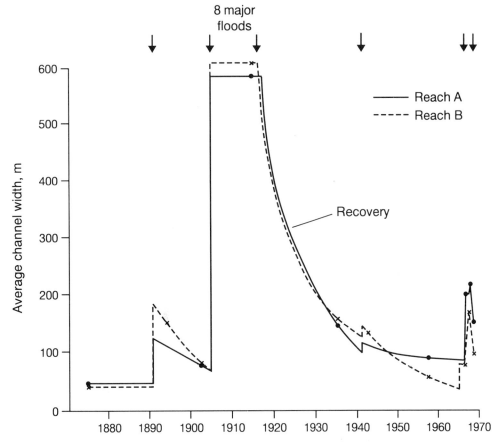

Figure 10.6 *Historical changes in channel width, 1875–1970, Gila River, Arizona (after Burkham 1972). The vertical arrows indicate major floods*

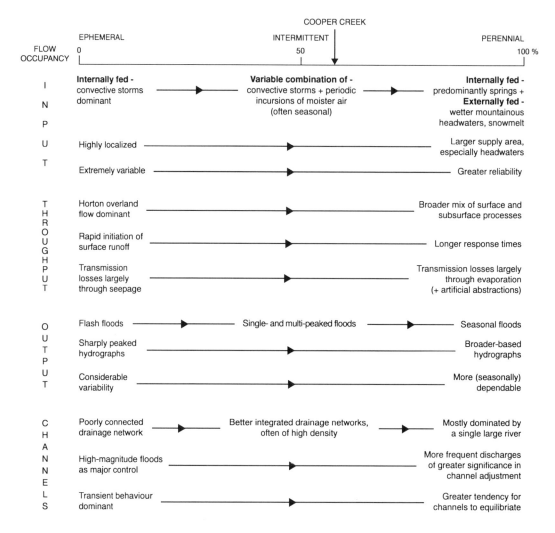

Figure 10.7 *Diversity in hydrologic input, throughput and output, and in channel characteristics within the arid zone, expressed in terms of a linear flow occupancy scale*

The rapid generation of surface runoff through Hortonian overland flow is still likely to be dominant across a wide spectrum of input conditions, but as surface vegetation cover increases so does infiltration. As a result subsurface processes assume more importance towards the right hand end of the scale in Figure 10.7. Groundwater contributions probably remain small except where alluvial aquifers are forced toward the surface. The downstream decrease in discharge is, however, a more consistent feature of the arid zone fluvial environment, although the main cause of transmission loss may vary. The survival length of a flow depends on such factors as the permeability and depth of the channel substrate (the data for the Luni Basin in Table 10.1 show how variable this effect can be), the magnitude of the input discharge, and the distance between tributary inputs, if any exist. As regards the second of these factors, small- and medium-sized streams activated by local convective storms will lose water largely through seepage into the channel boundary since flow will rarely be maintained long enough for evaporation to have much effect. On the other

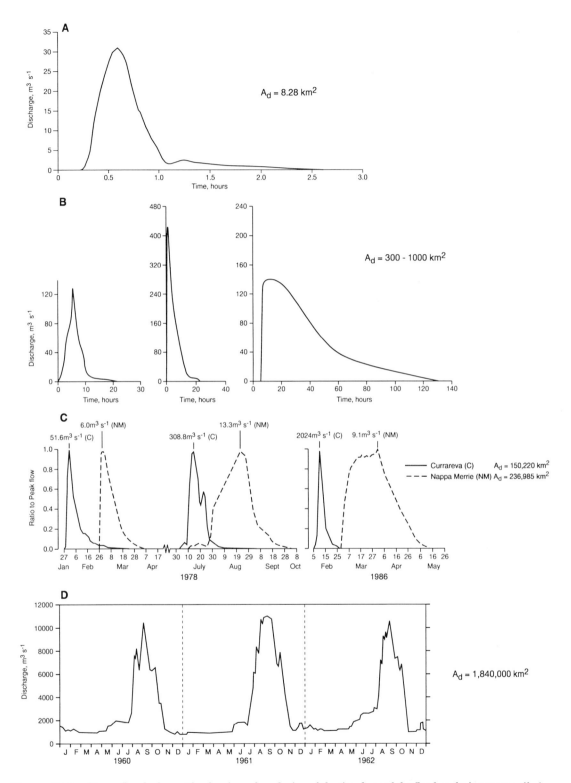

Figure 10.8 *Streamflow hydrographs showing a broadening of the time base of the flood as drainage area (A_d) increases: (A) Walnut Gulch, Arizona (after Lane et al. 1971); (B) Luni basin, northwest India (after Sharma et al. 1994); (C) Cooper Creek, east-central Australia; (D) flow of the River Nile at Aswan before construction of the Aswan High Dam (after Beaumont 1989)*

hand, large through-flowing rivers are likely to experience increasing evaporative loss as they cross arid areas, with artificial abstractions being a further cause of discharge depletion in very large catchments. Where such catchments cross political boundaries, a not uncommon situation, the potential for conflict over the use of water resources can become considerable.

Graf (1988) identifies four types of flood in dryland rivers — flash-floods, single peak events, multiple peak events, and seasonal floods. Although not mutually exclusive, they are partly scale dependent, with flash floods produced by convectional precipitation in basins of $\leqslant 100$ km^2 and seasonal floods associated with major perennial rivers as the two end members (Figure 10.7). Single and multiple-peak floods generated by tropical storms or frontal systems cover a wide range of intermediate conditions. In progressing from one end of the scale to the other, there is a corresponding reduction in the steepness of the rising limb of the hydrograph and a broadening of the time base of the flood (Figure 10.8). Despite this progression, Farquharson *et al.* (1992) demonstrate a strong similarity in the shape of regional flood frequency curves across many arid areas, although externally fed rivers are expressly excluded. On this basis they suggest that all such areas could be treated as a single homogenous region, which has potential value from engineering and predictive perspectives. However, considerable interregional differences do exist. In terms of the relation between specific peak annual discharge ($\bar{q}$) and mean annual runoff (*MAR*), McMahon (1979) shows that eastern Mediterranean streams produce higher and Australian streams considerably lower mean annual floods than those in other arid areas (Figure 10.9). In addition, peak discharges in North American streams tend to be much less variable than elsewhere. The diversity of hydrological conditions across the arid zone implies that no single strategy is likely to be effective for either runoff modelling or the management of already meagre water resources.

The peakedness of sediment yield curves in the mean annual precipitation range of 0–600 mm (Figure 10.3) underlines that diversity by indicating how widely susceptibility to surface water erosion varies within drylands. Small changes in total precipitation can lead to

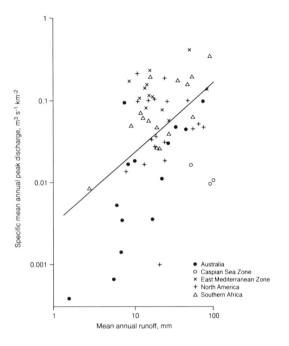

Figure 10.9 *Relationship between specific mean annual peak discharge and mean annual runoff for given arid regions (after McMahon 1979)*

large differences in erosion potential, temporally as well as spatially. Peak yields are commonly associated with semi-arid areas where drainage densities also tend to be at a maximum (Gregory 1976), largely because in drier areas there is insufficient runoff to cut and maintain channels and in wetter ones there is more vegetation to inhibit channel development. Thus, networks tend to be poorly connected in deserts and become better integrated as flow occupancy increases (Figure 10.7), although through-flowing rivers often only have occasional large tributaries beyond the headwaters. Despite the downstream decrease in discharge, those rivers may show elements of behaviour more akin to those of perennial rivers in humid regions (Cooke *et al.* 1992), both in terms of channel morphology and the tendency to approach some form of equilibrium. Also, the overall influence of infrequent, high-magnitude floods in controlling channel morphology is likely to diminish. Whether or not those rivers are treated as a separate category, there is a sense of progression along the flow occupancy scale as regards channel stability, responsiveness to extreme floods and ability to recover from flood damage, related in no small way to the effects of vegetation. Riparian vegetation, for example, can increase bank resistance to erosion (Smith 1976) and lead to significantly reduced channel widths, bringing them more in line with humid zone expectation. However, in so far as it is vegetation related, the progression is unlikely to be simple because of the existence of thresholds (Moss 1979), but it serves to emphasise the broad range of conditions in the arid zone fluvial environment.

Uniqueness: the Australian experience

Excepting Antarctica, Australia occupies a distinctive position as the world's driest continent with more than 70% of its land area classified as arid or semi-arid. The surrounding oceans provide extensive sources of moisture for weather systems which, given the absence of high topographic barriers across prevailing wind directions, can occasionally penetrate far into the arid zone. The sporadic nature of that penetration contributes to the extreme variability of the runoff record, manifest in the very high coefficients of variation (standard deviation divided by mean) of annual flows and the large values of $Q_{100}/Q_{2.33}$ (the 100-year recurrence interval flood divided by the mean annual flood) which is a measure of flood variability (Finlayson and McMahon 1988). Only southern Africa has comparable values. Such high variability in Australian streams, about twice that observed in northern hemisphere streams, increases the difficulty of detecting change in streamflow volumes which might result from global warming (Chiew and McMahon 1993), and influences the decadal-scale adjustment of river channel form (Erskine and Warner 1988).

The arid zone of Australia is dominated by three physiographic regions of which the Lake Eyre Basin is the most extensive (Figure 10.10). Large areas of the Western Plateau lack integrated drainage because of the extreme aridity. Mean annual rainfall in the Lake Eyre Basin varies from 400–500 mm along the northern and eastern margins to less than 100 mm in the Simpson Desert, with occasional incursions of moist tropical air from the north being the major influence on streamflow. However, this source is erratic and all streams are characterised by extreme discharge variation. The mean annual runoff of 4 km^3, or 3.5 mm depth equivalent across the basin, is the lowest of any major drainage system in the world (Kotwicki 1987). Despite the aridity and generally low relief, the drainage network is surprisingly well-developed, mainly because of the well-watered uplands on the periphery. Lake Eyre can be filled in three main ways — by Cooper Creek, by the Diamantina River, or by the western, desert tributaries (Figure 10.10). Major flooding episodes which occur on average once in 20 years seem to be most often associated with La Niña phases of the El Niño Southern Oscillation (ENSO) when Australia experiences enhanced summer monsoon activity (Allan 1988). Such quasi-periodicities could influence not only contemporary hydrological conditions but also palaeohydrological sequences which have been shown to include an alternating pattern of pluvial and arid episodes broadly correlated with late Quaternary interglacials and glacials, respectively (Nanson *et al.* 1992).

COOPER CREEK

In one sense every river is unique but the rivers which make up the Channel Country of east-central Australia are indeed very unusual when compared with arid zone rivers elsewhere. Rising as the Barcoo and Thomson Rivers on the southwestern side of the Great Dividing Range at an elevation of only 230 m, Cooper Creek maintains a well-defined system of channels for much of its 1523 km length (Figure 10.10). Between the gauging stations at Currareva and Nappa Merrie, a distance of ~420 km, the flow moves over channel gradients which rarely exceed 0.0002 m m^{-1} and can expand to exceptional widths of up to 60 km during large floods. In the absence of major tributaries, much of the outflow at Nappa Merrie can be attributed to inflow at Currareva with a distinctive anastomosing system of channels providing the hydrological connection (Knighton and Nanson 1993; 1994a).

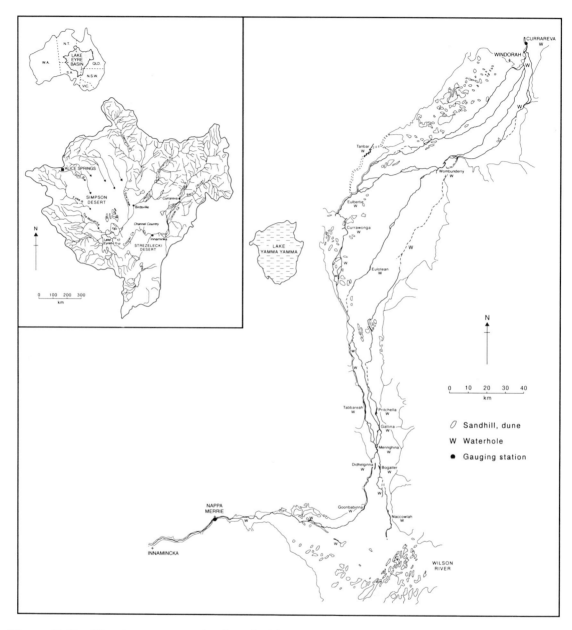

Figure 10.10   *The primary anastomosing channels of Cooper Creek between Currareva and Nappa Merrie. The inset map shows the position of Cooper Creek within the Lake Eyre basin*

Flow occurs during most years at Currareva. There is on average no flow for 44% of the time so Cooper Creek occupies an intermediate position in Figure 10.7. River regime, defined in terms of the ratios between mean monthly discharges and the overall mean, indicates a pronounced mid- to late-summer dominance (Figure 10.11), reflecting southerly penetrations of the north Australian monsoon. That the March and April ratios at Nappa Merrie are higher than those at Currareva is a consequence of the long transmission times which average 30 days. However, despite the long river distance,

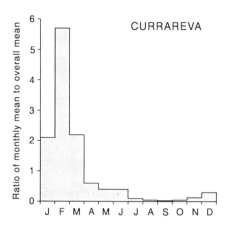
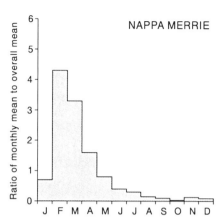

Figure 10.11 *Flow regime at the Currareva and Nappa Merrie gauging stations expressed by the ratio of each monthly mean discharge to the overall mean of the 12 monthly values*

hydrograph shape can be maintained over a wide range of discharges (Figure 10.12).

In common with other arid zone rivers mean annual discharge decreases downstream, from 3.05 km³ at Currareva to 1.26 km³ at Nappa Merrie, but it is much more variable than elsewhere. Also, annual flows become more variable downstream, a trait which appears to be unique to Australian rivers (Finlayson and McMahon 1988). Despite this increased variability and the long intervening distance, average annual discharge at Nappa Merrie is well correlated ($\rho = 0.967$) with that at Currareva, emphasising the strong dependence of outflow on inflow. As well as showing the variability and interstation correspondence of flow, the plot of annual data (Figure 10.13) indicates the dominance of 1974 in the flow record when the largest known flood occurred, probably the largest for over 100 years (Kotwicki 1986; 1987). Indeed, mean annual discharge and annual peak discharge are highly correlated at both Currareva and Nappa Merrie, underlining the important contribution of individual floods to total runoff.

Flood flows have a significant influence on the channel morphology of dryland rivers. The 1974 flood with a maximum daily discharge of 24 974 m³ s⁻¹ at Currareva was exceptionally large and yet, in common with Australian arid zone streams as a whole (Figure 10.9), mean annual flood is relatively small. Undoubtedly that contributes to the extreme steepness of the relative flood magnitude ($Q/Q_{2.33}$) curve

(Figure 10.2). However, the most prominent feature of Cooper Creek hydrology is the average five-fold decrease in discharge magnitude between Currareva and Nappa Merrie, a level of loss which is to be expected where the upstream river is the main source of runoff and where the flow undergoes considerable transformation during its downstream passage.

Transmission losses along the Cooper can be attributed to three main causes (Knighton and Nanson 1994a):

1. Evaporation/evapotranspiration, the effectiveness of which is probably increased here by the large surface area and relative shallowness of floodplain flows.
2. Infiltration into the channel boundary and floodplain surface, which may be significant only during the early stages of inundation since the clay-rich muds of the floodplain surface swell after wetting.
3. Drainage diffusion, wherein large areas of flat, low-lying land act as sumps for floodplain waters and thereby enhance loss through evaporation; water is also lost to Lake Yamma Yamma (Figure 10.10) during very large floods.

To assess how these losses vary with the state of flow, the ratio of output (V_{NM}) to input (V_C) runoff volumes is plotted against the input flow (V_C) for 34 periods when a hydrological connection between Currareva and Nappa Merrie could be reasonably well established (Figure 10.14). The mean value of V_{NM}/V_C is 0.23, suggesting that, of those input flows which do

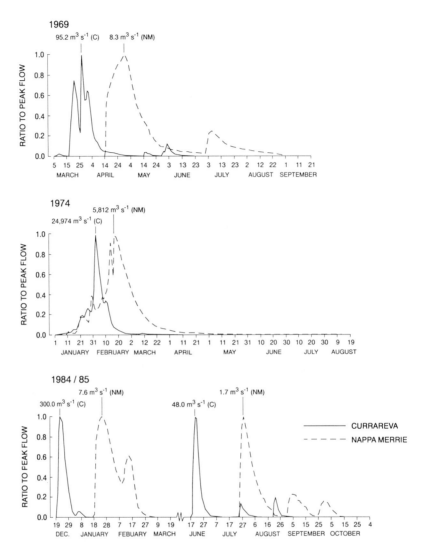

Figure 10.12   *Selected hydrographs from the flow records of the Currareva and Nappa Merrie gauging stations when there was a hydrological connection between the two*

reach Nappa Merrie, about 75–80% is lost on average, a not uncommon level of transmission loss in arid zone rivers (Walters 1990; Hughes and Sami 1992). That transmission losses are not higher given the long distance can probably be attributed to the presence of swelling clays which limit infiltration. However, the most surprising feature of the plot is the non-linear variation of V_{NM}/V_C. The initial peak at discharges in the 20–25% range is believed to reflect flow through the primary system of anastomosing channels whose good connectivity

and limited surface area ensure relatively efficient throughput and reduced losses. With an increase in input volume and activation of lesser channels, the output/input ratio falls to a fairly constant level of about 0.12 as evaporation and infiltration increase from an enlarging area of surface flow. Not until input volumes become very large ($>2 \times 10^9$ m^3) does transmission efficiency begin to improve again. Above 10×10^9 m^3 when the floodplain is fully inundated, transmission losses level off at about 30–40%. While the magnitude of transmission

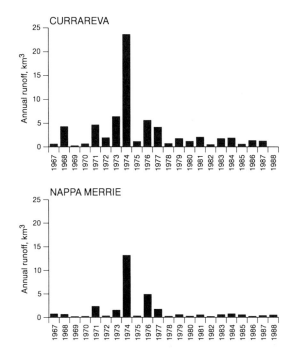

Figure 10.13 *Mean annual runoff at the Currareva and Nappa Merrie gauging stations*

losses is not atypical of arid zone rivers, the pattern of variation may well be unique to Cooper Creek, related to the exceptional width to which the river can expand during floods and to the varied system of channels present.

In the broad, shallow valley occupied by Cooper Creek, a hierarchy of channel forms can be recognised: the dominant system of anastomosing channels; a related system of floodplain channels (floodways); and reticulate drainage in backswamps. The anastomosing channels vary considerably in size and connectivity, from well-integrated primary channels to poorly connected and ill-defined tertiary channels. One of the most distinctive features of this system is the preponderance of 'waterholes' which represent relatively wide and deep sections of the channels (Figure 10.15). There are more than 300 waterholes between Currareva and Nappa Merrie, ranging in length from 100 m to over 20 km, some of which are inherited from large palaeochannels and some of which are contemporary in origin, a consequence of localised scour (Knighton and Nanson 1994b). Floodways are channel-like forms of high width–depth ratio (commonly about 60) cut into the floodplain surface and

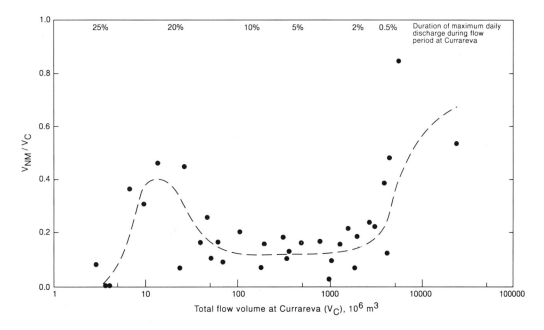

Figure 10.14 *Plot of output/input ratio against input value for total volumes of flow over definable periods when there was a hydrological connection between Currareva and Nappa Merrie (after Knighton and Nanson 1994a)*

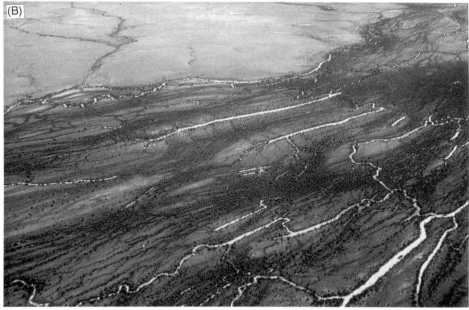

Figure 10.15 *(A) Meringhina Waterhole and associated anastomosing channels (see Figure 10.10 for location). (B) Linear waterholes* en echelon *and linking channels; the edge of the floodplain can be seen in the middle background.*

having the appearance of a braided pattern (Nanson *et al.* 1986). They tend to follow more direct routes than the anastomosing channels, often cutting across them at high angles. Away from the main flow systems are the backswamps which act as sumps during floods and support intricate networks of small reticulate channels. These three channel patterns interact at times of high flow and the non-linear variation of transmission losses (Figure 10.14) depends in large part on which of the various elements are active during the filling phase of a flood.

The unique character of the Channel Country rivers is due in no small way to their unusual sediment load, which is dominated by mud aggregates that move as bedload during floods through the deeper anastomosing and shallower floodplain channels (Nanson *et al.* 1986; 1988; Rust and Nanson 1986; 1989). The siltstones and labile sandstones in the catchment areas weather in the semi-arid/monsoon climate to form shrinking–swelling clays which pedogenically rework the soil into aggregates that behave as low-density sand-sized particles. After a rainstorm the floodplain surface is stirred by the self-mulching properties of the characteristic vertisols to produce a dry, 'fluffy', 10–20-cm-thick horizon of aggregates readily entrained during the next overbank flow. With deposition and burial the aggregates compress into compacted muds, and it is into these cohesive sediments that the anastomosing channels are incised (Nanson *et al.* 1986). One of the conditions for the development of an anastomosing pattern appears to be low bank erodibility (Knighton and Nanson 1993).

The morphology of Channel Country rivers has, however, varied over time. Along the Cooper airphotos and shallow augering reveal remnant scroll-bars and palaeochannels scaled to river meanders far larger than any today (Rust and Nanson 1986; Nanson *et al.* 1988). These features and their associated sediments show that the last two interglacials (250–195 ka and 128–75 ka) were pronounced pluvial episodes during which extensive alluvial sands were deposited (Nanson *et al.* 1988; 1992). The intervening glacials are interpreted as being generally arid (see Chapters 25 and 26) and associated with mud deposition along the rivers. This alternation between relatively inactive

mud-dominated systems during more arid glacial periods and relatively energetic sand-dominated systems during the middle to later periods of humid interglacials might suggest a corresponding change in sediment supply. However, there is little opportunity for significant temporal variation in sediment provenance and it appears that abundant sand and mud are present in the floodplain all the time but that there is a shift in the river's ability to rework and transport both (Nanson *et al.* 1988). Aeolian dunes along the valley sides and floodplain contain palaeosols which also indicate alternating cycles of activity and stability (Nanson *et al.* 1988).

Clearly, the Channel Country of east-central Australia represents an extensive and very unusual fluvial system within the arid zone context. A unique combination of physiographic conditions gives rise to seasonally active and coexisting anastomosing, braided and reticulate channels which transport an abnormal traction load of sand-sized mud aggregates. The narrow but clearly-defined anastomosing channels are incised into cohesive floodplain muds and constitute the main planimetric form, while the wide and shallow floodplain braid-channels rework the less compact surface. Well-developed anastomosing channel systems are not unknown in dry environments (see, for example, Schumann 1989), but scale, constituent morphology and flow conditions set Cooper Creek apart.

Conclusion

The regime of arid zone rivers is characteristically unsteady. The uncertainties of hydrologic input coupled with the highly variable effects of transmission loss make their behaviour much more difficult to predict than that of their humid zone counterparts. Relationships between process and form are less clear-cut, partly because discharge generally decreases downstream and partly because high-magnitude events exert such a dominant influence, both hydrologically and geomorphologically. The effects of floods can be preserved for long periods of time, especially where vegetation is very sparse, so that channel form may never become completely adjusted to coexisting process. The crucial role of vegetation in determining the efficacy of erosional processes acting within and outside the channel/floodplain system is indicated by the sediment yield curves (Figure 10.3), whose form underlines how sensitive dryland environments can be to small changes in conditions.

The distinctiveness of arid and semi-arid regions argues for the adoption of modelling and management strategies which are different from those developed for humid regions. However, pandemic solutions to arid zone problems are unlikely to be successful given the inherent diversity outlined in Figure 10.7. Some of the characteristics listed there are probably better represented as discrete rather than continuous variables but there is no denying the wide range of conditions. Many Australian rivers, and Cooper Creek in particular, stand out as especially distinct, not least because of their extreme variability which is partly due to the fact that large floods are monsoon-related, an unusual circumstance within dryland rivers. Cooper Creek is not a major through-flowing river. Indeed it flows towards a more arid area and a hydrological sink, but it has nevertheless developed a well-defined channel system which maintains a hydrological connection over long river distances. The active flow system expands to exceptional widths during floods which tend to be of long duration, moving slowly through low-gradient multiple-channel patterns. Figure 10.7 can only hint at the diversity within the arid zone fluvial environment, of which Cooper Creek is but one illustration.

References

Allan, R.J. 1988. El Niño Southern Oscillation influences in the Australasian region. *Progress in Physical Geography*, 12: 4–40.

Baker, V.R. 1977. Stream-channel response to floods, with examples from central Texas. *Bulletin of the Geological Society of America*, 88: 1057–1071.

Beaumont, P. 1989. *Drylands. Environmental management and development*. Routledge, London.

Bell, F.C. 1979. Precipitation. In D.W. Goodall and R.A. Perry (eds), *Arid land ecosystems*. Cambridge University Press, Cambridge: 373–393.

Burkham, D.E. 1972. *Channel changes of the Gila River in Safford valley, Arizona, 1846–1970*. US Geological Survey Professional Paper 655G.

Chiew, F.H.S. and McMahon, T.A. 1993. Detection of trend or change in annual flow of Australian rivers. *International Journal of Climatology*, 13: 643–653.

Cooke, R., Warren, A. and Goudie, A.S. 1993. *Desert geomorphology*. UCL Press, London.

Cordery, I., Pilgrim, D.H. and Doran, D.G. 1983. Some hydrological characteristics of arid western New South Wales. In *Hydrology and Water Resources Symposium 1983, Institute of Australian Engineers*, National Conference Publication 83/13: 287–292.

Costa, J.E. 1987. Hydraulics and basin morphometry of the largest flash floods in the conterminous United States. *Journal of Hydrology*, 93: 313–338.

Dunkerley, D.L. 1992. Channel geometry, bed material, and inferred flow conditions in ephemeral stream systems, Barrier Range, western N.S.W., Australia. *Hydrological Processes*, 6: 417–433.

Dunne, T., Zhang, W. and Aubry, B.F. 1991. Effects of rainfall, vegetation and microtopography on infiltration and runoff. *Water Resources Research*, 27: 2271–2287.

Erskine, W.D. and Warner, R.F. 1988. Geomorphic effects of alternating flood- and drought-dominated regimes on NSW coastal rivers. In R.F. Warner (ed.), *Fluvial geomorphology of Australia*. Academic Press, Sydney: 223–244.

Farquharson, F.A.K., Meigh, J.R. and Sutcliffe, J.V. 1992. Regional flood frequency analysis in arid and semi-arid areas. *Journal of Hydrology*, 138: 487–501.

Finlayson, B.L. and McMahon, T.A. 1988. Australia v the World: a comparative analysis of streamflow characteristics. In R.F. Warner (ed.), *Fluvial geomorphology of Australia*. Academic Press, Sydney: 17–40.

Graf, W.L. 1983. Flood-related channel change in an arid-region river. *Earth Surface Processes and Landforms*, 8: 125–139.

Graf, W.L. 1988. *Fluvial processes in dryland rivers*. Springer-Verlag, Berlin.

Gregory, K.J. 1976. Drainage networks and climate. In E. Derbyshire (ed.), *Geomorphology and climate*. Wiley, Chichester: 289–315.

Hughes, D.A. and Sami, K. 1992. Transmission losses to alluvium and associated moisture dynamics in a semi-arid ephemeral channel system in southern Africa. *Hydrological Processes*, 6: 45–53.

Knighton, A.D. and Nanson, G.C. 1993. Anastomosis and the continuum of channel pattern. *Earth Surface Processes and Landforms*, 18: 613–625.

Knighton, A.D. and Nanson, G.C. 1994a. Flow transmission along an arid zone anastomosing river, Cooper Creek, Australia. *Hydrological Processes*, 8: 137–154.

Knighton, A.D. and Nanson, G.C. 1994b. Waterholes and their significance in the anastomosing channel system of Cooper Creek, Australia. *Geomorphology*, 9: 311–324.

Kotwicki, V. 1986. *Floods of Lake Eyre*. Engineering and Water Supply Department, Adelaide.

Kotwicki, V. 1987. On the future of rainfall–runoff modelling in arid areas — Lake Eyre case study. In J.C. Rodda and N.C. Matalas (eds), *Water for the future. Hydrology in perspective. Proceedings of the Rome Symposium*. IAHS Publication 164: 341–351.

Lane, L.J., Diskin, M.H. and Renard, K.G. 1971. Input–output relationships for an ephemeral stream channel system. *Journal of Hydrology*, 13: 22–40.

Langbein, W.B. and Schumm, S.A. 1958. Yield of sediment in relation to mean annual precipitation. *Transactions of the American Geophysical Union*, 39: 1076–1084.

Laronne, J.B. and Reid, I. 1993. Very high rates of bedload sediment transport by ephemeral desert rivers. *Nature*, 366: 148–150.

Lewin, J. 1989. Floods in fluvial geomorphology. In K. Beven and P. Carling (eds), *Floods*. Wiley, Chichester: 265–284.

McMahon, T.A. 1979. Hydrological characteristics of arid zones. In *The hydrology of areas of low precipitation*. Proceedings of the Canberra Symposium. IAHS–AISH Publication, 128: 105–123.

McMahon, T.A., Finlayson, B.L., Haines, A.T. and Srikanthan, R. 1992. *Global runoff*. Catena Verlag, Cremlingen.

Moss, A.J. 1979. Thin-flow transportation of solids in arid and non-arid areas: a comparison of process. In *The hydrology of areas of low precipitation*. Proceedings of the Canberra Symposium. IAHS–AISH Publication, 128: 435–445.

Nanson, G.C., Rust, B.R. and Taylor, G. 1986. Coexistent mud braids and anastomosing channels in an arid-zone river: Cooper Creek, central Australia. *Geology*, 14: 175–178.

Nanson, G.C., Young, R.W., Price, D.M. and Rust, B.R. 1988. Stratigraphy, sedimentology and Late Quaternary chronology of the Channel Country of western Queensland. In R.F. Warner (ed), *Fluvial geomorphology of Australia*. Academic Press, Sydney: 151–175.

Nanson, G.C., Price, D.M. and Short, S.A. 1992. Wetting and drying of Australia over the past 300 ka. *Geology*, 20: 791–794.

Pilgrim, D.H., Chapman, T.G. and Doran, D.G. 1988. Problems of rainfall–runoff modelling in arid and semiarid regions. *Hydrological Sciences Journal*, 33: 379–400.

Renard, K.G. and Keppel, R.V. 1966. Hydrographs of ephemeral streams in the Southwest. *Proceedings of the American Society of Civil Engineers, Journal of the Hydraulics Division*, 92: 33–52.

Rust, B.R. and Nanson, G.C. 1986. Contemporary and palaeo channel patterns and the Late Quaternary stratigraphy of Cooper Creek, southwest Queensland, Australia. *Earth Surface Processes and Landforms*, 11: 581–590.

Rust, B.R. and Nanson, G.C. 1989. Bedload transport of mud as pedogenic aggregates in modern and ancient rivers. *Sedimentology*, 36: 291–306.

Schumann, R.R. 1989. Morphology of Red Creek, Wyoming, an arid-region anastomosing channel system. *Earth Surface Processes and Landforms*, 14: 277–288.

Sharma, K.D., Murthy, J.S.R. and Dhir, R.P. 1994. Streamflow routing in the Indian arid zone. *Hydrological Processes*, 8: 27–43.

Smith, D.G. 1976. Effect of vegetation on lateral migration of anastomosed channels of a glacial meltwater river. *Bulletin of the Geological Society of America*, 86: 857–860.

UNESCO 1979. *Map of the world distribution of arid regions*. MAB Technical Notes 7, UNESCO, Paris.

Walling, D.E. and Kleo, A.H.A. 1979. Sediment yields of rivers in areas of low precipitation. In *The hydrology of areas of low precipitation*. Proceedings of the Canberra Symposium. IAHS-AISH Publication, 128: 479–493.

Walters, M.O. 1990. Transmission losses in arid region. *American Society of Civil Engineers, Journal of Hydraulic Engineering*, 116: 129–138.

Williams, G.E. 1970. The central Australian stream floods of February–March 1967. *Journal of Hydrology*, 11: 185–200.

Wolman, M.G. and Gerson, R. 1978. Relative scales of time and effectiveness of climate in watershed geomorphology. *Earth Surface Processes*, 3: 189–208.

11

Channel form, flows and sediments in deserts

Ian Reid and Lynne E. Frostick

Introduction

Water is a significant agent of erosion in arid lands. Upon entering a desert or semi-desert area, it soon becomes obvious that access is often facilitated or hindered by the river system. Yet, until recently, the role of rivers in shaping desert landscapes has generally been underestimated by geomorphologists. The reasons for this are several. First, there has been a tendency to concentrate attention on processes and forms that are thought to be more peculiar to drylands. Seminal texts, such as that of Bagnold (1941), and collections of papers such as those of McKee (1979) and Brookfield and Ahlbrandt (1983), have highlighted the action of wind and given prominence to windblown dunes. This has been reinforced by popular portrayals of the desert in novels and on film. Second, the infrequence of rainfall and runoff in drylands has made data acquisition an expensive proposition for the fluvial geomorphologist. Few have had either the financial resources or the patience to collect long-term records that consist largely of non-information and that are punctuated only spasmodically with a frenzy of relevant data. Notable exceptions are Renard and Keppel (1966) in Arizona, and Schick (1970) in Israel. Indeed, many of those who

have examined ephemeral streams have done so over shorter periods of time. But to do this, they have moved towards semi-arid areas where rivers are no less dry and where higher rainfall gives a greater likelihood of flash floods (see, for example, Leopold and Miller 1956; Thornes 1976; Frostick and Reid 1977).

There is, as a consequence, a dearth of information about processes in ephemeral rivers. It would be an understatement to declare that the explosive increase in both rainfall–runoff and sediment–transport data for perennial systems of temperate latitudes has not been matched by that collected in drylands. As a result, there is little basis for stochastic or other forms of modelling of flows in ephemeral rivers. Each piece of information, each measured flood, is fairly unique and has to be treasured as a gem that gives insight into the workings of these mysterious elements of these vast landscapes. For the geomorphologist, hydrologist and sedimentologist, this remains pioneering territory.

Notwithstanding the small total information base for ephemeral river systems, they have figured greatly in the development of process-oriented geomorphology through the incorporation of the results of work done in the southwestern United States and published in the timely text of Leopold *et al.* (1964).

Arid Zone Geomorphology: Process, Form and Change in Drylands, 2nd edition. Edited by David S. G. Thomas.
© 1997 John Wiley & Sons Ltd.

However, the impetus that this text might have given to work on desert streams *per se* was lost in its adoption as a general statement about river processes by a generation whose imagination it had caught. There was also the fact that the rash of process studies that broke out during the 1960s and 1970s inevitably focused on the backyards of those concerned. These backyards were (and still are) predominantly in humid environments where water catchments are (or could be) fully clothed with vegetation, where the soil fauna ensures a perforated medium that accommodates infiltering rainfall, and where the stream at the bottom of the hill has a perennial flow regime.

More recently, attempts have been made to redress the continuing imbalance in emphasis on windblown sediments in desert environ-ments by giving equal attention to fluvial pro-cesses (Frostick and Reid 1987; Cooke *et al.* 1993; Abrahams and Parsons 1994). This is important not only because fluvial processes are a cause of so many problems in desert areas, but also, because the peculiarities of dryland environments cause sufficient differences in river behaviour that the lessons learnt in humid areas are not reliably translated (Pilgrim *et al.* 1988).

Rainfall and river discharge

STORM CHARACTERISTICS

One way in which desert streams differ from their perennial counterparts is the generation

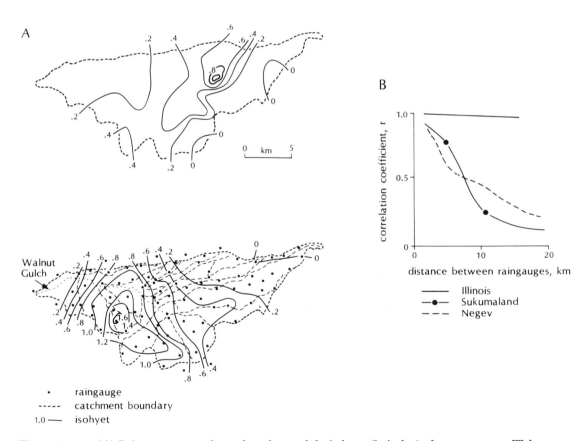

Figure 11.1 *(A) Raingauge net, catchment boundary and the isohyets (in inches) of two storms over Walnut Gulch, Arizona (modified and redrawn after Renard and Keppel 1966). (B) Correlogram of rainfall caught by point gauges spaced at distances up to 20 km for winter frontal rainfall (Illinois) and cellular convective storms (Negev Desert and Sukumaland, western Tanzania) (modified after Sharon 1974)*

and propagation of the flood wave. However, the peculiarities of flash floods are not entirely a function of processes on the ground. Of considerable significance for runoff is the fact that rain is more often than not associated with discrete convective cells (Sharon 1972; 1974). Data are few because the dense networks of gauges that would be required to measure such spotty rainfalls are extremely rare in drylands. Nevertheless, there is good evidence to suggest that these cells have a diameter that is generally less than 10–14 km (Renard and Keppel 1966; Diskin and Lane 1972; Figure 11.1A). This means that rainfall measured at one gauge cannot be used to predict rainfall even a few kilometres away, in contrast with temperate regions affected by ubiquitous frontal storms (Wheater *et al.* 1991a; Figures 11.1B and 11.2). Figure 11.1A illustrates the additional point, that because of the discrete nature of each convective cell, an individual storm is unlikely to affect the entire drainage net, while successive storms are more than likely to wet different parts of a river catchment (Schick and Lekach 1987).

This has implications for the flood hydrograph, because a different part of the drainage basin will contribute water during each event. Indeed, some tributaries have been noted first to run and then to remain dry during a period of successive floods in a trunk stream (Frostick *et al.* 1983). As a result of this and other variables, the shape of the flood hydrograph must change considerably. There is therefore less likelihood of identifying a *characteristic* outfall hydrograph for an ephemeral system, in contrast with perennial streams affected by widespread frontal rain (Wheater and Brown 1989). Besides this, the seemingly fixed maximum storm cell size means that the fraction of a river basin that will be affected by rain will fall as basin size increases. Although it is conceivable that several storms will wet a large basin more or less simultaneously, atmospheric dynamics will dictate that they are widely separated geographically. Sharon (1974) has suggested that the distance between storm centres may average 40–60 km.

One other factor that has considerable bearing on the nature of river discharge is that storm cells migrate as they deliver their rain (Sharon 1972; Frostick and Reid 1977; Frostick *et al.* 1983). Again, there are implications for the shape of the flood hydrograph, since a storm moving upstream across a drainage basin will cause runoff from lower tributaries to occur earlier than those of headwater regions. The fact that overland flow is generated within minutes in arid environments (Yair and Lavee 1976; Reid and Frostick 1986) encourages this

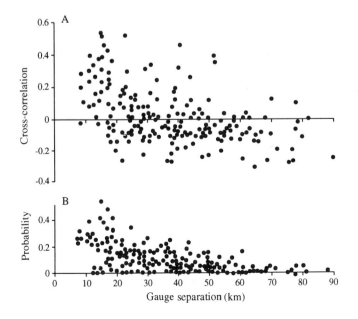

Figure 11.2 *(A) Cross-correlation of hourly rainfall as a function of gauge separation, Wadi Yiba, southwest Saudi Arabia. (B) Probability of simultaneous occurrence of hourly rainfall as a function of gauge separation, Wadi Yiba, southwest Saudi Arabia. Both reflect the spotty nature of rainfall in desert regions (modified after Wheater et al. 1991b)*

temporal separation of contributions from each tributary source. Indeed, Schick (1988) has speculated that the multi-peak nature of flash floods reflects, among other things, the piggy-backed contributions of individual tributary water catchments as each receives and disposes of rainfall. Ben-Zvi *et al.* (1991) have demonstrated for ephemeral streams in Israel that the downstream travel-time of flood crests is governed most significantly by crest discharge, as would be expected from hydraulic considerations of wave celerity. This would suggest that the contributions from individual tributaries would overrun each other in a manner dictated by the spatial inhomogeneity of storm rainfall, as much as by other factors. Because spatio-temporal rainfall patterns are highly unpredictable, this has to add to the complexity and changeability of the river outfall hydrograph.

In areas of pronounced topography, orographic heightening of convective storms can have a significant effect on the spatial pattern of rainfall, a factor that has to be taken into account when assessing probable stream discharge in basins that are not greatly separated, geographically. Wheater *et al.* (1991b) demonstrate reasonably well-defined direct relationships between both number of raindays and annual rainfall and altitude in the Asir escarpment of southwest Saudi Arabia. Martínez-Goytre *et al.* (1994) suggest a rain-shadow effect in the Santa Catalina Mountain range of southeastern Arizona after analysing the magnitude of palaeofloods in basins disposed at different locations around the massif.

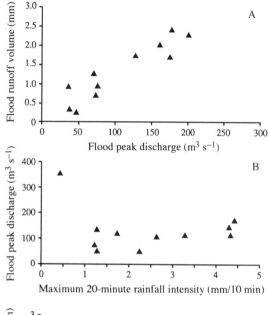

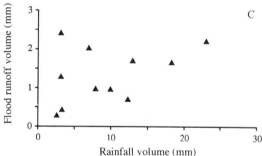

Figure 11.3 *(A) Surprisingly good relation between flash-flood volume and peak discharge, Wadi Ghat, southwest Saudi Arabia. (B and C) Lack of relations between flash-flood parameters and storm rainfall, Wadi Ghat, southwest Saudi Arabia (modified after Wheater and Brown 1989)*

FLASH-FLOOD HYDROGRAPH

The peculiarities of storms in desert areas mean that the relationships between rainfall and runoff are extremely complex, almost precluding hydrological modelling (Wooding 1966; Renard and Lane 1975; Srikanthan and McMahon 1980). Wheater and Brown (1989; Figure 11.3) have illustrated poor relations between flood peak discharge and maximum rainfall intensity and between flood runoff and storm rainfall volume in Wadi Ghat, a 597 km² basin draining towards the Red Sea in south-western Saudi Arabia. The most convincing

relation they show is, ironically, between flood runoff volume and flood peak discharge, suggesting some consistency between basic hydrograph shape and flood magnitude, though for modelling purposes the information base is extremely small.

However, there are three characteristics apparent in the flood hydrographs of streams in widely dispersed deserts. First, every one exhibits a steep rising limb that incorporates a bore (Figures 11.4 and 11.5). McGee (1897) gave a remarkable first description of a bore,

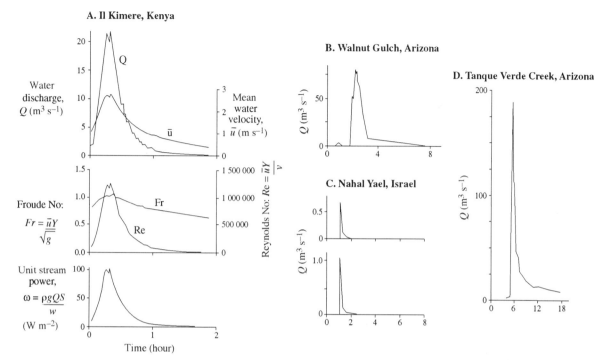

Figure 11.4 *Flood hydrographs and hydraulic parameters of ephemeral rivers in desert regions: (A) Il Kimere, Kenya (modified after Reid and Frostick 1987); (B) Walnut Gulch, Arizona (modified after Renard and Laursen 1975); (C) Nahal Yael, Israel (modified after Lekach and Schick 1982); (D) Tanque Verde Creek, Arizona (modified after Hjalmarson 1984)*

though for an unconfined sheetflood on a bajada fronting a mountain range in the Sonoran Desert. Since then the phenomenon has been noted in ephemeral channels in different deserts by various workers, including Hubbell and Gardner (1944), Leopold and Miller (1956), Renard and Keppel (1966), Gavrilovic (1969) and Frostick and Reid (1979). The bore is rarely the 'wall of water' often referred to in popular literature about flash floods. Hassan (1990a; Figure 11.6) has related the speed of advance and the depth of some bores observed in Israel, distinguishing between those travelling over antecedently dry or wet channel beds, and indicates remarkably modest velocities ≤ 1 m s⁻¹. This is corroborated by other observations made in Israel by Reid *et al.* (1994). Nevertheless, the arrival of a bore has often been fatal for people caught unawares (Hjalmarson 1984). This is undoubtedly because the time of rise of the flood hydrograph is short. Reid *et al.* (1994) have

measured rates of rise in water-stage as high as 0.25 m per minute and average water velocities approaching 3 m s⁻¹ within a few minutes of the passage of the bore. A person caught mid-channel in these circumstances can easily lose footing and be swept away. Reports of the interval between the arrival of the bore and peak discharge for catchments of small or moderate size range from 18 to 23 minutes (Renard and Keppel 1966), 10 minutes (Schick 1970) and 14 to 16 minutes (Reid and Frostick 1987).

The second characteristic of ephemeral stream hydrographs is the steepness of the flood recession (Figure 11.4). This may reflect the overriding importance of Hortonian overland flow in arid environments. Observations of the limited penetration of the wetting front in the soil (e.g. 210 mm, Reid and Frostick 1986) suggest no substantial routing of water that might sustain stream discharge (Pilgrim *et al.* 1988). Besides this, the extreme dryness of the soil always means a high gradient of matric potential that would

Figure 11.5   *(A) 'Normal' flood bore of an ephemeral stream, Il Kimere, Kenya. (B) Last phase of flood recession on the Il Eriet, Kenya, only three hours after peak discharge*

draw water downward into the soil profile and reduce the chance of achieving the positive pressures that would permit soil interflow.

This leads to the third characteristic that is peculiar to ephemeral systems. The flood is often extremely short-lived. The whole event from initial to final dry bed might have taken no more than a few hours (Figures 11.4 and 11.5). Because the number of floods that might be expected ranges from around six (Reid and Frostick 1987) to much less than one per year (Schick 1988) as one moves from semi-arid to arid environments, this means that the river system is active for much less than 1% of the time. The probability of being on-site to make

observations and to take measurements during an event is extremely small. This and the low rate of data acquisition are the chief reasons for the paucity of information that we have regarding desert flash floods.

TRANSMISSION LOSSES

Another factor which sets ephemeral systems aside from those that flow perennially or even intermittently is the loss of water to the stream bed. The process has been quantified by Renard and Keppel (1966), Burkham (1970) and Butcher and Thornes (1978) for streams

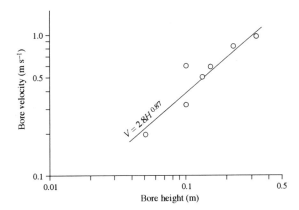

Figure 11.6 *Relation between flash-flood bore velocity and height observed in Nahal Og, Judean Desert, and Nahal Hebron, Negev Desert, Israel (modified after Hassan 1990a)*

in Arizona and Spain, and noted in many other instances (such as Murphey *et al.* 1972; Schick 1988; Reid *et al.* 1989). The rate of transmission losses will depend upon the porosity and depth of the channel fill, as well as upon the hydraulic conductivity of its least permeable layer, in relation to the length of the flood period (Parissopoulos and Wheater 1992). The fixed flow measuring flumes installed

in Walnut Gulch in Arizona (Renard and Keppel 1966) also indicate that an increase in wetted perimeter with increasing peak discharge is an important variable in reckoning transmission losses (Figure 11.7). Where flow exceeds channel capacity and spills onto the floodplain, transmission losses may increase dramatically (Knighton and Nanson 1994; see Chapter 10).

Thornes (1977; Figure 11.8) offers a conceptual model of the interaction of transmission losses and tributary discharge contributions for low- and high-magnitude floods, which alerts the geomorphologist to the complicated downstream changes in flow that can be expected in desert floods. Almost all statistical and mathematical models of transmission losses (e.g. Lane *et al.* 1971; Walters 1990; Sharma and Murthy 1995) have necessarily used channel reaches in which the outflow hydrograph is unaffected (as best as can be judged) by complications such as tributary inflow. Interestingly, they give tolerable first approximations, in some cases over channel lengths of up to 25 km. However, Knighton and Nanson (1994) draw attention to the fact that the pattern of behaviour evident in one reach of a river may not be repeated downstream, reinforcing the impression of unpredictability in ephemeral systems.

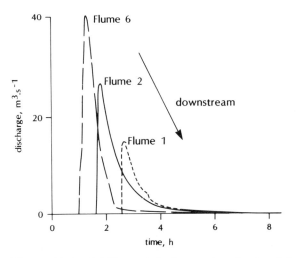

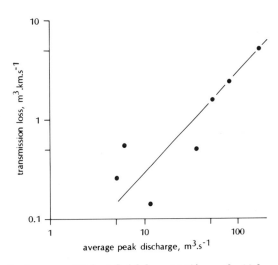

Figure 11.7 *(a) Transmission losses reduce the size of the flood wave on Walnut Gulch by c. 110% over the 10 km channel length between flumes 6 and 1. (b) Transmission losses, expressed in cubic metres per second per kilometre of channel, as a function of average peak discharge at flumes 6 and 2 on Walnut Gulch, Arizona (modified after Renard and Keppel 1966)*

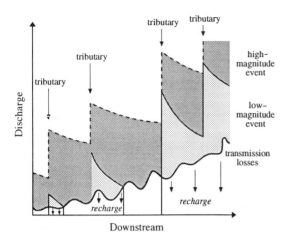

Figure 11.8 *Conceptual model of the interplay of tributary inflows and transmission losses during large and small flash-floods. The hachured areas represent net surface streamflow (modified after Thornes 1977)*

Transmission losses have two effects on the flood wave. They may steepen the bore (Butcher and Thornes 1978), though the effect will be highly variable depending upon the location of the storm in relation to the drainage net and, therefore, how dry the bed is before being overrun (Reid and Frostick 1987). They also ensure that flood discharge *decreases* downstream in the absence of significant tributary inflows (Figure 11.7). This is in complete contrast with perennial streams and has considerable implications for downstream changes in channel geometry and sediment transport.

DRAINAGE BASIN SIZE AND WATER DISCHARGE

The low annual rainfall of arid and semi-arid areas inevitably means that the annual discharge of ephemeral streams is low compared with perennial counterparts having the same size drainage basin. In fact, Wolman and Gerson (1978) demonstrated that the exponent in the log–log relation between mean annual runoff and drainage area of ephemeral streams in California is about half that of the same relation for perennial streams of Maryland (Figure 11.9). However, perhaps just as noteworthy in the relation is the fact that there is far greater

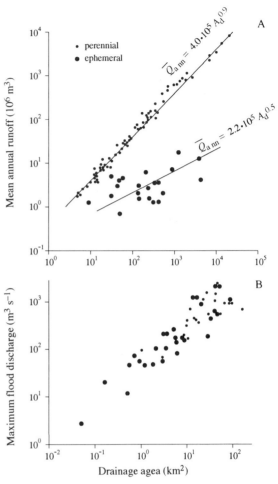

Figure 11.9 *(A) Annual runoff as a function of drainage basin size for perennial rivers in Maryland and ephemeral streams in California. (B) Peak flood discharge as a function of drainage area for ephemeral and perennial streams (modified after Wolman and Gerson 1978; data compiled from several sources)*

range in runoff values among ephemeral streams for any particular drainage area. It is tempting to speculate that this contrast with the perennial stream case is a function of the spatially discrete and highly variable rainfall pattern, together with wide-ranging transmission losses, and a host of other factors including lithologically and pedologically controlled runoff-to-rainfall ratios. Whatever the exact reasons, it highlights again the relative unpredictability of the ephemeral drainage system.

Despite the expected differences in gross annual discharge, Wolman and Gerson (1978) point to the unexpected similarities in flood peak discharge between ephemeral and perennial rivers having the same size drainage basin (Figure 11.9). This must be fortuitous from a process point of view given the now well-established role of soil interflow, the diminutive role of Hortonian overland flow and the general concept of partial-area contribution to streamflow in humid environments (Hewlett and Hibbert 1967) and the undoubted importance of Hortonian overland flow in deserts (see Chapter 5; Bonell and Williams 1986). Nevertheless, it is interesting that desert streams can match perennial counterparts at least in this one flood parameter despite the spottiness of the rainfall pattern, the transmission losses, and so on.

Ephemeral river channel geometry

CHANNEL WIDTH

Desert streams are notable for their great width (Leopold *et al.* 1966; Baker 1977; Frostick and Reid 1979; Graf 1983; Figure 11.10). When visiting a desert for the first time, those whose familiarity is rather with perennial streams of

a

b

Figure 11.10 *(A) Sand-bed ephemeral stream showing great width and planar bed, Laga Tulu Bor, Kenya. (B) Gravel-bed ephemeral showing great width and planar bed, unnamed stream, Sibilot National Park, Kenya*

humid regions might think that the ephemeral channels they come across represent much larger basins than is the case. Wolman and Gerson's (1978) compilation of data for different arid and semi-arid regions confirms this rapid increase in width with drainage area (see Chapter 10 and Figure 10.5), at least for small-to moderate-sized basins. However, their diagram also reveals a fascinating and fairly universal asymptote value of between 100 and 200 m for channel width once the drainage basin exceeds about 50 km². (This value is only achieved in perennial rivers when the drainage basin is about 10 000 km² in size!)

The large rate of increase of stream width in small- to moderate-sized basins might be attributable, in part, to the high drainage densities that are sustained by rapid runoff in drylands. The change to a fairly steady stream width once the water catchment exceeds *c.* 50 km² might result, in part, from the fact

that transmission losses from flows with such a high wetted perimeter compensate for any addition of tributary water. Indeed, in a situation where there are no significant tributary inputs, Dunkerley (1992; Figure 11.11) shows that severe transmission losses produce a commensurate decline in channel width and capacity over distances as small as a few kilometres. On the other hand, it could indicate that the finite size of convective rain cells imposes an upper limit on the discharge of an ephemeral drainage system regardless of its size. In other words, there would be no need for any further increase in channel capacity once the runoff of one rain cell is catered for. This is clearly an area where there is a need for greater information about both flows and channel geometry.

Superimposed on regional trends, Schumm (1961) provides us with an insight into the variability of channel width at a more local scale

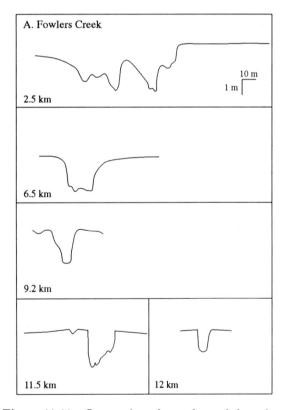
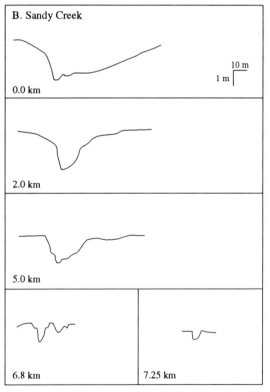

Figure 11.11 *Cross-sections of two ephemeral channels at specified distances beyond the reentrant they make in the scarp of the Barrier Range of New South Wales and as they flow towards a flood basin. Transmission losses ensure rapid reduction in channel capacity and width (modified after Dunkerley 1992)*

by examining a surrogate measure of bed and
bank material shear strength. He shows that the
lower the silt–clay content of the wettable
perimeter, the higher the ratio between bankfull
width and depth (Figure 11.12). Working in
another part of arid North America, Murphey *et
al.* (1972) have also noted downstream changes
in channel width that are related to the nature
of local bank materials. In their case, sandstones are more erodible than conglomerates,
so producing a less well-defined and less confined channel.

However, some reservations may be needed in
assessing the geometry of the arroyos of the
southwestern United States, if only because
land-use changes and changes in storm size and
frequency during the late nineteenth century are
thought to have been responsible for subsequent
channel incision and consequent confinement of
the flow (Cooke and Reeves 1976).

CHANNEL BED MORPHOLOGY

Not only are ephemeral streams wide, they are
noted for having a remarkably subdued bed
topography. In fact, the beds of single thread

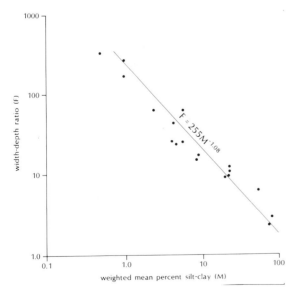

Figure 11.12 *Width-to-depth ratio of stable cross-
sections of four ephemeral streams in four western states of
the USA as a function of weighted mean percentage
silt–clay content of the stream bed and bank material
(after Schumm 1961)*

streams are often near-horizontal and planar;
any bar forms are also often flat-topped and
rise only 10–20 cm above the thalweg (Leopold
et al. 1966; Frostick and Reid 1977; 1979;
Figure 11.13). It is likely that channel width
and bed flatness are related through flow
depth, since great width spreads the flow, so
ensuring shallow depths. This in turn sup-
presses the secondary current cells that might
otherwise encourage the building of bars; but it
also maintains sediment transport efficiency by
maintaining relatively high values of grain
exposure (water depth: bed grain diameter)
throughout flood events. As a result, temporary
'highs' are planed off and 'lows' are filled in.

This does not mean that all ephemeral stream
channels are featureless. Where the bed material
is sand and the last bed-forming flow was of
appropriate strength, dunes and megadunes
may be found (Williams 1970). However,
exaggerated bar forms tend to occur only at
pronounced bends, as might be expected from
analogy with perennial rivers, or they occur
where the channel is diffused in a braid plan
(for example, Cooper Creek in the Lake Eyre
Basin of Australia; Rust and Nanson 1986).
There is a suggestion that single-thread ephemeral streams tend to low sinuosity (Schumm
1961). This straightness may be the reason why
flat channel beds are so characteristic. However, the number of examples that have been
documented and that can be categorised is too
small to make definitive comments.

Fluvial sediment transport

Ephemeral streams move vast quantities of sediment during each flood event, both as bedload
and as suspended load. In fact, they have produced the highest recorded values of suspended
sediment concentration, reaching 68% solids (by
weight) in the Rio Puerco, New Mexico (Bondurant 1951). Material is readily available for
transport (see also Chapter 10), not only from
the thinly vegetated or bare slopes of a catchment, but also from the bed of the stream.

However, before considering the entrainment
of sediment, attention should be drawn to
a characteristic of some ephemeral streams
that distinguishes them from their perennial
counterparts and has considerable effect on the

flow. The perched groundwater that is held in the channel-fill makes the channel bank a favoured, and sometimes only, location for the growth of large trees (Frostick and Reid 1979). Because flows of magnitude sufficient to sweep the channel in perennial streams are infrequent in ephemerals, trees are also able to establish themselves on the channels bed (Graeme and Dunkerley 1993), or trees that were once bankside may find themselves within-channel as the bank is eroded. Graeme and Dunkerley (1993; Figure 11.13) indicate that flow resistance can be greatly increased by the presence of the trunks. Indeed, because the flow may encounter more and more vegetative 'roughness' in the form of branches as discharge increases, they show that, instead of the 'usual' inverse relation between the roughness parameter and water depth, roughness values may increase with increasing depth. Where this is significant, it will obviously have an impact on the amount of energy available for sediment transport.

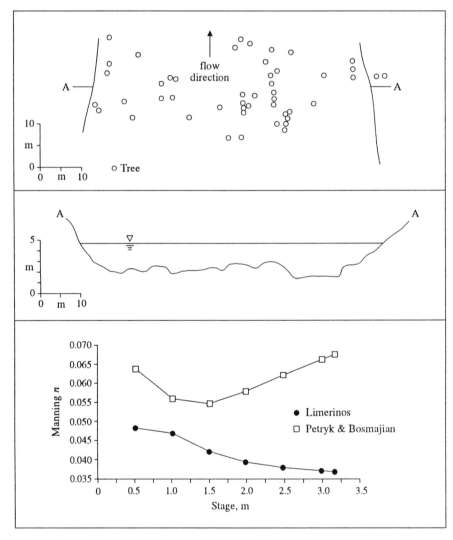

Figure 11.13   *(A and B) Plan and cross-section of a reach of ephemeral Fowlers Creek, New South Wales, showing the distribution of River Red Gum trunks in the channel. (C) Manning roughness coefficient as a function of water stage taking into account the bed material size characteristics only (following Limerinos) and, in addition, the effect of vegetation (following Petryk and Bosmajian). (Modified after Graeme and Dunkerley 1993)*

SCOUR AND FILL

The fact that ephemeral streams allow complete access to their bed sediments between floods without the need for pumping equipment, coffers, and so on, means that they were the first river type to reveal the process of scour and fill.

By emplacing scour chains in the small but elongate Arroyo de Los Frijoles in New Mexico, Leopold *et al.* (1966) were able to show that a single flood might incise the river bed by as much as 0.3 m. However, just as significant was the fact that this scour would be matched by a more or less equal amount of deposition during flood recession (Figure 11.14). Over and above this single flood pattern, it was also revealed that the entire channel was in grade, that is to say, in equilibrium,

since the bed was shown to be restored to more or less the same altitude over four successive flood seasons. Scour and fill has been shown subsequently to operate in other predominantly sandy (Foley 1978; Figure 11.15) and gravelly (Schick *et al.* 1987) streams. Leopold *et al.* were able to provide a loosely deterministic relation between a hydraulic parameter — unit discharge — and the depth of scour (Figure 11.14). Foley (1978) has gone further in his explanation by suggesting that the depth of scour is determined by the amplitude of migrating antidunes that form under upper flow regime conditions. Actual observations by Reid and Frostick (1987) indicate the short-lived nature of supercritical flow during a flash-flood. This suggests that, if scour were indeed due to antidune migration, it occurs in a short time interval of perhaps several tens of minutes, and

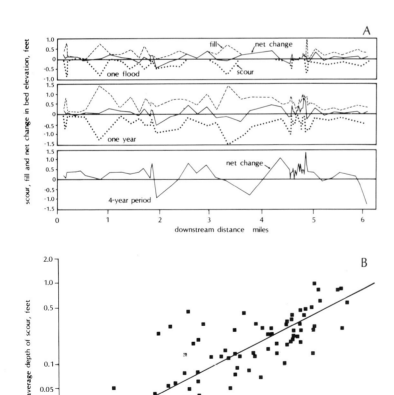

Figure 11.14 *(A) Downstream pattern of bed scour and fill for different periods on the Arroyo de Los Frijoles, New Mexico. (B) Average depth of scour as a function of flood peak discharge in the Arroyo de Los Frijoles. (Modified after Leopold et al. 1966)*

only near peak flow. In fact, in some way corroborating the restricted nature of this interval, the values of the Manning roughness coefficient that have been derived for the major part of flash-flood duration are consistent with plane bed conditions rather than with more complex bedforms (Nordin 1963; Frostick and Reid 1987). It seems likely, therefore, that if antidunes were to be implicated in the process, scour is achieved much quicker than subsequent fill. However, observations of standing waves suggest that the underlying antidunes are spatially discrete and extremely transient. The channel-wide scour and fill patterns indicated by the scour chains of Leopold *et al.* (1966) may require a different explanation, perhaps involving plane beds in the upper flow regime.

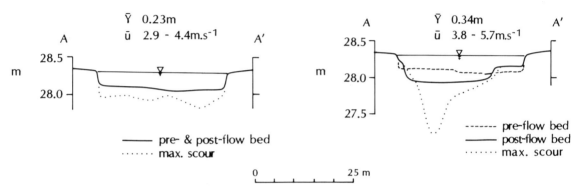

Figure 11.15   *The patterns of scour and fill at one cross-section of Quatal Creek, California, during two flash-floods of different magnitude. Y is calculated average flow depth; u is calculated average flow velocity (modified after Foley 1978)*

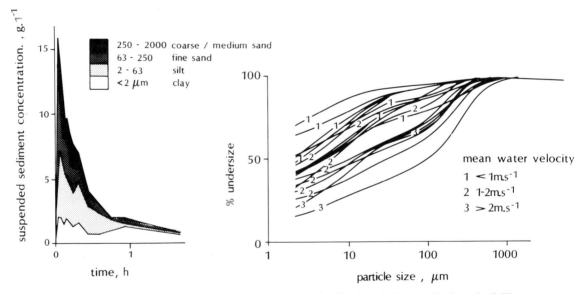

Figure 11.16   *(A) Concentration of suspended sediment by grain-size class during one flood on the Il Kimere, Kenya. (B) Size distribution of suspended sediment sampled at various flow velocities during one flood in the Il Kimere, Kenya. (Modified after Frostick et al. 1983; Reid and Frostick 1987)*

SEDIMENT TRANSPORT IN SUSPENSION

Irrespective of the point in a flood when bed scour takes place, the process provides an easily erodible source of sediment. Added to this is the material that is brought to the stream channel by overland flow. This produces sediment concentrations that have inspired local farmers to

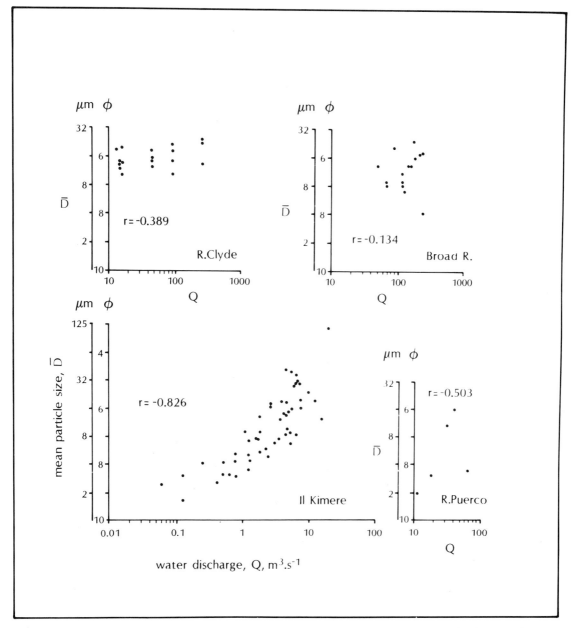

Figure 11.17 *Comparison of the well-defined relationship between the mean size of suspended sediment and water discharge in flash-floods on ephemeral rivers — Il Kimere and Rio Puerco — and the poor relationship established in perennial rivers — River Clyde and Broad River. (Modified after Reid and Frostick 1987; data compiled from several sources)*

describe the Colorado River and some of its tributaries as 'too thin to plough and too thick to drink!' (Beverage and Culbertson 1964).

From the few measurements of suspended sediment that have been taken for flash floods (e.g. Nordin 1963; Lekach and Schick 1982; Reid and Frostick 1987; Sutherland and Bryan 1990), it can be shown that the concentration rises along with water discharge, as with perennial streams for which there is, of course, vastly greater information. However, this is the only similarity. The constant a in the relationship $C = aQ^b$, where C is sediment concentration (measured in mg L^{-1}) and Q is water discharge (measured in $m^3 s^{-1}$), is anywhere between 6 and 4500 times higher in ephemeral streams. This reflects the fact that, even at low flows, sediment concentrations are often more than five times as high as at times of *high* flow in perennial systems. On the other hand, the exponent b is usually less than unity for ephemeral streams, but always greater than unity for perennial rivers. This suggests that perennial systems are more responsive to changes in discharge, but only in relative terms, since the range of concentrations that might be expected in an ephemeral stream will be anywhere from 35 to 1700 times higher than that expected in perennial systems (Frostick *et al.* 1983).

There is hysteresis in the relation between suspended sediment concentration and flow, as in perennial rivers. However, it appears to be less pronounced. This might be because particles of all sizes are available for transport, and transport is dictated by changes in the hydraulic environment rather than by sediment supply limitation, as has been suggested for perennial rivers. In fact, this is one area where ephemeral streams might prove to be more predictable than their perennial counterparts. It has been shown that the size distribution of suspended sediment varies systematically with water velocity or discharge in desert streams, in complete contrast with perennial rivers (Reid and Frostick 1987; Figures 11.16 and 11.17).

SEDIMENT TRANSPORT ALONG THE STREAM BED

In keeping with river sediment studies in all environments, much less is known about bedload transport in desert streams than about transport in suspension. In fact, in relative terms there is probably more information about bedload in desert streams because of recent field monitoring programmes, mainly in Israel. But this 'anomaly' only reflects the paucity of studies of any sort on fluvial sediment transport in the world's drylands.

Leopold *et al.* (1966) were able to show that large clasts of pebble to cobble size can be moved by as much as 3 km during a single flood event by overpassing a predominantly sand-bed channel. Schick *et al.* (1987) and Hassan (1990b) have extended this work to cover gravel-bed streams. They have traced tagged clasts as they move downstream in the Nahal Hebron in the Negev Desert. The use of a two-coil metal detector allowed them to locate buried clasts as well as those exposed on the surface of the stream bed. As a result, they have been able to highlight two important processes: for the first time, scour and fill has been shown to operate in a gravel-bed channel; besides, this, there is a clear pattern of exchange between buried and exposed clasts, flood by flood, which Schick *et al.* relate to the depth of scour in each event (Figure 11.18). In addition, using some of the same data, Church and Hassan (1992) have indicated the effects that clast interlock and

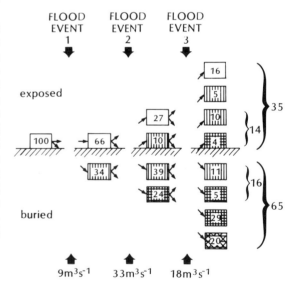

Figure 11.18 *Pattern of burial and exhumation of tagged gravel clasts during downstream transport by three successive floods in the Nahal Hebron, Israel. The tagged clasts were seeded **initially** on the surface of the gravel bed (after Schick* et al. *1987)*

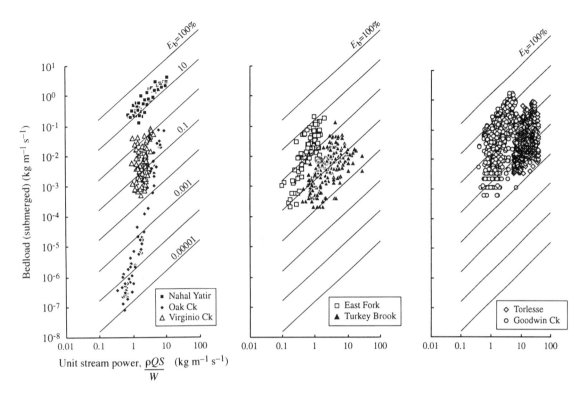

Figure 11.19 *Comparison of bedload transport rates measured by slot samplers in the ephemeral Nahal Yatir (Israel), the seasonal Goodwin Creek (Mississippi), and the perennial East Fork River (Wyoming), Oak Creek (Oregon), Torlesse Stream (New Zealand), Turkey Brook (England), and Virginio Creek (Italy). East Fork River has a gravelly sand-bed. The other streams have gravel beds. E_b is Bagnold's efficiency parameter (modified after Reid and Laronne 1995)*

exposure have on transport distance in the context of clast size and flood magnitude, suggesting that bed structure appears more important in controlling downstream displacement during moderate floods in which values of excess shear stress are comparatively small.

These studies provide valuable information about bedload, but the data are of a 'black box' nature in that they report displacement of clasts after the event. Indeed, in some cases (e.g. Schick and Lekach 1987) even the flood hydrograph has been reconstructed from slackwater deposit evidence of maximum water-stage. Live-bed information in ephemeral streams is inevitably rare given the dangers of wading in flash floods with portable bedload samplers, let alone the fact that the chance of being on-site during an event is extremely small (Lekach and Schick 1983).

This absence of information has been remedied in recent years by the installation of Birkbeck-type bedload samplers in two

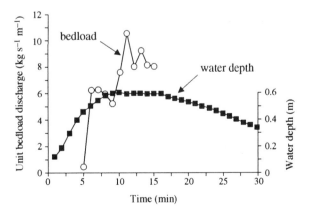

Figure 11.20 *Bedload transport rate and hydrograph of a flash-flood in the Nahal Yatir, Israel, showing the comparatively simple response of bedload to changes in flow (modified after Laronne et al. 1994)*

gravel-bed ephemeral streams in the northern Negev Desert, Israel (Laronne *et al.* 1992). These have revealed patterns of behaviour that, as with other characteristics, set desert streams apart from those of humid environments. First, bedload transport rates are much higher over the same range of shear stress. The differences might be as high as a million-fold at the threshold of entrainment and 10-fold at moderate levels of shear stress (Figure 11.19; Laronne and Reid 1993; Reid and Laronne 1995). Second, the relation between bedload transport rate and flow appears to be simpler and, therefore, more predictable (Figure 11.20; Reid *et al.*. 1996). These differences have been attributed to a lack of armour development and this is thought to arise because the sparsely vegetated slopes of the desert are capable of supplying large quantities of sediment. Indeed, Pickup (1991) draws attention to the effect of reducing vegetation cover through drought or grazing in the Todd River Basin, central Australia, pointing to a 10-fold increase in sediment transport over those rates measured during marginally wetter periods. The ready supply of sediment to the channel system counteracts the tendency for selective entrainment of finer clasts on the channel bed, so reducing armour development and permitting high bedload transport rates through an absence of protection

(Dietrich *et al.* 1989; Laronne *et al.* 1994). This has obvious implications for sediment yield, giving clarity to the signal that bedload cannot be ignored as a component in arid environments in the same way that it often and conveniently is in more humid environments.

Desert river deposits

It is perhaps curious that much of what has been written about the *dynamic* behaviour of desert streams is speculative, and comes from observation of *static* fluvial deposits. Perhaps even more curious is that there is more written about the dynamics of desert sediments that may be as much as 400 million years old (see, for example, Allen 1964; Schumm 1968; Tunbridge 1984; Olsen 1987; Frostick *et al.* 1988; Frostick *et al.* 1992) than about those of modern streams for which there is more information about the likely character of flows that were responsible, the nature of the channel, and so on. There are, however, a few studies of modern flash-flood deposits. They are often concerned with a high magnitude event that has achieved local notoriety (see, for example, McKee *et al.* 1967; Williams 1971; Sneh 1983), though some are less specific (such as Picard and High 1973).

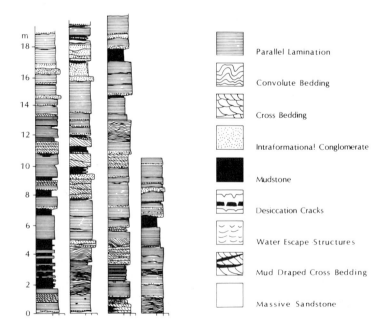

Figure 11.21 *Measured section of flash-flood deposits of the Middle Devonian Trentishoe Formation, exposed in the sea cliffs of north Devon, England (modified after Tunbridge 1984)*

A

B

C

Figure 11.22 *(A) Flash-flood deposits of the Triassic Burghead Beds on the Moray coast, Scotland. (B) Section in a modern ephemeral stream channel-fill, Kenya, showing couplets of horizontal laminae. (C) Thick mud drape in a modern ephemeral stream channel, Kenya, laid down in the last stages of flood flow*

Since the hydraulic processes that entrain a clast are universal regardless of the environmental setting, the question arises as to whether or not there are sedimentary characteristics that distinguish ephemeral from perennial stream deposits.

There appear to be at least three attributes in particular that might lead a sedimentologist to conclude that an ancient sedimentary sequence was laid down by an ephemeral stream.

THIN BEDS IN SANDS AND PEBBLY SANDS

There is no quantification which can be used to justify a claim that desert river deposits are composed characteristically of thin beds, that is, 0.1–0.3 m thick. However, deposits of widely differing age — Devonian (Tunbridge 1984, Figure 11.21) and Triassic (Frostick *et al.* 1988) — undoubtedly have an easily recognisable affinity because of the nature of their bedding (Figure 11.22). There is plenty of evidence of the minor incision that is associated with scour in the nested fill-sets of ancient deposits. The range in bed thickness is also consistent with the

depth of scour and fill as defined by field experiments in modern ephemeral streams (Leopold *et al.* 1966; Figure 11.14).

PREDOMINANCE OF HORIZONTAL LAMINATION IN SAND-BEDS

The planar surface of many single-thread ephemeral streams has already been remarked on. This surface reflects the apparent ubiquity (Picard and High 1973) and prevalence (McKee *et al.* 1967; Figure 11.23) of horizontal primary structures in ephemeral stream deposits. There is, however, some confusion over the conditions of flow responsible for their formation. This stems largely from the singular lack of observations of floods in ephemeral channels, but it is more than probable that they represent upper flow regime plane beds. They usually comprise a set of couplets, however, in which laminae of coarser and finer particles alternate (Figure 11.22). An examination of the deposits of floods associated with one storm but in four contiguous river basins of northern Kenya suggests that the separation of particles by size is related to small pulses of sediment-laden water that are superimposed on the main flood wave by

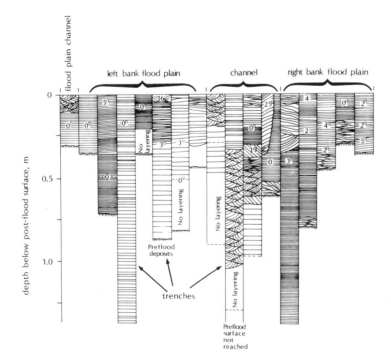

Figure 11.23 *Flash-flood deposits of Bijou Creek, Colorado, exposed in trenches cut in the floodplain and channel. Note the predominance of horizontal primary structures (modified after McKee* et al. *1967)*

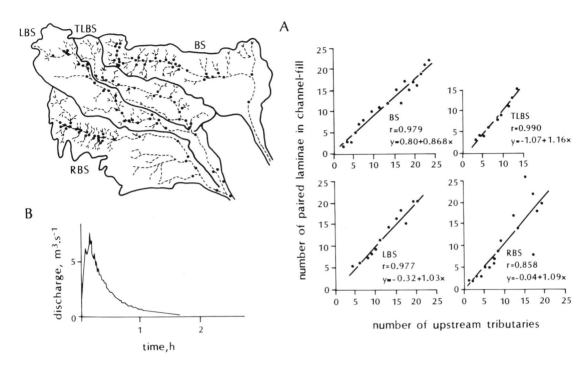

Figure 11.24 *(A) Drainage nets and catchments of four contiguous sand-bed ephemeral streams, northern Kenya; and the relationship between the number of coarse–fine couplets of horizontal laminae and the number of influent tributaries upstream of any sampling trench in the channel-fill of the most recent flash-flood (modified and redrawn after Frostick and Reid 1979). (B) Flood hydrograph of the Il Kimere, Kenya, showing the discharge pulses that arise from tributary contributions to the trunk stream (modified after Frostick* et al. *1983)*

contributions from tributary channels (Frostick and Reid 1977; Figure 11.24).

MUD DRAPES AND MUD INTERCLASTS

The fact that the concentration of suspended sediment is often in excess of several thousand milligrams per litre even at the last stages of flow, that much of this is of clay size, and that the channel dries out after each flood so that fines are not winnowed as they might be in a perennial system, mean that ephemeral stream sequences are often punctuated by comparatively thick (say 0.1 m) drapes of clay that have settled from the flow or from stagnant pools in the channel (Tunbridge 1984; Reid *et al.* 1989; Figure 11.22). However, not only are there intact clay drapes, but an abundance of clay intraclasts. These have started as mud-curls lying on the stream bed, have been entrained by a subsequent flood, and have ended by being incorporated in a superposed bed downstream.

Conclusion

Rivers play an important role in shaping the Earth's deserts despite the fact that they run for only a vanishingly small fraction of time as one moves towards the hyper-arid core regions. The thick sequences of fluviatile desert sediments that range in age from Precambrian (Williams 1969), through Devonian (Carruthers 1987) and Triassic (Steel and Thompson 1983), to the Pleistocene (Vondra and Bowen 1978) are more than adequate as a testimony.

When they run, desert rivers are more efficient erosional agent than their perennial counterparts. This is due to the fact that material is readily available for erosion on the poorly vegetated slopes of a water catchment, as well as from the channel bed. Sediment transport is much less supply-limited. As a result, suspended sediment concentrations reach record levels as do bedload transport rates. However, the consequences of such efficiency are far from beneficial. For instance, the life expectancy of water impoundment structures is shortened considerably as deltas prograde into dam lakes. Even where the dammed trunk river is perennial, it may be being supplied with sediment by its highly efficient ephemeral tributaries as it passes through a desert region. As a result, the half-life of the reservoirs — so essential to human existence in such regions — will be much reduced. This is the case with the Elephant Butte Reservoir on the Rio Grande (Vanoni 1975) and with the Tarbella Reservoir on the Indus where 6% of the capacity had been lost within five years of its construction (Ackers and Thompson 1987).

However, the spottiness of rainfall in desert regions, together with the highly variable effects of transmission losses, make ephemeral rivers highly unpredictable in comparison with perennial equivalents. As a result, there is less chance of achieving a stochastic model of flow and therefore less chance of predicting their nuisance. There is a great need for an increase in information about channel form, flood hydrology and sediment transport through an increase in fixed gauging stations, however unrewarding it may seem to document floods that often occur as infrequently as less than once per year.

References

Abrahams, A.D. and Parsons, A.J. (eds) 1994. *Geomorphology of desert environments.* Chapman & Hall, London.

Ackers, P. and Thompson, G. 1987. Reservoir sedimentation and influence of flushing. In C.R. Thorne, J.C. Bathurst and R.D. Hey (eds), *Sediment transport in gravel-bed rivers.* Wiley, Chichester: 845–868.

Allen, J.R.L. 1964. Studies in fluviatile sedimentation: 6. Cyclothems from the lower Old Red Sandstone, Anglo-Welsh Basin. *Sedimentology*, 3: 89–108.

Bagnold, R.A. 1941. *The physics of blown sand and desert dunes.* Methuen, London.

Baker, V.R. 1977. Stream channel response to floods, with examples from central Texas. *Bulletin of the Geological Society America Bulletin*, 88: 1057–1071.

Ben-Zvi, A., Massoth, S. and Schick, A.P. 1991. Travel time of runoff crests in Israel. *Journal of Hydrology*, 122: 309–320.

Beverage, J.P. and Culbertson, J.K. 1964. Hyper-concentrations of suspended sediment. *Proceedings of the American Society of Civil Engineers, Journal of the Hydraulics Division*, 90 (HY6): 117–128.

Bondurant, D.C. 1951. Sedimentation studies at Conchas Reservoir in New Mexico. *Transactions of the American Society Civil Engineers*, 116: 1292–1295.

Bonell, M. and Williams, J. 1986. The generation and redistribution of overland flow on a massive oxic soil in a eucalypt woodland within the semi-arid Tropics of north Australia. *Hydrological Processes*, 1: 31–46.

Brookfield, M.E. and Ahlbrandt, T.S. (eds) 1983. *Eolian sediments and processes.* Developments in Sedimentology, 38. Elsevier, Amsterdam.

Burkham, D.E. 1970. *Depletion of streamflow by infiltration in the main channels of the Tucson Basin, southeastern Arizona.* US Geological Survey, Water Supply Paper 1939.

Butcher, G.C. and Thornes J.B. 1978. Spatial variability in runoff processes in an ephemeral channel. *Zeitschrift für Geomorphologie*, Supplementband 29: 83–92.

Carruthers, R.A. 1987. Aeolian sedimentation from the Galtymore Formation (Devonian), Ireland. In L.E. Frostick and I. Reid (eds), *Desert sediments: ancient and modern.* Geological Society of London Special Publication, 35. Blackwell Scientific, Oxford: 251–268.

Church, M. and Hassan, M.A. 1992. Size and distance of travel of unconstrained clasts on a streambed. *Water Resources Research*, 28: 299–303.

Cooke, R.U. and Reeves, R.W. 1976. *Arroyos and environmental change in the American Southwest.* Clarendon Press, Oxford.

Cooke, R.U., Warren, A. and Goudie, A.S. 1993. *Desert geomorphology.* UCL Press, London.

Dietrich, W.E., Kirchner, J.W., Ikeda, H. and Iseya, F. 1989. Sediment supply and the development of the coarse surface layer in gravel-bedded rivers. *Nature*, 340: 215–217.

Diskin, M.H. and Lane, L.J. 1972. A basinwide stochastic model of ephemeral stream runoff in southeastern Arizona. *International Association Scientific Hydrologists Bulletin*, 17: 61–76.

Dunkerley, D.L. 1992. Channel geometry, bed material, and inferred flow conditions in ephemeral stream systems, Barrier Range, Western N.S.W., Australia. *Hydrological Processes*, 6: 417–433.

Farquharson, F.A.K., Meigh, J.R. and Sutcliffe, J.V. 1992. Regional flood frequency analysis in arid and semi-arid areas. *Journal of Hydrology*, 138: 487–501.

Foley, M.G. 1978. Scour and fill in steep, sand-bed ephemeral streams. *Bulletin of the Geological Society of America*, 89: 559–570.

Frostick, L.E. and Reid, I. 1977. The origin of horizontal laminae in ephemeral stream channel-fill. *Sedimentology*, 24: 1–9.

Frostick, L.E. and Reid, I. 1979. Drainage-net control of sedimentary parameters in sand-bed ephemeral streams. In A.F. Pitty. (ed.), *Geographical approaches to fluvial processes*. Geoabstracts, Norwich: 173–201.

Frostick, L.E. and Reid, I. (eds) 1987. *Desert sediments: ancient and modern.* Geological Society of London Special Publication, 35. Blackwell Scientific, Oxford.

Frostick, L.E., Reid, I. and Layman, J.T. 1983. Changing size distribution of suspended sediment in arid-zone flash floods. *Special Publication International Association Sedimentologists*, 6: 97–106.

Frostick, L.E., Reid, I., Jarvis, J. and Eardley, H. 1988. Triassic sediments of the Inner Moray Firth, Scotland: early rift deposits. *Journal of the Geological Society of London*, 145: 235–248.

Frostick, L.E., Linsey, T.K. and Reid, I. 1992. Tectonic and climatic control of Triassic sedimentation in the Beryl Basin, northern North Sea. *Journal of the Geological Society of London*, 149: 13–26.

Gavrilovic, D. 1969. Die Überschwemmungen im Wadi Bardague im Jahr 1968 (Tibesti, Rep. du Tchad). *Zeitschrift für Geomorphologie*, NF, 14: 202–218.

Graeme, D. and Dunkerley, D.L. 1993. Hydraulic resistance by the River Red Gum, *Eucalyptus camaldulensis*, in ephemeral desert streams. *Australian Geographical Studies*, 31: 141–154.

Graf, W.L. 1983. Flood-related channel change in an arid-region river. *Earth Surface Processes and Landforms*, 8: 125–139.

Hassan, M.A. 1990a. Observations of desert flood bores. *Earth Surface Processes and Landforms*, 15: 481–485.

Hassan, M.A. 1990b. Scour, fill and burial depth of coarse material in gravel bed streams. *Earth Surface Processes and Landforms*, 15: 341–356.

Hewlett, J.D. and Hibbert, A.R. 1967. Factors affecting the response of small watersheds to precipitation in humid areas. In W.E. Sopper and H.W. Lull (eds), *Forest hydrology*. Pergamon, Oxford: 267–290.

Hjalmarson, H.W. 1984. Flash flood in Tanque Verde Creek, Tucson, Arizona. *Journal of Hydraulic Engineering, ASCE*, 110: 1841–1852.

Hubbell, D.S. and Gardner, J.L. 1944. Some edaphic and ecological effects of water spreading on rangelands. *Ecology*, 55: 27–44.

Knighton, A.D. and Nanson, G.C. 1994. Flow transmission along an arid zone anastomosing river, Cooper Creek, Australia. *Hydrological Processes*, 8: 137–154.

Lane, L.J., Diskin, M.H. and Renard, K.G. 1971. Input–output relationships for an ephemeral stream channel system. *Journal of Hydrology*, 13: 22–40.

Laronne, J.B. and Reid, I. 1993. Very high bedload sediment transport in desert ephemeral rivers. *Nature*, 366: 148–150.

Laronne, J.B., Reid, I., Yitshak, Y. and Frostick, L.E. 1992. Recording bedload discharge in a semiarid channel, Nahal Yatir, Israel. In J. Bogen, D.E. Walling and T.J. Day (eds), *Erosion and sediment monitoring programmes in river basins*. International Association Hydrological Sciences, Publication, 210: 79–86.

Laronne, J.B., Reid, I., Yitshak, Y. and Frostick, L.E. 1994. The non-layering of gravel streambeds under ephemeral flow regimes. *Journal of Hydrology*, 159: 353–363.

Lekach, J. and Schick, A.P. 1982. Suspended sediment in desert floods in small catchments. *Israel Journal of Earth Sciences*, 31: 144–156.

Lekach, J. and Schick, A.P. 1983. Evidence for transport of bedload in waves: analysis of fluvial sediment samples in a small upland stream channel. *Catena*, 10: 267–279.

Leopold, L.B. and Miller, J.P. 1956. *Ephemeral streams — hydraulic factors and their relation to the drainage net.* US Geological Survey, Professional Paper 282A.

Leopold, L.B., Wolman, M.G. and Miller, J.P. 1964. *Fluvial processes in geomorphology*. Freeman & Co., San Francisco.

Leopold, L.B., Emmett, W.W. and Myrick, R.M. 1966. *Channel and hillslope processes in a semiarid area of New Mexico.* US Geological Survey, Professional Paper 352G.

Martínez-Goytre, J., House, P.K. and Baker, V.R. 1994. Spatial variability of small-basin paleoflood magnitudes for a southeastern Arizona mountain range. *Water Resources Research*, 30: 1491–1501.

McGee, W.J. 1897. Sheetflood erosion. *Bulletin of the Geological Society of America*, 8: 87–112.

McKee, E.D. (ed.) 1979. *A study of global sand seas.* US Geological Survey, Professional Paper 1052.

McKee, E.D., Crosby, E.J. and Berryhill, H.L. 1967. Flood deposits of Bijou Creek, Colorado, June 1965. *Journal of Sedimentary Petrology*, 37: 829–851.

Murphey, J.B., Lane, L.J. and Diskin, M.H. 1972. Bed material characteristics and transmission losses in an ephemeral stream. *Hydrology and water resources in Arizona and the Southwest.* Proceedings 1972 Meeting Arizona Section, American Water Association and the Hydrology Section, Arizona Academy of Sciences, Prescott, Arizona, 2: 455–472.

Nordin, C.F. 1963. *A preliminary study of sediment transport parameters, Rio Puerco, near Bernardo, New*

Mexico. US Geological Survey, Professional Paper 462C.

Olsen, H. 1987. Ancient ephemeral stream deposits: a local terminal fan model from the Bunter Sandstone Formation (L. Triassic) in the Tønder-3, -4 and -5 wells, Denmark. In L.E. Frostick and I. Reid (eds), *Desert sediments: ancient and modern*. Geological Society of London, Special Publication, 35. Blackwell Scientific, Oxford: 69–86.

Parissopoulos, G.A. and Wheater, H.S. 1992. Experimental and numerical infiltration studies in a wadi stream bed. *Hydrological Sciences Journal*, 37: 27–37.

Picard, M.D. and High, L.R. 1973. *Sedimentary structures of ephemeral streams*. Developments in Sedimentology, 17. Elsevier, Amsterdam.

Pickup, G. 1991. Event frequency and landscape stability on the floodplain systems of arid central Australia. *Quaternary Science Reviews*, 10: 463–473.

Pilgrim, D.H., Chapman, T.G. and Doran, D.G. 1988. Problems of rainfall-runoff modelling in arid and semiarid regions. *Hydrological Sciences Journal*, 33: 379–400.

Reid, I. and Frostick, L.E. 1984. Particle interaction and its effect on the thresholds of initial and final bedload motion in coarse alluvial channels. In E.H. Koster and R.J. Steel (eds), *Sedimentology of gravels and conglomerates*. Canadian Society of Petroleum Geologists, Memoir. 10: 61–68.

Reid, I. and Frostick, L.E. 1986. Slope process, sediment derivation and landform evolution in a rift valley basin, northern Kenya. In L.E. Frostick, R.R. Renaut, I. Reid and J.J. Tiercelin (eds), *Sedimentation in the African rifts*. Geological Society of London, Special Publication, 25, Blackwell Scientific, Oxford: 99–111.

Reid, I. and Frostick, L.E. 1987. Flow dynamics and suspended sediment properties in arid zone flash floods. *Hydrological Processes*, 1: 239–253.

Reid, I. and Laronne, J.B. 1995. Bedload sediment transport in an ephemeral stream and a comparison with seasonal and perennial counterparts. *Water Resources Research* 31, 773–781.

Reid, I., Best, J.L. and Frostick, L.E. 1989. Floods and flood sediments at river confluences. In K. Beven and P.A. Carling (eds), *Floods*. British Geomorphological Research Group Symposium Series. Wiley, Chichester: 135–150.

Reid, I., Laronne, J.B., Powell, D.M. and Garcia, C. 1994. Flash floods in desert rivers: studying the unexpected. *EOS, Transactions American Geophysical Union*, 75: 452.

Reid, I., Powell, D.M. and Laronne, J.B. 1996. Prediction of bedload transport by desert flash-floods. *Journal of Hydraulic Engineering, ASCE*, 122: 170–173.

Renard, K.G. and Keppel, R.V. 1966. Hydrographs of ephemeral streams in the Southwest. *Proceedings of the American Society of Civil Engineers, Journal of the Hydraulics Division*, 92 (HY2): 33–52.

Renard, K.G. and Lane, L.J. 1975. Sediment yield as related to a stochastic model of ephemeral runoff. *Present and prospective technology for predicting sediment yields and sources*. Proceedings Sediment-yield Workshop, USDA Sedimentation Laboratory, Oxford, MS, 1972: 253–263.

Renard, K.G. and Laursen, E.M. 1975. Dynamic behaviour model of ephemeral streams. *Proceedings of the American Society Civil Engineers, Journal of the Hydraulics Division*, 101 (HY5): 511 528.

Rust, B.R. and Nanson, G.C. 1986. Contemporary and palaeochannel patterns and the late Quaternary stratigraphy of Cooper Creek, southwest Queensland, Australia. *Earth Surface Processes and Landforms*, 11: 581–590.

Schick, A.P. 1970. Desert floods. In *Symposium on the results of research on representative experimental basins*. International Association Scientific Hydrologists/Unesco: 478–493.

Schick, A.P. 1988. Hydrologic aspects of floods in extreme arid environments. In V.R. Baker, R.C. Kochel and P.C. Patton (eds), *Flood geomorphology*. Wiley, New York: 189–203.

Schick, A.P. and Lekach, J. 1987. A high magnitude flood in the Sinai Desert. In L. Mayer and D. Nash (eds), *Catastrophic flooding*. Allen & Unwin, Boston: 381–410.

Schick, A.P., Lekach, J. and Hassan, M.A. 1987. Vertical exchange of coarse bedload in desert streams. In L.E. Frostick and I. Reid (eds), *Desert sediments: ancient and modern*. Geological Society of London Special Publication, 35. Blackwell Scientific, Oxford: 7–16.

Schumm, S.A. 1961. *Effect of sediment characteristics on erosion and deposition in ephemeral stream channels*. US Geological Survey, Professional Paper 352C.

Schumm, S.A. 1968. *River adjustment to altered hydrologic regimen — Murrumbidgee River and paleochannels, Australia*. US Geological Survey, Professional Paper 598.

Sharma, K.D. and Murthy, J.S.R. 1995. Hydrological routing of flow in arid ephemeral channels. *Journal of Hydraulic Engineering, ASCE*, 121: 466–471.

Sharon, D. 1972. The spottiness of rainfall in a desert area. *Journal of Hydrology*, 17: 161–175.

Sharon, D. 1974. The spatial pattern of convective rainfall in Sukumaland, Tanzania — a statistical analysis. *Arch. Met. Geoph. Bookl.*, Ser. B, 22: 201–218.

Sneh, A. 1983. Desert stream sequences in the Sinai peninsula. *Journal of Sedimentary Petrology*, 53: 1271–1279.

Srikathan, R. and McMahon, T.A. 1980. Stochastic

generation of monthly flows for ephemeral streams. *Journal of Hydrology*, 47: 19–40.

Steel, R.J. and Thompson, D.B. 1983. Structures and textures in Triassic braided stream conglomerates ('Bunter' Pebble Beds) in the Sherwood Sandstone Group, north Staffordshire, England. *Sedimentology*, 30: 341–367.

Sutherland, R.A. and Bryan, R.B. 1990. Runoff and erosion from a small semiarid catchment, Baringo District, Kenya. *Applied Geography*, 10: 91–109.

Thornes, J.B. 1976. *Semi-arid erosional systems: case studies from Spain*. London School of Economics Geographical Paper 7.

Thornes, J.B. 1977. Channel changes in ephemeral streams: observations, problems and models. In K.G. Gregory (ed.) *River channel changes*. Wiley, Chichester: 317–335.

Tunbridge, I.P. 1984. Facies model for a sandy ephemeral stream and clay playa complex: the Middle Devonian Trentishoe Formation of north Devon, UK. *Sedimentology*, 31: 697–715.

Vanoni, V.A. (ed.) 1975. *Sedimentation engineering. American Society of Civil Engineers Manual on Sedimentation*. American Society of Civil Engineers, New York.

Vondra, C.F. and Bowen, B.E. 1978. Stratigraphy, sedimentary facies and palaeoenvironments, East Lake Turkana, Kenya. In W.W. Bishop (ed.), *Geological background to fossil Man*. Scottish Academic Press, Edinburgh: 395–414.

Walters, M.O. 1990. Transmission losses in arid region. *Journal of Hydraulic Engineering, ASCE*, 116: 129–138.

Wheater, H.S. and Brown, R.P.C. 1989. Limitations of design hydrographs in arid areas — an illustration from south west Saudi Arabia. In *Proceedings of the British Hydrological Society National Symposium*, Sheffield, September 1989: 3.49–3.56.

Wheater, H.S., Butler, A.P., Stewart, E.J. and Hamilton, G.S. 1991a. A multivariate spatial–temporal model of rainfall in southwest Saudi Arabia. I. Spatial rainfall characteristics and model formulation. *Journal of Hydrology*, 125: 175–199.

Wheater, H.S., Onof, C., Butler, A.P. and Hamilton, G.S. 1991b. A multivariate spatial–temporal model of rainfall in southwest Saudi Arabia. II. Regional analysis and long-term performance. *Journal of Hydrology*, 125: 201–220.

Williams, G.E. 1969. Characteristics and origins of a Precambrian pediment. *Journal of Geology*, 77: 183–207.

Williams, G.E. 1970. The central Australian stream floods of February–March 1967. *Journal of Hydrology*, 11: 185–200.

Williams, G.E. 1971. Flood deposits of the sand-bed ephemeral streams of central Australia. *Sedimentology*, 17: 1–40.

Wolman, M.G. and Gerson, R. 1978. Relative scales of time and effectiveness of climate in watershed geomorphology. *Earth Surface Processes*, 3: 189–208.

Wooding, R.A. 1966. A hydraulic model for the catchment-stream problem. *Journal of Hydrology*, 4: 21–37.

Yair, A. and Lavee, H. 1976. Runoff generative process and runoff yield from arid talus mantled slopes. *Earth Surface Processes*, 1: 235–247.

12

The role of alluvial fans in arid zone fluvial systems

Adrian M. Harvey

The occurrence and role of alluvial fans

OCCURRENCE

Alluvial fans are depositional landforms which occur where confined streams emerge from mountain catchments into zones of reduced stream power. An abrupt reduction in stream power commonly occurs at mountain-front locations (Bull 1979), and here fans may be built up by progressive deposition, especially of the coarse fraction of the stream's sediment load. The gross topography of mountain fronts may be tectonically or erosionally controlled. At faulted mountain fronts, alluvial fan sediments may form part of a thick sequence of sedimentary basin-fill deposits, but at erosionally controlled mountain fronts the fan deposits may be thinner, burying former erosional pediment surfaces. In both situations the alluvial fans form a transitional environment between mountains and plains (Figure 12.1a). Fans may also occur within mountain regions, in intermontane basins and at valley junctions (Figures 12.1b and 12.2a); in these cases they may be confined at their margins by valley walls, whereas mountain-front fans tend to be unconfined or confined only by neighbouring fans.

Alluvial fans generally have a conical surface form with slopes radiating away from an apex at the point where the stream issues from the mountain catchment (Figure 12.2b). Some fans, especially along active faults, may have a single stream source at the apex; others, especially where pedimentation has occurred prior to burial by the fan deposits, or where the fan sediments have backfilled into the mountain catchment (Figure 12.2c) may have more complex multiple sources (Figure 12.1b). Away from the mountain front the fans may coalesce to form an alluvial apron or *bajada* (Figure 12.2d). Distally, fan deposits may merge with river, lake or windblown deposits, and may form part of a continuous suite from mountain source to basin centre or main drainage. Fans that terminate in standing water are known as fan-deltas. The long-profiles of alluvial fans tend to be slightly concave upwards and cross profiles convex upwards (Hooke 1967; Bull 1977).

The relationships between the stream channels and the fan surface reflect the erosional/depositional behaviour of the system as a whole. Within the mountain source area the stream network has a conventional tributary pattern of channels which issue as one channel onto the fan surface. Where deposition begins

Arid Zone Geomorphology: Process, Form and Change in Drylands, 2nd edition. Edited by David S. G. Thomas.

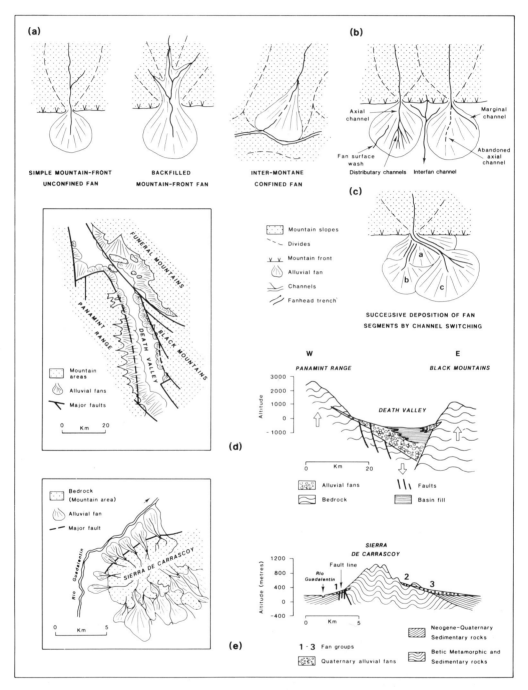

Figure 12.1 *Topographic and locational characteristics of alluvial fans. (a) Topographic relationships. (b) Fan/channel relationships. (c) Switching and successive fan surface age (modified from Denny 1967). (d) Death Valley, California: fan location and geological section (modified from Hunt 1975). (e) Sierra de Carrascoy, Murcia, Spain: fan location and geological section (modified from Harvey 1988)*

Figure 12.2   *Fan topography. (a) Tributary junction fan, Wadi Al Bih, Oman. (b) Hanaupah Canyon fan, Death Valley, west-side. A mountain-front fan: note apex (x), fanhead trench (f) and intersection point (i). (c) Roy fan, Sierra de Carrascoy, Murcia, Spain: note backfilling into mountain catchment. (d) Death Valley, east-side fans; coalescence to form a* bajada. *(e) Deep fanhead trench at Nijar, Almeria, southeast Spain. Trench cuts through fan deposits (f) into underlying bedrock (b)*

to dominate, this channel may become a multi-thread distributary system (Figure 12.1b). On fully aggrading, non-dissected fans, this transition will occur at the fan apex, but on many fans the stream at the apex may be incised below the fan surface in a fanhead trench (Figure 12.2e), and emerge onto the fan surface at a mid-fan intersection point (Hooke 1967). On such fans active deposition is restricted to distal locations. On some fans the channel system may be trenched below the fan surface throughout the fan environment, in which case the fan surface is essentially a fossil surface receiving no further deposition. The position of the main channel may vary. It might be an axial channel running more or less down the centre of the fan but may switch to run down the steeper flanks of the fan to become or join an interfan channel, draining the depression between two adjacent fans (Figure 12.1b). As the result of such switching the focus of deposition will shift from one part of the fan surface to another (Denny 1967) (Figure 12.1c). The former channel then becomes abandoned. Other small channels may form on the fan surface as fan surface washes (Figure 12.1b), draining the fan surface itself but having no mountain source area.

Alluvial fans are characteristic landforms of arid and semi-arid mountain areas. Much of the classic research on alluvial fans has been based on the southwest United States (see, for example, Blissenbach 1954; Bluck 1964; Denny 1965; Hooke 1967; Bull 1977), but fans have also been described in other mountainous dry regions; for example, in Australia (Wasson 1974; 1979), Israel (Bowman 1978; Gerson *et al.* 1993), elsewhere in the Middle East (Al-Sarawi 1988; Maizels 1990; Freytet *et al.* 1993), Crete (Nemec and Postma 1993), Tunisia (White 1991; White and Walden 1994) and Spain (Harvey 1984a; 1990). Alluvial fans are not restricted to dry regions, and indeed they are common in many mountain regions (see examples in Rachocki and Church 1990), in arctic environments (see Leggett *et al.* 1966), in alpine environments (see Kostaschuk *et al.* 1986; Derbyshire and Owen 1990), including paraglacial environments (Ryder 1971a; 1971b), in humid temperate regions (see Harvey *et al.* 1981; Harvey and Renwick 1987; Kochel 1990), and even in the humid tropics

(see Kesel 1985; Kesel and Spicer 1985). In all these areas several conditions favour alluvial fan development, particularly the combination of a high rate of sediment supply from the mountain source areas, with the sudden downstream decrease in stream power at mountain-front locations. This combination is particularly favoured in dry-region mountain environments for four main reasons. First, as the result of sparse vegetation cover, intense storm rainfall and the dominance of overland flow processes on the hillslopes, desert mountains have high rates of storm sediment production. Second, steep, flashy, desert mountain streams have high rates of sediment transport and delivery to mountain-front locations. Third, the episodic nature of sediment transport is accentuated in arid regions as the result of the dominance of rare high-magnitude storm events in sediment transport (Baker 1977; Beaty 1974; 1990; Wolman and Gerson 1978; Laronne and Reid 1993). Fourth, arid region fluvial systems are characterised by spatial discontinuity (Brunsden and Thornes 1979), resulting from high evaporation rates and high transmission losses, that accentuate down-channel decreases in stream power through alluvial fans.

THE ROLE OF ALLUVIAL FANS IN DRYLAND FLUVIAL SYSTEMS

With their location between sediment-source areas and either enclosed depositional basins or main arterial river systems, alluvial fans have an important role in either coupling or buffering dry-region fluvial systems. In the long term, alluvial fans act as major sediment stores, trapping the coarse fraction of the incoming sediment. Locally, deposition will occur if the sediment supply rate is greater than the transporting capacity, in other words, if the actual stream power on the fan is less than that needed to carry sediment through the system, a threshold defined by Bull (1979; 1991) as the threshold of critical stream power. Spatial variations in stream competence will cause variations in deposition on the fan, in relation to the stream power behaviour (Figure 12.3). Temporal variations will cause variations in sedimentation rate or even switches between net aggradation on, and dissection of, the fan surface.

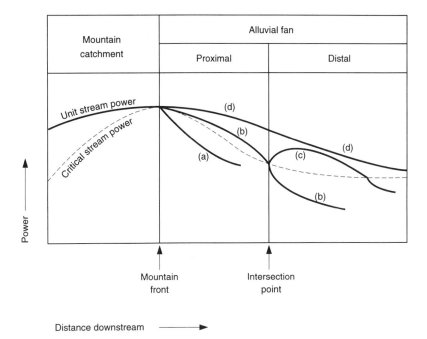

Figure 12.3   *Theoretical stream power relationships through alluvial fans. Amended from Harvey (1992b), based on the concept of the threshold of critical stream power (Bull 1979). (a) aggrading fans, (b) proximally trenched fans, (c) mid-fan trenched fans (see page 251), (d) through-trenched fans*

Aggrading fans, whether untrenched and proximally aggrading, or fanhead-trenched and distally aggrading, will act as buffers within the fluvial system, breaking the continuity of sediment movement between source area and either sedimentary basin or main channel system. The implications are important not only in the context of sediment movement through the system but also in the context of the system's response to environmental change. Alluvial fans may absorb such changes by varying their rates or patterns of sedimentation (Harvey 1987b), and inhibit the downstream transmission of the effects of such changes. For this reason alluvial fan sedimentary sequences may preserve a more sensitive record of environmental change than might main river alluvial sequences. Indeed, in drainage basins with large alluvial fans there may be very little correlation between the stratigraphy of fan sequences and that of arterial river or basinal depositional sequences. Similarly, the presence of aggrading alluvial fans may prevent, or at least delay, the upstream transmission of changes related to arterial river base-level

change. On the other hand, where fans have become dissected throughout their length there may be continuity of coarse-sediment movement from source area to arterial drainage, permitting an increased sensitivity of the main river, with aggradation and dissection occurring in response to source-area environmental change.

Factors affecting arid zone alluvial fan development

GENERAL CONTROLS

It has been argued above that fan development and dynamics, the temporal and spatial variations in aggradation/dissection relationships, are controlled by the rate of sediment supply to the fan relative to transport competence. These in turn respond to three sets of controlling factors, gross morphological and tectonic factors, factors controlling water and sediment supply *to* the fan, and those controlling sediment transport *through* the fan system.

TECTONIC FACTORS

Long-term tectonic and regional denudational histories create the gross topography within which alluvial fans may form, such as the juxta-position of mountain and basin, the major valleys and the form and position of mountain fronts. Many texts emphasise the importance of fault-controlled mountain-front/basin-margin locations. Over long timescales and in tectonically active regions tectonism is often seen as the major control over fan location and sedimentary style. Many ancient alluvial fan sequences, preserved in the geological record, are interpreted primarily in a tectonic context (see, for example, Mack and Rasmussen 1984; Nichols 1987). Within modern tectonically active areas, tectonism may create the gross framework for fan deposition; however, tectonism is not essential for fan development. Fans occur in both tectonically active and tectonically stable regions, and, in both, the detailed locations may be erosionally controlled, at mountain/pediment junctions, below erosional escarpments and at valley junctions. Indeed, in areas of very rapid uplift, base-level-induced incision may cause dissection of alluvial fans, and in extreme cases stream power may be so increased as to prevent alluvial fan accumulation. For example, in Almeria, southeast Spain, a tectonically active semi-arid region, Quaternary, dry-climate alluvial fans formed only in areas *away* from zones of maximum uplift, zones where fan development was inhibited by tectonically induced dissection (Harvey 1987a; 1990).

In another study of Spanish Quaternary fans, Silva *et al.* (1992) demonstrate the influence of the tectonic context on the spatial geometry of mountain-front fans. Two other examples illustrate the influence of tectonic style on the patterns of fan development.

Death Valley, California (Figure 12.1d), is a deep sedimentary basin bounded by active strike-slip and normal faults. Alluvial fans, issuing from the Funeral Mountains, Black Mountains and the Panamint Range, fringe the basin margins. The Panamint Range is tilting up towards the west and very large fans occur along its eastern side (Figure 12.2b). Although this margin is followed by active antithetic faults the fans bury the faults, backfilling into

the mountain catchments. There are minor faults too, along the Funeral Mountains, but this mountain front is largely an erosional form whose base is blanketed by extensive fans. Only along the mountain front of the Black Mountains do alluvial fans coincide with an active fault line. Here small steep fans interdigitate with the basin-fill sediments (Hunt 1975).

The Sierra de Carrascoy, Murcia, southeast Spain (Figure 12.1e), is bounded to the north by a major reverse fault. The southern margin is an extensive unfaulted pediment surface. Alluvial fans occur in three groups (Harvey 1988). Group 1 (Figure 12.1e) are simple mountain-front fans along the fault line. Faulting was active in the mid-Pleistocene, during early phases of fan deposition (Harvey 1988; 1990; Silva *et al.* 1992). It appears to have influenced fan location and spatial geometry but has had relatively little continuing influence on modern fan processes or morphology. These fans have backfilled into their mountain catchments (Figure 12.2c). Group 2 are intermontane fans occupying small basins and burying pediment surfaces (Figure 12.1c), and Group 3 are mountain-front fans burying pediment surfaces. Neither group are directly related to tectonic activity. Interestingly, despite the overall uplift of this mountain range during the Pleistocene, and the faulting along the northern flank, the Group 1 fans are all distally aggrading fans and not directly connected to the incised Guadalentin drainage (Figure 12.1e), an indication of the weakness of base-level as a control of alluvial fan processes.

Active tectonism may influence fan development in two ways. First, it may directly influence fan gradients, base-levels, and may involve deformation of the fan sediments themselves, either during or immediately after deposition (Gerson *et al.* 1993; Yoshida 1994). Second, tectonism may act indirectly, through its influence on erosion rates. Bull (1961; 1964) has demonstrated the influence of pulsed uplift on fan-profile segmentation in the California Coast Ranges. The influence of active tectonism on erosion rates may be through the continuing elevation of the source area, maintaining active erosion and a high rate of sediment supply to the fan. Tectonic factors are most important in the long term, controlling fan location, the overall context of fan development and spatial geometry, but in the shorter term in most situations other factors become more important in controlling fan dynamics and development.

FACTORS CONTROLLING WATER AND SEDIMENT SUPPLY TO THE FAN ENVIRONMENT

The factors which govern the rates of water and sediment supply, and the calibre of the sediment supplied are essentially those controlling the hydrology and erosion rates within the mountain catchment. They include relatively invariant geological and topographic factors, and more dynamic climatic and vegetation factors.

Source-area geology influences the rate of sediment supply through the inverse relationship between rock resistance and erosion rate. It is also important in controlling dominant clast size, itself an important factor in depositional processes and fan morphology (Bull 1962a). Indeed, since large alluvial fans are composed dominantly of coarse sediments, major fans are likely to occur only where, on weathering, the source rocks yield coarse clasts. In the Vera basin of southeast Spain, despite rapid erosion during the Quaternary (Harvey 1987a), alluvial fans are restricted to locations immediately proximal to the resistant rocks of the surrounding mountains. Elsewhere in the basin, erosion of weak marls has created pediment surfaces with little or no veneer of fan sediments. The marls weather to very fine material, which is transported away as suspended sediment. Bedrock lithology can also influence the ratio of coarse to fine sediments, and therefore the likelihood of transport by fluvial or debris-flow processes.

Topographic factors influence the rate of sediment supply, in that erosion rates tend to be higher from steeper slopes. They also influence the relative importance of fluvial and debris-flow processes. Fluvial processes are favoured over debris flows where streamside as opposed to hillslope sediment sources are dominant. Therefore fluvial processes tend to characterise fans supplied by large and less steep basins, whereas debris flows tend to dominate fans supplied by small steep drainage basins (Harvey

1984b; 1992a; Kostaschuk *et al.* 1986; Wells and Harvey 1987).

Climatic factors are important both directly and indirectly. High rates of coarse sediment supply are directly associated with high-magnitude, rare storm events. Sediment transport in arid areas is dominated by such events (Beaty 1974; Baker 1977; Wolman and Gerson 1978) rather than by frequent events more typical in humid regions (Wolman and Miller 1960). Climatic factors are indirectly also important, through their effects on vegetation cover. Sparse vegetation cover, together with high-intensity storm rains, both characteristics of arid and semi-arid regions, favour the rapid generation of runoff, the dominance of overland flow and a high rate of sediment supply. On arid-region fans, in comparison to those in more humid regions, fluvial processes tend to be relatively more important than debris-flow processes (Harvey 1992a). The contrast between so-called 'wet' and 'dry' fans (Schumm 1977), is not so much a climatic as a topographic contrast (see also Blair and McPherson 1994a, b).

The importance of climatic factors lies not only in their relation to short-term sediment supply, but also in the context of changes in aggradational or dissectional behaviour in response to longer-term climatic change. In the southwest Mediterranean region, Quaternary climates, corresponding to the glacial phases of northern Europe, appear to have been cold and arid with too little precipitation for tree growth (Amor and Florschutz 1964), but with high seasonal hillslope erosion rates resulting from intense seasonal storms (Butzer 1964; Rhodenburg and Sabelberg 1973; 1980; Sabelberg 1977). In north Africa (Sabelberg 1977; Rhodenburg and Sabelberg 1980) and southeast Spain (Harvey 1984a; 1990) these periods appear to have been times of major alluvial fan aggradation, followed during the Holocene by fan dissection. In Australia, however, both Williams (1973) and Wasson (1979) have associated fan aggradation with more moist phases and fan dissection with drier phases. In the Mojave Desert of the southwestern United States the late Pleistocene was a period of major fan aggradation (Wells *et al.* 1987; Dorn 1988; 1994; Harvey and Wells 1994), coincident with a climate cooler and/or wetter than that of today (Brackenridge 1978; Galloway 1983; Van

Devender and Spaulding 1983; Spaulding 1990). In the arid conditions of the Holocene, fanhead trenching and distal progradation have been the dominant trends (Harvey and Wells 1994).

FACTORS CONTROLLING SEDIMENT TRANSPORT WITHIN THE FAN ENVIRONMENT

Sediment transport *through* the fan depends on the transport mechanism and on transporting power. Transport mechanisms range from debris flows to fluvial transport, either as unconfined sheetfloods over the surface of the fan or as fluvial transport within channels (Blissenbach 1954; Hooke 1967). For fluvial processes, unit stream power, the power available at a point (Richards 1982), depends directly on discharge and slope, and inversely on channel width. It is therefore influenced by the gradient of the channel or fan surface, itself the product of erosion or deposition, and by the hydraulic geometry. If the unit power locally falls below that required to transport the sediment supplied, or in other words, falls below the threshold of critical stream power (Bull 1979), deposition will take place. The gradient at which deposition takes place varies between processes, being steepest for debris flows. For fluvial transport, as the result of greater depths and therefore greater unit power (Richards 1982), this gradient is lower for channel flows than for sheetfloods. Fan morphology, as expressed by the gradient (slope) of the channel or of the fan surface, results from an internal adjustment of form to process, which responds to the externally controlled water and sediment supply rates and dominant transport mechanism.

Changes in fan morphology, and especially in fan slope, can result from three types of cause: externally induced (extrinsic) changes in sediment and water supply; internally induced (intrinsic) changes related to short-term random variations in process/form relationships within the fan environment (Schumm 1979); and long-term progressive changes related to either long-term source-area erosion or long-term fan deposition.

Externally controlled tectonic, or, more commonly in the shorter term, climatic or vegetation changes, may cause variations in sediment and water supply to the fan environment (see above). Particularly important are changes influencing sediment supply rate, sediment calibre and the water-to-sediment ratio, especially if a change of transport mechanism, debris flow to fluvial transport or vice versa, is involved. The process would then no longer be that to which the fan slopes are adjusted and the result would be either modification of the depositional slope by burial, or dissection of the previous slope, resulting in channel slope reduction by the formation of a fanhead trench (Figure 12.1c).

Because the fan morphology may adjust, in the short term, to the dominant sediment transport and depositional processes, and as these processes on many fans may fluctuate between debris flow and fluvial processes, the fanhead environment may undergo short-term random fluctuations between erosion and deposition (Hooke 1967). Erosion might produce shallow fanhead trenches initiating one of two feedback mechanisms (Chorley and Kennedy 1971): a negative feedback, whereby slope is reduced thus enhancing later deposition; or a positive feedback where the increased unit stream power enhances further erosion. Which of the two occurs may depend on the succeeding sequence of processes, though it is clear that a potential or intrinsic threshold (Schumm 1979) exists, where fan trenching may be initiated simply as the result of random fluctuations between processes within the fan system, without any major external change. Similarly, incision *may* be initiated by random switching of channels across the fan surface (Figure 12.1b and 12.1c) from an axial to a steeper marginal position (Hooke and Rohrer 1979).

Although short-term equilibrium can be identified between processes and morphology, over the long term fan morphology is subject to progressive change. This may be brought about by continued deposition on the fan, extending the fan distally or causing backfilling into the mountain catchment (Figure 12.2c). Furthermore, progressive, long-term erosion of the source area may cause progressive changes in the rate of sediment supply, the calibre of the sediment and the water-to-sediment ratio. Eckis (1928) was the first to recognise long-term diminution in sediment availability as a possible cause of fan dissection. Harvey (1984a; 1987b; 1990) has suggested similar reasons for progressive changes in sedimentation, and for a long-term trend from aggradation to dissection on Spanish Quaternary alluvial fans.

Fanhead trenches are common, and numerous explanations have been put forward for their formation. Schumm *et al.* (1987) list 16 possible causes but these can be grouped in relation to the three sets of causes identified above: extrinsic, intrinsic and ageing-related causes. Excluding the results of base-level change and 'toe-trimming' which initially influence only the distal fan areas, all produce dissection at the fan apex, or at least in the proximal part of the fan. Dissection resulting from intrinsic causes may be short-lived and create only shallow fanhead or intermittent trenches (Hooke 1967). Dissection resulting from extrinsic or ageing-related causes tends to produce a deep and permanent fanhead trench (Figure 12.2e), within which the channel slope is less than that of the fan surface, the two profiles converging towards an intersection point (Hooke 1967), where the channel emerges onto the fan surface. Here, rapid channel widening often occurs, causing a decrease in unit stream power (Harvey 1984c), and even if slope increases, deposition usually occurs (Wasson 1974). In some cases (see later, page 251) environmental change may trigger incision at the intersection point in mid-fan (Harvey 1987b). Successive phases of either fanhead trenching or midfan incision may lead to the downfan migration of intersection points (Bowman 1978), and ultimately to the total dissection of the fan surface.

Alternatively, fan dissection may be induced by a local fall in base-level in which case incision of the toe of the fan may occur and work progressively upfan (Harvey 1987a). In most cases, even in areas where tectonic uplift of the mountains has occurred, base-level-induced dissection is rare, and occurs only where an adjacent river undergoes deep incision, or in the special case of coastal locations, in response to sea-level change (Nemec and Postma 1993; Yoshida 1994). In the Carrascoy example, described above, deep dissection of the Guadalentin River has not caused dissection of the fan toes. In many studies considering the relative

importance of proximal and distal controls on fan geomorphology, it is the proximal controls, in the longer term, mountain uplift, and in the shorter term climatically induced sediment supply, that have an overwhelming influence on fan dynamics (Bowman 1988; Frostick and Reid 1989).

Alluvial fan sediments

SEDIMENT TRANSPORT MECHANISMS

Sediment transport processes on arid region fans range from debris flows to fluvial processes. Arid region fans are often composite, with a range from fluvial or sheetflood dominance, as in the large Death Valley fans (Hunt 1975), to debris-flow dominance (Beaty 1963; 1970; 1990). Fluvial deposition takes place either by unconfined sheetflow, or in wide, shallow, usually braided channels (Beaumont and Oberlander 1971; Bull 1977).

Fluvial depositional mechanisms are described in many geomorphology and sedimentology texts (e.g. Schumm 1977; Nilsen 1982; Richards 1982; Reinech and Singh 1983), but until recently there has been less work on debris-flow deposition (see Johnson and Rodine 1984; and Blair and McPherson 1994a; 1994b for discussions). Debris flows on arid region fans have been described by Blackwelder (1928) and Beaty (1974; 1990), but much of the more recent work on debris flows has dealt with other climatic environments (see, for example, Pierson 1980; 1981; Suwa and Okuda 1983; Pierson and Scott 1985; McArthur 1987; Wells and Harvey 1987; Rickenmann and Zimmerman 1993; DeGraff 1994).

The dominant controls over transport mechanisms appear to be the water-to-sediment ratio of the mix entering the fan environment, and the availability of fine, mud-sized sediment. High water-to-sediment ratios, especially with a low concentration of fines, lead to fluvial transport of coarse clasts by traction and the deposition of fluvial bar and sheet forms (Wells and Harvey 1987). Increasing proportions of sediment enable movement to take place as a mass flow, and with increasing fine content there is sufficient internal strength for the matrix to support the clasts and for debris-flow movement by plastic flow to take place (Pierson 1981). On deposition this process produces the matrix-supported gravels, characteristic of debris flows (facies Gms of Miall 1978). Recent work has identified the importance of hyperconcentrated flows, which move as mass flows but which drain on deposition to produce structureless, matrix-free gravels (Pierson and Scott 1985; Wells and Harvey 1987; facies Gm of Miall 1978).

On any one fan the water-to-sediment ratio may vary during a storm (Wells and Harvey 1987), and may vary downfan (Pierson and Scott 1985), resulting in a complex assemblage of depositional facies, but with the overall pattern and the differences between fans reflecting the gross catchment controls over water and sediment production. As indicated earlier, fans issuing from small steep catchments, especially if the geology is conducive to the production of fine sediment, tend to be debris-flow dominated, while those issuing from larger less steep catchments tend to fluvial dominance.

DEPOSITIONAL FACIES

Three main groups of alluvial fan sediments — debris-flow, sheetflood and channel deposits — have long been recognised (Blissenbach 1954; Hooke 1967). More recently, transitional types such as sieve deposits (Hooke 1967; Wasson 1974) and those resulting from hyperconcentrated flows (Wells and Harvey 1987) have been described. Each group may be recognised by sedimentary structure, clast size and shape characteristics, fabric and clast-to-matrix relationships, and each may be described by characteristic facies assemblages (after Miall 1977; 1978; Rust 1978).

True debris-flow deposits are massive, poorly sorted, matrix-supported gravels (facies Gms, after Rust 1978), which may show shear-induced clast alignment parallel to the base of the flow, and a 'push' fabric of near-vertical clasts towards the front of the flow (Figure 12.4a) (Wells and Harvey 1987). Alternatively, if the flow is more fluid, such deposits would show little internal structure. Subaerial debris flows, characteristic of alluvial fan environments, have a greater tendency to exhibit internal shear-related and 'push' structures than do subaqueous debris flows that are common in fan-delta environments. Debris flows occur in thick beds, each equivalent to individual flows (Figure 12.4b), and may show a weak within-bed vertical clast sorting. They may grade laterally into massive, clast-supported gravels (facies Gm, after Miall 1977), a facies which may have one of several origins. Wells and Harvey (1987) recognise such a gravel as the product of deposition by hyperconcentrated flows. On deposition the matrix flushes out and the deposit settles, adopting a clast-supported framework, but usually little or no internal structure or fabric except for a closely packed 'collapse' fabric (Wells and Harvey 1987). It is possible that many of the thicker beds of poorly sorted, structureless, clast-supported gravels, recognised in Quaternary or ancient fan sediments, and described as fanglomerates, may have been deposited by such flows.

As indicated above, fluvial deposits within alluvial fan sediments include sheetflood, sieve and channel deposits. Sheetflood deposits comprise thin sheets of gravels (facies Gm after Miall 1977), and sand or silt often forming couplets (Blair and McPherson 1994a). Sieve deposits (Hooke 1967) have been described as lobate forms often deposited at the fan intersection point and show vertical and lateral fining-up as the coarser material traps and 'sieves' the fines (Figure 12.3b). In section they may appear as Gm facies (after Miall 1977) but may show imbrication of the coarser frontal clasts and a marked vertical and front-to-rear fining.

Fluvial channel deposits are easily recognisable by their channelled bases, their internal bedding and lensed sedimentary structures, often by clast imbrication and by better within-bed sorting. They include a variety of sand and gravel microfacies (for example Gm, Gt, Gp, St, Sp, Sh facies of Miall 1977). Multiple-channel structures inset one within another, characteristic of braided channel sediments, are common (Figure 12.4c).

In addition to purely depositional facies, arid-area fans are often characterised by other sedimentary features, related to arid region weathering and pedogenic processes (see Bull 1991). A desert pavement is developed on many

Figure 12.4 *Fan sedimentology: facies types. (a) Bouldery debris-flow deposits, Zzyzx fans California: note clast fabric. (b) Stratified fan deposits (including debris flows) Cayola fan, Alicante, southeast Spain. d = debris flow. c = channel gravels; hammer is 30 cm. (c) Stratified fan deposits (channel deposits), Nudos fan, Almeria, southeast Spain. Scale bar is 50 cm. (d) Intersection-point sieve lobe on a small aggrading fan near Vera, Almeria, Spain. (e) Ceporro Fan, Almeria, Spain. Fanhead trench section showing early debris-flow deposition (d), followed by channel gravel deposits (c), capped by a calcrete crust. Height of section c. 10 cm*

fan surfaces, comprising a fractured and sorted stone layer at the surface over an underlying silty layer (McFadden *et al.* 1987). There has been much debate (see Chapters 9, 11 and 28) on the origins of desert pavement surfaces, but what is clear for fan pavements is that the stone pavement develops by sorting processes concurrently with the accumulation of the fines. The clasts themselves are subject to mechanical weathering processes by fracture and in some cases to chemical weathering, so that the pavement characteristics are age dependent. In a recent study in the Middle East, Al Faraj and Harvey (in preparation) have demonstrated relative age relationships of fan and terrace surfaces through differential pavement development.

The clasts on desert pavement surfaces commonly carry a rock varnish (Chapter 6) of precipitated iron and manganese salts, which increases in thickness and darkens with age. Again, there is debate concerning the formation of desert varnishes (see Dorn and Oberlander 1981; 1982), but their age-related characteristics have allowed the relative dating of fan surfaces, for example in the Death Valley region (Hunt and Mabey 1966). More recently, it has been demonstrated that their trace carbon content may allow radiocarbon dating of desert surfaces (Dorn *et al.* 1989), and other analytical techniques may not only reveal age information (Dorn 1983; Hooke and Dorn 1992), but also give other palaeoenvironmental signals (Dorn 1984; Dorn *et al.* 1987; see Chapters 26 and 27). Much of this recent work has concentrated on the Death Valley fans (Dorn 1988; 1994).

Many studies in both arid and semi-arid regions have used soil characteristics as indicators of surface age (e.g. Birkeland 1985). Some have used characteristics of the soil profile as a whole (Harden 1982; Harden and Taylor 1983; McFadden *et al.* 1987; 1989); others have focused on the B-horizon, dealing with soil colour (for example Hurst 1977), iron oxide behaviour (Alexander 1974) or magnetic mineral behaviour (White and Walden 1994).

Perhaps the most well-known aspect of arid and semi-arid soils is the accumulation of pedogenic carbonate. The stages of carbonate accumulation defined by Gile *et al.* (1966) and elaborated by Machette (1985) have been used in many studies as aids to relative dating and correlation of geomorphic surfaces, including alluvial fan surfaces. On exposure, pedogenic carbonate indurates to form calcrete (Chapter 6), and undergoes a complex sequence of brecciation and recementation. Many geomorphic surfaces in arid and semi-arid regions are crusted by various types of calcrete (Butzer 1964; Lattman 1973; Goudie 1983). Calcrete has been used to assess relative ages of pediments (Dumas 1969) and alluvial fans (Harvey 1978), and has potential for providing absolute dates (Ku *et al.* 1979), though there are many problems inherent in dating calcretes (see Chapters 26 and 27).

A palaeoclimatic implication of the widespread occurrence of Pleistocene calcretes in today's arid and semi-arid regions is that dryness must have occurred at least at times in the Pleistocene for calcium carbonate accumulation to occur. Once formed, a calcrete crust influences fan surface processes. It may cause a reduction in infiltration into the fan surface but, most importantly, creates a surface resistent to erosion, one which may influence channel morphology and channel erosional behaviour (Van Arsdale 1982; Harvey 1987b).

Quaternary and older fan sediments often preserve former soils, either red B-horizons or fossil pedogenic calcretes or both, as palaeosols (see, for example, Wright and Alonso Zarza 1990).

VERTICAL AND SPATIAL FACIES VARIATIONS

Assemblages of depositional facies produce characteristic vertical and spatial facies variations within alluvial fan deposits. The most distinctive vertical sedimentary sequence for arid region fans is that described by Miall (1978) as the 'Trollheim' type, named after the Trollheim fan in California, northwest of Death Valley (Hooke 1967). It comprises a sequence dominated by alternating debris-flow deposits, massive (sheet?) gravels and channel gravels (Figure 12.5a, showing facies Gms, Gm, Gt; after Miall 1978). This type of sequence is also shown on Figure 12.4b, and is characteristic of proximal fan environments. Further downfan, or on fans dominated by fluvial processes, the sequence may be of the 'Scott' type (Figure 12.5b), dominated by sheet and channel gravels (facies Gm, Gp, Gt and various sand facies, after Miall 1978), similar to the sequence shown on Figure 12.4c.

Short-term or intrinsic variations in sedimentary processes produce facies assemblages of these simple sequences, however, major climatic or tectonic changes, or the long-term progressive erosion of the source area ('ageing') may cause a progressive vertical trend in sedimentary style. Such progressive changes have been identified in Quaternary fan sequences in Australia (Williams 1973; Wasson 1979), Nevada (Bluck 1964) and Spain (Harvey 1978;

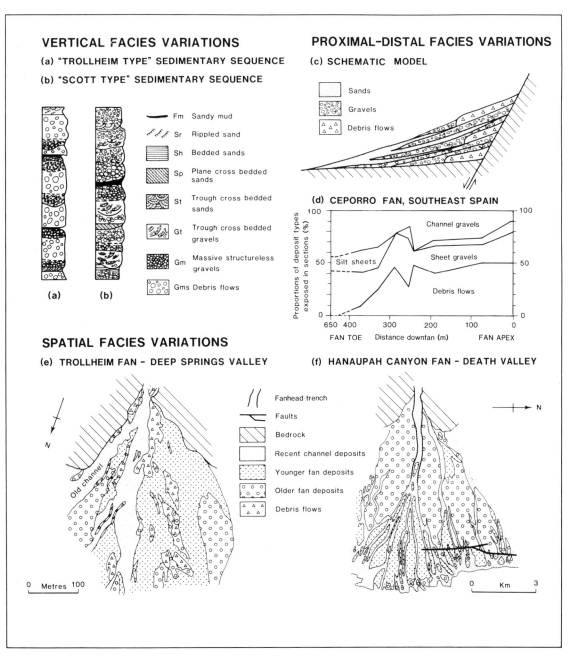

Figure 12.5 *Fan sedimentology: facies variations.*
Vertical facies variations: (a) 'Trollheim' type; (b) 'Scott' type sedimentary sequences (after Miall 1978).
Proximal–distal facies relations: (c) Schematic model (after Rust 1979). (d) Downfan variations in proportions of facies exposed in fan sections on Ceporro fan; southeast Spain (after Harvey 1984b).
Spatial facies variations: (d) Trollheim fan, Deep Springs Valley, California, apex area of a proximally aggrading fan (after Hooke 1967). (e) Hanaupah Canyon fan, Death Valley, California, a proximally trenched distally aggrading fan (after Hunt and Mabey 1966)

1984a; 1990). In the Spanish case many fans show early dominantly debris-flow deposition, followed by later sheet and channel gravel deposition (Figure 12.4e), and finally by dissection becoming dominant over aggradation. A similar sequence has been observed at Zzyzx, California, but with a strong climatic signal (Harvey and Wells 1994). Late Pleistocene fan aggradation dominantly by debris-flow deposition was followed by Holocene fanhead trenching and fan progradation under fluvial processes (Figure 12.6).

Ancient alluvial fan sequences often show progressive changes in sedimentary environment, sometimes involving upwards coarsening of megasequences (Steel 1974; Steel *et al.* 1977; Rust 1979; Nilsen, 1982; Mack and Rasmussen 1984). In most cases the progressive changes are seen largely as tectonically or 'ageing' controlled. The coarsening-up model, often assumed for ancient sequences, does not seem to be borne out by Quaternary examples. In many Quaternary cases fan surface aggradation is associated with backfilling; hence specific locations effectively become more 'distal' (Harvey 1990). In this case, fining-up would be expected.

Within the geological record, phases of fan sedimentation are often only temporary stages in the evolution of sedimentary basins

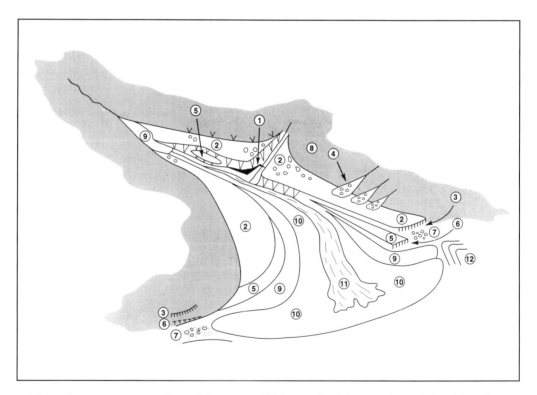

Figure 12.6   *Summary sequence of Late Pleistocene and Holocene alluvial fan erosion and deposition: Zzyzx, California.*
1. Mid-Pleistocene fan deposits (heavily calcreted). 2. Main phase of late Pleistocene aggradation, debris-flow deposits in proximal zones and from small tributary fans, fluvial deposits in main fan. 3. Late Pleistocene shoreline of Lake Mojave. 4. Hillslope debris flows, contemporaneous with phases 2 and 5. 5. Latest Pleistocene fan deposits resulting from limited fanhead trenching and distal progradation. 6. Early Holocene shoreline of Lake Mojave. 7. Lake shoreline sediments relating to phases 3 and 6. 8. Hillslopes now stabilised during the Holocene. 9. Mid-Holocene fan sedimentation, limited deposition at fan apex, fanhead trenching in midfan and distal progradation by fluvial deposition. 10. Further fanhead trenching and distal progradation by fluvial processes during the later Holocene. 11. Active fan sediments. 12. Modern playa, salt flats (modified after Harvey and Wells 1994)

associated with basin uplift and inversion (Mather 1993). In the Neogene basins of southeast Spain a phase of Plio-Pleistocene alluvial fan sedimentation followed earlier marine, then low-energy terrestrial, environments. Basin-wide fan deposition ceased following uplift and dissection of the basin (Mather 1993; Mather and Harvey 1995). Smaller fans developed around the basin margins later in the Pleistocene, in response to climatic change (Harvey 1987a; 1990).

Spatial variations in fan deposition reflect proximal to distal variations in depositional process and sorting mechanism. Dominance by debris flows in proximal areas may give way downfan to sheetflood deposits in distal locations (Figure 12.5c and 12.5d), as 'Trollheim'-type sequences give way to 'Scott'-type sequences. Channel deposits, too, may give way distally to sheetflood deposits as the flow spreads over the distal fan surface. Where this transition takes place, often in midfan, sieve deposits may be common. In distal environments sand and silt sheets may be dominant over gravel deposits. These trends are apparent both in Quaternary and ancient fan sequences (Harvey 1978; 1990; Gloppen and Steel 1981). Within the sediments as a whole and within individual sedimentary units there is normally a downfan decrease in clast size and an increase in sorting (Bull 1962a; 1963; Bluck 1964; 1987; Lustig 1965).

Channel switching and fanhead trenching will produce detailed variations in the distribution of sedimentary units over the fan surface, resulting in a mosaic of fan deposits and of depositional surfaces of varying ages. There is a contrast in the spatial pattern of sedimentary units between the random pattern produced by channel switching near the apex of an undissected fan (Figure 12.5e), and the more ordered age-progression of deposits 'younging' downfan and telescoped into the fanhead trench of trenched fans (Figure 12.5f).

Alluvial fan morphology

FAN MORPHOMETRY

Depositional processes create the surface morphology of alluvial fans. Since these processes are controlled by sediment supply, which in turn reflects the source-area characteristics, it is reasonable to suppose that for fans of similar age, climatic and tectonic history, fed by areas of similar geology, that the gross morphology of each fan would reflect the topographic characteristics of the source area. This hypothesis has provided the basis for much research into the geomorphology of dry-region alluvial fans. The fan property that has been most thoroughly investigated is fan area (Bull 1962b; 1977; Denny 1965; Hooke 1968; Hooke and Rohrer 1977; Silva *et al.* 1992), which has been demonstrated to have a simple relationship to drainage area that may be expressed as:

$$F = pA^q \qquad (12.1)$$

where F and A are fan area and drainage area (measured in km^2), respectively. Numerous studies have demonstrated similarity between different fan groups for the exponent q of between *c.* 0.7 and *c.* 1.1, but values for p show a wide range of between *c.* 0.1 and *c.* 2.1, reflecting regional differences in fan age and history. Hooke and Rohrer (1977) and Lecce (1991) have argued that in any one region the differences reflect different erosion rates, and therefore can give an indication of rock resistance. Figure 12.7 illustrates some of the available data and demonstrates that the California Coast Range fans (after Bull 1962b; 1977) have much larger fan areas per unit drainage area than do fans in the Death Valley region (after Denny 1965; Hooke 1968; Hooke and Rohrer 1977). The relationships for the Spanish fans (based in part on data from Silva *et al.* 1992, and in part on hitherto unpublished information for 68 fans) lie midway between the other two groups. The data for the 68 Spanish fans yield a regression equation as follows:

$$F = 0.72A^{0.82} \qquad (12.2)$$

with a correlation coefficient of 0.89 (standard error of the estimate 0.297 log units). The data from Silva *et al.* (1992) give separate regression lines for individual mountain fronts.

Adjacent drainage basins, of differing size along a mountain front, will produce a mosaic of coalescent fans of differing size (and slope, see below; and Figure 12.11a). Such relationships also have implications for the

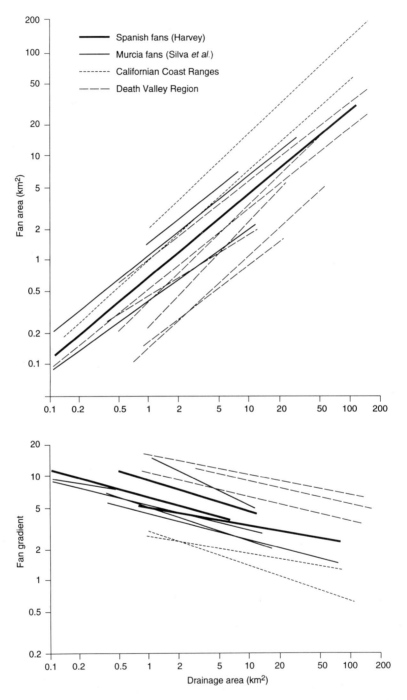

Figure 12.7   *Summary of regression relationships of drainage area to fan area and fan gradient. Data for Spanish fans, fan area regression see text, fan gradient regressions from Harvey (1987b). Data for Murcia (Spain) fans from Silva* et al. *(1992); data for Californian Coast Ranges from Bull (1962b; 1977); data for Death Valley region from Denny (1965), Hooke (1968) and Hooke and Rohrer (1977)*

interpretation of ancient erosional topography from the spatial geometry of ancient fan sequences (Hirst and Nichols 1986).

Other studies, notably Blissenbach (1952), Bluck (1964; 1987), Denny (1965), Lustig (1965) and Hooke and Rohrer (1979), have demonstrated relationships between fan sediment properties (see above) and drainage area and fan morphometry, particularly downfan decreases in clast size. However, perhaps the most important morphometric studies that have implications for fan processes and morphological development are those dealing with fan gradient, channel slope and channel geometry, in that these variables influence stream power on the fan. The relation of fan gradient (G), usually taken as the axial fan surface slope in the upper part of the fan, to drainage area can be expressed by:

$$G = aA^b \qquad (12.3)$$

Values of the exponent b show a fairly narrow range, from -0.15 to -0.35 but values for a show a greater range from $c.$ 0.03 to $c.$ 0.17, reflecting different sedimentary processes. Figure 12.7 shows these relationships for the same regional groups as before. According to data from Bull (1962b) and Hooke (1968) the Death Valley fans are much steeper than the California Coast Range fans. The Spanish plots again lie between those for the other two groups. The overall regression equation for 77 Spanish fans (Harvey 1987b; 1990) is:

$$G = 0.066A^{-0.20} \qquad (12.4)$$

with a correlation coefficient of -0.64 (standard error of the estimate 0.170 log units). The Spanish data shown on Figure 12.7 are presented as three separate regression lines, for three broad geographical/lithological groupings of fans. In two of the three cases correlation coefficients improve and standard errors decrease in relation to the overall regression (Harvey 1987b; 1990). Correlation coefficients improve still further to between 0.74 and 0.89 when multiple correlation coefficients are calculated, taking into account drainage basin relief as well as drainage area (Harvey 1987b). Also shown as separate plots on Figure 12.7 are the relationships from Silva et al. (1992) for fans at individual mountain fronts in Murcia, southeast Spain.

One of the main causes for different plotting positions on Figure 12.7 is the influence of depositional mechanism on fan gradient. In a study of fan gradients in a variety of dry regions, including the southwestern United States, Greece and Spain, Harvey (1992b) demonstrated that debris-flow fans not only have higher gradients per drainage area than fluvial fans (Figure 12.8), but also tend towards lower values for the exponent b in the above equation, indicating less sensitivity in this relationship for debris-flow processes. The overall spatial morphometry of alluvial fans may reflect overall tectonic evolution as well as climatic controls over fan processes. Silva et al. (1992) demonstrate four different categories of fan in the Guadalentin depression, southeast Spain, based on fan area and gradient to drainage area relationships, governed by the sedimentary and tectonic contexts and expressed by differences in the spatial geometry of the fans (Figure 12.9).

The geometry of fan channels can be examined in the same way as were fan area and fan gradient. The slope of the channel within the fanhead trench (S) can be expressed in relation to drainage area by the equation:

$$S = fA^g \qquad (12.5)$$

For the Spanish examples the exponent g ranges from $c.$ -0.2 to $c.$ -0.4 and f between $c.$ 0.02 and $c.$ 0.06, but there have been too few equivalent American studies for direct comparison. The data that exist suggest that American fanhead trench channels tend to be braided and marginally steeper than the Spanish channels (Harvey 1987b).

For channel cross-sectional variables, width (W, in metres) is the most readily measurable, and useful, as it has an inverse relationship with unit stream power. Its relationship to drainage area can be expressed by the equation:

$$W = hA^k \qquad (12.6)$$

For the Spanish examples the exponent k ranges from 0.17 to 0.34 and h between 6 and 13. Again, there are few comparable American studies, but for the Death Valley fans Denny (1965) quotes the following:

$$W = 23.3A^{0.5} \qquad (12.7)$$

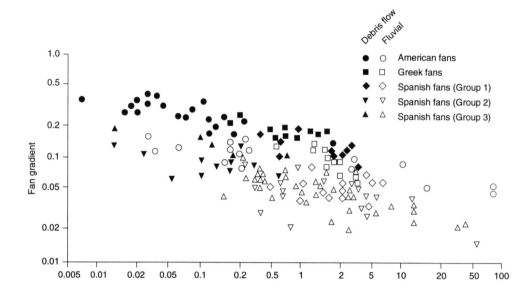

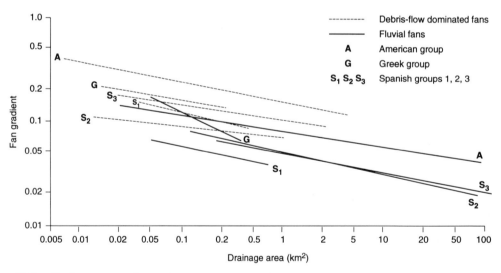

Figure 12.8 *Drainage area to fan gradient relationship for selected Spanish, Greek and American fans, subdivided into dominantly debris-flow and dominantly fluvial fans: above — scatter diagram, below — regression lines (for details see Harvey 1992b)*

This is in marked contrast to the Spanish channels. These, and other American fan channels are much wider than the Spanish channels, especially on larger fans. On individual American fans width tends to increase along the fanhead trench, in contrast to the more or less constant widths within Spanish fanhead trenches (Harvey 1987b). One characteristic common to both American and Spanish channels is the tendency for narrower channels to occur on crusted fans (Harvey 1987b; Van Arsdale 1982), reflecting the greater resistance to bank erosion.

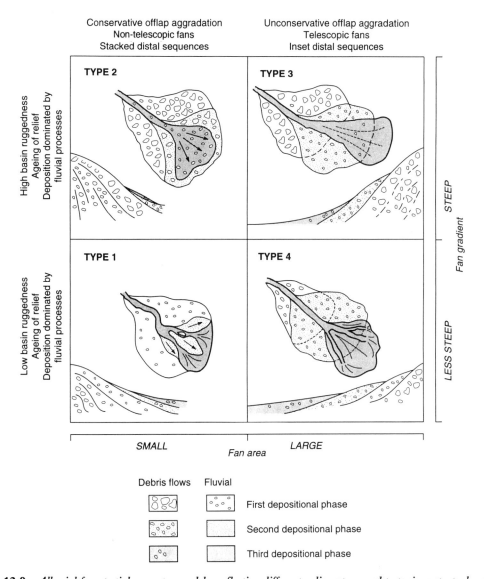

Figure 12.9   *Alluvial fan spatial geometry models, reflecting different sedimentary and tectonic contexts, based on Murcia fans (Spain) (modified after Silva* et al. *1992)*

FAN DYNAMICS

Alluvial fans are dynamic landforms showing progressive morphological change during their development. As the morphology changes, so too does its influence on the pattern of erosion and deposition. A simple schematic model can be devised (Figure 12.10a) to represent the progressive morphological

development of alluvial fans, incorporating headstream incision, fanhead dissection and fanhead trench formation, and the progradation of the distal portion of the fan. As fanhead entrenchment occurs successive intersection points are formed further and further downfan and the focus of deposition progressively shifts distally. This model is loosely based on the west-side Death Valley

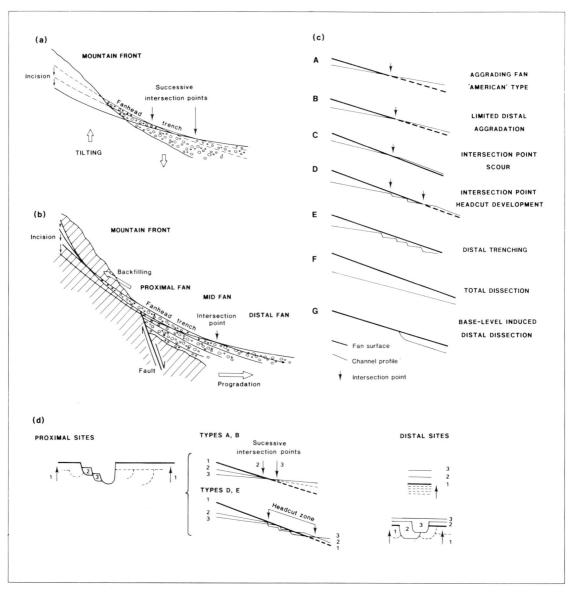

Figure 12.10 *Fan development models. (a)Simple schematic fan model for headstream incision; fanhead trench development and fan progradation, based on west-side Death Valley fans, where this sequence is enhanced by simple tectonic tilting (based on field observations, and on Hunt 1975). (b) Modification of the simple model (a) to incorporate early faulting followed by backfilling, then incision and progradation as in (a). (c) Model fan and channel profile relationships and associated mid-fan aggradation/dissection behaviour (after Harvey 1987b). (d) Stratigraphic implications of profile types illustrated in (c) (after Harvey 1987b)*

fans, but appears to be broadly applicable to most proximally trenched distally aggrading fans. Tectonic tilting would accelerate the sequence but is not an essential feature of the general model. The sequence would be modified by contemporaneous tectonism or by climatically induced changes in sediment supply. Figure 12.10b depicts the sequence that would be produced by early faulting along the mountain front, followed by later sediment excess causing backfilling, then by the sequence as depicted in Figure 12.10a.

This type of model, however, cannot account for all types of fan development sequence. Many of the fans in southeast Spain exhibit intersection-point headcut development and midfan-to-distal trenching, sometimes involving continuous trenching throughout the distal zone. Figure 12.10c presents a simple schematic model of fan-surface and channel-profile relationships for all types of dissected fan (after Harvey 1987b), omitting any profile for proximally aggrading fans. Types A and B are conventional, proximally trenched, distally aggrading fans. Type A represents the profiles often described in the literature for American fans (see Bull 1964), with continuity of channel slope through the fanhead trench onto the aggrading distal zone, but a segmented fan surface profile as a whole. In reality, many American fans, as well as distally aggrading Spanish fans (Figure 12.12, Honda and Grotto Canyon), exhibit profiles of Type B, with an increase in channel slope at the intersection point (Harvey 1987b). In Type C (Figure 12.12, Ceporro), scour, rather than deposition takes place at the intersection point, but with distal aggradation (Figure 12.11b). In Types D and E (Figure 12.12, Corachos) this scour is sufficiently effective to create a headcut (Figure 12.11c) or a series of headcuts which ultimately link together as a midfan to distal trench, dissecting the fan throughout its length (Type F). Type G is rather different and occurs where base-level-induced dissection trenches the fan upwards from the fan 'toe' (Figure 12.12, Sierra).

All types of profile occur on Spanish fans (Harvey 1992b); types C D and E are especially common on crusted fans, but I am not aware of any American fans with interesion-point headcuts, even on crusted fans, except on tributary fans where the incision is really local-base-level-induced. Through-fan trenches of Type F, however, do occur (Figure 12.11d).

The causes of intersection-point headcut development may be explained in relation to the threshold of critical stream power, within the context of fan morphometry and the history of fan development. The channel within the fanhead trench is adjusted to sediment transport by fluvial processes. At the intersection point it emerges onto the fan surface, whose slope is inherited from past processes, either

Figure 12.11 *Alluvial fan morphology. (a) Nested mountain-front alluvial fans, Panamint Valley, California. Note: faulted mountain front; fan size and gradients that reflect drainage basin size; fan surface washes dissecting distal surface of the largest fan.*
(b) Intersection-point and distal aggradation, Ceporro fan, Almeria, Spain. (c) Intersection-point headcut, Torre fan, Murcia, Spain. (d) Through-fan trenches, Virgin Mountain fans, Nevada

debris-flow or sheetflood, processes other than within-channel processes, and is therefore steeper than the adjusted channel slope. In most cases the potential increase in unit stream power, due to increased slope, would be compensated for by a rapid increase in channel width. However, under conditions of low sediment transport and low channel widths, unit stream power may increase because of the increase in slope. Incision into the fan surface may then result (Harvey 1987b). These conditions would be favoured by a climatically induced reduction in sediment supply, the presence of crusted fan surfaces inhibiting channel widening, and a major inherited contrast between fan and channel slopes. Such conditions occur on many Spanish fans. Holocene sediment supply rates appear to be low compared with excess sediment supply during fan aggradation in the Pleistocene, and many fans have seen a progression from dominantly debris-flow deposition during fan aggradation, to fluvial dominance during dissection (see Harvey 1984a). This sequence has accentuated the contrasts between fan and channel slopes to a greater degree in Spain than on the American fans. Finally, channel widths on crusted fans tend to be less than on non-crusted fans, and on Spanish fans generally to be less than on American fans (Harvey 1987b). All these conditions make the Spanish fans more prone to intersection-point headcut development and mid-fan trenching than the American fans.

Figure 12.12 illustrates fan profiles for several Spanish and American examples, and Table 12.1 summarises their morphometric data. Note that on all fans shown except La Sierra, distal slopes are greater than channel slopes within the fanhead trench. Note, too, that Corachos (Type E), is a strongly crusted fan, has the greatest $G:S$ ratio and a relatively narrow channel, all properties conducive to the formation of intersection-point headcuts and distal trenches. Both Type B fans shown, Honda and Grotto Canyon, have had relatively simple developmental histories; the terraces shown within the fanhead trenches are erosional only. The other three fans have had more complex erosional and dissectional histories. The terraces in these fans are aggradational terraces formed by cut and fill during the period of overall dissection.

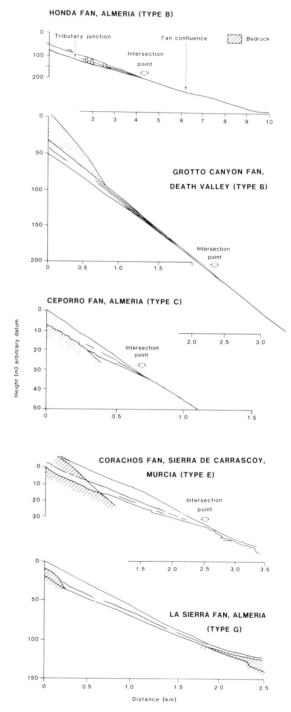

Figure 12.12 *Example fan profiles for fan types identified on Figure 12.10c. Morphometric data for these fans are given on Table 12.1*

Table 12.1 *Morphometric data for the sample fans shown on Figure 12.12*

	Drainage area (km^2)	Fan slope (*G*)	Channel slope (*S*)	Distal slope	Width range (*m*)	*G/S*
Honda	27.0	0.032	0.020	0.029	22–30	1.6
Grotto Canyon	10.6	0.085	0.064	0.075	60–120	1.3
Ceporro[*]	0.4	0.081	0.043	0.063	2.0–2.5	1.9
Corachos[*]	4.6	0.033	0.014	0.020	7.0–10.0	2.4
La Sierra	3.3	0.055	0.045	0.045	20–25	1.2

[*] Strongly crusted fans.

Throughfan trenching, whether initiated by headcut development in mid-fan or by base-level-induced distal trenching, has two important implications for the geomorphology and sedimentology of dry region fluvial systems. First, it provides continuity of transport for coarse sediment between mountain source areas and main drainages and this has importance for sediment transport and channel adjustment throughout the fluvial system. This also means a connectivity through the drainage net, whereby the whole system may respond to erosional changes within the source area, enhancing the likelihood of correlation between upland and lowland alluvial sedimentary sequences. Second, it has implications for the stratigraphy and sedimentology of alluvial fan deposits themselves (Figure 12.10d). Sedimentary sequences in alluvial fan proximal sites may be horizontally bedded if there has been no history of fanhead trenching, but show vertical junctions and inset stratigraphy if there has been fanhead trenching. In non-trenched distal sites, stacked horizontal stratigraphy will predominate, but on fans with a history of distal trenching, inset stratigraphy may occur in these locations too (Harvey 1987b).

The significance of dryland alluvial fans

Understanding the geomorphology of dry-region alluvial fans is important for three main reasons. First, in the context of the human use of arid lands, alluvial fans may be important for their groundwater resources, for their potential under irrigation for agricultural development, and for their potential as sites for settlement

and communications. They are, however, hazardous environments for settlements, subject to temporally and spatially unpredictable flash flooding, erosion and sedimentation. Second, alluvial fans are critical zones in the coupling or buffering of arid region fluvial systems, and therefore understanding their geomorphology is central to the understanding of arid region fluvial geomorphology as a whole. Third, alluvial fans are important landforms and sedimentary environments in their own right, in the contexts of modern sedimentation, as important Quaternary landforms, and as important sedimentary sequences in the geological record.

Despite the rather separate development of our knowledge of alluvial fans into, on the one hand, studies of their geomorphology and short-term erosion and deposition and, on the other hand, of their geology as represented by their sedimentary sequences, the two themes are closely linked and each contributes to the understanding of the other. It is important to recognise that the relative importance of the three groups of factors identified above — tectonic, climatic and morphological — varies with timescale. Over long-term, geological timescales, appropriate for the study of ancient sequences, tectonics play a major role, but climatic and 'ageing' factors may also be important. Over medium-term, Quaternary, timescales, climatic factors perhaps become the most important and tectonics less so, and over the short term the problem is often one of differentiating between extrinsically and intrinsically induced morphological and sedimentological change.

One final comment, that applies to other arid region landforms as well, is that there is debate about the extent to which the landforms of today's arid regions are truly arid area forms. Much of the 'classic' literature on alluvial fans stresses arid and semi-arid regions. More recent literature, some of it referred to above, stresses the importance of fans in other climatic regions, and identifies similarities between fans in arid and non-arid areas. It is open to question whether the particular characteristics of arid region fans, for example crusted surfaces, process dominance by rare events, and the distal loss of discharge, make the arid area fan distinct from other fans.

Furthermore, many of the fans in today's dry regions, especially extra-tropical dry regions, are Pleistocene landforms produced under climatic conditions very different from those of the Holocene. For the two regions extensively discussed in this chapter there has been debate over the nature of Pleistocene climates. Both regions today have semi-arid to arid climates, with mean average rainfall totals of *c.* 200–350 mm for southeast Spain (Geiger 1970), and *c.* 50–250 mm for much of the southwestern United States. Traditionally both areas were thought to have had wetter climates during the Pleistocene, but it is now generally accepted that full-glacial Pleistocene climates in the southwest Mediterranean region were cold and arid, but subject to intense storms (Sabelberg 1977). Recent work has begun to question the validity of Pleistocene 'pluvial' climates in the American southwest (Galloway 1983). Whatever the precipitation totals, both regions had dominantly dry climates through much of the Pleistocene, sufficiently dry to allow carbonate-rich soil development and calcrete formation, but both also experienced massive sediment production. High sediment production may have primarily been the result of higher weathering rates during Pleistocene cold phases (Harvey and Wells 1994). This factor may account for the existence of alluvial fans in both regions.

Of overriding importance in controlling both the occurrence and location and the erosional/dissectional dynamics of alluvial fans is the relationship between sediment generation and sediment transport capacity. Fans develop where sediment generation is high in relation to transport capacity. This is the case, above all, in arid mountain regions.

References

Alexander, E.B. 1974. Extractable iron in relation to soil age in terraces along the Truckee River, Nevada. *Soil Science Society of America, Proceedings*, 38: 121–124.

Al Faraj, A. and Harvey, A.M., in preparation. Differential desert pavement development on terrace and fan surfaces, Wadi Al Bih, UAE and Oman.

Al Sarawi, A.M. 1988. Morphology and facies of alluvial fans in Kadhmah Bay, Kuwait. *Journal of Sedimentary Petrology*, 58: 902–907.

Amor, J.M. and Florschutz, F. 1964. Results of the preliminary palynological investigation of samples from a 50 m boring in Southern Spain. *Boletin de la Real Sociedad Espanola de Historia Natural (Geologia)*, 62: 251–255.

Baker, V.R. 1977. Stream channel response to floods, with examples from central Texas. *Bulletin of the Geological Society of America*, 88: 1057–1071.

Beaty, C.B. 1963. Origin of alluvial fans, White Mountains, California and Nevada. *Annals of the Association of American Geographers*, 53: 516–535.

Beaty, C.B. 1970. Age and estimated rate of accumulation of an alluvial fan, White Mountains, California, U.S.A. *American Journal of Science*, 268: 50–77.

Beaty, C.B. 1974. Debris flows, alluvial fans and a revitalized catastrophism. *Zeitschrift für Geomorphologie*, Supplementband, 21: 39–51.

Beaty, C.B. 1990. Anatomy of a White Mountains debris flow — the making of an alluvial fan. In A.H. Rachocki and M. Church (eds), *Alluvial fans: a field approach*. Wiley, Chichester: 69–89.

Beaumont, P. and Oberlander, T.M. 1971. Observations on stream discharge and competence at Mosaic Canyon, Death Valley, California. *Bulletin of the Geological Society of America*, 82: 1695–1698.

Birkeland, P.W. 1984. *Soils and geomorphology*. Oxford University Press, Oxford.

Birkeland, P.W. 1985. Quaternary soils of the western United States. In J. Boardman (ed.), *Soils and Quaternary Landscape Evolution*. Wiley, Chichester: 303–324.

Blackwelder, E. 1928. Mudflow as a geologic agent in semi-arid mountains. *Bulletin of the Geological Society of America*, 39: 465–484.

Blair, T.C. and McPherson, J.G. 1994a. Alluvial fan processes and forms. In A.D. Abrahams and A.J. Parsons (eds), *Geomorphology of Desert Environments*. Chapman & Hall, London: 354–402.

Blair, T.C. and McPherson, J.G. 1994b. Alluvial fans and their natural distinction from rivers based on morphology, hydraulic processes, sedimentary processes and facies assemblages. *Journal of Sedimentary Research*, A64: 450–489.

Blissenbach, E. 1952. Relation of surface angle

distribution to particle size distribution on alluvial fans. *Journal of Sedimentary Petrology*, 22: 25–28.

Blissenbach, E. 1954. Geology of alluvial fans in semi-arid regions. *Bulletin of the Geological Society of America*, 65: 175–190.

Bluck, B.J. 1964. Sedimentation of an alluvial fan in Southern Nevada. *Journal of Sedimentary Petrology*, 34: 395–400.

Bluck, B.J. 1987. Bed forms and clast size changes in gravel-bed rivers. In K. Richards (ed.), *River Channels: Environment and Process*. Blackwell, Oxford: 159–178.

Bowman, D. 1978. Determination of intersection points within a telescopic alluvial fan complex. *Earth Surface Processes*, 3: 265–276.

Bowman, D. 1988. The declining but non-rejuvenating base-level — the Lisan Lake, the Dead sea, Israel. *Earth Surface Processes and Landforms*, 13: 239–249.

Brackenridge, G.R. 1978. Evidence for a cold, dry full-glacial climate in the American southwest. *Quaternary Research*, 9: 22–40.

Brunsden, D. and Thornes, J.B. 1979. Landscape sensitivity and change. *Transactions of the Institute of British Geographers*, NS, 4: 463–484.

Bull, W.B. 1961. Tectonic significance of radial profiles of alluvial fans in western Fresno County, California. *US Geological Survey*, Professional Paper 424B: 182–184.

Bull, W.B. 1962a. Relation of textural (CM) patterns to depositional environment of alluvial fan deposits. *Journal of Sedimentary Petrology*, 32: 211–216.

Bull, W.B. 1962b. Relations of alluvial fan size and slope to drainage basin size and lithology, in western Fresno County, California. *US Geological Survey*, Professional Paper 430B: 51–53.

Bull, W.B. 1963. Alluvial fan deposits in western Fresno County, California. *Journal of Geology*, 71: 243–251.

Bull, W.B. 1964. Geomorphology of segmented alluvial fans in western Fresno County, California. *US Geological Survey*, Professional Paper 352E: 89–129.

Bull, W.B. 1977. The alluvial fan environment. *Progress in Physical Geography*, 1: 222–270.

Bull, W.B. 1979. Threshold of critical power in streams. *Bulletin of the Geological Society of America*, 90: 453–464.

Bull, W.B. 1991. *Geomorphic responses to climatic change*. Oxford University Press, New York. 326 pp.

Butzer, K.W. 1964. Climatic–geomorphologic interpretation of Pleistocene sediments in the Eurafrican sub-tropics. In F.C. Howell and F. Bouliere (eds), *African ecology and human evolution*. Methuen, London: 1–25.

Chorley, R.J. and Kennedy, B.A. 1971. *Physical geography: a systems approach*. Prentice-Hall, London: 370 pp.

DeGraff J.V. 1994. The geomorphology of some debris flows in the southern Sierra Nevada, California. *Geomorphology*, 10: 231–252.

Denny, C.S. 1965. *Alluvial fans in Death Valley region, California and Nevada*. US Geological Survey, Professional Paper 466, 59 pp.

Denny, C.S. 1967. Fans and pediments. *American Journal of Science*, 265: 81–105.

Derbyshire, E. and Owen, L.A. 1990. Quaternary alluvial fans in the Karakoram Mountains. In A.H. Rachocki and M. Church, (eds), *Alluvial fans: a field approach*. Wiley, Chichester: 27–53.

Dorn, R.I. 1983. Cation-ratio dating: a new rock varnish age-determination technique. *Quaternary Research*, 20: 49–73.

Dorn, R.I. 1984. Cause and implications of rock varnish microchemical laminations. *Nature*, 310: 767–770.

Dorn, R.I. 1988. A rock varnish interpretation of alluvial-fan development in Death Valley, California. *National Geographic Research*, 4: 56–73.

Dorn, R.I. 1994. The role of climatic change in alluvial fan development. In A.D. Abrahams and A.J. Parsons (eds), *Geomorphology of desert environments*. Chapman & Hall, London: 593–615.

Dorn, R.I. and Oberlander, T.M. 1981. Rock varnish origin, characteristics and usage. *Zeitschrift für Geomorphologie*, NF, 25: 420–436.

Dorn, R.I. and Oberlander, T.M. 1982. Rock varnish, *Progress in Physical Geography*, 6: 317–367.

Dorn, R.I., De Niro, M.J. and Ajie, H.O. 1987. Isotopic evidence for climatic influence on alluvial fan development in Death Valley, California. *Geology*, 15: 108–110.

Dorn, R.I., Jull, A.J.T., Donahue, D.J., Linick, T.W. and Toolin, L.T. 1989. Accelerator mass spectrometry radiocarbon dating of rock varnish *Bulletin of the Geological Society of America*, 101: 1363–1372.

Dumas, M.B. 1969. Glacis et croutes calcaires dans le levant espanol. *Association des Geographes Français, Bulletin*, 375: 553–561.

Eckis, R. 1928. Alluvial fans of the Cucamunga district, southern California. *Journal of Geology*, 36: 225–247.

Freytet, P., Baltzer, F. and Conchon, O. 1993. A Quaternary piedmont on an active rift margin: the Egyptian coast of the Red Sea. *Zeitschrift für Geomorphologie*, 37: 215–236.

Frostick, L.E. and Reid, I. 1989. Climatic versus tectonic controls of fan sequences: lessons from the Dead Sea, Israel. *Journal of the Geological Society of London*, 146: 527–538.

Galloway, R.W. 1983. Full-Glacial southwestern United States: mild and wet or cold and dry? *Quaternary Research*, 19: 236–248.

Geiger, F. 1970. Die ariditat in sudostspanian. *Stuttgarter Geographische Studien*, 77: 173 pp.

Gerson, R., Grossman, S., Amit, R. and Greenbaum, N. 1993. Indicators of faulting events and periods of quiescence in desert alluvial fans. *Earth Surface Processes and Landforms*, 18: 181–202.

Gile, L.H., Peterson, F.F. and Grossman, R.B. 1966. Morphological and genetic sequence of carbonate accumulation in desert soils. *Soil Science*, 101: 347–360.

Gloppen, T.G. and Steel, R.J. 1981. The deposits, internal structure and geometry in six alluvial fan–fan delta bodies (Devonian–Norway): A study in the significance of bedding sequence in conglomerates. *Society of Economic Palaeontologists and Mineralogists*, Special Publication, 31: 49–69.

Goudie, A.S. 1983. Calcrete. In A.S. Goudie and K. Pye (eds), *Chemical sediments and geomorphology*. Academic Press, London: 93–131.

Harden, J.W. 1982. A quantitative index of soil development from field descriptions: examples from a chronosequence in central California. *Geoderma*, 28: 1–28.

Harden, J. W. and Taylor, E.M. 1983. A quantitative comparison of soil development in four climatic regimes. *Quaternary Research*, 20: 342–359.

Harvey, A.M. 1978. Dissected alluvial fans in southeast Spain. *Catena*, 5: 177–211.

Harvey, A.M. 1984a. Aggradation and dissection sequences on Spanish alluvial fans: influence on morphological development. *Catena*, 11: 289–304.

Harvey, A.M. 1984b. Debris flows and fluvial deposits in Spanish Quaternary alluvial fans: implications for fan morphology. In E.H. Koster and R. Steel (eds), *Sedimentology of gravels and conglomerates*. Canadian Society of Petroleum Geologists, Memoir 10: 123–132.

Harvey, A.M. 1984c. Geomorphological response to an extreme flood: a case from southeast Spain. *Earth Surface Processes and Landforms*, 9: 267–279.

Harvey, A.M. 1987a. Patterns of Quaternary aggradational and dissectional landform development in the Almeria region, southeast Spain: a dry-region tectonically-active landscape. *Die Erde*, 118: 193–215.

Harvey, A.M. 1987b. Alluvial fan dissection: relationships between morphology and sedimentation. In L. Frostick and I. Reid (eds), *Desert sediments, ancient and modern*. Geological Society of London, Special Publication, 35. Blackwell, Oxford: 87–103.

Harvey, A.M. 1988. Controls of alluvial fan development: The alluvial fans of the Sierra de Carrascoy, Murcia, Spain. In A.M. Harvey and M. Sala (eds), *Geomorphic processes in environments with strong seasonal contrasts — Volume II: Geomorphic systems*. Catena, Supplement 13: 123–137.

Harvey, A.M. 1990. Factors influencing Quaternary alluvial fan development in southeast Spain. In A.H. Rachocki and M. Church (eds), *Alluvial fans: a field approach*. Wiley, Chichester: 247–269.

Harvey, A.M. 1992a. Controls on sedimentary style on alluvial fans. In P. Billi, R.D. Hey, C.R. Thorne and P. Tacconi (eds), *Dynamics of gravel-bed rivers*. Wiley, Chichester: 519–535.

Harvey, A.M. 1992b. The influence of sedimentary style on the morphology and development of alluvial fans. *Israel Journal of Earth Sciences*, 41: 123–137.

Harvey, A.M. and Renwick, W.H. 1987. Holocene alluvial fan and terrace formation in the Bowland Fells, northwest England. *Earth Surface Processes and Landforms*, 12: 249–257.

Harvey, A.M. and Wells, S.G. 1994. Late Pleistocene and Holocene changes in hillslope sediment supply to alluvial fan systems: Zzyzx, California. In A.C. Millington and K. Pye (eds), *Environmental change in drylands: biogeographical and geomorphological perspectives*. Wiley, Chichester: 66–84.

Harvey, A.M., Oldfield, F., Baron, A.F. and Pearson, G.W. 1981. Dating of post-glacial landforms in the central Howgills. *Earth Surface Processes and Landforms*, 6: 401–412.

Hirst, J.P.P. and Nichols, G.J. 1986. Thrust tectonic controls on Miocene alluvial distribution patterns, southern Pyrennees. *International Associoation of Sedimentologists, Special Publication*, 8: 247–258.

Hooke, R. le B. 1967. Processes on arid region alluvial fans. *Journal of Geology*, 75: 438–460.

Hooke, R. le B. 1968. Steady state relationships on arid-region alluvial fans in closed basins. *American Journal of Science*, 266: 609–629.

Hooke, R. le B. and Dorn R.I. 1992. Segmentation of alluvial fans in Death Valley, California: New insights from surface-exposure dating and laboratory modelling. *Earth Surface Processes and Landforms*, 17: 557–574.

Hooke, R. le B. and Rohrer, W.L. 1977. Relative erodibility of source area rock types from second order variations in alluvial fan size. *Bulletin of the Geological Society of America*, 88: 1177–1182.

Hooke, R. le B. and Rohrer, W.L. 1979. Geometry of alluvial fans: effect of discharge and sediment size. *Earth Surface Processes*, 4: 147–166.

Hunt, C.B. 1975. *Death Valley: geology, archaeology, ecology*. University of California Press, Berkeley. 234 pp.

Hunt, C.B. and Mabey, D.R. 1966. *Stratigraphy and structure, Death Valley, California*. US Geological Survey, Professional Paper 494A, 162 pp.

Hurst, V.J. 1977. Visual estimates of iron in saprolite. *Bulletin of the Geological Society of America*, 88: 174–176.

Johnson, A.M. and Rodine, J.R. 1984. Debris flows. In D. Brunsden and D.B. Prior (eds), *Slope instability*. Wiley, New York: 257–361.

Kesel, R.H. 1985. Alluvial fan systems in a wet-tropical environment, Costa Rica. *National Geographic Research*, 1: 450–469.

Kesel, R.H. and Spicer, B.E. 1985. Geomorphic relationships and ages of soils on alluvial fans in the Rio General valley, Costa Rica. *Catena*, 12: 149–166.

Kochel, R.C. 1990. Humid fans of the Appalachian Mountains. In A.H. Rachocki, and M. Church (eds), *Alluvial fans: a field approach*. Wiley, Chichester: 109–129.

Kostaschuk, R.A., MacDonald, G.M. and Putnam, P.E. 1986. Depositional processes and alluvial fan – drainage basin morphometric relationships near Banff, Alberta, Canada. *Earth Surface Processes and Landforms*, 11: 471–484.

Ku, T.-L., Bull, W.B., Freeman, S.T. and Knauss, K.G. 1979. $Th^{230} - U^{234}$ dating of pedogenic carbonates in gravelly desert soils of Vidal Valley, southeastern California. *Bulletin of the Geological Society of America*, 90: 1063–1073.

Laronne, J.B. and Reid, I. 1993. Very high rates of bedload sediment transport by ephemeral desert rivers. *Nature*, 366: 148–150.

Lattman, L.H. 1973. Calcium carbonate cementation of alluvial fans in southern Nevada. *Bulletin of the Geological Society of America*, 84: 3013–3028.

Lecce, S.A. 1991. Influence of lithologic erodibility on alluvial fan area, western White Mountains, California and Nevada. *Earth Surface Processes and Landforms*, 16: 11–18.

Leggett, R.F., Brown, R.J.E. and Johnston, G.H. 1966. Alluvial fan formation near Aklavik, Northwest Territories, Canada. *Bulletin of the Geological Society of America*, 77: 15–30.

Lustig, L.K. 1965. Clastic sedimentation in Deep Springs Valley, California. *US Geological Survey, Professional Paper*, 352F: 131–192.

Macheete, M.N. 1985. Calcic soils of the southwestern United States. In D.L. Weide (ed.), *Soils and Quaternary Geology of the Southwestern United States*. Geological Survey of America Special paper 203: 1–21.

Mack, G.M. and Rasmussen, K.A. 1984. Alluvial fan sedimentation of the Cutler formation (Permo-Pennsylvanian), near Gateway, Colorado. *Bulletin of the Geological Society of America*, 95: 109–116.

Maizels, J. 1990. Long-term palaeochannel evolution during episodic growth of an exhumed Plio-Pleistocene alluvial fan, Oman. In A.H. Rachocki and M. Church (eds), *Alluvial fans: a field approach*. Wiley, Chichester: 271–304.

Mather, A.E. 1993, Basin inversion: Some consequences for drainage evolution and alluvial architecture. *Sedimentology*, 40: 1069–1089.

Mather, A.E. and Harvey, A.M. 1995. Controls on drainage evolution in the Sorbas basin, southeast Spain. In J. Lewin, M.G. Macklin and J. Woodward (eds), *Mediterranean Quaternary river environments*. Balkema, Rotterdam: 65–76.

McArthur, J.L. 1987. The characteristics, classification and origin of late Pleistocene fan deposits in the Cass Basin, Canterbury, New Zealand. *Sedimentology*, 34: 459–471.

McFadden, L.D., Wells, S.G. and Jercinavich, M.J. 1987. Influences of eolian and pedogenic processes on the origin and evolution of desert pavements. *Geology*, 15: 504–508.

McFadden, L.D., Ritter, J.B. and Wells, S.G., 1989. Use of multiparameter relative-age methods for age estimation and correlation of alluvial fan surfaces on a desert piedmont, eastern Mojave Desert, California. *Quaternary Research*, 32: 276–290.

Miall, A.D. 1977. A review of the braided river depositional environment. *Earth Science Reviews*, 13: 1–62.

Miall, A.D. 1978. Lithofacies types and vertical profile models in braided river deposits: a summary. In A.D. Miall (ed.), *Fluvial sedimentology*. Canadian Society of Petroleum Geologists, Memoir 5: 597–604.

Nemec, W. and Postma, G. 1993. Quaternary alluvial fans in southwest Crete: sedimentation processes and geomorphic evolution. *International Association of Sedimentologists*, Special Publication, 17: 235–276.

Nichols, G.J. 1987. Structural control of fluvial distributing systems, Luna system, northern Spain. *Society of Economic Palaeontologists and Mineralogists*, Special Publication, 39: 269–277.

Nilsen, T.H. 1982. Alluvial fan deposits. In P.A. Scholle and P. Spearing (eds), *Sandstone depositional environments*. American Association of Petroleum Geologists, Memoir 31: 49–86.

Pierson, T.C. 1980. Erosion and deposition by debris flows at Mt. Thomas, North Canterbury, New Zealand. *Earth Surface Processes*, 5: 226–247.

Pierson, T.C., 1981. Dominant particle support mechanisms in debris flows at Mt. Thomas, New Zealand, and implications for flow mobility. *Sedimentology*, 28: 49–60.

Pierson, T.C. and Scott, K.M. 1985. Downstream dilution of a lahar: Transition from debris flow to hyperconcentrated streamflow. *Water Resources Research*, 21: 1151–1524.

Rachocki, A.H. and Church, M. (eds) 1990. *Alluvial fans: a field approach*. Wiley, Chichester: 391 pp.

Reinech, H.E. and Singh, I.B. 1983. *Depositional sedimentary environments with reference to terrigenous classics*, 3rd edn. Springer-Verlag, Berlin: 549 pp.

Rhodenburg, H. and Sabelberg, U. 1973. Quartare Kleinzyklen im Westlichen Mediterrangebiet und ihre Auswirkungen auf die Relief- und Bodenentwicklung. *Catena*, 1: 71–80.

Rhodenburg, H. and Sabelberg, U. 1980. Northwest Sahara Margin: terrestrial stratigraphy of the Upper Quaternary and some palaeoclimatic implications. In E.M. Van Sinderen Bakker Sr and J.A. Coetsee (eds), *Palaeoecology of Africa and the surrounding islands*, 12: 267–276.

Richards, K.S. 1982. *Rivers, form and process in alluvial channels*. Methuen, London: 358 pp.

Rickenmann, D. and Zimmerman, M. 1993. The 1987 debris-flows in Switzerland: documentation and analysis. *Geomorphology*, 8: 175–189.

Rust, B.R. 1978. Depositional models for braided alluvium. In A.D. Miall (ed.), *Fluvial sedimentology*. Canadian Society of Petroleum Geologists, Memoir 5: 605–625.

Rust, B.R. 1979. Facies models 2: Coarse alluvial deposits. In R.G. Walker (ed.), *Facies models*. Geoscience Reprint Series 1, Kitchener, Ontario, Canada: 9–21.

Ryder, J.N. 1971a. The stratigraphy and morphology of paraglacial alluvial fans in south central British Columbia. *Canadian Journal of Earth Sciences*, 8: 279–298.

Ryder, J.N. 1971b. Some aspects of the morphometry of paraglacial alluvial fans in south central British Columbia. *Canadian Journal of Earth Sciences*, 8: 1252–1264.

Sabelberg, U. 1977. The stratigraphic record of late Quaternary accumulation series in southwest Morocco and its consequences concerning the pluvial hypothesis. *Catena*, 4: 204–215.

Schumm, S.A. 1977. *The fluvial system*. Wiley, New York: 338 pp.

Schumm, S.A. 1979. Geomorphic thresholds: the concept and its application. *Transactions of the Institute of British Geographers*, NS, 4: 485–515.

Schumm, S.A., Mosley, M.P. and Weaver, W.E. 1987. *Experimental fluvial geomorphology*. Wiley, New York: 413 pp.

Silva, P.G., Harvey, A.M., Zazo, C. and Goy, J.L. 1992. Geomorphology, depositional style and morphometric relationships of Quaternary alluvial fans in the Guadalentin depression (Murcia, southeast Spain). *Zeitschrift für Geomorphologie*, NF, 36: 325–341.

Spaulding, W.G. 1990. Vegetation dynamics during the last deglaciation, southeastern Great Basin, USA. *Quaternary Research*, 33: 188–203.

Steel, R.J. 1974. New Red Sandstone floodplain and piedmont sedimentation in the Hebridean Province, Scotland. *Journal of Sedimentary Petrology*, 44: 336–357.

Steel, R.J., Moehle, S., Nilsen, H., Roe, S.L. and Spinnangr, A. 1977. Coarsening upwards cycles in the alluvium of Homelen Basin (Devonian), Norway: Sedimentary response to tectonic events. *Bulletin of the Geological Society of America*, 88: 1124–1134.

Suwa, H. and Okuda, S. 1983. Deposition of debris flows on a fan surface, Mt. Yakedale, Japan. *Zeitschrift für Geomorphologie*, Supplementband, 46: 79–101.

Van Arsdale, R. 1982. Influence of calcrete on the geometry of arroyos near Buckeye, Arizona. *Bulletin of the Geological Society of America*, 93: 20–26.

Van Devender, T.R. and Spaulding, W.G. 1983. Development of vegetation and climate in the south western United States. In S.G. Wells and D.R. Haragan (eds), *Origin and evolution of deserts*. University of New Mexico Press, Albuquerque: 131–156.

Wasson, R.J. 1974. Intersection point deposition on alluvial fans: an Australian example. *Geografiska Annaler*, 56A: 83–92.

Wasson, R.J. 1979. Sedimentation history of the Mundi Mundi alluvial fans, western New South Wales. *Sedimentary Geology*, 22: 21–51.

Wells, S.G. and Harvey, A.M. 1987. Sedimentologic and geomorphic variations in storm generated alluvial fans, Howgill Fells, northwest England. *Bulletin of the Geological Society of America*, 98: 182–198.

Wells, S.G., McFadden, L.D. and Dohrenwend, J.C. 1987. Influence of late Quaternary climatic change on geomorphic and pedogenic processes on a desert peidmont, eastern Mojave Desert, California. *Quaternary Research*, 27: 130–146.

White, K.H. 1991. Geomorphological analysis of piedmont landforms in the Tunisian Southern Atlas using ground data and satellite imagery. *Geographical Journal*, 157: 279–294.

White, K.H. and Walden, J. 1994. Mineral magnetic analysis of iron oxides in arid zone soils from the Tunisian Southern Atlas. In A.C. Millington and K. Pye (eds), *Environmental change in drylands: biogeographical and geomorphological perspectives*. Wiley, Chichester: 43–65.

Williams, G.E. 1973. Late Quaternary piedmont sedimentation, soil formation and palaeoclimates in arid South Australia. *Zeitschrift für Geomorphologie*, 17: 102–125.

Wolman, M.G. and Gerson, R. 1978. Relative scales of time and effectiveness of climate in watershed geomorphology. *Earth Surface Processes*, 3: 189–208.

Wolman, M.G. and Miller, J.P. 1960. Magnitude and frequency of forces in geomorphic processes. *Journal of Geology*, 68: 64–74.

Wright, V.P. and Alonso Zarza, A.M. 1990. Pedostratigraphic models for alluvial fan deposits: a tool for interpreting ancient sequences. *Journal of the Geological Society of London*, 147: 8–10.

Yoshida, F. 1994. Interaction between alluvial fan sedimentation, thrusting, and sea-level changes: an example from the Komeno Formation (Early Pleistocene), southwest Japan. *Sedimentary Geology*, 92: 97–115.

13

Badlands and badland gullies

Ian A. Campbell

Introduction

Geomorphologists often face the problem of defining the nature of the spatial and temporal relationships which exist between the observed, measurable rates of geomorphological processes and the landforms that evolve, or are assumed to evolve, as a result of those processes. The time available for such observations, and the slow rates at which most processes operate, is usually too restricted either for discernible changes to occur in the landforms or for sufficient, accurate data to be collected which will allow valid theories about process/form relationships to be formulated and rigorously tested. Furthermore, the complexity and the interactive nature of most geomorphological systems precludes other than an often highly selective approach to studying the multivariate relationships between processes and forms.

These interrelated spatial and temporal linkage problems require geomorphologists to use a wide variety of approaches in their attempts to devise theories of landform evolution (Thorn 1988). They include inductive theories of landform development, geometrical and mathematical models, physically based predictive equations, laboratory simulations and, increasingly, computer models which often incorporate elements of several techniques. Other researchers have studied natural or artificial landscapes which exemplify rapid responses to certain geomorphic processes, and which may represent simpler landscapes with fewer complex process/form relationships. Such landscapes, which include badlands, favour both direct observations and the collection of detailed measurements. Badlands, moreover, may act as analogues or surrogates for more slowly evolving, larger-scale landforms (Schumm 1956a; Scheidegger *et al.* 1968).

The word badland is likely derived from the French '*mauvais terres à traverser*' — land that is difficult to cross — as the French were among the first Europeans to enter the northern Great Plains of western North America where there are extensive areas of badlands, for example Badlands National Monument (*c.* 98 000 ha) in South Dakota (Scheidegger *et al.* 1968). Badlands have little or no intrinsic economic value due to their barren nature, rugged terrain and difficulty of access. Several badlands, however, have become important tourist attractions (Figure 13.1).

Badlands develop where soft, predominantly horizontally bedded, relatively impermeable rocks are exposed to rapid fluvial erosion; they have been called the drylands' most

Arid Zone Geomorphology: Process, Form and Change in Drylands, 2nd edition. Edited by David S. G. Thomas.

Figure 13.1　*Badlands, Dinosaur Provincial Park, Alberta, Canada, a UNESCO World Heritage Site. Relief is c. 30 m from the alluvial-covered slope to the butte crest. Interbedded units of shales, mudstones, channel sandstones and thin, discontinuous lenses of concretionary clay ironstone dominate the lithology. Clay ironstone fragments form a colluvial armour over portions of the lower and midslopes and alluvial fines cover the basal pediments. Lenticular outcrops of indurated channel sandstone form incipient hoodoos in the lower slopes. Small pipes emerge at the shale–sandstone interface channelling subsurface flow into rill heads. Slopes are etched by rills, collapsed pipes and small gullies whose density and size reflect lithology and variations in runoff and erosion. Miniature mudflows and creep processes occur mainly on the dark bentonitic shale members*

distinctive fluvial landform (Warren 1984). Gilbert (1877), in his authoritative work on the Henry Mountains, was fascinated by the stark beauty of '*bad-lands*' and used them for much of his reasoning on the relationship between slope form, erosional processes and landscape evolution.

While badlands are customarily associated with drylands they form minor parts of deserts. In an admittedly selective survey, Clements *et al.* (1957) found badlands comprised 2.6% of the southwestern United States, 8% of the Libyan Desert, 2% of the Sahara and 1% of the Arabian Desert. Nevertheless, badlands have attracted a research interest out of all proportion to their restricted areal extent, partly because badlands, and related landscapes such as spoil heaps, etc., are not limited to arid regions.

Badlands form in a wide range of materials and climates (Table 13.1), from glaciomarine silts in the high arctic (Christie 1967) to deeply weathered granite in the humid tropics (Lam 1977). Badlands-like terrain results from accelerated erosion in devegetated areas caused by smelter fumes (Strahler 1956a), strip-mining, waste or spoil heap deposition (Schumm 1956b), or construction and unwise agricultural

practices (Aghassy 1973). Localised badlands formed in Greece in Classical times because of human misuse of land, destruction of vegetation, and a general, climatically induced, erosional phase of severe gullying (Harris and Vita-Finzi 1968). In the Göreme valley in Cappadocia, Turkey, large-scale excavation for troglodytic settlements in volcanic tuffs has contributed to badlands development (UNESCO 1983). The principal concern here is with badlands as natural components of arid zone landscapes.

Their widespread geographic distribution results in an equally wide variety of terms being used to describe, in particular, the fluvially eroded channels which often characterise badlands and deserts. Such words as gully, arroyo, wadi and donga are often used synonymously, along with many others, depending on language and regional usage (Stone 1967); but there is controversy over terminology and little agreement on exact definitions, especially in the case of gullies (Schumm *et al.* 1984; Graf 1988, p. 218).

Regardless of their origin badlands are barren and usually intricately dissected by rills and gullies (Figure 13.1). Steep-sided residuals rise above gently sloping alluvial or pediment surfaces with intervening slope angles often reflecting the erosional resistance of different lithologies. Microrelief is complex; deep desiccation cracks, pipes, rills, knife-edge divides and vertical faces alternate abruptly with rounded forms. Some surfaces may be mantled with a variable thickness of fragmental debris or weathered aggregates while others show little evidence of weathering (Figure 13.1).

Badlands as model landscapes

Badlands resemble some aspects of desert landscapes in miniature and certain geomorphological analogies have been drawn between them especially with reference to pediment development (Johnson 1932; Bradley 1940; Smith 1958; Schumm 1962; Engelen 1973). Mandelbrot (1977) examined the way in which forms may transcend scale differences and exhibit 'self-similarity' such that in the absence of a scale it is difficult to tell by geometry alone the size of a feature. Fractal geometry and morphologic measures offer some promising methods of analysis in comparative landscape studies (see several papers in Snow and Mayer 1992). However, the extent to which such scale-independent observations are valid in terms of *processes* is limited by dimensional and gravitational constraints on particle and hydraulic characteristics (Bloom 1978, p. 280), especially in physical models.

The effects of magnitude and frequency of processes, and the effects of different processes on convergence or equifinality of forms requires particular attention when considering transfer of data across spatial and temporal scales. Extrapolation from small temporal and spatial-scale studies in badlands to larger landforms also presents special difficulties (Campbell and Honsaker 1982). Small-scale features in badlands often have short reaction times to erosional processes and may adjust their forms quite rapidly. These landforms have 'short memories' of past geomorphic events. The larger-scale badlands landscape may preserve the effects of formative events for protracted periods (e.g. Alexander *et al.* 1994) and may show comparatively little response except to major changes (de Boer 1992b). Badlands are continually at, or close to, threshold conditions

but the erosional response of their surfaces or lithologies varies greatly with respect to similar external stresses.

The dramatic character of badlands, which apparently reveal clear morphological relationships resulting from the interaction of geomorphic processes with highly erodible materials, has long stimulated geomorphic speculation and investigation. Gilbert (1877), while declining to give an explanation, stressed the significance of convex crestal divides on badlands slopes as an exception to his 'law' of concave divides. He indicated that at positions near the crest, where the flow of water was very small (discharge increasing proportionately downslope), some other process must be more important than wash in creating slope form. In response, and in one of his few definitive statements on slope processes, Davis (1892) suggested that the convexity was due to surface creep; expansion and contraction induced by variations in moisture and thermal conditions caused differential movements which produced a slow, downslope movement of debris. Slope form reflected the extent to which creep or wash predominated. Gilbert (1909) evidently accepted Davis' explanation for he included it in his paper on the convexity of hillslope form.

Davis' (1892) statements presaged Schumm's (1956a) study on the relative roles of creep and wash in the evolution of badland slopes. Schumm poured water on two different lithologies in Badlands National Monument, South Dakota, to show that the loose, clay aggregate surface of the Chadron shales, with their high infiltration rates, would slide *en masse* in a saturated state. The less permeable, densely rilled Brule Formation responded to the same amount of wetting by almost instantaneous runoff. Slope geometry thus reflected lithological control in terms of a varied response via infiltration rates to moisture input; Chadron slopes lowered, Brule slopes retreated — with little lowering. Schumm (1956a) concluded that the different geomorphic responses might be a basis for comparing humid and arid slope forms in general.

Despite their obvious suitability in terms of geomorphological examples with respect to drainage system development, runoff generation and rill formation (Figure 13.2), Horton (1945), in his seminal work on those topics,

Figure 13.2   *A once-smooth, pediment surface is now furrowed by rills and pipe-controlled small gullies (grass-covered mounds in the distance give scale). The narrow, rounded divides have been used as examples of Horton's (1945) 'belt of no erosion'. Note the even spacing and alignment of the small tributary rills — especially in the right foreground and centre — controlled by small joint-sets and associated fractures in this outcrop of channel sandstone*

makes no reference to badlands and only passing reference to arid landscapes. Brief mention (Horton 1945, p. 331) is made that new road cuts, spoil banks and mine dumps are excellent sites for rilling. Horton cites Gilbert's Henry Mountain report in his references but he does not refer to it in his text, nor does he make it explicit that he recognised Gilbert's contribution to rill development yet he echoes Gilbert's geometrical approach: cf. Gilbert's (1877; his figure 58) badland ridge, and Horton's (1945; his figures 37 and 38) interior cross divide. Horton also ignores the role of creep and uses a 'belt of no erosion' to explain divide convexity (Figure 13.2).

Theories of drainage system development (Strahler 1950; Schumm 1956b; Leopold *et al.* 1966) and landscape textural differences (Smith 1950) were considerably influenced by observations made in badlands. Formative ideas on erosional processes and slope evolution by Strahler (1950; 1956b), during the rise of quantitative geomorphology in the 1950s, used examples of badlands topography to rigorously test concepts such as the theory of dynamic equilibrium. Only recently, however (Bryan and Yair 1982; Howard 1994), has badlands geomorphology been comprehensively reviewed. Badlands are ideal sites for geomorphological study. Their barren, often sharply defined slopes, and their rapid rates of erosion, encourage detailed measurements of slope forms and the effects and rates of erosional processes (Campbell 1974; Haigh 1975; Toy 1983). Erosion rates of 2.00–20.00 mm per year have been reported (Haigh 1975), though differences in materials, measurement techniques and other factors can make it difficult to arrive at meaningful comparisons among the different studies (Yair *et al.* 1980). In a 10-year study of erosion rates in the Alberta badlands, Campbell (1981) found that mean (1969–79) erosion rates were 38.00 mm, but that they ranged from 83.60 to 7.40 mm with no clear evidence that the variations reflected lithology, slope angle or topographic position. In the absence of natural badlands, geomorphic process studies have utilised anthropogenic or geotechnical badlands for short- and long-term studies (e.g. Schumm 1956b; Bridges and Harding 1971; Haigh 1978).

The stark, highly geometrical form of the badlands, and their clearly defined channels, favours computer-based analysis of drainage systems (Howard 1990; 1994) in which simulated waterflows can be controlled to produce models of drainage evolution. As Howard (1994) notes, however, the establishment of the boundary conditions in such models (for erosion rates, types of erosion, lithological variations, and other natural factors) renders exact comparison between the model and actual badlands drainage systems and landforms uncertain.

The characteristically high rates of erosion associated with badlands, and the large and rapid runoff that they generate, can produce exceptionally high sediment yields; values in excess of 38 000 t km^{-2} yr^{-1} are reported from the Loess Plateau badlands in the headwaters of the Huangho, China (Chen 1983), which records one of the highest sediment concentrations in the world (Long and Xiong 1981). In the Cheyenne River badlands, South Dakota, Hadley and Schumm (1961) estimated yields of about 13 400 t km^2 between July and October 1954, and similar values were obtained by Leopold *et al.* (1966) from badlands-type terrain in New Mexico. Campbell (1970) found potential average sediment yields equivalent to 9976 t km^{-2} yr^{-1} in a runoff plot study of badlands in Alberta, Canada. Such findings are of fundamental importance in drainage basin studies because they show, albeit as extreme examples, that the majority of the sediment yield can be derived from a comparatively minor portion of a watershed. Badlands, with their often massive potential sediment yield, epitomise the 'partial area' effect (Campbell 1985). In the 44 800 km^2 Red Deer River basin in Alberta, Canada, Campbell (1977; 1992) showed that some 800 km^2 of badlands (about 2% of the basin) could supply almost 80% of the river's mean annual suspended sediment load of *c.* 1.5×10^6 tonnes.

Variations in the kinds of measurements, differences in lengths of records, measurement techniques, geology, suspended sediment bulk density conversion values, and spatial representativeness of the data can make it difficult to compare these studies. Nevertheless, badlands are some of the most erosionally active areas on Earth with both surface erosion and subsurface erosion being of great local significance (Jones 1981).

Scale problems and badlands evolution

Most research in badlands has focused on the nature and effects of geomorphic processes at micro- and meso-scales of operation. Little attention has been given to the large-scale components of badlands, particularly their long-term evolution. Concern with detailed process studies has dominated modern geomorphic research (Slaymaker 1980). Thornes and Brunsden (1977, p. 185) suggested that it may reflect a desire to equate landforming processes and landscapes with anthropocentric spatial and temporal scales. However, if badland process studies are to have wider application in terms of understanding the way badlands evolve, this detailed research must be put into the context of regional scales and long time spans. A key problem in badlands studies is how to satisfactorily resolve what are usually short-term, highly site-specific data with much longer-term, often inferential observations, related to much larger areas. Graf (1988) noted that space can be arranged in a hierarchical sense in which the system variables assume different causal relationships where the value of any given variable is influenced by at least three factors — regional, local and specific. Schumm and Lichty (1965) considered temporal effects in much the same way.

Badland landscapes, with their high rates of erosion, and the large sediment yields they produce, are controlled by regional, geological conditions. However, not all the exposed surfaces erode at the same rate or even in the same manner; there are wide local variations which reflect site-specific controls on erosion (Figure 13.1). Very minor variations in the lithology, slope, aspect (Figure 13.3), or other factors can greatly influence the behaviour of each badland surface. Regional badland landscapes cannot be understood without knowledge of the local and specific variations (Campbell 1987; de Boer and Campbell 1989).

Figure 13.3   *Aspect can have a major effect on slope form and processes. In the Dinosaur badlands, Alberta, on the same lithological sequence, north-facing slopes (left) are more densely rilled and have a lower slope angle than their steeper, mass-movement dominated, south-facing counterparts. The surface characteristics also differ — smooth, wash slopes contrast with 'popcorn' aggregate-covered surfaces*

In his badlands-based analysis of the role of spatial scale in geomorphic process studies, de Boer (1992a; 1992b) emphasised that the hierarchical structure inherent in each geomorphic system means that as the spatial size increases the amounts of information which can be validly extrapolated from small-scale studies diminishes. While the dominant characteristics of the landscape system remain identifiable across scale boundaries, detailed data must be compared at compatible spatial and temporal scales. For example, the present mean surface erosion rates in the Dinosaur badlands,which average about 4.00 mm a year (Campbell 1974; Campbell and Honsaker 1982) could, if extrapolated uniformly over the 12–14 000 years since the area was deglaciated (Bryan *et al.* 1987), reasonably closely account for the present landscape in terms of the volume of eroded material. Dated deposits (Bryan *et al.* 1987), however, show that comparatively little erosion has occurred since the mid-Holocene, and field evidence indicates that most erosion occurred very quickly during and immediately following deglaciation

(Campbell and Evans 1990; O'Hara and Campbell 1993).

Other studies have also provided important long time perspectives on badland development. Wells (1983) and Wells and Guiterrez (1982) documented the Quaternary evolution of badlands in New Mexico. Linkages between stripping of the Quaternary deposits coupled to base-level lowerings of the Chaco River exposed the Cretaceous-aged deposits in which the badlands formed. Palaeoenvironmental data and calculated erosion rates based on instrumented catchments provided a chronology of the rate of badland development.

Wise *et al.* (1982) examined badlands in Almeria, southeastern Spain, where infrequent, high-magnitude events of great spatial variability, when averaged over long time spans, account for a landscape which may date from the Pliocene. Alexander *et al.* (1994), in the same region, show that episodic development since the late Pleistocene appears to be punctuated (stabilised) by gravel lag deposition and biotic (lichen) crusts which retard surface erosion. Their study indicates that several centuries may be required for complete stabilisation to occur; interruptions may be due to either long-term effects acting through base-level changes (tectonic or climatically induced) or shorter disturbances reflecting land use, climatic variations or effects of major storms (Alexander *et al.* 1994). In the northern Negev, Israel (Yair *et al.* 1982), the badlands evolved rapidly some 40–70 000 years ago, but erosion has been minimal over the last 20 000 years when drier conditions began to affect the area.

Badlands convey strong visual impressions of rapid, often spectacular erosion, but not all the parts of any one badland show these dramatic changes. Understanding their forms requires a combination of viewpoints from different temporal and spatial scales. Badlands are complex, intricate landscapes which reflect a wide variety of surface and subsurface adjustments and responses to delicately poised external and internal thresholds (Campbell and Honsaker 1982).

Badland materials

The climate characteristics associated with arid and semi-arid environments — prolonged dryness punctuated by sometimes intense rainstorms — occur in many badlands, but not all landscapes with such climates evolve into badlands and, as discussed above, badlands form outside arid regions. Fundamental attention must be given to the nature of the badland materials. Whereas climate is the forcing mechanism that drives the geomorphic processes, the primary landform control is the generally easily erodible, yet highly variable, character of the materials which form the base for the interaction of weathering and erosion processes. Badlands exemplify structural and lithological control at all scales (Figure 13.1).

Erosion in badlands is frequently so rapid that a regolith or soil suitable for the establishment of a dense, or any, vegetation cover is inhibited or prohibited. Frequent deep desiccation cracking hinders plant root development and the badlands materials are often infertile or inhospitable for plant establishment. Rapid erosion limits vegetation colonisation and, in a positive feedback, the lack of vegetation enhances erosion. Aridity exacerbates these conditions. The landscape becomes an expanse of barren, edaphically unprotected surfaces which respond differentially only in terms of *rates* of badland evolution which are controlled by the efficiency of the processes and the rapidity with which slopes are eroded or stripped of weathered debris, i.e. the balance between weathering-limited and transport-limited processes.

Culling (1963) noted that on bare rock slopes, erosion is controlled by the rate of weathering, whereas on soil-mantled slopes the thickness of the soil, and the rate of weathering, depends on the rate of erosion. Carson and Kirkby (1972) refer to weathering-limited and transport limited-processes in which critical limits are set by the comparative rates of weathering and erosion. Where weathering predominates, a weathered mantle forms; such slopes or surfaces are then transport limited. Where the rate of erosion is greater, weathered debris does not accumulate and a weathering-limited condition exists.

In drylands, slopes are generally weathering limited and erosion controlled with shallow or no weathered mantles. Badlands slopes, however, present an interesting transport-limited case. Because badlands form mostly in poorly consolidated rock, large quantities of *pre-weathered* debris is readily available. Erosion is limited only by the frequency and magnitude of fluvial events of sufficient energy to transport an apparently endless supply of sediment.

A review of representative badland and related landscapes (Table 13.1) shows that most form in fine-grained argillaceous sediments which come under the general generic heading of shale. Shale is a widely distributed rock and the term is not always used precisely. Shale, and its lithologically associated sediments, have a variety of names (e.g. bedded silts, clay muds, siltstone, mudshale, clayshale, claystone, mudstone, etc.) depending upon the degree of induration, lamination or bedding characteristics and percentage of clay-size constituents (Potter *et al.* 1980). The role of cementing agents (iron oxide, calcite, etc.) and various clay minerals have a considerable effect on the behaviour of the sediments when these are exposed to erosion and alternating cycles of wetting and drying (Fleming *et al.* 1970). Additionally, many shales are over consolidated

Table 13.1   *Representative badlands locations and materials in arid, semi-arid and other environments*

References	Mean annual precipitation (mm)	Location	Materials
1. Badlands in arid and semi-arid areas			
Alexander (1982)	450	Agri Basin, Basilicata, Italy	Plio-Pleistocene marine clays, silt–clays, interbedded sands, soft shales and mudstones. Clays mostly impure montmorillonite
Bergstrom and Schumm (1981)	450	Kraft Badlands, Lusk, Wyoming, USA	Tertiary siltstones of the Brule Formation
Brown (1983)	135	Borrego Springs. California, USA	Mid to late Pleistocene poorly consolidated sands and gravels with interdigitated lacustrine clays and silts
Campbell (1970)	350	Red Deer Valley, Alberta, Canada	Upper Cretaceous highly bentonitic shales interbedded with clay-ironstone and sandstone
Christie (1967)	100	Clements Markham Inlet, Ellesmere Island, Canada	Late Quaternary glacio-marine sands, silts and clays
Clayton and Tinker (1971)	450	Little Missouri badlands. North Dakota, USA	Palaeocene Sentinal Butte and Tongue River formations. Montmorillonite and mica-rich sediments with minor chlorite and kaolinite clays
De Ploey (1974)	350	Kasserine area, central Tunisia	Clays, loams and sandy lithosoils developed on Cretaceous marls. Illite, kaolinite and montmorillonite clays
Hadley and Schumm (1961)	355	Cheyenne River basin, Nebraska, USA	Oligocene clays, sandstones, mudstones, siltstones and shales interbedded with bentonite
Harvey (1982)	170–350	Almeria–Alicante region, Spain	Cenozoic (mostly Tertiary) and Triassic marls, silts, shales and sandstones
Imeson et al. (1982)	300	Rif Mountains, Morocco	Villafranchian, Pliocene and subrecent marine sediments, alluvial and colluvial deposits. Clays mainly illite and kaolinite
Liu et al. (1985)	500	Central Huanghe (Yellow River) valley, China	Pleistocene loess, mainly silts with illite, montmorillonite and kaolinite being main clay minerals
Schumm (1964)	450	Western Colorado, USA	Cretaceous Mancos Shale, marine shales with thin coal and limestone lenses containing montmorillonite, illite, chlorite and mica
Smith (1958)	380	Badlands National Monument, South Dakota, USA	Oligocene White River Group (Brule and Chadron formations) of poorly consolidated clays and silts with channel sandstones. Bentonitic with illite and montmorillonite

continued

Table 13.1 *Continued*

References	Mean annual precipitation (mm)	Location	Materials
Wells and Gutierrez (1982)	220	Chaco River basin, New Mexico, USA	Cretaceous friable sandstones, thin coal beds and thick mudstones of the Kirtland and Fruitland formations
Yair *et al.* (1982)	90	Northern Negev, Israel	Palaeocene marls and soft shales
2. Badlands in other environments			
Bridges and Harding (1971)	1100	Lower Swansea Valley, Wales, UK	Infertile, acidic sandy-clay loam soils derived from solifluction redistributed glacial material on industrial wasteland
Haigh (1978)	1100	Blaenavon, Wales, UK	Colliary spoil mounds and open-pit fill. Fine coal washings of gravel- to clay-size particles
Lam (1977)	1900	Tai Lam Chung region, Hong Kong	Deeply weathered (80 m) Jurassic granite regolith of sands, gravels and clays with kaolinite and illite
Schumm (1956b)	1090	Perth Amboy, New Jersey, USA	Backfilled clay-pit, homogenous mix of sands, silts and clays
Segerstrom (1959)	2000	Paricutin, Michoacán, Mexico	Recent (1943) basaltic andesite volcanic ash deposits of gravel- to dust-size particles

in that the present load is less than when the sediments underwent diagenesis. Unloading of over consolidated shales causes reduction in shear strength and enhances physico-chemical weathering by intraparticle rehydration of expandable clay minerals. Together with surface erosion this weathering process can be a major factor affecting slope stability (Taylor and Cripps 1987) and stresses caused by deformation during diagenesis or unloading may create fractures facilitating piping and triggering massive landslides (de Lugt and Campbell 1992).

Detailed consideration has been given to the role and effects of the physical and clay-mineralogical constituents of badland materials (e.g. Alexander 1982; Hodges and Bryan, 1982; Imeson *et al.* 1982; Bryan *et al.* 1984; Campbell 1987). Particularly significant is the presence of bentonite, a rock composed dominantly of the montmorillonite–smectite group of clay minerals (Table 13.2). In the Dinosaur badlands, most of the clay minerals are sodium montmorillonite (bentonite or smectite) with secondary amounts of kaolinite and illite (Table 13.2). Fifty per cent of the clays may be <0.04 μm in diameter (Bryan *et al.*, 1984). The different behaviour of these clays, and their relative abundance, has profound effects in desiccation cracking, surface crusting, infiltration rates and aggregate stability. Barendregt and Ongley (1977) regarded the presence of montmorillonite as an *a priori* factor in piping. Montmorillonite gels, secreted by dispersion during weathering (Van Olphen 1977), act as almost impenetrable surface seals and encourage rapid runoff. The clay content and the clay chemistry therefore determine a wide range of different responses by closely adjacent surfaces to identical rainfall amounts (Hodges 1982; Bryan *et al.* 1984; de Boer and Campbell 1990). Badlands are not the simple models of landform development they were often assumed to be.

Bentonite, while formerly regarded as a weathering derivative of volcanic ash or tuff,

Table 13.2   *Granulometry and clay mineralogy of representative badland lithologic units, Dinosaur Provincial Park, Alberta* [*]

Unit	Sand (%)	Silt (%)	Clay (%)	Clay fraction		
				Monmorillonite (%)	Illite (%)	Kaolinite (%)
Pediment†						
surface	26.3	57.5	16.2	100	0	0
5 cm depth	31.3	65.1	3.6	70	30	0
Mudstone						
surface	8.7	31.8	59.5	5	80	15
5 cm depth	7.5	36.1	56.4	15	60	25
Channel sandstone						
surface	59.5	37.7	2.8	10	70	20
5 cm depth	59.2	24.8	16.0	100	0	0
Dark grey shale						
surface	15.0	14.5	70.5	80	20	0
5 cm depth	19.8	15.4	64.8	65	30	5
Yellow shale						
surface	8.9	49.7	41.4	100	0	0
5 cm depth	17.6	28.6	53.8	90	5	5
Reddish shale						
surface	15.9	50.0	34.1	70	25	5
5 cm depth	2.8	57.8	44.4	70	25	5
Grey shale						
surface	18.2	24.4	57.4	5	65	30
5 cm depth	14.4	46.0	39.6	10	75	15

[*] Clay mineralogy by Alberta Research Council.
† A deposit, not a lithologic unit.
Source: Campbell (1987).

has various origins (Grim and Güven 1978). Bentonitic deposits are widespread in the Cretaceous rocks of the North American prairies where badlands are common; bentonites are also widely distributed globally, usually in sediments of marine origin. Recycling of bentonites by fluvial or aeolian processes may introduce considerable concentrations of clay minerals, such as montmorillonite, into resedimented deposits (Teruggi 1957; Liu *et al.* 1985). Such redeposition may favour local badland formation, perhaps in areas where they would not normally occur. The bentonitic clay minerals tend to slake, swelling when water is introduced into or between the particles (Taylor and Cripps 1987, 416–417), and the volumetric expansion and contraction involved by inter- and intra-particle hydration and desiccation produces severe disruptive effects on aggregate stability. The clays can also form almost impermeable surface seals and crusts. Percolation regulation by crusts and other clay layers may be as or more important in piping than the hitherto more emphasised role of clay mineral interparticle bonds (Jones 1981). Deep, wide desiccation cracks which are characteristic of bentonitic shales favour moisture penetration. Cracks have a key role in near-surface piping processes (Bryan *et al.* 1978; Hogg 1978; Hodges and Bryan 1982; Guiterrez 1983; Bryan and Harvey 1985) and may affect deeper infiltration. De Ploey (1974) found that desiccation in Tunisian badlands formed polygonal crack networks which penetrated the subsoil for several metres completely controlling pluvial erosion and generating extensive piping.

The processes of surface and subsurface fluvial activity are greatly influenced by the physical and chemical characteristics of the badlands materials. The association of high percentages of exchangeable sodium with swelling clays causes dispersion and encourages deeper moisture penetration into the regolith and bedrock (Bryan *et al.* 1984). The importance of these processes, while long recognised in piping studies, requires more attention in order to gain a comprehensive understanding of their effects on badlands geomorphology (Stocking 1984).

The response of badlands surfaces to erosional processes, especially raindrop impact and runoff generation, are greatly affected by infiltration characteristics, degree of impermeability, cohesion and aggregate stability (Schumm 1956a; Parker 1963; Yair *et al.* 1980; Hodges 1982; Bryan *et al.* 1984). In Colorado, Schumm and Lusby (1963) and Schumm

(1964) showed frost action affected permeability, changing a less permeable, rilled surface to a highly permeable, non-rilled, aggregate-covered surface. Frost action in Alberta assists erosion by breaking down aggregates and preparing surfaces for rapid stripping by runoff during snowmelt and spring rainfall (Campbell 1974; Harty 1984). Seasonal effects of erosional stripping and rill obliteration are also found in the Ebro badlands in Spain (Benito *et al.* 1991).

Variations in topography, particularly as they affect aspect and exposure, in addition to lithological contrasts, appear to influence the effects of frost and other processes. In the South Dakota badlands, Churchill (1981) found south-facing slopes significantly shorter, steeper and generally straighter in profile than their north-facing counterparts. These

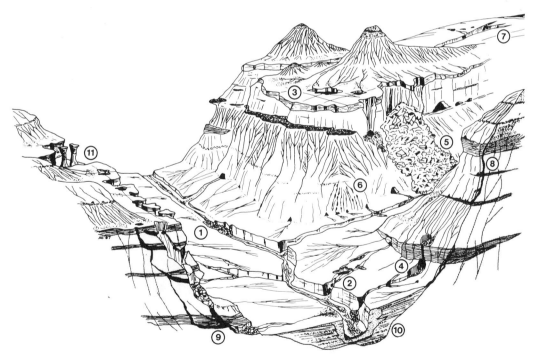

Figure 13.4 *Composite of badlands lithogeomorphic characteristics. (1) Headcutting by valley-bottom gully in alluvial fill with accompanying slump and collapse by basal sapping. (2) Pipe-induced collapse initiates discontinuous gully. (3) Surface drainage into sinks and pipe shafts controlled by structural discontinuities. (4) Valley-side gullies fed by convergent rill flow. (5) Percoline-controlled large-scale piping generates slope collapse and triggers mass movements. (6) Near-surface micropiping produces honeycomb erosion of slope surfaces by collapsed rills. (7) Gully heads expanding into undissected surface. (8) Early stage of meso-scale pipe development. (9) Mature meso-scale piping initiating slope collapse, large gullies and tributary valley formation. (10) Multiple cut-and-fill gully deposits in bedrock-floored valley. (11) Hoodoos formed by dissection of resistant caprock*

morphological differences were attributed to type and rate of weathering with densely rilled north-facing slopes indicating more intense fluvial action. In Dinosaur Provincial Park, Alberta, Harty (1984) also found north-facing slopes retained snow longer and had moister regoliths. Statistical analysis of 110 paired slopes on identical lithological units showed that south-facing slopes are steeper, by an average of 10.4°, than their north-facing counterparts (Figure 13.3). Yair *et al.* (1980) recorded aspect-related infiltration differences in badlands in the Negev, Israel. Northern-facing slopes developed significantly deeper regoliths than their southern-oriented counterparts showing that microclimatic factors affected by exposure influence the character of the surficial materials. In the Little Missouri badlands of North Dakota, seepage-step erosion (creep) occurred mainly on northeast-facing slopes while rilling and wash processes dominated slopes with a southwest aspect (Clayton and Tinker 1971).

On a larger scale, the effects of aspect on valley asymmetry, slope length, steepness, degree of dissection and other morphological, vegetational and soil characteristics is a widely debated topic (Toy 1977). Aspect effects, where these occur, should be more pronounced in regions where general aridity promotes extreme variations in regolith moisture content (Churchill 1981), and perhaps these variations are of special importance in controlling slope form in badlands. Alexander *et al.* (1994) note that in the Tabernas badlands of Almeria, Spain, the bare, south-facing hills are dominated by overland flow and rilling; there is comparatively little swelling, cracking, piping, or mass movements. On north-facing slopes the regolith is thicker, there is more vegetation, and erosion, including rilling, is less evident and there are only small mass movements. The frequent and abrupt changes in slope facets, the often deeply incised landscape and the rapid alternations in slope orientation and exposure favour pronounced microtopoclimatic contrasts. Moreover, because the buffering effects of vegetation cover are absent, small lithological differences are often sharply etched by weathering and erosion processes.

Sites favouring snow accumulation (in mid- or high-latitude badlands), shading, pipe-induced

Figure 13.5 *Piping is ubiquitous in many badlands. The size, form and location of the pipe shafts and outlets is a function of the rate and amount of infiltration and lateral water movement. Complex hydraulic gradients reflect microlithological and clay mineralogical differences. Control of lateral flow by less permeable layers is a key factor in the process with collapse, seen here, occurring below permeable sandstones. Pipes originate rills or gullies, or form from them by percolation. Large pipes induce collapse of slopes and channel segments initiating valley formation. R. Bryan, at right, gives scale*

seepage planes or other controls which influence slope moisture conditions — either annually, seasonally or over shorter periods — can be more important in controlling slope processes than large, regional climatic influences (de Lugt and Campbell 1992).

Badlands are often multi-process landscapes (Figure 13.4). Even where different badlands share the same general assemblage of geomorphological processes the *process-mix* differs spatially and temporally. Some badlands, or parts of them, may be dominated by rills, others by creep, while in others piping may be the key process. The sequence and mechanisms of slope formation and the processes by which sediment is derived and removed from the slopes in terms of timing, rates and areal distribution, has profound effects on badland development.

Piping

Piping, sometimes called tunnel erosion or pseudokarst (Jones 1981), is a subsurface erosion process in which percoline-controlled flowing water removes rock and soil particles. A subterranean drainage network forms which varies from simple systems of one or a few conduits to extensive, integrated networks of great complexity. Pipes may be a few millimetres in diameter to several metres and they can extend for scores of metres (Figure 13.5). The hydrological significance of pipes is that they form a separate drainage system which often transgresses topographic divides.

Piping has been recognised globally as an important geomorphic process since the late nineteenth century and there are many studies on pipe origin, the controlling physico-chemical soil/rock conditions, its hydrological significance, and its relationship to other landforming processes. Pipes, rills and gullies can produce a geomorphological kaleidoscope of interchanging processes and forms (Figure 13.6). Jones (1981), in a comprehensive review, regarded the papers by Parker (1963) and Dalrymple *et al.* (1968) as key works in which piping processes were placed in the general context of landscape models.

Figure 13.6 *Rills beget pipes which beget gullies ... or pipes beget rills which beget gullies or ...? The complexity of the relationships between surface and subsurface flow shows on this 40-m-long badlands slope. Rills dominate the upper portion, pipes and gullies the lower. Incipient hoodoos, capped by thin lenses of clay ironstone, form above now exposed vertical pipe shafts*

While piping is common in arid regions (Parker 1963; Jones 1990; Parker *et al.* 1990) its role in badlands, despite its almost ubiquitous presence, has only been given detailed attention (e.g. Mears 1963; Bell 1968; Barendregt and Ongley 1977; Bryan *et al.* 1978; Yair *et al.* 1980; Drew 1982; Harvey 1982; Bryan and Harvey 1985; Beaty and Barendgregt 1987) corresponding to the growth in interest in badlands since the 1950s.

Many badlands have the basic conditions necessary for piping (Parker 1963; Barendregt and Ongley 1977; Jones 1981). These include, not necessarily in priority: strongly alternating wet and dry climatic patterns (preferably with at least some intense rainstorms to generate large volumes of surface runoff and prolonged periods of dry weather); deep desiccation cracking (or other fractures for entry of moisture); steep slopes (providing sufficient hydraulic head to assist subsurface flow); presence of swelling clay minerals such as montmorillonite or illite, and large amounts of exchangeable sodium (to develop strong expansion and contraction stresses and to cause deflocculation and dispersion of the clay aggregates along moisture routes); alternating layers of permeable and impermeable strata (to facilitate lateral flow of water in some rock units); and an outlet for the pipe system — such as a gully floor or a porous layer.

Badlands dramatically express the dominance of surface erosion and the importance of subsurface processes is often disguised (Figure 13.7). However, many of the minor and major surface erosional forms, including rills, gullies,

Figure 13.7 *A 2-m-deep, pipe-controlled gully system dissects a once extensive pediment. The pipe shaft entrance (lower right) connects to the outlets shown in Figure 13.5*

mass movements and valleys (Figure 13.8), owe their origin to piping erosion (Barendregt and Ongley 1977; Harvey 1982; Beaty and Barendregt 1987; de Lugt and Campbell 1992). While piping is an important erosional process in most badlands, the collection of reliable data on the subsurface distribution of pipes, and their contributions to erosional processes in terms of suspended sediment yields and solute concentrations during runoff, is extremely difficult. Complex, three-dimensional pipe systems are almost impossible to accurately trace and map, and the dynamic nature of pipe erosion causes frequent variations in their sizes and their location. Access to experimental sites, especially during rainstorms, can be hazardous, and the sporadic

character of runoff-generating rainfall poses major problems of instrumentation, monitoring and data collection. Bryan and Harvey (1985) studied two pipe (tunnel) systems in the Dinosaur badlands in Alberta during summer rainstorms. Suspended sediment concentrations were as high as 97.1 gl^{-1} — much higher than surface channel loads — and very high values for solute loads were recorded. It was estimated that about 10% of the total catchment discharge was pipe-routed. However, measurements taken in surface channels do not discriminate between pipe-derived materials, and pipe- and non-pipe-routed surface inputs, so that the actual significance of pipe contributions is difficult to assess (Drew 1982; de Boer and Campbell 1990).

Figure 13.8 *Sequential stages of valley formation by piping and associated gullying activities. This series of pictures, taken from a kilometre-long section of one side of a valley in the badlands of Dinosaur Provincial Park, Alberta, Canada, shows tributary valley evolution. (A) Initial pipe-dominated slope; (B) collapsed pipe and major gully stage; (C) lateral valley slope development; (D) final valley form with micropiping and minor surface wash and gullies. Compare with schematic Figure 13.12*

Runoff generation

Because badlands demonstrate so vividly the results of fluvial erosion, the processes by which running water affects badlands slopes assumes a paramount significance. Major research advances have taken place concerning the generation of overland flow, especially with respect to rill development, since Horton's (1945) elegantly stated hypotheses on the roles of infiltration capacity, rainfall intensity and surface runoff; these advances have been extensively reviewed by Bryan (1987; 1990). The relationships which Horton (1945) identified concerning threshold velocities, critical shear stresses, particle detachment and rill initiation have undergone considerable revision and refinement as a result of many detailed studies in laboratory flumes and field runoff plots.

Much attention is focused on the hydraulics of rill and interrill flow (e.g. Kirby 1990; Benito *et al.* 1991; Abrahams *et al.* 1993; Abrahams and Parsons 1994) and the effect rock fragments have on surface flow, erosion and infiltration (Poesen and Lavee, 1994). While most rill research is not directed at badlands *per se*, it has profound implications on understanding the way runoff is generated in badlands.

Despite wide regional and local lithological variations, it is possible to present a basic model of the hydrological responses of representative types of badlands surfaces (Figure 13.9). Three characteristic slope units are shown: a basal pediment surface; a mid-slope, developed on a bentonitic mudstone with its typical 'popcorn' surface, and an upper slope of mud-cemented sandstone with a dense rill network. While the model sets specific morphological conditions,

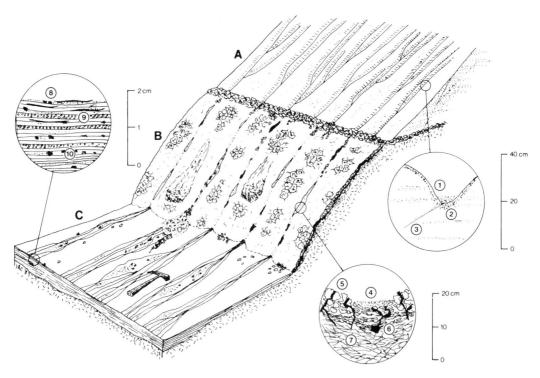

Figure 13.9 *Slope schematic showing three contrasting lithologic units, erosional processes and forms. (A) Densely rilled fine-grained sandstone, silty stringers, (1) line rill floor allowing moisture retention and deeper weathering (2) below rill channel which here is slightly asymmetric reflecting structural control by (3) fractures probably induced during diagenesis. (B) Bentonitic grey mudstone with characteristic desiccated surface, scarred by minor mudflows, pipes and discontinuous rills (4) lined with silty deposits below popcorned (5) interfluves and with small pipes (6) perched above unweathered (7) shard-like bedrock. (C) Basal pediment capped by microlaminar silt layers and clay galls (8) with vesicular deposits (9) and scattered clay ironstone fragments (10)*

and the arrangement and association of units is arbitrary, it reflects the results from a wide range of field observations and experimental studies (e.g. Schumm 1956a; 1956b; 1962; Smith 1958; Clayton and Tinkler 1971; Engelen 1973; Bryan *et al.* 1978; 1984; Hogg 1978; Yair *et al.* 1980; Hodges and Bryan 1982; Hodges 1982; Guiterrez 1983; Bowyer-Bower and Bryan 1986; Dunne and Aubry 1986; Campbell 1987; de Boer 1992a).

Given a hypothetically uniform rainfall the units (Figure 13.9) show markedly different responses in infiltration rates and associated regolith changes, surface conditions (e.g. sealing), runoff generation and the yield of sediment and solutes. Antecedent moisture conditions and the presence of swelling clays within the units are important variables. In the Negev badlands, Yair *et al.* (1980) found that in the presence of other prevalent ions, such as calcium, the swelling process may be suppressed or reduced though the exact nature of the processes involved were not completely understood. Hydraulic conductivity of the materials, macroporosity, hydraulic gradient and crack density are key factors which also greatly influence the rate of internal drainage (Gerits *et al.* 1987).

Runoff on the sandstone slopes occurs almost instantaneously under even light rainfall conditions. Typical threshold amounts for runoff in the Alberta badlands are about 3 mm of rain (Bryan *et al.* 1978), almost regardless of intensity. The compact lithology, which may consist of smectite-cemented fine sands, has extremely low permeability. Surface sealing is rapid and moisture rarely penetrates more than 2–3 mm below the thin, 1–3 mm, weathering rind which develops at the surface (Bowyer-Bower and Bryan 1986). Raindrop impact and splash erosion provide effective mechanisms for particle detachment, though the indurated sandstones produce significantly different rill and interrill environments than the soils that are normally investigated (De Ploey and Poesen 1985). Both concentrated and unconcentrated slopewash are generated. Rill networks, which may be partly controlled by small discontinuities or stress-release-related fractures, show deeper wetting below the rill channels though this is rarely more than a few centimetres (Figure 13.9).

Rills may be propagated by piping and desiccation-crack-controlled seepage from overlying shale caprocks (Gerits *et al.* 1987). Bowyer-Bower and Bryan (1986) suggested that in the absence of such initial flow-concentrating conditions rills may not form. Dunne and Aubry (1986) regard rilling as depending on the balance between sediment transport (which causes channel incision) and rainsplash (which obliterates irregularities); they confirm Horton's (1945) model which suggests that rill cutting may be a function of slope length and steepness. Shear velocity, which affects transport capacity of the flow, and physico-chemical interactions between the water and the surface also are important as these affect aggregate stability and grain cohesion (Gerits *et al.* 1987).

The reaction of the mudstone members to rainfall reflects clay-rich properties as expressed in the dense network of deep (10–15 cm) desiccation cracks, and a 'popcorn' crusted surface of individual, highly irregularly shaped aggregates about 2–3 cm in diameter (Figure 13.9). The mudstones, typically with a weathered regolith 15–20 cm thick, are much less compacted than the sandstones. In the Alberta badlands mudstones may consist of up to 60% clay of which 5–15% is montmorillonite and 60–80% is illite (Table 13.2). Similarly, in North Dakota shales, montmorillonite and illite clays predominate (Engelen 1973). In the Negev, badland units contain up to 75% clay with about 50% montmorillonite; montmorillonite may form 80–100% of the upper regolith (Yair *et al.* 1980).

The surface sealing effects due to wetting can be profound. Hogg (1978) and Hodges and Bryan (1982) describe a sequential pattern of runoff initiation and surface changes. On wetting, some aggregates swell and minor desiccation cracks close. The initially dry surfaces provide a multitude of microchannels (Haigh 1978), many of which remain open during the early wetting phase. As rainfall continues micro-piping occurs with incipient flow along major cracks. Schumm (1956a) found that water poured on a dry shale surface flowed beneath the aggregates and reappeared further downslope. As the surface seals, water trapped during early rainfall flows beneath the sealed surfaces, and some water moves deeper within the unit along fractures. Prolonged rain causes almost complete surface sealing and overland flow in shallow rills. Once saturation occurs, mudflows may occur (Schumm 1956a; Engelen 1973; Yair *et al.* 1980) in a series of liquid surges and microslumps (Bryan *et al.* 1978). Piping continues below the saturated layer. Complex networks of voids forming both macro and micropores act as conduits for subsurface flow and present a wide range of downslope routes interchanging surface and subsurface moisture (Jones 1978; Gerits *et al.* 1987).

Under varied conditions of surface sealing with different clays, clay contents and regolith structures, depths of wetting vary widely. Some regoliths consist of multilayered clay-dominated units which greatly complicate the infiltration process, i.e. the position of the most impermeable layer dictates infiltration depths, rates and regolith behaviour. Yair *et al.* (1980) recorded wetting depths varying from 2 to 25 cm on experimental plots in calcareous shales. Moisture penetration was greater beneath rills and depressions, and shallower elsewhere. Typical moisture penetration was about 5–6 cm. Bryan *et al.* (1978) found that wetting depths rarely exceeded 4–5 cm. Such differences represent wide variations in runoff-generating capacity and the rate of drainage even within a single shale unit. As a result runoff varies enormously both spatially and temporally even under controlled conditions (Hodges 1982; Bryan *et al.* 1984; Campbell 1987), reflecting dynamic and static variations in permeability and other properties (de Boer 1992a).

The basal, pedimented slopes consist of deposits of thin (1–2 mm) microlaminae of silt-rich sediments over bedrock (Schumm 1962; Engelen 1973; Hodges 1982). Their hydrological and hydraulic significance lies in the near-uniform sheetflow which occurs under almost all rainfall conditions. The silts, which often contain lenses of thin clay sheets deposited as gels (Bryan *et al.* 1984), are compact and present a smooth, gently sloping plane (2–10% slope is typical — Schumm 1962), which favours sheetwash through a dense network of shallow (2–3-mm-deep) anastomosing channels. Clay galls or mud curls representing evaporation residues line the channel floors.

When flattened and interbedded with later silt deposits they further reduce permeability (Figure 13.9).

Vesicular layers, about 1 mm thick, form in the near-surface sediments of the pediments (Figure 13.9). Such layers are common in fine-grained desert deposits forming by air entrapment in advance of a downward moving wetting front (Springer 1958) or by entrapped air expanding during heating after summer rains (Evanari *et al.* 1974). Engelen (1973) studied badlands in North Dakota and stated that vesicles could result from meltout of ice granules, but this process, if it occurs at all, is restricted to frost-affected areas. A fine-grained layer appears necessary for vesicle development and the vesicular layers are progressively destroyed, apparently by eluviation of clay and solute plasma (McFadden *et al.* 1986).

Runoff generation from these different surfaces, its timing and amount, is highly variable. Variations in intensity, duration and spatial patterns of rainfall, produce complex hydrological responses as the surfaces yield widely differing amounts of water, sediment and solutes (Bryan *et al.* 1984; Campbell 1987; de Boer and Campbell 1990; de Boer 1992a; 1992b). Badlands, like arid regions generally, are extreme examples of the partial or variable source-area model of runoff and sediment generation (Bryan and Campbell 1980; 1986; de Boer and Campbell, 1989). High-yielding surfaces vary spatially and temporally depending entirely on their locational coincidence with precipitation and not, as is typical in humid regions, with seasonal fluctuations of near-channel contributing areas. This discontinuous pattern of surface runoff is also reflected in the distribution of the badland gully systems.

Badland gully systems

Gullies are ubiquitous phenomena and an enormous amount of international literature has dealt with the mechanics of gully formation, the hydraulic and hydrological processes involved in gully erosion, and their relationship to features such as rills and pipes (Figure 13.7). The principal concern that has directed this research focuses on the often serious effects gullying has on agricultural lands. One particular area of investigation has considered gully (arroyo) formation in the US southwest. There, gully erosion has been variously ascribed to climatic changes, land use, or variations in geomorphological processes involving extrinsic and intrinsic threshold conditions (Cooke and Reeves 1976; Graf 1983; 1988). Graf (1983) regarded gullies as distinct from arroyos with the latter occurring in alluvial-filled, through-flowing channels. Cooke and Reeves (1976) make no such distinction; they call arroyos 'valley-bottom gullies' and build a case for equifinality in gully formation. Whatever the name or the cause, the end result looks the same (Figure 13.10).

Little research on badlands gullies *per se* has been done. Until comparatively recently geomorphologists largely ignored badlands. Badland gullies involve no special processes of formation and, more importantly, because badlands are agriculturally useless, their gullies are of no economic significance. As with their other landforms however, badland gullies are often highly visible being unmasked by vegetation. Furthermore, gullying, especially in its pipe-related forms, can be a major mechanism in badlands formation when allied with mass movements (Beaty and Barendregt 1987; de Lugt and Campbell 1992).

Brice (1966) classified gullies in Nebraska on the basis of gully-head location and topographic position, into valley-bottom, valley-side and valley-head gullies; there was often a smooth transition between each type. The classification used here follows Brice's system and identifies badland gullies on the basis of: (1) gullies in alluvial-fill in existing valleys; (2) gullies cut in bedrock on slopes or valley sides; and (3) gullies forming as headcut extensions of a drainage system on undissected upland surfaces (Figure 13.4). While these gully systems may be related, the various modes of formation and the different materials involved can produce variations in gully morphology.

GULLIES IN ALLUVIAL VALLEY FILLS

Gully-trenched valleys in arid lands have been the subject of more geomorphological research than pediments or alluvial fans (Graf 1983). In badlands, where the effects of land use are rarely relevant to gully incision, the role of climatic variations or geomorphological threshold effects (both extrinsic and intrinsic) are of paramount importance. Complex interactions between these two factors produce a vast range of gully-initiating situations. In Kenya, Oostwoud Wijdenes and Bryan (1991) identified three forms of instability which are associated with active gullying: intense storms, and/or climatic changes, or changes in catchment hydrology, all of which increase peak discharges; reductions in surface resistance by devegetation or variations in soil physicochemical properties; and changes in hydraulic conditions such as base-level lowering or channel scouring in a trunk stream. Gullies may be initiated, reactivated, or stabilised depending on the nature and extent of the disturbance and the related feedback conditions. As Bull (1979; 1980) notes, the threshold conditions between cutting and filling modes of stream behaviour

Figure 13.10 *Valley-floor gullying in silty alluvium, Red Deer valley, Alberta, Canada. Gully here is c. 2.5 m deep, up to 5 m wide, and is typical of gullies in western North America. Piping is absent and head and sidewall recession is mainly controlled by undercutting and loss of cohesion by saturation of basal sediments*

operate within very narrow margins (Figure 13.11). Oostwoud Wijdenes and Bryan (1991) found that seasonal variations affected gully development; early in the wet season there was rapid headcutting and sediment aggradation — this was followed by reduced headcutting and flushing of the stored sediment. This episodic pattern varied from year to year and gully to gully.

A particular example of this situation is seen in gullies in alluvial-filled bedrock valleys (O'Hara and Campbell 1993). Where bedrock valleys have been both cut and filled by larger streams than the present gullies, as has long been recognised in arroyos in the western US (Bryan 1928), the valley gradient is predetermined for a protracted time. Present gully

systems in such valleys accommodate to this slope as best they may (Leopold and Miller 1956). The result is an epicyclic alternation in gully cutting and filling (Figure 13.11) which occurs entirely as a result of slope-driven internal adjustments (Patton and Schumm 1975) which may be superimposed upon larger externally derived cycles of alluviation and erosion (Schumm and Hadley 1957; Schumm 1988). Watson *et al.* (1984) discussed reactivation of drainage channels causing dramatic gullies or 'dongas' in the widespread valley colluvial deposits of southeastern Africa. These have produced ravine-like badlands in coarse-grained, sodium-rich slope deposits that were deposited between about 30 000 and 12 000 years BP. In South Africa, Botha *et al.* (1994)

Figure 13.11   *Sequence of down-gully photographs showing form and width–depth relationship changes characteristic of discontinuous gullies. (A) Near the gullyhead (see Figure 13.10), typical steep-sided (c. 2.5 m), flat-bottomed form. (B) 100 m downstream shows increased sediment infill (partly vegetated) and increased width to depth ratio with banks here about 1.5 m high. (C) 100 m downstream banks are lower, channel is shallow and increasingly vegetated. (D) 100 m downstream, channel is almost obscured by vegetation, banks are low (<1.0 m) and graded*

also found a series of palaeosols indicative of episodic, non-synchronous, gully cut-and-fill successions which have occurred over at least the last 135 000 years. Cut-and-fill patterns in the gullies in the Dinosaur badlands may be controlled by relative base-level changes in the Red Deer River induced by meander migration (O'Hara 1986). McPherson (1968) showed up to 200 m of lateral migration between 1879 and 1960. Meander shifts could cause equivalent base-level variations of ± 60–70 cm in tributary valleys as the gully was shortened or lengthened by up to 70 m. Detailed sequential aerial photography analysis from 1937 to 1977 (O'Hara 1986) reveals multiple discontinuous head-cutting of the gully in valley-fill deposits which are dated at about 300 years BP (Bryan *et al.* 1987; O'Hara and Campbell 1993).

Leopold and Miller (1956) discuss the geomorphic implications of continuous and discontinuous gully systems. Discontinuous gullies resulting from headcutting caused by local, often depositional-induced, changes of gradient or other disturbances eventually develop into continuous trench-like forms (Schumm and Hadley 1957). Heede (1974) identified continuous gullies with rill coalescence and discontinuous gullies with single headcuts but it is evident that all combinations of rills and gullies exist across the complete spectrum of badland drainage systems. In narrow badland valleys, local injections of sediment-rich discharge from valley-side drainage (sheetwash, rills, pipes, gullies and channels) together with mass movement debris, provide numerous sites for alluvial fans and debris cones both in and adjacent to the main channel (Butcher and Thornes 1978). These accumulations cause gradient changes in the main channels which are independent of those formed by the main channel's own discharge of sediment and water.

Small badland catchments in Wyoming (Bergstrom and Schumm 1981) respond in similar fashion as headwaters (erosion zones) contribute sediment and water to intermediate zones (pulsatory storage and flushing) and thence to downstream zones forming sites of local aggradation and cutting. In the Borrego Springs badlands of southern California, Brown (1983) found that irregular channel profiles were influenced by entry of tributary streams.

Decreased gradients in the main channel occurred above the tributary entry with increased channel gradients downstream. Each zone or length of channel responds differently to intrinsic and extrinsic conditions and is frequently out of phase with its neighbour zones (Graf 1988), so that each individual case of gully cutting is constrained by the sequential and spatial pattern of events (Figure 13.11).

These alternating erosional and depositional conditions affect gully morphology. Along the banks and at the head, where these exist, gully form constantly varies in response to slumping caused by local bank oversteepening, undercutting and, in some circumstances, piping or tunnel erosion. Stocking (1978) explained morphological variations in headcut forms on the basis of erosivity factors, arguing that precipitation patterns, depth of percolation and erosion at different levels within the gully head determined the type of headcut and the principal mechanism (slumping, piping or surface erosion) which was responsible. Blong (1985) found that gully sidewall evolution may be as or more significant in gully erosion than headcut erosion. Oostwoud Wijdenes and Bryan (1991) found that headcutting dominates. Gully morphology is dynamic and it often is impossible to separate cause from effect and probably no one process operates entirely independently (Jones 1981).

While piping is a major factor in certain types of gullying it is not a universal phenomenon in gullies cut in valley-fill deposits. Evidently one or more of the various piping requirements may be absent (Jones 1981). Stocking (1979) emphasises the importance of soils with high exchangeable sodium percentages which cause greater dispersity and hence piping; vegetation cover and steep local relief may play a subsidiary role (Barendregt and Ongley 1977). Where valley-fill deposits consist almost entirely of silt and fine-sand size sediments, clay layers may not be present — or may not form layers sufficiently continuous or thick to generate lateral pipeflow. In the extensive Maama gully system, Lesotho, soil shrinkage limits were zero and piping was insignificant in comparison to other gullies (Faber and Imeson 1982); in Morocco Imeson *et al.* (1982) found that piping occurred in wadis only where a relatively impermeable layer was present.

No piping occurs in Holocene filled intra-basin alluvial channels and valley floors in the New Mexico badlands (Wells and Gutierrez 1982), but adjacent Cretaceous shale and mudstone units are extensively piped. Gullied loess deposits in Missouri (Roloff *et al.* 1981) exhibit no piping despite widespread through-flow conditions and predominantly montmorill-onitic clays. While the environmental conditions are not those of badlands, the widespread nature of gullying and the type of materials are analogous to those of many badland alluvial-filled valleys. In Iowa, Bradford and Piest (1980) identified gully-head retreat mechan-isms in loess-derived alluvium. Headcut processes included arc failure, slab failure and basal 'popout' which triggered slumping. While seepage and associated loss of soil cohesion occur, piping is not in evidence nor suggested as an erosional mechanism.

Pipe-induced gullying and sinkhole develop-ment are widespread in glacio-lacustrine valley-fill sediments of interior British Columbia, where extensive deposits of silts create local badlands (Slaymaker 1982b). The steep hydraulic gradient is the only factor normally associated with piping that is present, but the closely jointed and varied nature of the silts serves the same purpose for water flow control as desiccation cracks and impermeable horizons elsewhere. Harvey (1982), for example, shows the relationship between piping and gullying as highly variable, depending on the presence of differential porosity within the trenched deposits and deep tension cracks.

VALLEY-SIDE AND SLOPE GULLIES

On steep badlands bedrock slopes the regolith is thin and discontinuous (Figure 13.6). Fluvial processes are almost entirely in direct contact with bedrock and variations in erosional resist-ance, permeability and thickness of lithologic units determine the rates and type of gully development. Piping, together with pipe-induced slope collapse and mass movement, is probably the most significant mode of valley-wall retreat and slope formation. These pro-cesses operate in conjunction with gullying at all landscape scales from single pipe-controlled gullies on individual slopes to macro-scale

valley evolution. The same basic erosion princi-ples and sequence of events is common in many badlands and results in a series of stages of gully-related valley formation.

A sequence of events (Figure 13.12) shows the pattern of meso-scale gully and valley-side evolution in parts of the Red Deer badlands. Local relief is of the order of 20 m. A three-stage sequence can be modelled:

1. Piping shafts and horizontal tunnels form in response to locally steep hydraulic gradients along the edge of a structurally controlled escarpment bordering a main channel. Centripetal drainage concentrates upper surface runoff into the pipe entrance. Downward-percolating water, following local structural discontinuities which may reflect unloading/overconsolidation stresses, causes sodium-induced dispersal and deflocculation of the montmorillonite-rich clays. Impeded drainage layers encourage lateral flow and initiate horizonal shafts below more porous units. The limit to percolation is determined by local base-level of the main channel. Progressive enlargement of the pipes leads to partial collapse.

2. Here piping is fully developed and blocks of unsupported roof and side materials collapse and partially plug some pipes. Continued subsurface erosion processes extend the existing pipes and remove the collapsed debris. Headwater develop-ment of the pipe-contributing surface drainage system eventually ceases at the local watershed divide and the channel gradient from divide to base-level declines. Channel gradient reduction and comminution of collapsed blocks continues until stage three conditions prevail.

3. At this stage of valley development gullies domi-nate the surface. Retreat of the basin headwall by runoff, weathering and seepage-induced basal sapping may extend the drainage system, and small-scale piping may form in response to local slope hydraulic conditions. If base-level con-ditions in the main channel are fixed the cycle of major pipe-induced valley development ceases. The reduced hydraulic head and water supply at the surface no longer sustain deep piping.

A similar sequence of events is documented by Harvey (1982) for badlands in southeastern Spain where gully and pipe erosion continue until finally the badlands become dominated by surface and near-surface processes. This sequence appears generally to reflect lowering of relief, and stripping of surface materials which favour piping may also be involved.

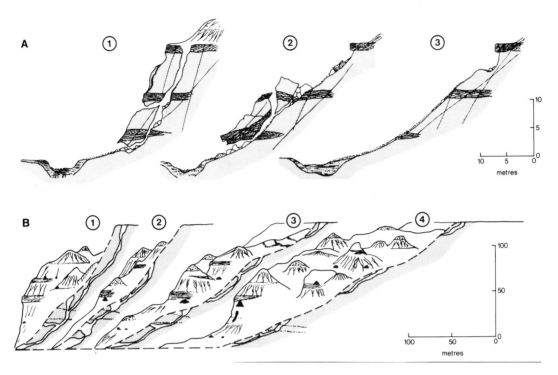

Figure 13.12 *Slope and valley development by combination piping, collapse and gullying. (A) Three-stage meso-scale evolution of slope and valley formation from initial piping (1) through pipe-induced collapse at (2), to final valley form at (3), where hydraulic gradients are no longer steep enough to generate extensive large pipe systems. (B) Macrovalley development from the early deep incision of trunk streams at (1) and extensive pipe formation with steep topographic and hydraulic gradients, through (2) and (3) gradually diminishing as badlands extend into upper surface with large pipe collapse triggering mass movements, to the stage at 4, where piping ceases to play the major role in badland valley evolution because of reduced hydraulic and topographic gradients*

The extensive areas of badlands in western North America are frequently *peripheral* to major river valleys. Along the Missouri and its tributaries and in Alberta along the Red Deer and the Milk rivers, badlands formed initially along the valley walls where local relief favoured rapid incision in the weak, shale-dominated bedrock. In the Red Deer badlands of Alberta, deep valley incision by glacial meltwaters and isostatic readjustment in the early Holocene preceded badland formation (Bryan *et al.* 1987). Similar events occurred in the Milk River Canyon (Beaty and Barendregt 1987) and in the Big Muddy badlands of Saskatchewan (Drew 1982). In New Mexico, badlands flank the Chaco River (Wells and Gutierrez 1982) apparently forming in response to the onset of arid climatic conditions in late Holocene times *c.* 3000 years BP. In the Okanagan silts, pipes

and gullies followed a rapid draining of glacial Lake Penticton and subsequent base-level lowering (Slaymaker 1982b). Topographic conditions, high local relief, and steep hydraulic gradients in combination with piping-prone materials, form an ideal environment for gullying and piping in a circuitous cause and effect relationship (Jones 1971).

Slope and valley-side pipe–gully systems occur on a massive scale in the Red Deer badlands and the Milk River Canyon (Barendregt and Ongley 1977) and have contributed substantially to the width of the badlands and the growth of their valleys (Figure 13.10). Beaty and Barendregt (1987) estimate that about 80 m of Milk River badland valley widening has occurred since deglaciation and that the greater volume of this material is pipe-supplied. In Saskatchewan, Drew (1982) found a close link between gullies and pipes. Tunnels up to 30 m long and 2 m in diameter were mapped and flow events traced by dye; over 50% of the local drainage is via piping. As discussed above, a cyclical pattern of percolation, pipe enlargement, subsurface drainage integration and collapse into surface gullies is repeated until the system is graded to the lowest base-level.

Valley-side and slope pipes and gullies in badlands have a complex, intermeshed relationship. Episodic cutting and filling characterises valley-bottom gullies and a repetitive pattern of alternating piping, collapse and gullying, dominates the valley walls (Figure 13.12). As long as free headward retreat and sufficient hydraulic gradient (Jones 1971) permit continual replacement of the channel at a more or less constant rate and slope, as is seen in some gullies on the uplands, the cycle will continue. Expansion of the lower, alluvial-covered surfaces (Barendregt and Ongley 1977; Drew 1982), together with a gradual lowering of the entire badlands surface gradient, eliminates the major, preferred, pipe locations but small-scale interactive piping and gullying continue until the last vestiges of the badlands disappear (Harvey 1982).

GULLIES ON UPLAND SURFACES

Drainage network growth on undissected surfaces has been approached in various ways including theoretical, mathematical–geometrical analysis in which stochastic or random events may be incorporated to add physical realism; physical models, which may include runoff-plot devices or use evolutionary analogues based on rapidly forming natural drainage systems; combinations of techniques using field examples to develop explanatory models of network growth; and historic analysis or stratigraphic-geomorphic investigation based on field evidence to interpret past events.

Each technique has its uses and certain disadvantages. While theoretical approaches of necessity omit some processes and concentrate, it is hoped, on the most dominant ones, significant single or combinational processes may be excluded (Howard 1994). Field-based investigations concerned with the location of erosion in specific sites depends on sequences of past events which are impossible to reconstruct with certainty (Brice 1966). Natural systems almost inevitably possess topographic, lithologic or other constraints which impart often subtle controls on the expression of the drainage network and on its development.

The headward extension of gully systems (or coulees) into the flat, prairie landscape of southern Alberta provides numerous examples of the roles of various gully-activating processes. Major gully development can, in many cases, be directly traced to scouring and incision by glacial lake drainage into the Red Deer River meltwater channel following late Wisconsin ice retreat about 14 000 years BP (Bryan *et al.*1987; Campbell and Evans 1990). Drainage direction of these gully systems was controlled initially by the general topographic slope towards the Red Deer. Incision was rapid, valleys over 60 m deep and several kilometres long were probably cut in the first few hundred years following deglaciation (Bryan *et al.* 1987). Contemporaneous large-scale piping accompanied this cutting phase, together with major mass movements (Campbell and Evans 1990).

Control of gully forms and location by both topography and structure are clearly evident in many locations (Figure 13.13). Joint-controlled drainage networks are a feature of southern Alberta. In the Dinosaur badlands, many channels follow a NE–SW and NW–SE orientation, possibly reflecting orthogonal joint sets which are roughly parallel and normal to the main structural trend of the Rocky Mountains (Babcock 1973; Koster 1983). Beaty (1975) has described coulee systems showing a striking northeastern alignment that he attributes to fluvial erosion along chinook wind-eroded furrows. These furrows channel snowdrifts, and meltwater erosion further aids the process of coulee formation.

Although little theoretical attention has been given to the growth of badland upland gullies, Faulkner (1974) developed an allometric growth rate model for badland gully formation on an undissected prairie surface. Faulkner proposed that the evolutionary pattern of gullies that she identified could provide the basis for an erogodic approach to gully development. Her model defined thresholds of gully basin size beyond which the gully attains self-sustaining growth as it expands freely into the prairie

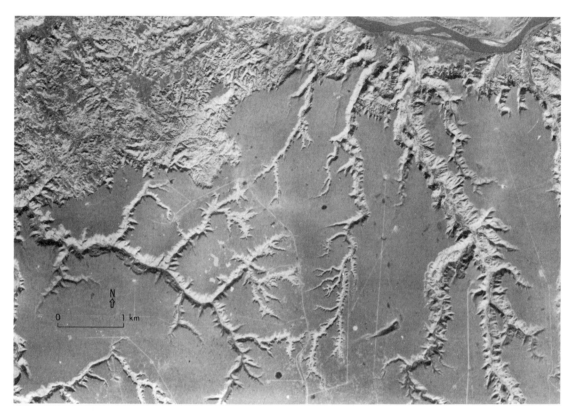

Figure 13.13   *Portion of Dinosaur Provincial Park badlands, Alberta, Canada (top left); Red Deer River (top right) and prairie surface with extending gully (coulee) systems. Gullies on left of photograph show pronounced structural control, possibly by deep bedrock fractures reflecting through thin glacio-lacustrine sediments. Main gullies on right evolved principally by glacial lake drainage into rapidly incising Red Deer meltwater channel c. 14 000 years BP. Photograph, AS 3423-12, reproduced by permission of Government of Alberta*

surface. Under such conditions, she argued that ergodic transformations were valid. Schumm's (1956a) study of badland slopes assumes a similar set of relationships i.e. sampling spatial patterns of hillslope form in a rapidly evolving system is equivalent to sampling through time (Schumm 1973).

Conclusion

Most badlands research has focused on the nature and effects of geomorphological processes at micro- and meso-scales of operation. There is a need to integrate the results of these process studies in the context of the longer-term spatial and temporal variations that characterise these fascinating landscapes. Fractal analysis

and digital terrain studies could provide a powerful technique for identifying and quantifying the precise nature of the spatial patterns of badlands; it could be an important step in considering whether the ergodic hypothesis (e.g. Bull 1975; Church and Mark 1980; Craig 1982; Paine 1985) is a valid approach.

References

Abrahams, A.D. 1972. Environmental constraints on the substitution of space for time in the study of natural channel networks. *Bulletin of the Geological Society of America*, 83: 1523–1530.

Abrahams, A.D. and Parsons, A.J. 1994. Hydraulics of interrill overland flow on stone-covered desert surfaces. *Catena*, 23: 111–140.

Abrahams, A.D., Howard, A.D. and Parsons, A.J.

1993. Rock mantled slopes. In A.D. Abrahams and A.J. Parsons (eds), *Geomorphology of desert environments*. Chapman & Hall, London: 173–212.

Aghassy, J. 1973. Man-induced badlands topography. In D.R. Coates (ed.), *Environmental geomorphology and landscape conservation. Volume III. Non-urban*. Dowden Hutchinson and Ross, Stoudsburg, PA: 124–136.

Alexander, D. 1982. Difference between 'calanchi' and 'biancane' badlands in Italy. In R.B. Bryan and A. Yair (eds), *Badland geomorphology and piping*. GeoBooks, Norwich: 71–88.

Alexander, R.W., Harvey, A.M., Calvo, A., James, P.A. and Cerda, A. 1994. Natural stabilisation mechanisms on badland slopes: Tabernas, Almeria, Spain. In A.C. Millington and K. Pye (eds), *Environmental change in drylands: bigeoographical and geomorphological perspectives*. Wiley, Chichester: 85–111.

Babcock, A.E. 1973. Regional jointing in southern Alberta. *Canadian Journal of Earth Sciences*, 10: 1769–1781.

Barendregt, R.W. and Ongley, E.D. 1977. Piping in the Milk River Canyon, southeastern Alberta: A contemporary dryland geomorphic process. In *Erosion and solid matter transport in inland waters*. Proceedings of the Paris Symposium, July. International Association of Hydrological Sciences, Publication 122: 233–243.

Beaty, C.B. 1975. Coulee alignment and the wind in southern Alberta Canada. *Bulletin of the Geological Society of America*, 86: 119–128.

Beaty, C.B. and Barendregt, R.W. 1987. The Milk River Canyon, Alberta: superposition and piping in a semiarid environment. *Geological Society of America Centennial Field Guide — Rocky Mountain Section* 29–32.

Bell, G.L. 1968. Piping in the Badlands of North Dakota. In *Proceedings of the 6th Annual Engineering Geology and Soils Engineering Symposium*, Boise, Idaho: 242–257.

Benito, G., Gutierrez, M. and Sancho, C. 1991. Erosion patterns in rill and interrill areas in badlands zones of the middle Ebro basin (NE-Spain). In M. Sala, J.L. Rubio and J.M. García-Ruiz (eds), *Soil erosion studies in Spain*. Geoforma Ediciones, Longroño: 41–54.

Bergstrom, F.W. and Schumm, S.A. 1981. Episodic behaviour in badlands. In T.R.H. Davies and A.J. Pearce (eds), *Erosion and sediment transport in Pacific Rim steeplands*. Proceedings of the Christchurch Symposium, January. International Association of Hydrological Sciences, Publication 132: 478–492.

Blong, R.J. 1985. Gully sidewall development in New South Wales, Australia. In S.A. El-Swaify, W.C. Moldenhauer and A. Lo (eds), *Soil erosion and conservation*. Soil Conservation Society of America, Ankeny, IA: 574–584.

Bloom, A.L. 1978. *Geomorphology: a systematic analysis of late Cenozoic landforms*. Prentice-Hall, Englewood Cliffs, NJ.

Botha, G.A., Wintle, A.G. and Vogel, J.C. 1994. Episodic late Quaternary palaeogully erosion in northern KwaZulu–Natal, South Africa. *Catena* 23: 327–340.

Bowyer-Bower, T.A.S. and Bryan, R.B. 1986. Rill initiation: concepts and experimental evaluation on badland slopes. *Zeitschrift für Geomorphologie*, Supplementband, 59: 161–175.

Bradford, J.M. and Piest, R.F. 1980. Erosional development of valley-bottom gullies in the upper midwestern United States. In D.R. Coates and J.D. Vitek (eds), *Thresholds in geomorphology*. George Allen and Unwin, London: 75–102.

Bradley, W.H. 1940. Pediments and pedestals in miniature. *Journal of Geomorphology*, 3: 244–254.

Brice, J.C. 1966. Erosion and deposition in the loess-mantled Great Plains, Medicine Creek drainage basin, Nebraska. *US Geological Survey, Professional Paper* 352H: 255–335.

Bridges, E.M. and Harding, D.M. 1971. Micro-erosion processes and factors affecting slope development in the Lower Swansea Valley. *Institute of British Geographers Special Publication* 3: 65–79.

Brown, A.J. 1983. Channel changes in arid badlands, Borrego Springs, California. *Physical Geography*, 4: 82–102.

Bryan, K. 1928. Historic evidence on changes in the channel of the Rio Puerco, a tributary of the Rio Grande in New Mexico. *Journal of Geology*, 36: 265–282.

Bryan, R.B. 1987. Processes and significance of rill erosion. In R.B. Bryan (ed.), *Rill erosion*. Catena Supplement, 8, Catena Verlag, Cremlingen, Germany: 1–15.

Bryan, R.B. (ed.) 1990. *Soil erosion — experiments and models*. Catena Supplement, 17, Catena Verlag, Cremlingen, Germany.

Bryan, R.B. 1991. Surface wash. In O. Slaymaker (ed.), *Field experiments and measurement programs in geomorphology*. University of British Columbia Press, Vancouver: 107–167

Bryan, R.B. and Campbell, I.A. 1980. Sediment entrainment and transport during local rainstorms in the Steveville badlands, Alberta. *Catena*, 7: 51–65.

Bryan, R.B. and Campbell, I.A. 1982. Surface flow and erosional processes in semiarid mesoscale channels and basins. In D.E. Walling (ed.), *Recent developments in the explanation and prediction of erosion and sediment yield*. Proceedings of the Exeter Symposium, 1982, International

Association of Hydrological Sciences, Publication, 137: 123–133.

Bryan, R.B. and Campbell, I.A. 1986. Runoff and sediment discharge in a semiarid ephemeral drainage basin. *Zeitschrift für Geomorphologie, Supplementband,* 58: 121–143.

Bryan, R.B. and Harvey, L.E. 1985. Observations on the geomorphic significance of tunnel erosion in a semi-arid ephemeral drainage system. *Geografiska Annaler,* 67A: 257–272.

Bryan, R.B. and Yair, A. (eds), 1982. *Badlands geomorphology and piping.* GeoBooks, Norwich.

Bryan, R.B., Yair, A. and Hodges, W.K. 1978. Factors controlling the initiation of runoff and piping in Dinosaur Provincial Park badlands, Alberta, Canada. *Zeitschrift für Geomorphologie, Supplementband,* 29: 151–168.

Bryan, R.B., Imeson, A.C. and Campbell, I.A. 1984. Solute release and sediment entrainment on microcatchments in the Dinosaur Park badlands, Alberta, Canada. *Journal of Hydrology,* 71: 79–106.

Bryan, R.B., Campbell, I.A. and Yair, A. 1987. Postglacial geomorphic development of the Dinosaur Provincial Park badlands, Alberta. *Canadian Journal of Earth Sciences,* 24: 135–146.

Bull, W.B. 1975. Allometric change of landforms. *Bulletin of the Geological Society of America,* 86: 1489–1498.

Bull, W.B. 1979. The threshold of critical power in streams. *Bulletin of the Geological Society of America,* 90: 453–464.

Bull, W.B. 1980. Geomorphic thresholds as defined by ratios. In D.R. Coates and J.D. Vitek (eds), *Thresholds in geomorphology.* Allen & Unwin, London: 259–263.

Butcher, G.C. and Thornes, J.B. 1978. Spatial variability in runoff processes in an ephemeral channel. *Zeitschrift für Geomorphologie, Supplementband,* 29: 83–92.

Campbell, I.A. 1970. Erosion rates in the Steveville badlands, Alberta. *The Canadian Geographer,* 14: 202–216.

Campbell, I.A. 1974. Measurements of erosion on badland surfaces. *Zeitschrift für Geomorphologie, Supplementband,* 21: 122–137.

Campbell, I.A. 1977. Stream discharge, suspended sediment and erosion rates in the Red Deer River basin, Alberta, Canada. In *Erosion and solid matter transport in inland waters.* Proceedings of the Paris Symposium, July. International Association of Hydrological Sciences, Publication, 122: 244–259.

Campbell, I.A. 1981. Spatial and temporal variations in erosion measurements. In *Symposium on erosion and sediment transport measurement.* Proceedings of the Florence Symposium. International Association of Hydrological Sciences, Publication, 133: 447–456.

Campbell, I.A. 1985. The partial area concept and its application to problems of sediment source areas. In S.A. El-Swaify, W.C. Moldenhauer and A. Lo (eds), *Soil erosion and conservation.* Soil Conservation Society of America, Ankeny, IA: 128–138.

Campbell, I.A. 1987. Infiltration characteristics of badlands surfaces and storm runoff. In Y.-S. Fok (ed.), *Infiltration development and application.* Proceedings of the International Conference on Infiltration Development and Application, 1987, Water Resources Center, University of Hawaii, Honolulu: 251–261.

Campbell, I.A. 1992. Spatial and temporal variations in erosion and sediment yield. In *Erosion and sediment transport monitoring programmes in river basins.* Proceedings of the International Symposium, Oslo. International Association of Hydrological Sciences, Publication, 210: 244–259.

Campbell, I.A. and Evans, D.J.A. 1990. Glaciotectonism and landsliding in Little Sandhill Creek, Alberta. *Geomorphology,* 4: 19–36.

Campbell, I.A. and Honsaker, J.L. 1982. Variability in badlands erosion; problems of scale and threshold identification. In C.E. Thorn (ed.), *Space and time in geomorphology.* Allen & Unwin, London: 59–79.

Carson, M.A. and Kirkby, M.J. 1972. *Hillslope form and process.* Cambridge University Press, Cambridge.

Chen, Y. 1983. A preliminary analysis of the processes of sediment production in a small catchment on the Loess Plateau. *Geographical Research* (China), 2: 35–47.

Christie, R.L. 1967. Reconnaissance of the surficial geology of northeastern Ellesmere Island, Arctic Archipelago. *Geological Survey of Canada Bulletin,* 138.

Church, M. and Mark, D.M. 1980. On size and scale in geomorphology. *Progress in Physical Geography,* 4: 342–390.

Churchill, R.E. 1981. Aspect-related differences in badlands slope morphology. *Annals of the Association of American Geographers,* 71: 374–388.

Clayton, L. and Tinker, J.R. Jr 1971. *Rates of hillslope lowering in the badlands of North Dakota.* North Dakota Water Resources Research Institute Report WI-221-012-71, Fargo, ND.

Clements, T. *et al.* 1957. *A study of desert surface conditions.* US Army Environmental Protection Research Division Technical Report EP-33, Quartermaster Research and Development Center, Natick, MA.

Cooke, R.U. and Reeves, W.R. 1976. *Arroyos and environmental change in the American South-West.* Clarendon Press, Oxford.

Craig, R.G. 1982. The ergodic principle in erosional models. In C.E. Thorn (ed.), *Space and Time in Geomorphology*. Allen & Unwin, London: 81–115.

Culling, W.E.H. 1963. Soil creep and the development of hillside slopes. *Journal of Geology*, 71: 127–161.

Dalrymple, J.B., Blong, R.J. and Conacher, A.J. 1968. An hypothetical nine unit land surface model. *Zeitschrift für Geomorphologie*, 12: 60–76.

Davis, W.M. 1892. The convex profile of bad-land divides. *Science*, 20: 245.

de Boer, D.H. 1992a. Constraints on spatial transference of rainfall–runoff relationships in semiarid basins drained by ephemeral streams. *Hydrological Sciences Journal*, 37: 491–504.

de Boer, D.H. 1992b. Hierarchies and spatial scale in process geomorphology: a review. *Geomorphology* 4: 303–318.

de Boer, D.H. and Campbell, I.A. 1989. Spatial scale dependence of sediment dynamics in semi-arid drainage basins. *Catena*, 16: 277–290.

de Boer, D.H. and Campbell, I.A. 1990. Runoff chemistry as an indicator of runoff sources and routing in semi-arid, badland drainage basins. *Journal of Hydrology*, 121: 379–394.

de Lugt, J. and Campbell, I.A. 1992. Mass movements in the badlands of Dinosaur Provincial Park, Alberta, Canada. In K.H. Schmidt and J. De Ploey (eds), *Functional geomorphology*. Catena, Supplement, 23: 75–100.

De Ploey, J. 1974. Mechanical properties of hillslopes and their relation to gullying in central semi-arid Tunisia. *Zeitschrift für Geomorphologie*, Supplementband, 21: 177–190.

De Ploey, J. and Poesen, J. 1985. Aggregate stability, runoff generation and interrill erosion. In K.S. Richards, R.R. Arnett and S. Ellis (eds), *Geomorphology and soils*. Allen & Unwin, London: 99–120.

Drew, D.P. 1982. Piping in the Big Muddy badlands, southern Saskatchewan, Canada. In R.B. Bryan and A. Yair (eds), *Badland geomorphology and piping*. GeoBooks, Norwich: 293–304.

Dunne, T. and Aubry, B.F. 1986. Evaluation of Horton's theory of sheetwash and rill erosion on the basis of field experiments. In A.D. Abrahams (ed.), *Hillslope processes*. Allen & Unwin, London: 31–53.

Engelen, G.B. 1973. Runoff processes and slope development in Badlands National Monument, South Dakota. *Journal of Hydrology*, 18: 55–79.

Evanari, J., Yaalon, D.H. and Gutterman, Y. 1974. Note on soils with vesicular structures in deserts. *Zeitschrift für Geomorphologie*, 18: 162–172.

Faber, T. and Imeson, A.C. 1982. Gully hydrology and related soil properties in Lesotho. In D.E. Walling (ed.), *Recent developments in the explanation and prediction of erosion and sediment yield*. Proceedings of the Exeter Symposium, July. International Association of Hydrological Sciences, Publication, 137: 135–144.

Faulkner, H. 1974. An allometric growth model for competitive gullies. *Zeitschrift für Geomorphologie*, Supplementband, 21: 76–87.

Fleming, R.W., Spencer, G.S. and Banks, D.C. 1970. *Empirical study of behavior of clay shale slopes*. US Army Engineer Nuclear Cratering Group Technical Report 15, Vol. 1. Defense Documentation Centre, Alexandria, VA.

Gerits, J., Imeson, A.C., Verstraten, J.M. and Bryan, R.B. 1987. Rill development and badland regolith properties. In R.B. Bryan (ed.), *Rill erosion*. Catena Supplement 8, Catena Verlag, Cremlingen, Germany: 141–160.

Gilbert, G.K. 1877. *Report on the geology of the Henry Mountains*. US Geographical and Geological Survey of the Rocky Mountain Region, US Government Printing Office, Washington, DC.

Gilbert, G.K. 1909. The convexity of hilltops. *Journal of Geology*, 17: 344–351.

Graf, W.L. 1983. The arroyo problem — palaeohydrology and palaeohydraulics in the short term. In K.J. Gregory (ed.), *Background to palaeohydrology*. Wiley, Chichester: 279–302.

Graf, W.L. 1988. *Fluvial processes in dryland rivers*. Springer-Verlag, Berlin.

Grim, R.E. and Güven, N. 1978. *Bentonites*. Developments in Sedimentology, 24. Elsevier, Amsterdam.

Guitierrez, A.A. 1983. Geomorphic processes and sediment transport in badland watersheds, San Juan County, New Mexico. In S.G. Wells, D.W. Love and T.W. Garner (eds), *Chaco Canyon Country, American Geomorphological Field Group, Field Trip Guidebook*. Adobe Press, Albuquerque, NM: 113–120.

Hadley, R.F. and Schumm, S.A. 1961. Sediment sources and drainage-basin characteristics in upper Cheyenne River basin. *US Geological Survey, Water Supply Paper* 1531B: 137–198.

Haigh, M.J. 1975. The use of erosion pins in the study of slope evolution. In *Shorter Technical Methods II*. British Geomorphological Research Group Technical Bulletin, 18: 31–49.

Haigh, M.J. 1978. *Evolution of slopes on artificial landforms — Blaenaven, U.K.* Department of Geography Research Paper, 183, University of Chicago.

Harris D. and Vita-Finzi, C. 1968. Kokkinopilos — a Greek badland. *Geographical Journal*, 134: 537–546.

Harty, K.M. 1984. The geomorphic role of snow in a badlands watershed. Unpublished MSc thesis, University of Alberta.

Harvey, A. 1982. The role of piping in the development of badlands and gully systems in south-east Spain. In R.B. Bryan and A. Yair (eds), *Badland geomorphology and piping*. GeoBooks, Norwich: 317–336.

Heede, B.H. 1974. Stages of development of gullies in western United States of America. *Zeitschrift für Geomorphologie*, Supplementband, 18: 260–271.

Hodges, W.K. 1982. Hydraulic characteristics of a badland pseudo-pediment slope system during simulated rainfall experiments. In R.B. Bryan and A. Yair (eds), *Badland geomorphology and piping*. GeoBooks, Norwich: 127–152.

Hodges, W.K. and Bryan, R.B. 1982. The influence of material behaviour on runoff initiation in the Dinosaur Badlands, Canada. In R.B. Bryan and A. Yair (eds), *Badland geomorphology and piping*. GeoBooks, Norwich: 12–46.

Hogg, S.E. 1978. The near surface hydrology of the Steveville Badlands, Alberta. Unpublished MSc thesis, University of Alberta.

Horton, R.E. 1945. Erosional development of streams and their drainage basins: hydrophysical approach to quantitative morphology. *Bulletin of the Geological Society of America*, 56: 275–370.

Howard, A.D. 1990. Theoretical model of optimal drainage networks. *Water Resources Research*, 26: 2197–2217.

Howard, A.D. 1994. Badlands. In A.D. Abrahams and A.J. Parsons (eds), *Geomorphology of desert environments*. Chapman & Hall, London: 213–242.

Imeson, A.C., Kwaad, F.J.P.M. and Verstraten, J.M. 1982. The relationship of soil physical and chemical properties to the development of badlands in Morocco. In R.B. Bryan and A. Yair (eds), *Badland geomorphology and piping*. GeoBooks, Norwich: 47–70.

Johnson, D. 1932. Miniature rock fans and pediments. *Science,* 76: 546.

Jones, J.A.A. 1971. Soil piping and stream channel initiation. *Water Resources Research*, 7: 602–610.

Jones, J.A.A. 1978. Soil pipe networks: distribution and discharge. *Cambria*, 5: 1–21.

Jones, J.A.A. 1981. *The nature of soil piping — a review of research*. British Geomorphological Research Group Research Monograph 3. GeoBooks, Norwich.

Jones, J.A.A. 1989. The initiation of natural drainage networks. *Progress in Physical Geography*, 11: 207–245.

Jones, J.A.A. 1990. Piping effects in drylands. *Geological Society of America, Special Paper*, 252: 111–138.

Kirkby, M.J. 1980. The stream head as a significant geomorphic threshold. In D.R. Coates and J.D. Vitek (eds), *Thresholds in geomorphology*. Allen & Unwin, London: 53–73.

Kirkby, M.J. 1990. A one-dimensional model for rill–inter-rill interactions. In R.B. Bryan (ed.), *Rill Erosion*, Catena Supplement 8, Catena Verlag, Cremlingen, Germany: 133–145.

Koster, E.H. 1983. *Sedimentology of the Upper Cretaceous Judith River (Belly River) Formation, Dinosaur Provincial Park, Alberta*. Alberta Geological Survey Field Trip Guidebook. Alberta Research Council, Edmonton.

Lam, K.C. 1977. Patterns and rates of slopewash on the badlands of Hong Kong. *Earth Surface Processes*, 2: 319–232.

Leopold, L.B. and Miller, J.P. 1956. *Ephemeral streams — hydraulic factors and their relation to the drainage net*. United States Geological Survey, Professional Paper, 282A.

Leopold, L.B., Emmett, W.E. and Myrick, R.M. 1966. Channel and hillslope processes in a semi-arid area of New Mexico. *United States Geological Survey, Professional Paper* 352G: 193–253.

Liu, T., An, Z., Yuan, B. and Han, J. 1985. The loess–paleosol sequence in China and climatic history. *Episodes*, 8: 21–28.

Long, Y. and Xiong, G. 1981. Sediment measurement in the Yellow River. In *Erosion and sediment transport measurement*. Proceedings of the Florence Symposium, International Association of Scientific Hydrology, Publication, 133: 275–285.

Mandelbrot, B.B., 1977. *Fractals: form, chance and dimension*. W.H. Freeman & Co, San Francisco.

McFadden, L.D., Wells, S.G. and Dohrenwand, J.C. 1986. Influences of Quaternary climatic changes on processes of soil development on desert loess deposits of the Cima volcanic field, California. *Catena*, 13: 361–389.

McPherson, H.J. 1968. Historical development of the Lower Red Deer Valley, Alberta. *The Canadian Geographer*, 12: 227–240.

Mears, B. Jr 1963. Karst-like features in badlands of the Arizona Painted Desert. *Wyoming University Contributions to Geology*, 2: 101–104.

O'Hara, S.L. 1986. Postglacial Geomorphic Evolution and Alluvial Chronology of a Valley in the Dinosaur Badlands, Alberta. Unpublished MSc thesis, University of Alberta.

O'Hara, S.L. and Campbell, I.A. 1993. Holocene geomorphology and stratigraphy of the lower Falcon Valley, Dinosaur Provincial Park, Alberta, Canada. *Canadian Journal of Earth Sciences*, 30: 1846–1852.

Oostwoud Wijdenes, D.J. and Bryan, R.B. 1991. Gully development on the Njemps Flats, Baringo, Kenya. In H.R. Bork, J. DePloey and A.P. Schick (eds), *Erosion, transport and deposition processes — theories and models*. Catena Supplement 19, Catena Verlag, Cremingen, Germany: 71–90.

Paine, A.D.M. 1985. 'Ergodic' reasoning in

geomorphology: time for a review of the term? *Progress in Physical Geography*, 9: 1–15.

Parker, G.G. 1963. Piping, a geomorphic agent in landform development of the drylands. *International Association for Scientific Hydrology Publication* 65: 103–113.

Parker, G.G. Sr, Higgins, C.G. and Wood, W.W. 1990. Piping and pseudokarst in drylands. *Geological Society of America Special Paper*, 252: 77–110.

Patton, P.C. and Schumm, S.A. 1975. Gully erosion, northwestern Colorado: a threshold phenomenon. *Geology*, 3: 88–90.

Poesen, J.W. and Lavee, H. (eds) 1994. *Rock fragments in soils: surface dynamics*. Special Issue, Catena 23, Catena Verlag, Cremlingen, Germany.

Potter, P.E., Maynard, J.B. and Pryor, W.A. 1980. *Sedimentology of shale*. Springer-Verlag, New York.

Roloff, G., Bradford, J.M. and Schrivents, C.L. 1981. Gully development in Deep Loess Hills region of central Missouri. *Soil Science Society of America Journal*, 45: 119–123.

Scheidegger, A.E., Schumm, S.A. and Fairbridge, R.W. 1968. Badlands. In R.W. Fairbridge (ed.), *The encyclopedia of geomorphology*. Reinhold, New York: 43–48.

Schumm, S.A. 1956a. The role of creep and rainwash on the retreat of badland slopes. *American Journal of Science*, 254: 693–706.

Schumm, S.A. 1956b. Evolution of drainage systems and slopes at Perth Amboy, New Jersey. *Bulletin of the Geological Society of America*, 67: 597–646.

Schumm, S.A. 1962. Erosion on miniature pediments in Badlands National Monument, South Dakota. *Bulletin of the Geological Society of America*, 73: 719–724.

Schumm, S.A. 1964. Seasonal variations of erosion rates and processes on hillslopes in Western Colorado. *Zeitschrift für Geomorphologie*, Supplementband, 5: 215–238.

Schumm, S.A. 1973. Geomorphic thresholds and complex response of drainage systems. In M. Morisawa (ed.), *Fluvial geomorphology*. Publications in Geomorphology, State University of New York, Binghampton, NY: 299–310.

Schumm, S.A. 1980. Some applications of the concept of geomorphic thresholds. In D.R. Coates and J.D. Vitek (eds), *Thresholds in geomorphology*. Allen & Unwin, London: 473–485.

Schumm, S.A. 1988. Variability of the fluvial system in space and time. In T. Rosswall, R.G. Woodmawee and P.G. Risser (eds), *Scales and global change: spatial and temporal variability in biospheric and geospheric processes*. Wiley, Chichester: 225–250.

Schumm, S.A. and Hadley, R.F. 1957. Arroyos and the semiarid cycle of erosion. *American Journal of Science*, 255: 161–174.

Schumm, S.A. and Lichty, R.W. 1965. Time, space, and causality in geomorphology. *American Journal of Science*, 263: 110–119.

Schumm, S.A. and Lusby, G.C. 1963. Seasonal variation of infiltration capacity and runoff on hillslopes in western Colorado. *Journal of Geophysical Research*, 68: 3655–3666.

Schumm, S.A., Harvey, M.D. and Watson, C.C. 1984. *Incised channels: morphology, dynamics and control*. Water Resources Publications, Littleton, CO.

Segerstrom, K. 1950. *Erosion studies at Paricutin, State of Michoacán, Mexico*. US Geological Survey Bulletin, 965-A.

Slaymaker, O. 1982a. Geomorphic field experiments: inventory and prospect. *Zeitschrift für Geomorphologie*, Supplementband, 35: 183–194.

Slaymaker, O. 1982b. The occurrence of piping and gullying in the Penticton glacio-lacustrine silts, Okanagan Valley, B.C. In R.B. Bryan and A. Yair (eds), *Badland geomorphology and piping*. GeoBooks, Norwich: 305–316.

Slaymaker, O. (ed.) 1991. *Field experiments and measurement programs in geomorphology*. University of British Columbia Press, Vancouver.

Smith, K.G. 1950. Standards for grading texture of erosional topography. *American Journal of Science*, 248: 655–668.

Smith, K.G. 1958. Erosional processes and landforms in Badlands National Monument, South Dakota. *Bulletin of the Geological Society of America*, 69: 975–1008.

Snow, R.S. and Mayer, L. (eds), 1992. *Fractals in geomorphology*. Geomorphology (Special Issue), 5.

Springer, M.E. 1958. Desert pavement and vesicular layers of some soils of the desert of the Lahonton Basin, Nevada. *Proceedings of the Soil Science Society of America*, 37: 323–324.

Stocking, M.A. 1978. The measurement, use and relevance of rainfall energy in investigations into erosion. *Zeitschrift für Geomorphologie*, Supplementband, 29: 141–150.

Stocking, M.A. 1979. Catena of sodium-rich soils in Rhodesia. *Journal of Soil Science*, 30: 139–146.

Stocking, M.A. 1984. Review of Bryan, R.B. and Yair, A. (eds), Badland geomorphology and piping. *Progress in Physical Geography*, 8: 624–625.

Stone, R.O. 1967. A desert glossary. *Earth Science Reviews*, 3: 211–268.

Strahler, A.N. 1950. Equilibrium theory of erosional slopes approached by frequency distribution analysis. *American Journal of Science*, 248: 673–696, 800–814.

Strahler, A.N. 1956a. The nature of induced erosion and aggradation. In W.L. Thomas Jr. (ed.), *Man's role in changing the face of the earth*. University of Chicago Press, Chicago: 621–638.

Strahler, A.N. 1956b. Quantitative slope analysis.

Bulletin of the Geological Society of America, 67: 571–596.

Taylor, R.K. and Cripps, J.C. 1987. Weathering effects: slopes in mudrocks and over-consolidated clays. In M.G. Anderson and K.S. Richards (eds), *Slope stability*. Wiley, Chichester.

Teruggi, M.E. 1957. The nature and origin of Argentine loess. *Journal of Sedimentary Petrology*, 27: 322–332.

Thorn, C.E. 1988. *An introduction to theoretical geomorphology*. Unwin Hyman, London.

Thornes, J.B. and Brunsden, D. 1977. *Geomorphology and time*. Methuen, London.

Toy, T.L. 1977. Hillslope form and climate. *Bulletin of the Geological Society of America*, 88: 16–22.

Toy, T.L. 1983. A comparison of the LEMI and erosion pin techniques. *Zeitschrift für Geomorphologie*, Supplementband, 46: 25–34.

UNESCO 1983. *Structural conservation of Gôreme*. Ministry of Culture and Tourism, General Directorate of Antiquities and Museums, Ankara, Turkey.

Van Olphen, H. 1977. *An introduction to clay colloid chemistry*. Wiley, New York.

Warren, A. 1984. Arid geomorphology. *Progress in Physical Geography*, 8: 399–420.

Watson, A., Price Williams, D. and Goudie, A.S. 1984. The palaeoenvironmental interpretation of colluvial sediments and palaeosols of the late Pleistocene hypothermal in southern Africa. *Palaeogeography, Palaeoclimatology, Palaeoecology*, 45: 225–229.

Wells, S.G. 1983. Regional badland development and a model of late quaternary evolution of badland watersheds, San Juan Basin, New Mexico. In S.G. Wells, D.W. Love and T.W. Garner (eds), *Chaco Canyon Country*. American Geomorphological Field Group, Field Trip Guidebook. Adobe Press, Albuquerque, NM: 121–132.

Wells, S.G. and Gutierrez, A.A. 1982. Quaternary evolution of badlands in the southeastern Colorado Plateau, USA. In R.B. Bryan and A. Yair (eds), *Badland geomorphology and piping*. GeoBooks, Norwich: 239–258.

Wise, S.M., Thornes, J.B. and Gilman, A. 1982. How old are the badlands? A case study from south-east Spain. In R.B. Bryan and A. Yair (eds), *Badland geomorphology and piping*. GeoBooks, Norwich: 259–277.

Yair, A., Bryan, R.B., Lavee, H. and Adar, E. 1980. Runoff and erosion rates in the Zin valley badlands, northern Negev, Israel. *Earth Surface Processes*, 5: 205–225.

Yair, A., Goldberg, P. and Brimer, B. 1982. Long term denudation rates in the Zin-Havarim badlands, northern Negev, Israel. In R.B. Bryan and A. Yair (eds), *Badland geomorphology and piping*. GeoBooks, Norwich: 279–291.

14

Pans, playas and salt lakes

Paul A. Shaw and David S.G. Thomas

The nature and occurrence of pans, playas and salt lakes

Given the nature of hydrological inputs in arid and semi-arid regions, it is not surprising that many are characterised by endoreic (internal) drainage or lack integrated surface drainage altogether. Under these circumstances surface depressions become important local and regional foci for the accumulation of water in lakes. Only a few of these lakes are perennial; examples such as the Caspian Sea, the Dead Sea and the Aral Sea invariably have inflowing rivers rising in fringing uplands or distant humid areas. Frequently, but not exclusively, these water bodies are highly saline and, in some cases, supersaturated with salts. Such salt lakes, which have a minimum salinity of 5000 $mg\,L^{-1}$, in turn lie at one end of a spectrum of otherwise ephemeral and often relict closed basins of varying scales and origins frequently termed *playas* or *pans*.

Pans and playas have been identified throughout the world's arid lands. In Africa they have been described from the Sahara (Boulaine 1954; Coque 1962; Smith 1972), Senegal (Tricart 1953), the Kalahari (Passarge 1911; Lancaster 1978a) and neighbouring regions (de Bruiyn 1971). Holm (1960),

Powers *et al.* (1966) and others have described Arabian sabkhas. Playas in many Asian arid environments have been discussed: in Mongolia (Cotton 1942); in Iran (Krinsley 1970); in India (Godbole 1972); in China (Chen and Bowler 1986); and in Russia (Zemljanitzyna 1973). There is a considerable literature on the Australian basins extending from the early works such as Woodward (1897) through to the results of detailed research programmes (Torgersen *et al.* 1986; Chivas and de Dekker 1991; Chivas 1995). Playas and pans of the United States received the early attentions of geomorphological pioneers (Russell 1885; Gilbert 1895) and have been subject to considerable interest ever since, especially those in the southwestern deserts and the Texas High Plains, while those in South America have been given less attention (Bowman 1924; Tricart 1969). Pans have also been described in areas which experience cold climates, for example, Highway Lake in Antarctica (Bird *et al.* 1991) and the numerous salt lakes of the Pallisers Triangle region of the Canadian Prairies (Last 1994).

The identification of pans and playas within arid environments has suggested to some authors that there may be climatic controls on their development (e.g. Tricart 1954, reported in Cooke and Warren 1973). The majority of

Arid Zone Geomorphology: Process, Form and Change in Drylands, 2nd edition. Edited by David S. G. Thomas.

southern Kalahari and peri-southern Kalahari pans, for example, occur on the arid side of the 500 mm mean annual isohyet and the 1000 mm free evaporation isoline (Goudie and Thomas 1985). However, comparable features are found beyond the limits of modern aridity, for example the *Bays* of Carolina, southeast USA (Prouty 1952; Gamble *et al.* 1977), the *Plains* of Zambia (Goudie and Thomas 1985; Williams 1987) and the *Wokanu* of the Darwin region of Australia (van den Broek 1975).

Playas and pans vary in size from the frequently very small depressions of a few tens of square metres in the Kalahari, Western Australia and Texas (Killigrew and Gilkes 1974; Goudie and Thomas 1985; Osterkamp and Wood 1987) to massive tectonic basins such as Lake Eyre, south-central Australia, and Lake Uyuni, Bolivia, which may exceed 10 000 km^2 in area (Lowenstein and Hardie 1985).

These landforms occupy a limited proportion of the dryland area, perhaps 1% of the total landscape. However, they are important and often numerous features. In parts of southern Africa pans attain densities of up to 1.14 pans km^2 (Goudie and Thomas 1985) and occupy 20% of the surface area (Goudie and Wells 1995), while there are an estimated 30 000 to 37 000 basins on the southern High Plains of northwest Texas and adjoining New Mexico (Reddell 1965; Osterkamp and Wood 1987).

Playas and pans have been ascribed a great variety of origins, ranging from the simple desiccation of depressions filled by seasonal rainfall, particularly under wetter palaeoclimatic conditions, to tectonic subsidence, aeolian deflation or the excavatory activity of large mammals. Given such a range of origins and scales these basins display a bewildering degree of variability in characteristics such as morphology, hydrology and sedimentology. The playa environment is both complex and dependent on local conditions, making generalisation about the landform a difficult task.

Playas and pans have been important to human populations from prehistoric times to the present day as sources of water and minerals. In modern times they have been used for urban development (Cooke *et al.* 1982), for airfields and racetracks, and in the case of Lop Nor, China, for testing nuclear weapons. Regrettably these uses conflict with the inherent value of pans and playas as extreme and unusual habitats, while development itself is not without difficulties, as the flooding of Salt Lake City in the 1980s shows (Atwood 1994). Scientifically they are also important for elucidation of palaeoenvironmental conditions from their sediments and landforms. It is not surprising that they have become the focus of multidisciplinary scientific research in which geomorphology has played a large part, as exemplified by the SLEADS (Saline Lakes, Evaporites and Aeolian Deposits) project in Australia, now in its second decade.

PLAYA AND PAN TERMINOLOGY

Arid and semi-arid lacustrine basins have a rich terminology. The most commonly used terms, pan and playa, are interchangeable. There is, none the less, an increasing tendency to use pan for small basins formed by arid zone geomorphological, rather than geological processes (Goudie and Wells 1995), and playa for a depression with a saline surface (Rosen 1994). There are also many descriptors in regional usage and with approximately equivalent meaning. However, attempts to restrict the application of some terms, such as *sabkha*, to features with specific characteristics or to specific parts of closed basins have led to confusion in the literature, as specific usage has not been widely adhered to. More recent attempts to introduce broad terms such as *hemiarid basin* (Currey 1994) have done little to clarify the issue. Some of the general and specific terms are given in Table 14.1.

Under these circumstances we assume the terms in Table 14.1 to be interchangeable and synonymous with *playa* and *pan*, while recognising that local interpretations can occur. Taking all these considerations into account, pans or playas can be defined as arid zone basins of widely varying size and origin which, although generally above the present groundwater table, are subject to ephemeral surface-water inundation of variable periodicity and extent. Their basal and marginal sediments often display evidence of evaporite accumulation, aeolian deflation and accumulation, and/or lacustrine activity.

The term *coastal sabkha* (Glennie 1970) specifically applies to saline flats in arid areas

Table 14.1 *Playa and pan terminology*

Name	Usage	Example
(a) General		
Playa	Widespread	Reeves (1966)
Playa lake	Australia	Killigrew and Gilkes (1974)
	China	Bowler *et al.* (1986)
	North America	Osterkamp and Wood (1987)
Pan	Southern Africa	Rogers (1940)
	Australia	Bowler and Wasson (1984)
Saline or salt lake	USA	Hardie *et al.* (1978)
Dried lake	Australia	Woodward (1897)
Sebkha (sabkha, sebkra, etc.)	North Africa	Smith (1969)
	Arabia	Glennie (1970)
Mamlahah	Arabia	Holm (1960)
Kavir	Iran	Krinsley (1970)
Salar	Peru	Bowman (1924)
Mier	South Africa	Du Toit (1926)
(b) Specific		
Sebkha (Arabia)	Coastal salt flats	Holm (1960)
Mamlahah (Arabia)	Inland basins	Holm (1960)
Sabkhah (Arabia)	Coastal or inland flats or basins	Powers *et al.* (1966)
Mamlahah (Arabia)	Flats excavated for salt	Powers *et al.* (1966)
Clay pan (Australia)		
Takir (Central Asia)	Playa with silt or clay surface	Neal (1969)
Khabra (Arabia)		
Qu (Jordan)		
Salt pan (South Africa)	Playa with saline surface	Hugo (1974)
Salt pan (USA)	Playa with saline surface	Lowenstein and Hardie (1985)
Salina (USA)		
Kavir (Iran)	Playa with saline surface	Neal (1969)
Salar (China)		
Tsaka (Mongolia)		
Salt pan (USA)	The portion of a playa subject to ephemeral flooding	Hardie *et al.* (1978)
Grassed pan (Kalahari)	Sandy clay surface with grass cover	Boocock and Van Straten (1962)

that occur above the level of high tide, but nevertheless receive periodic marine incursions and associated sediments. They have many of the features and processes of inland playas; indeed there is a recognised continuum between coastal flats and inland playas which receive sea water (Bye and Harbison 1991). However, sebkhas are not considered in this chapter.

GENERAL CHARACTERISTICS

Although there is considerable variability among pans and playas, a number of common characteristics emerge (Table 14.2). The most

obvious of these is that pans occupy topographic lows, though not necessarily the lowest areas in enclosed drainage basins, because small pans can develop almost anywhere in relatively flat arid landscapes, for example along deranged drainage lines (Osterkamp and Wood 1987), in the lee of strandlines in fossil lakes (Cooke and Verstappen 1984), and along interdune hollows (Bond 1948). With larger playas there is inevitably a strong geological control on form (e.g. Salama 1994). Topographic position and geological framework may both influence groundwater regime, and hence the formative processes. This is nowhere more apparent than in the *Boinkas* (groundwater

Table 14.2   *General characteristics of playas and pans*

1. Occupy regional or local topographic lows
2. Lack surface outflows
3. Occupied by ephemeral water bodies
4. Usually have flat surfaces
5. Have a hydrological budget in which evaporation greatly exceeds inputs
6. Have vegetation-free surfaces, or, where vegetation occurs, distinct vegetation associations

discharge areas) of southeast Australia (Jacobson *et al.* 1994), where contemporary playas are nested in the topographic lows of larger groundwater-controlled landscapes and geologically older lake basins.

In terms of surface hydrology they are essentially 'closed' systems, having no surface outflow. The dominance of potential evaporation over precipitation and other inputs is the essential contributory factor to their closed status. Hydrological inputs may be by direct precipitation, surface or subsurface inflow, or any combination of the three. Standing surface water is inevitably an ephemeral feature, and accounts for the distinct morphological and sedimentological differences between arid basins and those of more humid areas, where permanent water bodies are the norm. The extent, frequency and length of time of surface water occupancy is dependent on climatological and hydrological regimes, and is a major source of pan and playa variability (Table 14.3). Overall, arid and humid closed basins can be viewed as part of a climate-based spectrum ranging from 0 to 100% surface water occupancy at its extremes, each with a distinct morphology and processes (Bowler 1986).

Table 14.3   *Major sources of pan and playa variability*

1. Origin (geology, structure)
2. Size
3. Morphology
4. Frequency of surface inundation by water (climate, catchment)
5. Relative importance of surface water and groundwater regimes
6. Sedimentology

Much research has focused on the role of groundwater in pan formation and function (see Chapter 15 for a discussion of the general role of groundwater in drylands), particularly in terms of interaction with surface processes (Friedman *et al.* 1982; Torgersen 1984; Torgersen *et al.* 1986; Osterkamp and Wood 1987; Fryberger *et al.* 1988) and of the origins of the brines themselves (Jankowski and Jacobson 1989; Bye and Harbison 1991; Herczeg and Lyons 1991). Not only is the role of groundwater important in pan and playa formation, but there is again a spectrum of conditions and subsequent effects present. These range from pans and playas where the groundwater table intersects the basin surface (Figure 14.1), accompanied by evaporite accumulation and evaporative effects, as in the *Chotts* of Tunisia

Figure 14.1   *Badwater, the lowest point in the playa system of Death Valley, California, exhibits shallow brine pools due to groundwater seepage*

(e.g. Roberts and Mitchell 1987), to those where the water table lies at depth. These latter features are usually clay-floored, and percolating groundwater plays a major role in deep weathering and eluviation. Such variations are thus a function of topography and geology rather than climate.

Pans and playas usually have vegetation-free surfaces, particularly at their lowest elevations. Episodic flooding, vertisol or solonchak formation and salt accumulation discourage vegetation growth, although halophytic plants and shallow-rooting grasses may be established. Grassed pans exist alongside bare clay surfaces, as in the Kalahari (Boocock and Van Straten 1962), suggesting small variations in soil alkalinity and wind action. Butterworth (1982) has proposed a cycle of development linking grassed and clay pans.

PANS AND PLAYAS AS AGGRADATIONAL FEATURES

Playas are sometimes regarded as receptacles of sediments, derived through episodic inflows or aeolian inputs. The sediments which are received are almost exclusively fine grained, which can be explained in three ways. First, where playas represent drainage terminals, only fine sediments are likely to be transported to the basin as discharge generally decreases downstream in arid channels. Second, where playas occur in a tectonic basin closely bordered by uplands, alluvial fans at the basin–mountain interface usually act as buffers in the fluvial system, trapping sediment. Third, aeolian sediments are, by their very nature, no coarser than sand.

Surface evaporation plays a major role in pan evolution, together with the complex processes involved in salt and water transfer at the groundwater–surface-water interface, leading to the accumulation of evaporite deposits on and within near-surface deposits. Where bedrock is close to, or outcrops at, the playa surface, high rates of evaporation may favour rapid breakdown by salt weathering (Goudie and Thomas 1985, see Chapter 3). The various geomorphic and hydrologic depositional processes which operate in the basins are neither mutually exclusive in space

nor time (Bowler 1986). Consequently a series of depositional subenvironments may be present (Hardie *et al.* 1978, Figure 14.2), including concentric zonation of salts and sediments (e.g. Rosen 1991) that may be identifiable as facies (Magee 1991). Any individual basin may possess only some of these subenvironments at any given time (Eugster and Kelts 1983).

The aggradational attributes of playas contributes to their usually flat, horizontal surfaces, especially in the subenvironments subject to inundation. Given the fine-grained nature of the sediments, any irregularities, including those derived from evaporite growth (see Lowenstein and Hardie 1985) are smoothed out by water movement and dissolution when surface water occupies the basin. Playas with highly infrequent (possibly not recorded in historical times) surface-water inundation may develop uneven surfaces through evaporite growth or sand dune development. The extension of dunes (Bowler 1986) and fluvial distributaries (Townshend *et al.* 1989) onto playa surfaces from surrounding areas may lead to uneven margins.

PAN AND PLAYA SURFACE DEGRADATION

Surface degradation, particularly by deflation, is a significant factor in pan development, and is essential for the maintenance of small basins in the landscape. Although the development of evaporite and clay crusts can protect surfaces from deflation, the lack of a protective vegetation cover, high sodium concentrations (Le Roux 1978), the development of wind-susceptible clay desiccation curls and pellets (Bowler 1973; Bowler and Wasson 1984), surface breakdown by salt weathering (Goudie and Watson 1984) and presence of fine sediments all favour the operation of deflation from playa and pan floors when they are not inundated by surface water. Deflation from basin surfaces, which can be controlled by the depth of the groundwater table (see Fryberger *et al.* 1988), may result in the presence of fringing dunes (Bowler 1973; Lancaster 1978b; Goudie and Thomas 1986), though particles may be transported further afield (Young and Evans 1986).

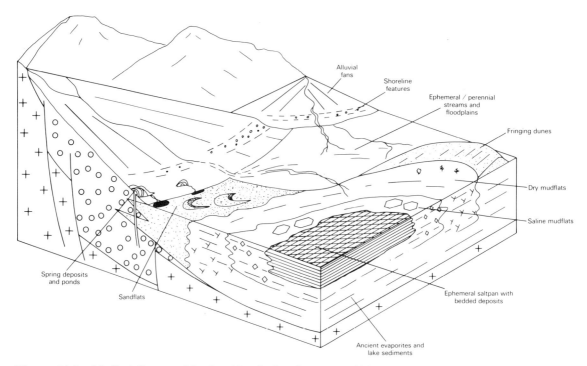

Figure 14.2 *Idealised diagram of the depositional subenvironments which can occur in closed arid zone playa basins (after Hardie* et al. *1978)*

The origins and development of pans and playas

Pans and playas may arise from a variety of causes, which can be classified into *structural, erosional* and *ponding* controls (Table 14.4, Figure 14.3). A few pans have more dramatic origins: Pretoria Saltpan (South Africa), Zuni Salt Lake (New Mexico) and Meteor Crater (Arizona) have been attributed to the impact of meteors or volcanic crater development (Wellington 1955; Mabbutt 1977; Goudie and Thomas 1985).

STRUCTURAL CONTROLS

Faulting and downwarping have led to the development of major regional basins of interior drainage in some arid environments. For example, Cenozoic tectonic activity, including block faulting, has been responsible for the concentration of the intermontane basins in the 'Great Basin' (parts of California, Utah and

Table 14.4 *Origins of closed arid zone basins*

(a) Structural controls
Faulting and rifting
Downwarping
Fracture lines
Intrusions
Differential weathering of adjacent rock types

(b) Erosional controls
Deflation
Subsurface solution and karstic development
Animal scouring

(c) Ponding
Ephemeral or abandoned drainage lines
Floodplain depressions
Interdune troughs
Interstrandline troughs
Coastal sedimentation
Drainage disrupted by tilting

(d) Dramatic
Meteor impact
Volcanic crater development

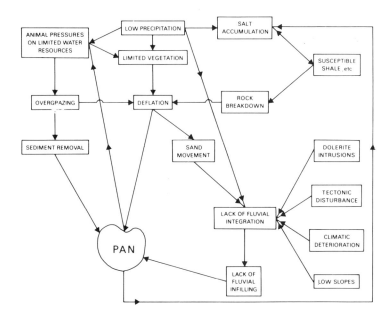

Figure 14.3 *Flow diagram showing complex linkages of factors favouring pan and playa initiation and development by erosion (in this case, predominantly by deflation) (after Goudie and Thomas 1985)*

Oregon) and 'Basin and Range' desert regions of the southwestern United States (Smith and Street-Perrott 1983). Gentler Cenozoic tectonism has contributed to the development of the Etosha and Makgadikgadi basins in southern Africa (Wellington 1955; Thomas and Shaw 1988) and the Eyre Basin in Australia (Johns 1989).

Under current arid conditions these basins contain closed pan and playa features. Their closed status, and the fact that climates have fluctuated substantially in the Quaternary, has meant that large basins such as the Great Salt Lake Basin, Utah, and Makgadikgadi, Botswana, have responded to major local and regional hydrological inputs by developing massive lakes. This record of climatic fluctuation may extend to millions of years, as at Uinta, Utah (Eugster and Kelts 1983). Changes in the status of the hydrologic budget are preserved in the sedimentary record, or in former shorelines (strandlines) present in the basins, and have been termed *playa-lake complexes* by some authors (e.g. Eugster and Hardie 1975; Eugster and Kelts 1983). In some cases the lakes have lost their closed status and overflowed, as in the Bonneville and Lahontan lakes of the Great Basin (Benson *et al.* 1990).

Fracture patterns also influence pan and playa development in two ways. First, lineaments may mark the boundaries of major fault- or warping-controlled basins. Second, lineaments, by acting as conduits for groundwater movement, are the preferential sites for the development of smaller pans, as suggested for some of the features of the Texas High Plain (Osterkamp and Wood 1987) and the south and southeast Kalahari (Arad 1984; Shaw and De Vries 1988), where groundwater activity has occurred since the Tertiary. Intruded bodies at depth found in association with lineaments may also influence pan location, as indicated by the geophysical studies of Lokgware and Mogatse pans in the Kalahari (Farr *et al.* 1982) and the influence of dolerite sills on the pans of the Lake Chrissie complex in the eastern Transvaal (Wellington 1955).

EROSIONAL CONTROLS

Erosional processes, such as deflation, contribute to the formation of the larger structural basins, as in the Qattara and Siwa depressions, Egypt (Gindy 1991), but are especially important in the genesis of smaller local or subregional-scale features. Both aeolian deflation and removal of material by solution during deep weathering have been proposed as erosional mechanisms, but as the debate on small depressions in, for example, Texas and New Mexico shows (Reeves 1966; Carlisle and Marrs 1982; Osterkamp and Wood 1987; Wood and Osterkamp 1987), there is a strong case for a polygenetic origin for many small pans.

Deflation has been cited as an originator or contributor in pan development in many arid and formerly arid environments: for example, Egypt (Haynes 1980), the Kalahari (Lancaster 1978a), Australia (Hills 1940; Bowler 1973), Texas (Gilbert 1895; Reeves 1966), Carolina (Thom 1970), the Argentine Pampas (Tricart 1969) and Zaire (de Ploey 1965). For deflation to be effective the criteria necessary for aeolian entrainment must be satisfied. Of special importance is the susceptibility of surfaces to deflation (Goudie and Thomas 1985), both in terms of material susceptibility, and in the absence of a protective vegetation cover. The latter may be effected by concentration of salts (Le Roux 1978) or seasonal surface inundation (Bowler 1986). In this respect Osterkamp and Wood's (1987) observation that any slight depression in an otherwise flat surface has the potential to develop into a pan or playa should be noted.

A near-surface groundwater table in a playa can act as a base-level control on the depth of deflation. In areas where playa development has taken place in unconsolidated, highly permeable sediments, the presence of impermeable strata may provide this control, for example in the southwest Kalahari, where calcrete horizons are known to be associated with perched local water tables (Bruno 1985). Coastal sabkhas and near-coast playas may have groundwater tables influenced by sea level (Fryberger *et al.* 1988).

The role of deflation in playa and pan development may be indicated by the presence of one or more fringing transverse or *lunette* (Hills 1940) dunes on the downwind margin of the depression, or, indeed, by orientation of the pan transverse to prevailing winds (see Le Roux 1978; Goudie and Thomas 1985; Bowler 1986) (Figure 14.4). Some authors (e.g. Wood and Osterkamp 1987) have opposed the deflationary hypothesis on the grounds that the volume of sand in the fringing dune does not represent the volume removed from the pan. However, deflated playa sediment can be transported beyond the margins of the depression. A more serious objection is the observation that, in many Kalahari pans, the material comprising the lunette may be of different grain size to the pan surface (Goudie and Thomas 1986).

Solution, piping and subsurface karstic collapse may be locally important mechanisms in areas underlain by carbonate and other sedimentary lithologies, and has been proposed for the development of some depressions in the Texas and New Mexico High Plains (Baker 1915; Judson 1950), in Morocco and other parts of the Sahara (Mitchell and Willimot 1974) and in some parts of arid Australia. The contribution of deep weathering and carbonate dissolution to the development of playas in the High Plains has been summarised as a hydrological model (Wood and Osterkamp 1987; see next section)

The excavations and trampling of animals were seen as important factors in forming depressions by early investigators in Texas (Gilbert 1895) and the Kalahari (Allison 1899; Passarge 1904). While clearly inapplicable to the evolution of larger basins, animal activity has been observed to contribute to depression development in areas of seasonally limited water supplies (Weir 1969; Ayeni 1977; see Thomas 1988 for review). Overgrazing around water-points in semi-arid areas can allow deflation to operate more efficiently (Goudie and Thomas 1985). Termites have also been implicated in the formation of small, highly saline pans on islands in swamp ecosystems, such as the Okavango Delta (McCarthy *et al.* 1986).

The mechanisms proposed require suitably susceptible surfaces. In southern Africa, pans are preferentially found on lithologies which readily breakdown to fine-grained sediments, or which are generally poorly consolidated (Goudie and Wells 1995). Susceptibility may be enhanced in lithologies which contain significant amounts of sodium sulphate, which enhances salt weathering and retards plant growth, or clays such as bentonite which have high coefficients of expansion on hydration.

Extensive low-relief terrain also seems to favour pan development, as in Texas and southern Africa. Such surfaces limit the potential to develop integrated drainage, and promote the concentration of both moisture and fine-grained clastic material into surface depressions.

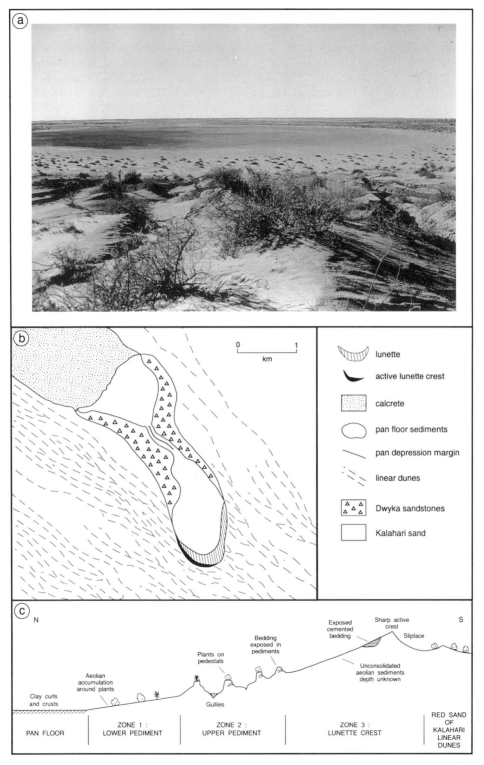

Figure 14.4 *Witpan, southwest Kalahari. (a) View northwards from the lunette dune onto the pan surface, (b) geomorphological map, (c) idealised cross-section and zonation of the lunette dune. (a) Shows the main features of the pediment zones 1 and 2 (after Thomas* et al. *1993)*

PONDING CONTROLS

Some pans in parts of the Kalahari, Western Australia, the Canadian Prairies and Texas are associated with relict or ephemeral drainage lines. In ephemeral channels, pans may simply occupy depressions in the channel floor, though the operation of deflation on channel alluvium or the extension of dunes across channels, as cited with reference to Western Australia (Gregory 1914) and in the Namib Sand Sea (Rust and Wieneke 1974), can lead to subsequent flow being disrupted and the creation of a closed basin by the back-ponding of water. Some channels which contain pans may be relict features inherited from earlier, more humid, climates (e.g. Van der Graaff *et al.* 1977; Nash *et al.* 1994).

Linear depressions between longitudinal dunes are frequently the site of pan formation, as in parts of the Kalahari (Mallick *et al.* 1981); less commonly they may occur between strandlines of palaeolakes, as in the Dautsa Ridge sequence of Lake Ngami (Shaw 1985).

Coastal processes, particularly the construction of beaches and bars, has led to the ponding of coastal sabkhas on susceptible coasts, such as north Africa (Smith 1969) and Namibia. Playa development through ponding can also be effected by tectonic movements, while in specific cases, for example Alkali Dry Lake, California (Mabbutt 1977), lava flows may be responsible.

Playa sedimentation and hydrology

Several sedimentary subenvironments exist in playas and pans. The processes involved can be grouped into those resulting in deposition on the basin floor, the basin subsurface and the basin margins. Deposition in basin margin locations is not necessarily directly related to the processes operative in the basin itself. Depending on the setting of an individual basin, marginal sedimentation can be achieved by the activity of ephemeral rivers, alluvial fans, sand seas or, more rarely, by mass-movement processes, and the resulting landforms are discussed elsewhere.

The dominant sediment types encountered within the basin are fine sediments brought in by surface flow or aeolian action, organic materials and solutes. The main ions encountered are SiO_2, Ca^{2+}, Mg^{2+}, K^+, Na^+, Cl^-, HCO_{3-}, CO_{3-} and SO_{4-}, which are derived from both the surface and groundwater catchments. Hardie *et al.* (1978) note that weathering reactions and catchment lithology are the first determinants of the types of salts precipitated within the basin, although airborne salts may be important in coastal locations (Jack 1921), and have, in the long-term, contributed to inland playas as well (Chivas *et al.* 1991; Jones *et al.* 1994). Precipitation, in turn, is controlled by salt composition and concentration, and the relative influence of the surface and groundwater regimes (Figure 14.5).

GROUNDWATER–SURFACE-WATER INFLUENCES

The hydrological regime affects pan development and morphology in two ways. First, in relation to external factors such as climate and catchment; second, in the relationship between the surface-water and groundwater inputs within the basin.

Bowler (1986) addresses this first control in his model of a six-stage hydrological sequence for closed basins in Australia. The stages range from a lake with permanent surface water at one end of the continuum to an ephemeral terminal sink totally controlled by groundwater at the other (Figure 14.6). An index of disequilibrium, Δ_{eL},

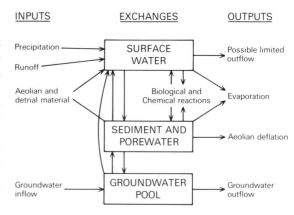

Figure 14.5 *Diagrammatic representation of the possible exchange pathways for water and salts in a pan system (after Torgersen* et al. *1986)*

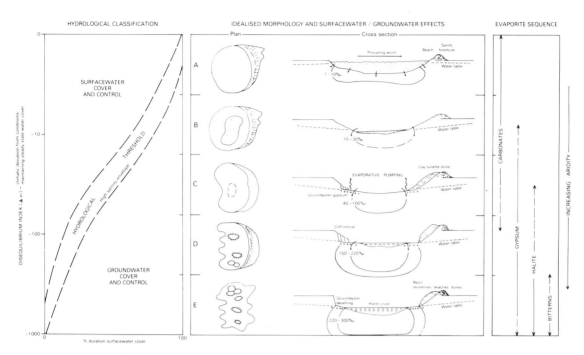

Figure 14.6 *Hydrological classification, idealised morphology and groundwater–surface-water interaction of evaporative basins (after Bowler 1986)*

calculated from hydrological and climatic data, was also used to relate present conditions in a basin to those necessary to maintain a steady-state water cover, with values ranging from 0 in presently perennial basins to –1000 for those currently in the driest locales. While the results, in ignoring many of the other basin variables, are not universally applicable, they serve to emphasise the difference between surface-water and groundwater processes, both in terms of the nature of the waters and sedimentation, and in their interrelationships, particularly within the flooding–desiccation cycle. In the first instance, groundwater may be highly saline, whereas surface waters generally have low solute loads but appreciable suspended sediment. Basins with a large surface inflow component will develop significant evaporite deposits only in the long term; Hardie *et al.* (1978) note that the flood in Lake Eyre in 1950, which covered 8000 km^2, evaporated two years later to leave only a thin layer of halite over an area of about only one-tenth of the original flood, while Holser (1979) estimated the evaporation of a 200 m depth of surface inflow would be necessary to produce a 3 m thickness of the same mineral.

The interval between episodes of surface inundation is also important for sedimentation, as surface water halts evaporation from subsurface water and leads to resolution of salts; Eugster and Kelts (1983) point out that the Great Salt Lake, Utah, has only deposited major halite beds in historic times on two occasions, 1930–35 and 1960–64, coincident with periods of drought.

The position of the water table has been used as the basis of classification of playas by Mabbutt (1977), Wood and Sanford (1990) and Rosen (1994). In pans where the watertable lies at depth, features defined as *recharge playas* by Rosen (1994), there will be little interaction between surface water and groundwater, and sedimentation will occur in a shallow surface and subsurface layer, with overall transfer of water towards the water table. These pans usually have clay or vegetated surfaces, with little sign of evaporite accumulation. Conversely, saline basins with near-surface water tables have complex transfers of water and salts along physical and chemical pathways in three dimensions leading to the accumulation of surface crusts and displacive evaporites.

CLAY-FLOORED BASINS

Clay-floored pans are characteristic of regimes with low groundwater input, or where the surface lies above the influence of capillary rise from the water table, a depth of usually about 3 m (Rosen 1994). Usually they are composed of a flat clay or sandy clay surface, either as the base of a pan, or as a higher surround to a more saline basin (Mabbutt 1977). The dominant sediment is clastic material deposited from suspension during inundation, although lenses of sand may be deposited under higher-energy conditions. The clay surface, in turn, forms an impervious layer to groundwater recharge at the pan centre. Salt input is generally low to basins of this type. Precipitation of calcium and magnesium is common, producing a range of carbonates from calcite to dolomite, dependent upon the Ca/Mg ratio (Müller *et al.* 1972), as cements, laminates, crusts or other structures (Eugster and Kelts 1983). Gypsum efflorescence may follow the drying of the clay surface, while silica mobility is also apparent; silcretes as well as calcretes are found in both the bed and periphery of pans in semi-arid environments (Summerfield 1982).

The Wood and Osterkamp (1987) model (Figure 14.7) for the formation of small clay-floored pans is based on studies of the Texas High Plains, where pan development has taken place in post-Pliocene times under arid to semi-arid conditions in a variety of sedimentary strata which overlie, or are part of, the Ogallala Formation. The Ogallala Formation itself is composed of remnant alluvial silts and clays, with calcrete (caliche) lenses and a thick calcrete caprock. Early stages of development also saw a lowering of the regional water table as margins of the Plains underwent fluvial incision. During initial development stages depressions originate by various means, including deflation, drainage ponding and along structural lineaments. The protobasins act as sites of seasonal runoff concentration on the relatively flat plains surface, and through which groundwater recharge occurs. This results in sub-basin locations in the unsaturated zone becoming foci for oxidation and carbonate dissolution, leading to piping development and the disintegration of the calcrete, thus contributing to basin enlargement. The lateral

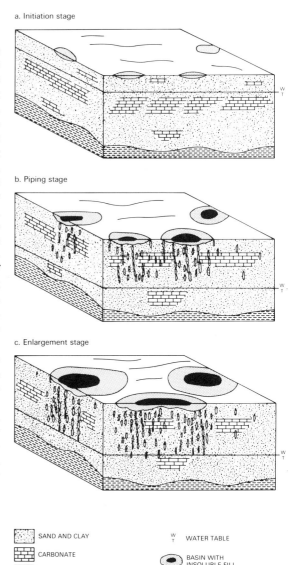

a. Initiation stage

b. Piping stage

c. Enlargement stage

	SAND AND CLAY		W T	WATER TABLE
	CARBONATE		●	BASIN WITH INSOLUBLE FILL
	LOWER CONFINING LAYER			

Figure 14.7 *A model of playa evolution in the High Plains of Texas and New Mexico. (a) Initiation stage. (b) Development of subsurface piping due to concentration of seepage to groundwater from playa depressions. (c) Playa enlargement through concentration of seepage from basin perimeters (after Wood and Osterkamp 1987)*

growth of basins is encouraged by the accumulation of insoluble clays in the lowest, usually central areas, favouring the concentration of seepage to groundwater, and therefore carbonate solution, around the basin margins. This bulk loss is counteracted by the accumulation of lacustrine deposits during more humid phases, when percolation is prevented by a high water table and solutional development ceases. However, this may be partly offset by deflational loss from the basin when it is dry.

This model accords well with other regions of small pans unaffected by groundwater inputs, such as the Kalahari, although here percolating water rarely reaches the water table under present climatic conditions, leading to the precipitation of fresh calcrete at depth. The contention of Wood and Osterkamp (1987) that pans are capable of enlargement by peripheral weathering is backed by the observation of Farr *et al.* (1982) that some pans in the Kalahari are capable of migration over a long period of time.

SALINE BASINS

Increasing near-surface salinity, resulting from either climatic or hydrological factors, will result in evolution from clay-floored pans to salt pans containing evaporites. As already noted, the chemistry of inflowing water is largely dependent on solute output from the catchment; the subsequent evolution of evaporites will be dependent on the ratios of solutes present and the precipitation gradient of the salts involved. Hardie *et al.* (1978) identified four main brine types resulting from a series of evaporative concentration steps on undersaturated inflow; a model which has been subsequently modified by others (see Jankowski and Jacobson 1989; Rosen 1994). These represent a set of geochemical pathways along which evaporites develop (Figure 14.8).

Initial evaporation and degassing (Eugster and Kelts 1983) leads to the progressive precipitation of low-Mg calcite, aragonite and high-Mg calcite (protodolomite), followed by gypsum ($CaSO_4.2H_2O$) at concentrations of $40-100 \, g \, L^{-1}$ (Bowler 1986), dependent on the type and duration of processes in the evaporation zone. Gypsum precipitation is also dependent on the degree of carbonate depletion. Halite (NaCl) saturation occurs at around $200-350 \, g \, L^{-1}$, while other common salts include (Table 14.5) trona ($NaHCO_3.Na_2CO_3.2H_2O$), thenardite (Na_2SO_4), epsomite ($MgSO_4.7H_2O$) and burkeite ($Na_2CO_3.2Na_2SO_4$). Where sodium-rich brines come into contact with deposits of gypsum or calcite, double salts, such as glauberite ($CaSO_4.Na_2SO_4$) or gaylussite ($Na_2CO_3.CaCO_3.5H_2O$) may be formed. Less common evaporites are potassium and magnesium chlorides, which are found in the Qaidam Basin of China (Yuan *et al.* 1983; Chen and Bowler 1986) and nitrates, as in the saltpetre-rich Matsap Pan of South Africa (Wellington 1955).

The precipitation of this range of salts inevitably requires increasing concentration of the brine by evaporation, resulting in zonation of the evaporites by solubility. In individual salt pans this leads to a 'bulls eye' effect of lateral zonation of facies from carbonates at the edge, through sulphates to chlorides in the sump (Jones 1965; Hunt *et al.* 1966). This zonation will also be apparent in the texture of the pan surface, with a transition from the peripheral clay floor to a soft mud with surface efflorescence, described as 'self-raising ground' (Mabbutt 1977), which represents the capillary fringe of the groundwater. This, in turn, gives way to a salt crust whose thickness is dependent on the frequency of surface flooding and groundwater characteristics (see next section), and to a brine layer if present. The concentric surface zonation of salts may be mirrored by a vertical zonation as a result of variations in solubility, with the most soluble minerals at the surface as a result of variations in solubility, or as a response to subsurface processes, particularly reduction (Neev and Emery 1967; Rosen 1991).

On a regional scale, given a standard groundwater regime, evaporite zonation between basins will be apparent along a climatic gradient, with highly soluble salts associated with high levels of groundwater control and few episodes of surface inundation (Bowler 1986).

PROCESSES IN THE SALINE ENVIRONMENT

Where the water table lies close to, but does not intersect, the pan surface, three zones may be identified: a saturated zone; a porewater zone in which capillary rise, enhanced by surface evaporation, occurs; and the surface crust. These zones

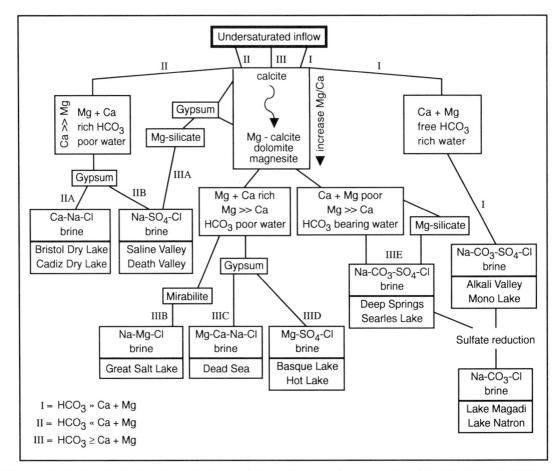

Figure 14.8 *Geochemical pathways for brine evolution with progressive evaporation (after Hardie et al. 1978; Rosen 1994). Common evaporative minerals are shown in Table 14.5*

will change in extent, laterally and vertically, with changes in the water table, leading to corresponding change in the surface sediments (see previous section). Evaporative concentration through this system, whether the groundwater intersects the surface or lies below it, is controlled not only by rates of evaporation, but also groundwater salinity and density, hydraulic conductivity of the aquifer, and the depth of the porewater zone (Bowler 1986). Changes within the porewater zone, termed 'shallow interstitial waters' by Bowler, have profound effects on the ultimate character of the playa. For example, evaporation within this zone will cause variations in water density, often to great depth, enhancing vertical and horizontal transfers of groundwater to balance the salinity gradient.

Salts precipitate within this system as surface crusts, or by interstital crystallisation within existing sediments, or as subaqueous evaporites in brinepools. Salt emplacement can arise by direct crystallisation from the brine, or by reaction between the brine and surrounding sediments and organisms.

An influx of surface water halts evaporative loss from groundwater, reverses many of the reactions in the interstital zone, and sets up new gradients between the surface and groundwater bodies. It also introduces clastic material from inflow and from the atmosphere, which settles on the lake bed, and provides an environment for organisms which play a part in overall sedimentation. Diatoms, which store SiO_2 (Neev and Emery 1967), and algal mats, involved in

Table 14.5 *Common evaporite minerals found in different brine types in playas (after Eugster and Hardie 1978; Rosen 1994)*

Brine type	Saline mineral	
Ca−Mg−Na−(K)−Cl	Antarcticite	$CaCl_2.6H_2O$
	Bischofite	$MgCl_2.6H_2O$
	Carnallite	$KCl.MgCl_2.6H_2O$
	Halite	$NaCl$
	Sylvite	KCl
Na−(Ca)−SO$_4$−Cl	Glauberite	$CaSO_4.NaSO_4$
	Gypsum	$CaSO_4.2H_2O$
	Halite	$NaCl$
	Mirabilite	$NaSO_4.10H_2O$
	Tachyhydrite	$CaCl_2.2MgCl_2.12H_2O$
	Thenardite	Na_2SO_4
Mg−Na−(Ca)−SO$_4$−Cl	Bischofite	$MgCl_2.6H_2O$
	Bloedite	$Na_2SO_4.MgSO_4.4H_2O$
	Epsomite	$MgSO_4.7H_2O$
	Glauberite	$CaSO_4.Na_2SO_4$
	Gypsum	$CaSO_4.2H_2O$
	Halite	$NaCl$
	Hexahydrite	$MgSO_4.6H_2O$
	Kieserite	$MgSO_4.H_2O$
	Mirabilite	$NaSO_4.10H_2O$
	Thenardite	$NaSO_4$
Na−CO$_3$−Cl	Halite	$NaCl$
	Nahcolite	$NaHCO_3$
	Natron	$Na_2CO_3.10H_2O$
	Thermonatrite	$Na_2CO_3.H_2O$
	Trona	$NaHCO_3.Na_2CO_3.2H_2O$
Na−CO$_3$−SO$_4$−Cl	Burkeite	$Na_2CO_3.2Na_2SO_4$
	Halite	$NaCl$
	Mirabilite	$NaSO_4.10H_2O$
	Nahcolite	$NaHCO_3$
	Natron	$Na_2CO_3.10H_2O$
	Thenardite	$NaSO_4$
	Thermonatrite	$Na_2CO_3.H_2O$

the precipitation of carbonates and the formation of kerogen-rich organic layers (Brock 1979; Grant and Tindall 1985), are particularly important in this respect. Larger organisms, such as the brine shrimp *Artemia*, cause bioturbation and sediment reworking (Eardley 1938).

As surface waters evaporate they become increasingly brackish, leading to precipitation of salts at the periphery of the water body, and on the surface of the brine as precipitation thresholds are reached. These crystals, initially held by surface tension, sink to the lake floor, and become nuclei for further, distinctive patterns of crystal growth (Lowenstein and Hardie 1985). They may also be concentrated by wind action into arcuate bands known as *salt ramps* (Millington *et al.* 1995), which persist as minor landforms after evaporation of the brine.

The desiccation of the pan surface will lead to further interstitial crystal growth and dissolution, and a return to the groundwater-dominated regime. As surface flooding is

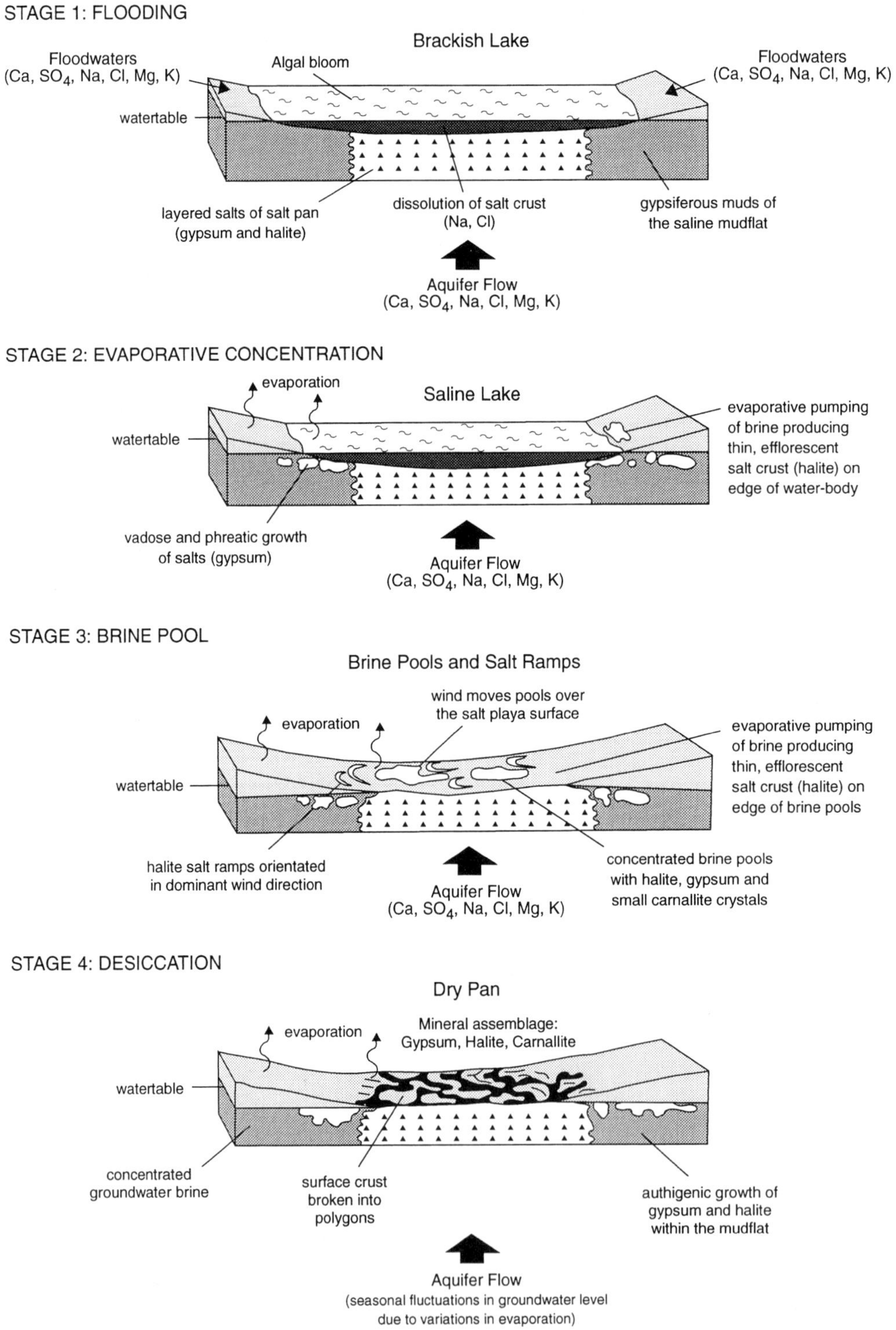

STAGE 1: FLOODING

Brackish Lake

Floodwaters
(Ca, SO_4, Na, Cl, Mg, K)

Algal bloom

Floodwaters
(Ca, SO_4, Na, Cl, Mg, K)

watertable

layered salts of salt pan
(gypsum and halite)

dissolution of salt crust
(Na, Cl)

gypsiferous muds of
the saline mudflat

Aquifer Flow
(Ca, SO_4, Na, Cl, Mg, K)

STAGE 2: EVAPORATIVE CONCENTRATION

evaporation

Saline Lake

watertable

evaporative pumping
of brine producing
thin, efflorescent
salt crust (halite) on
edge of water-body

vadose and phreatic growth
of salts (gypsum)

Aquifer Flow
(Ca, SO_4, Na, Cl, Mg, K)

STAGE 3: BRINE POOL

Brine Pools and Salt Ramps

wind moves pools over
the salt playa surface

evaporation

watertable

evaporative pumping
of brine producing
thin, efflorescent
salt crust (halite) on
edge of brine pools

halite salt ramps orientated
in dominant wind direction

concentrated brine pools
with halite, gypsum and
small carnallite crystals

Aquifer Flow
(Ca, SO_4, Na, Cl, Mg, K)

STAGE 4: DESICCATION

Dry Pan

Mineral assemblage:
Gypsum, Halite, Carnallite

evaporation

watertable

concentrated
groundwater brine

surface crust
broken into
polygons

authigenic growth of
gypsum and halite
within the mudflat

Aquifer Flow
(seasonal fluctuations in groundwater level
due to variations in evaporation)

Figure 14.9   *The saline pan cycle (after Lowenstein and Hardie 1985; Bryant et al. 1994)*

episodic it is possible to represent these processes as a saline pan cycle, characterised by the stages of flooding, evaporative concentration and desiccation (Lowenstein and Hardie 1985; Bryant *et al.* 1994, Figure 14.9)

Given the continual reworking of surface crusts and sediments within the pan cycle, it is not surprising that evaporite deposits do not persist as sedimentary strata in many playas, and, when they do so, may take thousands of years to accumulate. Many of the larger salt lakes owe their extensive evaporite deposits, usually in the form of a series of mud–salt (Hardie *et al.* 1978) or protodolomite–gypsarenite (Dutkiewicz and von der Borch 1995) couplets, to the desiccation of larger water bodies by climatic change or tectonic activity, as in the case of Lake Magadi (Kenya), Lake Bonneville (USA), the Makgadkgadi Pans (Botswana) and the Dead Sea (Israel–Jordan).

PAN SURFACE FEATURES

The surface morphology of the pan is the product of periodic flooding and desiccation, including the effects of rainfall onto the pan surface, crystal growth and dissolution, and aeolian deflation. The surface features themselves are among the most ephemeral of geomorphological phenomena, some lasting no longer than the interval between one rainfall event and the next.

Haloturbation is an important process, usually involving gypsum or halite. In the clay-flat environment surface cracking is apparent, usually forming polygons up to 25 cm across. Thin hard crusts of carbonates or puffy crusts of more soluble minerals may appear on drying (Hardie *et al.* 1978). Surface flaking, with gypsum precipitation, is also common.

Salt crusts have smooth surfaces only while above the level of capillary action, as at Bonneville Flats (Eugster and Kelts 1983), or when wet; on drying crystal expansion leads to the formation of salt blisters and salt polygons, the latter up to 10 m in diameter (Krinsley 1970; Figures 14.10 and 14.11). Plate boundaries become foci for evaporation and precipitation, producing thrust surfaces up to 50 cm above the pan floor, and capable of lifting gravel-size material. Extrusion at the plate edges may lead to the formation of mud and salt pinnacles. Subsequent inundation and desiccation leads to a fresh cycle of polygon development.

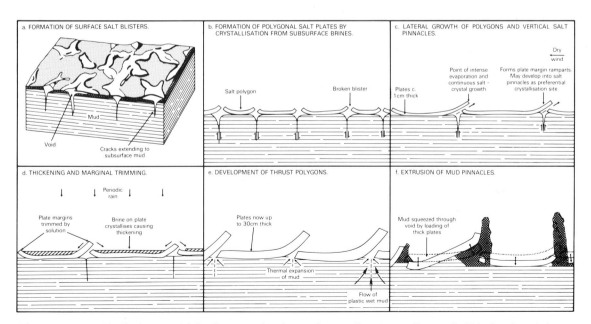

Figure 14.10 *Idealised sequential development of a playa salt crust. Based on a diagram in Krinsley (1970) from investigations in Iran*

Figure 14.11 *Polygons developed on a playa surface in the vicinity of a surface-water inflow, Iran*

Under artesian conditions groundwater effluents may be marked by the growth of spring mounds, known collectively as *aioun* in North Africa (Roberts and Mitchell 1987), and tufa deposits, partly organic in origin, as in many of the playas of the southwestern United States (Neal 1965). Algae may also be preserved as calcareous or siliceous stromatolites towards pan margins, as at Urwi Pan, southern Kalahari (Lancaster 1977), and the East African salt lakes (Casanova and Hillaire-Marcel 1992).

Aeolian processes in pan environments

The presence of extensive unvegetated surfaces, composed mostly of unconsolidated materials, provides ideal conditions for aeolian activity. Pan surfaces experience deflation during dry periods, with transport in the dominant wind direction to form dunes on the pan and its margins.

WIND ACTION ON THE PAN SURFACE

The pan surface, composed of sands, clays and salts, is vulnerable to wind erosion, although crusting reduces its effect. Wind scour removes material to the level of the near-surface water table, creating an unconformity known as a *Stokes surface* (Stokes 1968), present in a number of depositional subenvironments (Fryberger *et al.* 1988). Differential erosion of horizontal sediments may lead to the development of *yardang* topography, with a relief of up to 20 m (Mabbutt 1977). The *kalut* landscape of the Kerman Basin in Iran, a series of parallel ridges and troughs with a relief of 60 m (Dresch 1968; Krinsley 1970) is an extreme example of this landform suite.

On a regional scale, playas with falling water tables may be lowered by deflation and groundwater weathering to form pan and escarpment landscapes, as in the Oasis depressions of north Africa, including the Qattara Depression at 134 m below sea level.

On the pan surface wind action entrains surface materials, mainly fine sands and small pellets of clay of equivalent dimension, the latter produced by salt efflorescence or desiccation (Bowler 1973). Winnowing of fine material from fractures separating surface plates also occurs. Removal of fines may lead to the formation of lag deposits composed of gravels, silcrete fragments or remnant crusts.

Depositional forms will include the formation of sand ripples on salt crusts, which may become accentuated by incorporation into the edges of polygon structures. Sediments will accumulate around plant stems to form phreatophytic mounds, which, in turn, may lead to the formation of *nebkha* dunes (see Chapters 7 and 17 for discussion). On a larger scale parabolic (see Chapter 17) or *lunette* dunes (Hills 1940) form on the downwind wide of the pan, particularly where wind direction is strongly controlled by basin structure (Hardie *et al.* 1978). These dunes may be modified by later inundation, as in the Makgadikgadi Basin of Botswana (Cooke 1980). In large basins with ample sediment supply, such as Lake Eyre, Australia, a range of dune forms may occur. Playas are also important sources for suspended fines carried beyond the pan margin (Gillette 1981).

The wind may also be implicated in the movement of larger rocks, called *sliding stones* or *playa scrapers*, across the pan surface under low frictional conditions (Sharp and Carey 1976). Although the hydraulic energy of surface runoff has been suggested as a contributory mechanism (Wehmeier 1986), recent experimental work (Reid *et al.* 1995) indicates that the formation of a thin ice sheet is necessary for scraper movement. If this is widely applicable it clearly has implications for the environments in which playa scrapers are likely to occur.

LUNETTE DUNES

Pan-margin dunes are important landscape features in many regions, including southern Australia (Bowler 1973), the southern Kalahari (Goudie and Thomas 1986), Tunisia (Coque 1979) and Texas (Huffman and Price 1979). Although commonly between 10 and 50 m high, one example in Tunisia rises almost 150 m above the basin floor (Goudie and Thomas 1986). The surfaces of lunettes are frequently vegetated, which contributes to their development through encouraging sediment accretion. Dune size is a function of a range of factors, but basin size, morphology and sediment supply are important. Cyclical episodes of lunette formation have been identified (Thomas *et al.* 1993; Dutkiewicz and von der Borch 1995) on the basis of depositional hiatuses and palaeosol formation.

Individual pans and playas can possess more than one fringing dune, each representing a separate deflational period, with as many as three identified on the margins of some southern Kalahari pans. Differences in morphology, orientation and sedimentology between dunes on the margins of individual basins have been interpreted as indicators of changing palaeoenvironmental (wind regime and hydrological) conditions (e.g. Lancaster 1978b). In the southern Kalahari some pans

possess an outer quartz sand lunette and an inner form which has a higher silt and clay content of between 12 and 20% by weight (Lancaster 1978b). The former are interpreted as the outcome of deflation in the initial stages of pan development from the sandy Kalahari floor, under relatively dry conditions in an arid environment, in a manner comparable to parabolic dune development from partially vegetated surfaces. Conversely, in Australia, the orientation of quartz-rich fringing dunes reflects wet-season winds. They have therefore been seen as the outcome of deflation from pan and playa beach sediments during periods of high or seasonal lakes, in a manner comparable to coastal dune development (see, for example, Twidale 1972).

Great importance has been attached to the deflation of clay pellets from seasonally dry pan surfaces in the development of the clay lunettes of Australia (e.g. Bowler 1973; 1986), and this is also the mechanism which Lancaster (1978b) ascribes to the development of the inner sandy-clay dunes found on pan margins in the southern Kalahari. As pellet formation is dependent on the breakdown of basin-floor clays by salt efflorescence (Australian lunettes also contain high percentages of gypsum as well as clay), clay lunette development is unlikely in extremely dry or surface-water-dominated environments (Bowler 1986).

Up to 80% of the material in Australian clay lunettes is in the form of clay pellets (Bowler 1973), though overall dune sediments range from sandy clay, comparable to that forming the inner Kalahari lunettes, to almost pure gypsum (Bowler 1976). Analyses of the Australian clays show that the material persists in pelletal form after deposition (Bowler and Wasson 1984), whereas in Nevada, Young and Evans (1986) report that clay pellets break down after deposition on the dunes when the binding salts are dissolved down in subsequent rainfall events. The resultant landform is termed a *mud dune*.

Pans and playas as palaeoenvironmental indicators

Pans and playas have become important sources of palaeoenvironmental reconstruction in arid environments, particularly in Australia (e.g. Harrison 1993), in Africa (Street-Perrott and Roberts 1983) and the southwest USA (Benson *et al.* 1990), even though the evidence they preserve is often discontinuous, and absolute dating remains problematic. The evidence is based upon three aspects of playa research. First, by mapping and dating shorelines, it is possible to re-create past water budgets on the assumption that most playas are 'amplifier lakes' (Street 1980) in that, lacking surface outflow, they tend to emphasise the influence of precipitation on the water budget. The evidence usually takes the form of a lake of successively decreasing volumes represented by a suite of strandlines, each controlled by a threshold within the hydrological system, such as lake morphology, or sometimes an overflow.

Second, recent advances in the multidisciplinary study of playas as three-dimensional features, with distinct hydrological, sedimentary, chemical and organic budgets, has allowed the identification of distinct facies associated with the saline pan cycle, and thus allowed the interpretation of their sedimentary record, even where it is discontinuous. Geochemical studies have contributed greatly to this; in particular the recognition that common minerals such as gypsum may take on different crystal forms in lacustrine, groundwater, aeolian and pedogenic environments (Magee 1991; Magee *et al.* 1995). Although organic materials do not display the same degree of preservation that occurs in temperate environments, studies involving pollen, diatoms, stromatolites (Casanova and Hillaire-Marcel 1992), ostracods (Lister *et al.* 1991) and even ostrich shells (Miller *et al.* 1991) have added to the debate.

The third aspect is the study of landforms associated with pans, especially the stratigraphy of lunette dunes (Chen 1995; Dutkiewicz and von der Borch 1995), which supports the contention that many playas have cycles of erosion and sedimentation with a distinct local signature.

Within the past decade absolute dating has moved on from radiocarbon to the application of Th/U isotopes (e.g. Herczeg and Chapman 1991), amino acid racemisation (Miller *et al.* 1991) and luminescence techniques (e.g. Chen *et al.* 1993). It is these advances, alongside the understanding of pans and playas as dynamic,

three-dimensional landforms involving the interface of aeolian, groundwater and surface water processes, which will contribute much to our understanding of the arid zone in the future.

References

Allison, M.S. 1899. On the origin and formation of pans. *Transactions of the Geological Society of South Africa*, 4: 159–161.

Arad, A. 1984. Relationship of salinity of groundwater to recharge in the southern Kalahari Desert. *Journal of Hydrology*, 71: 225–238.

Atwood, G. 1994. Geomorphology applied to flooding problems of closed-basin lakes — specifically Great Salt Lake, Utah. In M. Morisawa (ed.), *Geomorphology and natural hazards*. Proceedings of the 25th Binghampton Symposium: 197–220.

Ayeni, J.S.O. 1977. Waterholes of the Tsavo National Park. *Journal of Applied Ecology*, 14: 369–378.

Baker, C.L. 1915. Geology and underground waters of the northern Llano Estacado. *University of Texas Bulletin* 57.

Benson, L.V., Currey, D.R., Dorn, R.I., Lajoie, K.R., Oviatt, C.G., Robinson, S.W., Smith, G.I. and Stine, S. 1990. Chronology of expansion and contraction of four Great Basin lake systems during the past 35,000 years. *Palaeogeography, Palaeoclimatology, Palaeoecology*, 78: 241–286.

Bird, M.I., Chivas, A.R., Radnell, C.J. and Burton, H.R. 1991. Sedimentological and stable-isotope evolution of lakes in the Vestfold Hills, Antarctica. *Palaeogeography, Palaeoclimatology, Palaeoecology*, 84: 109–130.

Boocock, C. and Van Straten, J.J. 1962. Notes on the geology and hydrology of the central Kalahari region, Bechuanaland Protectorate. *Transactions of the Geological Society of South Africa*, 65: 130–132.

Bond, G. 1948. The direction of origin of the Kalahari Sand of Southern Rhodesia. *Geological Magazine*, 85: 305–313.

Boulaine, J. 1954. La sebka de Ben Ziane et sa 'lunette' ou Bourrelet exemple de complexe morphologique formé par la dégradation éolienne des sols salés. *Revue de Géomorphologie Dynamique*, 4: 102–123.

Bowler, J.M. 1973. Clay dunes, their occurrence, formation and environmental significance. *Earth Science Reviews*, 12: 279–310.

Bowler, J.M. 1976. Aridity in Australia: age, origins and expression in aeolian landforms and sediments. *Earth Science Reviews*, 12: 279–310.

Bowler, J.M. 1986. Spatial variability and hydrological evolution of Australian lake basins: analogue for Pleistocene hydrological change and evaporite formation. *Palaeogeography, Palaeoclimatology, Palaeoecology*, 54: 21–41.

Bowler, J.M. and Wasson, R.J. 1984. Glacial age environments of inland Australia. In J. Vogel (ed.), *Late Cenozoic palaeoclimates of the southern hemisphere*. Balkema, Rotterdam: 183–208.

Bowler, J.M., Huang, Q., Chen, K., Head, M.J. and Yuan, B. 1986. Radiocarbon dating of playa-like hydrologic changes: examples from northwest China and central Australia, *Palaeogeography, Palaeoclimatology, Palaeoecology*, 54: 241–260.

Bowman, I. 1924. *Desert trails of Atacama*. American Geographical Society, Special Publication 5.

Brock, T.D. 1979. Environmental biology of living stromatolites. In: M.R. Walter (ed.), *Stromatolites*. Developments in Sedimentology 20, Elsevier, Amsterdam.

Bruno, S.A. 1985. Pan genesis in the southern Kalahari. In: D.G. Hutchins and A.P. Lynam (eds), *The proceedings of a seminar on the mineral exploration of the Kalahari*. Geological Survey Department, Botswana, Bulletin 29.

Bryant, R.G., Sellwood, B.W., Millington, A.C. and Drake, N.A. 1994. Marine-like potash evaporite formation on a continental playa: case study from Chott el Djerid, southern Tunisia. *Sedimentary Geology*, 90: 269–291.

Butterworth, J.S. 1982. *The chemistry of Mogatse Pan, Kgalagadi District*. Botswana Geological Survey Department, Report JSB/14/82.

Bye, J.A.T. and Harbison, I.P. 1991. Transfer of inland salts to the marine environment at the head of the Spencer Gulf, South Australia. *Palaeogeography, Palaeoclimatology, Palaeoecology*, 84: 357–368.

Carlisle, W.J. and Marrs, R.W. 1982. Eolian features of the Southern High Plains and their relationship to windflow. *Geological Society of America, Special Paper*, 192: 89–105.

Casanova, J. and Hillaire-Marcel, C. 1992. Chronology and palaeohydrology of Late Quaternary high lake levels in the Manyara Basin (Tanzania) from isotopic data (^{18}O, ^{13}C, ^{14}C, Th/U) on fossil stromatolites. *Quaternary Research*, 38: 205–226.

Chen, K. and Bowler, J.M. 1986. Late Pleistocene evolution of salt lakes in Qaidam Basin, Quinghai Province, China. *Palaeogeography, Palaeoclimatology, Palaeoecology*, 54: 87–104.

Chen, X.Y. 1995. Geomorphology, stratigraphy and thermoluminescence dating of the lunette dune at Lake Victoria, western New South Wales. *Palaeogeography, Paleoeclimatology, Palaeoecology*, 113: 69–86.

Chen, X.Y., Bowler, J.M. and Magee, J.W. 1993. Late Cenozoic stratigraphy and hydrologic history

of Lake Amadeus, a central Australian playa. *Australian Journal of Earth Sciences*, 40: 1–14.

Chivas, A.R. 1995. Preface. *Palaeogeography, Palaeoclimatology, Palaeoecology*, 113: 1.

Chivas, A.R. and de Dekker, P. 1991. Introduction. *Palaeogeography, Palaeoclimatology, Palaeoecology*, 84: 1–2.

Chivas, A.R., Andrew, A.S., Lyons, W.B., Bird, M.I. and Donelly, T.H. 1991. Isotopic constraints on the origin of salts in Australian playas. 1. Sulphur. *Palaeogeography, Palaeoclimatology, Palaeoecology*, 84: 309–332.

Cooke, H.J. 1980. Landform evolution in the context of climatic change and neo-tectonism in the Middle Kalahari of north central Botswana. *Transactions of the Institute of British Geographers*, 5: 80–99.

Cooke, H.J. and Verstappen, H. 1984. The landforms of the western Makgadikgadi Basin in northern Botswana, with a consideration of the chronology of the evolution of Lake Palaeo-Makgadikgadi. *Zeitschrift für Geomorphologie*, NF, 28: 1–19.

Cooke, R.U. and Warren, A. 1973. *Geomorphology in deserts*. Batsford, London.

Cooke, R.U., Brunsden, D., Doornkamp, J.C. and Jones, D.K.C. 1982. *Urban geomorphology in drylands*. Oxford University Press, Oxford.

Cooke, R.U., Warren, A. and Goudie, A.S. 1993. *Desert geomorphology*. UCL Press, London.

Coque, R. 1962. *La Tunisie pré-Saharienne, étude géomorphologique*. Colin, Paris.

Coque, R. 1979. Sur la place du vent dans l'érosion en milieu aride. L'exemple des lunettes (bourrelets éoliens) de la Tunisie. *Mediterranée*, 1/2: 15–21.

Cotton, C.A. 1942. *Climatic accidents in the making*. Whitcombe and Tombs, Christchurch, NZ.

Currey, D.R. 1994. Hemiarid lake basins: hydrographic patterns. In A.D. Abrahams and A.J. Parsons (eds), *Geomorphology of desert environments*. Chapman & Hall, London: 405–421.

de Bruiyn, H. 1971. Pans in the western Orange Free State. *Annals of the Geological Society of South Africa*, 9: 121–124.

de Ploey, J. 1965. Position géomorphologique, genèse et chronologie de certains dépôts superficiels au Congo occidental. *Quaternaria*, 7: 131–154.

Dresch, J. 1968. Reconnaissance dans le Lut (Iran). *Bulletin, Association Géographie Français*, 362/3: 143–153.

Dutkiewicz, A. and von der Borch, C.C. 1995. Lake Greenly, Eyre Peninsula, South Australia; sedimentology, paleoclimatic and paleohydrologic cycles. *Palaeogeography, Palaeoclimatology, Palaeoecology*, 113: 43–56.

Du Toit, A.L. 1926. The Mier country. *South African Geographical Journal*, 9: 21–26.

Eardley, A.J. 1938. Sediments of the Great Salt Lake, Utah. *Bulletin of the American Association of Petroleum Geologists*, 22: 1305–1411.

Eugster, H.P. and Hardie, L.A. 1975. Sedimentation in an ancient playa-lake complex: the Wilkins Peak Member of the Green River Formation of Wyoming. *Bulletin of the Geological Society of America*, 86: 319–334.

Eugster, H.P. and Hardie, L.A. 1978. Saline lakes. In A. Lerman (ed.), *Chemistry, geology and physics of lakes*. Springer-Verlag, New York: 237–239.

Eugster, H.P. and Kelts, K. 1983. Lacustrine chemical sediments. In A.S. Goudie and K. Pye (eds), *Chemical sediments and geomorphology*, Academic Press, London: 321–368.

Farr, J.J., Peart, R.J., Nellisse, C. and Butterworth, J.S. 1982. *Two Kalahari pans: a study of their morphometry and evolution*. Botswana Geological Survey Department Report GS10/10.

Friedman, I., Smith, G.I. and Matsou, S. 1982. Economic implications of the deuterium anomaly in the brine and salts in Seales Lake, California. *Economic Geology*, 77: 694–702.

Fryberger, S.G., Schenk, C.J. and Krystinik, L.F. 1988. Stokes surfaces and the effects of near-surface groundwater-table on aeolian deposition. *Sedimentology*, 35: 21–41.

Gamble, E.E., Daniels, R.B. and Wheeler, W.H. 1977. Primary and secondary rims of Carolina Bays. *Southeastern Geology*, 18: 199–212.

Gilbert, C.K. 1895. Lake basins created by wind erosion. *Journal of Geology*, 3: 47–49.

Gillette, D.A. 1981. Production of dust that may be carried great distances. In T.L. Péwé (ed.), *Desert dust*. Geological Society of America Special Paper, 186: 11–26.

Gindy, A.R. 1991. Origin of the Qattara Depression, Egypt — a discussion. *Bulletin of the Geological Society of America*, 103: 1374–1376.

Glennie, K.W. 1970. *Desert sedimentary environments*. Elsevier, Amsterdam.

Godbole, N.N. 1972. Theories on the origin of salt lakes in Rajasthan, India. *24th International Geological Congress*, 10: 354–357.

Goudie, A.S. and Thomas, D.S.G. 1985. Pans in southern Africa with particular reference to South Africa and Zimbabwe. *Zeitschrift für Geomorphologie*, NF, 29: 1–19.

Goudie, A.S. and Thomas, D.S.G. 1986. Lunette dunes in southern Africa. *Journal of Arid Environments*, 10: 1–12.

Goudie, A.S. and Watson, A. 1984. Rock block monitoring of rapid salt weathering in southern Tunisia. *Earth Surface Processes and Landforms*, 9: 95–98.

Goudie, A.S. and Wells, G.L. 1995. The nature, distribution and formation of pans in arid zones. *Earth Science Reviews*, 38: 1–69.

Grant, W.D. and Tindall, B.J. 1985. The alkaline environment. In R.A. Herbert and G.A. Codd (eds), *Microbes in extreme environments*. Society of General Microbiology, Special Publication 17.

Gregory, J.W. 1914. The lake system of Westralia. *Geographical Journal*, 43: 656–664.

Hardie, L.A., Smoot, J.P. and Eugster, H.P. 1978. Saline lakes and their deposits: a sedimentological approach. In A. Matter and M. Tucker (eds), *Modern and ancient lake sediments*. International Association of Sedimentologists, Special Publication, 2: 7–41.

Harrison, S.P. 1993. Late Quaternary lake-level changes and climates of Australia. *Quaternary Science Reviews*, 12: 211–231.

Haynes, C.V. 1980. Geological evidence of pluvial climates in the Nabta area of the Western Desert, Egypt. In F. Wendorf and R. Schild (eds), *Prehistory of the Eastern Sahara*. Academic Press, New York: 353–371.

Herczeg, A.L. and Chapman, A. 1991. Uranium-series dating of lake and dune deposits in southeastern Australia: a reconnaissance. *Palaeogeography, Palaeoclimatology, Palaeoecology*, 84: 285–298.

Herczeg, A.L. and Lyons, W.B. 1991. A chemical model for the evolution of Australian sodium chloride lake brines. *Palaeogeography, Palaeoclimatology, Palaeoecology*, 84: 43–53.

Hills, E.S. 1940. The lunette: a new landform of aeolian origin. *Australian Geographer*, 3: 1–7.

Holm, D.A. 1960. Desert geomorphology in the Arabian Peninsula. *Science*, 132: 1369–1379.

Holser, W.T. 1979. Mineralogy of evaporites. In R.G. Burns (ed.), *Marine minerals: reviews in mineralogy*. Mineralogical Society of America, 6: 211–294.

Huffman, G.W. and Price, W.A. 1979. Clay dune formation near Corpus Christi, Texas. *Journal of Sedimentary Petrology*, 19: 118–127.

Hugo, P.J. 1974. Salt in the Republic of South Africa. *Memoirs of the Geological Society of South Africa*, 65: 105.

Hunt, C.B., Robinson, T.W., Bowles, W.A. and Washburn, A.I. 1966. *Hydrologic Basin, Death Valley, California*. US Geological Survey, Professional Paper 494B.

Jack, R.L. 1921. The salt and gypsum resources of South Australia. *Bulletin of the Geological Survey of South Australia*, Bulletin 8.

Jacobson, G., Ferguson, J. and Evans, W.R. 1994. Groundwater-discharge playas of the Mallee Region, Murray Basin, southeast Australia. In M.R. Rosen (ed.), *Paleoclimate and basin evolution of playa systems*. Geological Society of America, Special Paper, 289: 81–96.

Jankowski, J. and Jacobson, G. 1989. Hydrochemical evolution of regional groundwaters to playa brines in central Australia. *Journal of Hydrology*, 108: 123–173.

Johns, R.K. 1989. The geological setting of Lake Eyre. In C.W. Bonython and A.S. Fraser (eds), *The great filling of Lake Eyre in 1974*. Royal Geographical Society of Australasia, Adelaide: 60–66.

Jones, B.F. 1965. *The hydrology and mineralogy of Deep Springs Lake, Inyo County, California*. US Geological Survey, Professional Paper. 502A.

Jones, B.F., Hanor, J.S. and Evans, W.R. 1994. Sources of dissolved salts in the central Murray Basin, Australia. *Chemical Geology*, 111: 135–154.

Judson, S. 1950. Depressions of the Nu portion of the Southern High Plains of eastern New Mexico. *Bulletin of the Geological Society of America*, 61: 253–274.

Killigrew, L.P. and Gilkes, R.J. 1974. Development of playa lakes in south western Australia. *Nature*, 247: 454–455.

Krinsley, D.P. 1970. *A geomorphological and palaeoclimatological study of the playas of Iran*. US Geological Survey, Final Scientific Report, CP 70–800.

Lancaster, I.N. 1977. Pleistocene lacustrine stromatolites from Urwi Pan, Botswana. *Transactions of the Geological Society of South Africa*, 80: 283–285.

Lancaster, I.N. 1978a. The pans of the southern Kalahari. *Geographical Journal*, 144: 80–98.

Lancaster, I.N. 1978b. Composition and formation of southern Kalahari pan margin dunes. *Zeitschrift für Geomorphologie*, NF, 22: 148–169.

Last, W.M. 1994. Palaeohydrology of playas in the northern Great Plains: perspectives from Palliser's Triangle. In M. R. Rosen (ed.), *Paleoclimate and basin evolution of playa systems*. Geological Society of America Special Paper, 289: 69–80.

Le Roux, J.S. 1978: The origin and distribution of pans in the Orange Free State. *South African Geographer*, 6: 167–176.

Lister, G.S., Kelts, K., Zao, C.K., Yu, J.-Q. and Niessen, F. 1991. Lake Qinghai, China: closed-basin lake levels and the oxygen isotope record for ostracoda since the latest Pleistocene. *Palaeogeography, Palaeoclimatology, Palaeoecology*, 84: 141–162.

Lowenstein, T.K. and Hardie, L.A. 1985. Criteria for the recognition of salt pan evaporites. *Sedimentology*, 32: 627–644.

Mabbutt, J.A. 1977. *Desert landforms*. MIT Press, Cambridge, MA.

Magee, J.W. 1991. Late Quaternary lacustrine, groundwater, aeolian and pedogenic gypsum in the Prungle Lakes, southeastern Australia. *Palaeogeography, Palaeoclimatology, Palaeoecology*, 84: 3–42.

Magee, J.W., Bowler, J.M., Miller, G.H. and

Williams, D.L.G. 1995. Stratigraphy, sedimentology, chronology and palaeohydrology of Quaternary lacustrine deposits at Madigan Gulf, Lake Eyre, South Australia. *Palaeogeography, Palaeoclimatology, Palaeoecology*, 113: 3–42.

Mallick, D.I.J., Habgood, F. and Skinner, A.C. 1981. *A geological interpretation of LANDSAT imagery and air photography of Botswana*. Overseas Geological and Mining Research, 56. HMSO, London.

McCarthy, T.S., McIver, J. and Cairncross, B. 1986. Carbonate accumulation on islands in the Okavango Delta. *South African Journal of Science*, 82: 588–91.

Miller, G.H., Wendorf, F., Ernst, E., Schild, R., Close, A.E., Friedman, I. and Schwarcz, H.P. 1991. Dating lacustrine episodes in the eastern Sahara by the epimerization of isoleucine in ostrich eggshells. *Palaeogeography, Palaeoclimatology, Palaeoecology*, 84: 175–189.

Millington, A.C., Drake, N.A., White, K. and Bryant, R.G. 1995. Salt ramps: wind-induced depositional features on Tunisian playas. *Earth Surface Processes and Landforms*, 20: 105–113.

Mitchell, C.W. and Willimot, S.G. 1974. Dayas of the Moroccan Sahara and other arid regions. *Geographical Journal*, 140: 441–453.

Müller, G., Irion, G. and Forstner, U. 1972. Formation and diagenesis of inorganic Ca–Mg carbonates in the lacustrine environment. *Naturwissenschaften*, 59: 158–164.

Nash, D.J., Shaw, P.A. and Thomas, D.S.G. 1994. Duricrust development and valley evolution: process–landform links in the Kalahari. *Earth Surface Processes and Landforms*, 19: 299–317.

Neal, J.T. 1965. Playa variation. In W.G. McGinnies and B.J. Goldman (eds), *Arid lands in perspective*, University of Arizona Press, Tucson: 14–44.

Neev, D. and Emery, K.O. 1967. The Dead Sea — depositional processes and environments of evaporites. *Bulletin of the Israel Geological Survey*, 41.

Osterkamp, W.R. and Wood, W.W. 1987. Playa-lake basins on the southern High Plains of Texas and New Mexico: Part 1 — Hydrologic, geomorphic, and geologic evidence for their development. *Bulletin of the Geological Society of America*, 99: 215–223.

Passarge, S. 1904. *Die Kalahari*. Dietrich Riemer, Berlin.

Passarge, S. 1911. Die pfannenformigen Hohlformen der Südafrikanischen Steppen. *Petermann's Geographische Mitteilungen*, 57: 130–135.

Powers, R.W., Ramirez, L.F., Redmond, C.D. and Elberg, E.L. 1966. *Geology of the Arabian Peninsula*. US Geological Survey, Professional Paper, 560D: 147 pp.

Prouty, W.F. 1952. Carolina Bays and their origin. *Bulletin of the Geological Society of America*, 63: 167–224.

Reddell, D.L. 1965. Water resources of playa lakes. *Cross Section*, 12: 1.

Reeves, C.C. 1966. Pluvial lake basins of west Texas. *Journal of Geology*, 74: 269–291.

Reid, J.B., Bucklin, E.P., Copenagle, L., Kidder, J., Pack, S.M., Pollissor, P.J. and Williams, M.L. 1995. Sliding rocks at the Racetrack, Death Valley: What makes them move. *Geology*, 23: 819–822.

Roberts, C.R. and Mitchell, C.W. 1987. Spring mounds in southern Tunisia. In L.E. Frostick and I. Reid (eds), *Desert sediments: ancient and modern*. Geological Society of London, Special Publication, 35: 321–336.

Rogers, A.W. 1940. Pans. *South African Geographical Journal*, 22: 55–60.

Rosen, M.R. 1991. Sedimentologic and geochemical constraints on the evolution of Bristol Dry Lake Basin, California, USA. *Palaeogeography, Palaeoclimatology, Palaeoecology*, 84: 229–257.

Rosen, M.R. 1994. The importance of groundwater in playas: a review of playa classifications and the sedimentology and hydrology of playas. In M.R. Rosen (ed.), *Paleoclimate and basin evolution of playa systems*. Geological Society of America, Special Paper, 289: 1–18.

Russell, U.C. 1885. *Playa lakes and playas*. US Geological Survey Monograph, 11: 81–86.

Rust, U. and Wieneke, F. 1974. Studies on the grammadulla formation in the middle part of the Kuiseb River, South West Africa. *Madoqua*, 3: 5–15.

Salama, R.B. 1994. The Sudanese buried saline lakes. In M.R. Rosen (ed.), *Paleoclimate and basin evolution of playa systems*. Geological Society of America, Special Paper, 289: 33–47.

Sharp, R.P. and Carey, D.L. 1976. Sliding stones, Racetrack Playa, California. *Geological Society of America Bulletin*, 87: 1704–1717.

Shaw, P.A. 1985. Late Quaternary landforms and environmental change in northwest Botswana: the evidence of Lake Ngami and the Mababe Depression. *Transactions of the Institute of British Geographers*, NS, 10: 333–346.

Shaw, P.A. and De Vries, J.J. 1988. Duricrust, groundwater and valley development in the Kalahari of southeast Botswana. *Journal of Arid Environments*, 14: 245–254.

Smith, G.I. and Street-Perrott, F.A. 1983. Pluvial lakes of the western United States. In H.E. Wright (ed.), *Late Quaternary environments of the United States*, Vol. 1. Longman, London: 63–87.

Smith, H.T.U. 1969. *Photo-interpretation studies of desert basins in North Africa*. US Air Force Cambridge Research Laboratories Final Report AF19(628)2486.

Smith, H.T.U. 1972. Playas and related phenomena in the Saharan region. In C.C. Reeves (ed.), *Playa Lake Symposium Proceedings*. ICASALS Publication 4, Texas University, Lubbock: 63–87.

Stokes, W.L. 1968. Multiple parallel truncation bedding planes — a feature of wind-deposited sandstone formations. *Journal of Sedimentary Petrology*, 38: 510–515.

Street, F.A. 1980. The relative importance of climate and local hydrogeological factors in influencing lake-level fluctuations. *Palaeoecology of Africa*, 12: 137–158.

Street-Perrott, F.A. and Roberts, N. 1983. Fluctuations in closed-basin lakes as an indicator of past atmospheric circulation patterns. In F.A. Street-Perrott (eds), *Variations in the global water budget*. Reidel, Dortmund: 331–345.

Summerfield, M.A. 1982. Distribution, nature and probable genesis of silcrete in arid and semi-arid southern Africa. In D.H. Yaalon (ed.), *Aridic soils and geomorphic processes*. Catena Supplement. 1: 37–56.

Thom, B.G. 1970. Carolina Bays in Horry and Marion Counties, South Carolina. *Bulletin of the Geological Society of America*, 81: 783–814.

Thomas, D.S.G. 1988. The biogeomorphology of arid and semi-arid environments. In H.A. Viles (ed.), *Biogeomorphology*. Blackwell, Oxford: 193–221.

Thomas, D.S.G. and Shaw, P.A. 1988. Late Cainozoic drainage evolution in the Zambezi Basin: evidence from the Kalahari Rim. *Journal of African Earth Sciences*, 7: 611–618.

Thomas, D.S.G. and Shaw, P.A. 1991. *The Kalahari environment*. Cambridge University Press, Cambridge.

Thomas, D.S.G., Nash, D.J., Shaw, P.A. and Van der Post, C. 1993. Present day lunette sediment cycling at Witpan in the arid southwestern Kalahari Desert. *Catena*, 20: 515–527.

Torgersen, T. 1984. Wind effects on water and salt loss in playa lakes. *Journal of Hydrology*, 74: 137–149.

Torgersen, T., De Dekker, P., Chivas, A.R. and Bowler, J.M. 1986. Salt lakes: a discussion of the processes influencing palaeoenviromental interpretations and recommendations for future study. *Palaeogeography, Palaeoclimatology, Palaeoecology*, 54: 7–19.

Townshend, J.R.G., Quarmby, N.A., Millington, A.C., Drake, N., Reading, A.J. and White, K.H. 1989. Monitoring playa sediment transport systems using Thematic Mapper data. *Advances in Space Research*, 9: 177–183.

Tricart, J. 1953. Influence des sols salés sur la

déflation éolienne en basse Mauritaine et dans la delta du Senegal. *Revue de Géomorphologie Dynamique*, 4: 124–132.

Tricart, J. 1954. Une forme de relief climatique: les Sebkhas. *Revue de Géomorphologie Dynamique*, 5: 97–101.

Tricart, J. 1969. Actions éoliennes dans la Pampa Deprimada (Republique Argentine). *Revue de Géomorphologie Dynamique*, 19: 178–189.

Twidale, C.R. 1972. Evolution of sand dunes in the Simpson Desert, central Australia. *Transactions of the Institute of British Geographers*, 56: 77–110.

Van den Broek, P.H. 1975. *The urban and engineering geology of the proposed Darwin East urban development area, N.T.* Bureau of Mineral Resources, Australia, Record, 1975/171.

Van der Graaff, W.J.E., Crowe, R.W.A., Bunting, J.A. and Jackson, M.J. 1977. Relict early Cenozoic drainage in arid Western Australia. *Zeitschrift für Geomorphologie*, NF, 21: 379–400.

Wehmeier, E. 1986. Water induced sliding of rocks on playas: Alkali Flat in Big Smoky Valley, Nevada. *Catena*, 13: 197–209.

Weir, J.S. 1969. Chemical properties and occurences on Kalahari Sands of salt licks created by elephants. *Journal of Zoology*, 138: 292–310.

Wellington, J.H. 1955. *Southern Africa*, Vol. 1. Cambridge University Press, Cambridge.

Williams, G.J. 1987. A preliminary LANDSAT interpretation of the relict landforms of western Zambia. In G.J. Williams and A.P. Wood (eds), *Geographical perspectives on development in Southern Africa*. Commonwealth Geographical Bureau, James Cook University, Queensland: 23–33.

Wood, W.W. and Osterkamp, W.R. 1987. Playa-lake basins on the southern High Plains of Texas and New Mexico: part 2 — a hydrologic model and mass-balance arguments for development. *Bulletin of the Geological Society of America*, 99: 224–230.

Wood, W.W. and Sanford, W.E. 1990. Groundwater control of evaporite deposition. *Economic Geology*, 85: 1226–1235.

Woodward, H.P. 1897. The dry lakes of Western Australia. *Geological Magazine*, 4: 363–366.

Young, J.A. and Evans, R.A. 1986. Erosion and deposition of fine sediments from playas. *Journal of Arid Environments*, 10: 103–116.

Yuan, J., Chengyu, H. and Keqin, C. 1983. Characteristics of salt deposits in the dry salt lake. In *Abstracts of the 6th International Symposium on Salt*. Northern Ohio Geological Society, Cleveland.

Zemljanitzyna, L.A. 1973. Inflow to lakes of the semi-arid zone of USSR from groundwater. In *Hydrology of Lakes*. International Association of Hydrological Science, Special Publication, 109, 185–190.

15

Groundwater as a geomorphological agent in drylands

David J. Nash

Introduction: groundwater and dryland landscapes

The role of groundwater as an agent influencing the geomorphology of environments *beyond* the arid zone has received considerable attention in the geological, geographical and engineering literature. Subsurface water has long been recognised as an important factor in the processes of weathering (particularly carbonate dissolution), soil development and hillslope stability, and as a component of river discharge. However, the influence of groundwater in sculpting the landscape in areas where carbonate rocks are not present has been either underestimated or ignored and remains one of the least understood factors in landform genesis (Higgins 1984). The lack of understanding of the role of groundwater processes is especially pronounced in dryland environments where landscapes are frequently viewed as being a product of the long-term interaction of wind and surface water operating under different structural and tectonic settings, with little reference to subsurface activity. That is not to say that the importance of groundwater in arid zones has been completely overlooked — many of the pioneering studies of piping and tunnel scour (e.g. Bryan and Yair 1982; Parker and

Higgins 1990; see Chapter 13), salt weathering (e.g. Cooke *et al.* 1982 and Chapter 3) and groundwater seepage erosion (e.g. Peel 1941) were based upon observations made in dryland environments. However, it would appear that unless the geomorphological impact of subsurface water is manifest at relatively short timescales and has an impact upon engineering structures it is largely overlooked. This can, in part, be explained by the difficulties in identifying the long-term role of groundwater in drylands, particularly where groundwater processes contributed to early landscape development and have been subsequently overwritten by more easily observed surface-water and aeolian processes (Nash *et al.* 1994a).

Groundwater plays an important role in dryland geomorphology, which is reflected in other chapters; in surface processes and characteristics (Section 2), in the geomorphology of dryland slopes (Chapter 8), drainage networks (Chapter 10), badland gullies (Chapter 13) and in pans and playas (Chapter 14). In addition to these various direct roles as an agent of weathering and erosion, groundwater may also act as an important control on the operation of specific processes, such as where the maximum extent of aeolian deflation is limited by the

Arid Zone Geomorphology: Process, Form and Change in Drylands, 2nd edition. Edited by David S. G. Thomas.

depth to the regional water table (Chapter 17). This chapter will identify key areas of ground-water influence in drylands, complementing discussion in other chapters where reference is made to specific landform suites. The water resource implications arising from the influence of geomorphology upon groundwater availability in drylands are not discussed as they have been considered elsewhere (e.g. Berger 1992; Carter 1994; Carter *et al.* 1994). Karst processes and landscapes in arid environments are also not considered as, while dissolution processes operate on limestone surfaces even under conditions of limited available moisture (Smith 1988; 1994), many dryland karst landscapes are largely relict at a macro-scale (Smith 1987; Palmer 1990). Within the phreatic zone (the zone beneath the water table in which all voids are completely filled with water), groundwater processes and morphologies in karst terrain are virtually independent of climate (Palmer 1990) although there may be reduced dissolution in dryland regions owing to the higher salinity, low levels of recharge and subsequent antiquity of many desert groundwater sources (Lowry and Jennings 1974). For a wider discussion of the role of groundwater in geomorphology in general, readers are referred to the excellent summaries provided in the volumes edited by La Fleur (1984), Higgins and Coates (1990) and Brown (1995).

Groundwater processes in valley and scarp development

There are two main ways in which ground-water can act as a factor in dryland valley and scarp development; first, through the processes of tunnel scour and seepage erosion associated with subsurface water emerging at a free face or along a hillside or scarp or valley floor; and, second, through the operation of *in situ* deep-weathering processes (principally chemical and biochemical corrosion) and associated erosion due to the lateral and vertical movement of groundwater along preferential subsurface flow paths such as fractures and faults. Neither of these processes is unique to arid zone environments, but some of the best-documented resultant landforms occur within drylands.

EROSION BY EXFILTRATING WATER: DEFINITIONS AND MECHANISMS

In general, subsurface flow will discharge from the ground surface where either a water table in an unconfined aquifer intersects the landscape, or where the land lies below the piezometric surface of a confined aquifer which is linked to the surface by a fracture or fault in the aquiclude (the impermeable rock stratum which prevents the upward passage of groundwater). Erosion by exfiltrating (i.e. emerging) subsurface water can operate in three ways. First, near-surface groundwater flow may apply stress to the walls of a pre-existing macropore, commonly within a partially or fully consolidated material, which may have originated by a variety of means (e.g. as a result of subsurface flowing water, as a shrinkage crack or from burrowing animals or plant roots). Second, sufficient drag force may be generated as water seeps through and exfiltrates from a porous, usually semi- or unconsolidated material, to entrain particles, cause failure or liquify the material (Dunne 1990). Third, groundwater outflow may, through the operation of biological, chemical and resultant physical weathering processes, exert stress on the walls of pores, weaken the material and ultimately lead to mass wasting (e.g. by providing a moist micro-environment for algal growth or through the precipitation of salts in pore spaces; Laity 1983).

The first of these three processes is termed tunnel scour while the second and third both contribute to the process of seepage erosion (Dunne 1990). Of the two, seepage erosion appears to be the most significant in terms of scarp and valley formation (Uchupi and Oldale 1994) with tunnel scour being most effective at a smaller scale (Dunne 1980). There has been considerable confusion over the terminology utilised to describe these processes, with terms such as piping, pipe formation, tunnel erosion (Bennett 1939), sapping, spring sapping (Bates and Jackson 1980), spring erosion, artesian sapping (Milton 1973), basal sapping and seepage erosion (Hutchinson 1968) often used imprecisely and interchangeably. In the following discussion the terminology suggested by Dunne (1990) is adopted.

The formation of subsurface pipes may result from both tunnel scour or seepage erosion, and, if such pipes collapse, may lead to channel initiation and ultimately valley development (Dunne 1980). Both tunnel scour and seepage erosion may also lead to sapping, in its simplest sense 'the undermining of the base of a cliff, with the subsequent failure of the cliff face' (Bates and Jackson 1980, p. 556). If groundwater flow is sufficiently focused to emerge as a spring then spring sapping may occur, while seepage erosion may lead to weakening and collapse along a more diffuse seepage zone. The processes of tunnel scour in dryland drainage development have been discussed in Chapter 13 in the context of badland development. As such the following section focuses predominantly upon the role of seepage erosion in scarp and valley evolution.

SEEPAGE EROSION AND VALLEY FORMATION

The earliest reference to the role of seepage erosion in dryland valley formation can be traced to Peel (1941) arising from observations made as part of Major R.A. Bagnold's expedition to the Gilf Kebir plateau of Libya in 1938. In this region Peel identified wadis with flat floors and steep sides which terminated in a headward cliff with little or no evidence for fluvial activity in the plateau region surrounding the valley head. This led him to suggest that the wadis appeared to have been 'cut out from *below* rather than "let down from above"' (Peel 1941, p. 13, italics author's original). This description neatly summarises the main difference between scarp and valley development by groundwater seepage erosion processes as opposed to surface incision by rivers — seepage erosion and sapping effectively undermine valley heads and sides due to enhanced weathering and erosion within a zone of groundwater emergence, while erosion by flowing water operates from the surface downwards (Laity and Malin 1985).

Since Peel's observations, seepage erosion has been identified as an important factor in the formation of scarps (see Chapter 8), canyons and drainage systems in a variety of terrestrial and extraterrestrial settings (Table 15.1). These include submarine canyons (e.g. Robb *et al.* 1982; Robb 1990), erosion cirques (Issar 1983)

and valleys in environments ranging from some of the Earth's driest (e.g. southwestern Egypt; Maxwell 1979) to its wettest (the Hawaiian Islands; Kochel and Piper 1986; Baker 1990). It was, however, the identification of vast valley systems on remotely sensed images of Mars from the Mariner 9 mission in the mid-1970s which generated most interest in groundwater as a factor in valley formation (Baker 1982; Baker *et al.* 1992). Despite the obvious difficulties in ground-truthing these images and identifying evidence for the operation of seepage erosion processes, not to mention the dangers of circular argument, many Martian valleys were suggested to have formed by groundwater 'sapping' by analogy with terrestrial valley networks (Pieri *et al.* 1980; Higgins 1982).

The use of terrestrial analogues to explain the origin of Martian valleys highlights one of the major problems of many studies of the role of exfiltrating water in valley development, namely that seepage erosion is often invoked purely on the basis of morphological and morphometric properties rather than by direct observation of processes. This arises, in part, from the difficulties of making direct field observations of groundwater seepage erosion processes in operation, primarily due to the lack of accessibility at headwalls of active gullies and streams. With the notable exceptions of Laity (1983) and Onda (1994), most quantitative assessments of the role of seepage erosion in valley development have come from experimental work using stream table simulations (e.g. Kochel and Piper 1986; Howard and McLane 1988; Baker 1990; Gomez and Mullen 1992) and associated computer modelling (Howard and Selby 1994). These experimental approaches use unconsolidated or semi-consolidated sediments to allow the rapid development of drainage features and are, as such, relatively limited in terms of their applicability to valley development in bedrock settings. Differences in experimental technique, particularly in the variety of initial conditions used in stream table experiments, limit the detailed conclusions of these investigations. Field studies must also be treated with caution as much work has concentrated upon 'model' landscapes such as the Colorado Plateau where the observation of groundwater erosion processes is relatively free from other influences which would mask the

Table 15.1   *Valleys and canyons suggested to have formed by groundwater seepage erosion processes*

Landform	References
(a) Terrestrial valley networks	
(i) Drylands	
Australia	Jennings (1979); Baker (1980); Young (1986)
Botswana	Shaw and de Vries (1988), Nash (1992; 1995), Nash *et al.* (1994a; 1994b)
Colorado Plateau (USA)	Ahnert (1960), Laity (1980; 1983), Laity *et al.* (1980), Pieri *et al.* (1980), Laity and Malin (1985), Howard and Kochel (1988), Howard *et al.* (1988), Baker (1990),
Egypt	Maxwell (1979), El-Baz *et al.* (1980), El-Baz (1982)
Libya	Peel (1941)
Morocco	Smith (1987)
(ii) Non-drylands	
Florida (USA)	Schumm *et al.* (1995)
Hawaii (USA)	Hinds (1925), Baker (1980; 1990), Kochel and Piper (1986)
Japan	Onda (1994)
Massachusetts (USA)	Uchupi and Oldale (1994)
New Zealand	Schumm and Philips (1986)
UK	Sparks and Lewis (1957–58), Small (1964), Nash (1996)
(iii) Beach microdrainage networks	
	Higgins (1982; 1984)
(iv) Submarine canyons	Robb *et al.* (1982), Robb (1990)
(v) Experimental drainage networks	
	Kochel *et al.* (1985), Kochel and Piper (1986), Sakura *et al.* (1987), Howard (1988), Howard and McLane (1988), Baker (1990), Gomez and Mullen (1992)
(b) Extraterrestrial valley networks	
Mars	Milton (1973), Sharp (1973), Baker and Kochel (1979), Carr (1980), Laity *et al.* (1980), Pieri (1980), Pieri *et al.* (1980), Baker (1982; 1983; 1985; 1990), Higgins (1982; 1983; 1984), Belderson, El-Baz (1982), (1983), Mars Channel Working Group (1983), Stiller (1983), Kochel *et al.* (1985), Sharp and Malin (1985), Kochel and Piper (1986), Howard *et al.* (1988), Gulick and Baker (1990), Baker *et al.* (1992), Tanaka and Chapman (1992), Craddock and Maxwell (1993)

effect of seepage erosion (such as fluvial activity and extensive mass movement; Dunne 1990). Nonetheless, some generalisations can be made on the operation of seepage erosion processes. These observations, taken from studies in a number of different environments, appear to hold for a wide range of drainage network scale from beach microdrainage networks (Higgins 1982) up to mega-scale Martian valleys (Baker 1982; 1990).

As noted in the preceding section, the process of seepage erosion in bedrock involves at least some intergranular flow, with weathering proceeding by the slow release of grains within the zone of groundwater emergence leading to spalling and mass wasting of the rock about the seepage zone. Experimental studies show that the main method of drainage network development is by headward erosion, which proceeds rapidly by headwall collapse during the early stages of valley formation (Kochel *et al.* 1985; Baker 1990; Gomez and Mullen 1992). Tributary growth occurs as a result of permeability variations (Howard *et al.* 1988) and disturbances in subsurface flow (Dunne 1980) and may also be influenced by joints and geological structures

Figure 15.1 *Scanning electron micrographs of weathering features associated with groundwater seepage within the Navajo Sandstone, Colorado Plateau. (a) Macropore development within sandstone, (b) small calcite crystal adhering to a gypsum crystal precipitated within a sandstone pore (micrographs courtesy of Julie Laity)*

(Laity *et al.* 1980; Pieri *et al.* 1980; Laity, in Baker 1990). Headward erosion has been found to occur most effectively in gently dipping lithologies with an overall regional dip of 1 to 4°, with erosion of the valley head proceeding in an up-dip direction (Howard *et al.* 1988). In cohesionless sediment, seepage forces at the site of emergence of subsurface flow are most important controls on headward erosion (Howard and McLane 1988), whereas in cohesive bedrock mechanical and chemical weathering are likely to be the dominant displacive processes (Laity *et al.* 1980). Mechanical weathering processes are significant factors in the development of dryland canyons in the Colorado Plateau. SEM analyses of sandstone within the 20 to 25-m-thick zone of

groundwater emergence identify macropore development (Figure 15.1a) together with algal growth and the precipitation of calcite and salt efflorescences within pore spaces as agents of rock weakening (Figure 15.1b) (Laity 1983; Laity, in Baker 1990). In dryland environments the presence of salts within pores may contribute to spalling through pressures exerted by expansion and contraction of minerals due to thermal expansion and rehydration of minerals (Cooke and Smalley 1968). Other processes leading to rock wasting within Colorado Plateau sapping valleys are the dissolution of cements by exfiltrating groundwater and the weathering of fine shale layers within the sandstone formations (see Chapter 8).

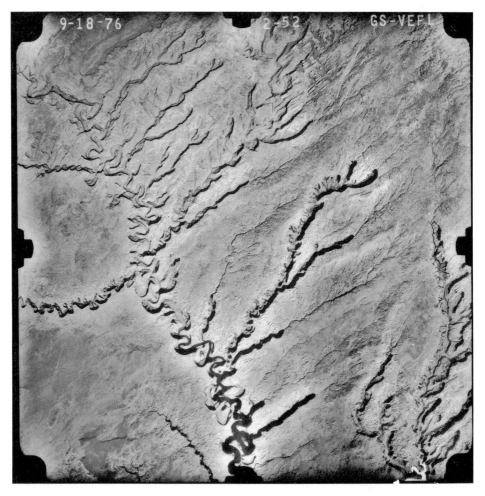

Figure 15.2   *Aerial photograph of the Escalante River (Colorado Plateau) together with Long, Bowns and Explorer Canyons which developed by groundwater seepage erosion (photograph courtesy of Julie Laity)*

MORPHOLOGICAL CHARACTERISTICS OF DRAINAGE NETWORKS DEVELOPED BY GROUNDWATER SEEPAGE EROSION

Valleys developed predominantly by seepage erosion have a number of distinctive morphological features which, to a certain extent, may be diagnostic of the operation of groundwater processes in their formation (Howard *et al.* 1988; Baker 1990). In short, the flow generating these valleys is thought to differ significantly from the processes described in Chapter 9, and valley network behaviour to differ from that discussed in Chapters 9 and 10. These diagnostic features are best illustrated by consideration of the characteristics of the most intensively investigated of the dryland seepage erosion valley systems, those of the Colorado Plateau (Figure 15.2). The most prominent morphometric properties of these (and other non-dryland) systems are summarised in Table 15.2. They include the abrupt initiation of valleys at amphitheatre headwalls, little evidence of surface flow above the valley head, the presence of alcoves and springs in the headward region (Figure 15.3a), steep valley flanks with an abrupt change in slope angle to a flat valley floor (Figure 15.3b), a long valley with a constant valley width, short first-order tributaries with possible hanging valleys and a paucity of downstream tributaries. However, not all valleys influenced by seepage erosion will exactly

Table 15.2 *Diagnostic morphometric features of terrestrial valley networks developed by groundwater seepage erosion or sapping processes (Howard* et al. *1988). See text for discussion*

Abrupt channel initiation, possibly with amphitheatre valley headwalls
Alcove development with springs or seepage zones in headwater regions
Steep valley walls with an abrupt angle to a flat valley floor
Small basin area to canyon area ratio
Low drainage density
Long main valley with constant valley width
Short first-order tributaries with a paucity of downstream tributaries
Possible parallelism of tributaries
Structurally controlled tributary asymmetry
Flat or stepped longitudinal profile

match all of these criteria (Dunne 1990). In particular, the amphitheatre valley head and near-vertical valley sides often considered an essential feature of groundwater outflow networks may not be present (Figure 15.4) if erosion by surface flow and hillslope processes exceeds erosion by groundwater seepage (Sakura *et al.* 1987). This will partly depend upon local climatic characteristics but will also be greatly affected by the angle of internal resistance of the material within which the valley has developed. Valleys such as those with near-vertical canyon walls in the Colorado Plateau (which have developed predominantly by seepage erosion of well-cemented sandstones) must be viewed as forming one end of a process spectrum, with systems produced entirely by surface incision as the opposite end member (Nash 1996). Seepage erosion networks in drylands are often more readily identified than their temperate counterparts as there is less likelihood of extensive contemporary surface flow to mask the influence of groundwater processes. It is, however, possible in such systems that surface flows have been more prevalent during former periods of wetter climate with the balance between groundwater and surface erosion processes fluctuating with time (Nash *et al.* 1994a).

PARAMETERS PROMOTING THE OPERATION OF GROUNDWATER SEEPAGE EROSION PROCESSES

In addition to many common morphometric characteristics, there also appear to be a number of parameters common to valley systems at a variety of scales which influence the effectiveness of groundwater seepage processes. Howard *et al.* (1988) suggest five factors necessary for the operation of seepage erosion (Table 15.3). These include the need for a permeable aquifer of a transmissive rock type, a rechargeable groundwater system (ideally of a large areal extent), a free face at which water can emerge, a structural or lithological inhomogeneity to increase local hydrologic conductivity and a means of transporting material from the free face. In the case of the Colorado Plateau these conditions are ideally met (Figure 15.5) with a low overall annual rainfall (130 to 380 mm)

Figure 15.3   *The morphology of valleys within the Colorado Plateau developed by groundwater seepage erosion. (a) Tributaries of Long Canyon, (b) the amphitheatre valley head of Explorer Canyon showing the zone of seepage emergence (photographs courtesy of Julie Laity)*

but seasonal recharge, and permeable bedrock (the Navajo Sandstone) overlying a series of impermeable sedimentary units (the Kayenta Formation) (Laity, in Baker 1990). Numerous minor seeps with associated alcove development also occur in parts of the Navajo above the main seepage zone where thin shales interbedded with the aeolian sandstone act as aquicludes (Laity 1988).

The operation of seepage erosion processes is affected by a variety of factors which vary in significance according to scale and time

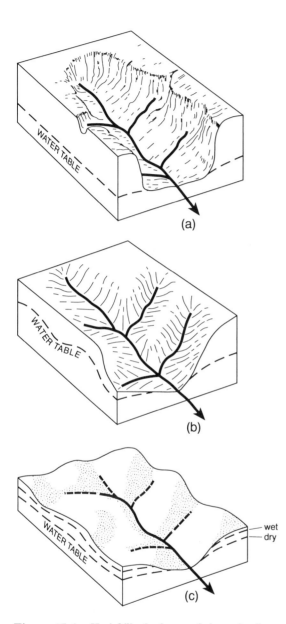

Figure 15.4 *Variability in the morphology of valley heads produced by groundwater seepage erosion. (a) Amphitheatre-headed valley produced by seepage erosion in rocks with a high angle of internal friction, (b) valley head formed by seepage erosion in rocks which either have a lower angle of internal friction or are more susceptible to mass wasting, (c) valley heads where seepage erosion makes only a minor contribution to valley morphology (after Dunne 1990)*

Table 15.3 *Prerequisites for the operation of groundwater seepage erosion in valley development (Howard* et al. *1988; Baker 1990)*

A permeable aquifer of a transmissive rock type
A rechargeable groundwater system, preferably of a large areal extent
A free face at which water can emerge
A structural or lithological inhomogeneity to increase local hydraulic conductivity
A means of transporting material from the free face

(Figure 15.6). These include megascale characteristics, where climate and regional geology may determine whether seepage is perennial, ephemeral or unlikely to occur. Regional water tables will also affect process operation, as will the gradual development of any valley system which will progressively change the distribution and foci of groundwater flow paths by feedback processes. At meso- and micro-scales, the scales at which the actual processes of seepage erosion occur, there are a variety of complex feedback mechanisms, with, for example, the amount of surface-water flow influencing the slope morphology and hence the operation of seepage erosion processes.

GROUNDWATER SEEPAGE EROSION AND ENVIRONMENTAL CHANGE

Finally, the environmental significance of seepage erosion in dryland valley development should not be overlooked (Nash *et al.* 1994a). While theatre-headed forms in some arid regions are relict features, the fact that groundwater erosion is an ongoing process in many valleys within the Colorado Plateau suggests that, given ideal geological conditions (Table 15.3), seepage erosion may be an extremely important landforming process under semi-arid conditions. The process of valley development by seepage erosion does, however, appear to have inherent thresholds, with different modes of operation under different lithological and climatic settings. Laity (in Baker 1990) suggests that present-day groundwater discharge and rates of cliff retreat in the Colorado Plateau may be less than during previous wetter periods, implying that a wetter climate will promote

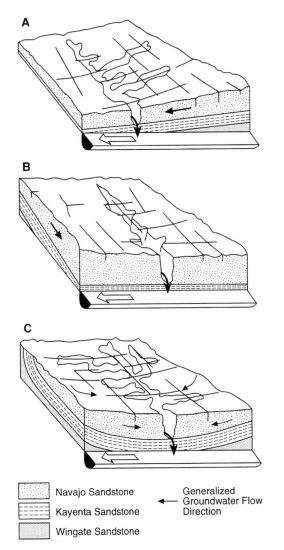

Navajo Sandstone
Kayenta Sandstone
Wingate Sandstone

Generalized
← Groundwater Flow
Direction

Figure 15.5　*Diagrammatic representations of valleys developed by seepage erosion in the Colorado Plateau. The relationship between geological structure, valley morphology, drainage pattern and groundwater flow is clearly visible, with the drainage pattern and form controlled by the direction of groundwater flow. Network length, tributary length and tributary asymmetry vary in response to geological structure (after Laity and Malin 1985)*

High rates of groundwater outflow do not, however, appear to be a hindrance to the development of seepage erosion valleys in the Hawaiian Islands (Kochel and Piper 1986) where chemical erosion and basal sapping of basalts occurs around spring sites with considerably higher discharges than those of the Colorado Plateau (Laity and Malin 1985; Kochel and Baker, in Baker 1990). Clearly, further investigation of the variation in weathering processes associated with ephemeral and perennial groundwater seepage acting upon varying lithologies under different climatic regimes is required.

IN SITU DEEP-WEATHERING AND VALLEY DEVELOPMENT

Another method by which groundwater can contribute to dryland valley formation is through the operation of deep-weathering processes as groundwater moves vertically and laterally along preferential subsurface flow paths. Such flow paths include geological faults and fractures which may act as linear aquifers due to their enhanced permeability (Buckley and Zeil 1984), or igneous dykes which may form barriers to groundwater movement and compartmentalise aquifers (Bromley *et al.* 1994). In bedrock, subsurface flow will be greatest along fractures and joint planes which are approximately normal to the force equipotential surface of the regional water table. Any leaching concentrated along these zones will gradually lower valley bases (Newell 1970; Buckley and Zeil 1984).

This process has been identified as a possible factor in the formation of a number of valleys in dryland and subhumid regions, most notably in the development of some African dambos (also termed *fadamas*, *vleis*, *bas-fonds* and *bolis*). Dambos are broad, shallow, seasonally waterlogged, grassed depressions without a marked stream channel, occupying valley floors, which commonly occur at the headwaters of stream networks (Mäckel 1974; Acres *et al.* 1985; Boast 1990; Bullock 1992). They are found in areas with strongly seasonal rainfall regimes, with present-day total annual precipitation in the range 600 to 1500 mm, and, like valleys attributed to formation by seepage erosion

greater seepage erosion. Conversely, Howard *et al.* (1988) hypothesise that greater rainfall and groundwater outflow in this region would not lead to an increased rate of erosion as it would hinder the accumulation of minerals and salts which act as important heave mechanisms.

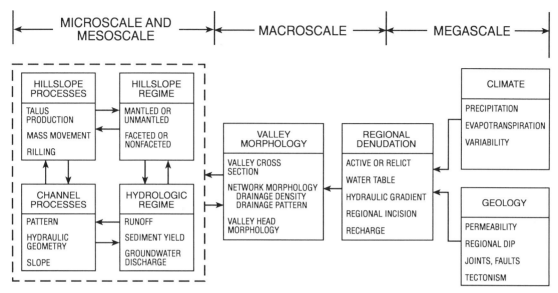

Figure 15.6 *Factors influencing valley development by groundwater seepage erosion at a variety of scales (after Baker 1990)*

processes, are not restricted to semi-arid or dry-subhumid environments. Climate does not appear to be the overriding control of contemporary dambos, otherwise many hydrologically active dambos would be relict features (Whitlow 1985). Geology and the influence of lithology upon soil characteristics are important in determining drainage, while a flat, gentle relief appears to be prerequisite for formation (Mäckel 1974).

There are two schools of thought regarding the origin of dambos; one sees fluvial erosion and slope transport processes dominating development (e.g. Mäckel 1974) while the other implies that dambos have developed by pseudo-karstic *in situ* deep-weathering during drier climates, largely independent of the fluvial network (e.g. McFarlane 1989; 1990; McFarlane and Whitlow 1990). Under the latter theory of formation it is envisaged that Zimbabwean dambos evolved without the action of rivers. Chemical and biochemical corrosion due to vertically and laterally moving water operating along lines of geological weakness such as concentrations of fractures, joints and faults is suggested to have led to gradual surface lowering by solute leaching and ultimately to the formation of a valley (McFarlane 1990). This theory is based upon the identification of a

number of dambos displaying strong structural control and features untypical of a fluvial valley, most notably where dambos cross drainage divides.

The coincidence of surface lowering associated with dissolution of non-carbonate bedrock along lines of fractures has also been noted in semi-arid eastern Botswana, where fractured aquifers are characterised by shallow surface depressions (Gieske and Selaola 1988). The alignment of many structurally controlled Kalahari valley systems (Figure 15.7) has been attributed to the same process with similar evidence for bedrock dissolution identified from boreholes drilled within valley floors (Von Hoyer *et al.* 1985; Nash *et al.* 1994a; Nash 1995). The significance of this process for dryland valley development is that surface lowering is suggested to occur in the absence of fluvial activity and does not necessarily require seasonal recharge. However, it should be noted that the model proposed by McFarlane (1989; 1990) does not totally preclude river action in dambo formation, rather, it suggests an alternation between leaching and fluvial incision. Leaching would dominate drier phases but be interrupted by periods of river activity due to rejuvenation of the fluvial system, with associated incision, removal of dambo floor

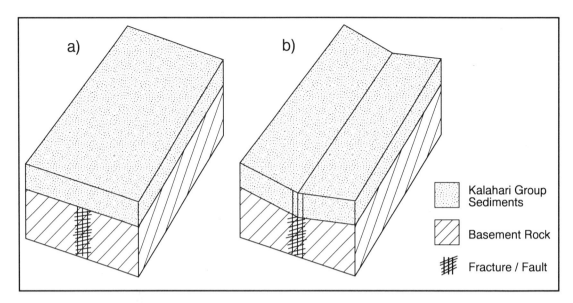

Figure 15.7 *A conceptual diagram of Kalahari valley development by* in situ *deep-weathering processes.
(a) Following the deposition of the Jurassic to Recent Kalahari Group sediments but prior to groundwater movement
along subsurface faults and fractures, and (b) once chemical and biochemical corrosion has lowered the ground surface
by bedrock dissolution (modified after Nash 1995)*

sediments and a concomitant lowering of
regional water tables. There still remains,
however, the as yet unresolved 'chicken and
egg' question of which came first, the valley or
the deep-weathering.

Groundwater and pan/playa development

In a recent reassessment of the classification of
pans and playas, Rosen (1994) notes that the
balance between surface-water and groundwater inputs, as reflected by the position of the
water table beneath the playa surface, is one of
the most important determinants influencing
the development, morphology and sedimentology of playas within arid zones (see Chapter
14). Playas in which the water table lies at
depth beneath the surface are termed *recharge
playas*. Such basins experience little interaction
between surface and groundwater since the
playa floor is frequently above the zone of
capillary rise, have minimal accumulation of
salts at their surface and tend to be clay-
floored. In contrast, more saline *discharge
playas*, where the water table is close to or

outcrops at the surface, experience appreciable
seasonal groundwater outflow and are commonly associated with accumulations of
evaporite deposits (e.g. Clarke 1994).

The role of groundwater as a factor in pan
formation and development has been most
closely considered through studies in Australia
(e.g. Bowler 1986; Torgersen *et al.* 1986;
Jacobson *et al.* 1994), the United States (e.g.
Osterkamp and Wood 1987; Paine 1994; Rosen
1994) and in the Kalahari (Butterworth 1982;
Farr *et al.* 1982; Thomas *et al.* 1993), with
other basins fed by groundwater seepage
described in Tunisia by Coque (1962), in
Algeria by Boulaine (1954) and in many other
locations in North Africa by Glennie (1970).

Groundwater can operate as a factor in playa
development in three main ways. First, percolating groundwater can lead to pan-floor
subsidence by direct dissolution processes
(Baker 1915; Judson 1950; Osterkamp and
Wood 1987), as discussed in more detail in
Chapter 14. Second, a long-term reduction in
groundwater head levels can lead to a change in
the status of a playa, for example, from discharge to recharge as groundwater flow is
focused towards other lower-lying playas in a

region. Such a change may be associated with a concomitant decrease in the surface salinity, in turn promoting an increase in the vegetation cover and ultimately leading to a change in the distribution of discharge playas by a process of playa capture (Jacobson and Jankowski 1989). Third, the subsurface water table can also act as a base-level for wind deflation (Bowler 1986; Thomas *et al.* 1993). Groundwater dissolution and playa capture will now be considered, with the links between groundwater and aeolian deflation discussed later in this chapter.

The processes of deep-weathering and bedrock dissolution outlined in the previous section have also been proposed as possible mechanisms (along with deflation, biogenic activity, volcanism, tectonism and meteorite impact; Goudie and Thomas 1985; 1986; Goudie and Wells 1995) in the formation of playas. Osterkamp and Wood (1987) and Wood and Osterkamp (1987) have proposed a lithologically specific groundwater solution model for the development of clay-floored basins based upon observations in the Southern High Plains of Texas and New Mexico, substantiated by mass-balance calculations (see Chapter 14). These authors suggest that deepening and expansion of a playa-floor area occurs essentially by dissolution and removal of material beneath the playa surface. The infiltration, weathering and downward transport of solutes by percolating groundwater (Zartman *et al.* 1994), along with the removal to the

subsurface of clastic material along solutional pipes, is suggested to lead to the gradual subsidence of the playa surface. Subsidence is a particularly important mechanism in the Southern High Plains in areas where many playas are underlain by evaporite-bearing Permian bedrock (Paine 1994). The dissolution of the Permian strata is suggested to have been continuous throughout the deposition of later formations during the Neogene and may be occurring today.

In addition to operating directly as a factor in playa floor dissolution, groundwater may also be an important control on the hydrological and sedimentological characteristics of a playa through time. The depth of the water table beneath a playa surface is likely to vary in response to seasonal and longer-term drought and also due to regional climatic change. A long-term reduction in the level of groundwater head has been suggested to cause the migration of playas by a process termed *playa capture* (Jacobson and Jankowski 1989). This process, illustrated by Figure 15.8, is broadly analogous to river capture and results from a shift in subsurface groundwater flow arising from the combination of a fall in regional groundwater tables and the preferential deepening of one playa floor relative to adjacent basins. The model is based upon studies of discharge playas near Curtin Springs, central Australia, an area where groundwater head is known to have decayed over several thousand years (Jacobson

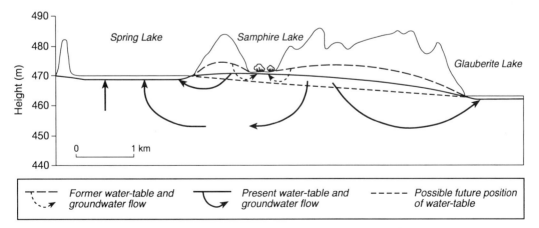

Figure 15.8 *A schematic representation of the process of 'playa capture' or playa abandonment due to groundwater head decay. Samphire Lake formerly had a strong groundwater discharge but is now vegetated, with Glauberite Lake progressively capturing the groundwater flow system in this region (after Jacobson and Jankowski 1989)*

et al. 1988). In this region, Samphire Lake contains thick deposits of groundwater-derived gypsum, indicating that it was a previously active discharge playa. Lowering of the water table has reduced the levels of saline groundwater outflow onto the playa surface, allowed the encroachment of vegetation onto the playa, and focused groundwater discharge into the neighbouring Spring and Glauberite lakes. The dry playa surfaces are thus rendered susceptible to both aeolian sediment deflation as well as alluviation by sediments from around the playa periphery. This may, in part, explain the occurrence of groundwater-discharge derived playa sediments now buried by aeolian material. The decay in groundwater head therefore results in a variation in the spatial distribution of active and abandoned groundwater discharge playas through time (Jacobson and Jankowski 1989). Glauberite Lake will become the primary focus of groundwater activity in this region, ultimately forming a regional groundwater sump.

Groundwater and aeolian processes

Groundwater has also been recognised as an important factor in the control of aeolian sediment deflation and deposition within dunefields and ergs (see Chapters 16 and 17), particularly where the water table is close to the surface as is the case in many coastal sand seas, playas and sabkha environments. Aeolian entrainment and deposition in dunefields is a function of sediment availability and transportability, both of which are controlled by factors such as sediment size, sediment dryness, the degree of surface cementation, the wind energy and the degree of vegetation or lag cover (Kocurek and Nielson 1986; Kocurek 1988; 1991). The presence of a high water table can substantially modify ground surface conditions and limit the extent to which the wind can act as an erosional agent.

The concept of the water table acting as a base-level for wind scour was first suggested by Stokes (1968) and has been more recently described by Fryberger *et al.* (1988). A near-surface water table or *Stokes surface* (Fryberger *et al.* 1988) can act as a factor influencing aeolian processes in three ways. The first, as originally described by Stokes (1968), arises from the higher cohesivity of damp sand in proximity to the water table, primarily as a result of increased intergranular surface tension due to the presence of porewater. Damp or wet sand is therefore less easily entrained by the wind and the boundary between dry and wet sediment may act as an erosional disconformity. Stokes surfaces in modern arid environments (Table 15.4) vary in extent from extremely localised zones within interdune areas up to extensive planar surfaces of at least $25 \, km^2$ (Fryberger *et al.* 1988). Modern Stokes surfaces have a variety of features and associated sedimentary structures, the most typical of which (in aeolian systems) is the extremely sharp truncation of underlying cross-beds with only a thin veneer of overlying sediments. Although erosional in nature, Stokes surfaces also act as important controls upon deposition, and are commonly associated with thin layers of sediment indicating the role of the water table in

Table 15.4   *Examples of modern Stokes surfaces*

Location	Setting	References
Sabkha Matti, Arabian Gulf	Continental sabkha	Kinsman (1969)
White Sands dunefield, New Mexico, USA	Alkali flats in continental sand sea	Loope (1984); Fryberger *et al.* (1988)
Jafurah Sand Sea, Saudi Arabia	Coastal offshore prograding sand sea	Fryberger *et al.* (1988)
Guerrero Negro, New Mexico, USA	Coastal onshore prograding sand sea	Fryberger *et al.* (1988; 1990)
Great Salt Lake Desert, USA	Continental sand sea and salt flats	Stokes (1968)

aeolian sedimentation. The erosional topography of Stokes surfaces is not always completely planar, partly due to the water table mimicking the dunefield topography, and may also exhibit surface irregularities; the capillary tension within damp sand can assist in the formation of a variety of sedimentary adhesion structures due to wind sculpturing, including adhesion ripples, laminations, warts and other steep-sided, irregular bedforms (Kocurek 1981a).

The second influence of near-surface groundwater as a limit to wind scour is through the cementation of sediment in the vicinity of the water table. High levels of evaporation combined with a high water table may promote the precipitation of evaporite cements in the phreatic zone, or of salcretes nearer the surface at the top of the capillary fringe zone (see next section). Schenk and Fryberger (1988) and Fryberger *et al.* (1988) record massive phreatic gypsum cementation in the White Sands dune field, New Mexico, which has acted as a basal limit to aeolian deflation and created extensive planar surfaces. Cementation in proximity to the water table may also produce an irregular surface topography with Fryberger *et al.* (1988) recording 'mini-yardangs', non-streamlined eroded bumps, salt ridges and scour pits in halite-cemented sediments in the Jafurah Sand Sea, Saudi Arabia. In addition to cementing the

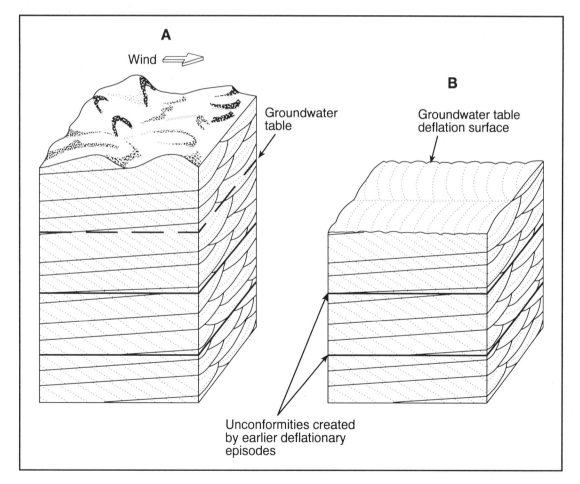

Figure 15.9 *The formation of flat bedding planes by aeolian deflation to the level of the groundwater table. (a) Bedform climbing leads to the accumulation of sand in trough cross-beds. (b) Subsequent deflation to the water table occurs due to either a change in wind regime or a decrease in sand supply, with eroded sand deposited far downwind (after Loope 1984)*

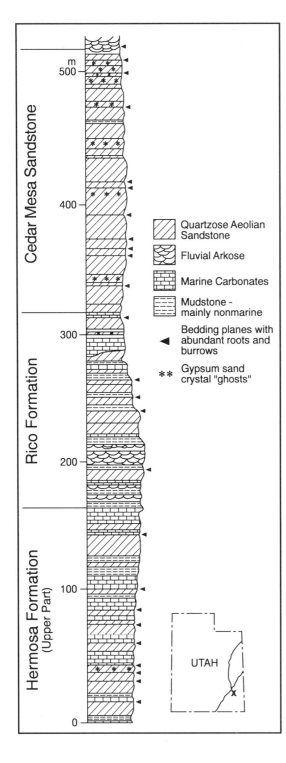

sand surface, the presence of salts also raises the threshold velocity of sand (Pye 1980).

The third link between groundwater and aeolian processes in dryland regions occurs where shallow water tables enable the establishment of a permanent vegetation cover as opposed to acting as an erosional base-level. Fixed dunes in central Niger have had their surfaces stabilised by the packing of fines due to increased rainfall and the action of cyanophytes and fungi acting as binding agents (Talbot 1980). Former water tables are suggested by Talbot (1985) to have influenced vegetation colonisation of the dunefield and hence stabilised the dune surface.

While near-surface groundwater tables have been widely recognised in present-day arid zones (e.g. Loope 1984; Fryberger *et al.* 1988; 1990), there is considerable debate over whether extensive planar first-order bounding surfaces (Brookfield 1977) forming bedding planes in ancient aeolian sandstones, such as the Jurassic age Entrada and Navajo sandstones in Utah and Colorado, were generated in the same way (e.g. Kocurek 1981a; 1981b; 1984; 1988; 1991; Rubin and Hunter 1982; 1984; Loope 1984). An indicative model for the role of the water table in the creation of bedding planes is shown in Figure 15.9, although this model is not accepted as an explanation for all major bounding surfaces. The original concept of a water table acting as a control of deflation as put forward by Stokes (1968) is now viewed as one possible scenario for the creation of bounding surfaces, with interdune migration and bedform climbing (e.g. Rubin and Hunter 1982) viewed as less site-specific alternatives (Mader and Yardley 1985).

Despite these reservations, there is widespread evidence for sediment cohesion, cementation and salt encrustation associated with the groundwater table in a number of ancient aeolian sandstones (e.g. Loope 1985; 1988; Chan and Kocurek 1988; Gaylord 1990; Bromley 1992). The Permian age Cedar Mesa Sandstone in southeast Utah is one example of an ancient

Figure 15.10 *A stratigraphic section of Upper Pennsylvanian (Carboniferous) and Lower Permian rocks exposed in Canyonlands National Park, Utah (after Loope 1985).*

aeolian system where a near-surface ground-water table is suggested to have played a major role in the sandstone architecture (Loope 1985; 1988). The sandstone comprises thin sand bodies separated by extensive planar erosion surfaces attributed to deflation to a former water table (Figure 15.10). Rhizoliths are abundant within the sandstone beds directly beneath these deflation surfaces suggesting a close proximity to the water table. The presence of replaced gypsum sand crystals in a number of horizons provides further supporting evidence that the water table was sufficiently close to the surface to allow evaporite formation (Loope 1988).

Groundwater and duricrust formation

As has been noted in Chapter 6, groundwater may also play a very important role in determining the nature of arid zone surficial materials. The range of duricrusts which occur in drylands can be broadly classified into pedogenic and non-pedogenic varieties, according to their hydrological setting (Carlisle 1980; 1983). There is, however, some confusion over the application of the word 'pedogenic' in many classifications. 'Non-pedogenic' calcretes are defined by Goudie (1983, p. 113), in the strictest sense, as those where authigenic cement has been introduced into the host sediment or soil from an external source, i.e. by absolute accumulation. This is in contrast to pedogenic duricrusts where cementation of sediments occurs due to relative accumulation within a profile. This general distinction can also be applied to other duricrust types such as silcrete and gypcrete, although the extent to which supplies of the cementing agent have been brought in from elsewhere or originated within the host material may be difficult to identify.

While the level of the water table may be important in limiting processes of pedogenic duricrust development, groundwater is at its most significant in the formation of non-pedogenic duricrusts where it may have three, strongly interlinked, roles. First, groundwater may act as an important source of silica or carbonate as a result of water movement through bedrock and weathered material. Second, subsurface water may also act as a

transporting agent from a source of silica, Ca- or Mg-bicarbonate to a site of precipitation. Finally, groundwater fluctuations combined with processes such as evaporation, pH changes and the common ion effect may act as an important component in the precipitation of a duricrust.

Non-pedogenic duricrusts are referred to by a range of terminologies with, for example, groundwater-, phreatic-, valley- and channel-calcretes frequently being used synonymously. Within this confusing nomenclature there are essentially three main subdivisions or models of non-pedogenic duricrust formation, all of which involve the role of subsurface water within either the vadose or phreatic zone. These are duricrusts produced in playa or pan environments (lacustrine models), those produced in riverine environments (fluvial or sheetflood models) and those produced at or below the water table (groundwater models) (Summerfield 1982; 1983; Goudie 1983). Despite operating in a range of geomorphological settings, groundwater plays a potentially important role in all three of these models with the boundaries between different types of model being frequently indistinct, as will be discussed below.

LACUSTRINE MODELS OF DURICRUST FORMATION

Lacustrine models of duricrust formation (Figure 15.11) have been discussed by, among others, Ambrose and Flint (1981) and Summerfield (1982; 1983). These models are especially applicable in arid and semi-arid environments where the high potential evaporation rates and fluctuating pH conditions associated with playa lakes or pans which contain standing water on a seasonal basis are conducive to duricrust formation (see Chapter 6). These conditions may combine to produce an environment which promotes the weathering, leaching and neoformation of pan-floor clay minerals which may act as a source material for duricrust formation (Summerfield 1983; Nash *et al.* 1994b). Additionally, playa- and pan-dwelling organisms such as diatoms have been suggested as an important source of silica for duricrusts developed in lacustrine settings (Du Toit 1954).

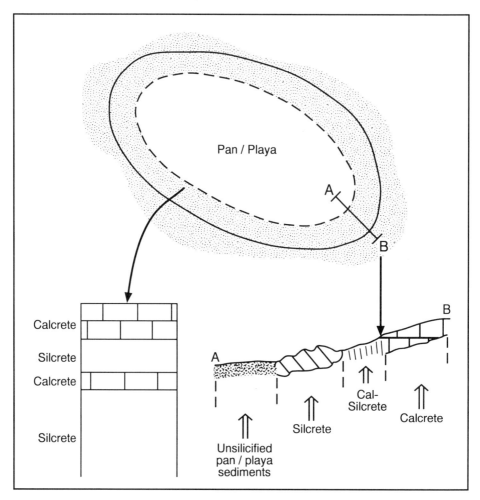

Figure 15.11 *The spatial relationships between different duricrust types associated with a pan/playa environment in the Kalahari (after Summerfield 1982)*

The association of lacustrine environments and duricrusts in drylands has been described from a number of regions. For example, Alley (1977) notes how silcrete surfaces in South Australia are best developed within the vicinity of lake basins, where high water tables occur, there is a focus of drainage, relatively slow movement of groundwater and high alkalinity. Many pans in the Kalahari are associated with subsurface duricrusts (Figure 15.12) which have formed at or near the water table due to the combination of carbonate- or silica-saturated surface waters interacting with groundwater (Butterworth 1982; Farr *et al.* 1982; Holmgren 1995). The landforms and chemical sediments associated with many playas in the USA and within the central Australian groundwater discharge zone have been shown to be directly related to the transmission and emergence of groundwater beneath and onto the playa surface (e.g. Arakel 1986; 1991; Jacobson *et al.* 1988; Arakel *et al.* 1989). Groundwater discharge onto playa surfaces is frequently associated with the deposition of carbonates as spring mounds (Pentecost and Viles 1994). Spring mounds are predominantly submerged features formed where Ca-rich exfiltrating waters and saline lake waters mix, leading to the precipitation of $CaCO_3$ (Bischoff *et al.* 1993).

Figure 15.12   *Partially silicified calcretes developed in a small pan on the northern edge of the Makgadikgadi Depression, Botswana*

Groundwater within a playa setting may also act as an agent in the diagenetic alteration of pre-existing duricrusts as a result of the movement of subsurface water through a duricrust mass. The most detailed studies of diagenetic alteration of duricrusts have been described from Australian palaeodrainage playas by Arakel (1986) and Arakel *et al.* (1989), where calcretes have been progressively silicified in association with fluctuating groundwater tables.

GROUNDWATER AND FLUVIAL MODELS OF DURICRUST FORMATION

Groundwater, fluvial and sheetflood models of duricrust formation have been suggested to explain the nature and distribution of many occurrences of silcrete and calcrete, including studies in Australia (e.g. Stephens 1971; Mann and Deutscher 1978; Mann and Horwitz 1979; Carlisle *et al.* 1978; Arakel and McConchie 1982; Carlisle 1983; Twidale and Milnes 1983; Arakel 1986; Arakel *et al.* 1989; Thiry and Milnes 1991; Milnes and Thiry 1992), southern Africa (e.g. Netterburg 1975; 1982; Carlisle *et al.* 1978; Carlisle 1983; Nash *et al.* 1994b; McCarthy and Ellery 1995), France (e.g. Thiry and Millot 1987; Thiry *et al.* 1988; Milnes and Thiry 1992) and Oman (Maizels 1987; 1988; 1990). In many cases it is difficult to distinguish between purely 'groundwater' duricrusts and 'fluvial/sheetflood' varieties since the two sets of models have considerable overlap. Duricrust accumulation by fluvial processes can involve deposition within channels or valleys, deposition from floods, and/or the lateral seepage of throughflow water or groundwater (Goudie 1983). In many cases, the mechanisms of duricrust formation are identical to the pan/lacustrine models described above with the fluvial system simply acting as a transfer mechanism for Si, Ca- and Mg-bicarbonate-bearing waters.

The close links between fluvial/sheetflood and groundwater models are clearly illustrated by considering three models put forward to explain the formation and silicification of groundwater calcretes in Australian palaeodrainages (e.g. Mann and Horwitz 1979), the occurrence of silcretes within sandstones in the Paris Basin (e.g. Thiry *et al.* 1988) and the development of calcretes and silcretes within the Okavango Delta of Botswana (McCarthy and Ellery 1995).

Valleys and drainage systems, like playas and pans, are potentially important sites for the formation of duricrusts in arid and semi-arid environments since water tables are generally closer to the surface in the vicinity of depressions. Mann and Horwitz (1979) describe a model for the development of groundwater calcretes within alluvially filled drainage lines with shallow water tables (Figure 15.13). In this model, carbonate deposition occurs within the phreatic zone with the progressive precipitation of calcium carbonate leading to the development of pods and domes of calcrete which may divert surface and subsurface drainage and outcrop above the valley floor. Such mounds have been widely noted in Australian valley calcretes (e.g. Mabbutt *et al.* 1963; Sanders 1973) and are commonplace in Kalahari palaeo-drainages (Nash 1992). They are considered to represent loci of upwelling groundwater, forming in a similar way to carbonate domes associated with playas (Carlisle 1983). Cementation occurs preferentially where basement irregularities (highs) bring groundwaters nearer to the surface and thus increase the potential for evaporation, evapotranspiration and degassing (Wright and Tucker 1991).

Where renewed fluvial activity has occurred following a period of calcretisation, pre-existing duricrusts may be dissected and isolated from the valley floor (Mann and Horwitz 1979; Nash *et al.* 1994b). This is a similar situation to that described by Thiry *et al.* (1988) for dissected silcrete lenses within the Fontainebleau Sand of the Paris Basin (Figure 15.14). While these silcretes are not thought to have developed under arid conditions, the model of their formation may be widely applicable to duricrusts developed in valley settings. The silcrete occurs as discrete lenses exposed within valley flanks

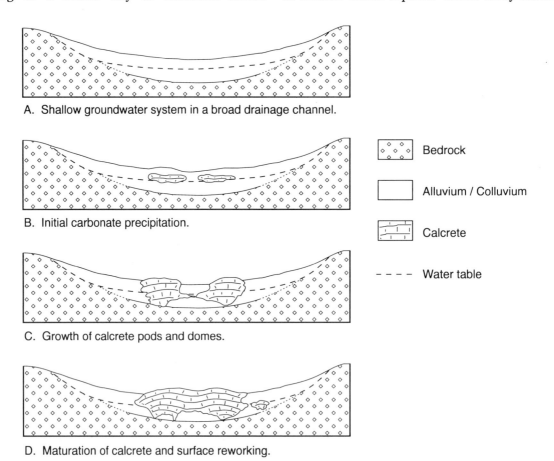

A. Shallow groundwater system in a broad drainage channel.

B. Initial carbonate precipitation.

C. Growth of calcrete pods and domes.

D. Maturation of calcrete and surface reworking.

Bedrock

Alluvium / Colluvium

Calcrete

- - - - Water table

Figure 15.13   *The stages of development of a groundwater calcrete (after Mann and Horwitz 1979)*

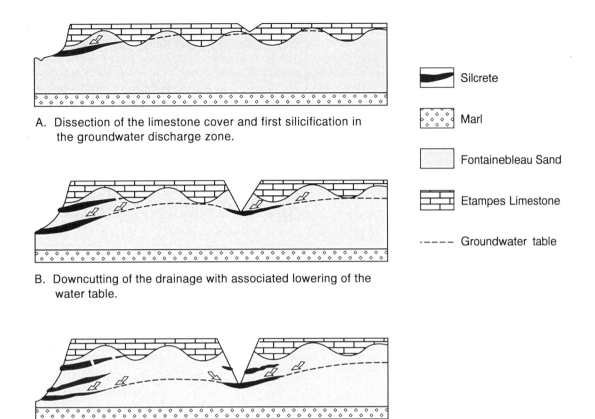

A. Dissection of the limestone cover and first silicification in the groundwater discharge zone.

B. Downcutting of the drainage with associated lowering of the water table.

C. The process is repeated, leading to a thicker leached profile above the water table and a deeper silcrete level. Note that uppermost (oldest) silcrete lenses show evidence of partial dissolution.

■ Silcrete

Marl

Fontainebleau Sand

Etampes Limestone

----- Groundwater table

Figure 15.14 *A model for the formation of groundwater silcretes based upon successive stages in the development of silcrete lenses in the Fontainebleau Sand of the Paris Basin (after Thiry et al. 1988)*

with the body of silcrete 'pinching out' within a short distance of the valley side, suggesting formation in association with the groundwater table. Silcrete lenses are superposed up the valley flanks as a result of progressive down-cutting due to dissection of the landscape, with the oldest lenses uppermost. The older lenses show the greatest signs of dissolution and weathering (Thiry and Millot 1987) while those lower down, closer to the valley floor, are comparatively fresh and unweathered. This is similar to superposed groundwater silcretes in South Australia described by Thiry and Milnes (1991) where former silicified horizons, stranded because of a lowering water table, experience dissolution by percolating water. The mobilised silica is then recrystallised and

incorporated into new groundwater silcretes at deeper levels associated with the lower water table.

The final example of a fluvial model of duricrust development is provided by the work of McCarthy and Ellery (1995) in the Oka-vango Delta (or Delta-fan) of northwest Botswana. The Okavango Delta, which has a total area of over 20 000 km^2 of permanent and seasonal swamps, is flooded between March and August as a result of rain falling over the Angolan Highlands. The Delta is in a region with a hot, dry climate (rainfall of 500 mm yr^{-1} and potential evapotranspiration of 1860 mm yr^{-1}; Wilson and Dincer 1976) but, as a result of the passage of the annual flood the dry sandy sediments at the distal margin of the fan receive an annual influx of silica-saturated water. High evaporation rates lead to increases in groundwater salinity levels within floodplain sediments which promotes the precipitation of amorphous silica (McCarthy and Metcalfe 1990) as surface floodwater gradually permeates down to the water table. Trees and other plants such as sedges are also suggested to play an important role in controlling groundwater salinity levels (McCarthy *et al.* 1991) and the precipitation of silica, primarily due to the production of silica phytoliths by

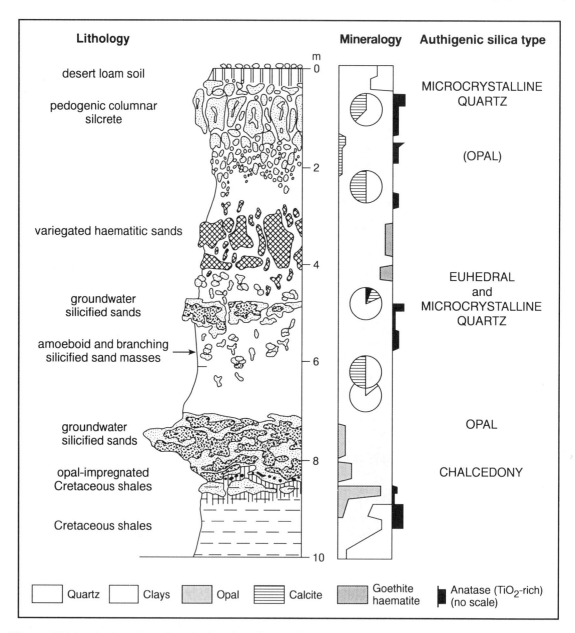

Figure 15.15   *A schematic section of pedogenic and non-pedogenic silcretes exposed in the Stuart Creek opal field, South Australia. The uppermost pedogenic silcretes exhibit a columnar structure as opposed to the more massive, lens-shaped non-pedogenic varieties (after Thiry and Milnes 1991; Milnes and Thiry 1992)*

absorption of silica from the groundwater into plant tissue. The processes leading to silcrete formation are the progressive illuviation of phytolithic silica into the sediment profile and the precipitation of silica from within the groundwater. The silica is introduced into the sediment by direct precipitation and by plants, with subsequent diagenetic alteration leading to the cementation of sediment. While sediments beneath the Okavango Delta are often only partly consolidated, the models suggested by McCarthy and Ellery (1995) may provide the first detailed explanation of the link between fluvial systems and groundwater in the generation of duricrusts.

CHARACTERISTICS OF GROUNDWATER DURICRUSTS

On the basis of the preceding discussion, it may seem likely that groundwater duricrusts would be clearly distinguishable from their pedogenic counterparts on the basis of their different modes of formation. In practice, however, this is not always the case. Groundwater calcretes and dolocretes are frequently difficult to distinguish from their pedogenic counterparts, although there are some macro- and micro-scale differences. Groundwater calcretes and dolocretes are typically massive, composed of micritic carbonate, are commonly silicified, (e.g. Jacobson *et al.* 1988; Arakel *et al.* 1989), are frequently deformed (Mann and Horwitz 1979) and are unlikely to appear as a typical mature pedogenic carbonate profile with peloidal structures (Wright and Tucker 1991). One possible method for distinguishing between pedogenic and non-pedogenic carbonates is through the analysis of the isotopic values in carbonates deposited in these different settings. Wright and Tucker (1991) suggest that carbonates derived from evolved groundwater which has interacted with local rock should have a heavier ^{18}O value compared with a pedogenic carbonate. However, it may also be the case that rapid infiltration of rainwater during storm events may make groundwater carbonates isotopically lighter than pedogenic varieties which have experienced strong evaporation.

It may also be possible to distinguish between groundwater and pedogenic silcretes

by analysing the silcrete structure (Figure 15.15). From studies of the Stuart Creek opal field in South Australia, Thiry and Milnes (1991) and Milnes and Thiry (1992) identify clear differences between silcretes formed near the land surface under the influence of the prevailing climate and varieties formed at depth in association with the water table. Pedogenic silcretes at Stuart Creek exhibit a distinctive profile, with columnar and illuviation structures suggesting a sequence of infiltration processes associated with alternating wet and dry conditions over a considerable period of development. In contrast, groundwater silcretes appear much more massive and uniform, with many of the primary sedimentary structures of the original host sediment preserved at least partially.

One of the best methods for distinguishing between pedogenic and non-pedogenic duricrusts on a broad scale is through analysis of the overall geometry of the duricrust body. This will, however, only be possible if a duricrust is exposed over a large area in a landscape with sufficient dissection to allow detailed investigation of considerable lengths of outcrop. As has been noted, non-pedogenic or groundwater duricrusts are commonly associated with certain geomorphological features. In the case of many calcretes in palaeodrainages, the result of this close association is that the duricrusts occur as narrow, elongate bodies. Mallick *et al.* (1981), Summerfield (1982) and Nash *et al.* (1994b) note that duricrusts within the Kalahari Group sediments are typically more extensively developed in the vicinity of former drainage lines and around lakes and pans, and this can be explained by the occurrence of local shallow water tables within these landscape components.

References

Acres, B., Blair Rains, A., King, R., Lawton, R., Mitchell, A. and Rackham, L. 1985. African dambos: their distribution, characteristics and use, *Zeitschrift für Geomorphologie*, Supplementband, 52: 63–86.

Ahnert, F. 1960. The influence of Pleistocene climates upon the morphology of cuesta scarps on the Colorado Plateau. *Annals of the Association of American Geographers*, 50: 139–156.

Alley, N.F. 1977. Age and origin of laterite and

silcrete duricrust and their relationship to episodic tectonism in the mid-north of South Australia, *Journal of the Geological Society of South Australia*, 24: 107–116.

Ambrose, G.J. and Flint, R.B. 1981. A regressive Miocene lake system and silicified strandlines in northern South Australia: implications for regional stratigraphy and silcrete genesis, *Journal of the Geological Society of Australia*, 28: 81–94.

Arakel, A.V. 1986. Evolution of calcrete in palaeo-drainages of the Lake Napperby area, central Australia. *Palaeogeography, Palaeoclimatology, Palaeoecology*, 54: 283-303.

Arakel, A.V. 1991. Evolution of Quaternary duricrusts in Karinga Creek drainage system, central Australian groundwater discharge zone. *Australian Journal of Earth Sciences*, 38: 333–347.

Arakel, A.V. and McConchie, D. 1982. Classification and genesis of calcrete and gypsite lithofacies in palaeo-drainage systems of inland Australia and their relationship to carnotite mineralisation. *Journal of Sedimentary Petrology*, 52: 1149–1170.

Arakel, A.V., Jacobson, G., Salehi, M. and Hill, C.M. 1989. Silicification of calcrete in palaeodrainage basins of the Australian arid zone. *Australian Journal of Earth Sciences*, 36: 73–89.

Baker, C.L. 1915. *Geology and underground waters of the northern Llano Estacado*. University of Texas Bulletin, 57.

Baker, V.R. 1980. Some terrestrial analogs to dry valley systems on Mars. *NASA Technical Memo*, TM 81776: 286–288.

Baker, V.R. 1982. *The channels of Mars*. University of Texas Press, Austin.

Baker, V.R. 1983. Large scale fluvial paleohydrology. In K.J. Gregory (ed.), *Background to Palaeohydrology*. Wiley, Chichester: 453–478.

Baker, V.R. 1985. Models of fluvial activity. In M. Woldenberg (ed.), *Models in geomorphology*. Allen & Unwin, London: 287–312.

Baker, V.R. 1990. Spring-sapping and valley network development, with case studies by R.C. Kochel, V.R. Baker, J.E. Laity and A.D. Howard. In C.G. Higgins and D.R. Coates (eds), *Groundwater geomorphology; the role of subsurface water in Earth-surface processes and landforms*. Geological Society of America, Special Paper, 252, Boulder, CO: 235–265.

Baker, V.R. and Kochel, R.C. 1979. Martian channel morphology; Maja and Kasei Vallis. *Journal of Geophysical Research*, 84: 7961–7983.

Baker, V.R., Carr, M.H., Gulick, V.C., Williams, C.R. and Marley, M.S. 1992. Channels and valley networks. In H.H. Kiefer, B.M. Jakosky, C.W. Snyder and M.S. Mathews (eds), *Mars*. University of Arizona Press, Tucson: 493–522.

Bates, R.L. and Jackson, J.A. (eds) 1980. *Glossary of geology*, 2nd edn. American Geological Institute, Falls Church, VA.

Belderson, R.H. 1983. Comment: drainage systems developed by sapping on Earth and Mars. *Geology*, 11: 55.

Bennett, H.H. 1939. *Soil conservation*. McGraw-Hill, New York.

Berger, D.L. 1992. Ground-water recharge through active sand dunes in northwestern Nevada. *Water Resources Bulletin*, 28: 959–965.

Bischoff, J.L., Fitzpatrick, J.A. and Rosenbauer, R.J. 1993. The solubility and stabilisation of ikaite $(CaCO_3 \cdot 6H_2O)$ from 0–25 °C – environmental and paleoclimatic implications for thinolite tufa. *Journal of Geology*, 101: 21–33.

Boast R. 1990. Dambos: a review. *Progress in Physical Geography*, 14: 153–177.

Boulaine J. 1954. La sebkha de Ben Ziane et sa 'lunette' ou Bourellet exemple de complex morphologique formé par la dégradation éolienne des sols salés. *Revue de Géomorphologique Dynamique*, 4: 102–123.

Bowler, J.M. 1986. Spatial variability and hydrological evolution of Australian lake basins: an analogue for Pleistocene hydrologic change and evaporite formation. *Palaeogeography, Palaeoclimatology, Palaeoecology*, 54: 21–41.

Bromley, J., Mannstrom, B., Nisca, D. and Jamtlid, A. 1994. Airborne geophysics — application to a groundwater study in Botswana. *Ground Water*, 32: 79–90.

Bromley, M. 1992. Topographic inversion of early interdune deposits, Navajo Sandstone (Lower Jurassic), Colorado Plateau, USA. *Sedimentary Geology*, 80: 1–25.

Brookfield, M.E. 1977. The origin of bounding surfaces in ancient aeolian sandstones. *Sedimentology*, 24: 303–332.

Brown, A.G. (ed.) (1995). *Geomorphology and groundwater*. Wiley, Chichester.

Bryan, R.B. and Yair, A. (eds), 1982. *Badland geomorphology and piping*. GeoBooks, Norwich.

Buckley, D.K. and Zeil, P. 1984. The character of fractured rock aquifers in eastern Botswana. In *Challenges in African hydrology and water resources*. Proceedings of the Harare Symposium July 1984. International Association of Hydrological Sciences Publication, 144: 25–36.

Bullock, A. 1992. Dambo hydrology in southern Africa — review and reassessment. *Journal of Hydrology*, 134: 373–396.

Butterworth, J.S. 1982. *The chemistry of Mogatse Pan — Kgalagadi District*. Botswana Department of Geological Survey, Unpublished Report, JSNB/14/82.

Carlisle, D. 1980. *Possible variation in the calcrete-gypcrete uranium model*. US Department of Energy Open-File Report GJBX-53(80).

Carlisle, D. 1983. Concentration of uranium and vanadium in calcretes and gypcretes. In R.C.L. Wilson (ed.), *Residual deposits: surface related weathering processes and materials.* Geological Society Special Publication 11, Blackwell Scientific, London: 185–195.

Carlisle, D., Merifield, P., Orme, A.R and Kolke, O. 1978. *The distribution of calcretes and gypcretes in the southwestern United States. Based on a study of deposits in Western Australia and South West Africa.* US Department of Energy Open File Report, GJBX-29 (78), subcontract 76-022-E, Grand Junction, Colorado.

Carr, M.H. 1980. Survey of Martian fluvial features. *NASA Technical Memo,* TM 81776: 265–267.

Carter, R.C. 1994. The groundwater hydrology of the Manga Grasslands, northeast Nigeria: importance to agricultural development strategy for the area. *Quarterly Journal of Engineering Geology,* 27: S73–S83.

Carter, R.C., Morgulis, E.D., Dottridge, J. and Agbo, J.U. 1994. Groundwater modelling with limited data: a case study in a semi-arid dunefield of northeast Nigeria. *Quarterly Journal of Engineering Geology,* 27: S85–S94.

Chan, M.A. and Kocurek, G. 1988. Complexities in eolian and marine interactions: processes and eustatic controls on erg development. *Sedimentary Geology,* 56: 283–300.

Clarke, J.D.A. 1994. Lake Lefroy, a palaeodrainage playa in Western Australia. *Australian Journal of Earth Sciences,* 41: 417–427.

Cooke, R.U. and Smalley, I.J. 1968. Salt weathering in deserts. *Nature,* 220: 1226–1227.

Cooke, R.U., Brunsden, D., Doornkamp, J.C. and Jones, D.K.C. 1982. *Urban geomorphology in drylands.* Oxford University Press, Oxford.

Coque, R. 1962. *La Tunisie pré-Saharienne, étude géomorphologique.* Colin, Paris.

Craddock, R.A. and Maxwell, T.A. 1993. Geomorphic evolution of the Martian Highlands through ancient fluvial processes. *Journal of Geophysical Research,* 98 (E2): 3453-3468.

Dunne, T. 1980. Function and control of channel networks. *Progress in Physical Geography,* 4: 211–239.

Dunne, T. 1990. Hydrology, mechanics, and geomorphic implications of erosion by subsurface flow. In C.G. Higgins and D.R. Coates (eds), *Groundwater geomorphology; the role of subsurface water in Earth-surface processes and landforms.* Geological Society of America, Special Paper, 252, Boulder, Colorado: 1–28.

Du Toit, A.L. 1954. *The Geology of South Africa.* Oliver and Boyd, Edinburgh.

El-Baz, F. 1982. *Desert landforms of southwest Egypt: a basis for comparison with Mars.* National Air and Space Museum Report, NASA-CR-3611.

El-Baz, F., Boulos, L., Breed, C., Dardir, A., Dowidar, H., El-Etr, H., Embabi, N., Grolier, M., Haynes, V., Ibrahim, M., Issawi, B., Maxwell, T., McCauley, J., McHugh, W., Moustafa, A. and Yousif, M. 1980. Journey to the Gilf Kebir and Uweinat, southwest Egypt. *Geographical Journal,* 146: 51–93.

Farr, J.L., Peart, R.J., Nelisse, G. and Butterworth, J.S. 1982. *Two Kalahari pans: a study of their morphology and evolution.* Botswana Department of Geological Survey Unpublished Report, GS 10/10.

Fryberger, S.G., Schenk, C.J. and Krystinik, L.F. 1988. Stokes surfaces and the effects of near-surface groundwater-table on aeolian deposition. *Sedimentology,* 35: 21–41.

Fryberger, S.G., Krystinik, L.F. and Schenk, C.J. 1990. Tidally flooded back-barrier dunefield, Guerrero Negro area, Baja California, Mexico. *Sedimentology,* 37: 23–43.

Gaylord, D.R. 1990. Holocene paleoclimatic fluctuations revealed from dune and interdune strata in Wyoming. *Journal of Arid Environments,* 18: 123–138.

Gieske, A. and Selaola, E. 1988. A proposed study of recharge processes in fractured aquifers of semi-arid Botswana. In I. Simmers (ed.), *Estimation of natural groundwater recharge.* NATO ISI Series C, Volume 222, Reidel, Dordrecht: 117–124.

Glennie, K.W. 1970. *Desert sedimentary environments.* Developments in Sedimentology 14, Elsevier, Amsterdam.

Gomez, B. and Mullen, V.T. 1992. An experimental study of sapped drainage network development. *Earth Surface Processes and Landforms,* 17: 465–476.

Goudie, A.S. 1983. Calcrete. In A.S. Goudie and K. Pye (eds), *Chemical sediments and geomorphology.* Academic Press, London: 93–132.

Goudie, A.S. and Thomas, D.S.G. 1985. Pans in southern Africa with particular reference to South Africa and Zimbabwe. *Zeitschrift für Geomorphologie,* 29: 1–19.

Goudie, A.S. and Thomas, D.S.G. 1986. Lunette dunes in southern Africa. *Journal of Arid Environments,* 10: 1–12.

Goudie, A.S. and Wells, G.L. 1995. The nature, distribution and formation of pans in arid zones. *Earth Science Reviews,* 38: 1–69.

Gulick, V.C. and Baker, V.R. 1990. Origin and evolution of valleys on Martian volcanoes. *Journal of Geophysical Research,* 95: 14325–14344.

Higgins, C.G. 1974. Model drainage networks developed by groundwater sapping. *Geological Society of America Abstracts with Programs,* 6: 794–795.

Higgins, C.G. 1982. Drainage systems developed by sapping on Earth and Mars. *Geology*, 10: 147–152.

Higgins, C.G. 1983. Reply: drainage systems developed by sapping on Earth and Mars. *Geology*, 11: 55–56.

Higgins, C.G. 1984. Piping and sapping: development of landforms by groundwater outflow. In R.G. La Fleur (ed.), *Groundwater as a geomorphic agent*. Allen & Unwin, London: 18–58.

Higgins, C.G. 1990. Seepage-induced cliff recession and regional denudation, with case studies by W.R. Osterkamp and C.G. Higgins. In C.G. Higgins and D.R. Coates (eds), *Groundwater geomorphology; the role of subsurface water in Earth-surface processes and landforms*. Geological Society of America, Special Paper, 252, Boulder, Colorado: 291–318.

Higgins, C.G. and Coates, D.R. (eds) 1990. *Groundwater geomorphology; the role of subsurface water in Earth-surface processes and landforms*. Geological Society of America, Special Paper, 252, Boulder, Colorado.

Hinds, N.E.A. 1925. Amphitheatre valley heads. *Journal of Geology*, 33: 816–818.

Holmgren, K. 1995. *Late Pleistocene climatic and environmental changes in central southern Africa*. Department of Physical Geography, Stockholm University, Dissertation Number 3.

Howard, A.D. 1988. Groundwater sapping experiments and modelling at the University of Virginia. In A.D. Howard, R.C. Kochel and H. Holt (eds), *Sapping features of the Colorado Plateau — a comparative planetary geology fieldguide*. NASA Publication, SP-491: 71–83.

Howard, A.D. and Kochel, A.D. 1988. Introduction to cuesta landforms and sapping processes on the Colorado Plateau. In A.D. Howard, R.C. Kochel and H. E. Holt (eds), *Sapping features of the Colorado Plateau — a comparative planetary geology fieldguide*. NASA Publication, SP-491: 6–56.

Howard, A.D. and McLane, C.F. 1988. Erosion of cohesionless sediment by groundwater seepage. *Water Resources Research*, 24: 1659–1674.

Howard, A.D. and Selby, M.J. 1994. Rock slopes. In A.D. Abrahams and A.J. Parsons (eds), *Geomorphology of desert environments*. Chapman & Hall, London: 123–172.

Howard, A.D., Kochel, R.C. and Holt, H.E. 1988. *Sapping features of the Colorado Plateau — a comparative planetary geology fieldguide*. NASA Publication, SP-491.

Hutchinson, J.N. 1968. Mass movement. In R.W. Fairbridge (ed.), *Encyclopedia of geomorphology*, Reinhold, New York.

Issar, A. 1983. Emerging groundwater, a triggering factor in the formation of the Makhteshim (erosion cirques) in the Negev and Sinai. *Israel Journal of Earth Sciences*, 32: 53–61.

Jacobson, G. and Jankowski, J. 1989. Groundwater-discharge processes at a central Australian playa. *Journal of Hydrology*, 105: 275–295.

Jacobson, G., Arakel, A.V. and Chen, Y.J. 1988. The central Australian groundwater discharge zone — evolution of associated calcrete and gypcrete deposits. *Australian Journal of Earth Sciences*, 35: 549–565.

Jacobson, G., Ferguson, J. and Evans, W.R. 1994. Groundwater-discharge playas of the Mallee Region, Murray Basin, southeast Australia. In M.R. Rosen (ed.), *Paleoclimate and basin evolution of playa systems*. Geological Society of America, Special Paper, 289: 81–96.

Jennings, J.N. 1979. Arnhem Land, city that never was. *Geographical Magazine*, 51: 822–827.

Judson, S. 1950. Depressions of the northern portion of the Southern High Plains of eastern New Mexico. *Bulletin of the Geological Society of America*, 61: 253–274.

Kinsman, D.J.J. 1969. Modes of formation, sedimentary associations, and diagnostic features of shallow-water and supratidal evaporites. *Bulletin of the American Association of Petroleum Geologists*, 53: 830–840.

Kochel, R.C. and Piper, J.F. 1986. Morphology of large valleys on Hawaii: evidence for groundwater sapping and comparisons with Martian valleys. *Journal of Geophysical Research*, 91: e175–e192.

Kochel, R.C., Howard, A.D. and McLane, C.F. 1985. Channel networks developed by groundwater sapping in fine-grained sediments: analogs to some Martian valleys. In M.J. Woldenberg (ed.), *Models in geomorphology*. Allen & Unwin, London: 313–341.

Kocurek, G. 1981a. Significance of interdune deposits and bounding surfaces in aeolian dune sands. *Sedimentology*, 28: 753–780.

Kocurek, G. 1981b. Erg reconstruction: the Entrada Sandstone (Jurassic) of Northern Utah and Colorado. *Palaeogeography, Palaeoclimatology, Palaeoecology*, 36: 125–153.

Kocurek, G. 1984. Reply — origin of first-order bounding surfaces in aeolian sandstones. *Sedimentology*, 31: 125–27.

Kocurek, G. 1988. First-order and super bounding surfaces in eolian sequences — bounding surfaces revisited. *Sedimentary Geology*, 56: 193–206.

Kocurek, G. 1991. Interpretation of ancient aeolian sand dunes. *Annual Review of Earth and Planetary Science*, 19: 43–75.

Kocurek, G. and Nielson, J. 1986. Conditions favourable for the formation of warm climate aeolian sand sheets. *Sedimentology*, 33: 795–816.

La Fleur, R.G. (ed.) 1984. *Groundwater as a geomorphic agent.* Allen & Unwin, London.

Laity, J.E. 1980. Groundwater sapping on the Colorado Plateau. In *Reports of Planetary Geology Programme 1980.* NASA Technical Memo, TM 82385: 358–360.

Laity, J.E. 1983. Diagenetic controls on groundwater sapping and valley formation, Colorado Plateau, as revealed by optical and electron microscope. *Physical Geography*, 4: 103–125.

Laity, J.E. 1988. The role of groundwater sapping in valley evolution on the Colorado Plateau. In A.D. Howard, R.C. Kochel and H.E. Holt (eds), *Sapping features of the Colorado Plateau — a comparative planetary geology fieldguide.* NASA Publication, SP-491: 63–70.

Laity, J.E. and Malin, M.C. 1985. Sapping processes and the development of theater-headed valley networks in the Colorado Plateau. *Bulletin of the Geological Society of America*, 96: 203–217.

Laity, J.E., Pieri, D.C. and Malin, M.C. 1980. Sapping processes in tributary valley systems. *NASA Technical Memo*, TM 81776: 295–297.

Loope, D.B. 1984. Origin of extensive bedding planes in aeolian sandstones: a defence of Stokes' hypothesis. *Sedimentology*, 31: 123–125.

Loope, D.B. 1985. Episodic deposition and preservation of eolian sands: a late Paleozoic example from southeastern Utah. *Geology*, 13: 73–76.

Loope, D.B. 1988. Rhizoliths in ancient eolianites. *Sedimentary Geology*, 56: 315–339.

Lowry, D.C. and Jennings, J.N. 1974. The Nullarbor karst Australia. *Zeitschrift für Geomorphologie*, 18: 35–81.

Mabbutt, J.A., Litchfield, W.H., Speck, N.H., Sofoulis, J., Wilcox, D.G., Arnold, J.M., Brookfield, M. and Wright, R.L. 1963. *General report on lands of the Wiluna–Meekatharra area, Western Australia, 1958.* Australia CSIRO Land Research Series No. 7.

Mäckel, R. 1974. Dambos: a study in morphodynamic activity on the plateau regions of Zambia. *Catena*, 1: 327–365.

Mader, D. and Yardley, M.J. 1985. Migration, modification and merging in aeolian systems and the significance of the depositional mechanisms in Permian and Triassic dune sands of Europe and North America. *Sedimentary Geology*, 43: 85–218.

Maizels, J. 1987. Plio-Pleistocene raised channel systems of the western Sharquiya (Wahiba), Oman. In L.E. Frostick and I. Reid (eds), *Desert sediments, ancient and modern.* Geological Society Special Publication, 35, London: 31–50.

Maizels, J. 1988. Palaeochannels: Plio-Pleistocene raised channel systems of the Western Sharqiyah. *Journal of Oman Studies*, Special Report, 3: 95–112.

Maizels, J. 1990. Raised channel systems as indicators of palaeohydrologic change: a case study from Oman. *Palaeogeography, Palaeoclimatology, Palaeoecology*, 76: 241–277.

Mallick, D.I.J., Habgood, F. and Skinner, A.C. 1981. *A geological interpretation of Landsat imagery and air photography of Botswana.* Institute of Geological Sciences, Overseas Geology and Mineral Resources, 56, London, HMSO.

Mann, A.W. and Deutscher, R.L. 1978. Genesis principles for the precipitation of carnotite in calcrete drainages in Western Australia. *Journal of the Geological Society of Australia*, 26: 293–303.

Mann, A.W. and Horwitz, R. 1979. Groundwater calcrete deposits in Australia: some observations from Western Australia. *Journal of the Geological Society of Australia*, 26: 293–303.

Mars Channel Working Group 1983. Channels and valleys on Mars. *Bulletin of the Geological Society of America*, 94: 1035–1054.

Maxwell, J.A. 1979. *Field investigation of Martian canyonlands in southwestern Egypt.* NASA Conference Publication, 2072.

McCarthy, T.S. and Ellery, W.N. 1995. Sedimentation on the distal reaches of the Okavango Fan, Botswana, and its bearing on calcrete and silcrete (ganister) formation. *Journal of Sedimentary Research*, A65: 77–90.

McCarthy, T.S. and Metcalfe, J. 1990. Chemical sedimentation in the Okavango Delta, Botswana. *Chemical Geology*, 89: 157–178.

McCarthy, T.S., McIver, J.R. and Verhagen, B.Th. 1991. Groundwater evolution, chemical sedimentation and carbonate brine formation on an island in the Okavango Delta swamp, Botswana. *Applied Geochemistry*, 6: 577–595.

McFarlane M.J. 1989. Dambos — their characteristics and geomorphological evolution in parts of Malawi and Zimbabwe, with particular reference to their role in the hydrogeological regime of surviving areas of African surface. *Proceedings of the Groundwater Exploration and Development in Crystalline Basement Aquifers Workshop*, Harare, Zimbabwe, 15–24 June 1987, Commonwealth Science Council 1 (Session 3): 254–308.

McFarlane, M.J. 1990. A review of the development of tropical weathering profiles with particular reference to leaching history and with examples from Malawi and Zimbabwe. *Proceedings of the Groundwater Exploration and Development in Crystalline Basement Aquifers Workshop*, Harare, Zimbabwe, 15–24 June 1987, Commonwealth Science Council 1 (Session 8): 93–145.

McFarlane, M.J. and Whitlow, R. 1990. Key factors affecting the initiation and progress of gullying in dambos in parts of Zimbabwe and Malawi. *Land Degradation and Rehabilitation*, 2: 215–235.

Milnes, A.R. and Thiry, M. 1992. Silcretes. In I.P. Martini and W. Chesworth (eds), *Weathering, soils and palaeosols*. Developments in Earth Surface Processes 2, Elsevier, Amsterdam: 349–377.

Milton, D.J. 1973. Water and processes of degradation in the Martian landscape. *Journal of Geophysical Research*, 78: 4037–4047.

Nash, D.J. 1992. The development and environmental significance of the dry valleys (mekgacha) in the Kalahari, central southern Africa. Unpublished PhD thesis, University of Sheffield.

Nash, D.J. 1995. Structural control and deep-weathering in the evolution of the dry valley systems of the Kalahari, central southern Africa. *Africa Geoscience Review*, 2: 9–23.

Nash, D.J. 1996. Groundwater sapping and valley development in the Hackness Hills, North Yorkshire, England. *Earth Surface Processes and Landforms*, 21: 781–795.

Nash, D.J., Thomas, D.S.G. and Shaw, P.A. 1994a. Timescales, environmental change and dryland valley development. In A.C. Millington and K. Pye (eds), *Environmental Change in Drylands*. Wiley, Chichester: 25–41.

Nash, D.J., Shaw, P.A. and Thomas, D.S.G. 1994b. Duricrust development and valley evolution: process–landform links in the Kalahari. *Earth Surface Processes and Landforms*. 19: 299–317.

Netterberg, F. 1975. Calcretes and silcretes at Sambio, Okavangoland. *South African Archaeological Bulletin*, 29: 83–88.

Netterberg, F. 1982. Calcretes and their decalcification around Rundu, Okavangoland, South West Africa. *Palaeoecology of Africa*, 15: 159–169.

Newell, W.L. 1970. Factors influencing the grain of the topography along the Willoughby Arch in northeastern Vermont. *Geografiska Annaler*, 52A: 103–112.

Onda, Y. 1994. Seepage erosion and its implication to the formation of amphitheatre valley heads: a case study at Obara, Japan. *Earth Surface Processes and Landforms*, 19: 627–640.

Osterkamp, W.R. and Wood, W.W. 1987. Playa lake basins on the Southern High Plains of Texas and New Mexico: Part 1. Hydrologic, geomorphic and geologic evidence for their development. *Bulletin of the Geological Society of America*, 99: 215–223.

Paine, J.G. 1994. Subsidence beneath a playa basin on the Southern High Plains, U.S.A.: evidence from shallow seismic data. *Bulletin of the Geological Society of America*, 106: 233–242.

Palmer, A.N. 1990. Groundwater processes in karst terranes. In C.G. Higgins and D.R. Coates (eds), *Groundwater geomorphology; the role of subsurface water in Earth-surface processes and landforms*. Geological Society of America, Special Paper, 252, Boulder, CO: 177–210.

Parker, G.G. Sr and Higgins, C.G. 1990. Piping and pseudokarst in drylands, with case studies by G.G. Parker, Sr. and W.W. Wood. In C.G. Higgins and D.R. Coates (eds), *Groundwater geomorphology; the role of subsurface water in Earth-surface processes and landforms*. Geological Society of America, Special Paper, 252, Boulder, CO: 77–110.

Peel, R.F. 1941. Denudational landforms of the central Libyan desert. *Journal of Geomorphology*, 4: 3–23.

Pentecost, A. and Viles, H. 1994. A review and reassessment of travertine classification. *Géographie physique et Quaternaire*, 48: 305–314.

Pieri, D.C. 1980. Martian valleys: morphology, distribution, age and origin. *Science*, 210: 895–897.

Pieri, D.C., Malin, M.C. and Laity, J.E. 1980. Sapping: network structure in terrestrial and Martian valleys. *NASA Technical Memo*, TM-81979: 292–293.

Pye, K. 1980 Beach salcrete and aeolian sand transport, evidence from North Queensland. *Journal of Sedimentary Petrology*, 50: 257–261.

Robb, J.M. 1990. Groundwater processes in the submarine environment. In C.G. Higgins and D.R. Coates (eds), *Groundwater geomorphology; the role of subsurface water in Earth-surface processes and landforms*. Geological Society of America, Special Paper, 252, Boulder, CO: 267–281.

Robb, J.M., O'Leary, D.W., Booth, J.S. and Kohout, F.A. 1982. Submarine spring sapping as a geomorphic agent on the East Coast Continental Slope. *Geological Society of America, Abstracts and Programs*, 14: 600.

Rosen, M.R. 1994. The importance of groundwater in playas: a review of playa classifications and the sedimentology and hydrology of playas. In M.R. Rosen (ed.), *Paleoclimate and basin evolution of playa systems*. Geological Society of America, Special Paper, 289: 1–18.

Rubin, D.M. and Hunter, R.E. 1982. Bedform climbing in theory and nature. *Sedimentology*, 29: 121–138.

Rubin, D.M. and Hunter, R.E. 1984. Reply. *Sedimentology*, 31: 128–132.

Sakura, Y., Mochizuki, M. and Kawasaki, I. 1987. Experimental studies on valley headwater erosion due to groundwater flow. *Geophysical Bulletin of Hokkaido University*, 49: 229–239.

Sanders, C.C. 1973. Hydrogeology of a calcrete deposit on Paroo Station, Wiluma, and surrounding areas. *Western Australia Geological Survey Annual Report 1972*: 15–26.

Schenk, C.J. and Fryberger, S.G. 1988. Early diagenesis in eolian dune and interdune sands at White Sands, New Mexico. *Sedimentary Geology*, 55: 109–120.

Schumm, S.A. and Phillips, L. 1986. Composite channels of the Canterbury Plains, New Zealand: a Martian analogue. *Geology*, 14: 326–330.

Schumm, S.A., Boyd, K.F., Wolff, C.G. and Spitz, W.J. 1995. A ground-water sapping landscape in the Florida Panhandle. *Geomorphology*, 12: 281–297.

Sharp, R.P. 1973. Mars: fretted and chaotic terrains. *Journal of Geophysical Research*, 78: 4063–4072.

Sharp, R.P. and Malin, M.C. 1985. Channels on Mars. *Bulletin of the Geological Society of America*, 86: 593–609.

Shaw, P.A. and de Vries, J.J. 1988. Duricrust, groundwater and valley development in the Kalahari of south-east Botswana. *Journal of Arid Environments*, 14: 245–254.

Small, R.J. 1964. The escarpment dry valleys of the Wiltshire Chalk. *Transactions of the Institute of British Geographers*, 34: 33–52.

Smith, B.J. 1987. An integrated approach to the weathering of limestone in an arid area and its role in landscape evolution: a case study from southeast Morocco. In V. Gardiner (ed.), *International Geomorphology 1986 Part 2*. Wiley, Chichester: 637–657.

Smith, B.J. 1988. Weathering of superficial limestone debris in a hot desert environment. *Geomorphology*, 1: 355–367.

Smith, B.J. 1994. Weathering processes and forms. In A.D. Abrahams and A.J. Parsons (eds), *Geomorphology of desert environments*. Chapman & Hall, London: 39–63.

Sparks, B.W. and Lewis, W.V. 1957–58. Escarpment dry valleys near Pegsdon, Hertfordshire. *Proceedings of the Geologists Association*, 68: 26–38.

Stephens, C.G. 1971. Laterite and silcrete in Australia. *Geoderma*, 5: 5–52.

Stiller, D. 1983. Comment: drainage systems developed by sapping on Earth and Mars. *Geology*, 11: 54–55.

Stokes, W.L. 1968. Multiple parallel truncation bedding planes — a feature of wind-deposited sandstone formations. *Journal of Sedimentary Petrology*, 38: 510–515.

Summerfield, M.A. 1982. Distribution, nature and probable genesis of silcrete in arid and semi-arid southern Africa. In D.H. Yaalon (ed.), *Aridic soils and geomorphic processes*. Catena Supplement, 1: 37–65.

Summerfield, M.A. 1983. Silcrete. In A.S. Goudie and K. Pye (eds), *Chemical sediments and geomorphology*. Academic Press, London: 59–91.

Talbot, M.R. 1980. Environmental responses to climatic change in the West African Sahel over the past 20 000 years. In M.A.J. Williams and H. Faure (eds), *The Sahara and the Nile*. A.A. Balkema, Rotterdam: 37–62.

Talbot, M.R. 1985. Major bounding surfaces in aeolian sandstones — a climatic model. *Sedimentology*, 32: 257–265.

Tanaka, K.L. and Chapman, M.G. 1992. Kasei Valles, Mars: interpretation of canyon materials and flood sources. *Proceedings of Lunar and Planetary Science*, 22: 73–83.

Thiry, M. and Millot, G. 1987. Mineralogical forms of silica and their sequence of formation in silcretes. *Journal of Sedimentary Petrology*, 57: 343–352.

Thiry, M. and Milnes, A.R. 1991. Pedogenic and groundwater silcretes at Stuart Creek opal field, South Australia. *Journal of Sedimentary Petrology*, 61: 111–127.

Thiry, M., Ayrault, M.B. and Grisoni, J. 1988. Ground-water silicification and leaching in sands: example of the Fontainebleau Sand (Oligocene) in the Paris Basin. *Bulletin of the Geological Society of America*, 100: 1283–1290.

Thomas, D.S.G., Nash, D.J., Shaw, P.A. and Van der Post, C. 1993. Present day lunette sediment cycling at Witpan in the arid southwestern Kalahari Desert. *Catena*, 20: 515–527.

Torgersen, T., De Deckker, P., Chivas, A.R. and Bowler, J.M. 1986. Salt lakes: a discussion of processes influencing palaeoenvironmental interpretations and recommendations for future study. *Palaeogeography, Palaeoclimatology, Palaeoecology*, 54: 7–19.

Twidale, C.R. and Milnes, A.R. 1983. Aspects of the distribution and disintegration of siliceous duricrusts in arid Australia. *Geologie en Mijnbouw*, 62: 373–382.

Uchupi, E. and Oldale, R.N. 1994. Spring sapping origin of the enigmatic relict valleys of Cape Cod and Martha's Vineyard and Nantucket Islands, Massachusetts. *Geomorphology*, 9: 83–95.

Von Hoyer, M., Keller, S. and Rehder, S. 1985. *Core borehole Lethlakeng 1*. Botswana Department of Geological Survey, Unpublished report MVH/4/85, Lobatse.

Whitlow J.R. 1985. Dambos in Zimbabwe. A review. *Zeitschrift für Geomorphologie*, Supplementband, 52: 115–146.

Wilson, B.H. and Dincer, T. 1976. An introduction to the hydrology and hydrography of the Okavango Delta. In *Proceedings of a Symposium on the Okavango Delta and its future utilisation*. Botswana Society, Gaborone: 33–48.

Wood, W.W. and Osterkamp, W.R. 1987. Playa lake basins on the Southern High Plains of Texas and New Mexico: Part 2. A hydrologic model and mass-balance arguments for their development. *Bulletin of the Geological Society of America*, 99: 224–230.

Wright, V.P. and Tucker, M.E. 1991. Calcretes: an

introduction. In V.P. Wright and M.E. Tucker (eds), *Calcretes*. International Association of Sedimentologists Reprint Series Volume 2, Blackwell Scientific Publications, Oxford: 1–22.

Young R.W. 1986. Sandstone terrain in a semi-arid littoral environment: the lower Murchison valley, Western Australia. *Australian Geographer*, 17: 143–153.

Zartman, R.E., Evans, P.W. and Ramsey, R.H. 1994. Playa lakes on the Southern High Plains in Texas: reevaluating infiltration. *Journal of Soil and Water Conservation*, 49: 299–301.

Section 4
The Work of the Wind

16

Sediment mobilisation by the wind

Giles F.S. Wiggs

Introduction

Wind can be a particularly effective medium for sediment movement in drylands where a relatively sparse vegetation cover and thin soils combine to create highly erosive conditions on highly erodible surfaces. Despite most desert winds being no stronger than winds in more humid climates (see Pye and Tsoar 1990), Cooke *et al.* (1993) suggest that these winds carry more sediment than any other geomorphological agent. The power of desert winds is underlined by the inexorable movement of sand in such massive dunefields as the Rub'al Khali (Empty Quarter) in Arabia, which covers 560 000 square kilometres, and the vast deposits of aeolian loess in China reaching thicknesses of 300 m (Derbyshire 1983; see Chapter 21).

To interpret the evolution of landscapes such as those mentioned above, the processes of aeolian erosion, transportation and deposition of sediment in drylands have to be understood. Understanding of aeolian processes has been inspired by the work of Bagnold (1941; 1953; 1956). Indeed, no discussion of aeolian processes is complete without reference to his north African research and for a comprehensive fundamental understanding it is necessary to

consult these works. There has been a great deal of research in the intervening 50 years, particularly in the 1960s and 1970s when the burgeoning oil industry in the Middle East required an understanding of dune movement for the protection of installations and communication links from sand burial (Livingstone 1990).

The study of aeolian processes, however, is not simply an attempt to predict dune movement for the sake of protecting arable land, oases and communications, although this is certainly one branch of the study (Watson 1985; Cooke and Doornkamp 1990). In recent years, an understanding of aeolian processes has been used to clarify a wide range of natural and environmental questions and problems. While improved understanding of aeolian processes is used in modelling the dynamics and distribution of dunes (Howard *et al.* 1977; Lancaster 1985; 1987; 1988; 1989; Livingstone 1989; Shehata *et al.* 1992; Bullard *et al.* 1995; Wiggs *et al.* 1995a, 1996), it has also proved vital in understanding and interpreting the significance of palaeosandstones (Havholm and Kocurek 1994). Furthermore, as well as informing the debates concerning desertification and land degradation today (Khalaf 1989; Thomas and Middleton 1994; Williams and Balling 1996),

Arid Zone Geomorphology: Process, Form and Change in Drylands, 2nd edition. Edited by David S. G. Thomas.

process studies are also important in understanding relict dune systems (see Chapter 26) in terms of their palaeoenvironmental significance and susceptibility to future environmental changes, including the potential effects of greenhouse warming (Ash and Wasson 1983; Thomas and Shaw 1991; Livingstone and Thomas 1993; Muhs and Maat 1993; Gaylord and Stetler 1994).

Following on from investigations of soil erosion and depletion from arable land, particularly in the United States (Chepil 1945; Chepil and Woodruff 1963; Lyles 1977; Fryrear and Saleh 1993), recent aeolian process research has focused on the transport and deposition of aeolian dust (Pye 1987; Goosens and Offer 1990; Nickling and Gillies 1993; Lee *et al.* 1994; Goosens 1995) as well as of sand. Further from home, studies of extraterrestrial sediment transport by White (1979) and Greeley and Iversen (1985) have improved our understanding of environmental conditions on other planets (see Chapter 29).

Sediment movement is essentially a function of both the power of the wind (*erosivity*) and grain characteristics holding particles in place (*erodibility*). In general terms, where erosivity is greater than erodibility, particle dislodgement and erosion will take place. This simple equation is complicated by a variety of surface characteristics (such as roughness, slope and moisture content) which affect both erosivity and erodibility. This chapter investigates factors influencing erosivity and erodibility in terms of their effect on sediment transport, erosion and deposition. Recent advances in our understanding of complicating factors such as vegetation growth and wind turbulence are also discussed.

The nature of windflow in deserts

As air moves over the surface of the Earth it is retarded by friction at its base and a *velocity profile* develops. The zone of flow where air is affected by surface friction is called the *boundary layer* and within this layer there is a gradation from zero velocity in a very thin layer at the surface to free stream velocity at a height beyond the effects of surface friction. The atmospheric boundary layer (ABL) is approximately one to two kilometres thick.

The structure of the velocity profile within the boundary layer is highly dependent on the type of airflow: laminar or turbulent. In *laminar* flow, there is little mixing between the different layers of fluid and faster layers slip over slower layers, while at the surface itself the air is stationary. With this type of flow, momentum transfer between layers of fluid is accomplished by means of molecular transfer. Slower moving molecules of air drift into faster moving layers, thus causing a drag on the faster airflow. Such a transfer of momentum produces shearing forces between layers of air.

Turbulent flow also comprises zero flow velocity at the surface, but the exchange of momentum in the boundary layer is achieved through the action of gusts and turbulent eddies mixing between layers. Such momentum exchange is far more efficient than the molecular exchange seen in laminar flow and this is represented by differing velocity profiles. The greater mixing in turbulent flow results in a steeper velocity gradient at the surface and hence higher shearing stresses (Figure 16.1).

Laminar and turbulent flows can be distinguished from each other mathematically by the Reynolds number (*Re*):

$$Re = \frac{\rho h v}{\nu} \qquad (16.1)$$

where: ρ = fluid density, h = flow depth, v = flow velocity and ν = viscosity.

An increasing Reynolds number signifies an increase in turbulent intensity and a predomination of inertial forces over viscous forces

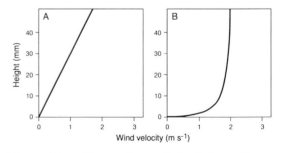

Figure 16.1　*Near-surface vertical velocity profiles showing the smaller shear stresses in laminar flow (A) when compared with turbulent flow (B). After Bagnold (1941)*

such that $Re < 500-2000 =$ laminar flow, and $Re \geqslant 2000 =$ turbulent flow.

In the atmosphere, airflow is nearly always turbulent because air has a low viscosity and boundary layer depths are normally quite high. Only in very viscous, slow or thin flows do laminar characteristics develop.

THE TURBULENT VELOCITY PROFILE

Under normal atmospheric conditions on flat, unvegetated surfaces and in the absence of intense solar heating, the turbulent velocity profile plots as a straight line on a semi-logarithmic chart (Figure 16.2). The gradient of the semi-logarithmic profile is a result of the *surface roughness* producing a drag on the overlying airflow. Hence, if the gradient of the velocity profile is known, the *shear stress* at the surface can be determined. In sedimentological research, a common method for describing the gradient of the velocity profile is in terms of the *shear velocity* (u_*). It is the value of u_* which is normally used in calculations for determining rates of sediment transport (Oke 1987). The shear velocity is proportional to the velocity profile gradient (Figure 16.2) and can be calculated simply from two velocities at known heights (Bagnold 1941). The shear velocity (u_*) is related to the actual surface shear stress (τ_0) by the following expression:

$$u_* = \sqrt{\tau_0/\rho} \qquad (16.2)$$

In Figure 16.2a the semi-logarithmic turbulent velocity profile does not reach the surface. This means that very close to the surface the wind velocity is zero. The height of this zero-velocity region is termed the *aerodynamic roughness length* (z_0) and it is an important parameter for it is a function of the surface roughness and it partly controls the gradient of the velocity profile, and hence u_*. For example, over a rougher surface (Figure 16.2b), the aerodynamic roughnesss (z_0) is larger, and with all other parameters remaining constant, the velocity profile gradients become steeper. Hence, shear velocities (u_*) increase and, consequent upon this, sediment transport would be likely to rise.

The relationship between aerodynamic roughness (z_0), shear velocity (u_*) and wind

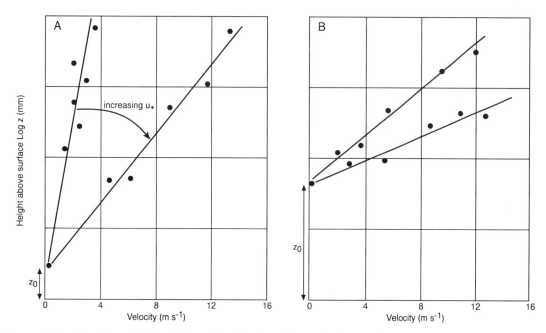

Figure 16.2 *Semi-logarithmic velocity profiles showing a focus at height z_0, the aerodynamic roughness. The aerodynamic roughness in (A) is less than that in (B), hence shear velocities (u_*) in (B) are greater. See text for details*

velocity (u) at a height (z) are described by the Karman–Prandtl velocity distribution, otherwise known as the 'law-of-the-wall':

$$\frac{u}{u_*} = \frac{1}{\kappa} \ln \frac{(z-d)}{z_0} \qquad (16.3)$$

where: κ = von Karman's constant (≈ 0.4) and d = zero-plane displacement. The *zero-plane displacement* (d) is discussed on page 347 and can largely be ignored on smooth and unvegetated desert surfaces.

MEASURING SHEAR VELOCITY (u_*)

Shear velocity (u_*) is commonly determined from regression analysis of time-averaged velocity measurements at several known heights. Wilkinson (1983/1984) reports that at least five velocity data points are required to determine u_* and z_0 with any statistical significance. Alternatively, if z_0 is known, then u_* can be determined from a velocity measurement at one known height (u_z) following the arguments of Bagnold (1941):

$$u_* = \frac{\kappa}{\ln z/z_0} u_z \qquad (16.4)$$

Such an approach has been widely used (e.g. Mulligan 1988) but has been questioned by Wiggs (1993) who showed that the method relied too heavily on changes in wind velocity alone and assumed a constant value of z_0. Widespread concern with this approach has also been published with circumstances where sand is being transported in the air (e.g. Gerety 1985) and these are discussed on page 364. Furthermore, use of equation 16.4 to calculate u_* relies on being able effectively to determine z_0, a far from easy task.

Further problems with measuring values of u_* occur where the effects of topography (e.g. dune slopes) produce a non-log-linear velocity profile above the surface. In such cases, the coupling between shear velocity measured at a known height above the surface and sediment transport occurring on that surface is not clear (see Mulligan 1988; Wiggs 1993; Wiggs *et al.* 1995b and Chapter 17).

Corrections to measured values of u_* also have to be made for the effect of thermal

stability or instability, which can significantly alter the shape of velocity profiles. If strong thermal heating or atmospheric stratification occurs, both of which are common in arid environments, then wind turbulence may be driven by buoyant forces rather than mechanical forces, and the velocity profile may become non-log-linear (Rasmussen *et al.* 1985; Greeley *et al.* 1991; Lancaster *et al.* 1991). Hence, shear velocity measured above the surface may not be representative of the near-surface shear velocity (Frank and Kocurek 1994). Rasmussen *et al.* (1985) found that omitting stability corrections could lead to errors in the measurement of u^* of the order of 15–25%, with subsequent estimates of sand transport in error by as much as 15 times (Frank and Kocurek 1994). Correction factors have been proposed by Businger *et al.* (1971), Dyer (1974) and Högström (1988) and involve the calculation of the Monin–Obukhov length or Richardson number, both of which require knowledge of the local atmospheric temperature gradient.

While correcting velocity profiles for the effects of stability is clearly important, Rasmussen *et al.* (1985) suggested that errors in measured velocity due to such instability was a function of height, and that at elevations below 0.5 m the correction required was unimportant. From their measurements in White Sands, New Mexico, Frank and Kocurek (1994) also found that correction factors were not necessary at wind speeds higher than about 10 m s^{-1}.

MEASURING AERODYNAMIC
ROUGHNESS (z_0)

The aerodynamic roughness (z_0) of desert surfaces varies widely both temporally and spatially. Typical values for z_0 may be 0.0007 m for stationary sand surfaces (Stull 1988), 0.003 m for surfaces with moving sand (Rasmussen *et al.* 1985) to 0.2 m and greater for vegetated or semi-vegetated desert surfaces (Wiggs *et al.* 1994). Owing to its importance in determining aeolian processes, it is a vital parameter to calculate correctly (Blumberg and Greeley 1993). Bagnold (1941) found a relationship between z_0 and surface roughness. He noted that $z_0 = d/30$ where d is the mean surface particle diameter. However, this relationship assumes a homogeneous surface and well-sorted sediment. Furthermore, it ignores the effect of roughness element spacing and scale where the aerodynamic roughness is determined perhaps by larger particles on a stone pavement. For example, Greeley and Iversen (1985) noted that z_0 may reach a maximum value of $d/8$ for widely spaced elements, before returning to a $d/30$ ratio as element spacing increased further (see Figure 16.3). Lancaster *et al.* (1991) found that z_0 varied in relation to changes in microtopography, suggesting that mean particle size was not a good determinant of z_0 for surfaces with a mix of particle sizes or large spatial variations in particle size.

Aerodynamic roughness has often been determined from the focus of velocity profiles as they intercept the *y*-axis, as shown in Figure 16.2. Such an approach has been shown to work well for flat and non-complex surfaces (Wiggs *et al.* 1994), but, as mentioned earlier, there is concern about the difficulty in defining such a focus when sand is being transported by the wind (Gerety 1985; McEwan 1993).

Blumberg and Greeley (1993) note that there is no clear or agreed method for calculating z_0, and that this is a response to the fact that surface roughnesses are extremely complex, difficult to describe and vary considerably spatially. They regard one of the most difficult hurdles as being able to take account of the varying scales of roughness found in nature. The problem is to combine the effects of a surface roughness change into the development of a velocity profile, for as there is a transition from a smooth to rough surface (or rough to smooth), a boundary layer grows downwind in response to that transition (Figure 16.4). Hence, different parts of the velocity profile are responding to different surface roughnesses (Panofsky 1984; Panofsky and Dutton 1984), with many subsidiary boundary layers responding to roughnesses provided at different scales by individual grains, surface ripples, or sand dunes.

In the example shown in Figure 16.4, the wind passes from a smooth surface (perhaps a flat sand sheet) to a rougher surface (such as coarse-grained ripples). At the smooth–rough transition, airflow close to the surface is immediately decelerated by the additional drag and the aerodynamic roughness (z_0) and surface shear velocity (u_{*0}) increase rapidly in response. As the internal boundary layer responding to the rougher surface grows downwind, shear velocity steadily decreases to reach an equilibrium level. It should be noted that such a rise in u_* may not necessarily lead to an increase in sediment transport or erosion because the increase in surface roughness may result in more grains lying below the height of z_0 and hence in a zone of zero wind velocity. However, there is some evidence that increased turbulence around large, non-erodible roughness elements may be important in sediment entrainment (Robert *et al.* 1992), at least in fluvial environments.

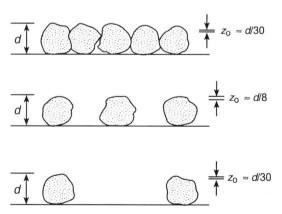

Figure 16.3 *Aerodynamic roughness height (z_0) as a function of roughness element spacing (after Greeley and Iversen 1985)*

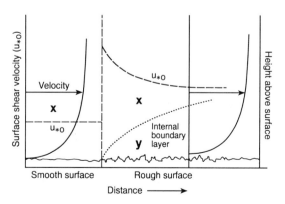

Figure 16.4 *The growth of an internal boundary in response to a changing surface roughness. See text for details (after Greeley and Iversen 1985)*

From Figure 16.4, it can be seen that at points marked X the wind is still in equilibrium with the smooth upstream surface, while at point Y the wind is responding to the rougher downstream surface. Hence, z_0 and u_* measured at any particular point in a velocity profile are functions not only of the size and spacing of surface roughness elements in the immediate locality, but also of changes in surface roughness within the fetch of the wind (Panofsky 1984; Blumberg and Greeley 1993).

Clearly, although z_0 is a critical value to determine in terms of assessing aeolian sediment transport, there are many difficulties in its successful calculation. There has been some recent success in interpreting values and spatial variations in z_0 from remote sensing. Blumberg and Greeley (1993) found correlations between regional variations in z_0 and radar backscatter in the Mojave Desert. A similar approach has been taken in Death Valley by Lancaster *et al.* (1991) and Greeley *et al.* (1991) who, in an ongoing experiment, are investigating aerodynamic roughness using 10-m-high anemometer towers (Figure 16.5), roughness element laser profiling technology and NASA Shuttle Imaging Radar.

Figure 16.5 *A 10-m-high anemometer array being used to assess the aerodynamic roughness of a stone surface in Death Valley, California (see Lancaster* et al. *1991; Greeley* et al. *1991)*

THE EFFECT OF VEGETATION ON VELOCITY PROFILES

Vegetation on desert surfaces acts as a surface roughness, providing significant drag on overlying airflow and considerably altering velocity profile parameters (Wiggs *et al.* 1994; 1995b). In such cases, vegetation growth may have a considerable protective role in terms of sediment transport (Ash and Wasson 1983; Wasson and Nanninga 1986; Wolfe and Nickling 1993; Wiggs *et al.* 1994; 1995a; see also Chapter 7).

Figure 16.6 shows the effect that a vegetated surface can have on velocity profiles. The additional drag on the airflow provided by the vegetation has displaced z_0 upwards, resulting in a higher shear velocity (u_*) above the vegetation canopy than would be expected over an unvegetated surface. From measurements on vegetated and bare dunes in the Kalahari Desert, Wiggs *et al.* (1994) found three-fold increases in above-canopy u_* on vegetated dunes. Vegetated surfaces are therefore seen to establish higher shear velocities than bare surfaces, at least above the vegetation canopies. However, Wiggs *et al.* (1994) also noted reductions in near-surface wind velocity by 200% on vegetated dunes. A question therefore arises as to the height above the surface at which the additional shear stresses are absorbed. It is in these cases that the zero-plane displacement height (d) (equation 16.3 and Figure 16.6) becomes important.

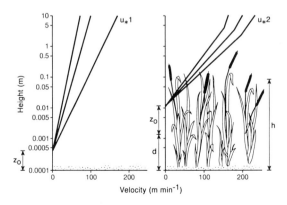

Figure 16.6 *Vegetation effect on velocity profile structure.* d = *zero-plane displacement,* h = *mean vegetation canopy height.* $u_{*2} > u_{*1}$ *(after Chepil and Woodruff 1963)*

The *zero-plane displacement* height is the level of the mean momentum sink and is the elevation at which the mean drag appears to act (Thom 1971; Jackson 1981). It represents the origin of the velocity profile and so all height measurements taken from the surface should have d subtracted from them (see equation 16.3). When velocity profiles are presented graphically (e.g. Figure 16.2), it is often assumed that the relevant zero-plane displacement has already been subtracted from the height data. Hence, the y-axis (height above the surface) in Figure 16.2 should properly be labelled 'log $z - d$ (mm)'. On many surfaces the value of d is too small to be concerned with. On vegetated surfaces, however, it can have a significant effect on velocity profiles and should be properly accounted for. For a large range of flows, the approximation $d/h = 0.7$ (where h is defined as the mean vegetation canopy height) is considered reasonably good (Jackson 1981). The distinction between d and z_0 can be understood more clearly in the following example. Where roughness elements (vegetation in this case) are widely spaced, one would expect the z_0 to be determined by the geometry of the roughness elements and d to be small, having a value just above the base of the roughness elements. At the opposite extreme, where roughness elements are so dense that their tops are touching, the flow would skim across the gaps, the apparent z_0 would decrease (being controlled only by the roughness of the tops of the canopies), d would approach the scale of the roughness elements, and u_* in the fluid trapped beneath the canopy would be zero (Jackson 1981; Lee 1991).

The real difficulty from a sediment-transport perspective is in determining the proportion of u_* which acts on the surface. Methods for partitioning the shear stress between that acting on the vegetation canopy and that affecting the surface have been proposed by Gillette and Stockton (1989), Musick and Gillette (1990), Stockton and Gillette (1990) and Wolfe and Nickling (1993). Many of these approaches, however, rely on mean time-averaged velocity measurements, while it has been shown that the majority of momentum transfer beneath vegetation canopies is accounted for by gusts of highly intermittent turbulence (Raupach 1989; 1991). Calculating the proportion of above-canopy shear stress which reaches the surface (and is thus available for sediment transport) under a variety of vegetation types, is therefore still an area of active research.

Further complications arise due to the pliable nature of many vegetation stands. The effective height of vegetation stands may vary in response to changing wind speeds. Greater wind speeds can bend grass stems, and hence produce a constant variation in both aerodynamic roughness and zero-plane displacement.

Sediment in air

GRAIN ENTRAINMENT

Sediment is entrained into the airflow when forces acting to move a stationary particle overcome the forces resisting sediment movement. The relevant forces for the entrainment of dry, bare sand are shown diagrammatically in Figure 16.7.

Particles are subjected to three forces of movement: lift, surface drag and form drag. Lift is a result of the air flowing directly over the particle forming a region of low pressure (in contrast to relatively high pressure beneath the particle), hence there is a tendency for the particle to be 'sucked' into the airflow. Surface drag is the shear stress on the particle provided by the velocity profile, and the form drag is also related to pressure differences around the particle. When these forces overcome the forces

of particle cohesion, packing and weight, the particle tends to shake in place and then lift-off, spinning into the airstream.

It has been shown that aerodynamic entrainment is primarily a function of the mean grain size of the particles involved and the shear velocity of the wind (Williams *et al.* 1990a). Bagnold (1941) studied these relationships and derived values of *critical threshold shear velocity* (u_{*ct}) for a wide range of particle sizes, with the principal determinant as the square root of grain diameter:

$$u_{*ct} = A \sqrt{\left[\frac{(\sigma - \rho)}{\rho}\right] gd} \qquad (16.5)$$

where σ = particle density, g = acceleration due to gravity, d = grain diameter and A = constant dependent upon the grain Reynolds number (≈ 0.1).

Figure 16.8 shows the relationship between fluid threshold shear velocity and grain size. From this figure, it can be seen that, in general, larger particles have a higher threshold of entrainment. However, smaller particles (with diameters less than about 0.06 mm) also require higher shear velocities to entrain them. This is because particles in this size range and

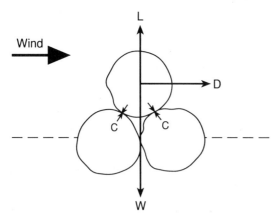

Figure 16.7 *Forces exerted on a grain by the wind.*
L = lift; D = drag; W = weight; C = inter-particle cohesion

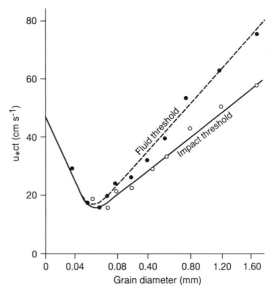

Figure 16.8 *The relationship between particle size* (d*)*
and threshold shear velocity (u$_{*ct}$*) showing the fluid and*
impact thresholds (after Chepil 1945). See text for details

smaller tend to have additional molecular and electrostatic forces of cohesion. They also have an affinity for the retention of moisture, often become protected from erosion by larger particles and frequently rest in the zero-velocity layer (beneath the height of z_0). The most susceptible grain size for entrainment is seen to be between about 0.04 and 0.40 mm, i.e. sand-sized particles. The heightened mobility of sediment in this size-range in drylands enables large volumes of sand to be moved and accumulate, resulting in the extensive dunefields present in many drylands (see Chapter 17).

The relationships shown in Figure 16.8 tend to result in a downwind fining of aeolian sediment both at the scale of the individual ripple and across whole sand seas. Such a downwind sorting can be used to identify process-form relationships; sand surfaces subject to aeolian erosion tend to consist of a coarse lag material (Nickling and McKenna-Neumann 1995), while those subject to aeolian deposition are commonly composed of fine material.

The discussion above refers to the value of the *static* or *fluid* critical threshold of entrainment where the grains are entrained only by the drag and lift forces of the moving air. However, one of the most significant influences on the entrainment of grains is the existence of grains already in motion. In particular, the impact of saltating grains on the bed surface imparts momentum to previously stationary grains, hence 'splashing' them into saltation (see later discussion). A second threshold of grain entrainment can therefore be identified; the *impact* (or *dynamic*) threshold which refers to the threshold of entrainment where other grains are already in motion.

Anderson and Haff (1988) note that the impact threshold is about 0.8 of the static threshold (see Figure 16.8) because saltating grains bring extra momentum from higher in the velocity profile to the surface. It is the action of impacting grains which drives the saltation process in this case, rather than the lift and drag forces of the wind.

Once entrainment of grains has been accomplished, however, the relative importance of the two processes of grain entrainment is still not entirely clear (Ungar and Haff 1987). Rice (1991) has shown that when impact dislodgement becomes established, direct fluid

entrainment still accounts for a significant proportion of entrainments. However, the consensus from empirical observations (Bagnold 1941; Willetts and Rice 1985a) and theoretical examinations (notably by Anderson and Haff 1988; Werner 1990; Haff and Anderson 1993; McEwan and Willetts 1994) is that grain impact is the principal mechanism by which saltation is maintained. Hence, once grains are entrained at the fluid threshold, a reduction in wind velocity will not necessarily result in a reduction in sand transport so long as the velocity remains above the impact threshold, as shown in Figure 16.8.

Surface modifications

While the relationships shown in Figure 16.8 have been found to be satisfactory for loose, dry, flat and homogeneous surfaces (Williams *et al.* 1994), the critical thresholds of motion on natural sediment beds are also influenced by variations in factors such as sediment mixtures, surface crusting, surface slope, moisture and vegetation. The relationships between these parameters are frequently complex and are not yet fully understood. For example, a surface crust will inhibit the entrainment of sediment into the wind. However, where a sediment bed contains a mixed grain-size population, a surface crust may readily be destroyed and abraded by the action of impacting sand grains, hence releasing smaller dust-sized particles into the airstream (Gillette 1978; 1981; 1986a; 1986b; Gillette *et al.* 1982; Nickling 1983). The effects of vegetation are dealt with elsewhere (see above and also Chapter 7). Two other particularly important modifications are provided by surface slope and moisture content.

BEDSLOPE

Despite the potential importance of surface slope on the threshold of entrainment of particles (some windward dune slopes may reach an angle of 30°), there has been relatively little empirical research in this area. Theoretical analyses have been presented by Allen (1982) and Dyer (1986), while Hardisty and

Whitehouse (1988) and Iversen and Rasmussen (1994) have used a portable field wind tunnel, and a tilting laboratory wind tunnel, respectively. Hardisty and Whitehouse (1988) found a good agreement between theory and practice. Figure 16.9 shows their data plotted against theoretical data, both indicating a relative increase in critical threshold on positive slopes (upslope) and a decrease on negative slopes (downslope). A similar, though more comprehensive data set is provided by Iversen and Rasmussen (1994), incorporating the effect of particle-size variation.

Surface slope may not only have an important effect on the *threshold* of sediment movement, but also on the *rate* of sediment transport. Despite much discussion of the potential effect (Hunt and Nalpanis 1985; Nalpanis 1985; Hardisty and Whitehouse 1988; Whitehouse and Hardisty 1988; Iversen and Rasmussen 1994), conclusions are still far from complete. Bagnold (1941) derived a simple geometric relationship to describe the effect of bedslope on sand transport rate:

$$q_s = q/\cos \theta (\tan \alpha + \text{tab } \theta) \qquad (16.6)$$

where q_s = sand transport rate on a sloping surface, q = sand transport rate on a flat surface, θ = bedslope angle and α = angle of repose of sand ($\approx 32-34°$).

Howard *et al.* (1977) studied the operation of this equation and found that it had only a minor influence on transport rate predictions and characterised actual sand transport rate no better than if bedslope was not taken into account. In contrast, Hardisty and Whitehouse (1988) found that sand transport rate was much more dependent upon surface slope than predicted by the Bagnold relationship (Figure 16.10).

MOISTURE CONTENT

Although the potential influence of moisture content on the threshold of entrainment has long been recognised, the exact physical nature of the relationship has still to be defined quantitatively (Sherman 1990; Arens 1994). The theoretical basis for the available models are that the critical shear velocity increases as a function of the increased surface tension associated with pore moisture (Kawata and Tsuchiya 1976). This is shown diagramatically in Figure 16.11. Both the Hotta *et al.* (1984) and Belly (1964) models

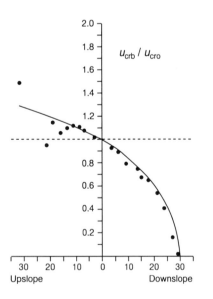

Figure 16.9 *The effect of bedslope on threshold of grain entrainment. Solid line = theoretical models of Allen (1982) and Dyer (1986). Circles = experimental data (from Hardisty and Whitehouse 1988)*

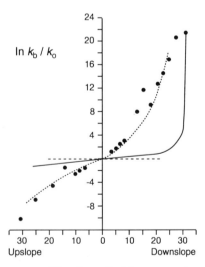

Figure 16.10 *The effect of bedslope on sand transport rate. Solid line = geometric relationship (Bagnold 1941). Circles = experimental data (from Hardisty and Whitehouse 1988)*

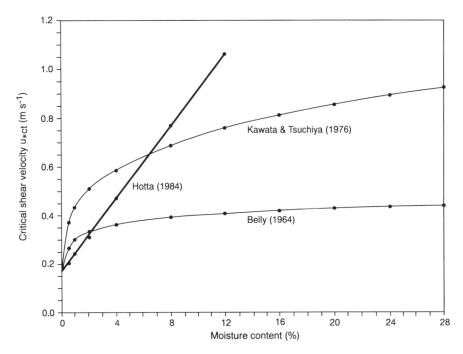

Figure 16.11   *Critical shear velocity for sediment entrainment as a function of sand moisture content (from Sherman 1990)*

are based on empirical wind tunnel data while that of Kawata and Tsuchiya (1976) is theoretical.

Figure 16.11 shows a great deal of scatter between the data of the different investigators. This is understandable as the action of water between individual grains is highly sensitive to the grain-size distribution of the sediment (McKenna-Neumann and Nickling 1989). However, from Figure 16.11 it can be seen that the models of Belly (1964) and Kawata and Tsuchiya (1976) both indicate large changes in critical threshold below about 8% moisture content. Once sediment has become entrained and is in transport, however, Sarre (1988; 1990) found that surface moisture contents of up to 14% had no appreciable effect on sand already in transport, stating that sand may saltate across significantly 'damp' surfaces without hindrance. Similar results have been reported by Svasek and Terwindt (1974). This is in contrast to Horikawa *et al.* (1984) who found that transport rates on sand with 3–4% moisture content were only 80% those of dry

sand.

Modes of sediment transport

Once entrained into the airflow, sediment may be transported by any one of four mechanisms, principally dependent upon the sediment grain size (Bagnold 1941 and Figure 16.12). The following modes of transport are not discrete classes and the transition from one to another may not be well-defined.

SUSPENSION

Small particles (of less than 0.06 mm) whose settling velocity may be very small in comparison to the combined effects of wind lift and drag might be transported in *suspension*. The turbulent motion of airflow can keep very fine sediment suspended for many days, high in the atmosphere, and it may ultimately be deposited as loess or dust (see Chapter 18). At the coarse end of this spectrum, some material may be transported in modified saltation (Hunt

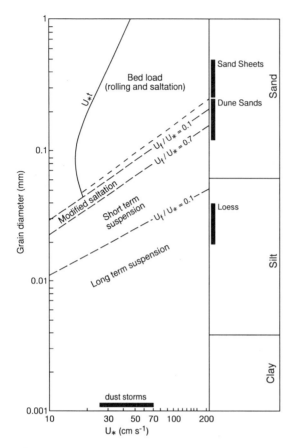

Figure 16.12 *The relationships between grain diameter, shear velocity and mode of sediment transport showing the distinction between the suspension of dust-sized sediment and the bedload transport of sand-sized sediment. U$_f$ is the particle fall velocity (after Tsoar and Pye 1987)*

and Nalpanis 1985) where saltation particle trajectories are affected by random wind turbulence (Figure 16.12).

CREEP

Larger particles tend to be transported by one of the three modes of bedload (or contact) transport; creep, reptation or saltation. Surface *creep* describes the rolling action of coarse particles (0.5–2.0 mm) as a result both of wind drag on the grain surfaces and the impact of high-velocity saltating grains. In the case of the finer particles in this size range, a creep

movement may become apparent immediately prior to the onset of saltation (Nickling 1983; Willetts and Rice 1985a). The difficulty in isolating creep in experimental observations has made the task of defining the relative importance of creep in terms of other transport mechanisms problematic. Willetts and Rice (1985b) estimate that creep accounts for approximately one-quarter of the bedload transport rate.

REPTATION

Recent research has identified an important transitional state between the modes of creep and saltation. The low-hopping of several grains consequent upon the high-velocity impact of a single saltating grain has been termed *reptation* (Anderson and Haff 1988). On impact and subsequent rebound, a saltating grain may lose 40% of its velocity (Willetts and Rice 1989; Anderson and Haff 1991; Haff and Anderson 1993). This energy imparted to the grain bed results in the ejection (or 'splashing') of perhaps 10 other grains (Werner and Haff 1988) with velocities at approximately 10% of the impact velocity, often too low to enter into saltation (Willetts and Rice 1989; Anderson and Haff 1991). Hence, each grain takes a single hop (Figure 16.13), the majority in a downwind direction (Haff and Anderson 1993). Much more research is required to fully understand this process, but Anderson *et al.* (1991) suggest

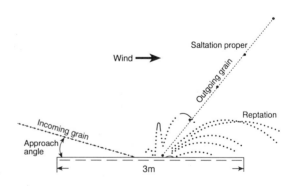

Figure 16.13 *The process of reptation where the impact of a high-velocity saltating grain ejects other grains into the airflow (after Anderson 1987)*

that it could be extremely important in near-surface aeolian transport.

SALTATION

The most intensively researched mode of aeolian transport is that of *saltation*. This is the characteristic ballistic trajectory of grains (≈ 0.06 mm-0.5 mm in diameter) as they are ejected from the grain bed, are given horizontal momentum by the airflow, descend to impact the grain bed and then continue 'leaping'

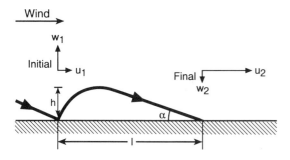

Figure 16.14 *The ballistic trajectory of a saltating sand grain.* w *and* u *represent vertical and horizontal velocities, respectively (after Bagnold 1941)*

downwind (see Figure 16.14). The impact of saltating grains on the surface often leads to the splash of other grains into the airflow which may then undergo reptation or, if sufficient momentum is transferred, saltation. Hence, saltation is a very efficient transporting mechanism whereby a few salting grains can rapidly induce mass transport of sediment in a cascading system (Nickling 1988). See Figure 16.15.

Bagnold (1936; 1941) was the first to appreciate the ballistic trajectory of saltating grains and a plethora of recent studies have attempted to model the subprocesses of grain entrainment, trajectory, bed collision and velocity profile modification, principally relying on wind tunnel and numerical modelling approaches (Owen 1964; Owen and Gillette 1985; Willetts and Rice 1986a; 1986b; 1989; Anderson and Hallett 1986; Werner and Haff 1986; 1988; Ungar and Haff 1987; Anderson and Haff 1988; Rasmussen and Mikkelsen 1988; 1991; Werner 1990; Anderson *et al.* 1991; McEwan *et al.* 1992; McEwan 1993; Haff and Anderson 1993; McEwan and Willetts 1994; Spies *et al.* 1995).

Bagnold (1941) and Chepil (1945) both

Figure 16.15 *Saltating sand being swept from the crest of an 80-m-high linear dune in the Namib Desert.*
Increasing wind speeds and saltation impacts result in progressively more intense saltation from dune foot to dune crest

identified steep initial take-off angles of grains at angles approaching 90°. However, more recent studies have recognised more normal lift-off angles of between 30 and 50°, while accepting that some grains are ejected at greater angles, back into the flow (White and Schultz 1977; Nalpanis 1985; Willetts and Rice 1985a; Anderson 1989). The actual grain trajectory depends very much on the height of bounce of the grain into the boundary layer. Wind speeds increase at a logarithmic rate away from the grain bed (see Figure 16.2), hence the higher a grain leaps, the more momentum it can extract from the airflow and the faster and longer is its saltating jump. Grain shape is also influential, with platy grains tending to have lower and longer paths than spherical grains (Willetts 1983). The length of jump is thought to be of the order of 12 to 15 times the height of bounce (Livingstone and Warren 1996) and the trajectory is also influenced if the grain starts to spin after a glancing impact. Reports of spin rates reaching over 400 r.p.s (White and Schultz 1977; White 1982) can induce a lift force (termed the Magnus effect) which may extend saltation trajectories.

Angles of descent of saltating particles have been shown to be of the order of 10 to 15°, with smaller impact angles being related to smaller grain sizes (Willetts and Rice 1985a; Anderson 1989). Ungar and Haff (1987) specified that the number of grains splashed up into the airstream by an impacting grain was proportional to the square of the impacting grain velocity, which may be up to five times its initial velocity (Anderson and Hallett 1986).

The height of the saltation layer is dependent on the value of u_*, the grain size in transport and surface characteristics. Bagnold (1941) recognised that saltation leaps were higher on a pebbly or hard rock surface than on a loose sand surface because hard surfaces are less absorbent of the grain momentum on each bounce. Pye and Tsoar (1990) quote a maximum saltation height on such surfaces of 3 m, with a mean saltation height of 0.2 m. However, the mass of saltating particles decreases very rapidly with height and has been described approximately by a declining exponential function (Chepil 1945; Rasmussen *et al.* 1985), with up to 80% of all transport taking place within 2 cm of the surface (Butterfield 1991).

Sand transport modification of the wind profile

The incorporation of sediment into the airstream has a complex and not yet fully understood effect on the shape and function of the velocity profile. Bagnold (1941) recognised that a saltation layer caused a near-surface kink in the velocity profile which shifted the velocity ray focus (shown in Figure 16.2) from z_0 to a higher, but consistent value of z_0'. Many investigators, however, have failed to identify such a focal point (Gerety 1985; Anderson and Haff 1991; McEwan 1993). Rather, it is now considered that the kink is caused by saltating grains extracting momentum from the air, carried in the form of a *grain-borne shear stress*, which acts as an additional roughness on the overlying boundary layer (McKenna-Neumann and Nickling 1994). The height of this kink increases as u_* increases because the height of the saltation layer grows. Gerety (1985) remarked that a rise in u_* must therefore be associated with a rise in z_0'.

In order to satisfy the concept of an increasing z_0' with increasing u_* when saltation is occurring, several workers (Rasmussen *et al.* 1985; Anderson and Haff 1991; McEwan 1993) have turned their attention to the relationship applied by Owen (1964):

$$z_0' = C_0 u_*^2 / 2g \qquad (16.7)$$

where: z_0' = aerodynamic roughness height during saltation and C_0 = constant (≈ 0.02).

In this case, the depth of saltation (associated with the aerodynamic roughness height, z_0') is related to the lift-off velocity of the individual sand grains, which in turn is governed by the shear velocity (u_*) and the force of gravity. While this relationship has been widely used to determine u_* in wind tunnel studies, it has yet to be fully utilised in field situations where the constant C_0 is difficult to define (McEwan and Willetts 1993).

The Owen (1964) equation has been used as the basis for constructing self-regulatory models of saltation (Anderson and Haff 1988; 1991; Werner 1990; McEwan and Willetts 1991; 1993). These steady-state models work on the premise that as u_* (and hence saltation) increases, so too does the effective aerodynamic roughness (z_0'). This increased aerodynamic roughness exerts an extra drag on the airflow and this effect propagates upwards through the velocity profile as an internal boundary layer. This results in near-surface winds being reduced, eventually to reach a steady-state value where as many grains are leaving the surface as are falling onto it (Anderson *et al.* 1991). An equilibrium is therefore established between u_*, z_0' and saltation load.

It has been shown both numerically and experimentally that the primary response time for saltating grains to saturate the airflow and for z_0' to rise to a peak is approximately one second (Anderson and Haff 1988; Gillette and Stockton 1989; Butterfield 1991; 1993) as shown in Figure 16.16. A secondary response time also exists, in which saltation flux decays in tandem with decreasing u_* and z_0'. This phase is of the order of tens of seconds and

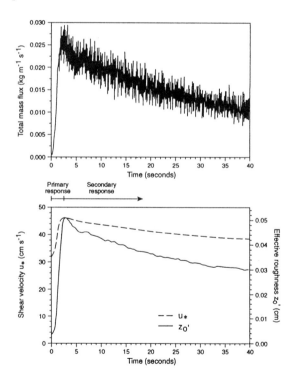

Figure 16.16 *The primary and secondary response of mass flux (top), shear velocity and effective surface roughness (bottom) to an increase in shear velocity at t=0–3 seconds. Calculated by the self-regulatory numerical saltation model of McEwan and Willetts (1991; 1993)*

has been explained as the period during which the three parameters find their steady-state equilibrium (McEwan and Willetts 1991; 1993; Butterfield 1993).

Prediction and modelling of bulk sediment transport

There are many equations to calculate mass sand flux from wind velocity data, nearly all derived from theoretical or wind tunnel experimental work (see Greeley and Iversen 1985 for a comprehensive review). All the expressions tend to the form of:

$$q = A u_*^3 \qquad (16.8)$$

The two types of relationship frequently used to calculate sand transport rate are typified by the expressions of Bagnold (1941) and Kawamura (1951):

$$q = C(d/D)^{0.5} u_*^3 \rho/g \quad \text{(Bagnold 1941)} \quad (16.9)$$

$$q = K_k (u_* - u_{*ct})(u_* + u_{*ct})^2 \rho/g$$
$$\text{(Kawamura 1951)} \quad (16.10)$$

where: q = sand transport rate $(g\,m^{-1}\,s^{-1})$, C = constant (1.8 for naturally graded dune sand), d = grain diameter (mm), D = standard grain diameter (0.25 mm), ρ = air density, K_k = constant (2.78) and u_{*ct} = threshold of grain entrainment.

Well-known problems with the Bagnold (1941) expression include the fact that it predicts sand movement below the threshold of entrainment and it commonly predicts rates which are considered too low at high values of shear velocity (Sarre 1987). Owing to the inclusion of a threshold term in the Kawamura (1951) calculation, it should be more accurate at lower levels of shear velocity. However, this expression only incorporates the effect of grain size in the threshold term (u_{*ct}) despite the fact that it is also likely to have an important effect on the transfer of momentum on grain impact.

Zingg (1953) followed a similar argument to Bagnold (1941) but used a $\frac{3}{4}$ power function:

$$q = C_2 (d/D)^{3/4} u_*^3 \rho/g \qquad (16.11)$$

where C_2 = constant (0.83).

Owen (1964) found that this expression was more accurate over a wider range of particle sizes than the Bagnold (1941) formula, but it still omits a term for the threshold of entrainment. Such a term was incorporated into a later analysis of Bagnold (1956) which was subsequently refined by Lettau and Lettau (1978) into a very popularly used expression:

$$q = C_3 (d/D)^{0.5} (u_* - u_{*ct}) u_*^2 \rho/g \quad (16.12)$$

where C_3 = constant (4.2).

There has been very little empirical testing of these relationships in the field, and that which has been accomplished (Sarre 1987; Sherman 1990; Wiggs 1992) has shown a great deal of variation between the observed rates and those predicted as a function of u_* (Figure 16.17). This is not surprising when one considers the complex nature of saltation which is only beginning to be unravelled and the fact that all of the expressions are based on numerical or wind-tunnel experimental observations with homogeneous sediment beds. No allowance is made for complexities arising from variations in terrain, vegetation, moisture levels or turbulence.

Further complications arise with the accurate measurement of sand transport rates in the field. In order to measure sand transport rate, a collecting device has to be inserted into the airflow for a known period of time, but because traps inevitably cause airflow disturbance they are unable to intercept all of the sand in transport. Two major problems with sand traps are those of back pressure and scouring of sand around the base of traps. To reduce these errors, the majority of traps are of a thin vertical design (hence presenting a minimum disturbance to the wind flow) and incorporate the bleeding of air from within the trap to reduce back-pressure. Most designs are based on Bagnold's collector (e.g. Chepil and Woodruff 1963; Gillette and Goodwin 1974) which collects particles through a vertical slot about 1 cm wide and up to 80 cm high. A modified version consists of vertical separators so that the vertical flux can also be established (Kawamura 1951).

The efficiencies (i.e. the proportion of moving sand collected) of the various types of trap vary considerably. That of the Bagnold trap has been reported to be as low as 20% (Knott and Warren 1981) and as high as 60% for the modified version of Gillette and Goodwin (1974). A more recent design by Leatherman (1978) reportedly achieves a 70% efficiency (Marston 1986). However, the determination of efficiencies is fraught with problems, and Jones and Willetts (1979) have shown that small differences in the operation of traps in the field may give rise to large contrasts in efficiency.

Experimental problems such as these have made it very difficult to define a simple relationship linking u^* to bulk sediment transport, despite advances in our understanding of the physics of sediment transport.

The role of turbulence

The majority of relationships discussed to this point concerning wind action on sediment beds have involved the calculation of 'average' conditions. The turbulent velocity profile, however, by definition, consists of momentum transfer through turbulent eddies with some frequencies of 10 Hz or greater. Furthermore, while Raupach (1989; 1991) has shown the importance of turbulence in providing momentum beneath vegetation canopies, Nickling (1978) has suggested that dust entrainment is highly susceptible to wind turbulence.

The difficulties encountered in measuring this 'gustiness' of the wind has necessitated that

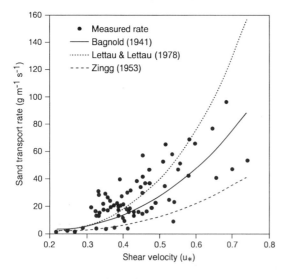

Figure 16.17 *Relationship between shear velocity, measured sand transport rate and that calculated from three sand transport rate formulae (after Wiggs 1992)*

field, laboratory and numerical modelling rely on statistical averages of wind over time, with many relationships between shear velocity and sediment transport being calculated over periods of between one minute (Wiggs 1993) and one hour (Tsoar 1985). Recent investigations modelling saltation dynamics (see above) and wind-tunnel studies of the interaction between fluctuating wind velocities and sediment flux (Butterfield 1991; 1993) have now begun to identify relevant timescales of the order of a second, as discussed earlier (see Figure 16.16).

Investigations by Williams *et al.* (1990b), Willetts *et al.* (1991) and McEwan and Willetts (1993) have shown that the entrainment of grains into saltation is greatly influenced by peak values of instantaneous shear stress, driven by near-surface burst and sweep flow structures. Butterfield (1993) found that gusts of wind could induce 'spikes' in sediment transport several times the magnitude of the mean mass flux. It is these peaks in turbulent energy which determine the threshold of entrainment.

Butterfield (1991; 1993) also used a wind tunnel to investigate the response of sand transport to fluctuations in velocity above the entrainment threshold. Some of his results are shown in Figure 16.18. Gust sequences produced a rapidly varying shear velocity with parallel responses from both mass flux (q) and effective aerodynamic roughness (z_0'). Butterfield found that mass flux and z_0' responded within about one second to an acceleration in flow, but that the response was several seconds longer in a decelerating flow due to the momentum of the grains. Furthermore, as discussed above, the secondary response of q, z_0' and u_* to a change in wind velocity in which a boundary layer propagates upwards in response to a changing effective aerodynamic roughness, may take tens of seconds, by which time further velocity variations may have taken place. These findings led Butterfield (1993) to conclude that with naturally fluctuating winds over a sand bed, the grain-laden boundary layer is in constant adjustment, and may rarely achieve an equilibrium state.

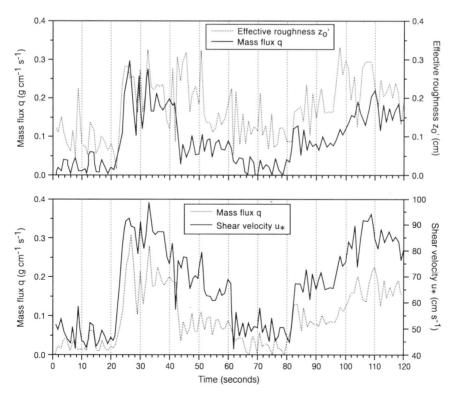

Figure 16.18   *The response of mass flux* (q)*, effective aerodynamic roughness* (z_0') *and shear velocity* (u_*) *to wind velocity gust sequences (from Butterfield 1993)*

An understanding of the complex role of wind turbulence in the entrainment and transportation of sediment is still in its infancy. Butterfield (1993) notes that too little is currently known about turbulence within the saltation layer and about the occurrence of near-bed flow structures, although they undoubtedly affect grain movement. Without further research in these areas, it is uncertain whether improvements in sediment transport rate predictions can be accomplished.

Conclusions

Great advances have been made in the understanding of the processes of sediment entrainment and transportation in the last 50 years. With the reductionist nature of much recent research, a future goal is to be able to satisfactorily link small-scale studies of individual grain movements to large-scale geomorphic questions and problems. Meanwhile, investigations concerning aeolian processes can only become more important. An appreciation of such processes is vital for the prediction and understanding of the effects of human settlement and activities on wind erosion, particularly in marginal and susceptible tracts of arid and semi-arid land.

References

Allen, J.R.L. 1982. Simple models for the shape and symmetry of tidal sand waves: (1) statically stable equilibrium forms. *Marine Geology*, 48: 31–49.

Anderson, R.S. 1987. Eolian sediment transport as a stochastic process: the effects of a fluctuating wind on particle trajectories. *Journal of Geology*, 95: 497–512.

Anderson, R.S. 1989. Saltation of sand: a qualitative review with biological analogy. In C.H. Gimmingham, W. Ritchie, B.B. Willetts and A.J. Willis (eds), *Symposium: coastal sand dunes*. Royal Society of Edinburgh, Proceedings, B96, 149–165.

Anderson, R.S. and Haff, P.K. 1988. Simulation of eolian saltation. *Science*, 241: 820–823.

Anderson, R.S. and Haff, P.K. 1991. Wind modification and bed response during saltation of sand in air. *Acta Mechanica*, Supplement, 1: 21–51.

Anderson, R.S. and Hallett, B. 1986. Sediment transport by wind: toward a general model.

Bulletin of the Geological Society of America, 97: 523–535.

Anderson, R.S., Sørensen, M. and Willetts, B.B. 1991. A review of recent progress in our understanding of aeolian sediment transport. *Acta Mechanica*, Supplement, 1: 1–19.

Arens, S.M. 1994. *Aeolian processes in the Dutch foredunes*. Landscape and Environmental Research Group, University of Amsterdam, Amsterdam.

Ash, J.E. and Wasson, R.H. 1983. Vegetation and sand mobility in the Australian desert dunefield. *Zeitschrift für Geomorphologie*, Supplementband, 45: 7–25.

Bagnold, R.A. 1936. The movement of desert sand. *Proceedings of the Royal Society of London*, A157: 594–620.

Bagnold, R.A. 1941. *The physics of blown sand and desert dunes*. Methuen, London.

Bagnold, R.A. 1953. *The surface movement of blown sand in relation to meteorology*. Research Council of Israel, Special Publication, 2: 89–93.

Bagnold, R.A. 1956. The flow of cohesionless grains in fluids. *Philosophical Transactions of the Royal Society of London*, A249: 29–297.

Belly, P.Y. 1964. *Sand movement by wind*. US Army Coastal Engineering Research Center, Technical Memo 1.

Blumberg, D.G. and Greeley, R. 1993. Field studies of aerodynamic roughness length. *Journal of Arid Environments*, 25: 39–48.

Bullard, J.E., Thomas, D.S.G., Livingstone, I. and Wiggs, G.F.S. 1995. Analysis of linear sand dune morphological variability, southwestern Kalahari Desert. *Geomorphology*, 11: 189–203.

Businger, J.A., Wyngaard, J.C., Izumi, Y. and Bradley, E.F. 1971. Flux-profile relationships in the atmospheric surface layer. *Journal of Atmospheric Sciences*, 28: 181–189.

Butterfield, G.R. 1991. Grain transport rates in steady and unsteady turbulent airflows. *Acta Mechanica*, Supplement, 1: 97–122.

Butterfield, G.R. 1993. Sand transport response to fluctuating wind velocity. In N.J. Clifford, J.R. French and J. Hardisty, (eds), *Turbulence, perspectives on flow and sediment transport*, Wiley Chichester: 305–335.

Chepil, W.S. 1945. Dynamics of wind erosion: 1. Nature of movement of soil by wind. *Soil Science*, 60: 305–320.

Chepil, W.S. and Woodruff, N.P. 1963. The physics of wind erosion and its control. *Advances in Agronomy*, 15: 211–302.

Cooke R.U. and Doornkamp, J.C. 1990. *Geomorphology in environmental management*, 2nd edn. Oxford University Press, Oxford.

Cooke R.U., Warren, A. and Goudie, A.S. 1993. *Desert geomorphology*. UCL Press, London.

Derbyshire, E. 1983. On the morphology, sediments and origin of the Loess Plateau of central China. In R.A.M. Gardner and H. Scoging (eds), *Megageomorphology*. Oxford University Press, Oxford: 172–194.

Dyer, A.J. 1974. A review of flux–profile relationships. *Boundary-Layer Meteorology*, 7: 363–372.

Dyer, K. 1986. *Coastal and estuarine sediment dynamics*. Wiley, Chichester.

Frank, A. and Kocurek, G. 1994. Effects of atmospheric conditions on wind profiles and aeolian sand transport with an example from White Sands National Monument. *Earth Surface Processes and Landforms*, 19: 735–745.

Fryrear, D.W. and Saleh, A. 1993. Field wind erosion. *Soil Science*, 155 (4): 294–300.

Gaylord, D.R. and Stetler, L.D. 1994. Aeolian-climatic thresholds and sand dunes at the Hanford Site, south-central Washington, USA. *Journal of Arid Environments*, 28: 95–116.

Gerety, K.M. 1985. Problems with determination of u^* from wind-velocity profiles measured in experiments with saltation. *Proceedings of International Workshop on Physics of Blown Sand*, Memoirs 8, Department of Theoretical Statistics, Aarhus University, Denmark.

Gillette, D.A. 1978. A wind tunnel simulation of the erosion of soil; effect of soil moisture, sand-blasting, wind speed, and soil consolidation on dust production. *Atmospheric Environments*, 12: 1735–1743.

Gillette, D.A. 1981. Production of dust that may be carried great distances. *Geological Society of America, Special Paper*, 186: 11–26.

Gillette, D.A. 1986a. Production of dust. In F. El-Baz and M.H.A. Hassan (eds), *Physics of desertification*. Martinus Nijhoff, The Hague: 251–260.

Gillette, D.A. 1986b. Threshold velocities for dust production. In F. El-Baz and M.H.A. Hassan (eds), M.H.A. *Physics of desertification*. Martinus Nijhoff, The Hague: 321–326.

Gillette, D.A. and Goodwin, P.A. 1974. Microscale transport of sand-sized soil aggregates eroded by wind. *Journal of Geophysical Research*, 7927: 4080–4084.

Gillette, D.A. and Stockton, P.A. 1989. The effects of non-erodible particles on wind erosion of erodible surfaces. *Journal of Geophysical Research*, 94 (D10), 12885–12893.

Gillette, D.A., Adams, J., Muhs, D. and Kihl, R. 1982. Threshold friction velocities and rupture moduli for crusted desert soils for the input of soil particles to the air. *Journal of Geophysical Research*, 87C: 9003–9015.

Goosens, D. 1995. Field experiments of aeolian dust accumulation on rock fragment substrata. *Sedimentology*, 42: 391–402.

Goosens, D. and Offer, Z.I. 1990. A wind tunnel simulation and field verification of desert dust deposition (Avdat Experimental Station, Negev Desert). *Sedimentology*, 37: 7–22.

Greeley, R. and Iversen, J.D. 1985. *Wind as a geomorphological process*. Cambridge University Press: Cambridge.

Greeley, R., Dobrovolskis, A., Gaddis, L., Iversen, J., Lancaster, N., Leach, R., Rasmussen, S., Saunders, S., VanZyl, J., Wall, S., White, B. and Zebker, H. 1991. *Radar-Aeolian Roughness Project*, NASA Contractor Report, 4378.

Haff, P.K. and Anderson, R.S. 1993. Grain scale simulations of loose sedimentary beds: the example of grain-bed impacts in aeolian saltation. *Sedimentology*, 40: 175–198.

Hardisty, J. and Whitehouse, R.J.S. 1988. Evidence for a new sand transport process from experiments on Saharan dunes. *Nature*, 332: 532–534.

Havholm, K.G. and Kocurek, G. 1994. Factors controlling aeolian sequence stratigraphy: clues from super bounding surface features in the Middle Jurassic Page Sandstone. *Sedimentology*, 41: 913–934.

Högström, U. 1988. Non-dimensional wind and temperature profiles in the atmospheric surface layer: a re-evaluation. *Boundary-Layer Meteorology*, 42: 55–78.

Horikawa, K., Hotta, S., Kubota, S. and Katori, S. 1984. Field measurement of blown sand transport rate by trench trap. *Coastal Engineering (Japan)*, 27: 214–232.

Hotta, S., Kubota, S., Katori, S. and Horikawa, K. 1984. Blown sand on a wet sand surface. *19th Coastal Engineering Conference, Proceedings*. American Society of Civil Engineers, New York: 1265–1281.

Howard, A.D., Morton, J.B., Gad-el-Hak, M. and Pierce, D.B. 1977. *Simulation model of erosion and deposition on a barchan dune*. NASA Contractor Report, CR-2838.

Hunt, J.C.R. and Nalpanis, P. 1985. Saltating and suspended particles over flat and sloping surfaces. i. modelling concepts. *Proceedings of International Workshop on Physics of Blown Sand*, Memoirs 8, Department of Theoretical Statistics, Aarhus University, Denmark.

Iversen, J.D. and Rasmussen, K.R. 1994. The effect of surface slope on saltation threshold. *Sedimentology*, 41: 721–728.

Jackson, P.S. 1981. On the displacement height in the logarithmic velocity profile. *Journal of Fluid Mechanics*, 111: 15–25.

Jones, J.R. and Willetts, B.B. 1979. Errors in measuring uniform aeolian sandflow by means of an adjustable trap. *Sedimentology*, 26: 463–468.

Kawamura, R. 1951. *Study of sand movement by wind* (in Japanese). Report of the Institute of Science and Technology, University of Tokyo. Translated to English in NASA Technical Transactions, F14.

Kawata, Y. and Tsuchiya, Y. 1976. Influence of water content on the threshold of sand movement and the rate of sand transport in blown sand. *Proceedings of the Japanese Society of Civil Engineers*, 249: 95–100. (in Japanese).

Khalaf, F.I. 1989. Textural characteristics and genesis of aeolian sediments in the Kuwait desert. *Sedimentology*, 36: 253–271.

Knott, P. and Warren, A. 1981. Aeolian processes. In A. Goudie (ed.), *Geomorphological techniques*. Allen & Unwin, London: 226–246.

Lancaster, N. 1985. Variations in wind velocity and sand transport on the windward flanks of desert sand dunes. *Sedimentology*, 32: 581–593.

Lancaster, N. 1987. Reply: variations in wind velocity and sand transport on the windward flanks of desert dunes. *Sedimentology*, 34: 511–520.

Lancaster, N. 1988. Controls of eolian dune size and spacing. *Geology*, 16: 972–975.

Lancaster, N. 1989. The dynamics of star dunes: an example from the Gran Desierto, Mexico. *Sedimentology*, 36: 273–289.

Lancaster, N., Greeley, R. and Rasmussen, K.R. 1991. Interaction between unvegetated desert surfaces and the atmospheric boundary layer: a preliminary assessment. *Acta Mechanica*, Supplement, 2: 89–102.

Leatherman, S.P. 1978. A new aeolian sand trap design. *Sedimentology*, 25: 303–306.

Lee, J.A. 1991. The role of desert shrub size and spacing on wind profile parameters. *Physical Geography*, 12: 72–89.

Lee, J.A., Allen, B.L., Peterson, R.E., Gregory, J.M. and Moffett, K.E. 1994. Environmental controls on blowing dust direction at Lubbock, Texas, USA. *Earth Surface Processes and Landforms*, 19: 437–449.

Lettau, K. and Lettau, H.H. 1978. Experimental and micrometeorological field studies on dune migration. In H.H. Lettau and K. Lettau (eds), *Exploring the world's driest climate*, University of Wisconsin-Madison, Institute for Environmental Studies, Report 101: 110–147.

Livingstone, I. 1989. Monitoring surface change on a Namib linear dune. *Earth Surface Processes and Landforms*, 14: 317–332.

Livingstone, I. 1990. Desert sand dune dynamics: review and prospect. In M.K. Seely (ed.), *Namib ecology: 25 years of Namib research*. Transvaal Museum Monograph, 7: 47–53.

Livingstone, I. and Thomas, D.S.G. 1993. Modes of linear dune activity and their palaeoenvironmental significance: an evaluation with reference to southern African examples. In K. Pye (ed.), *The dynamics and environmental context of aeolian sedimentary systems*. Special Publication of the

Geological Society of London, 72. Geological Society, London: 91–101.

Livingstone, I. and Warren, A. 1996. *Aeolian geomorphology: an introduction*. Longman, London.

Lyles, L. 1977. Wind erosion: processes and effects on soil productivity. *Transactions of the American Society of Agricultural Engineering*, 20: 880–884.

Marston, R.A. 1986. Maneuver-caused wind erosion impacts, south-central New Mexico. In W.G. Nickling (ed.), *Aeolian geomorphology*. Proceedings of the 17th Annual Binghampton Symposium: 273–306.

McEwan, I.K. 1993. Bagnold's kink: a physical feature of a wind velocity profile modified by blowing sand. *Earth Surface Processes and Landforms*, 18: 145–156.

McEwan, I.K. and Willetts, B.B. 1991. Numerical model of the saltation cloud. *Acta Mechanica, Supplement*, 1: 53–66.

McEwan, I.K. and Willetts, B.B. 1993. Sand transport by wind: a review of the current conceptual model. In K. Pye (ed.), *The dynamics and environmental context of aeolian sedimentary systems*. Geological Society of London, Special Publication, 72: 7–16.

McEwan, I.K. and Willetts, B.B. 1994. On the prediction of bed-load sand transport rate in air. *Sedimentology*, 41: 1241–1251.

McEwan, I.K., Willetts, B.B. and Rice, M.A. 1992. The grain/bed collision in sand transport by wind. *Sedimentology*, 39: 971–981.

McKenna-Neumann, C. and Nickling, W.G. 1989. A theoretical and wind tunnel investigation of the effect of capillary water on the entrainment of soil by wind. *Canadian Journal of Soil Science*, 69: 79–96.

McKenna-Neumann, C. and Nickling, W.G. 1994. Momentum extraction with saltation: implications for experimental evaluation of wind profile parameters. *Boundary-Layer Meteorology*, 68: 35–50.

Muhs, D.R. and Maat, P.B. 1993. The potential response of eolian sands to greenhouse warming and precipitation reduction on the Great Plains of the USA. *Journal of Arid Environments*, 25: 351–361.

Mulligan, K.R. 1988. Velocity profiles measured on the windward slope of a transverse dune. *Earth Surface Processes and Landforms*, 13: 573–582.

Musick, H.B. and Gillette, D.A. 1990. Field evaluation of relationships between a vegetation structural parameter and sheltering against wind erosion. *Land Degradation and Rehabilitation*, 2: 87–94.

Nalpanis, P. 1985. Saltating and suspended particles over flat and sloping surfaces ii. experiments and numerical simulations. In *Proceedings of International Workshop on Physics of Blown Sand*,

Memoirs 8, Department of Theoretical Statistics, Aarhus University, Denmark: 37–66.

Nickling, W.G. 1978. Eolian sediment transport during dust storms: Slims River Valley, Yukon Territory. *Canadian Journal of Earth Sciences*, 15: 1069–1084.

Nickling, W.G. 1983. Grain size characteristics of sediment transported during dust storms. *Journal of Sedimentary Petrology*, 53: 1011–1024.

Nickling, W.G. 1988. The initiation of particle movement by wind. *Sedimentology*, 35: 499–511.

Nickling, W.G. and Gillies, J.A. 1993. Dust emission and transport in Mali, West Africa. *Sedimentology*, 40: 859–868.

Nickling, W.G. and McKenna-Neumann, C. 1995. Development of deflation lag surfaces. *Sedimentology*, 42: 403–414.

Oke, T.R. 1987. *Boundary layer climates*. Methuen, New York.

Owen, P.R. 1964. Saltation of uniform grains in air. *Journal of Fluid Mechanics*, 20: 225–242.

Owen, P.R. and Gillette, D. 1985. Wind tunnel constraint on saltation. In *Proceedings of International Workshop on Physics of Blown Sand*, Memoirs 8, Department of Theoretical Statistics, Aarhus University, Denmark: 253–269.

Panofsky, H.A. 1984. Vertical variation of the roughness length at the Boulder Atmospheric Observatory. *Boundary-Layer Meteorology*, 28: 305–308.

Panofsky, H.A. and Dutton, J.A. 1984. *Atmospheric turbulence: models and methods for engineering applications*. Wiley, New York.

Pye, K. 1987. *Aeolian dust and dust deposits*. Academic Press, London.

Pye, K. and Tsoar, H. 1990. *Aeolian sand and sand dunes*. Unwin Hyman, London.

Rasmussen, K.R. and Mikkelsen, H.E. 1988. Aeolian transport in a boundary layer wind tunnel. *Geoskrifter*, 29, Geologisk Institut Aarhus Universitet, Denmark.

Rasmussen, K.R. and Mikkelsen, H.E. 1991. Wind tunnel observations of aeolian transport rates. *Acta Mechanica*, Supplement, 1: 135–144.

Rasmussen, K.R., Sørensen, M. and Willetts, B.B. 1985. Measurement of saltation and wind strength on beaches. In *Proceedings of International Workshop on Physics of Blown Sand*, Memoirs 8, Department of Theoretical Statistics, Aarhus University, Denmark: 301–325.

Raupach, M.R. 1989. Stand overstorey processes. *Philosophical Transactions of the Royal Society of London*, 324B: 175–190.

Raupach, M.R. 1991. Saltation layers, vegetation canopies and roughness lengths. *Acta Mechanica*, Supplement, 1: 83–96.

Rice, M.A. 1991. Grain shape effects on aeolian sediment transport. *Acta Mechanica*, Supplement, 1: 159–166.

Robert, A., Roy, A.G. and De Serres, B. 1992. Changes in velocity profiles at roughness transitions in coarse grained channels. *Sedimentology*, 39: 725–735.

Sarre, R.D. 1987. Aeolian sand transport. *Progress in Physical Geography*, 11: 157–181.

Sarre, R.D. 1988. Evaluation of aeolian sand transport equations using intertidal-zone measurements, Saunton Sands, England. *Sedimentology*, 35: 671–679.

Sarre, R.D. 1990. Evaluation of aeolian sand transport equations using intertidal-zone measurements, Saunton Sands, England — reply. *Sedimentology*, 37: 389–392.

Shehata, W., Bader, T., Irtem, O., Ali, A., Abdallah, M. and Aftab, S. 1992. Rate and mode of barchan dune advance in the central part of the Jafurah sand sea. *Journal of Arid Environments*, 23: 1–17.

Sherman, D.J. 1990. Discussion: evaluation of aeolian sand transport equations using intertidal-zone measurements, Saunton Sands, England — discussion. *Sedimentology*, 37: 385–392.

Spies, P.J., McEwan, I. and Butterfield, G.R. 1995. On wind velocity profile measurements taken in wind tunnels with saltating grains. *Sedimentology*, 42: 515–521.

Stockton, P.H. and Gillette, D.A. 1990. Field measurement of the sheltering effect of vegetation on erodible land surfaces. *Land Degradation and Rehabilitation*, 2: 77–85.

Stull, R.B. 1988. *An introduction to boundary layer meteorology*. Kluwer, Dordrecht.

Svasek, J.N. and Terwindt, J.H.J. 1974. Measurements of sand transport by wind on a natural beach. *Sedimentology*, 21: 311–322.

Thom, A.S. 1971. Momentum absorption by vegetation. *Quarterly Journal of the Royal Meteorological Society*, 97: 414–428.

Thomas, D.S.G. and Middleton, N.J. 1994. *Desertification: exploding the myth*. Wiley, Chichester.

Thomas, D.S.G. and Shaw, P. 1991. *The Kalahari Environment*. Cambridge University Press, Cambridge.

Tsoar, H. 1985. Profiles analysis of sand dunes and their steady state signification. *Geografiska Annaler*, 67A: 47–61.

Tsoar, H. and Pye, K. 1987. Dust transport and the question of desert loess formation. *Sedimentology*, 34: 139–154.

Ungar, J.E. and Haff, P.K. 1987. Steady state saltation in air. *Sedimentology*, 34: 289–299.

Wasson, R.J. and Nanninga, P.M. 1986. Estimating wind transport of sand on vegetated surfaces. *Earth Surface Processes and Landforms*, 11: 505–514.

Watson, A. 1985. The control of wind blown sand and moving dunes: a review of the methods of sand control in deserts, with observations from Saudi Arabia. *Quarterly Journal of Engineering Geology*, 18: 237–252.

Werner, B.T. 1990. A steady-state model of wind-blown sand transport. *Journal of Geology*, 98 (1): 1–17.

Werner, B.T. and Haff, P.K. 1986. A simulation study of the low energy ejecta resulting from single impacts in eolian saltation. In *Proceedings ASCE Conference on Advancements in Aerodynamics, Fluid Mechanics and Hydraulics*: 337–345.

Werner, B.T. and Haff, P.K. 1988. The impact process in eolian saltation: two dimensional simulations. *Sedimentology*, 35: 189–196.

White, B.R. 1979. Soil transport by winds on Mars. *Journal of Geophysical Research*, 84 (B9): 4643–4651.

White, B.R. 1982. Two-phase measurements of saltating turbulent boundary layer flow. *International Journal of Multiphase Flow*, 9: 459–473.

White, B.R. and Schultz, J.C. 1977. Magnus effect in saltation. *Journal of Fluid Mechanics*, 81: 497–512.

Whitehouse, R.J.S. and Hardisty, J. 1988. Experimental assessment of two theories for the effect of bedslope on the threshold of bedload transport. *Marine Geology*, 79, 135–139.

Wiggs, G.F.S. 1992. Sand dune dynamics: field experimentation, mathematical modelling and wind tunnel testing. Unpublished PhD thesis, University of London.

Wiggs, G.F.S. 1993. Desert dune dynamics and the evaluation of shear velocity: an integrated approach. In K. Pye (ed.), *The dynamics and environmental context of aeolian sedimentary systems*. Geological Society of London, Special Publication, 72: 37–46.

Wiggs, G.F.S., Livingstone, I., Thomas, D.S.G. and Bullard, J.E. 1994. The effect of vegetation removal on airflow structure and dune mobility in the southwest Kalahari. *Land Degradation and Rehabilitation*, 5: 13–24.

Wiggs, G.F.S., Thomas, D.S.G., Bullard, J.E. and Livingstone, I. 1995a. Dune mobility and vegetation cover in the southwest Kalahari desert. *Earth Surface Processes and Landforms*, 20 (6): 515–529.

Wiggs, G.F.S., Livingstone, I., Thomas, D.S.G. and Bullard, J.E. 1995b. Airflow characteristics over partially-vegetated linear dunes in the southwest Kalahari desert. *Earth Surface Processes and Landforms*, 21: 19–34.

Wiggs, G.F.S., Livingstone, I. and Warren, A. 1996. The role of streamline curvature in sand dune dynamics: evidence from field and wind tunnel measurements. *Geomorphology*, 21: 19–34.

Wilkinson, R.H. 1983/1984. A method for evaluating statistical errors associated with logarithmic velocity profiles. *Geo-Marine Letters*, 3: 49–52.

Willetts, B.B. 1983. Transport by wind of granular materials of different grain shapes and densities. *Sedimentology*, 30: 669–679.

Willetts, B.B. and Rice, M.A. 1985a. Inter-saltation collisions. In *Proceedings of International Workshop on the Physics of Blown Sand, Memoirs 8*, Department of Theoretical Statistics, Aarhus University, Denmark: 83-100.

Willetts, B.B. and Rice, M.A. 1985b. Wind tunnel tracer experiments using dyed sand. In *Proceedings of International Workshop on the Physics of Blown Sand, Memoirs 8, Department of Theoretical Statistics*, Aarhus University, Denmark: 225–242.

Willetts, B.B. and Rice, M.A. 1986a. Collisions in aeolian saltation. *Acta Mechanica*, 63: 255–265.

Willetts, B.B. and Rice, M.A. 1986b. Collisions in aeolian transport: the saltation/creep link. In W.G. Nickling (ed.), *Aeolian geomorphology*. Proceedings of the 17th Annual Binghampton Symposium, September 1986: 1–18.

Willetts, B.B. and Rice, M.A. 1989. Collisions of quartz grains with a sand bed: the influence of incidence angle. *Earth Surface Processes and Landforms*, 14: 719–730.

Willetts, B.B., McEwan, I.K. and Rice, M.A. 1991. Initiation of motion of quartz sand grains. *Acta Mechanica*, Supplement, 1: 123–134.

Williams, A.J. and Balling, R.C. 1996. *Interactions of desertification and climate*. Arnold, London.

Williams, J.J., Butterfield, G.R. and Clark, D. 1990a. Aerodynamic entrainment thresholds and dislodgement rates on impervious and permeable beds. *Earth Surface Processes and Landforms*, 15: 255–264.

Williams, J.J., Butterfield, G.R. and Clark, D. 1990b. Rates of aerodynamic entrainment in a developing boundary layer. *Sedimentology*, 37: 1037–1048.

Williams, J.J., Butterfield, G.R. and Clark, D.G. 1994. Aerodynamic entrainment thresholds: effects of boundary layer flow conditions. *Sedimentology*, 41: 309–328.

Wolfe, S.A. and Nickling, W.G. 1993. The protective role of sparse vegetation in wind erosion. *Progress in Physical Geography*, 17 (1): 50–68.

Zingg, A.W. 1953. Wind tunnel studies of movement of sedimentary material. *Proceedings of 5th Hydraulic Conference Bulletin*, 34: 111–134.

17

Sand seas and aeolian bedforms

David S. G. Thomas

Introduction

Aeolian sand deposits, excluding non-arid coastal dune systems, cover approximately 5% of the global land area. In comparison, loess (aeolian silt) extends over double this area (see, for example, Pecsi 1968), mainly in the mid-latitudes but also on the margins of some deserts. In total, about 20% of the world's drylands are covered by aeolian sands, though their importance varies significantly between deserts, covering less than 1% of those of the Americas (Lancaster 1995) but almost half of the area of the Australian deserts (Mabbutt 1977).

Arid zone aeolian deposits have long attracted the attention of geographers, geomorphologists and allied scientists. Bagnold's (1941) seminal work established a depth of understanding and rigour which set the framework for research in the ensuing half-century. In more recent decades, at markedly different scales, both the availability of remotely sensed data (e.g. Breed and Grow 1979; Paisley *et al.* 1991) and a growing number of studies involving the detailed monitoring of windflow and sandflow over dunes (e.g. Weng *et al.* 1991; Wiggs 1993) have enhanced our understanding of the distribution of aeolian bedforms and the processes which mould them.

Over 99% of arid zone aeolian sand deposits are found in extensive sand seas or *ergs* (Wilson 1973), the largest landform unit of aeolian deposition (Table 17.1). The form and size of dunes can vary within and between sand seas, through the interaction of and variations in sand supply, vegetation cover and wind regime. Sand seas or parts of sand seas which are largely devoid of dunes are called *sand sheets, stringers* or *streaks. Ripples* are the smallest bedform unit and can occur on any larger-scale aeolian bedform.

The global distribution of sand seas

Sand seas are neither clearly nor consistently defined in the literature. Following Wilson (1971), Fryberger and Ahlbrandt (1979) and others, an *active sand sea* may be defined as an area of at least 125 km^2 where no less than 20% of the ground surface is covered by windblown sand. Lancaster (1995) has defined sand seas as 'dynamic sedimentary bodies that form part of local- and regional-scale sand transport systems in which sand is moved from source zones to depositional sinks'. Dune forms are commonly present, but in areas where sand throughflow dominates over sand accumulation, sand sheets

Arid Zone Geomorphology: Process, Form and Change in Drylands, 2nd edition. Edited by David S. G. Thomas.

Table 17.1 *Scales of aeolian depositional landforms*

Scale	Description	Name*
I	Deposits of regional extent	Sand seas or ergs
II	Deposits commonly devoid of scale III bedforms	Sand sheets, streaks and stringers
IIIa	Bedforms superimposed on scale I deposits	Complex, compound and megadunes
IIIb	As IIIa or independently	Simple dunes
IV	Bedforms superimposed on scale I, II or III deposits	Normal, fluid drag and megaripples

* Many deposits have alternative names. These are given in the tables of Breed and Grow (1979).

occur and dune forms are normally absent. When vegetation is present, it is not so dense as to significantly inhibit aeolian sand transport.

Figure 17.1 shows the global distribution of major active and fixed sand seas. Fixed sand seas (discussed in Chapter 26) contain aeolian bedforms which are presently inactive or have limited surface activity, because current environmental factors do not favour the significant operation of aeolian processes. Most of the major sand seas that are active lie within the 150 mm mean annual isohyet (Wilson 1973), where given a supply of sediment and suitable wind conditions, aeolian transport is uninhibited by plants.

SEDIMENT SUPPLY IN SAND SEAS

Aeolian transportation sorts sediment effectively, so that usually over 90% of particles in active ergs are sand-sized. Finer silt is

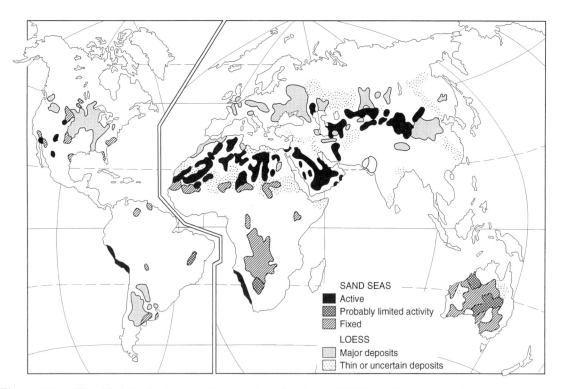

Figure 17.1 *The global distribution of aeolian deposits. After Snead (1972), modified and supplemented with data from numerous sources*

commonly transported out of the desert aeolian system in suspension (see Chapter 18). Clay is normally only transported by the wind in pellets, which then saltate and creep like the more familiar quartzose sands (Bowler and Wasson 1984; Al Janabi *et al.* 1988); Dare-Edwards (1984) has called such deposits *sand loess*.

Sand may move between source areas and sand seas along preferred transport pathways. In some areas, such as the Mojave Desert in California, these pathways may cross drainage basins and divides prior to accumulation in sand seas (Zimbelman *et al.* 1995). The highest rates of sand transport occur across the duneless sand sheets and stringers which often lead into areas of sand seas with dunes (Sarnthein and Walger 1974 give a range of $62.5-162.5$ m^3 m^{-1} width yr^{-1} for Mauritania). Mobile barchan dunes represent bulk transport rates of up to 3.49 m^3 m^{-1} width yr^{-1}. Substantial quantities of sand may be transported huge distances across sand sheets before accumulating into dunes, as in the case of the Makteir erg in Mauritania (Fryberger and Ahlbrandt 1979). Sand transport pathways may be sinuous, showing sensitivity to local topography and being analogous to 'rivers of sand' (Zimbelman *et al.* 1995).

Dune sediments can reach sand seas along non-aeolian pathways. In the Taklamakan Desert, China (Zhu Zenda *et al.* 1987), central Australia (Twidale 1972) and other areas, so-called 'source bordering dunes' have been deflated from dry river valleys. Wadi sediments may be particularly important local aeolian sediments as they dry out through drainage rather than evaporation and are therefore uncemented (Glennie 1970). It is in fact likely that alluvial sources have made direct or indirect contributions to the sediments in parts of many sand seas, especially in low-relief situations (Cooke *et al.* 1993).

Namib Desert dune sands have arrived at their destination through the combined effects of river, sea and wind transport (Lancaster and Ollier 1983). In the Wahiba Sands of Oman, hyperaridity has precluded fluvial contributions to Holocene sand sea accumulation (Warren 1988). Instead, sediment supply is attributed to coastal erosion of soft sandy materials (Figure 17.2). The extensive dune systems of the Kalahari represent periods of aeolian

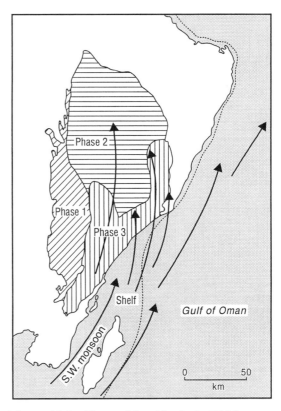

Figure 17.2 *Phases of deposition in the Wahiba Sands, Oman. Sands have been derived by the inland transport of coastal carbonate sands in at least three phases; in the third, Holocene, phase, cemented cliff sands have been deflated and transported inland. Redrawn after Cooke* et al. *(1993)*

reworking of pre-existing continental basin sediments which have accumulated by various means since the mid-Jurassic (Thomas 1987). These have been supplemented in the southwestern Kalahari by aeolian inputs operating along thin sand sheets (Fryberger and Ahlbrandt 1979; Figure 17.3). Such *intraergal* and *extraergal* sands can sometimes be distinguished by grain size and mineralogical differences (see Bowler and Wasson 1984). Finally, it can be noted that for many sand seas there is evidence for multiple phases of accumulation (e.g. Figure 17.2). One reason for hiatuses in accumulation is climate change. It has been suggested that during the last 30 000 years active accumulation of sediment in sand seas may have been restricted to less than a third of the time (Lancaster 1990), with episodicity of accumulation

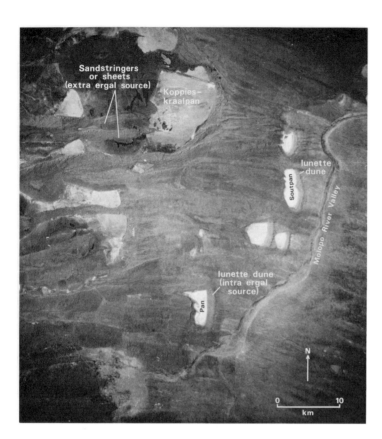

Figure 17.3 *Space Shuttle hand-held camera photograph of part of the southwestern Kalahari dunefield, southern Africa. Linear dunes dominate this sand sea, but larger, more amorphous sand sheets or stringers, also trending west–east can be identified, too. These may represent paths of extraergal sand inputs. Linear dunes also extend downwind from local sand sources, notably pan-margin lunette dunes and dry river valleys*

even affecting sand seas in the driest regions (Haney and Grolier 1991). Other reasons for disruptions in accumulation are tectonic activity and, in coastal situations, base-level changes.

SANDFLOW CONDITIONS AND SAND SEA DEVELOPMENT

Sand sea development requires both favourable sediment budgets and wind conditions. Where net sand transport rates are high, throughflow exceeds deposition and bedform development is limited to sheets or stringers with ripples or highly mobile dune types such as barchans (Wilson 1971). In such throughflow-dominated locations, sandflow is usually unsaturated and ground sand cover incomplete. If surface conditions are favourable, however, isolated bedforms can develop so long as the sand received by the bedform is equal to that which is lost downwind. Wilson (1971) termed this

situation *metasaturated flow*. Sand seas proper accrue where sandflow is saturated and accumulation exceeds net transport, with consequential bedform development and the vertical or lateral accumulation of sand.

Wilson (1971) provided theoretical models of erg development in terms of wind regime and sandflow variations, demonstrating how sand seas should grow in locations of convergent windflow and wind-speed deceleration, for example in inter-montane basins (Figure 17.4a). An empirical study (Fryberger and Ahlbrandt 1979) has supported Wilson's major assertions and identified the synoptic and topographical conditions favouring sand sea development (Figure 17.4b).

Fryberger (1979) classified sand seas into low-, intermediate- and high-energy environments. Most sand seas experience less than $27 \text{ m}^3 \text{ m}^{-1}$ width yr^{-1} potential sandflow (Table 17.2), though caution should be exercised in the interpretation of estimated sandflow values

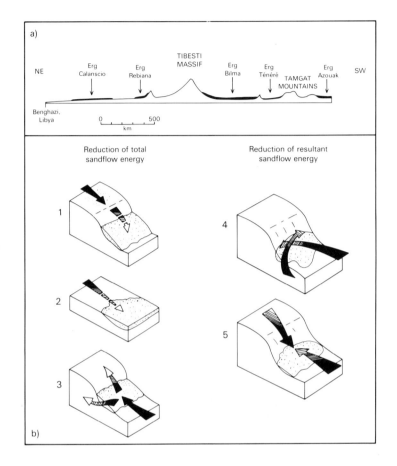

Figure 17.4 *(a) The relationships of ergs in part of north Africa to topography (after Wilson 1973). (b) Models of topographic influences on sand sea development (after Fryberger and Ahlbrandt 1979). (1) In the shadow of a topographic barrier; (2) in shallow desert depressions and (3) by the direct blocking of wind, all leading to a total reduction in sand transporting energy. Resultant energy may be reduced (4) when surface winds are deflected leading to sites of favourable accumulation and (5) when katabatic winds off a highland oppose a dominant wind. Sand seas can also develop where water bodies interrupt regional sandflow patterns*

(see Chapter 16). Many low-energy sand seas occur near the centre of semi-permanent high- and low-pressure cells, while the high-energy sand seas fall in the Trade wind zones near the margins of these pressure systems.

The transference of sand from high- to low-energy locations within sand seas has been demonstrated by studies in the Jafurah erg, Saudi Arabia (Fryberger *et al.* 1984) the Namib Desert (Lancaster 1985a) and the Gran Desierto, Mexico (Lancaster *et al.* 1987). Clearly, therefore, the general classification of wind environments by Fryberger and Ahlbrandt (1979) masks considerable intraregional variations in sandflow potential. This may account for the difference between their classification of the Simpson Desert (Table 17.2) and Ash and Wasson's (1983) assertion that the major limitation on sand movement there today is the low wind speeds.

SAND SHEETS

Sand sheets can develop in aeolian environments where conditions do not favour dune development, though they may exhibit low-relief aeolian features such as ripples and *zibar* (see pages 380 and 390). They can be small, local features of a few square kilometres, often on the margins of dunefields (Kocurek 1986), or major regional landscape components, such as the vast Selima sand sheet of the eastern Sahara, with an area of about 100 000 km^2 (Breed *et al.* 1987).

Kocurek and Nielson (1986) recognised five major controls on aeolian sand sheet development. *Vegetation*, especially grasses, may encourage the accretion of low-angle laminae while limiting the construction of dunes. *Coarse sand*, which is not readily formed into dunes other than zibar, can characterise sand sheets,

Table 17.2   *Sand sea wind environments*

Sand sea	Rate of sand drift ($m^3\ m^{-1}$ width yr^{-1})
1. High-energy environments Northern Arabian sand seas Northwestern Libya	25–40
2. Intermediate-energy environments Simpson Desert, Australia Western Mauritania Peski Karakumy, former USSR Peski Kyzylkum, former USSR Erg Oriental, Algeria Erg Occidental, Algeria Namib Sand Sea, Namibia Rub'al Khali, Saudi Arabia	15–25
3. Low-energy environments SW Kalahari, southern Africa Sahelian zone sand seas Gobi Desert, China Thar Desert, India Taklamakan Desert, China	<15

sometimes upwind of a dunefield where it remains as a surface lag deposit, as at the Algodones dunefield, California (Kocurek and Nielson 1986). A coarse sand sheet can therefore sometimes be regarded as a deflational remnant, rather like some desert pavements or *regs*. A *near-surface groundwater table* (Stokes 1968) or 'Stokes surface' (Fryberger *et al.* 1988) can be an important control on sand sheet development, acting as a base-level to the action of wind scour. This can occur in locations deprived of upwind aeolian sand supply, or playa, sabkha or coastal sand sea situations (Fryberger *et al.* 1988), where *periodic* or *seasonal* flooding can also prevent dune development. Langford (1989) identified a number of ways in which interactions between aeolian and fluvial processes could lead to different sedimentary deposits in areas with fluctuating groundwater tables. The formation of *surface crusts*, for example salcretes (see Chapter 6), or the growth of algal mats (Fryberger *et al.* 1988) can also inhibit deflation and contribute to the formation of sand sheets, as in the White Sands, New Mexico.

With the exception of the presence of coarse sand, the controls identified by Kocurek and Nielson (1986) are probably inapplicable to the development of the hyper-arid Selima sand sheet, because their effectiveness requires either the presence of some moisture or proximity to a dunefield. Rather, Breed *et al.* (1987) ascribe its development to the blanketing of an ancient fluvial landscape by aeolian redistribution of the sand fraction contained within widespread alluvial deposits. The absence of major topographic barriers able to inhibit long-distance wind transport accounts for the great size of this sand sheet, while the coarseness of the sand has probably caused the general lack of dune development and the formation of gently undulating subhorizontal tabular deposits 1–10 cm thick.

SAND RAMPS

Where aeolian sand transport paths cross undulating terrain, sand ramps may accumulate on both upwind and downwind sides of topographic obstacles (Tchakerian 1991; Zimbelman *et al.* 1995). In the Mojave Desert, these ramps may exceed 100 m in thickness (Smith 1982). Sand ramps commonly have shallow gradients, and interdigitated aeolian, slope and fluvial deposits (reflecting the consequences of climatic change). During periods of aeolian activity they act as shallow slopes along which saltating grains can progress in the general direction of sand transport (Zimbelman *et al.* 1995).

Aeolian bedforms

Scale III and IV features (Table 17.1) are *bedforms*, a term which applies to any regularly repeated erosional or depositional surface feature, whether aeolian or subaqueous (Wilson 1972a). This section considers general aspects of aeolian bedform development, while ripples and dunes are discussed in greater detail in subsequent sections.

SCALE EFFECTS IN AEOLIAN BEDFORM DEVELOPMENT

Contrary to early opinion (Cornish 1914), ripples do not develop over time into dunes.

Table 17.3 *Wilson's hierarchy of aeolian bedforms*

Order	Name	Wavelength (m)	Height (m)	Origin
1	Draas or megadunes	300–5500	20–450	Aerodynamic instability
2	Dunes	3–600	0.1–100	Aerodynamic instability
3	Aerodynamic ripples	0.015–0.25	0.002–0.05	Aerodynamic instability
4	Impact ripples	0.05–2.0	0.0005–0.1	Impact mechanism

Source: Wilson (1972a).

Ripples and dunes develop independently, at different scales, in response to different factors. Wilson (1972a; 1972b) developed the concept of bedform scale into one of a bedform hierarchy with four components: ripples (two orders), dunes and draas (Table 17.3), draas being large whaleback sand bodies, or megadunes.

The spacing of the three highest-order bedforms was attributed to secondary atmospheric flows of differing sizes. As bigger sand grains require higher shear velocities for their mobilisation, particle size was deemed to set the scale of lower-atmospheric eddy currents (Wilson 1972a; 1972b). Grain size would therefore be coarser in the higher-order, more widely spaced, bedforms.

The universal applicability of this 'granulometric control' hypothesis has not survived the scrutiny of subsequent investigations. Draas can not be separated from dunes in terms of grain size (Wasson and Hyde 1983a), probably being compound or complex dune forms that have developed in areas of high net sand accumulation, complex wind regimes and probably long-term aridity (see, for example, McKee 1979; Lancaster 1985a).

MECHANISMS OF AEOLIAN SAND DEPOSITION

Bedform development takes place when mobile particles come to rest in a sheltered position on the ground surface, through *accretion* (Bagnold 1941) or *tractional deposition* (Kocurek and Dott 1981). It also occurs when wind velocities decrease causing a reduction in shear stress and a fall in transportational potential, termed *grain-fall deposition*, and when particles come to rest after rolling down a slope under the effects of gravity, called *avalanching or*

grainflow deposition, which is important for particle movement down dune slip faces (see Tsoar 1982).

Dune initiation is still one of the least understood components of aeolian geomorphology, though it is clear that a nucleus is required to concentrate particle accumulation by tractional or grainfall mechanisms. A disruption in windflow, causing a decrease in u_* (see Chapter 16) is necessary for deposition. Although this can be caused by major topographic obstacles, small changes in surface elevation, including the presence of shallow hollows and small plant patches (Table 17.4), also contribute to deposition, by reducing aerodynamic roughness by up to 86%. The ensuing expansion of flow is followed by flow separation as sand accumulates, further enhancing deposition, as proposed in a five-stage model by Kocurek *et al.* (1990) (Figure 17.5). After initiation, dune development occurs through the movement of sand up the windward slope towards the crest as the developing dune increasingly modifies its own boundary layer environment. As sand crosses the *brink*, which separates the windward and lee sides, a *slip face* which is near to the angle of repose for sand develops in the zone where windflow separates from the dune surfaces.

Table 17.4 *Possible nuclei for sand dune development*

1. Topographical obstacles
2. Smaller obstacles: pebble or boulder
   plant or bush
   animal carcass, e.g. dead camel
   human artefact, e.g. oil drum
3. Remnant of previous bedform
4. Surface irregularities and hollows

Source: After Wilson (1972b) with modifications.

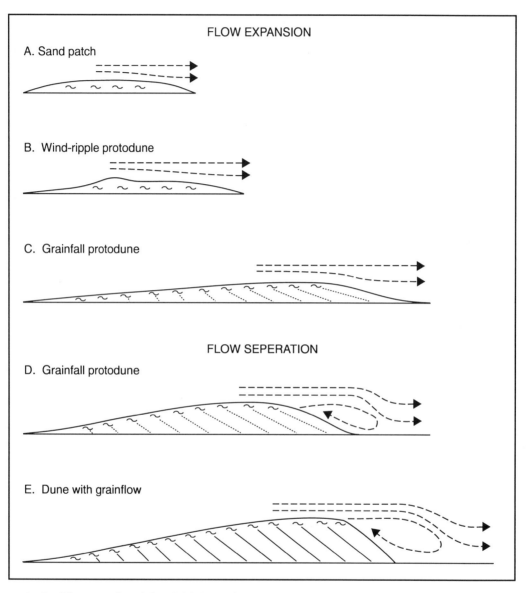

Figure 17.5 *The stages of sand dune initiation and development, and associated modifications to boundary layer airflow (after Kocurek* et al. *1990)*

Ripples

Bagnold (1941) and others recognised three types of aeolian ripple which have various names: *normal* or *ballistic*; *granule, sand ridge* or *mega*; and *fluid drag* or *aerodynamic* ripples. The last of these three categories, according to Wilson (1972a), can form longitudinally or transverse to wind direction, whereas normal and megaripples only develop transversely and commonly in regularly repeated patterns, which in the case of smaller ripples can adjust from minute to minute in response to windflow changes. Pattern wavelengths can also vary abruptly in spatial terms due to grain-size differences in the parent sediment (see Ellwood *et al.* 1975).

Normal ripples commonly have wavelengths of 1–25 cm and heights of 0.5–1.0 cm (see Sharp 1963) and are asymmetrical in profile with windward slopes of about 10° (Mabbutt 1977, Figure 17.4b). Megaripples have crest-to-crest wavelengths which may exceed 20 m and are more symmetrical in cross-profile (Greeley and Iversen 1985). This symmetry is probably a response to shifts in wind directions, suggesting that the form and spacing of larger ripples are less dynamic than those of smaller forms. Sharp (1963) considered very high wavelengths to be unrepresentative of megaripples formed on homogeneous sandy surfaces. Wavelengths in excess of a few metres were derived from ripples superimposed on wet substrate and not directly developed from bed material: figures of up to about 3 m may therefore be more appropriate (see Wilson 1972a; Mabbutt 1977).

Theoretical and empirical studies by Ellwood *et al.* (1975) led to the conclusion that mega and normal ripples do not form distinct populations, but are the upper and lower limits of an impact mechanism ripple continuation. Wavelength variations are determined by differences in grain size and windspeed.

IMPACT MECHANISM RIPPLES

Bagnold's (1941) impact mechanism theory has been widely used to explain ripple formation, but has its critics (Sharp 1963; Folk 1977; Brugmans 1983; Walker and Southard, reported in Warren 1984). In this theory, ripple wavelength is related to a characteristic saltation path length, which since Bagnold's (1941) original publication has been interpreted as the mean saltation path length. Surface irregularities act as erosional and depositional nuclei for moving particles (Figure 17.6a). The windward side of an irregularity is bombarded by more saltating grains per unit area than the sheltered lee side (Bagnold 1941). Collision causes a transference of energy, with new grains set in motion and net erosion on the preferentially bombarded windward side.

The preferential loss of particles from the windward side of the original irregularity creates a new preferential zone of bombardment downwind, at a distance equivalent to the mean saltation path length, which forms the focal point for erosion. Hence, the ripple pattern advances downwind as a series of alternating zones of erosion and protection. Coarser grains which are not set into saltation by the impact of descending particles creep forward to accumulate in the less bombarded crestal zone.

However, ripple spacing increases with time and is therefore unlikely to be related to the mean saltation path length under steady wind conditions (Sharp 1963). Sharp suggested that the angle of incidence of descending particles and ripple height determined ripple spacing. Height can perhaps be seen as crucial: given the

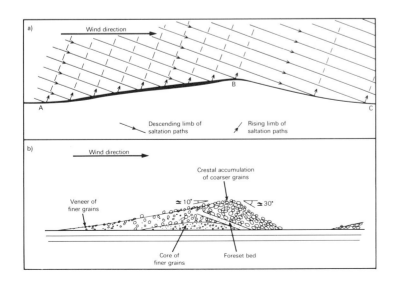

Figure 17.6 *(a) Ripple development (after Bagnold 1941). AB is preferentially bombarded by descending grains compared with BC (see text for full explanation). (b) Laminae and grain-size distribution within a ripple (from diagrams in Sharp 1963)*

narrow range of ripple slope angles which have been recorded, it must geometrically affect the length of the windward and lee slopes. Thus, the minimum wavelength must increase with greater ripple height (Brugmans 1983); correspondingly, ripple wavelength would increase as ripple growth proceeds.

As a ripple grows it protrudes into zones of higher wind speed in the atmospheric boundary layer. Ripple height is limited by this as ultimately the coarser crestal grains are more readily moved forward under the direct effect of the wind (Greeley and Iversen 1985). Once an equilibrium form is established, when as many grains reach the crest as are lost, the ripples migrate in the direction of sand transport as erosion is concentrated on the windward flanks. If local wind speeds increase ripples tend to flatten out (see Glennie 1970), perhaps because larger particles are brought into saltation (Jensen and Sorensen 1986). Under these conditions saltation lengths also increase, so that while ripple height might set the *minimum* possible spacing, it would seem more probable that saltation lengths play an important part in determining the *actual* wavelength.

Wind-tunnel and theoretical studies have indicated how the population of moving grains can affect ripple wavelengths. Saltating particles require vertical lift-off momentum while creeping grains require forward momentum. Willetts and Rice (1986a; 1986b) have calculated that bombarding grains transfer relatively more vertical energy on steeper slopes than on gentler slopes. Conversely, more forward momentum is transferred on gentler slopes. Saltation is therefore favoured from the middle of the windward side of ripples and creep activity increases towards the crest where the profile flattens out, favouring coarse grain accumulation (Willetts and Rice 1986a; Figure 17.6b). This effect would narrow the zone of saltation paths downwind, suggesting how a ripple field could be initiated from a single surface irregularity.

Anderson (1987) has suggested that while ripple wavelength is affected by grain trajectory lengths, it is not the same distance as the mean saltation path length. Importantly, Anderson's (1987) model does not require a narrow range of saltation path lengths, which was a major criticism levelled at the impact mechanism by

Folk (1977) and others. The grains ejected from the surface by the impact of a descending high-energy saltating grain have a range of trajectory lengths and speeds. The low-energy grains (creeping particles and saltating grains with short path lengths), called the *reptating* population by Anderson (1987), outnumber, by about nine to one, grains travelling on high-energy trajectories. Unlike high-energy grains, reptating grains do not travel forward on infinite successive saltations, but come to rest on the ground surface. The high-energy grains therefore power the mechanism of ripple formation, causing ripple growth by an impact mechanism, through the net accumulation of reptating grains. Ripple spacing, however, is a function of the probability distribution of the total trajectory population and the ejection rate of the reptating grains, and is roughly six times the mean reptation path length.

Anderson's (1987) model of reptation control of ripple development is supported by the wind-tunnel experiments of Willetts and Rice (1986b), while numerical modelling by Forrest and Haff (1992) lends some support for saltation influencing some characteristics of ripple development. Modelling also suggests that ripples develop a regular spacing through controls exerted by particle–bed impacts and interactions (Anderson 1990; McEwan et al. 1992). Clearly, as further studies have been conducted, the original saltation length model of Bagnold (1941) has been modified. Rather than viewing these developments as leading to a simple rejection of Bagnold's (1941) thesis (as suggested by Cooke et al. 1993), it is possible to see the outcomes of modelling, wind tunnel and field investigations as leading to progressive development of the impact mechanism model. Other aspects of ripple development, notably Folk's (1977) observation of the common development of Y-junctions in ripple patterns, still await explanation.

GRAIN-SIZE EFFECTS

Small (or 'normal') ripples probably develop in unimodal sands, and ripples with a greater wavelength in bimodal sands (Figure 17.7), where the coarse and fine fraction respond differently to aeolian action (Ellwood et al.

saltation paths are at any given wind strength (Willetts and Rice 1986b). This would presumably lead to a corresponding increase in ripple spacing.

Dunes

Over 60% of the total area of the Earth's sand seas are covered by dunes of different forms and sizes (Table 17.5 and 17.6). The vast and bewildering nomenclature used in dune description (see Breed and Grow 1979) can be simplified into a classification scheme which recognises the most pertinent aspects of form and structure and also the relationships between juxtaposed and superimposed forms (McKee 1979).

DUNE CLASSIFICATION

The morphological classification of dunes is based on their relationship to formative winds and on the number of slip faces present (McKee 1979). *Linear or longitudinal* dunes have long axes which form parallel to the resultant sand drift direction (RDD — see the next subsection). *Traverse* or *crescentic* forms are aligned normal to the RDD while *star* dunes have no less than three arms radiating from a central peak and develop in complex multidirectional wind regimes (Figure 17.8). These relationships determine the number of slip faces which dunes possess, with in general, one for each modal direction of sand moving winds (Figure 17.8).

This relatively simple classification scheme is complicated by a number of important considerations:

1. As Rubin and Hunter (1985) note, dune axes are not always perfectly parallel or normal to the RDD. Strictly speaking, therefore, a linear dune may be *longitudinal* or *oblique* to the RDD. In practice, it is generally accepted that linear dune development can be parallel or subparallel and transverse dune development normal or subnormal to the RDD. Nevertheless *hybrid* or *oblique* dune forms have been recognised in some areas, for example by Carson and MacLean (1986) in the Williams River Desert dunefield, Saskatchewan.

2. There are, therefore, considerable variations in the forms of the general dune types, so that such a classification is only a first approximation. For

Figure 17.7   *Variations in ripple wavelength on the flank of a linear sand ridge, SW Kalahari. Low (10–25 cm) wavelength ripples occur on unimodal dune sand (mean grain size at crest is 2.15φ). Dune plinth sand is bimodal (modes in 2.5 and 3.0φ classes), and possesses larger ripples with a wavelength of 0.75–1.5 m. In the foreground, smaller ripples are present on unimodal interdune sediments*

1975; Fryberger *et al.* 1992). For a bimodal sand, if winds are strong enough to cause the coarse fraction to saltate, ripple wavelength is determined by the impact mechanism, a function of the interaction between grain size, wind speed and particle trajectory lengths. If only the finer particles saltate, however, wavelength responds to the greater momentum which particles achieve through impact with the coarsest grains (Ellwood *et al.* 1975).

Grain shape probably also affects the nature of the ripple bed. Several studies (for example Willetts *et al.* 1982; Anderson and Hallet 1986) have shown that particle shape affects the length and height of saltation paths. Essentially the less spherical particles become, the longer

Table 17.5   *Sand sea characteristics*

	Sahara				Southern Africa		Asia				N America
	West ergs	South ergs	North ergs	NE ergs*	Namib	SW Kalahari	Arabian ergs	Thar	Takla-makan	Ala Shan	Gran Desierto
Area (10³ km²)	207	447	306	161.5	34	100	743	214	261	169	5.5
Dunes (% of area)	54.7	52.5	64.1	56.3	54.6	86.5	70.1	68.3	66.4	82.2	70.0
Sand sheets (% of area)	45.3	47.5	35.9	39.3	45.4	13.6	23.2	31.8	33.6	67.8	30.0
Uncertain (% of area)	–	–	4.5	–	–		6.7	–	–	–	–

*Excluding Egyptian and Lybian ergs.
Source: Fryberger and Goudie (1981); Lancaster *et al.* (1987).

Table 17.6   *Relative occurrence of major dune types*

	Sahara				Southern Africa		Asia				N America	
	West ergs	South ergs	North ergs	NE ergs*	Namib	SW Kalahari	Arabian ergs	Thar	Takla-makan	Ala Shan	Gran Desierto	MEAN
Linear (total)	65	46	36	30	60	99	71	20	33	5	0**	42.7
simple and compound complex: transverse	65	46	15	6	34	99	43	20	28	5	–	32.8
imposed	–	–	6	13	–	–	–	–	5	–	–	2.2
star imposed	–	–	15	11	26	–	28	–	–	–	–	7.3
Transverse (total)	35	54	52	26	22	0	21	38	56	84	70	41.6
simple	1	8	1	–	22	–	1	13	6	27	40	10.8
compound	–	–	11	3	–	–	–	–	–	–	18	2.9
complex	34	46	40	23	–	–	20	24	50	57	8	27.5
Star	0	0	12	43	18	0	8	0	0	9	33	11.1
Parabolic	0	0	0	0	0	1	0	42	0	0	0	43.9
Dome	0	0	0	1	0	0	0	0	11	3	0	1.3

*Excluding Egyptian and Lybian ergs.
** But see Figure 17.12 where Lancaster (1995) suggests some linear forms are present.
Source: As Table 17.5.

example, three broad subdivisions of transverse dunes are widely recognised (Figure 17.8), while detailed analysis of the southwest Kalahari dune-field (Bullard *et al.* 1995) shows considerable variation in form and pattern (Figure 17.9) within the general category of linear dune (Table 17.6). Given the wide range of wind environments in which desert dunes occur (Fryberger 1979), where the relative importance of the components of the wind regime for sand movement can be significantly or subtly different, it is perhaps important to recognise that dunes probably occur on a continuum ranging from transverse to star forms, rather than in totally discrete classes.

Dune type		Number of slip faces	Major control on form	Formative wind regime	Nature of movement
Zibar		0	Coarse sand	Various	Limited
Dome dune		0	"	"	"
Blow out		0	Disrupted vegetation cover	"	May extend down wind
Parabolic dune		1	"	Transverse. Unimodal	Slow, nose migration
TRANSVERSE DUNES					
Barchan dune		1	Wind regime and sand supply	"	Forward migratory
Barchanoid ridge		1	"	"	"
Transverse ridge	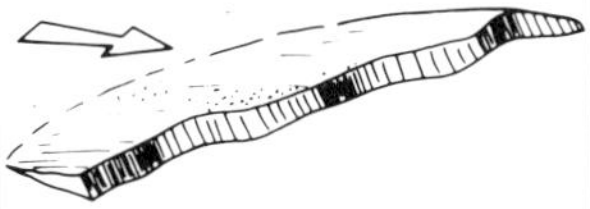	1	"	More directional variability than for barchans	"
LINEAR DUNES					
Linear ridge		1 - 2	"	Bimodal / wide unimodal	Extending
Seif dune		2	"	Bimodal	"
Reversing dune		2	"	Opposing bimodal	May migrate if one mode dominant
Star dune	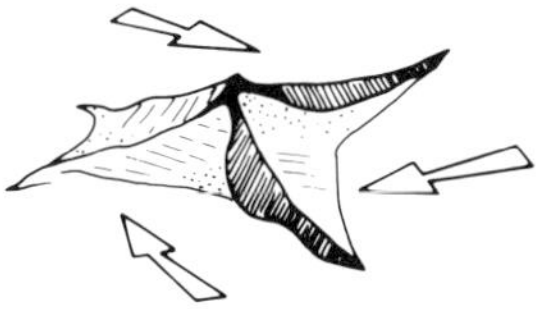	3+	"	Complex. Multimodal	Vertical accretion

Figure 17.8 *A classification of basic dune types*

	General Description	Planimetric Pattern		
		Example 1	Example 2	Example 3
1	Dunes parallel/sub-parallel, discontinuous occuring as short lengths (< 2km). Y-junctions uncommon and there are no transverse elements.			
2	Dunes parallel/sub-parallel and continuous for several km, few Y-junctions and no (or very rare) transverse elements. Those Y-junctions which do occur tend to form at the junction of two long dunes rather than as short spurs at the side of a dune. Occasional slip faces on crests but < 2m².			
3	Dunes parallel/sub-parallel and continuous for several km. Y-junctions common, both as parallel dunes merge and as short spurs < 600m on either side of dune. No slip faces on undisturbed dunes but may be common where grazing occurs.			
4	Linear dune network comprising large steep dunes and smaller gently sloping dunes. Larger dunes have broadly linear trend but are very sinuous. Small dunes tend to be orientated perpendicular to this trend. Small pans occur in deep interdune areas. Both types of Y-junction occur, Y-junctions and terminii are common.			
5	No obvious linear trend and a chaotic hummocky appearance. Dune slopes shallow with very rounded crests. Dunes have low relief but occasional dunes up to 7 - 8m high. Very little interaction between dunes.			

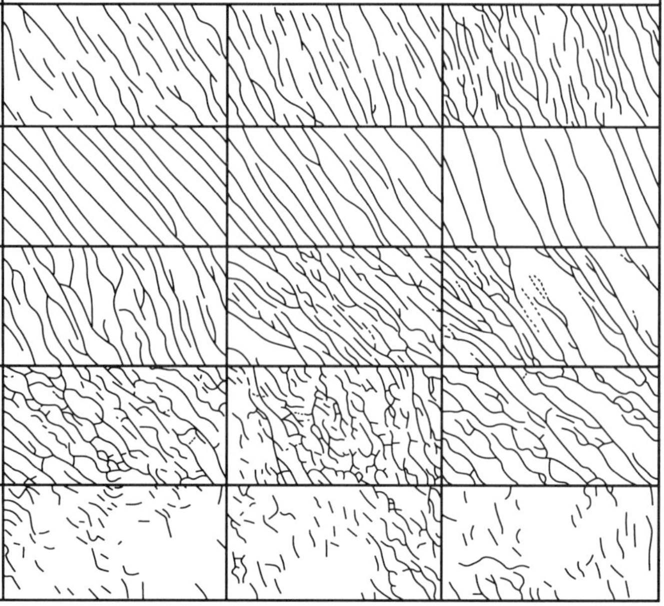

Figure 17.9   *Variations in 'linear' dune patterns and morphologies identified in the southwestern Kalahari Desert by Bullard* et al. *(1995)*

3. An RDD can be derived from a range of overall wind regimes (see Fryberger 1979).
4. Factors other than RDD and wind regime assume considerable importance for the development of some dune types, for example *zibar* and *parabolic* dunes.
5. Dune morphologies can result from the combination of more than one individual dune form. This led McKee (1979) to introduce a further level of classification, dividing dunes into *simple, compound* and *complex* forms. Simple dunes are basic forms which, although often occurring as part of the same basic type coalesce or merge into compound forms in many sand seas (Thomas and Martin 1987). When dunes of different basic types are superimposed, complex forms result. The wide range of complex forms which occur in the world's sand seas are comprehensively described by Breed and Grow (1979) and Breed *et al.* (1979) with the aid of satellite imagery.

Although the development of compound and complex forms is not fully understood, potentially important factors include a change in wind regime, leading to the superimposition of dunes of different orientations or types; the effect of dune size, which can cause differential rates of movement and consequently dune superimposition in a field of mobile dunes; the merging of dunes towards the centre of sand seas in areas of convergent circulation; and the modification of windflow on the flanks of large dunes, leading to the development of superimposed smaller dunes.

Table 17.6 shows that while linear and transverse forms account for the majority of desert sand dunes when simple, compound and complex dunes are considered together, there are notable differences between sand seas. The extremes range from the Kalahari and Australian dunefields which are overwhelmingly composed of linear dunes (Table 17.6; Wasson 1986), to the Gran Desierto in Mexico which is almost devoid of them. The Thar Desert, India, has many parabolic dunes while star dunes are especially dominant in the Grand Erg Oriental, northeastern Sahara (Mainguet and Callot 1978; Breed and Grow 1979). These and other differences are a response to the varying impact of the factors which influence dune development.

a. EXAMPLES OF ANNUAL SANDFLOW REGIMES

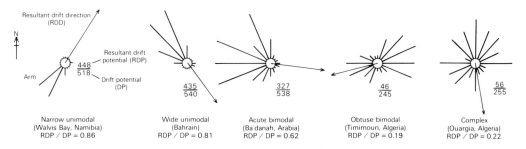

| Narrow unimodal (Walvis Bay, Namibia) RDP / DP = 0.86 | Wide unimodal (Bahrain) RDP / DP = 0.81 | Acute bimodal (Ba'danah, Arabia) RDP / DP = 0.62 | Obtuse bimodal (Timimoun, Algeria) RDP / DP = 0.19 | Complex (Ouargia, Algeria) RDP / DP = 0.22 |

b. EXAMPLES OF ANNUAL SANDFLOW REGIMES FOR MAJOR DUNE TYPES

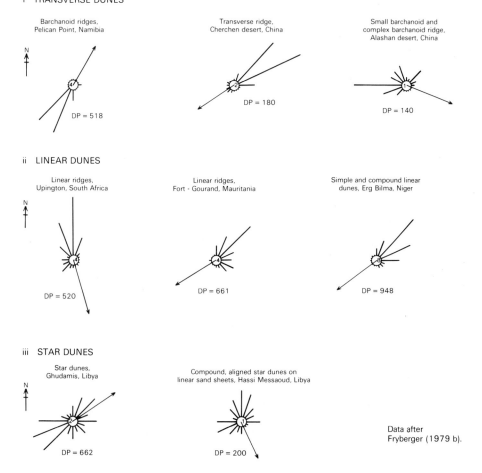

i TRANSVERSE DUNES

ii LINEAR DUNES

iii STAR DUNES

Data after
Fryberger (1979 b).

Figure 17.10 *Sand transport regimes represented by sandflow roses (after diagrams in Fryberger 1979). (a) Different types of sandflow regime. (b) Examples of sandflow roses from environments with three major dune types*

Controls on the development and form of major dune types

Fryberger (1978; 1979) used Lettau's equation and wind data to construct sandflow roses for environments possessing the major desert dune types. The length of the arms on these diagrams represents the relative amount of sand which would be transported by wind blowing from different directions (Figure 17.10). From these the total sand drift potential (DP), resultant drift potential (RDP) and the resultant drift direction (RDD) can be calculated. In addition the ration RDP/DP provides a measure of the directional variability of sand-transporting winds.

Wasson and Hyde (1983b) found that if a measure of the volume of sand in dunes was considered together with the RDP/DP ratio, statistically significant separations could be achieved for the major dune types (Figure 17.11). Transverse forms develop in wind environments which display the least directional variability (Figure 17.10b). Barchans occur where there is little directional variability in sand-moving winds and limited sand supply — Lettau and Lettau (1969) found that the sand contained in a field of barchans in Peru represented a spread out sand thickness of only 2 cm. It is therefore unsurprising that barchan dunes are most common in sand transport pathways between source and depositional areas and on the margins of sand seas (Lancaster

1995). More sand and more variability results in transverse ridges developing. Linear dunes occur when sand supply is restricted, but wind direction more variable, and star dunes where sand is abundant and wind direction unstable (RDP/DP > 0.5). Contrary to some opinions (such as Glennie 1986), there does not appear to be any evidence to support the view that differences in wind strength can contribute to the determination of dune type (Fryberger 1979; Wasson and Hyde 1983b; Steele 1986).

The precise significance of sand volume for dune formation is unclear. Wasson and Hyde's (1983b) findings could indicate the amount of sand necessary before a particular dune type develops. On the other hand, they could indicate the ability of different wind conditions to transport sand. For example, barchan dunes may not contain as much sand as star dunes because compared with complex wind regimes, those with little directional variability do not lead to sand concentration, but favour its continual movement in one direction.

DUNE ACTIVITY

In terms of overall dune activity and movement, Thomas (1992) has noted that transverse dunes are *migratory* forms, linear dunes are *extending* or *sandpassing* forms and star dunes are essentially *sand accumulating* forms. Dune activity does not therefore necessarily have to incorporate

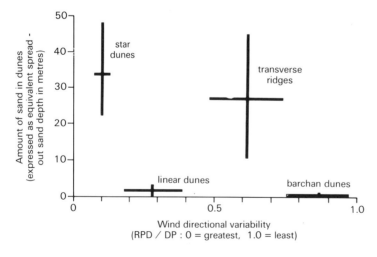

Figure 17.11   *The occurrence of major dune types in relation to sand volume and wind directional variability (after Wasson and Hyde 1983a)*

significant physical movement of the total body of a dune, but can be reflected in other attributes of surface change. This point is particularly appropriate in expressing the relationship between sand supply and wind regimes.

This is borne out by studies of the distribution of different dune types within some sand seas (Mainguet and Callot 1978; Lancaster 1985a; Lancaster *et al.* 1987), with sand-accumulating forms occurring preferentially towards the centres of dunefields (Figure 17.12). The progression from migratory to sand-accumulating dune forms is also reflected in the spatial variability of dune heights in the Namib Sand Sea (Figure 17.13).

In some circumstances the juxtaposition of dunes of different forms may not be due to such a spatial progression but to the development of different ages of dunes (Lancaster 1995). In such situations, differences in dune alignments, grain-size compositions and degrees of post-depositional modification may allow for the

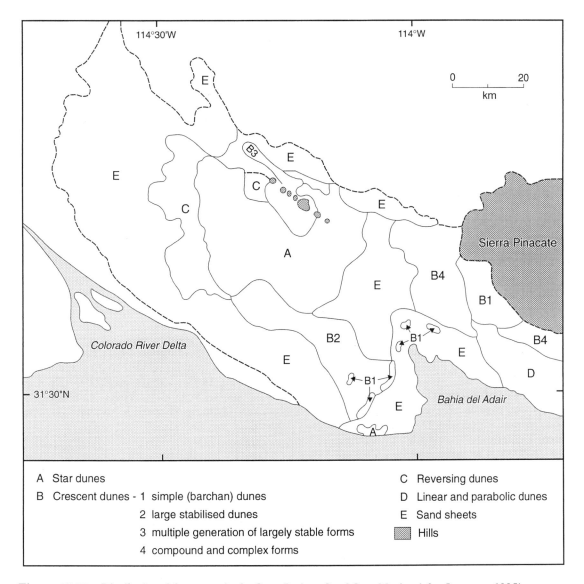

Figure 17.12 *Distribution of dune types in the Gran Desierto Sand Sea, Mexico (after Lancaster 1995)*

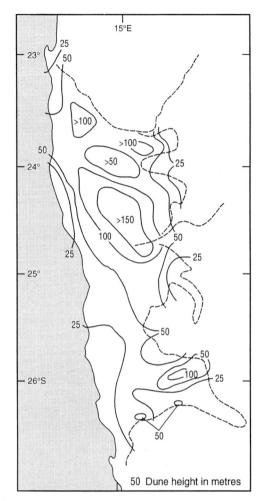

Figure 17.13 *The height of dunes in the Namib Sand Sea, with the highest dunes towards the centre and lowest on the margins of the sand sea (after Lancaster 1989)*

identification of different generations of dunes. This is the case in, for example, the Wahiba Sands, Oman, Simpson Desert, Australia, and Kelso Dunefield, California (Wasson 1983; Warren 1988, Lancaster 1993). Different dune generations or different phases of aeolian deposition may also be separated by 'bounding surfaces' which indicate hiatuses in the operation of aeolian processes (Kocurek 1988; Kocurek and Havholm 1994). The general tendency is for sand-moving forms to occur on sand sea margins, followed by sand-passing dunes and finally sand-concentrating forms in the central areas.

Other controls and resulting dune types

GRAIN SIZE AND ZIBAR DUNES

Zibars, dunes distinguished by their sedimentary characteristics, are perhaps more abundant than barchan and star dunes (Cooke *et al.* 1993) but have received little research attention. They are low transverse dunes, sometimes with a parabolic tendency (e.g. Anton and Vincent 1986; Goudie *et al.* 1987). Zibars may be up to about 5 m high, composed of coarse sand and lacking slip-face development (Holm 1960; McKee and Tibbitts 1964; Tsoar 1978). Grain-size characteristics probably affect their development in several ways. Vertical growth is restricted as only stronger winds generate the shear velocities needed to mobilise the large particles. The consequent steeper vertical velocity gradient may restrict upward growth (Cooke and Warren 1973), which also prevents slip-face development.

When wind velocities are lower, smaller particles move rapidly across zibar surfaces because greater saltation lengths are favoured on the more compact surface of the coarser grains (Bagnold 1941; Tsoar 1978). Erosion in the crestal zone is also likely to occur under these conditions, as wind velocity speed-up effects provide enhanced wind shear towards the top of the windward slope, allowing entrainment of the coarse particles to occur (Tsoar 1986). This acts as a further restriction on zibar height. Some dome dunes, which also lack morphological development (see Jawad Ali and Al-Ani 1983) may be related to zibars (Greeley and Iversen 1985). Zibars may be active even with a sparse vegetation cover (Nielson and Kocurek 1986), perhaps due to their response only to stronger winds.

THE ROLE OF VEGETATION

There has been a recent upsurge of interest in the role of vegetation in aeolian processes and dune development (Thomas and Tsoar 1990; Thomas and Shaw 1991; Livingstone and Thomas 1993; see Chapter 7). It can also be recalled, however, that over 50 years ago Hack (1941) considered vegetation cover to be a major determinant of dune type. Plants are certainly an important element of dune surfaces in the less arid sand seas, while the surfaces of the less mobile dune types may in fact provide ideal conditions for plant growth in arid environments because of the water-retention properties of dune sand (Tsoar and Møller 1986). A partial plant cover may also play an important role in modifying windflow (Ash and Wasson 1983; Wasson and Nanninga 1986) and trapping sediment on dune surfaces (Flenley *et al.* 1987), while the degree of dune surface vegetation cover has been identified as an important control determining the level of surface activity on linear dunes in the southwest Kalahari (Wiggs *et al.* 1994; 1995a; 1995b).

SPECIFIC DUNE TYPES LINKED TO VEGETATION

Parabolic dunes are often associated with blowouts on vegetated sandy surfaces (Hack 1941; Verstappen 1968; Eriksson *et al.* 1989). The arms of the dune are anchored by vegetation, limiting major sand movement to the central corridor where the cover has been disturbed. Parabolics may occur as individual forms or as nested patches, for example in India and South Africa (Wasson *et al.* 1983; Eriksson *et al.* 1989; Figure 17.14), and migrate downwind at reported rates of up to 13 m yr^{-1} (Hesp *et al.* 1989).

Vegetation can directly cause the accretion of sand and consequently the development of small dunes that are referred to by a variety of names including *coppice* or *nebkha dunes* (Boucart 1928; Melton 1940; Thomas 1987). The term *bush mound*, also applied to such dunes, may lead to confusion as it is also used to refer to features called *mima* in North America (Cooke *et al.* 1993). Mima are formed by non-aeolian processes, including rainsplash, burrowing mammal activity or even minor seismic action (Price 1949; Cox and Gakalm 1986; Berg 1990). Nebkha dunes vary markedly in size, may be up to 10 m high, and are streamlined in the direction of sand transport. Nebkha are often found on the downwind margins of playas, but if sediment delivery rates are sufficient and sand-trapping vegetation is abundant, more extensive *lunette dunes* may develop (Goudie and Thomas 1985; 1986; Thomas *et al.* 1993).

Figure 17.14 *Parabolic dune patch within the southwestern Kalahari linear dune field (photograph courtesy of Professor J. du P. Bothma)*

a)

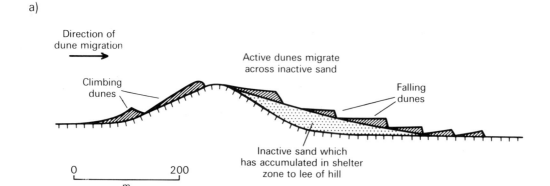

b)

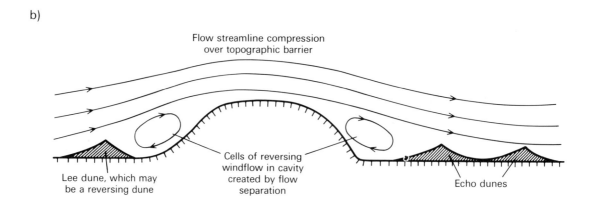

c)

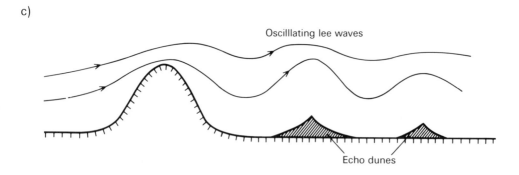

Figure 17.15   *(a) Climbing and falling dunes, Juniper Butte, Arizona (after diagrams in Greeley and Iversen 1985). (b) The development of obstacle dunes through boundary layer flow separation. This may occur on the windward, lee or both sides of a hill. (c) Echo dune development caused by boundary layer oscillating flow, analogous to a hydraulic jump, on the lee of a topographic barrier rising above a flat plain*

TOPOGRAPHIC EFFECTS

Topography modifies windflow and sand transport, and can therefore influence dune development. When a sand-laden wind or migrating dune encounters a topographic barrier, such as a hill or cliff face, *climbing* or *wrap-around dunes* may develop on the windward side (see Verstappen 1970; Figure 17.15). If sand is transported across the obstacle, *lee* or *falling dunes* can develop (Melton 1940; Grolier *et al.* 1974; Koscielniak in Greeley and Iversen 1985; Figure 17.15a). There is clearly potential for confusion in terminology between sand ramps, discussed earlier, and climbing and falling dunes. However, as Figure 17.15a shows, climbing and falling dunes are more likely to be mobile forms that in some circumstances may migrate across sand ramps.

Cells of reversed flow and oscillating waves can develop in the airflow close to a topographic barrier, contributing to the formation of *reversing* and *echo dunes* (Figures 7.15b and 7.15c; Merk 1960; Queney and Dubief 1943; Clos-Arceduc 1969). Topographic gaps can funnel windflow, so that in some situations elongated aeolian bedforms have been observed to mirror the orientation of the jet-like winds which often accompany flow concentration (Gaylord and Dawson 1987).

Transverse dunes

Barchan dunes develop where sand through-flow rates are high or where sand supply is limited, for example on stone pavements. They are usually small, simple dunes (Table 17.7) and, according to Mabbutt (1977), maximum height is almost consistently one-tenth of width. Small forms are not simply miniatures of larger barchans, being relatively flatter and with reduced angles between the windward flank and the desert floor (Hastenrath 1987). The actual size may be a function of a number of factors, including sand supply, wind strengths, atmospheric motion and age (Howard *et al.* 1978).

Megabarchans, which are compound forms consisting of superimposed barchans of different sizes, have been recorded in some dunefields (see Simons, 1956, Norris 1966; Figure 17.16d). In the Algodones, California, small barchans develop from the sand stringers which extend downwind from the horns of megabarchans. The small forms migrate forward relatively rapidly until they either merge with or ride up on to the next megabarchan downwind (Norris and Norris 1961; Howard *et al.* 1978).

Where sand supply increases, individual crescentic dunes may link up to form compound barchanoid ridges or transverse ridges

Table 17.7 *Dimensions of simple barchan dunes*

	Maximum height (m)	Total length (A–D)* (m)	Windward slope length (A–B)* (m)	Court length (C–D)* (m)	Horn to horn width (m)	Intersect angle windward slope to floor (°)
Salton Dunes, California	2.7–12.2†	62.5–193.6	33.5–108.2	–	41.1–251.5	–
Tulewash barchan, Algodones‡	9.4–11.0	–	54.6–77.4	–	158.5–210.4	–
Pampa de la Joya, southern Peru	2.1–3.9	31.6–58.1	14.8–25.5	13.5–26.5	20.5–42.0	8–10

* See Figure 17.16 for explanation.
† Recorded to highest point on brink, not crest.
‡ Represents minimum and maximum values recorded over period 1955–69.
Source: Long and Sharp (1964); Norris (1966); Hastenrath (1987).

(Figure 17.16), which are probably more common than individual barchans (Breed and Grow 1979). In practice, distinguishing between the two ridge types can be difficult, though the individual segments of barchanoid ridges display greater curvature than more amorphous transverse ridges (Cooper 1967). Barchanoid and transverse ridges range in width (tip to tip) from about 500 m to nearly 10 km (Table 17.8). A dunefield consisting of barchanoid ridges can form a fishscale pattern (Wilson 1971).

Table 17.8 *Dimensions of barchanoid and transverse ridges (dune types undifferentiated)*

	Dune length* (km)		Dune width (km)	
	Range	Mean	Range	Mean
Rub'al Khali, Saudi Arabia	0.3–1.1	0.67	0.9–2.2	1.43
NE Sahara	0.3–1.5	0.65	0.5–3.0	1.43
Taklamakan, China	1.1–3.4	2.20	2.0–5.2	3.24
Thar, India	0.8–2.0	1.30	1.0–2.5	1.50
Algodones, California	0.5–2.3	0.88	0.5–3.5	1.61

*From foot of windward slope to foot of slip face.
Source: Breed and Grow (1979).

Table 17.9 *Dune activity rates, transverse forms*

Location		Dune height (m)	Net migration rate (m yr^{-1})	References
Barchan dunes				
Pampa de la Joya, Peru		1–7	32–9	Finkel (1959)
	($n = 75$) Mean	3.7	15.4	
Pampa de la Joya, Peru	1964	4.5–6.0		Hastenrath (1987)
	($n = 6$) 1983	2.1–3.9	28–16	
Algodones, California	($n = 1$)	0.5	5	Norris (1966)
	($n = 79$) Mean	6	20	Smith (1970)
Salton Sand Sea, CA		3.1–8.2	27–14	Long and Sharp (1964)
(22 year data)	($n = 47$) Mean	6.1	18.7	
Tulewash, Salton, CA	($n = 1$)	11	13.4	Norris (1966)
Abu Moharic, Egypt		6–14	8–5	Beadnell (1910)
El-Arish, Sinai		1.9	13.1	
(individual dunes)		3.0	6.4	Tsoar (1974)
		4.0	6.2	
Mauritania	($n = 44$)	3–17	18–63	Sarnthein and
	Mean	8.8	30	Walger (1974)
Jafurah Sand Sea, Saudi Arabia	($n = 16$)	<30	15	Fryberger *et al.* (1984)
Transverse dunes				
Namib Sand Sea		2.5	5	Lancaster (1985a)
Erg Oriental		35	0.3	
		240	0.16	Wilson (1972a)

Source: Thomas (1992).

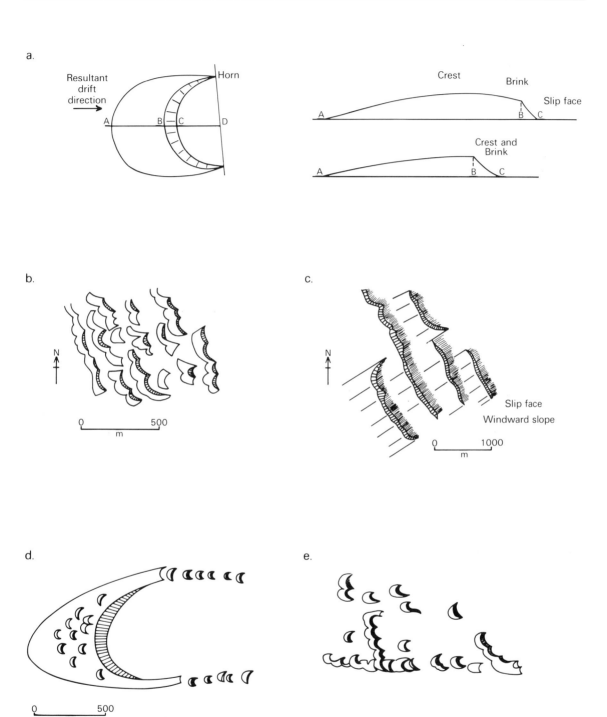

Figure 17.16 *(a) Barchan dune terminology. (b) Compound barchanoid ridges, White Sands, New Mexico and (c) transverse ridges (from aerial photographs in Breed and Grow 1979). (d) Compound megabarchan, Peru (after Simons 1956), (e) En echelon linked barchans, northern Sahara (after Clos-Arceduc 1969)*

THE DEVELOPMENT AND MOVEMENT OF BARCHANS

When sand saltating over a stone pavement encounters a patch of sand, the rate of forward movement is slowed because saltation paths are shorter on soft surfaces (Bagnold 1941). As the rate of particles arriving from upwind exceeds the rate of particle despatch on the downwind side, sand accumulation and dune development occur on the patch. Deposition will be greatest on the lee side where wind velocities are lower and flow patterns divergent, and a slip face develops as the angle of repose for sand is reached.

Sand accumulation is least at the edges of the original patch, where saltation rates are greatest, so the dune and slip face become highest in the middle. This leads to the development of the barchan horns. The slip face traps the same amount of sand regardless of position, but moves forward more rapidly at the edges where the dune is lowest, creating a concave profile (see Howard *et al.* 1978). Whether the horns are open or closed, long or short, depends on the rate of sand supply, degree of flow saturation and wind regime (Howard *et al.* 1978).

Migration rates of almost 40 m yr^{-1} have been recorded for small barchans (Finkel 1959), but as the volume of sand contained in a dune increases, the rate of movement declines (Table 17.9), especially in the case of isolated dunes moving over flat surfaces (Watson 1985). Barchan form and size usually remain constant during migration (Norris 1966) as dunes tend to develop a form which is in dynamic equilibrium or steady state with the controlling variables (Tsoar 1985; 1986; Lancaster 1987; Watson 1987). If sand supply decreases over time, dune size may shrink, too (Hastenrath 1987).

An equilibrium transverse dune profile is convex. Following the principle that boundary-layer flow compression occurs over hills (see Jackson and Hunt 1975; Walmsley *et al.* 1982), sand transport rates can be expected to increase towards the crest of a transverse form. At any point on the windward face of the dune, this can result in the continued movement of sand arriving from downwind as well as the erosion of surface sand, enabling forward momentum

to be maintained. The dune is not constantly lowered by this process because flow separation occurs in the crestal zone (Tsoar 1985; Lancaster 1987). As wind shear, responsible for particle entrainment, is also affected by factors such as slope angle and particle size as well as wind velocity (Walmsley and Howard 1985), this model probably oversimplifies the complex relationships which are likely to pertain between airflow and the surface of a dune (Watson 1987).

Particles arriving from upwind, together with newly entrained grains, progressively move up the windward slope as a transient climbing load. Deposition is temporary, usually in the form of ripples. Much deposition takes place on the lee-side slip face in steeply dipping foreset beds (32–33°, McKee 1957), through the effect of grainflow (see Kocurek and Dott 1981; Hunter 1985). These foresets re-emerge on the windward side as dune migration takes place (Hunter 1985) and bevelled remnants may remain on the desert floor after the passage of a dune (McKee 1966).

Some transverse dune forms have separate crests (highest point) and brinks (top of slip face) (Figure 17.16a; for discussion, see Tsoar 1985; and Watson 1987) as flow separation often occurs near the crest of dunes (Tsoar 1985) grainfall and tractional laminae are most likely to accrete in the zone between the crest and brink, or on both sides of the crest (Lai and Wu 1978; Watson 1986; McArthur 1987).

Linear dunes

Many linear dunes are simple forms, occurring in extensive dunefields, with individual dunes usually about 20 m high and up to 1 km apart (Goudie 1970; Twidale 1972), though they can attain considerable lengths and heights of up to 200 m. They occur in a relatively wide range of wind environments, displaying both directional variability and considerable differences in sand-moving capabilities (Fryberger 1979; Figure 17.10b).

Like transverse dunes, a considerable degree of morphological variation exists within the total population of linear dunes. Linear dunes typically consist of a plinth which is frequently vegetated, and a more active crestal zone. There are two broad classes of simple forms (Figure 17.8): seif dunes and linear ridges (see Breed and Grow 1979).

Seif dunes have sharp but sinuous crests with narrow reversing slip faces (Bagnold 1941; Tsoar 1982). Linear ridges, sometimes called *alâb* dunes (Monod 1958), are straighter, have convex asymmetrical crests and are generally more subdued forms. In the Simpson Desert, the steepest side always faces east, and in the southwest Kalahari to the southwest (Mabbutt 1968; Grove 1969). The crests may also be

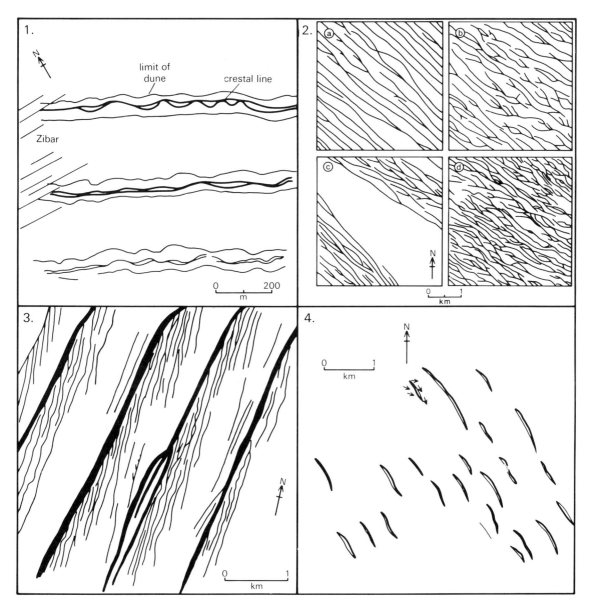

Figure 17.17 *Variations in linear dune morphology. 1. Seifs with zibar in interdune corridors: Sinai (after Tsoar 1983). 2. Different linear ridge patterns: SW Kalahari (after Thomas 1986). 3. Feathered linear dunes: Oman (after Glennie 1970). 4. Hybrid linear/transverse ('oblique') dunes: Williams River dunefield, Canada (after Carson and MacLean 1986)*

vegetated, though less than the plinths, and evidence of sand movement limited or localised (Grove 1969; Wasson and Hyde 1983a).

A notable characteristic of linear ridges in the Simpson and Kalahari deserts is that adjoining ridges often branch or merge at 'Y' junctions. In the Simpson Desert, the vast majority of the junctions are open to the wind (Mabbutt and Wooding 1983), though in the Kalahari over 15% point downwind (Thomas 1986). In the Kalahari, junctions are most common where ridges are closest together, probably occurring through the deflection of ridges under the action of winds from the different modal directions. Another important compound form, found for example, in the Wahiba Sands, Oman (Glennie 1970) and Sinai (Tsoar and Møller 1986), are feathered linear dunes (Figure 17.17), which have small secondary ridges extending from or intersecting with a larger major ridge.

Larger compound forms can have several parallel or subparallel crests developed on a single plinth (Livingstone 1986). In the Namib, such forms reach heights 160 m above interdune corridors (Lancaster 1981). Large complex forms with star dune peaks linked by sinuous crests can have smaller transverse forms on their flanks (Holm 1960; Lancaster 1981). It is a matter of conjecture whether these dunes are primarily linear or star dunes (see, for example, the classification of Monod 1958; Holm 1960; Mainguet and Callot 1978; Lancaster *et al.* 1987).

LINEAR DUNE FORMATION

Although linear dunes are the most common desert dune type (Table 17.6), they are probably the least understood in terms of formation

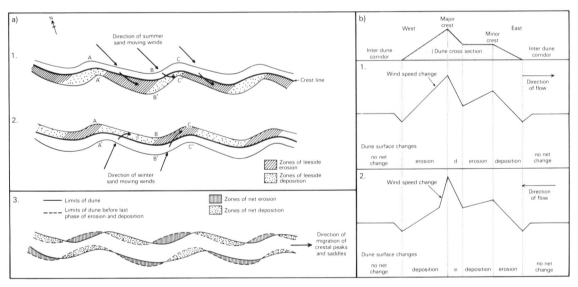

Figure 17.18   *(a) Tsoar's (1978) model of seif dune development, based on observations in the Sinai Desert, Egypt. Effective dominant winds are bimodal. (1) Summer winds approach from the northwest. Sand is eroded along the windward side of the dune. Winds approach the crest at <40° between A and B; the deflected lee-side winds erode sand in zone A'B'. Winds approaching section BC at <40° are also deflected after crossing the crestline but speeds are not enhanced: consequently deposition occurs in zone B'C'. (2) Winter winds from the southwest erode on the windward side of the crest and in zone BC on the leeside. (3) Overall, therefore, zones BC and A'B' are zones of net erosion and AB and B'C' are zones of net deposition. A pattern of peaks and saddles develop along the dune, which advances downwind under successive phases of sand movement. (b) Livingstone's (1986) model of linear dune development, based on observations in the Namib Sand Sea. (1) Shows changes in wind speed and dune surface conditions under winds blowing from the west, and (2) for winds blowing from the east. Although along-dune sand movement occurs because these winds blow obliquely to the crestline, there are no zones of overall net erosion and deposition, as the effects of winds from one modal direction are counteracted by those blowing from the opposition mode*

(McKee 1979) and certainly the most contentious (see, for example, Lancaster 1982). For many years linear dunes were widely believed to develop through the effects of parallel helical atmospheric vortices in unimodal wind regimes (Bagnold 1953; Hanna 1969; Glennie 1970; Folk 1971; 1976; Wilson 1972a; Warren 1979), though no direct relationship has ever been demonstrated been large vortices and dune development (Warren 1984; see also Leeder 1977 for an interesting discussion).

While some authorities persist as proponents of this hypothesis (see, for example, Glennie 1986; 1987; Tseo 1993), recent studies of actual windflow and sand movement over linear dunes (Tsoar 1978; Livingstone 1986) have provided justifiable grounds for its rejection. It has been demonstrated that linear dune development occurs in bidirectional wind regimes, which can presumably include the wide unimodal regimes also identified by Fryberger (1979), without the involvement of large helical vortices. Instead, the protrusion of the linear dune itself into the boundary layer creates windflow modifications which play a central role in dune dynamics.

Tsoar (1978; 1983) observed that winds passing obliquely across linear dune crests are deflected through the effect of flow separation at the crest to blow parallel to the dune axis on the lee side. Consequently, sand transported across the crest does not leave the dune, but is carried down its length. Because seif dune crests meander, the angle of wind approach fluctuates along the length of the dune, which affects whether the lee-side flow is erosional or depositional (Figure 17.18).

As formative winds are bidirectional, either seasonally (Tsoar 1978) or daily (McKee and Tibbitts 1964), zones of net erosion and net deposition develop on both flanks of the dune (Figure 17.18). Consequently, crestal peaks and saddles develop along the dune, which advance down dune in the resultant drift direction (Tsoar 1978).

Internal dune structure is affected by this process (Figure 17.19). When wind direction is acute to the crest, the zones of erosion on the lee side will probably have a rippled surface. Accretion in the lee-side deposition zones occurs through grainfall (as wind speed drops) and grainflow, giving rise to sets of steeply dipping laminae. As the zones of erosion and deposition advance down dune, deposition occurs on the former eroded zones, starting at angles of 10–20°, steepening as deposition progresses through the season, giving rise to an increase in grainflow structures (Tsoar 1982).

a)

b)

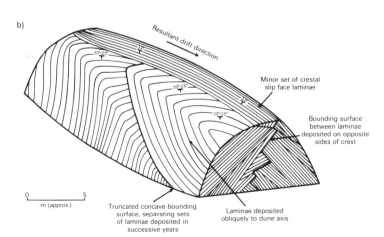

Figure 17.19 *(a) Internal structure of a linear dune (after Bagnold 1941). (b) Complex internal structure observed by Tsoar (1982)*

As the dune crest is sinuous, the bounding surfaces between sets of laminae are curved (Ahlbrandt and Fryberger 1982). Deposition on the plinth is almost always in the form of low-angle laminae (see Fryberger *et al.* 1979; Kocurek 1986).

The structures observed by Tsoar (1982) are complex, clearly differing from the simple model of Bagnold (1941). This consisted of alternate interdigitating units of grainflow deposits (Figure 17.19) and was supported by the observations of McKee and Tibbitts (1964) and McKee and Bigarella (1979). Differences may depend on whether dune formation is in an environment with seasonal or diurnal shifts in wind direction (Tsoar 1982).

Livingstone (1986) investigated the development of a large double-crested linear dune in the Namib Sand Sea, identifying atmospheric flow-line compression in the dune crestal zone as the most important factor for sand movement on the dune. Livingstone (1986) considered that wind-velocity changes caused by the protrusion of the dune into the atmosphere created net erosion on the windward side and net deposition to the lee of the crest (Figure 17.18), with zones of erosion and deposition alternating seasonally under the influence of winds from each of the modal directions. Observations by Lancaster (1987) also support this picture. On the dune studied by Livingstone (1986), flow deviation to the lee of the crest was neither persistent nor important for sand movement along the ridge, perhaps because the study dune was larger and more complex than the one investigated by Tsoar (1978).

Livingstone (1986), therefore, found that sand *did* leave the leeside of the dune, though this was balanced by sand arriving on the windward side. Interdune areas were zones of sand transport which did not experience net erosion or net deposition.

The angle between the components of the bimodal wind regime also affects dune development and morphology. When the angle between the dominant directions is large, dunes tend to build up into higher forms, as in the Namib. When it is narrower, along-dune sand transport will be greater and dune forms lower. Consequently, once an equilibrium dune form has developed, dune size is a function of the interaction between wind regime and sand supply (Livingstone 1986).

A seif dune has a sharp triangular crest, which is clearly not an equilibrium profile (see page 396). When the dune profile is considered parallel to each of the oblique wind directions, seif cross-sectional profiles are convex, comparable to the equilibrium forms of transverse dunes (Tsoar 1985).

On occasions when winds approach a seif perpendicular to the crest line, the profile is out of equilibrium. Under such conditions in the Namib, crestal erosion ensues (Lancaster 1985b), though this is only temporary as the form reverts under conditions of oblique flow. In the Sinai, extra erosion does not result because the triangular profile which perpendicular winds encounter causes enhanced flow separation in the crestal zone (Tsoar 1985).

OTHER ISSUES IN LINEAR DUNE DEVELOPMENT

The initiation of linear dunes still requires explanation. Bagnold (1941), McKee and Tibbitts (1964), Verstappen (1968) and Tsoar (1974) considered that this could result from the elongation of one arm of a barchan dune. Alternatively, development could be initiated through the blowing out of the nose of a parabolic dune (King 1960; Verstappen 1968). An alternative explanation for linear dune genesis is based on observed associations with coarse sand *zibar* in the Ténéré and Sinai deserts (Warren 1972; Tsoar 1978; see Figure 17.17). Saltation rates are high across the compact zibar surfaces, slowing down as particles move off the compact surface, creating favourable conditions for sand accumulation. Linear dunes then extend downwind from the zibar if this occurs in environments with bimodal winds.

These explanations appear less convincing for the origins of linear ridges, which usually occur in locations where other dune types are virtually absent. Linear ridges usually have at least a partial vegetation cover, which has often led to their being credited with an inactive, degraded and relict status. Tsoar and Møller (1986) have suggested that vegetation may be a relatively normal component of dunes in the less arid parts of sand seas, while, in the southwestern

Table 17.10 *Dune activity rates, linear forms*

Location	Dimensions (m)		Movement (m yr^{-1})			References
	Height	Width	Along-dune	Elongation	Lateral migration	
Seif dunes						
Sinai	12–14	50	8.4	14.5	Crest 5–7* Plinth 0	Tsoar (1978)
Namib	50	350			Crest 15* Plinth 0	Livingstone (1978)
Namib				>1.85		Ward (1984)
Linear ridges						
Strezlecki, Australia	10–20				50–100 m since Pleistocene	Rubin (1990)
William River, Canada†	12–30	120–160			0.2–0.6	Carson and MacLean (1986)

*Values are seasonal reversals, resulting in no net change.
†Dunes termed 'hybrid dunes' by authors, but regarded as linear ridges by Rubin (1990).
Source: Thomas (1992).

Kalahari, dunes with relatively densely vegetated plinths often display crestal activity (Grove 1969), with the level of activity determined by the interactions between wind velocities and vegetation density (Wiggs *et al.* 1995a; see Chapters 7 and 16). Such dunes should not, therefore, be simply considered as inactive forms on the basis of the presence of surface vegetation alone (Thomas and Shaw 1991). In fact, in terms of the ratio of the depth of dune surface change to dune cross-sectional area, the partially vegetated linear ridges of the southwestern Kalahari display a level of surface change comparable to that affecting the linear megadunes of the Namib Desert (Wiggs *et al.* 1995a; 1995b).

Overall, linear dunes display a range of activities and morphological behaviour, including elongation, along-dune sand transport, slow lateral migration (in situations where one component of the bimodal wind regime is slightly dominant) and seasonal surface relief changes (Table 17.10).

Star dunes

As Figure 17.11 shows, star dunes contain greater volumes of sand than other dune types and occur in environments with great directional variability in sand transporting wind directions. Nielson and Kocurek (1986) have shown that star dunes can develop where seasonal variations in dominant wind directions cause other dune types to merge and modify. If one wind direction has overall dominance, the incipient star form can be destroyed as the growth of one arm is favoured. However, if the incipient form is large enough, secondary windflows created over the other arms can counteract this, leading to their preservation and the ultimate growth of a star dune.

It is not surprising, given these characteristics, that star dunes often occur in the depositional centres of sand seas (e.g. Figure 17.12). They can also occur in association with topographic barriers (Breed and Grow 1979; Lancaster 1994), which may alter regional windflow and increase directional complexities. As star dunes grow in size they increasingly modify their own wind environments (Figure 17.20), leading to both the maintenance and growth of the overall star form and the development of other dune forms on their flanks and to many star dunes being complex forms. The position of star dunes in sand seas and their sand-accumulating behaviour (Thomas 1992) results in many of the world's largest reported dunes being of this type.

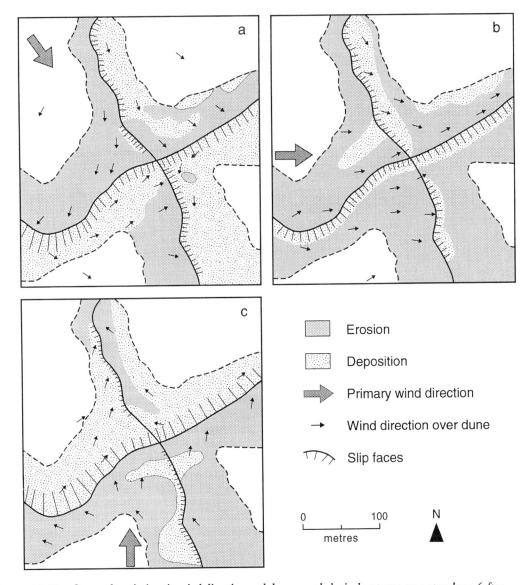

Figure 17.20   *Seasonal variations in wind direction and dune morphological response on a star dune (after Lancaster 1989; 1995)*

Star dunes of between 300 and 400 m high are reported from China, Iran, Namibia and Algeria (Wilson 1973; Mainguet and Chemin 1983; Walker *et al.* 1987).

Star dunes can occur in linked chains, for example in the Gran Desierto, Mexico (Lancaster 1995). Imbalances in the sand-transporting potential of different components of the overall wind regime can lead to migration of a star dune in one overall direction. Lancaster (1989) reports the annual displacement by $1-3 \text{ m yr}^{-1}$

of a star dune in the Gran Desierto, while seasonal displacement of individual dune arms of 10–20 m are reported from this dunefield and the Kelso dunes of the Mojave Desert (Sharp 1966; Lancaster 1989).

Dune form equilibrium and 'memory'

Migratory and sand-passing dunes do not grow indefinitely. If windflow and sand supply are constant, an equilibrium form will develop, and sand will be passed on at the same rate at which it is received, leading to dune migration and dune extension in the respective cases of transverse and linear dunes. Some dune forms are adapted to seasonal or even diurnal wind regimes (Tsoar 1978; 1985; Livingstone 1986): changes occur in the dune profile, especially in the crestal zone, but the basic form remains.

Total dune form will respond to quasipermanent changes in wind regime and sediment supply, such as those induced by a climatic change, or even to seasonal changes if the dune is small (see Nielson and Kocurek 1986). Larger dunes probably take longer to adjust to new environmental conditions that smaller dunes because the greater volume of sand which they contain requires more time to establish a new equilibrium form. The ability of a dune to retain its form when environmental conditions change has been termed 'dune memory' (Warren and Kay 1987). Thus 'dune systems with little sand have little memory, those with more sand "remember" the effect of the most dominant wind throughout the year: some dunes may have "mega memories"', remembering events from the Pleistocene' (Warren and Kay 1987, p. 205). Different memory scales of dunes are shown in Figure 17.21.

The concept of dune memory is closely related to the more widely applied idea of bedform reconstitution (Allen 1974). Reconstitution time is the time it takes a bedform to move one wavelength in the direction of net transport. Clearly an advantage of the memory concept over that of reconstitution is that it can be readily applied to dune forms for which migration is not a major characteristic. In the Namib Sand Sea, Lancaster (1988) notes that the average reconstitution time for crescentic dunes with a mean spacing of 90 m and a migration rate of 3 m yr^{-1} is 30 years. In terms of dune memory, it would take such dunes 30 years to adjust their form and orientation to a new wind regime — if other controlling variables remained constant. Consequently, a change in wind regime in an area of large dunes could result in the superimposition of a new dune pattern upon the pre-existing one (Fryberger 1980; Glennie 1986); this is therefore a mechanism which could account for the development of some compound and complex dunes.

DUNE SPACING

Mabbutt and Wooding (1983) and Thomas (1986) have shown that in areas of linear sand ridges in Australia and the Kalahari, new ridges begin when ridges extending from upwind either end or merge. Thus an equilibrium pattern tends to develop which maintains the number of linear ridges per unit area. Studies of linear dune morphometry have shown significant linear relationships between the height and spacing of sand ridges within spatial blocks of dunes, with dunes getting higher as spacing increases (Wasson and Hyde 1983b; Thomas 1988b). This is a relationship that appears to apply to all major dune types (Lancaster 1994; Figure 17.22).

Helical roll vortices have been suggested as determining the spacing of linear dunes by Folk (1971) and Tseo (1993), though as noted earlier the effect of these vortices on linear dune development is debatable. The fact that linear dune spacing can vary over quite short distances (Bullard *et al.* 1995) would appear to discount such a control, which is further supported by the fact that atmospheric roll vortices have a mean wavelength of 5 km (Kelly 1984), substantially greater than the maximum observed regular spacing of linear dunes patterns.

Wilson (1972b) proposed, on the basis of studies from Saharan sand seas, that a relationship existed between dune spacing and the coarse 20th percentile of crestal sand grain size.

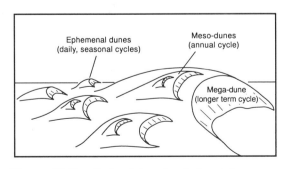

Figure 17.21 *Scales of dune in relation to the concept of dune memory. Although illustrated here in terms of transverse forms, the principles apply to all dune types (based on Cooke* et al. *1993)*

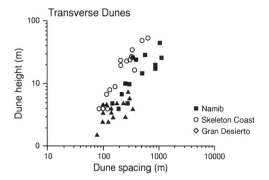

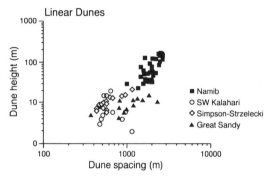

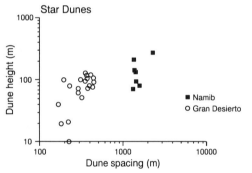

Figure 17.22 *Relationship between dune height and spacing for the principal dune types (after Lancaster 1994)*

Sedimentary variations across dune profiles

Dune sediments inherit some characteristics from the parent material (see McKee and Tibbitts 1964; Jawad Ali and Al-Ani 1983), upon which characteristics imparted by aeolian sorting are imposed. Aeolian processes can cause an increase in sorting and a reduction in mean grain size in the direction of sand transport within sand seas (Fryberger et al. 1984; Lancaster 1986), though there can be significant between-dune differences in grain-size characteristics (Livingstone 1987).

Grain-size variations also occur across dune profiles, in response to changes in windflow and sand transport characteristics (see the previous two sections). Consequently discrete sediment populations do not exist at different positions on a dune, but progressive changes take place across the profile (Watson 1986; Livingstone 1987). Seasonal and diurnal changes in surface sand characteristics can also be imparted by variations in wind regime.

Two distinct patterns of grain size variations have been observed across dune profiles (Table 17.11). The first is when dune crests are composed of coarser and better-sorted sand than the lower parts of dune flanks, as reported, for example by Lewis (1936) from the Kalahari, Folk (1971) from the Simpson Desert and Chandri and Khan (1981) from the Thar Desert. In the other case, crestal sand is finer but still better sorted than sediments elsewhere on the dune. This has been recorded from linear dunes in Libya, New Mexico, the Wahiba Sands, Oman, and Namib (Bagnold 1941; McKee and Tibbitts 1964; Glennie 1970; Lancaster 1981; Watson 1986) and from star dunes in the Gran Desierto in Mexico (Lancaster et al. 1987).

Both patterns result from the interactions between the parent sediment and aeolian transport processes. In each case the sediment fraction which is most readily transported by the wind, in the medium sand-size range (0.125–0.25 mm), dominates the upper dune positions. If the parent sediment is largely composed of coarser particles, the crestal sand will therefore appear to be finer; if the source sand has a finer mean grain size than the preferentially winnowed component, the opposite will occur.

Several tests of this relationship from linear dunefields (Wasson and Hyde 1983b; Lancaster 1988; Thomas 1988b) have not supported its validity. But for barchan and transverse dunes, spacing may become greater as the coarse 20th percentile grain size increases (Lancaster 1985a). This may be because transverse dunes formed from coarse sands tend to be lower and broader, giving a greater crest-to-crest spacing than those in finer sediments (Tsoar 1986; Lancaster 1995).

Table 17.11   *Examples of grain-size variations across linear dune profiles (phi units)* *

		Crest	Windward side			Slip face		
			a	b	c	a	b	c
Crests coarsest and best sorted								
Simpson (Folk 1971)†	M_z	2.53			2.75			
	σ_1	0.43			0.57			
SW Kalahari (Lancaster 1986)	M_z	2.16			2.21			2.26
	σ_1	0.49			0.62			0.59
Crests best sorted but slipfaces coarsest								
Sabha, Libya (Ahlbrandt 1979)‡	M_z	1.82			1.91			1.62
	σ_1	0.82			0.36			0.34
Namib Desert (Lancaster 1981)	M_z	2.44			2.41	2.32	2.49	
	σ_1	0.37			0.50	0.40	0.34	
Sinai Desert (Sneh and	M_z	–	2.21	2.40			2.08	2.28
Weissbrod 1983)	σ_1	–	0.60	0.50			0.35	0.50

Key:
a = lower slope position; b = upper slope position; c = position on slope undifferentiated.
M_z = mean grain size (Folk and Ward 1957).
σ_1 = sorting (one standard deviation) (Folk and Ward 1957).
* Larger values equal smaller grain diameters.
† Dune flanks undifferentiated.
‡ Uncertain whether all values are from the same dune.

Detailed investigations by Sneh and Weiss-brod (1983) and Watson (1986) have suggested a general model for grain-size variations across dune profiles. The sorting of linear dune windward flank sand increases towards the crest. On the other hand, slip-face sands always become better sorted downslope, under the influence of grainflow deposition. This picture must however be complicated for seif dunes because of the down-dune migration of the zones of net deposition and net erosion.

Transverse dunes tend to display the same pattern of grain-size changes in the slip face, but variations on the windward face are less consistent. Sorting also varies with changes in slope angle (Watson 1986) because of the effect of slope and morphology on wind shear. Some studies have found the coarsest and best-sorted sand at the foot of the windward face, due to the re-emergence of basal slipface deposits during the process of dune migration and the preferential removal of fine sand from the convex lower slope units (Watson 1986). However, other studies of transverse dunes have found the coarsest sand at the crest (Alimen

1953; Sharp 1966; McArthur 1987), perhaps due to the deposition of the coarsest sand in entrainment in the zone of reduced shear stress at the crest of dunes with convex profiles.

The apparently contradictory results of some studies of transverse dune sediment characteristics (see Watson 1986; McArthur 1987) may highlight the complex array of factors and interactions which can influence sediment characteristics on dunes (see Watson 1987). They may also, at least in part, result from the varying sensitivities of the different statistical techniques employed to describe sediment characteristics (see Flenley *et al.* 1987).

In some dunefields, the interdune areas are composed of a bimodal sand while the dune crests consist of better-sorted material (see Folk 1971; Warren 1971; 1972). The preferential removal of the most readily saltated component leaves behind a mix of the finest and coarsest grains. This moves more slowly as a creep load which contributes to the sands of dune plinths and interdune areas. Interdune areas may also accumulate thin, low-angle beds (Fryberger *et al.* 1979; Kocurek 1986), or lag deposits resulting from the passage of dunes. As even linear and star dunes can display some lateral migration (Rubin and Hunter 1985; Kocurek 1986), the more unstable, steeply bedded, upper components of dunes are perhaps less likely to be preserved in the very long term. This may explain why low-angle plinth and interdune deposits are well represented in the aeolian rock record (see Kocurek 1981; Rubin and Hunter 1985).

Conclusion

Understanding of the nature, distribution and dynamics of aeolian sand deposits and bedforms has been advanced substantially by the application of remote sensing techniques and the results gained through rigorous field measurement and monitoring programmes. The use of theoretical modelling and laboratory simulation to investigate the processes of aeolian entrainment and deposition is also contributing to our knowledge of bedform development. As human economic, agricultural and residential pressures have increased in the world's arid lands, geomorphological knowledge of the 20% covered by aeolian sands has taken on a new significance, too (see Cooke *et al.* 1982; Watson 1985), and will undoubtedly continue to do so.

References

Ahlbrandt, T.S. 1979. Textural parameters of aeolian deposits. *US Geological Survey Professional Paper*, 1052: 21–51.

Ahlbrandt, T.S. and Fryberger, S.G. 1982. Introduction to aeolian deposits. In P.A. Scholle and D. Spearing (eds), *Sandstone depositional environments*. American Association of Petroleum Geologists, Tulsa, OK: 11–47.

Al Janabi, K.Z., Jaward Ali, A. Al-Taie, F.H. and Jack, F.J. 1988. Origin and nature of sand dunes in the alluvial plain of southern Iraq. *Journal of Arid Environments*, 14: 27–34.

Alimen, M.-H. 1953. Variations granulometriques et morphoscopiques du sable le long de profils dunaires au Sahara occidental. *Actions Eoliennes. Centre Nationale de Recherche Scientifique, Paris*, 35: 219–235.

Allen, J.R.L. 1974. Reaction, relaxation and lag in natural sediment systems: general principles, examples and lessons. *Earth Science Reviews*, 10: 263–342.

Anderson, R.S. 1987. A theoretical model for aeolian impact ripples. *Sedimentology*, 34: 943–956.

Anderson, R.S. 1990. Eolian ripples as examples of self organisation in geomorphological systems. *Earth Science Reviews*, 29: 77–96.

Anderson, R.S. and Hallet, B. 1986. Sediment transport by wind: toward a general model. *Bulletin of the Geological Society of America*, 97: 523–535.

Anton, D. and Vincent, P. 1986. Parabolic dunes of the Jafurah Desert, Eastern Province, Saudi Arabia. *Journal of Arid Environments*, 11: 187–198.

Ash, J.E. and Wasson, R.J. 1983. Vegetation and sand mobility in the Australian desert dunefield. *Zeitschrift für Geomorphologie*, Supplementband, 45: 7–25.

Bagnold, R.A. 1941. *The physics of blown sand and desert dunes*. Methuen, London.

Bagnold, R.A. 1953. The surface movement of blown sand in relation to meterology. In *Desert Research: Proceedings of an International Symposium, Jerusalem, May 1952*. Research Council of Israel Special Publication, 2: 89–93.

Beadnell, H.J.L. 1910. The sand-dunes of the Libyan Desert. *Geographical Journal*, 35: 379–395.

Berg, A.W. 1990. Formation of mima mounds: a seismic hypothesis. *Geology*, 18: 281–284.

Boucart, J. 1928. L'Action du vent à la surface de la terre. *Revue Géomorphologie Physique et Geologie Dynamique*, 5: 102–123.

Bowler, J.M. and Wasson, R.J. 1984. Glacial age environments of inland Australia. In J.C. Vogel (ed.), *Late Cainozoic palaeoclimates of the southern hemisphere*. Balkema, Rotterdam: 183–209.

Breed, C.S. and Grow, T. 1979. Morphology and distribution of dunes in sand seas observed by remote sensing. *US Geological Survey, Professional Paper*, 1052: 253–302.

Breed, C.S., Fryberger, S.G., Andrews, S., McCauley, C., Lennartz, F., Gebel, D. and Horstman, K. 1979. Regional studies of sand seas using Landsat (ERTS) imagery. *US Geological Survey, Professional Paper*, 1052: 305–397.

Breed, C.S., McCauley, J.F. and Davis, P.A. 1987. Sand sheets of the eastern Sahara and ripple blankets on Mars. In L. Frostick and I. Reid (eds), *Desert sediments ancient and modern*. Geological Society of London Special Publication, 35. Blackwell, Oxford: 337–359.

Brugmans, F. 1983. Wind ripples in an active drift area in the Netherlands: a preliminary report. *Earth Surface Processes and Landforms*, 8: 527–534.

Bullard, J.E., Thomas, D.S.G., Livingstone, I. and Wiggs G.F.S. 1995 Analysis of linear sand dune morphological variability, southwestern Kalahari. *Geomorphology*, 11: 189–203.

Carson, M.A. and MacLean, P.A. 1986. Development of hybrid aeolian dunes: the William River dune field, northwest Saskatchewan, Canada. *Canadian Journal of Earth Sciences*, 23: 1974–1990.

Chandri, R.S. and Khan, H.M.M. 1981. Textural parameters of desert sediments — Thar Desert (India). *Sedimentary Geology*, 28: 43–62.

Clos-Arceduc, A. 1969. *Essai d'explication des formes dunaires sahariennes*. Etudes de photointerpretation, 4. Institute Geographie Nationale, Paris.

Cooke, R.U. and Warren, A. 1973. *Geomorphology in deserts*. Batsford, London.

Cooke, R.U., Brunsden, D., Doornkamp, J.C. and Jones, D.K.C. 1982. *Urban geomorphology in drylands*. Oxford University Press, Oxford.

Cooke, R.U., Warren, A. and Goudie, A.S. 1993 *Desert geomorphology*. UCL Press, London.

Cooper, W.S. 1967. Coastal dunes of California. *Memoir of the Geological Society of America*, 104.

Cornish, V. 1914. *Waves of sand and snow*. Fisher-Unwin, London.

Cox, G.W. and Gakalm, C.G. 1986. A latitudinal test of the fossorial rodent hypothesis of mima mound origin. *Zeitschrift für Geomorphologie*, 30: 485–501.

Dare-Edwards, A.J. 1984. Aeolian clay deposits of south-eastern Australia: parna or loessic clay? *Transactions of the Institute of British Geographers*, 9: 337–344.

Ellwood, T.M., Evans, P.D. and Wilson, I.G. 1975. Small scale aeolian bedforms. *Journal of Sedimentary Petrology*, 45: 554–561.

Eriksson, P.G., Nixon, N., Snyman, C.P. and Bothma, J. du P. 1989 Ellipsoidal parabolic dune patches in the southern Kalahari Desert. *Journal of Arid Environments*, 16: 111–124.

Finkel, H.J. 1959. The barchans of southern Peru. *Journal of Geology*, 67: 614–647.

Flenley, E.C., Fieller, N.R.J. and Gilbertson, D.D. 1987. The statistical analysis of 'mixed' grain size distributions from aeolian sands in the Libyan pre-desert using log skew Laplace models. In L. Frostick and I. Reid (eds), *Desert sediments ancient and modern*. Geological Society of London Special Publication, 35. Blackwell, Oxford: 271–280.

Folk, R.L. 1971. Longitudinal dunes of the north-western edge of the Simpson Desert, Northern territory. Australia. 1. Geomorphology and grain size relationships. *Sedimentology*, 16: 5–54.

Folk, R.L. 1976. Rollers and ripples in sand, streams and sky, rhythmic alteration of transverse and longitudinal vortices in three orders. *Sedimentology*, 23: 649–669.

Folk, R.L. 1977. Folk's bedform theory — reply. *Sedimentology*, 24: 864–847.

Folk, R.L. and Ward, A.C. 1957. Brazos river bar: a study in the significance of grain size parameters. *Journal of Sedimentary Petrology*, 27: 3–26.

Forrest, S.B. and Haff, P.K. 1992. Mechanics of wind ripple stratigraphy. *Science*, 225, 1240–1243.

Fryberger, S.G. 1978. *Techniques for the evaluation of surface wind data in terms of eolian sand drift*. US Geological Survey, Open File Report, 78–405.

Fryberger, S.G. 1979. Dune form and wind regime. *US Geological Survey, Professional Paper*, 1052: 137–169.

Fryberger, S.G. 1980. Duneforms and wind regime, Mauritania, West Africa: implications for past climate. *Palaeoecology of Africa*, 12: 79–96.

Fryberger, S.G. and Ahlbrandt, T.S. 1979. Mechanism for the formation of aeolian sand seas. *Zeitschrift für Geomorphologie*, NF, 23: 440–460.

Fryberger, S.G. and Goudie, A.S. 1981. Arid geomorphology. *Progress in Physical Geography*, 5: 420–428.

Fryberger, S.G., Ahlbrandt, T.S. and Andrews, S. 1979. Origin, sedimentary features and significance of low-angle eolian 'sand sheet' deposits, Great Sand Dunes, National Monument and vicinity, Colorado. *Journal of Sedimentary Petrology*, 49: 733–746.

Fryberger, S.G., Al-Sari, A.M., Clisham, T.J., Rizvi, S.A.R. and Al-Hinai, K.G. 1984. Wind sedimentation in the Jafurah sand sea, Saudi Arabia. *Sedimentology*, 31: 413–431.

Fryberger, S.G., Schenk, C.J. and Krystinik, L.F. 1988. Stokes surfaces and the effects of near surface groundwater-table on aeolian deposition. *Sedimentology*, 35: 21–41.

Fryberger, S.G., Hesp, P. and Hastings, K. 1992. Aeolian ripple deposits, Namibia. *Sedimentology*, 39: 319–331.

Gayloard, D.R. and Dawson, P.J. 1987. Airflow–terrain interactions through a mountain gap, with an example of eolian activity beneath an atmospheric hydraulic jump. *Geology*, 15: 789–792.

Glennie, K.W. 1970. *Desert sedimentary environments*. Elsevier, Amsterdam.

Glennie, K.W. 1986. Early Permian (Rotliegendes) palaeowinds of the North Sea - reply. *Sedimentary Geology*, 45: 297–313.

Glennie, K.W. 1987. Desert sedimentary environments, present and past — a summary. *Sedimentary Geology*, 50: 135–166.

Goudie, A.S. 1970. Notes on some major dune types in South Africa. *South African Geographical Journal*, 52: 93–101.

Goudie, A.S. and Thomas, D.S.G. 1985. Pans in southern Africa with particular reference to South Africa and Zimbabwe. *Zeitschrift für Geomorphologie*, NF, 29: 1–19.

Goudie, A.S. and Thomas, D.S.G. 1986. Lunette dunes in southern Africa. *Journal of Arid Environments*, 10: 1–12.

Goudie, A.S., Warren, A., Jones D.K.C. and Cooke, R.U. 1987. The character and possible origins of the aeolian sediments of the Wahiba Sand Sea, Oman. *Geographical Journal*, 153: 231–256.

Greeley, R. and Iversen, T.D. 1985. *Wind as a geological process on Earth, Mars, Venus and Titan*. Cambridge University Press, Cambridge.

Grolier, M., Ericksen, G.E., McCauley, T.F. and Morris, E.C. 1974. The desert landforms of Peru: a preliminary photographic atlas. *Astrogeology* (US Geological Survey Report), 57.

Grove, A.T. 1969. Landforms and climatic change in the Kalahari and Ngamiland. *Geographical Journal*, 135: 191–212.

Hack, J.T. 1941. Dunes of the western Navajo Country. *Geographical Review*, 31: 240–263.

Haff, P.K. and Presti, D.E. 1995. Barchan dunes of the Salton Sea region, California. In V.C. Tchakerian (ed.), *Desert aeolian processes*. Chapman & Hall, London: 153–177.

Haney, E.M. and Grolier, M.J. 1991. *Geologic map of major Quaternary aeolian features, northern and central coastal Peru*. US Geological Survey, Miscellaneous Investigation I-2162.

Hanna, S.R. 1969. The formation of longitudinal sand dunes by large helical eddies in the atmosphere. *Journal of Applied Meteorology*, 8: 874–883.

Hastenrath, S. 1987. The barchan sand dunes of Southern Peru revisited. *Zeitschrift für Geomorphologie*, NF, 31: 167–178.

Hesp, P.A., Hyde, R., Hesp, V. and Zhengyu, Q. 1989. Longitudinal dunes can move sideways. *Earth Surface Processes and Landforms*, 14: 447–452.

Holm, D.A. 1960. Desert geomorphology in the Arabian peninsula. *Science*, 132: 1329–1379.

Howard, A.D., Morton, T.B. Gad-El-Hak, M. and Pierce, D.B. 1978. Sand transport model of barchan dune equilibrium. *Sedimentology*, 25: 307–338.

Hunter, R.E. 1985. A kinetic model for the structure of lee-side deposits. *Sedimentology*, 32: 409–422.

Jackson, P.S. and Hunt, J.C.R. 1975. Turbulent wind flow over a low hill. *Quarterly Journal of the Royal Meteorological Society*, 101: 929–955.

Jawad Ali, A. and Al-Ani, R.A. 1983. Sedimentological and geomorphological study of sand dunes in the Western Desert of Iraq. *Journal of Arid Environments*, 6: 13–32.

Jensen, T.L. and Sorensen, M. 1986. Estimation of some aeolian saltation transport parameters: a re-analysis of William's data. *Sedimentology*, 33: 547–558.

Kelly, R.D. 1984. Horizontal roll and boundary-layer inter-relationships observed over Lake Michigan. *Journal of Atmospheric Sciences*, 41: 1811–1826.

King, D. 1960. The sand ridge deserts of South Australia and related aeolian landforms of Quaternary arid cycles. *Transactions of the Royal Society of South Australia*, 83: 99–108.

Kocurek, G. 1981. Significance of interdune deposits and bounding surfaces in aeolian dune sands. *Sedimentology* 28: 753–780.

Kocurek, G. 1986. Origins of low-angle stratification in aeolian deposits. In W.G. Nickling (ed.), *Aeolian geomorphology*. The Binghampton Symposia in Geomorphology, International Series, 17. Allen & Unwin, Boston: 177–193.

Kocurek, G. 1988. First order and super bounding surfaces in eolian sequences — bounding surfaces revisited. *Sedimentary Geology*, 56: 193–206.

Kocurek, G. and Dott, R.H. 1981. Distinctions and uses of stratification types in the interpretation of aeolian sand. *Journal of Sedimentary Petrology*, 51: 579–595.

Kocurek, G. and Havholm, K.G. 1994. Eolian sequence stratigraphy — a conceptual framework. In P. Weiner and H. Posamentier (eds), *Siliclastic sequence stratigraphy*. American Association of Petroleum Geologists, Tulsa: 393–409.

Kocurek, G. and Nielson, J. 1986. Conditions favourable for the formation of warm-climate eolian sand sheets. *Sedimentology*, 33: 795–816.

Kocurek, G., Townsley, M., Yeh, E., Sweet, M. and Havholm, K. 1990. Dune and dunefield development on Padre Island, Texas, with implications for interdune deposition and water table controlled accumulation. *Journal of Sedimentary Petrology* 62, 622–635.

Lai, R.J. and Wu, J. 1978. *Wind erosion and deposition along a coastal sand dune*. Sea Grant Program. University of Delaware, Report DEL-SG-10-78.

Lancaster, N. 1981. Grain size characteristics of Namib Desert linear dunes. *Sedimentology*, 24: 361–387.

Lancaster, N. 1982. Linear dunes. *Progress in Physical Geography*, 6: L475–504.

Lancaster, N. 1985a. Wind and sand movements in the Namib Sand Sea. *Earth Surface Processes and Landforms*, 10: 607–619.

Lancaster, N. 1985b. Variations in wind velocity and sand transport on the windward flanks of desert sand dunes. *Sedimentology*, 32: 501–593.

Lancaster, N. 1986. Grain size characteristics of linear dunes in the southwestern Kalahari. *Journal of Sedimentary Petrology*, 56: 395–400.

Lancaster, N. 1987. Reply. *Sedimentology*, 34: 516–520.

Lancaster, N. 1988. Controls on eolian dune size and spacing. *Geology*, 16: 972–975.

Lancaster, N. 1989. Star dunes. *Progress in Physical Geography*, 13: 67–91.

Lancaster, N. 1990. Palaeoclimatic evidence from

sand seas. *Palaeogeography, Palaeoecology, Palaeoclimatology,* 76: 279–290.

Lancaster, N. 1993. Development of Kelso Dunes, Mojave desert, California. *National Geographic Research and Exploration,* 9: 444–459.

Lancaster, N. 1994. Dune morphology and dynamics. In A.D. Abrahams and A.J. Parsons (eds), *Geomorphology of desert environments.* Chapman & Hall, London: 474–505.

Lancaster, N. 1995. *The geomorphology of desert dunes.* Routledge, London.

Lancaster, N. and Ollier, C.D. 1983. Sources of sand for the Namib sand sea. *Zeitschrift für Geomorphologie,* Supplementband, 45: 71–83.

Lancaster, N., Greeley, R. and Christensen, P.R. 1987. Dunes of the Gran Desierto, sand sea, Sonora, Mexico. *Earth Surface Processes and Landforms,* 12: 277–288.

Langford, I. 1989. Fluvial–aeolian interactions. Part 1 Modern systems. *Sedimentology,* 36: 273–289.

Leeder, M.R. 1977. Folk's bedform theory. *Sedimentology,* 24: 863–864.

Lettau, K. and Lettau, H. 1969. Transport of sand by the barchans of the Pampa La Toya in Southern Peru. *Zeitschrift für Geomorphologie,* NF, 13: 182–195.

Lewis, A.D. 1936. Sand dunes of the Kalahari within the Union. *South African Geographical Journal,* 19: 23–32.

Livingstone, I. 1986. Geomorphological significance of wind flow patterns over a Namib linear dune. In W.G. Nickling (ed.), *Aeolian geomorphology.* The Binghampton Symposium in Geomorphology International Series, 17. Allen & Unwin, Boston: 97–112.

Livingstone, I. 1987. Grain-size variation on a 'complex' linear dune in the Namib desert. In L. Frostick and I. Reid (eds), *Desert sediments ancient and modern.* Geological Society of London Special Publication, 35. Blackwell, Oxford: 281–91.

Livingstone, I. and Thomas, D.S.G. 1993. Modes of linear dune activity and their palaeoenvironmental significance: an evaluation with reference to southern African examples. In K. Pye (ed.) *The dynamics and context of aeolian sedimentary systems.* Geological Society of London, Special Publication 72. London: 91–101.

Long, J.T. and Sharp, R.P. 1964. Barchan-dune movement in Imperial Valley, California. *Bulletin of the Geological Society of America,* 75: 149–156.

Mabbutt, T.A. 1968. Aeolian landforms in central Australia. *Australian Geographical Studies,* 6: 139–150.

Mabbutt, J.A. 1977. *Desert landforms.* ANU Press, Canberra.

Mabbutt, J.A. and Wooding, R.A. 1983. Analysis of longitudinal dune patterns in the north western Simpson Desert, central Australia. *Zeitschrift für Geomorphologie,* Supplementband, 45: 51–70.

Mainguet, M. and Callot, Y. 1978. L'erg de FachiBilma (Tchad-Niger). *Memoire et Documents* CNRS, 18.

Mainguet, M. and Chemin, M.C. 1983. Sand seas of the Sahara and Sahel: an explanation of their thickness and sand dune type by the sand budget principle. In M.E. Brookfield and T.S. Ahlbrandt (eds), *Eolian sediments and processes.* Elsevier, Amsterdam: 353–364.

McArthur, D. 1987. Distinctions between grain-size distributions of accretion and encroachment deposits in an inland dune. *Sedimentary Geology,* 54: 147–163.

McEwan, I.K., Willetts, B.B. and Rice, M.A. 1992. The grain/bed collision in sand transport by wind. *Sedimentology,* 39: 971–981.

McKee, E.D. 1957. Primary structure in some recent sediments. *Bulletin of the American Association of Petroleum Geologists,* 41: 1704–1747.

McKee, E.D. 1966. Structures of dunes at White Sand National Monument (and a comparison with structure of dunes from other selected areas). *Sedimentology,* 7: 1–69.

McKee, E.D. 1979. Introduction to a study of global sand seas. *US Geological Survey, Professional Paper,* 1052: 1–19.

McKee, E.D. and Bigarella, J.T. 1979. Sedimentary structures in dunes. *US Geological Survey, Professional Paper,* 1052: 83–139.

McKee, E.D. and Tibbitts, G.C. Jr 1964. Primary structure of a seif dune and associated deposits in Libya. *Journal of Sedimentary Petrology,* 34: 5–17.

Melton, F.A. 1940. A tentative classification of sand dunes. *Journal of Geology,* 48: 113–173.

Merk, G.P. 1960. Great sand dunes of Colorado. In *Guide to the geology of Colorado.* Geological Society of America, Rocky Mountain Association of Geologists and the Colorado Science Society: 127–129.

Monod, T. 1958. Majabat Al-Koubra. *Memories de l'Institut Francais d'Alnigue Noire,* 52.

Nielson, J. and Kocurek, G. 1986. Surface processes, deposits and development of star dunes: Dumont dune field, California. *Bulletin of the Geological Society of America,* 99: 177–186.

Norris, R.M. 1966. Barchan dunes of Imperial Valley, California. *Journal of Geology,* 74: 292–306.

Norris, R.M. and Norris, K.S. 1961. Algodone dunes of southeastern California. *Bulletin of the Geological Society of America,* 72: 605–620.

Paisley, E.C.I., Lancaster, N., Gaddis, L.R. and Greeley, R. 1991. Discrimination of active and inactive sand from remote sensing: Kelso Dunes, Mojave Desert. *California Remote Sensing of the Environment,* 37: 153–166.

Pecsi, M. 1968. Loess. In R.W. Fairbridge (ed.), *Encyclopedia of geomorphology*. Reinhold, New York: 674–678.

Price, W.A. 1949. Pocket gophers as architects of mima (pimple) mounds of the western United States. *Texas Journal of Science*, 1: 1–17.

Queney, P. and Dubief, J. 1943. Action d'un obstacle oud d'un fosse sur un vent charge de sable. *Institute de Recheche Sahariennes, Travaiux*, 2: 169–176.

Rubin, D.M. 1990. Lateral migration of linear dunes in the Strzelecki Desert, Australia. *Earth Surface Processes and Landforms*. 15: 1–14.

Rubin, D.M. and Hunter, R.E. 1985. Why deposits of longitudinal dunes are rarely recognised in the geologic record. *Sedimentology*, 32: 147–158.

Sarnthein, M. and Walger, E. 1974. Der äolische Sandstrom aus der W-Sahara zur Atlantikküste. *Geologische Rundschau*, 63: 1065–1087.

Sharp, R.P. 1963. Wind ripples. *Journal of Geology*, 71: 617–636.

Sharp, R.P. 1966. Kelso dunes, Mohave Desert, California. *Bulletin of the Geological Society of America*, 77: 1045–1074.

Simons, F.S. 1956. A note on Pur-Pur Dune, Viru Valley, Peru. *Journal of Geology*, 64: 517–521.

Smith, J.D. 1970. Stability of a sand wave subject to a shear flow of a low Froude number. *Journal of Geophysical Research*, 75: 5928–5948.

Smith, R.S.U. 1982. Sand dunes in North American deserts. In W. Bender (ed.), *Reference handbook on the deserts of North America*. Greenwood Press, Westport. 481–526.

Snead, R.E. 1972. *Atlas of world physical features*. Wiley, New York.

Sneh, A. and Weissbrod, T. 1983. Size frequency distribution of longitudinal dune rippled flank sands compared to that of slip face sands of various dune types. *Sedimentology*, 30: 717–725.

Steele, R.P. 1986. Early Permian (Rotleigendes) palaeowinds of the North Sea — comments. *Sedimentary Geology*, 45: 293–297.

Stokes, W.L. 1968. Multiple parallel truncation bedding planes – a feature of wind-deposited sand-stone formations. *Journal of Sedimentary Petrology*, 38: 510–515.

Tchakerian, V.P. 1991. Late Quaternary aeolian geomorphology of the Dale Lake sand sheet, southern Mojave Desert, California. *Physical Geography*, 12: 347–369.

Thomas, D.S.G. 1986. Dune pattern statistics applied to the Kalahari Dune Desert, southern Africa. *Zeitschrift für Geomorphologie*, NF, 30: 231–242.

Thomas, D.S.G. 1987. Discrimination of depositional environments, using sedimentary characteristics, in the Mega Kalahari, central southern Africa. In L. Frostick and I. Reid (eds),

Desert sediments ancient and modern. Geological Society of London, Special Publication, 35. Blackwell, Oxford: 293–306.

Thomas, D.S.G. 1988a. Arid and semi-arid areas. In H.A. Viles (ed.), *Biogeomorphology*. Blackwell, Oxford: 193–221.

Thomas, D.S.G. 1988b. Analysis of linear dune sediment–form relationships in the Kalahari dune desert. *Earth Surface Processes and Landforms*, 13: 545–553.

Thomas, D.S.G. 1992. Desert dune activity: concepts and significance. *Journal of Arid Environments*, 22: 31–38.

Thomas, D.S.G. and Goudie, A.S. 1984. Ancient ergs of the southern hemisphere. In J.C. Vogel (ed.), *Late Cainozoic palaeoclimate of the southern hemisphere*. Balkema, Rotterdam: 407–418.

Thomas, D.S.G. and Martin, H.E. 1987. Grain size characteristics of linear dunes in the south western Kalahari — discussion. *Journal of Sedimentary Petrology*, 57: 572–573.

Thomas, D.S.G. and Shaw, P.A. 1991. Relict desert dune systems: interpretations and problems. *Journal of Arid Environments*, 20: 1–14.

Thomas, D.S.G. and Tsoar, H. 1990. The geomorphological role of vegetation in desert dune systems. In J.B. Thornes (ed.), *Vegetation and erosion*. Wiley, Chichester: 471–489.

Thomas, D.S.G., Nash, D.J., Shaw, P.A. and Van der Post, C. 1993. Present day sediment cycling at Witpan in the arid southwestern Kalahari Desert. *Catena*, 20: 515–527.

Tseo, G. 1993. Two types of longitudinal dune fields and possible mechanisms for their development. *Earth Surface Processes and Landforms*, 18: 627–645.

Tsoar, H. 1974. The formation of seif dunes from barchans — a discussion. *Zeitschrift für Geomorphologie*, NF, 28: 99–103.

Tsoar, H. 1978. *The dynamics of longitudinal dunes*. Final technical report, European Research Office, US Army.

Tsoar, H. 1982. Internal structure and surface geometry of longitudinal (seif) dunes. *Journal of Sedimentary Petrology*, 52: 823–831.

Tsoar, H. 1983. Dynamic processes acting on a longitudinal (seif) dune. *Sedimentology*, 30: 567–578.

Tsoar, H. 1985. Profiles analysis of sand dunes and their steady state significance. *Geografiska Annaler*, 67A: 47–59.

Tsoar, H. 1986. Two-dimensional analysis of dune profiles and the effect of grain size on sand dune morphology. In F. El-Baz and M.H.A. Hassan (eds), *Physics of desertification*. Martinus Nijhoff, Dordrecht: 94–108.

Tsoar, H. and Møller, J.T. 1986. The role of vegetation in the formation of linear dunes. In W.G.

Nickling (ed.), *Aeolian geomorphology*. Binghampton Symposia in Geomorphology, International Series, No. 17. Allen & Unwin, Boston: 75–95.

Twidale, C.R. 1972. Evolution of sand dunes in the Simpson Desert, central Australia. *Transactions of the Institute of British Geographers*, 56: 77–109.

Verstappen, H.Th. 1968. On the origin of longitudinal (seif) dunes. *Zeitschrift für Geomorphologie*, NF, 12: 200–220.

Verstappen, H.Th. 1970. Aeolian geomorphology of the Thar desert and palaeoclimates. *Zeitschrift für Geomorphologie*, Supplementband, 10: 104–120.

Walker, H.S., Olsen, J.W. and Bagen, J. 1987. The Badain Jaran Desert: remote sensing investigations. *Geographical Journal*, 153: 205–210.

Walmsley, J.R., Salmon, T.R. and Taylor, P.A. 1982. On the application of a model of boundary layer flow over low hills to real terrain. *Boundary-Layer Meteorology*, 23: 17–46.

Walmsley, J.R. and Howard, A.D. 1985. Application of boundary-layer model to flow over an eolian dune. *Journal of Geophysical Research*, 90: 10631–10640.

Ward, J.D. 1984. Aspects of the Cenozoic geology of the Kuiseb valley, central Namib Desert. Unpublished PhD thesis, University of Natal.

Warren, A. 1971. Dunes in the Ténéré Desert. *Geographical Journal*, 137: 458–461.

Warren, A. 1972. Observations on dunes and bimodal sands in the Ténéré Desert. *Sedimentology*, 19: 37–44.

Warren, A. 1979. Aeolian processes. In C. Embleton and J. Thornes (eds), *Process in geomorphology*. Edward Arnold, Sevenoaks, UK: 325–551.

Warren, A. 1984. Arid geomorphology. *Progress in Physical Geography*, 8: 399–420.

Warren, A. 1988. The dunes of the Wahiba Sands. *Journal of Oman Studies, Special Report*, 3: 131–160.

Warren, A. and Kay, S. 1987. The dynamics of dune networks. In L. Frostick and I. Reid (eds), *Desert sediments ancient and modern*. Geological Society of London Special Publication, 35. Blackwell, Oxford: 205–212.

Wasson, R.J. 1983. Dune sediment types, sand colour, sediment provenance and hydrology in the Strezlecki–Simpson dunefield, Australia. In M.E. Brookfield and T.S. Ahlbrandt (eds), *Eolian sediments and processes*. Elsevier, Amsterdam: 165–195.

Wasson, R.J. 1986. Geomorphology and Quarternary history of the Australian continental dunefields. *Geographical Review of Japan*, 59(B): 55–67.

Wasson, R.J. and Hyde, R. 1983a. A test of granulometric control of desert dune geometry. *Earth Surface Processes and Landforms*, 8: 301–312.

Wasson, R.J. and Hyde, R. 1983b. Factors determining desert dune type. *Nature*, 304: 337–339.

Wasson, R.J. and Nanninga, P.M. 1986. Estimating wind transport of sand on vegetated surfaces. *Earth Surface Processes and Landforms*, 11: 505–514.

Wasson, R.J., Rajaguru, S.N., Misra, V.N. Agrawal, D.P., Ohir, R.P., Singhvi, A.K. and Rao, K.K. 1983. Geomorphology, late Quaternary stratigraphy and palaeoclimatology of the Thar Desert. *Zeitschrift für Geomorphologie*, Supplementband, 45: 117–152.

Watson, A. 1985. The control of windblown sand and moving dune: a review of the methods of sand control in deserts, with observations from Saudi Arabia. *Quarterly Journal of Engineering Geology*, 18: 237–252.

Watson, A. 1986. Grain size variations on a longitudinal dune and a barchan dune. *Sedimentary Geology*, 46: 49–66.

Watson, A. 1987. Variations in wind velocity and sand transport on the windward flanks of desert sand dunes. *Sedimentology*, 34: 511–516.

Weng, W.S., Hunt, J.C.R., Carruthers, D.J., Warren, A., Wiggs, G.F.S., Livingstone, I. and Castro, I. 1991. Air flow and sand transport over dunes. *Acta Mechanica*, Supplement, 2: 1–22.

Wiggs, G.F.S., 1993. Desert dune dynamics and the evaluation of shear velocity: an integrated approach. In K.Pye (ed.), *The dynamics and environmental context of aeolian sedimentary systems*. Geological Society Special of London, Publication 72: 32–46.

Wiggs, G.F.S., Livingstone, I., Thomas, D.S.G. and Bullard, J.E. 1994. The effect of vegetation removal on airflow structure and dune mobility in the southwest Kalahari. *Land Degradation and Rehabilitation*, 5: 13–24.

Wiggs, G.F.S., Thomas D.S.G., Bullard, J.E. and Livingstone, I., Dune mobility and vegetation cover in the southwest Kalahari desert. *Earth Surface Processes and Landforms*, 20(6): 515–529.

Wiggs, G.F.S., Livingstone, I., Thomas, D.S.G. and Bullard, J.E. 1995b. Airflow characteristics over partially-vegetated dunes in the southwest Kalahari desert. *Earth Surface Processes and Landforms*, 21: 19–34.

Willetts, B.B. and Rice, M.A. 1986a. Collision in aeolian transport: the saltation/creep link. In W.G. Nickling (ed.), *Aeolian geomorphology*. Binghampton Symposia in Geomorphology, International Series, No. 17. Allen & Unwin, Boston: 1–17.

Willetts, B.B. and Rice, M.A. 1986b. Collisions in aeoian saltation. *Acta Mechanica*, 63: 255–265.

Willetts, B.B. and Rice, M.A. 1989. Collision of quartz grains with a sand bed: the influence of incidence angle. *Earth Surface Processes and Landforms*, 14: 719–730.

Willetts, B.B., Rice, M.A. and Swaine, S.F. 1982. Shape effects in aeolian grain transport. *Sedimentology*, 29: 409–417.

Wilson, I.G. 1971. Desert sand flow basins and a model for the development of ergs. *Geographical Journal*, 137: 180–197.

Wilson, I.G. 1972a. Aeolian bedforms — their development and origins. *Sedimentology*, 19: 173–210.

Wilson, I.G. 1972b. Universal discontinuities in bedforms produced by the wind. *Journal of Sedimentary Petrology*, 42: 667–669.

Wilson, I.G. 1973. Ergs. *Sedimentary Geology*, 10: 77–106.

Zhu, Z., Zou, B. and Yang, Y. 1987. The characteristics of aeolian landforms and the control of mobile dunes in China. In V. Gardiner (ed.), *International Geomorphology* 1986. Wiley, Chichester: 1211–1215.

Zimbelman, J.R., Williams, S.H. and Tchakerian, V.P. 1995. Sand transport paths in the Mojave Desert, southwestern United States. In V.P. Tchakarian (ed.), *Desert aeolian processes*. Chapman & Hall, London: 101–129.

18

Desert dust

Nick Middleton

Introduction

Wide appreciation of the significance of desert dust to the geomorphology of desert and peridesert zones has spawned considerable academic interest in recent decades (Morales 1979; Péwé 1981; Pye 1987). The entrainment, transport and deposition of dust also has importance for fields outside the geomorphologist's direct interest, including climatology, meteorology, ecology and environmental science in general. The range of environmental implications associated with desert dust is indicated in Table 18.1.

If Table 18.1 testifies to the wide-ranging relevance of desert dust study, a number of more specific areas of enquiry can be identified as being at the forefront of dust investigation. Among these are the importance of desert dust to the formation of the world's loess deposits (see page 427), its role as a palaeoenvironmental indicator in deep-sea and ice cores (see page 428), and, for applied geomorphology, dust's significance in various environmental problems faced by the increasing numbers of inhabitants of drylands.

The nature of desert dust

This chapter is concerned with desert dust: small particles, deflated from desert sediments, that are present in the air or have been deposited. Although desiccated sediments in any world environment can produce soil-derived atmospheric dust, the main sources are located in arid and semi-arid regions (Péwé 1981; Prospero 1981). While airborne particles in the world's atmosphere may be derived from a number of different sources, including cosmic dust, sea salt, volcanic dust and smoke particles from fire, these can be distinguished from soil dust by a variety of means such as grain size, mineralogy and chemical composition.

Dust particles have diameters of less than 0.08 mm according to Bagnold (1941), although many other workers prefer to define them according to the silt/sand boundary (i.e. less than 0.0625 mm). A subdivision can be made depending on the distance the material, once entrained, is transported. Dust that is carried a few kilometres to less than 100 km is generally between 0.005 and 0.05 mm. Material that can be transported over greater distances is generally smaller than 0.02 mm in diameter, according to Gillette (1979), or 0.002–0.01 mm according to Jackson *et al.* (1971). This material can remain suspended in the troposphere as an aerosol for many days, sometimes more than a week. Cumulative grain-size frequency curves of a number of local and far-travelled dust samples, together

Arid Zone Geomorphology: Process, Form and Change in Drylands, 2nd edition. Edited by David S. G. Thomas.
© 1997 John Wiley & Sons Ltd.

Table 18.1   *Some environmental consequences and hazards to human populations caused by dust storms*

Consequence	Example
Environmental	
Soil erosion	Kalma *et al.* (1988)
Soil nutrient gain	Swap *et al.* (1992)
Plant nutrient gain	Das (1988)
Ocean nutrient gain	Young *et al.* (1991)
Salt deposition and groundwater salinisation	Logan (1974)
Playa formation and relief inversion	Khalaf *et al.* (1982)
Sediment input to streams	Goudie (1978)
Case-hardening of rock	Conca and Rossman (1982)
Stone pavement formation	McFadden *et al.* (1987)
Desert varnish formation	Dorn (1986)
Loess formation	Zhang *et al.* (1991)
Silcrete development	Summerfield (1983)
Calcrete development	Coudé-Gaussen and Rognon (1988)
Bauxite metal enrichment	Brimhall *et al.* (1988)
Climatic change	Maley (1982)
Ocean sedimentation	Rea (1994)
Glacier mass budget alteration	Wake and Mayewski (1993)
Radiation budget alteration	Coakley and Cess (1985)
Ventifact sculpture	Whitney and Dietrich (1973)
Yardang sculpture	Breed *et al.* (1989)
Rock polish	Lancaster (1984)
Human-related	
Transport disruption	Burritt and Hyers (1981)
Disease transmission (human)	Leathers (1981)
Disease transmission (plants)	Clafin *et al.* (1973)
Radioactive dust transport	Bacon and Sarma (1991)
DDT transport	Riesborough *et al.* (1968)
Air pollution	Hagen and Woodruff (1973)
Radio communication problems	Martin (1937)
Animal suffocation	Choun (1936)
Animal madness	Choun (1936)
Rainfall acid neutralisation	Lôye-Pilot *et al.* (1986)
Machinery problems	Jones *et al.* (1986)
Warfare disruption	Agence France Press (1985)
Closing of business	Gillette (1981)
Reduction of property values	Gillette (1981)
Reduction of solar power potential	Reinking *et al.* (1975)
Car-ignition failure	Clements *et al.* (1975)
Drinking water contamination	Clements *et al.* (1975)
Electrical insulator failure	Kes (1983)

with a peridesert loess sample, are shown in Figure 18.1.

While the above comments on grain size are generally sound it should be noted that, contrary to classical theoretical considerations, many authors have found particles coarser than 0.08 mm in desert dust sampled far from its source. Glaccum and Prospero (1980), for example, found quartz grains up to 0.09 mm in diameter and mica flakes up to 0.35 mm in diameter in samples taken over Sal Island off west Africa, while Schroeder (1985) found aggregated dust particles up to 0.15 mm in diameter in samples taken on the coastal belt of

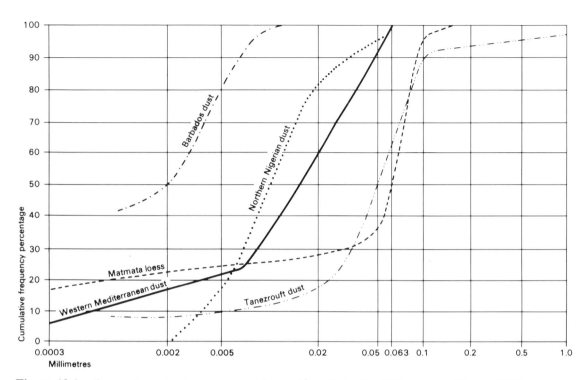

Figure 18.1 *Cumulative grain-size frequency of dust and loess (after Coudé-Gaussen and Rognon 1983)*

Sudan. Indeed, these so-called 'giant' dust parti-cles have been found at very large distances from source: Betzer *et al.* (1988) collected large num-bers of >0.075 mm diameter quartz particles in samples taken over the Pacific Ocean more than 10 000 km from where they were entrained in an Asian dust storm. Such large particles are unex-pected because of their high fall velocities, and their appearance at such great distances from source cannot be explained using currently acknowledged atmospheric transport mechanisms.

The dominant mineral in most desert dust is silica (SiO_2) mainly in the form of quartz (Goudie 1978). Other minerals found in desert dust include feldspars, calcite, dolomite, micas, chlorite, kaolinite, illite, smectite, mixed-layer clays, palygorskite, heavy oxide and silicate minerals, gypsum, halite, opal, amorphous inorganic material and organic material (Pye 1987). The precise composition of any dust will naturally depend upon the nature of the source region and the composition may vary with distance from the source due to differences in settling velocities of different minerals.

PROCESSES PRODUCING FINE PARTICLES IN THE DESERT ENVIRONMENT

Fine particles are commonly categorised into those of silt and clay sizes, with grain diameters of 0.004–0.0625 mm and less than 0.004 mm, respectively (Wentworth 1922). Whereas inor-ganic clay-size particles are generally agreed to be largely derived from chemical weathering, the processes responsible for silt formation in the desert environment remain a matter for debate. Table 18.2 shows some of the possible mechanisms that have been suggested.

The weathering of bedrock and quartz grains by salt processes has received much recent attention (see Pye and Sperling 1983; Smith *et al.* 1987 and also Chapters 3 and 4) and its ability to produce silt-size particles in the desert environment is now established. Naturally the relative importance of different mechanisms will vary according to such factors as climate, topo-graphy, relief and lithology. For example, it seems likely that the role of frost weathering is more dominant in high-latitude deserts such as

Table 18.2 *Possible mechanisms for silt particle formation through operation on larger particles or rock outcrops*

Mechanism	References
In situ	
Release from phyllosilicate parent rocks	Kuenen (1969)
Chemical weathering	Nahon and Trompette (1982)
Frost weathering	Zeuner (1949)
Salt weathering	Pye and Sperling (1983)
Biological origin	Wilding *et al.* (1977)
Clay pellet aggregation	Dare-Edwards (1984)
In transport	
Aeolian abrasion	Whalley *et al.* (1982)
Glacial grinding and crushing	Collins (1979)
Spalling during fluvial and fluvioglacial transport	Moss *et al.* (1973)

Source: Modified after Pye (1987).

the Gobi (Kowalkowski and Brogowski 1983) than in the Sahara, where its influence is relatively localised in areas such as the Tibesti and Ahaggar massifs. Even glacial grinding is likely to be important in certain arid areas such as the high Andes, the Tibetan Plateau and Central Asian mountain ranges, but earlier beliefs that this mechanism was the only one capable of generating large amounts of silt-size debris (Smalley 1966; Smalley and Vita-Finzi 1968) have now been rejected.

Dust entrainment

The entrainment of dust from the ground surface is controlled by the nature of the wind, the nature of the soil or sediment itself and the presence of any surface obstacles to windflow. To these three main variables may be added effects of topography both directly on windflow and indirectly by affecting the formation of certain meteorological systems. Human actions may affect any of these variables but are most important in disturbing earth surfaces and through effects on vegetation cover.

THRESHOLD VELOCITIES FOR DUST
ENTRAINMENT

Although different terrain types vary greatly according to their susceptibility to dust

entrainment, there are few reliable field data on threshold velocities for particle release into the air (see Chapter 16). However, the efforts of Gillette and his co-workers (Gillette *et al.* 1980; 1982) using portable wind tunnels mostly in the Mojave Desert have begun to fill this gap. They ranked different soil surface types according to the increase in threshold velocities from most erodible to least erodible: disturbed soils; sand dunes; alluvial and aeolian sand deposits; disturbed playa soils; skirts of playas; playa centres; and desert pavements (Figure 18.2).

GEOMORPHOLOGICAL UNITS
FAVOURABLE FOR DUST DEFLATION

Although silts and clays occur widely in soils and sediments, a number of factors influence their susceptibility to the deflation of dust. Among these factors are the ratios of clay-, silt- and sand-sized particles, the moisture content of the soil, the presence of particle cements such as salts or organic breakdown products, the compaction of sediments and the presence of crusts or armoured surfaces. The most favourable dust-producing surfaces are areas of bare, loose and mobile sediments containing substantial amounts of sand and silt but little clay. Terrains that satisfy these conditions are most commonly found in geomorphologically active landscapes where tectonic movements, climatic changes and/or human disturbance are

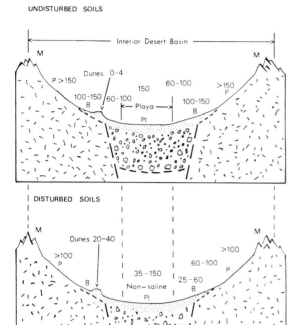

Figure 18.2 *Threshold friction velocity (in centimetres per second) relative to desert geomorphology (after Gillette 1982)*

responsible for rapid exposure, incision and reworking of sediment formations containing dust (Pye 1987).

Important sources of dust storm sediments are generally located in specific, small desert environments, according to research in the Sahara by Coudé-Gaussen (1984) and Yaalon (1987). These include flood-deposited debris

on floodplains, alluvial fans and in wadi beds, and terminal depressions that receive periodically removed material from desert slopes. Many other geomorphological terrain types can be identified globally as being significant dust sources, and they are summarised in Table 18.3.

The importance of water will be noted from Table 18.3 as the apparently dominant agent responsible for concentrating large amounts of dust-sized debris in deflatable environments. This observation is consistent with the findings of Goudie (1983), who found that markedly higher frequencies of dust storms are observed at meteorological stations in areas with 100–200 mm mean annual rainfall. This is in contrast to areas with less than 100 mm mean annual rainfall, where dust storm frequency appears to decline. In such areas, the apparent dearth of dust storm activity (apparent in that meteorological stations are relatively sparse in many hyper-arid regions) may be due to infrequent stream runoff which limits dust supply or because strong winds associated with fronts and cyclonic disturbances are rare in such areas (Goudie 1983). Although fluvial deposition will be more active in desert-marginal areas, thus in theory providing more regular deposits of fine material for deflation, Pye (1987) considers that recent cultivation of desert-marginal soils is a more important factor behind the peak of dust storm activity in the 100–200 mm mean annual rainfall zone.

In contrast to fluvial processes, aeolian action is directly responsible for just two of the 10 important sources shown in Table 18.3: loess deposits and dunes. Pleistocene loess regions

Table 18.3 *Geomorphological units from which substantial deflation occurs*

Geomorphological unit	Regional example
Floodplain	Lower Mesopotamia
Alluvial fan	Tibesti and Hoggar mtns, N Africa
Wadi	Negev
Glacial outwash plain	Slims River Valley, Yukon
Salt pan	Puna de Atacama, Argentina
Other desert depressions	Seistan Basin, Iran
Former lake bed	Lake Texcoco, Mexico City
Active dunes	Erg in Koussamene, Mali
Devegetated fossil dunes	Erg du Traza, Mauritania
Loess	Loess Plateau, China

may become important dust sources when protective vegetation cover is degraded by natural climatic fluctuations or trends towards aridity, or by human action. Deflation from cultivated fields in parts of the Chinese Central Loess Plateau is a serious problem for farmers in the region (see, for example, Pye 1987).

Although experimental work indicates that silt-sized fragments are readily produced from quartz grains during aeolian transport (Whalley *et al.* 1982; 1987), mobile dunes are not considered to be important dust sources. An assessment of an active dune area's dust output would require long-term monitoring of local atmospheric dust concentrations since the continual nature of the process would mean that large-scale high-magnitude events detectable on satellite imagery or by meteorological observers would be rare. Nonetheless, saltating grains are often important initiators of dust entrainment from non-dune sediments, by ballistic impact on smaller particles.

Stabilised dunes, by contrast, have been shown to contain up to 15–30% clay and silt (e.g. Thomas 1985) which can be attributed both to *in situ* weathering and to inputs of dust from elsewhere (Pye and Tsoar 1987). Many desert-marginal areas have experienced reactivation of formerly stable dunes by human and climatic influences, often producing increased dust output, as in the Sahel of Mauritania since the late 1960s (Middleton 1985; 1987).

GLOBAL GEOGRAPHY OF DUST STORM FREQUENCY

The major contemporary sources of desert dust have been identified using satellite imagery and analysis of standard meteorological observation data. Dust storms are recorded by observers using a 'target' method so that when visibility is reduced to below 1000 m by atmospheric dust being raised within sight of the observer a 'dust storm' is logged. The 1000-m visibility limit is an international standard, but different limits have been used by other authors: 700 m (Oliver 1945); 7 miles or 11.3 km (Orgill and Sehmel 1976); 800 m (Péwé *et al.* 1981), and the Japanese Meteorological Agency defines a dust storm as a dust haze or dustfall from Asian sources (known as 'Kosa') with a visibility of

less than 10 km (Arao and Ishizaka 1986). These inconsistencies among different authors and authorities have led to specific definitions of dust events (Middleton *et al.* 1986), and one geomorphically important distinction to be made separates a dust storm from 'dust haze'. The latter is dust that is suspended in the air but which is not actively being entrained within sight of the observer; it has been raised from the ground prior to the observation or at some distance away. Visibility in a dust haze may or may not be less than 1000 m. The other types of dust event are 'blowing dust', which is material being entrained within sight of the observer but not obscuring visibility to less than 1000 m, and the 'dust devil', which is a localised column of dust that neither travels far nor lasts long. McTainsh and Pitblado (1987) have further refined these definitions by including the relevant international meteorological observers' 'SYNOP' present weather (or 'ww') codes for each event type.

The highest frequencies of global dust storm occurrence are found in the arid and semi-arid regions of the northern hemisphere (Figure 18.3). In the regions in which dust storms are common features of the climate they have received a variety of local names; Middleton (1986c) lists over 70 worldwide. Table 18.4 shows those regions where on average more than 15 dust storm days per year have been recorded. The highest reported frequencies are in Asia, with Zabol in the Seistan Basin of Iran averaging 80.7 days a year (Middleton 1986b) and Takhiatash on the floodplain of the Amu Darya in Uzbekistan averaging 108 days a year (Sapozhnikova 1973). Other important high-frequency areas in Central Asia include the Karakum Desert, the steppes of Kazakhstan and the regions of Alma Alta and Altai, and in Europe the Rostov region between the Caspian and Black seas (Klimenko and Moskaleva 1979).

Apart from the Seistan Basin at the joint borders of Iran, Afghanistan and Pakistan, the plains of the upper Indus and the Thar Desert are centres of dust storm activity but most of the Indian subcontinent is relatively unaffected (Middleton 1986b). In the Middle East, the alluvial plains of Lower Mesopotamia and the desert areas of Syria, Jordan and northern Saudi Arabia experience the highest frequencies

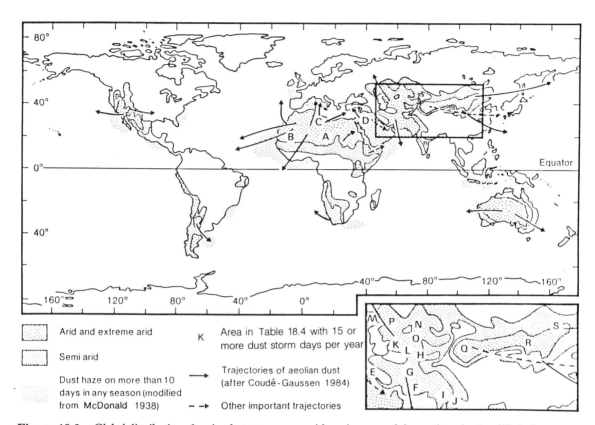

Figure 18.3　*Global distribution of major dust storm areas with main seasonal dust trajectories (modified after Middleton* et al. *1986)*

(Middleton 1986a). In northern China, the Taklamakan Desert of the Tarim Basin, and the Kansu Corridor are the most active regions (Middleton *et al.* 1986), while south of a line between Beijing and Xian fewer than three dust storms a year are recorded on average (Goudie 1983). Parts of the Gobi Desert in southern Mongolia average more than 30 dust storm days a year (Middleton 1991). In Africa, areas with annual frequencies between 10 and 25 days include the Bodélé Depression, southern Mauritania, northern Mali and central southern Algeria, northern Libya and central Algeria, and northern Sudan (Middleton *et al.* 1986).

In North America, long-term dust storm frequencies only exceed five days a year in small areas of Saskatchewan in the Canadian Prairies (Wheaton and Chakravarti 1990) and on the Great Plains of the USA (Changery 1983). In northern Mexico just one station averages more than 10 dust storm days a year (Middleton

1986c). In Australia more than five dust storms a year are recorded in the Simpson Desert (Middleton 1984; 1986c; McTainsh and Pitblado 1987). Little work has been done on dust-raising in the other southern hemisphere deserts of southern Africa and South America, but these areas are not thought to produce dust on the same scale as in the northern hemisphere, though sites such as dry lake beds may be locally very important.

METEOROLOGICAL SYSTEMS, SEASONALITY AND DIURNALITY

The meteorological systems capable of generating dust-raising winds can be divided broadly into synoptic-scale and mesoscale systems which are summarised in Table 18.5. The most important is the passage of low pressure fronts with intense baroclinic gradients that are

Table 18.4 *Major global dust source areas with key station annual dust storm day frequencies* (D)

Source area	Ref. to Fig. 18.3	Station	D	Data period
N and W Africa				
Bodele Depression	A	Maiduguri (Nigeria)	22.5	1955–79
S Mauritania/N Mali/C Algeria	B	Nouakchott (Mauritania)	27.4	1960–84
Libya/E Algeria	C	Sirte (Libya)	17.8	1956–77
Middle East				
N Saudi/Jordan/Syria	D	Abou Kamal	14.9	1959–79
Lower Mesopotamia	E	Kuwait International Airport	27.0	1962–84
SW Asia				
Makran coast	F	Jask (Iran)	27.3	1970–73
Seistan Basin	G	Zabol (Iran)	80.7	1967–73
Afghan Turkestan plains	H	Chardarrah (Afghanistan)	46.7	1974–80
Upper Indus plains	I	Jhelum (Pakistan)	18.9	1951–58
Thar Desert	J	Fort Abbas (Pakistan)	17.8	1951–59
Central Asia				
Turkmenistan	K	Repetek (Turkmenistan)	65.5	1936–60
Karakum Desert	L	Nebit Dag (Turkmenistan)	60.0	1936–60
Rostov	M	Zavetnoe (Russia)	23.3	1936–60
Altai	N	Rubtsovsk (Russia)	25.1	1936–60
Alma Alta	O	Bakanas (Kazakhstan)	47.7	1936–60
Kazakhstan	P	Dzhambeiti (Kazakhstan)	45.9	1936–60
NW Asia				
Taklamakan Desert	Q	Hotien (China)	32.9	1953–80
Gansu Corridor	R	Minqin (China)	37.3	1953–80
Gobi Desert	S	Zamiin Uud (Mongolia)	34.4	1956–86

Source: Modified after Middleton (1989b).

accompanied by very high-velocity winds. Such frontal passage is the dominant dust-generating mechanism in many of the world's dusty regions, including northern China and Mongolia (Ing 1972), Australia (Sprigg 1982), Central Asia (Nalivkin 1983), the Levant and northern Turkey (Middleton 1986a), the Mediterranean coast of north Africa (Yaalon and Ganor 1979), the Sahelian latitudes of West Africa (Tetzlaff and Peters 1986), the High Plains of the USA (Lee *et al.* 1993) and the plains of the Argentine Pampas (Woelcken 1951). Surface cyclones themselves may sweep out gyres of dust when circulation around the low pressure becomes very intense (Middleton *et al.* 1986). Other synoptic-scale dust-raising systems include winds generated in areas with steep pressure gradients such as in the Thar Desert of India and Pakistan (Middleton 1989a) and the northwesterly 'Shamal' that blows down the Arabian Gulf (Membery 1983).

More localised dust storms occur when katabatic winds deflate mountain-foot sediments such as at the northern slopes of Kopet Dag on the Iran–Turkmenistan border (Nalivkin 1983), at the Argentine foothills of the Andes (the 'Zonda' wind), or in California (the 'Santa Ana' wind). The high Andean Altiplano of Chile, northwest Argentina and southern Bolivia experiences strong dust-raising from the upper westerlies and similar upper airflow deflates sediments from the arid Tibetan Plateau.

The cold downburst wind of a dry thunderstorm is perhaps the most common mesoscale dust-raising system, as described for the classic 'Haboob' of southern Sudan by Freeman

Table 18.5 *Classification of dust-generating weather systems*

1. Dust devil	
2. Thunderstorm downdraft	Single cell
	Severe thunderstorm
3. Mountain downslope wind (katabatic/fohn type)	
4. Valley channelling (Venturi effect)	
5. Tropical cyclone	
6. Frontal	Pre-frontal
	Post-frontal
7. Cyclogenic	Low level jet
	Upper level jet
	Surface storm circulation (gyre)
	Mountain downslope
8. Pressure gradient (convergence zone)	Stationary systems
	Travelling systems
9. Upper westerlies	

(1952). Similar downdraft winds, which raise dust at the gust front some kilometres in advance of the towering convective clouds, occur in the Sahel (Tetzlaff and Peters 1986), northwestern India (Joseph *et al.* 1980) and Pakistan (Middleton and Chaudhary 1988), Arizona (Nickling and Brazel 1984), the Mongolian Gobi (Middleton 1991), central Mexico (Jauregui 1973) and Australia (Lindsay 1933), although often no distinction is made in the literature between the classic single cell 'Haboob' thunderstorm and a multi-cell severe thunderstorm.

Dust storms are typically highly seasonal in their occurrence (Littmann 1991). The important factors governing their seasonality include the occurrence of dust-raising meteorological systems, seasonal rainfall (Yu *et al.* 1993), vegetation cover and human activities such as harvesting and ploughing. The diurnal variability of dust-raising is also commonly marked, with the most intense activity occurring during the afternoon hours when solar heating is greatest and a turbulent boundary layer is present (Hinds and Hoidale 1975).

CHANGES IN DUST STORM FREQUENCY

Changes in the intensity and frequency of dust-raising have been observed and detected on a variety of timescales, quite apart from the seasonal and diurnal variations noted above. Through geological time these changes have been discerned from study of dust deposited in the world's ice caps (see, for example, Taylor *et al.* 1993) and in ocean sediments (see pages 428–429). These changes are related to large-scale global climatic variations, but, in the present century, analysis of standardised meteorological records and the monitoring of atmospheric dust loads has shown variations in which the human impact has also become important (Goudie and Middleton 1992).

Human activities that affect changes in dust storm frequencies can be divided broadly into those that break up naturally wind-resistant surfaces and those that remove protective vegetation cover from soils, in many cases the two effects occurring simultaneously. Such activities include a variety of agricultural impacts, construction, vehicle use, drainage and military movements.

Perhaps the best-known example of agricultural impacts is the 1930s Dust Bowl of the United States (Choun 1936; Martin 1937; Worster 1979). Figure 18.4A shows the variation in dust storm occurrence at Dodge City with a climatic index devised by Chepil *et al.* (1963). The dramatic rise in frequency during the late 1930s was the result of agricultural mismanagement and the application of inappropriate techniques which were exposed in drought years. This figure also indicates a secondary peak of activity in the 1950s, when more land was damaged by wind erosion in the Great Plains than in the 1930s, while soil loss in the 1970s was on a scale comparable to the 1930s (Lockeretz 1978). In all three periods, poor agricultural practices played a significant

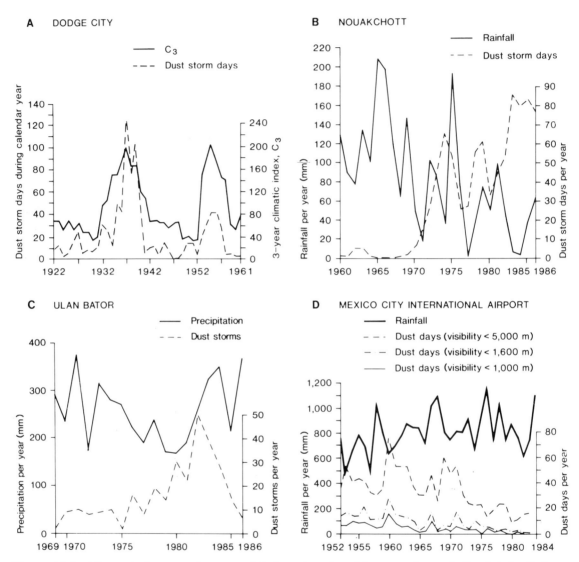

Figure 18.4   *Changes in dust storm frequency. (A) Annual number of dust storm days and climatic index C_3 for Dodge City, Kansas (after Chepil et al. 1963). (B) Annual number of dust storm days and annual rainfall for Nouakchott, Mauritania. (C) Annual number of dust storms and annual rainfall for Ulan Bator, Outer Mongolia. (D) Annual number of dust days and annual rainfall at Mexico City International Airport (after Middleton et al. 1986)*

role in exacerbating dust-raising during periods of drought (Ervin and Lee 1994).

Drought is commonly associated with an increase in dust-raising activity as vegetation cover dies off and soils dry out. The concentration of North African dust observed at Barbados was higher in 1972–74, when drought affected much of Sahelian Africa, than during the previous years (Prospero and Nees 1977) and higher still during 1983–84 when drought was more severe still in Sahelian latitudes (Prospero and Nees 1986). Analysis of meteorological station records in the Sahel of Mauritania and Sudan showed increases in dust storm frequencies by factors of five and six, respectively, during the 1970s (Middleton 1985). Both research programmes (Prospero and Nees 1986; Middleton 1987) have demonstrated a good correlation between increasing dust production and declining rainfall over periods of 20 years or so, and the high levels of dust production in the western Sahel of Mauritania have continued into the 1980s (Figure 18.4b). There is little doubt that as well as low rainfall, the actions of human populations in degrading vegetation cover in the Sahel have also had an important effect in the rise in dust storm activity.

Urban activities are also responsible for increasing dust production. A marked increase in dust storm activity in Ulan Bator, Mongolia (Figure 18.4c) is related to the rapid urban expansion which occurred in the vicinity of the meteorological station in the 1980s and the increasing use of off-road vehicles in and around the city which degrades the natural grass cover and breaks up soil surfaces. At the same time, it will be noted that the increase in dust storms throughout the 1970s is also mirrored by a gradual decline in rainfall over the same period (Middleton 1991).

Large-scale military movements can have a similar impact and have been reported during the North African campaign in the 1940s (Oliver 1945), in the USA during the Second World War when General Patton established a desert training camp at Parker, Arizona (Clements *et al.* 1975), and during the Gulf War in 1990–91.

The draining of water bodies, which exposes fine-grained lacustrine sediments to wind action, has been reported from a number of areas, including the fringes of the Aral Sea in Uzbekistan and Kazakhstan, the surface area of which has shrunk by about 50% since 1960 due to excessive offtakes for irrigation. More than 40 million tonnes of saline material is estimated to be deflated from the former sea bed annually causing problems of salinisation on cropland where it is deposited (Kotlyakov 1991). The demand of urban areas for water has caused the complete desiccation of lakes in other parts of the world, such as at Owens Lake in southern California (Reinking *et al.* 1975) and Lake Texcoco in Mexico City. The latter example has shown how a concerted effort to revegetate and stabilise the lake bed has been successful in eliminating what was a major dust source that caused environmental problems for one of the world's largest cities (Figure 18.4d).

Dust transport

Most atmospheric dust falls back to Earth a short time after entrainment and not far from its source. The distance travelled by dust particles depends upon many factors, including wind speed and turbulence, dust grain characteristics and settling velocities. The settling velocity (U_f) of a particle depends on its mass and shape and naturally plays an important role in determining the mode of transport of a grain. The force opposing the tendency of fine particles to settle is represented by the standard deviation of the shear velocity (u_*), and the ratio U_f/u_* has been found to be a suitable criterion for determining the degree of suspension for most particles (Chepil and Woodruff 1957; Gillette *et al.* 1974). The ratio does not, however, account for the presence of 'giant' particles found at considerable distances from source (see above), a phenomenon still to be explained adequately.

The giant particle conundrum reflects the fact that the dynamics of dust transport are not well understood. The influence of airborne dust on local meteorology and climate has been the subject of a major dust dynamics project conducted in Tajikistan, reported in a series of papers in *Atmospheric Environment* vol. 27A (16) (1993), but the best studied area in this respect is the Sahara, for which a number of dust transport models have been formulated (Prospero and Carlson 1972; Jaenicke and Schütz 1978; Westphal *et al.* 1987). A thorough understanding of the processes of long-range dust transport is important for a variety of reasons, including the need to attain clearer appreciation of the palaeoenvironmental signals contained in deep-sea dust deposits (see pages 248–249 and Chapter 26).

LONG-RANGE TRANSPORT

Dust that is carried to upper levels in unstable boundary layers may be transported for thousands of kilometres and remain in the atmosphere for a week or longer. The transport of dust over great distances has long been recognised in apparently bizarre depositional events such as 'blood rain' that are described in Homer's *Iliad* and in the works of numerous

Roman writers. Some of the earliest scientific observations were made by Darwin (1846) off the west coast of Africa and Ehrenburg (1849) in the same area and in southern Europe.

Today a wide variety of techniques are used to trace long-range transport of dust, with satellite monitoring being perhaps the clearest tool for detecting the densest events. Many techniques have been applied to trapped material in order to detect its source (Table 18.6). Ganor (1991), for example, analysed 28 heavy Saharan dustfalls over Israel in a 22-year period and used clay minerals to identify source areas which were corroborated with meteorological reports, satellite images and back-trajectory analysis. Some examples of long-range transport, often to areas well outside the desert realm, are shown in Table 18.7.

A number of authors have monitored dust events using measures of atmospheric turbidity, defined as the extinction of solar radiation by suspended particles that are large enough with respect to the wavelength of light, that is particles with radii of 0.0001–0.01 mm. Thus such a measure is one of atmospheric dust concentration. A number of turbidity studies have been undertaken over north Africa (see, for example, Jaenicke 1979) and across the northern equatorial Atlantic (Prospero *et al.* 1979). A network of Sun photometers established across the Sahara and Sahel has produced better estimates of the Sahara's source strength: in 1981 and 1982, 630 million tonnes and 710 million

tonnes were produced, respectively (d'Almeida 1986). McTainsh (1980), working in northern Nigeria, has demonstrated a relationship between dust deposition, solar radiation (as measured by turbidity) and horizontal visibility.

Atmospheric dust concentration during major dust storms may range from 100 to 100 000 $\mu g\,m^{-3}$ (Goudie 1978) while average concentrations over major world oceans are much lower (<1.0 $\mu g\,m^{-3}$) except those adjacent to large deserts where concentrations are an order of magnitude higher (Prospero 1979). High mean concentrations during dusty months have been recorded in desert areas: in Mali mean ambient air concentrations during April–June were 1176 $\mu g\,m^{-3}$, exceeding the recommended international health standard by an order of magnitude (Nickling and Gillies 1993). Similar concentrations can also occur during particularly severe long-range transport events as Chung (1992) has noted in South Korea: dust loadings of up to 1105 $\mu g\,m^{-3}$ have been recorded in Seoul, reducing visibility to <1 km, the material having been transported 2–3000 km from northern China and Mongolia.

ENVIRONMENTAL HAZARDS OF AIRBORNE DUST

Airborne dust presents a variety of problems to inhabitants of desert areas (Table 18.1),

Table 18.6   *Methods used for dust monitoring and identification of source areas*

Dust characteristic	Selected references
Mass size distribution	Prospero *et al.* (1970)
Mineralogy and elemental composition	Paquet *et al.* (1984)
Oxygen isotopes	Clayton *et al.* (1972)
Lead isotopes	Turekian and Graustein (1984)
Rubidium–strontium isotopes	Biscaye *et al.* (1974)
Thorium isotopes	Hirose and Sugimura (1984)
Radon-222	Prospero and Carlson (1970)
Magnetic mineral assemblages	Oldfield *et al.* (1985)
Aluminium concentration	Duce *et al.* (1980)
Aerosol–crust enrichment	Rahn *et al.* (1981)
Scanning electron microscopy of individual grain features	Prodi and Fea (1979)
Continentally derived lipids	Gagosian *et al.* (1981)
Pollen	Franzen *et al.* (1994)
Foraminifera	Ehrenburg (1849)

Table 18.7 *Examples of long-range dust transport*

Traced from	Traced to	Distance (km)	References
Africa			
West Africa	Barbados	6500	Delaney *et al.* (1967)
West Africa	Miami	8000	Prospero (1981)
West Africa	Cayenne	6500	Prospero *et al.* (1981)
Bilma/Faya Largeau	Gulf of Guinea	2000	Schütz (1980)
Tunisia/Libya	Central Caucasus	5000	Zamorskii (1964)
Algeria/Tunisia	Sweden/Finland	7000	Franzen *et al.* (1994)
Algeria	Britain	2500	Wheeler (1986)
Libya/Egypt	Negev	2000	Yaalon and Ganor (1979)
NE Sudan	E Mediterranean	2500	Middleton (1986c)
Horn of Africa	Makran coast	3000	Wells (pers. comm. 1985)
Namib Desert	S Atlantic	4000	Parkin *et al.* (1972)
Europe			
S Ukraine	Sweden	1500	Kes (1983)
Caucasus	E Europe	3500	Lisitzin (1972)
Middle East			
Lower Mesopotamia	Sharjah	1000	Middleton (1986c)
Arabian Peninsula	Turkmenistan	1500	Nalivkin (1983)
W Asia			
Thar Desert	SE Asia	4500	Prospero (1981)
W Kazakhstan	Gorky	1200	Kes (1983)
E Asia			
Loess Plateau	Japan	2500	Ishizaka (1973)
Taklamakan Desert	Oahu	10000	Betzer *et al.* (1988)
Gobi Desert	Barrow, Alaska	10000	Rahn *et al.* (1981)
Mongolian Gobi	Hawaii	10000	Shaw (1980)
Australia			
Central Australia	Singapore	3500	Durst (1935)
S Australia	New Zealand	3500	Kidson and Gregory (1930)
N America			
Great Plains	Mid-Atlantic	3500	McCauley *et al.* (1981)
Nebraska and Dakotas	Washington, DC	2500	Hand (1934)

hazards that have affected dryland peoples since time immemorial. Folk (1975), for example, suggests that the ancient Macedonian town of Stobi, which flourished between 400 BC and AD 400, was abandoned because of the severe affects of dust storms. Blowing dust and sand can cause serious damage to crops and natural vegetation by abrasion (Woodruff 1956), which is particularly critical for young shoots when fields are poorly protected by vegetation cover.

A severe dust storm in the San Joaquin Valley, California, in 1977 caused extensive abrasion damage to root crops, vineyards and citrus fruits, defoliation due to sandblasting, root exposure and the uprooting of trees (Wilshire *et al.* 1981). Dust storms may also transmit plant diseases (Rotem 1965).

A number of human medical conditions can be traced to the impact of dust in desert areas. Inhalation of fine particles can aggravate diseases such as bronchitis and emphysema, and high incidences of silicosis and pneumoconiosis have been reported in Bedouins in the Negev (Bar-Ziv and Goldberg 1974). Dust blown by the 'Irifi' wind of the Western (formerly Spanish) Sahara is responsible for the conjunctivitis that is common among the nomads of the country (Morales 1946). Organisms blown in dust may settle on the skin, be swallowed or inhaled into respiratory passages. In Arizona, 'Valley Fever' is caused by *Coccidioides immitis*, a common airborne fungus blown by dust storms, which causes more than 27 human deaths a year in the state (Leathers 1981).

The reduction in visibility caused by dust storms is a hazard to aviation and road transport. The severe pre-frontal storm of 7 November 1988 in South Australia, for example, caused road and airport closures all across the Eyre Peninsula (Crooks and Cowan 1993). Poor visibility has been the cause of some fatal traffic accidents, such as in January 1973 when a Royal Jordanian Airlines Boeing 707 crashed in thick 'Harmattan' dust haze at Kano airport in northern Nigeria, killing 176 passengers and crew (Pye 1987). Sudden loss of visibility associated with the arrival of a dry thunderstorm ('Haboob') dust wall was responsible for 32 multiple accidents between 1968 and 1975 on Interstate 10 in Arizona (Brazel and Hsu 1981), and the seriousness of the problem has inspired the development of a Dust Storm Alert System involving remote controlled road signs and special dust-alert messages broadcast on local radio (Burritt and Hyers 1981).

Applied geomorphologists have played a useful role in identifying dust sources and methods of dust prevention in arid zones. Hyers and Marcus (1981) highlighted abandoned farmland as a dust source near Arizona highways, while Jones *et al.* (1986) have suggested a general procedure for the assessment of dust hazards in urban areas after their work in the Middle East.

Dust deposition

The deposition of desert dust has a number of important implications for geomorphology, geology and ecology both within and beyond the arid realm. Dust contributes to the formation of desert surface coverings such as desert varnish and duricrusts (see Chapter 6) and case-hardening of rocks (Conca and Rossman 1982). Windblown salts can act as weathering agents (see Chapter 4) and increase the salinity of soils and water bodies. Dust deposition makes important contributions to alluvial fan sediments (Amit and Gerson 1986), stabilised sand dune sediments (Pye and Tsoar 1987), sabkhas, lakes, wadis and other fluvial sediments (Ali and West 1983), as well as soils in many parts of the world outside the arid realm: Saharan dust additions to soils have been investigated in southern Europe (Rapp 1984) and Bermuda (Syers *et al.* 1969), while Asian inputs have been studied in Japan (Mizota and Matsuhisa 1985) and Hawaii (Rex *et al.* 1969). Saharan dust is also thought to be of critical importance as a source of key trace elements to the Amazon Basin (Swap *et al.* 1992). In regions of snow and ice, inputs of aeolian dust affect the chemistry as well as the albedo and thus the freezing point (Wake and Mayewski 1993), while over the oceans desert dust has been shown to play an important role in primary productivity in some areas (Young *et al.* 1991). Dust deposited over long time periods is also used as a palaeoenvironmental indicator in sediment cores from lake and ocean beds, ice cores and in terrestrial loess profiles.

CONTEMPORARY TERRESTRIAL DEPOSITION RATES

Deposition in single dustfalls can be of very high magnitude (see Goudie 1978, Table 6). The 1903 fall of Saharan dust over England, for example, was estimated to have deposited 9.8×10^6 tonnes of sediment (Mill and Lempfert 1904), and when expressed per unit area rates may reach several hundred tonnes per square kilometre on desert margins. Quantities decline rapidly with distance from source of course, as indicated by data for annual deposition rates shown in Table 18.8. Rates of around $100 \, t \, km^{-2} \, a^{-1}$ have been calculated close to source areas. At best, however, most of these data on long-term deposition rates are rough estimates, as records are in most cases short and difficulties arise in distinguishing between

Table 18.8 *Modern dust deposition rates on land*

References	Location	Rate	
		t km^{-2} a^{-1}	mm 1000 a^{-1}*
Sahara			
Mediterranean region			
Löye-Pilot *et al.* (1986)	Corsica	14	16
Yaalon and Ganor (1975)	Israel	22–83	25–93
Goosens (1995)	Negev	15–30	10–20
Bücher and Lucas (1975)	Pyrenees	18–23	20–26
Pye (1992)	Crete	10–100	11–112
Harmattan plume			
Maley (1980)	S Chad	109	122
McTainsh and Walker (1982)	N Nigeria	137–181	154–203
Drees *et al.* (1993)	SW Niger	200	100–150
USA			
Smith *et al.* (1970)	High Plains	65–85	73–96
Péwé *et al.* (1981)	Arizona	54	61
Gile and Grossman (1979)	New Mexico	9.3–125.8	10–141
Muhs (1983)	California	24–31	27–35
Reheis and Kihl (1995)	California/Nevada	4.3–15.7	5–18
Middle East			
Safar (1985)	Kuwait	100	112
Behairy *et al.* (1985)	W Saudi Arabia	13–109	15–122
Miscellaneous			
Inoue and Naruse (1991)	Japan	3.5–6	4–7
Tiller *et al.* (1987)	SE Australia	5–10	6–11
Kukal (1971)	Caspian Sea	39.5	44

* Calculated on bulk density of dust of 0.89 g cm^{-3} where not derived in original reference.
Source: Modified after Goudie (1995).

deposited dust and amounts that are reworked. The presence of vegetation is thought to be important for trapping dust, while rock fragments also act as important traps, although such terrain probably retains less than 20% of settle dust (Goosens 1995). Similarly, estimates of dust contributed to ocean sediments based on extrapolation from terrestrial measurements may be invalid due to differences in particle scavenging by rainfall over land and at sea (Buat-Menard and Duce 1986).

PERIDESERT LOESS

Loess has been estimated to cover up to 10% of the world's land area (Pecsi 1968), and is by definition a wind-deposited dust with grain-size range in the region of 0.01–0.05 mm with a median size range of 0.02–0.03 mm (Tsoar and Pye 1987). Such deposits consist mainly of quartz, feldspar, mica, clay minerals and carbonate grains in varying proportions. Although the main areas of contemporary dust storm activity are in the world's arid and semi-arid regions, as seen above, peridesert loess deposits are poorly developed in comparison with those found in mid-latitudes which were formed when these areas experienced glaciation during the Pleistocene. Evidence from ocean and ice cores suggests that many of today's deserts produced more dust during the cold phases of the Pleistocene (see the next subsection) but loess formed only on certain desert margins.

The reasons for this apparent dichotomy have been the subject of some debate and are to date not clearly resolved. Former arguments were based on the premise that sufficient silt-sized material could only be produced in glacial environments (see above), but more recently attention has been focused on the mechanics of desert dust formation, transport and deposition (McTainsh 1987; Tsoar and Pye 1987; Yaalon 1987).

In a global appraisal Tsoar and Pye (1987) have examined dust dynamics and suggest that the absence of more widespread peridesert loess is largely due to a lack of available vegetation traps for dust. Coudé-Gaussen (1990), in comparing loess deposits north and south of the Mediterranean, also points to unfavourable dust-trapping terrain, as well as a relative lack of available sediment in north Africa. McTainsh (1987), however, has opted for a much more straightforward explanation, arguing that the apparent relative scarcity of peridesert loess deposits, compared with periglacial deposits, may simply be due to the use of inappropriate periglacial loess methodologies in the study of the desert and peridesert realms. The deposit that McTainsh cites as an example is located in northern Nigeria, but is not recognised as a loess by Tsoar and Pye. Four small areas of loess derived from the Sahara, the contemporary world's largest dust source, have been studied in some detail: in northern Nigeria, southern Tunisia (Coudé-Gaussen *et al.* 1982), in the central Sinai (Rögner and Smykatz-Kloss 1991) and in the Negev (Yaalon and Dan 1974). Other sparse deposits are catalogued by Coudé-Gaussen (1987): to the north of the Sahara in the Canary Islands, southern Morocco, Tripolitania and southwestern Egypt; and to the south in Guinea and northern Cameroon. It seems likely that this inventory is incomplete (Coudé-Gaussen 1987; Yaalon 1987), hence the uncertainty of peri-Saharan loess deposits shown in Figure 17.1. Elsewhere, it seems that some of the central Asian loess deposits may be at least partly derived from desert dust (see, for example, Lateef 1988 on northern Iran), although the relative roles of glacial and weathering processes in producing silt-sized material is yet to be established. Similarly in Argentina arid conditions have been suggested as being important for particle

formation in some of the Argentine loess deposits (Iriondo 1988).

Most of the world's thickest deposits, in China, are now considered to be of desert rather than glacial origin (Zhang *et al.* 1991), and the time of their greatest accumulation was probably during the maximum of the last glaciation around 22–18 000 years ago. Late Pleistocene accumulation rates at Lanzhou were 250–260 mm 1000 a^{-1}, while those at Luochuan were 50–70 mm 1000 a^{-1} (Pye 1987).

Loess can support vertical sections when dry but the shear strength is greatly reduced when wetted, making the material prone to subsidence, flow and sliding. Much geomorphological and engineering research has been conducted into the stability of slopes formed on loess in China and elsewhere (see, for example, Derbyshire *et al.* 1993; Rogers *et al.* 1994). The palaeoclimatic significance of loess is another active research area (see Chapter 21). Recent analysis of the Luochuan deposits, for example, has identified periods of grain-size maxima which correlate with episodes of massive iceberg release (Heinrich events) into the North Atlantic during the last glaciation, a reflection of a stronger East Asian winter monsoon circulation (Porter and Zhisheng 1995).

DEEP-SEA SEDIMENTS

Desert dust has made a significant contribution to ocean sediments in some parts of the world, and dust accumulations provide useful indicators of past global wind systems and environmental change on the continents (see Chapter 26). The ocean sediment record is relatively complete when compared with the continental record and fairly good stratigraphical and age control is usually provided by biogenic components.

Detailed analysis of cores from the Atlantic, Pacific and Indian oceans suggests that the northern hemisphere dominance of major dust sources (eastern and central Asia, northwest Africa, and Arabia) has characterised most of the Cenozoic. Aeolian flux data show that most of the northern hemisphere was more arid during Quaternary glacial maxima, with three to five times as much dust transported to the oceans during glacial stages than interglacials.

These results largely confirm those found from analysis of ice cores and loess profiles. Hovan *et al.* (1989), for example, used samples from a North Pacific core to construct a chronology of aeolian fluxes and grain sizes over the last 530 000 years, showing a clear relationship between dust flux maxima and glacial stages and suggesting a link of times of enhanced dust deposition to times of loess formation in China.

Only the northern Andes appear to have varied in the opposite sense, according to evidence from a core in the Equatorial Pacific, with less dust during glacials and more during interglacials (Rea *et al.* 1986). A comprehensive recent review of research into desert dust inputs to ocean sediments is given by Rea (1994) and further discussion of their arid zone implications can be found in Chapter 26.

Nevertheless, although there is no question that the deep-sea dust record relates to past climatic conditions over the continents, the interpretation of such records is not necessarily straightforward (Prospero 1985). The relationship between aridity and dust production is not a simple one. It has been noted above that modern dust storm occurrence is at a maximum in areas with 100–200 mm annual rainfall and that correlations of dust storm frequency with rainfall are weak. Further, the role of water in producing large quantities of fine particles has been hinted at, and the frequency of dust storm activity on the continents is not necessarily related directly to the rate of long-range transport and deposition over the oceans. Other findings also present challenges to those interpreting ocean sediments, such as Schroeder's (1985) identification of aggregated dust particles which become disaggregated in the ocean column and the presence of 'giant' aeolian dust particles at very large distances from source (Betzer *et al.* 1988), some of which may have been interpreted as evidence of iceberg rafting in previous analyses.

Figure 18.5 *Severe dust storm at Melbourne, Australia, 8 February 1983. This occurred during a period of intense drought, resulting in the erosion of topsoil from inland 'mallee' vegetation areas in Australian drylands (photo courtesy of the Australian Bureau of Meteorology)*

Conclusion

The realisation of the importance of dust to the geomorphology of arid zones is a relatively recent trend. Desert dust has significance for many geomorphological processes and affects the human inhabitants of desert and desert-fringe environments in a variety of ways (Figure 18.5). Research into desert dust proceeds apace, but there remain many basic questions to be answered: on the location of dust source areas, the effects of ground surface variables, and the processes of dust transport and deposition.

References

Agence France Press 1985. *Tchad – divers; fog de sable sur Ndjamena.* Ndjamena, 16 February.

Ali, Y.A. and West, I. 1983. Relationships between modern gypsum nodules in sabkhas of loess to compositions of brines in sediments in northern Egypt. *Journal of Sedimentary Petrology*, 53: 1151–1168.

Amit, R. and Gerson, R. 1986. The evolution of Holocene reg (gravelly) soils in deserts — an example from the Dead Sea region. *Catena*, 13: 59–79.

Arao, K. and Ishizaka, Y. 1986. Volume and mass of yellow sand dust in the air over Japan as estimated by atmospheric turbidity. *Journal of the Meteorological Society of Japan*, 64: 79–93.

Bacon, D.P. and Sarma, R.A. 1991. Agglomeration of dust in convective clouds initialized by nuclear bursts. *Atmospheric Environment*, 25A: 2627–2642.

Bagnold, R.A. 1941. *The physics of blown sand and desert dunes.* Methuen, London.

Bar-Ziv, J. and Goldberg, G.M. 1974. Simple siliceous pneumoconiosis in Negev Bedouins. *Archives of Environmental Health*, 29: 121–126.

Behairy, A.K.A., El-Sayed, M.K. and Rao, N.V. 1985. Eolian dust in the coastal area north of Jeddah, Saudi Arabia. *Journal of Arid Environments*, 8: 89–98.

Betzer, P.R., Carder, K.L., Duce, R.A., Merrill, J.T., Tindale, N.W., Uematsu, M., Costello, D.K., Young, R.W., Feely, R.A., Breland, J.A., Berstein, R.E. and Greco, A.M. 1988. Long-range transport of giant mineral aerosol particles. *Nature* 336: 568–571.

Biscaye, P.E., Chesselet, R. and Prospero, J.M. 1974. Rb–Sr, [87]r/[86]r isotope system as an index of the provenance of continental dusts in the open Atlantic Ocean. *Journal de Recherches Atmosphériques*, 8: 819–829.

Brazel, A. and Hsu, S. 1981. The climatology of hazardous Arizona dust storms. In T.L. Péwé (ed.), *Desert dust.* Geological Society of America, Special Paper, 186: 293–303.

Breed, C.S., McCauley, J.F. and Whitney, M.I. 1989. Wind erosion forms. In D.S.G. Thomas (ed.), *Arid zone geomorphology*, 1st edn. Belhaven, London: 284–307.

Brimhall, G.H., Lewis, C.J., Ague, J.J., Dietrich, W.E., Hampel, J., Teague, T. and Rix, P. 1988. Metal enrichment in bauxites by deposition of chemically mature aeolian dust. *Nature*, 333: 819–824.

Buat-Menard, P. and Duce, R.A. 1986. Precipitation scavenging of aerosol particles over remote regions. *Nature*, 321: 508–510.

Bücher, A. and Lucas, G. 1975. Poussières africains sur l'europe. *La Météorologie*, 5: 33, 53–69.

Burritt, B. and Hyers, A.D. 1981. Evaluation of Arizona's dust warning system. In T.L. Péwé (ed.), *Desert dust.* Geological Society of America, Special Paper, 186: 281–292.

Changery, M.J. 1983. *A dust climatology of the United States.* National Oceanic and Atmospheric Administration, Asheville, NC.

Chepil, W.S. and Woodruff, N.P. 1957. Sedimentary characteristics of dust storms, II. Visibility and dust concentration. *American Journal of Science*, 255: 104–114.

Chepil, W.S., Siddoway, F.H. and Armbrust, D.V. 1963. Climatic index of wind erosion conditions in the Great Plains. *Proceedings of the Soil Science Society of America*, 27: 449–451.

Choun, H.F. 1936. Dust storms in southwestern Plains area. *Monthly Weather Review*, 64: 195–199.

Chung, Y.-S. 1992. On the observations of yellow sand (dust storms) in Korea. *Atmospheric Environment*, 26A: 2743–2749.

Clafin, L.E., Stuteville, D.L. and Armburst, D.V. 1973. Windblown soil in the epidemiology of bacterial leaf spot of alfalfa and common blight of beans. *Phytopathology*, 63: 1417–1419.

Clayton, R.N., Rex, R.W., Syers, J.K. and Jackson, M.L. 1972. Oxygen isotope abundance in quartz from Pacific pelagic sediments. *Journal of Geophysical Research*, 77C: 3907–3915.

Clements, T., Stone, R.O., Mann, J.F. and Eymann, J.L. 1963. *A study of windblown sand and dust in desert areas.* US Army Natik Laboratories, Earth Sciences Division, Technical Report, ES-8. Natik, MA.

Coakley, J.A. and Cess, R.D. 1985. Response of the NCAR community climate model to radiative forcing by the naturally occurring tropospheric aerosol. *Journal of Atmospheric Sciences*, 42: 1677–1692.

Collins, D.N. 1979. Sediment concentration in melt waters as an indicator of erosion processes beneath an alpine glacier. *Journal of Glaciology*, 23: 247–257.

Conca, J.L. and Rossman, G.R. 1982. Case hardening of sandstone. *Geology*, 10: 520–523.

Coudé-Gaussen, G. 1984. Le cycle des poussières éoliennes désertiques actuelles et la sédimentation des loess peridésertiques quaternaires. *Bulletin Centre Researche et Exploration-Production Elf-Aquitaine*, 8: 167–182.

Coudé-Gaussen, G. 1987. The perisaharan loess: sedimentological characterisation and palaeoclimatic significance. *GeoJournal*, 15: 177–183.

Coudé-Gaussen, G. 1990. The loess and loess-like deposits along the sides of the western Mediterranean Sea: genetic and palaeoclimatic significance. *Quaternary International*, 5: 1–8.

Coudé-Gaussen, G. and Rognon, P. 1983. Les poussières sahariennes. *La Recherche*, 147: 1050–1061.

Coudé-Gaussen, G. and Rognon, P. 1988. Origine éolienne de certains encroûtements calcaires sur l'ile de Fuertebentura (Canaries Orientales). *Geoderma*, 42: 271–293.

Coudé-Gaussen, G., Mosser, C., Rognon, P. and Torenq, J. 1982. Une accumulation de loess Pleistocene supérieur dans le sud-Tunisien: coupe de Techine. *Bulletin de Société de Geologie Français*, 24: 283–292.

Crooks, G.A. and Cowan, G.R.C. 1993. Duststorm, South Australia, November 7th, 1988. *Bulletin of the Australian Meteorological and Oceanographic Society*, 6: 68–72.

d'Almeida, G.A. 1986. A model for Saharan dust transport. *Journal of Climate and Applied Meteorology*, 25: 903–916.

Dare-Edwards, A.J. 1984. Aeolian clay deposits of southeastern Australia: parna or loessic clay? *Transactions of the Institute of British Geographers*, 9: 337–344.

Darwin, C. 1846. An account of the fine dust which often falls in the Atlantic Ocean. *Quarterly Journal of the Geological Society*, 2: 26–30.

Das, T.M. 1988. Effects of deposition of dust particles on leaves of crop plants on screening of solar illumination and associated physiological processes. *Environmental Pollution*, 53: 421–422.

Delaney, A.C., Parkin, D.W., Goldberg, E.D., Riemann, B.E.F. and Griffin, J.J. 1967. Airborne dust collected at Barbados. *Geochimica et Cosmochimica Acta*, 31: 885–909.

Derbyshire, E., Dijkstra, T.A., Billard, A., Muxart, T., Smalley, I.J. and Li, Y.J. 1993. Thresholds in a sensitive landscape: the loess region of central China. In D.S.G. Thomas and R.J. Allison (eds), *Landscape sensitivity*. Wiley, Chichester: 97–127.

Dorn, R.I. 1986. Rock varnish as an indicator of environmental change. In W.G. Nickling (ed.), *Aeolian geomorphology*. Allen & Unwin, Winchester, MA: 291–307.

Drees, L.R., Manu, A. and Wilding, L.P. 1993. Characteristics of aeolian dusts in Niger, West Africa. *Geoderma* 59: 213–233.

Duce, R.A., Unni, C.K., Ray, B.J., Prospero, J.M. and Merrill, J.T. 1980. Long-range transport of soil dust from Asia to the tropical north Pacific: temporal variability. *Science*, 209: 1522–1524.

Durst, C.S. 1935. Dust in the atmosphere. *Journal of the Royal Meteorological Society*, 61: 81–89.

Ehrenburg, C.G. 1849. *Passat Staub und Blutregen*. Abhandlungen Akademie Wissenschaften, Berlin.

Ervin, R.T. and Lee, J.A. 1994. Impact of conservation practices on airborne dust in the southern High Plains of Texas. *Journal of Soil and Water Conservation*, 49: 430–437.

Folk, R.L. 1975. Geological urban hindplanning; an example from a Hellenistic Byzantian city, Stobi, Jugoslavian Macedonia. *Environmental Geology*, 1: 5–22.

Franzen, L.G., Hjelmroos, M., Kallberg, P., Brorstrom-Lunden, E., Juntto, S. and Savolainen, A.-L. 1994. The 'yellow snow' episode of northern Fennoscandia, March 1991 — a case study of long-distance transport of soil, pollen and stable organic compounds. *Atmospheric Environment*, 28 (22): 3587–3604.

Freeman, M.H. 1952. *Dust storms of the Anglo-Egyptian Sudan*. Meteorological Office Meteorological Reports, 11.

Gagosian, R.B., Peltzer, E.T. and Zafiriou, O.C. 1981. Atmospheric transport of continentally derived lipids to the tropical north Pacific. *Nature*, 291: 312–314

Ganor, E. 1991. The composition of clay minerals transported to Israel as indicators of Saharan dust emission. *Atmospheric Environment*, 25A: 2657–2664.

Gile, L.H. and Grossman, R.B. 1979. *The desert project soil monograph*. US Soil Conservation Service.

Gillette, D.A. 1979. Environmental factors affecting dust emission by wind erosion. In C. Morales (ed.), *Saharan dust*. Wiley, Chichester: 71–91.

Gillette, D.A. 1981. Production of dust that may be carried great distances. In T.L. Péwé (ed.), *Desert dust*. Geological Society of America, Special Paper, 186: 11–26.

Gillette, D.A. 1982. Threshold velocities for wind erosion on natural terrestrial surfaces (a summary). Unpublished manuscript.

Gillette, D.A., Bolivar, D.A. and Fryrear, D.W. 1974. The influence of wind velocity on the size distributions of aerosols generated by the wind erosion of soils. *Journal of Geophysical Research*, 79C: 4068–4075.

Gillette, D.A., Adams, J., Endo, A., Smith, D. and Khil, R. 1980. Threshold velocities for input of soil particles into the air by desert soils. *Journal of Geophysical Research*, 85C: 4068–4075.

Gillette, D.A., Adams, J., Muhs, D. and Khil, R. 1982. Threshold friction velocities and rupture moduli for crusted desert soils for the input of soil particles into the air. *Journal of Geophysical Research*, 87C: 9003–9015.

Glaccum, R.A. and Prospero, J.M. 1980. Saharan aerosols over the tropical north Atlantic. Mineralogy. *Marine Geology*, 37: 295–321.

Goosens, D. 1995. Field experiments of aeolian dust accumulation on rock fragment substrata. *Sedimentology*, 42: 391–402.

Goudie, A.S. 1978. Dust storms and their geomorphological implications. *Journal of Arid Environments*, 1: 291–310.

Goudie, A.S. 1983. Dust storms in space and time. *Progress in Physical Geography*, 7: 502–530.

Goudie, A.S. 1995. *The changing earth: rates of geomorphological processes*. Blackwell, Oxford.

Goudie, A.S. and Middleton, N.J. 1992. The changing frequency of dust storms through time. *Climatic Change*, 20: 197–225.

Hagen, L.J. and Woodruff, N.O. 1973. Air pollution from dust storms in the Great Plains. *Atmospheric Environment*, 7: 323–332.

Hand, I.F. 1934. The character and magnitude of the dense dust cloud which passed over Washington DC, 11 May 1934. *Monthly Weather Review*, 62: 156–157.

Hinds, B.D. and Hoidale, G.B. 1975. *Boundary layer dust occurrence, II. Atmospheric dust over the Middle East, Near East and North Africa*. Technical Report US Army Electronics Command, White Sands Missile Range, New Mexico.

Hirose, K. and Sugimura, Y. 1984. Excess [288]Th in the airborne dust: an indication of continental dust from the east Asian deserts. *Earth and Planetary Science Letters*, 70: 110–114.

Hovan, S.A., Rea, D.K., Pisias, N.G. and Shackleton, N.J. 1989. A direct link between the China loess and marine $_{18}$O records: aeolian flux to the North Pacific. *Nature* 340: 296–298.

Hyers, A.D. and Marcus, M.G. 1981. Land use and desert dust hazards in central Arizona. In T.L. Péwé (ed.), *Desert dust*. Geological Society of America, Special Paper, 186: 267–280.

Ing, G.K.T. 1972. A dust storm over central China, April 1969, *Weather*, 27: 136–145.

Inoue, K. and Naruse, T. 1991. Accumulation of Asian long-range eolian dust in Japan and Korea from the late Pleistocene to the Holocene. *Catena*, Supplement, 20: 25–42.

Iriondo, M. 1988. Map of the South American plains — its present state. *Quaternary of South America and Antarctic Peninsula*, 6: 297–308.

Ishizaka, Y. 1973. On materials of solid particles contained in snow and rain water: part 2. *Journal of the Meteorological Society of Japan*, 51: 325–336.

Jackson, M.L., Levett, T.W.M., Syers, J.K., Rex, R.W., Clayton, R.N., Sherman, G.D. and Uehara, G. 1971. Geomorphological relationships of tropospherically derived quartz in the soils of the Hawaiian Islands. *Proceedings of the Soil Science Society of America*, 35: 515–525.

Jaenicke, R. 1979. Monitoring and critical review of the estimated source strength of mineral dust from the Sahara. In C. Morales (ed.), *Saharan dust*. Wiley, Chichester: 233–242.

Jaenicke, R. and Schütz, L. 1978. Comprehensive study of physical and chemical properties of the surface aerosols in the Cape Verde Islands region. *Journal of Geophysical Research*, 83C: 3585–3599.

Jauregui, E. 1973. The urban climate of Mexico City. *Erdkunde*, 27: 298–307.

Jones, D.K.C., Cooke, R.U. and Warren, A. 1986. Geomorphological investigation, for engineering purposes, of blowing sand and dust hazard. *Quarterly Journal of Engineering Geology*, 19: 251–270.

Joseph, P.V., Raipal, D.K. and Deka, S.N. 1980. 'Andhi', the convective dust storm of northwest India. *Mausam*, 31: 431–442.

Kalma, J.D., Speight, J.G. and Wasson, R.J. 1988. Potential wind erosion in Australia: a continental perspective. *Journal of Climatology*, 8: 411–428.

Kes, A.S. 1983. Study of deflation processes and transfer of salts and dust. *Problems of Desert Development*, 1: 3–15 (in Russian).

Khalaf, F., Gharib, I.M. and Al-Khadi, A.S. 1982. Sources and genesis of the Pleistocene gravelly deposits in northern Kuwait. *Sedimentary Geology*, 3: 101–117.

Kidson, E. and Gregory, J.W. 1930. Australian origin of red rain in New Zealand. *Nature*, 125: 410–411.

Klimenko, L.V. and Moskaleva, L.A. 1979. Frequency of occurrence of dust storms in the USSR. *Meteorlogiia i Gidrologiia*, 9: 93–97.

Kotlyakov, V.M. 1991. The Aral Sea basin. *Environment* 33: 4–9, 36–38.

Kowalkowski, A. and Brogowski, Z. 1983. Features of cryogenic environment in soils of continental tundra and arid steppe on the southern Khangai slope under electron microscope. *Catena*, 10: 199–205.

Kuenen, P.H. 1969. Origin of quartz silt. *Journal of Sedimentary Petrology*, 39: 1631–1633.

Kukal, Z. 1971. *Geology of recent sediments*. Academic Press, London.

Lancaster, N. 1984. Characteristics and occurrence of wind erosion features in the Namib Desert. *Earth Surface Processes and Landforms*, 9: 469–478.

Lateef, A.S.A. 1988. Distribution, provenance, age

and palaeoclimatic record of the loess in Central North Iran. In D.N. Eden and R.J. Furkert (eds), *Loess: its distribution, geology and soils*. Balkema, Rotterdam: 93–101.

Leathers, C.R. 1981. Plant components of desert dust in Arizona and their significance for man. In T.L. Péwé (ed.), *Desert dust*. Geological Society of America, Special Paper, 186: 191–206.

Lee, J.A., Wigner, K.A. and Gregory, J.M. 1993. Drought, wind and blowing dust on the southern High Plains of the United States. *Physical Geography*, 14: 56–67.

Lindsay, H.A. 1933. A typical Australian line-squall dust storm. *Quarterly Journal of the Royal Meteorological Society*, 59: 350.

Lisitzin, A.P. 1972. *Sedimentation in the world ocean*. Society of Palaeontologists and Mineralogist Special Publication, 17.

Littmann, T. 1991. Dust storm frequency in Asia: climatic control and variability. *International Journal of Climatology*, 11: 393–412.

Lockeretz, W. 1978. The lessons of the Dust Bowl. *American Scientist*, 66: 560–569.

Logan, J. 1974. African dusts as a source of solutes in Gran Canaria ground waters. *Geological Society of America, Abstracts with Programs*, 6: 849.

Löye-Pilot, M.D., Martin, J.M. and Morelli, J. 1986. Influence of Saharan dust on the rainfall acidity and atmospheric input to the Mediterranean. *Nature*, 321: 427–428.

Maley, J. 1980. Études palynologques deans le bassin due Tchad et paléoclimatologie de l'Afrique nord tropical de 30,000 ans a l'époque actuelle. Unpublished PhD thesis, University of Montpellier, France.

Maley, J. 1982. Dust, clouds, rain types and climatic variations in tropical North Africa. *Quaternary Research*, 18: 1–16.

Martin, R.J. 1937. Duststorms of January–April 1937 in the United States. *Monthly Weather Review*, 65: 151–152.

McCauley, J.F., Breed, C.S., Grolier, M.J. and Mackinnon, D.J. 1981. The US dust storm of February 1977. In T.L. Péwé (ed.), *Desert dust*. Geological Society of America, Special Paper, 186: 123–147.

McDonald, W.F. 1938. *Atlas of climatic charts of the oceans*. Department of Agriculture, Weather Bureau, Washington DC.

McFadden, L.D., Wells, S.G. and Jercinovich, M.J. 1987. Influences of eolian and pedogenic processes on the origin and evolution of desert pavements. *Geology*, 15: 504–548.

McTainsh, G.H. 1980. Harmattan dust deposition in northern Nigeria. *Nature*, 286: 587–588.

McTainsh, G.H. 1987. Desert loess in northern Nigeria. *Zeitschrift für Geomorphologie*, NF, 26: 417–435.

McTainsh, G.H. and Pitblado, J.R. 1987. Dust storms and related phenomena measured from meteorological records in Australia. *Earth Surface Processes and Landforms*, 12: 415–424.

McTainsh, G.H. and Walker, P.H. 1982. Nature and distribution of Harmattan dust. *Zeitschrift für Geomorphologie*, NF, 26: 417–435.

Membery, D.A. 1983. Low level wind profiles during the Gulf Shamal. *Weather*, 38: 18–24.

Middleton, N.J. 1984. Dust storms in Australia: frequency, distribution and seasonality. *Search*, 15: 46–47.

Middleton, N.J. 1985. Effect of drought on dust production in the Sahel. *Nature*, 316: 431–434.

Middleton, N.J. 1986a. Dust storms in the Middle East. *Journal of Arid Environments*, 10: 83–96.

Middleton, N.J. 1986b. The geography of dust storms in south west Asia. *Journal of Climatology*, 6: 183–196.

Middleton, N.J. 1986c. The geography of dust storms. Unpublished DPhil thesis, University of Oxford.

Middleton, N.J. 1987. *Desertification and wind erosion in the western Sahel: the example of Mauritania*. University of Oxford, School of Geography Research Papers, 40.

Middleton, N.J. 1989a. Climatic controls on the frequency, magnitude and distribution of dust storms: examples from India and Pakistan, Mauritania and Mongolia. In M. Leinen and M. Sarnthein (eds), *Palaeoclimatology and palaeometeorology: modern and past patterns of global atmospheric transport*. NATO ASI Series C, Vol. 282: 97–132.

Middleton, N.J. 1989b. Desert dust. In D.S.G. Thomas (ed.), *Arid zone geomorphology*, 1st edn. Belhaven, London: 262–283.

Middleton, N.J. 1991. Dust storms of the Mongolian People's Republic. *Journal of Arid Environments*, 20: 287–297.

Middleton, N.J. and Chaudhary, Q.V. 1988. Severe dust storm at Karachi, 31st May 1986. *Weather*, 43: 298–301.

Middleton, N.J., Goudie, A.S. and Wells, G.L. 1986. The frequency and source areas of dust storms. In W.G. Nickling, (ed.), *Aeolian geomorphology*. Allen & Unwin, Boston: 237–259.

Mill, H.R. and Lempfret, M.A. 1904. The great dust-fall of February 1903, and its origin. *Quarterly Journal of the Royal Meteorological Society*, 30: 57–91.

Mizota, C. and Matsuhisa, Y. 1985. Eolian additions to soils and sediments of Japan. *Soil Science and Plant Nutrition*, 31: 369–382.

Morales, A.F. 1946. *El Sahara Español*. Alta Comisaria de España en Marruecos, Madrid.

Morales, C. (ed.) 1979. *Saharan dust*. Wiley, Chichester.

Moss, A.J., Walker, P.H. and Hutka, J. 1973. Fragmentation of granitic quartz in water. *Sedimentology*, 20: 489–511.

Muhs, D.R. 1983. Airborne dustfall on the Californian Channel Islands, USA. *Journal of Arid Environments*, 6: 223–228.

Nahon, D. and Trompette, R. 1982. Origin of siltstones: glacial grinding versus weathering. *Sedimentology*, 29: 25–35.

Nalivkin, D.V. 1983. *Hurricanes, storms and tornadoes*. Amerind, New Delhi.

Nickling, W.G. and Brazel, A.J. 1984. Temporal and spatial characteristics of Arizona dust storms (1965–1980). *Journal of Climatology*, 4: 645–460.

Nickling, W.G. and Gillies, J.A. 1993. Dust emission and transport in Mali, West Africa. *Sedimentology*, 40: 859–868.

Oldfield, F., Hunt, A., Jones, M.D.H., Chester, R., Dearino, J.A., Olsson, L. and Prospero, J.M. 1985. Magnetic differentiation of atmospheric dusts. *Nature*, 317: 516–518.

Oliver, F.W. 1945. Dust storms in Egypt and their relation to the war period, as noted in Maryut, 1939–45. *Geographical Journal*, 106: 26–49.

Orgill, M.M. and Sehmel, G.A. 1976. Frequency and diurnal variation of dust storms in the contiguous USA. *Atmospheric Environment*, 10: 813–825.

Paquet, H., Coudé-Gaussen, G. and Rognon, P. 1984. Étude minéralogique de poussières sahariennes le long d'un itinéraire entre 19° et 35° de latitude nord. *Revue de Geologie Dynamique et de Geographie Physique*, 25: 257–265.

Parkin, D.W., Phillips, D.R., Sullivan, R.A.L. and Johnson, L.R. 1972. Airborne dust collections down the Atlantic. *Quarterly Journal of the Royal Meteorological Society*, 98: 798–808.

Pecsi, M. 1968. Loess. In R.W. Fairbridge (ed.), *The encyclopedia of geomorphology*. Reinhold, New York: 674–678.

Péwé, T.L. (ed.) 1981. *Desert dust*. Geological Society of America, Special Paper, 186.

Péwé, T.L., Péwé, E.A., Péwé, R.H., Journaux, A. and Slatt, R.M. 1981. Desert dust: characteristics and rates of deposition in central Arizona. In T.L. Péwé (ed.), *Desert dust*. Geological Society of America, Special Paper, 186: 169–190.

Porter, S.C. and Zhisheng, A. 1995. Correlation between climate events in the North Atlantic and China during the last glaciation. *Nature*, 375: 305–308.

Prodi, F. and Fea, G. 1979. A case of transport and deposition of Saharan dust over the Italian Peninsula and southern Europe. *Journal of Geophysical Research*, 84C: 6951–6960.

Prospero, J.M. 1979. Mineral and sea-salt aerosol concentrations in various world ocean regions. *Journal of Geophysical Research*, 84: 725–731.

Prospero, J.M. 1981. Arid regions as sources of mineral aerosols in the marine atmosphere. In T.L. Péwé (ed.), *Desert dust*. Geological Society of America, Special Paper, 186: 71–86.

Prospero, J.M. 1985. Records of past continental climates in deep-sea sediments. *Nature*, 315: 279–280.

Prospero, J.M. and Carlson, T.N. 1970. Radon-222 in the North Atlantic trade winds: its relationship to dust transport from Africa. *Science*, 167: 974–977.

Prospero, J.M. and Carlson, T.N. 1972. Vertical and areal distribution of Saharan dust over the western equatorial North Atlantic Ocean. *Journal of Geophysical Research*, 77: 5255–5265.

Prospero, J.M. and Nees, R.T. 1977. Dust concentration in the atmosphere of the equatorial North Atlantic: possible relationship to the Sahelian drought. *Science*, 196: 1196–1198.

Prospero, J.M. and Nees, R.T. 1986. Impact of the North African drought and El Nino on mineral dust in the Barbados trade winds. *Nature*, 320: 735–738.

Prospero, J.M., Bonatti, E., Schuber, C. and Carlson, T.N. 1970. Dust in the Caribbean traced to an African dust storm. *Earth and Planetary Science Letters*, 9: 287–293.

Prospero, J.M. Savoie, D.L., Carlson, T.N. and Nees, R.T. 1979. Monitoring of Saharan aerosol transport by means of turbidity measurements. In C. Morales (ed.), *Saharan dust*. Wiley, Chichester: 171–186.

Prospero, J.M., Glaccum, R.A. and Nees, R.T. 1981. Atmospheric transport of soil dust from Africa to South America. *Nature*, 289: 570–572.

Pye, K. 1987. *Aeolian dust and dust deposits*. Academic Press, London.

Pye, K. 1992. Aeolian dust transport and deposition over Crete and adjacent parts of the Mediterranean Sea. *Earth Surface Processes and Landforms*, 17: 271–288.

Pye, K. and Sperling, C.H.B. 1983. Experimental investigation of silt formation by static breakage processes: the effect of temperature, moisture and salt on quartz dune sand and granitic regolith. *Sedimentology*, 30: 49–62.

Pye, K. and Tsoar, H. 1987. The mechanics and geological implications of dust transport and deposition in deserts, with particular reference to loess formation and dune sand diagenesis in the northern Negev, Israel. In L. Frostick and I. Reid (eds), *Desert sediments ancient and modern*. Geological Society of London, Special Publication, 35. Blackwell, Oxford: 139–156.

Rahn, K.A., Borys, R.D. and Shaw, G.E. 1981. Asian desert dust over Alaska: anatomy of an Arctic haze episode. In T.L. Péwé (ed.), *Desert dust*. Geological Society of America, Special Paper, 186: 37–70.

Rapp, A. 1984. Are terra rosa soils in Europe eolian deposits from Africa? *Geologiska Föreningens I Stockholm Förhendlingar*, 105: 161–168.

Rea, D.K. 1994. The paleoclimatic record provided by eolian deposition in the deep sea: the geologic history of wind. *Reviews of Geophysics*, 32: 159–195.

Rea, D.K., Chambers, L.W., Chuey, J.M., Janecek, T.R., Leinen, M. and Pisias, N.G. 1986. A 420,000-year record of cyclicity in oceanic and atmospheric processes from the eastern equatorial Pacific. *Paleoceanography*, 1: 577–586.

Reheis, M.C. and Kihl, R. 1995. Dust deposition in southern Nevada and California, 1984–1989: relations to climate, source area, and source lithology. *Journal of Geophysical Research*, 100 (D5): 8893–8918.

Reinking, R.F., Mathews, L.A. and St-Amand, P. 1975. Dust storms due to desiccation of Owens Lake. *International Conference on Environmental Sensing and Assessment*, 14–19 September. IEEE, Las Vegas, NV.

Rex, R.W., Syers, J.K., Jackson, M.L. and Clayton, R.N. 1969. Eolian origin of quartz in soils of Hawaiian Islands and in Pacific pelagic sediments. *Science*, 163: 277–279.

Riesborough, R.W., Hugger, R.J., Griffin, J.J. and Goldberg, E.D. 1968. Pesticides: transatlantic movements in the northeast trades. *Science*, 159: 1233–1236.

Rogers, C.D.F., Dijkstra, T.A. and Smalley, I.J. 1994. Hydroconsolidation and subsidence of loess: studies from China, Russia, North America and Europe. *Engineering Geology*, 37: 83–113.

Rögner, K. and Smykatz-Kloss, W. 1991. The deposition of eolian sediments in lacustrine and fluvial environments of Central Sinai (Egypt). *Catena*, Supplement, 20: 75–91.

Safar, M.I. 1985. *Dust and dust storms in Kuwait*. State of Kuwait, Kuwait.

Sapozhnikova, S.A. 1973. *Map diagram of the number of days with dust storms in the hot zone of the USSR and adjacent territories*. Report HT-23-0027. US Army Foreign and Technology Center, Charlottesville.

Schroeder, J.H. 1985. Eolian dust in the coastal desert of the Sudan: aggregates cemented by evaporites. *Journal of African Earth Sciences*, 3: 370–380.

Schütz, L. 1980. Long range transport of desert dust with special emphasis on the Sahara. *Annals of the New York Academy of Sciences*, 338: 515–532.

Shaw, G.E. 1980. Transport of Asian desert aerosol to the Hawaiian Islands. *Journal of Applied Meteorology*, 19: 1254–1259.

Smalley, I.J. 1966. The properties of glacial loess and the formation of loess deposits. *Journal of Sedimentary Petrology*, 36: 669–676.

Smalley, I.J. and Vita-Finzi, C. 1968. The formation of fine particles in sandy deserts and the nature of 'desert' loess. *Journal of Sedimentary Petrology*, 38: 766–774.

Smith, B.J., McGreevy, J.P. and Whalley, W.B. 1987. Silt production by weathering of a sandstone under hot arid conditions: an experimental study. *Journal of Arid Environments*, 12: 199–214.

Smith, R.M., Twiss, P.C., Krauss, R.K. and Bronn, M.J. 1970. Dust deposition in relation to site, season, and climatic variables. *Proceedings of the Soil Science Society of America*, 34: 112–117.

Sprigg, R.C. 1982. Alternating wind cycles of the Quaternary era and their influences on aeolian sedimentation in and around the dune deserts of south-east Australia. In R.J. Wasson (ed.), *Quaternary dust mantles of China, New Zealand and Australia*. Proceedings of a workshop at the Australian National University, Canberra. 3–5 December 1980: 211–240.

Summerfield, M.A. 1983. Silcrete. In A.S. Goudie and K. Pye, (eds), *Chemical sediments and geomorphology*. Academic Press, London.

Swap, R., Garstang, M., Greco, S., Talbot, R. and Kallberg, P. 1992. Saharan dust in the Amazon Basin. *Tellus*, 44B: 133–149.

Syers, J.K., Jackson, M.L., Berkeiser, V.E., Clayton, R.N. and Rex, R.W. 1969. Eolian sediment influence on pedogenesis during the Quaternary. *Soil Science*, 107: 421–427.

Taylor, K.C., Lamorey, G.W., Doyle, G.A., Alley, R.B., Grootes, P.M., Mayewskii, P.A., White, J.W.C. and Barlow, L.K. 1993. The 'flickering switch' of late Pleistocene climatic change. *Nature*, 361: 432–436.

Tetzlaff, G. and Peters, M. 1986. Deep-sea sediments in the eastern equatorial Atlantic off the African coast and meteorological flow patterns over the Sahel. *Geologische Rundschau*, 75: 71–79.

Thomas, D.S.G. 1985. Evidence of aeolian processes in the Zimbabwean landscape. *Transactions of the Zimbabwean Scientific Association*, 62: 45–55.

Tiller, K.G., Smith, L.H. and Merry, R.H. 1987. Accessions of atmospheric dust east of Adelaide, South Australia, and implications for pedogenesis. *Australian Journal of Soil Research*, 25: 43–54.

Tsoar, H. and Pye, K. 1987. Dust transport and the question of desert loess formation. *Sedimentology*, 34: 139–153.

Turekian, K.K. and Graustein, W.C. 1984. ^{210}Pb indicates source of mid-Pacific aerosols. *Eos*, 65: 837.

Wake, C.P. and Mayewski, P.A. 1993. The spatial

variation of Asian dust and marine aerosol contributions to glaciochemical signals in Central Asia. In G.J. Young (ed.), *Snow and glacier hydrology*. International Association of Hydrological Sciences, Publication, 218: 385–402.

Wentworth, C.K. 1922. A scale of grade and class terms for clastic sediments. *Journal of Geology*, 30: 377–392.

Westphal, D.L., Toon, O.B. and Carlson, T.N. 1987. A two-dimensional numerical investigation of the dynamics and microphysics of Saharan dust storms. *Journal of Geophysical Research*, 92D: 3027–3049.

Whalley, W.B., Marshall, J.R. and Smith, B.J. 1982; Origin of desert loess from some experimental observations. *Nature*, 300: 433–435.

Whalley, W.B., Smith, B.J., McAllister, J.J. and Edwards, A. 1987. Aeolian abrasion of quartz particles and the production of silt-size fragments. preliminary results and some possible implications for loess and silcrete production. In L. Frostick and I. Reid (eds), *Desert sediments ancient and modern*. Geological Society of London, Special Publication 35. Blackwell, Oxford: 129–138.

Wheaton, E.E. and Chakravarti, A.K. 1990. Dust storms in the Canadian prairies. *International Journal of Climatology*, 10: 829–837.

Wheeler, D.A. 1986. The meteorological background to the fall of Saharan dust, November 1984. *Meteorological Magazine*, 115: 1–9.

Whitney, M.I. and Dietrich, R.V. 1973. Ventifact sculpture by windblown dust. *Bulletin of the Geological Society of America*, 84: 2561–2582.

Wilding, L.R., Smeck, N.E. and Dress, L.R. 1977 Silica in soils: quartz, tridymite and opal. In J.B. Dixon and S.B. Weed (eds), *Minerals in soil environments*. Soil Science Society of America, Madison, WI: 471–552.

Wilshire, H.G., Nakata, J.K. and Hallet, B. 1981. Field observations of the December 1977 wind storm, San Joaquin Valley, California. In T.L. Péwé (ed.), *Desert dust*. Geological Society of America, Special Paper 186: 233–252.

Woelcken, K. 1951. Descripción de una violenta tempestad de polvo. *Meteoros*, 1: 211–226.

Woodruff, N.P. 1956. Windblown soil abrasive injuries to winter wheat plants. *Journal of Agronomy*, 48: 499–505.

Worster, D. 1979. *Dust bowl*. Oxford University Press, Oxford.

Yaalon, D.H. 1987. Saharan dust and desert loess: effect of surrounding soils. *Journal of African Earth Sciences*, 6: 569–571.

Yaalon, D.H. and Dan, J. 1974. Accumulation and distribution of loess-derived deposits in the semi-desert and desert fringe of Israel. *Zeitschrift für Geomorphologie*, Supplementband, 20: 91–105.

Yaalon, D.H. and Ganor, E. 1975. Rate of aeolian dust accretion in the Mediterranean and desert fringe environments of Israel. *19th International Congress of Sedimentology*: 169–174.

Yaalon, D.H. and Ganor, E. 1979. East Mediterranean trajectories of dust-carrying storms from the Sahara and Sinai. In C. Morales (ed.), *Saharan dust*. Wiley, Chichester: 187–193.

Young, R.W., Carder, K.L., Betzer, P.R., Costello, D.K., Duce, R.A., Ditullio, G.R., Tindale, N.W., Laws, E.A., Uematsu, M., Merrill, J.T. and Feely, R.A. 1991. Atmospheric iron inputs and primary productivity: phytoplankton responses in the North Pacific. *Global Biochemical Cycles*, 5: 119–134.

Yu, B., Hesse, P.P. and Neil, D.T. 1993. The relationship between antecedent rainfall conditions and the occurrence of dust events at Mildura, Australia. *Journal of Arid Environments*, 24: 109–124.

Zamorskii, A.D. 1964. Red snow. *Priroda*, 2: 127 (in Russian).

Zeuner, F.E. 1949. Frost soils on Mount Kenya and relation of frost soils to aeolian deposits. *Journal of Soil Science*, 1: 20–30.

Zhang, L., Dai, X. and Shi, Z. 1991. The sources of loess material and the formation of the Loess Plateau in China. *Catena*, Supplement, 20: 1–14.

19

Wind erosion in drylands

Carol S. Breed, John F. McCauley, Marion I. Whitney, Vatche P. Tchakerian and Julie E. Laity

Introduction

Landforms shaped by wind are produced where winds blow frequently at speeds sufficient to abrade and lift (deflate) particles from exposed surfaces. Vulnerable materials range in composition and texture from loose, sandy soils to sedimentary and even crystalline rocks. Wind erosion occurs most readily in areas where abrasive particles are blown against surfaces of easily disintegrated, granular material. Wind-erosion forms are therefore most widely distributed in arid regions, but they also occur elsewhere on beaches and in periglacial environments.

The shapes and textures produced by wind erosion vary with differences in material (hardness, homogeneity, grain size and cohesion), in shapes and orientations, and in structures (the presence of joints and bedding planes). The degree of streamlining and the arrangement of erosional markings (scores, pits, and flutes or grooves) in cohesive materials are influenced by the local topography, supply of abrasive particles, and evolving shape of the wind-eroded form itself (Whitney and Dietrich 1973). Wind-erosion patterns typically record not only the transport of material downslope, as in fluvial erosion, but also movement of material upslope

and in directions reverse, transverse and oblique to that of the primary wind. In areas of fairly homogeneous lithology and strongly directional winds, wind erosion produces streamlined, lemniscate forms that are commonly repeated in parallel arrays (Figure 19.1). Such streamlined hills (yardangs) and faceted stones (ventifacts: Figure 19.2) are readily recognized as the products of aerodynamic shaping. Less obviously wind eroded are the oddly shaped, pitted, grooved and fluted hillocks (demoiselles or hoodoos) and ventifacts found in localities where geology and windflow patterns are more complex (Figures 19.3 and 19.4).

After studies were published of wind-sculpted forms in arid to extremely arid regions of northwestern Asia (Hedin 1903; 1905; Stein 1909), north Africa (Ball 1900; Beadnell 1909), and southern Africa (Walther 1892; 1912; Passarge 1904), the idea of wind as a major agent of planation was applied to desert landforms too broadly (Keyes 1908; 1909). Subsequent challenges to this idea, particularly in regard to semi-arid regions, led to a long-prevailing view that wind as an agent of surface change is of only minor significance (among them Bryan 1923; 1925; 1927; Twidale 1976; Small 1978). A few geologists in the United States continued

Arid Zone Geomorphology: Process, Form and Change in Drylands, 2nd edition. Edited by David S. G. Thomas.
© 1997 John Wiley & Sons Ltd.

Figure 19.1   *Aerial view of 'marble yardangs' carved in silicified Thebes limestone that caps Limestone Plateau in Egypt. Yardangs have high, blunt heads facing into strong, prevailing north wind and taper to leeward. General lemniscate shapes and axial trends are regionally constant, and wind-erosion pattern cross-cuts relict stream channels and karst depressions. The largest yardangs are several tens of metres to hundreds of metres long. Ground views are shown in Figure 19.15*

Figure 19.2   *Series of streamlined, faceted, lemniscate ventifacts carved from beach pebbles of dolomitic limestone on shores of Lake Michigan, USA. Series runs from left to right in top row, right to left in bottom row, and is arranged according to degree of erosion. Most advanced stage is represented by classic, crestally keeled, triquetrous form at lower left. Long axis of this ventifact is aligned with the resultant of two primary winds, but others in this group formed in multidirectional wind regimes. In general, relations of long axes of streamlined ventifacts to wind directions are ambiguous, because they vary with original shapes and orientations of pebbles. Experimentally tested methods for determining impact and lee zones on ventifacts are outlined by Whitney (1983). Largest ventifact is approximately 6 cm long*

to study wind-eroded forms, mostly ventifacts and small yardangs (Blackwelder 1930; 1934; 1954; Maxson 1940; Sharp 1949). Meanwhile, soil physicists investigated erosion in farm fields of the American Midwest, which had suffered extreme losses of soil during the 'dust bowl' years of the 1930s (see, for example, Chepil 1945; Zingg 1949; 1953; Chepil and Woodruff 1963; Woodruff and Siddoway 1965; Skidmore and Woodruff 1968). In Britain the classic study of the movement of sand by wind was

published by R.A. Bagnold (1941). The work by Bagnold and by the soil scientists provided much of the theoretical framework for understanding aeolian processes on which geologists based their interpretations of aeolian phenomena. These studies, however, were concerned primarily with the physical mechanisms that govern the excavation and transport of loose particles, and only incidentally with the erosional effects that deflated particles might create by impinging on rocks. Such erosion was

a

b

Figure 19.3 *(a) Demoiselles and yardangs in the Farafra Depression, Egypt, carved in Tarawan Chalk of Palaeocene age and (b) lake-bed deposits of Quaternary age (bottom). Unusual shapes of demoiselles, relative to yardangs in the same localities, may be attributed to local topographic deflections of wind and to lithologic and structural control of erosional patterns. Figure of man on top of chalk yardang ((a), right) for scale; dog-shaped demoiselle (b) approximately 2 m long*

generally termed *sandblasting*; it was considered a minor process limited to abrasion by sand-size grains saltating to heights only a few centimetres to a metre or so above the ground (Hobbs 1917; Bagnold 1941; Sharp 1949; Simonett 1968; Twidale 1976; Small 1978).

A convergence of events during the 1960s and 1970s renewed geological interest in the capacity of wind not only to redistribute sediment, but also to shape large-scale landforms. Many poorly known parts of the Earth's deserts were photographed from space during the Mercury, Gemini and Apollo space missions of the 1960s. Their pictures provided compelling views of aeolian landforms, and later images from the Landsat earth-orbiting system (Figure 19.5) provided a basis for global surveys of

these features (McCauley *et al.* 1977a; Breed and Grow 1979; Breed *et al.* 1979). The views from space were especially influential in revising previous perceptions of wind as a geologic agent. As pointed out by Peel (1960), all of the United States would fit inside the Sahara, with room left over for parts of Europe, yet few textbooks at that time showed examples of desert landforms outside the United States. The pictures of Earth from space were matched in 1971 by pictures from the Mariner orbital mission to Mars (see Chapter 29), which first showed that planet shrouded in a global dust storm and later revealed fields of sand dunes and probable yardangs, indicating that wind was an active agent of erosion on Mars (McCauley 1973).

a

b

Figure 19.4   *Ventifacts from rocky plateau surfaces in the Western Desert of Egypt. (a) Wind-pitted and deflated orthoquartzite talus block (24 cm high) from Gilf Kebir Plateau. (b) Fluted ventifact (14 cm long) with projecting fingers ('dedos' of McCauley* et al. *1977a) in the silicified Thebes Limestone from Limestone Plateau. Both localities are about 1000 m above the adjacent sand plains, and sand is not abundant on the plateaux. Fingers on ventifact at right pointed into north wind; flutes on top surface have been cut by negative flow (reverse to the north wind), for they have Y-junctions that close in the downwind direction*

In search of Martian analogues, expeditions explored the hyper-arid deserts of Peru, Iran and southwestern Egypt (McCauley *et al.* 1977a; 1979). They investigated yardangs kilometres long and tens to hundreds of metres high, carved in bedrock and aligned in parallel arrays covering hundreds of square kilometres (Figure 19.6). The large-scale wind-erosion features demanded explanation, for abrasion by sandblasting had been considered incapable of significantly shaping hard bedrock in places high above the saltation zone and in lee zones not directly impacted by the primary wind. Some additional process seemed to be required, perhaps the removal of loosened granular particles by deflation (McCauley *et al.* 1977a).

Since the 1970s, characteristic shapes and textures of wind-eroded rocks have been documented in field settings worldwide, and some natural erosion patterns have been duplicated by laboratory studies (see, for example, McCauley *et al.* 1977a; 1977b; 1979; Whitney 1978; 1983; Whitney and Spletstoesser 1982). These studies suggested that wind-erosion forms develop in response not only to sandblasting and to deflation, but also to vorticity within systematic and complex air currents that flow over, around, and across rock obstacles. Aerodynamic erosion was found to occur when fine-particle abrasives were suspended within subsidiary wind currents and brought into contact with rock surfaces by vorticity (Whitney and Brewer 1968). This process offers an explanation for the formation of many yardangs, demoiselles and ventifacts, especially those located above the saltation zone or in sand-poor regions.

Figure 19.5 *Landsat image (30 October 1972) of very large yardangs in sandstone east of the Tibesti Plateau, Chad. Primary wind from the northeast, Unusually large single yardangs 20–30 km long are seen at A and B*

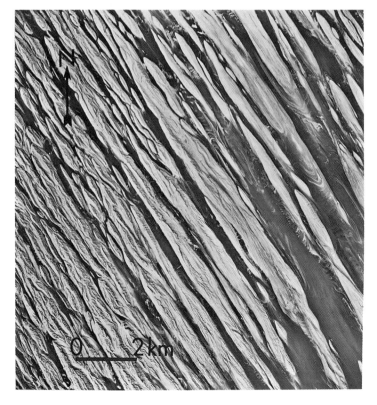

Figure 19.6 *Aerial view of yardangs carved in plateau capped by siltstones of Pleistocene age in Lut Desert, Iran. Intense gullying of yardang slopes (left side of photograph) indicates that running water is episodically effective in this region. Primary wind from north-northwest*

Major types of wind-erosion forms and their origins

DEFLATION HOLLOWS

Many early workers ascribed the excavation of desert depressions, including rock floors, to deflation: the removal of granular, poorly consolidated material by wind that lifts and carries finer grains in saltation and suspension and rolls coarser grains along the surface. Keyes (1908; 1909) proposed the extreme idea that desert rock floors have been planed to their present broad, low levels almost entirely by wind abrasion and deflation, even in the semi-arid American West: 'In the arid regions the wind is probably not only the most potent of the gradative agencies, but its efficiency is greater than all other geologic processes combined.' Such overstatements were vigorously refuted (for example, Tolman 1909; Bryan 1923). Although Hobbs (1917) and Peel (1941; 1960), argued eloquently for the recognition of wind as an agent of landform sculpture both emphasised that most broad, flat desert basins were formed in humid palaeoclimates and are fundamentally the products of erosion by running water.

The idea of wind as a powerful agent of excavation was widely applied to the origin of large closed bedrock depressions in Egypt's Western Desert (Ball 1900; 1927; Beadnell 1909; Walther 1912; Hobbs 1917; Blackwelder 1954; Peel 1960; Said 1962; Squyres and Bradley 1964; Smith 1972). The Qattara, Siwa, Farafra, Bahariya, Dakhla and Kharga depressions and several smaller basins have several physiographic elements in common: they are partly or entirely bounded by plateaux of limestone caprock, have been enlarged by cliff retreat, and have been severely deflated, some to depths below sea level. A common origin, by wind excavation to a base-level controlled by the groundwater table, has been the generally accepted explanation. Closer geological inspection, however, has suggested that solution and running water were involved in the origin of the excavations under more humid conditions during the Tertiary (Caton-Thompson and Gardner 1932; Stringfield *et al.* 1974; Brooks 1983; 1986).

Albritton *et al.* (1990) proposed a polygenetic origin for the Qattara Depression. The depression was originally excavated as a stream valley, subsequently modified by karstic activity, and further deepened and extended by mass-wasting, deflation, and fluviatile processes. Wind erosion was thought to play a major role only during arid phases of the Quaternary. Probably only during episodes of Quaternary aridity have the depressions undergone substantial deflation. The present similarity among the depressions may thus be ascribed to an equifinality of form resulting from different processes, with a strong aeolian overprint on all.

Wind abrasion and deflation of rock-floored depressions has thus been largely discounted, except for features such as the shallow, semi-circular to elongate hollows (pans) commonly found in semi-arid regions. These hollows (Figure 19.7) are common on the Great Plains of the western United States (Gilbert 1895; Bryan 1923; Price 1968; McCauley *et al.* 1981), in the Kalahari Desert of southern Africa (Grove 1969; Lancaster 1978), in central Australia (Mabbutt 1977) and in Mongolia (Berkey and Morris 1927). All of these regions consist largely of plains blanketed by aeolian and alluvial sediments, which include abundant materials in grain sizes vulnerable to mobilisation by wind. Deflation is initiated in spots where surface pavements or soil crusts are broken or the vegetation removed. The blow-outs often become playas by the collection of runoff or by excavation to the groundwater table (Blackwelder 1935; see Chapter 14).

The lack of vegetation on the pan enhances windflow and permits deflation of sediments made more susceptible as a result of salt weathering which produces fine-grained debris (Haynes 1982; Goudie 1989). As the depressions are enlarged, they become more attractive to grazing animals, which further disrupt the surface, rendering it more prone to erosion (Goudie 1989). According to Goudie and Wells (1995), pans are primarily the result of salt weathering and aeolian deflation, with the groundwater table both controlling the depth of deflation, and contributing to salt weathering processes. These deflation hollows are easily recognised because they represent localised wind erosion in otherwise vegetated regions. In extremely arid regions that lack vegetation, such as the Saharan wind plains, deflation occurs over a much broader area, and isolated deflation hollows are not obvious. A full account of

Figure 19.7   *Deflation hollow (about 200 m in diameter) eroded in sand sheet (with linear dunes) on the Moenkopi Plateau, northeastern Arizona, USA. Base of hollow is wind-scoured Navajo Sandstone of Mesosoic age, an aeolian sandstone that is friable and easily eroded by wind. This type of closed basin was described as a 'rock tank' by Bryan (1923). Dominant wind toward viewer from south-southwest*

playa and pan development is provided in Chapter 14.

DESERT PAVEMENTS

Many old spreads of alluvial gravel and residual talus slopes in arid regions have developed an armour of closely packed stones, one or two stones thick, over a layer of fine sediment several centimetres to metres thick (Springer 1958; Cooke 1970; McFadden *et al.* 1984; 1987; Wells *et al.* 1985). In the Sahara, such stony pavements are called *reg*, *serir* or *hamada*; in northern Asia they are the *gobi*

(Berkey and Morris 1927); in Australia they are known as *gibber plains*. Because they are remarkably level and smooth (except for micro-topographic variations from stone to stone) and because fine particles are conspicuously absent from the uppermost layer, the pavements were long believed to represent the residual surfaces of regions lowered and levelled almost entirely by deflation.

As pointed out by Peel (1960), French geomorphologists working in the Hoggar region of the central Sahara were the first to recognise desert pavements there as relics of Tertiary stream deposits preserved by Quaternary aridity (Birot *et al.* 1955). Similar gravel spreads in parts of the eastern Sahara, central Australia, deserts of the western United States and in the Hohsi Corridor and other inter-montane plains of northwestern China are also relict alluvial deposits. The lowering and levelling of the plains on which these pavements lie have clearly resulted from long-continued subaerial erosion by running water, not by wind.

How, then, did the alluvial deposits develop their characteristic flat, smooth desert pavements? This question was thoroughly addressed by Cooke (1970), who summarised evidence that deflation is only one among many processes that may produce desert pavements, depending on climate and topographic setting. These processes include not only weathering, which splits and disintegrates the pavement stones, but also freezing and thawing, wetting and drying of clayey soils, crystallisation of salts, and bioturbation — all of which may assist the upward migration of stones through the underlying soil to accumulate at the surface. Occasional rainfall drives dust particles (and airborne salts) into interstices among the stones and thus helps to form soil crusts, which enhance the cohesion of the fine surface particles between the stones and (until the crust is broken) protect them from deflation. Whereas all of these processes probably help to form desert pavements, they seem inadequate to account for the development, over hundreds of square kilometres, of the remarkably level mosaics of evenly distributed stones. The processes are particularly troublesome to invoke for the extremely hot, dry, stony plains of the eastern Sahara, where intervals between rainfall are measured in decades. Some other process seems to be at work, which operates in a highly uniform manner over very large areas — a characteristic of aeolian processes.

Another explanation has been offered (McFadden *et al.* 1984; 1987) that is applicable to any surface of loose rocks or gravels exposed to sediment-bearing winds. This explanation, the accretion hypothesis, is based on the discovery that soils beneath many stone pavements in the Mojave Desert (California and Nevada) have accumulated by the gradual, episodic infiltration of aeolian sand, silt and dust into lag gravels on alluvial fans and among loose fragments on the surfaces of Quaternary lava flows

(Figure 19.8). The progressive accumulation and pedogenesis of the infiltrated sediment result in the upward growth of a soil layer beneath the surface layer of stones, which settle evenly upon the fine material, called an *accretionary mantle* (Wells *et al.* 1985). The accretion mechanism can operate in concert with, and be assisted by, other processes in the formation of desert pavements, but accretion does not require the upward migration of stones through the soil layer, which was an especially troublesome aspect of the earlier explanations. The remarkably even accumulation of the soil layer, and the resulting level surfaces of the desert

a

b

Figure 19.8   *(a) Mosaic of basalt fragments is not original surface of lava flow but a stone pavement produced by aeolian infiltration and soil-forming processes. (b) Pavement one stone thick is underlain by more than 1 m of windblown silt, sand and clay that separates surface stones from lava bedrock below (Cima Volcanic Field, Mojave Desert, California, USA)*

pavement uplifted upon them, is explained by the uniform application of fine sediment by winds operating over broad areas.

The accretion hypothesis thus replaces deflation with another mechanism that also depends on wind for the formation of desert pavements. The accretion mechanism is consistent with other aeolian processes known to operate on stony desert surfaces. Wind carries clay minerals to the surfaces of individual stones to help form desert varnish (Potter and Rossman 1977), and it carries calcium carbonate dust to the soil to form caliche (calcrete) (Gile *et al.* 1966). These aeolian processes are depositional rather than erosional. Wind erosion plays a relatively minor role in the maintenance of desert pavements by deflation and by shaping some stones into ventifacts.

VENTIFACTS

Grooving and polishing (burnishing) of rocks by windborne abrasives was described quite early (Blake 1855), and the reshaping of stones by wind has been widely recognised (Wade 1910; Walther 1912). Evans (1911) proposed the generic term 'ventifacts' to include all varieties of 'wind-faced' stones, by which he meant pebbles or cobbles whose surfaces are flattened such that they intersect at sharp angles (Figure 19.2). Most early observers assumed that the faceting of such pebbles was accomplished entirely by sandblasting as particles of sand were driven against rock surfaces from a primary wind direction.

Abrasion of ventifact results from impact by either saltating or suspended material. Mass removal is greatest where large saltating grains impact a surface. Erosion by suspended grains is a much slower process, with particles commonly swept around an obstacle rather than striking it directly. The role of each of these agents of abrasion to the formation of ventifacts remains somewhat controversial, with Whitney and Dietrich (1973) arguing for a significant role for dust where fine-featured ventifacts are observed, and Laity (1994) considering the role of sand to be more important in ventifact formation. Arguments for each of these abrasive agents will be considered in turn.

Ventifact abrasion by dust

The puzzling occurrences of flattened surfaces on three or more sides of many ventifacts seemed to require rotation of the stones about a vertical axis; such rotations were explained by sequential shifting and overturning of the stones due to their undermining by deflation, or (in periglacial environments) by solifluction and frost-heaving (Bryan 1931; 1946; Schoewe 1932; Sharp 1949; 1964). A problem with this hypothesis, as Sharp (1949) noted, was that whatever process shifted the stones also ceased operating when the windcutting stopped.

Another problem was that periglacial processes could not be invoked to shift ventifacts in hot dry deserts. Furthermore, sandblasting could not account for the erosional markings (pits, flutes or grooves, helical scores), which may occur on all surfaces of ventifacts, and which appear too small and delicately textured (Figure 19.9) to have been made by the impact of sand-size grains (Whitney and Dietrich 1973). Sandblasting also could not explain the tapered forms of ventifacts, which typically have had more rock mass eroded from their lee areas than from their windward, frontal areas. Although sandblasting is clearly capable of abrading the slick, flattened surfaces seen on many faceted stones, it seems incapable of shaping ventifacts that are delicately fluted and fretted (Figure 19.4), like those in the Egyptian Sahara described by Hume (1925), McCauley *et al.* (1979) and others.

Many of the difficulties in accounting for ventifact shapes and textures are eliminated if flattening of the impact face by sandblasting is supplemented by fine-particle abrasion of all surfaces by subsidiary wind currents. The concept of wind erosion by vorticity was introduced by Maxson (1940) to explain the cutting of flutes on ventifacts. Cailleux (1942) also suggested that fine suspended grains could cut ventifacts. Sharp (1949) cited Maxson's (1940) attribution of flute-cutting to wind vortices, and he recognised that such cutting would require abrasion by 'particles fine enough to follow vortex currents'. Nonetheless, geologists generally remained reluctant to consider erosion of rock surfaces by suspended particles.

This reluctance may have been an unintended consequence of the use of dune sands as

Figure 19.9 *Vortex pits produced by aerodynamic erosion. (a) Pit on basal surface of ventifact carved in greywacke. Arrow points to helical score that is an incipient pit. (b) Scanning electron micrograph (SEM: 908×) of embossed pit in halite laboratory specimen, with myriads of secondary pits. (c) SEM (1400×) showing fine-textured surface of rayed vortex pit in halite. Both halite specimens were eroded by ordinary dusty air ((b) and (c) from Whitney 1979)*

abrasives in laboratory experiments by Bagnold (1941) and most others who studied physical controls on the movement of sand by wind (Kuenen 1928; 1960; Schoewe 1932; Suzuki and Takahashi 1981). The use of dune sand necessarily limited the results of these experiments to the effects of sandblasting, because particles capable of suspension were virtually excluded. As pointed out by Higgins (1956), dune sands are unusually well sorted and typically contain only a very small proportion of particles less than 0.1 mm in diameter. Many windblown particles in deserts, on the other hand, are derived from dry stream beds and playas, which have a high proportion of silt and clay. Except in the field experiments of Sharp (1964), the range of grain sizes used in wind-erosion studies was not representative of sediments in deserts, outside dune fields. Yet, in nature, ventifacts are rarely seen in or near fields of sand dunes, but they are common on the stony plains of sand-poor regions, particularly in localities where the topography channels windflow along preferred paths. Dust storms are frequent on bedrock plateaux and stony plains in many arid regions where wind-eroded features are common but sand is sparse.

Sand grains travel in a saltation curtain that generally rises no more than a metre or so above the ground, and the grains generally drop out of the moving airstream within a short distance of their source. On the other hand, silt- and clay-size particles are decoupled from their source areas and carried as dust in suspension, commonly for distances of tens to hundreds of kilometres; particles less than 5 µm in diameter can be transported thousands of kilometres (Péwé 1981; see Chapter 18). Dust storms over desert regions carry fine particles in clouds that range in density from 100 µg^{-3} to 176 000 µg^{-3} and that travel at altitudes of tens to thousands of metres (Goudie 1978). Suspended dust particles, unlike sand grains in saltation, travel with the same velocity as the air currents that move them, and thus they can be transported in local vortices that develop within subsidiary airflows. These abrasive currents (interfacial flows of Whitney 1978) can impinge on surfaces in the lee of obstacles and in other places beyond the reach of bouncing sand grains. Could dust be capable of erosion, beyond the mere burnishing of rock surfaces?

Ventifacts that developed while embedded in a cliff face were analysed by Higgins (1956). These ventifacts, although incapable of shifting their position, had developed multiple facets. Higgins concluded that the facets had been cut by suspended or 'partly suspended' particles travelling in wind currents deflected by local topography, and he suggested that 'currently accepted concepts of ventifact abrasion by saltating sand grains must be modified'. A decade later, Whitney and her associates began a series of wind-tunnel experiments to test the idea that dust might be capable of eroding rock surfaces (Whitney and Brewer 1968; Whitney *et al.* 1969; Whitney and Dietrich 1973). For two years, laboratory air (without artificially added particles) was continuously blown against the cleaved surfaces of mineral specimens that ranged in hardness from 2 to 6 (Mohs scale). Surfaces bearing the resulting erosional patterns (vortex pits, flutes and helical scores, see Figures 19.9 and 19.10) were compared with the original mirror surfaces of the cleaved specimens, and the results were recorded by scanning electron micrography (Whitney 1979). All specimens subjected to blown air showed erosional markings on all of their surfaces (Figures 19.9 and 19.10), and certain types of markings were consistently found in preferred locations. Fluting was rarely developed on windward surfaces but was common on surfaces out of the range of direct wind impact. Lee areas typically showed 'a great deal' more erosion than windward areas. All surfaces were pitted, although pits were rare on windward portions. The interiors of the pits were intricately rayed and embossed with delicate structures (Figure 19.9); these pits do not much resemble those formed by percussion (Marshall, in Greeley and Iverson 1985), but rather they appear to have formed by lift and abrasion due to vorticity. Many pits were aligned in chains that had evolved into U-shaped flutes of the type commonly seen on ventifacts and yardangs (Figure 19.4b; 19.17).

These experimental results led Whitney and Dietrich (1973) to conclude that 'plain old dusty air' is capable of erosion when blown against hard materials from constant directions, even at moderate speeds, over sufficient periods of time. Further testing, with dust (silt) particles added to the airstream, produced erosional

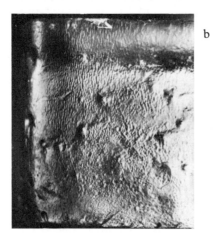

Figure 19.10   *Fluting produced by blowing dusty air at about 5 ms⁻¹ on halite block for 10 months (Whitney 1979). (a) Surface of block before erosion (3×). (b) Vortex pits. (c) Flutes (70×). Compare with pits and flutes on 'marble yardangs' (Figure 19.6)*

markings in a much shorter time, that is the increase in sediment load significantly increased the rate of erosion (Whitney and Dietrich 1973; Dietrich 1977; Whitney 1979). The experiments also demonstrated that all rock surfaces can be eroded simultaneously, but at different rates, when subjected to prolonged impingement by directed airstreams; this finding eliminated a factor that had been thought a necessity: successive shifts in position of ventifacts with time in strongly directional wind regimes. The erosional patterns obtained experimentally (Figures 19.9 and 19.10) were compared with the pits, flutes, and helical scores on ventifacts collected worldwide from natural settings (Whitney 1978; 1979). As a consequence of this work, Whitney and Dietrich (1973) expanded the definition of a ventifact from a stone or pebble that has been faceted, cut, shaped, or worn by sandblasting to include 'any rock or mineral fragment which has been shaped, worn, or polished by wind-promoted activities'.

To test relations of erosional markings on ventifacts to patterns of windflow that might have produced them, further experiments were conducted with directed air currents on ventifacts in field and laboratory settings (Whitney 1978) and in a wind tunnel (McCauley *et al.* 1977b; 1979). The ventifacts ranged from well-streamlined, faceted forms (Figure 19.2) to irregularly shaped, pitted and fluted rocks (Figure 19.4). Visualisation of low-speed airflow in the wind tunnel was provided by a bubble-generating device that simulated, in a general way, the behaviour of suspended particles as they moved over and around each rock. Airflow separation and negative flow (reverse flow in the upwind direction) was observed behind rock faces inclined 35–50° to windward. Secondary flow (transverse to the tunnel flow) was also observed. Vorticity normal to specimen faces occurred in the lee areas and on the sides of irregular and apparently non-streamlined rocks. Bubbles impacted on the windward faces or were pulled from separated airflow and then moved in a reverse direction, where some hovered behind or spiralled into pits, flutes, and irregular erosional concavities on the lee surfaces of the specimens. When samples were orientated so that the flow direction in the tunnel matched the dominant wind

in the field, bubble motions could be related to the erosional pattern on the rocks, but when the specimens were turned at angles of 45°, 90°, 135° and 180° to the flow direction, the bubbles lost contact with the specimen and their pattern became disorganised and unrelated to the surface morphology. From these experiments, McCauley *et al.* (1977b; 1979) concluded that the pits and flutes on all sides of these rocks could be explained by wind from a single direction. Although sandblasting undoubtedly played an episodic role in the development of these ventifacts, their small, fragile and complicated features were attributed to abrasion by suspended dust particles.

Laboratory experiments by Suzuki and Takahashi (1981) and Greeley *et al.* (1982) and numerical analyses by Anderson (1986) indicate that the mass of rock removed by impact is roughly proportional to the kinetic energy of the impacting particle. Anderson (1986) modelled the kinetic-energy flux produced by both saltating and suspended particles around a vertical cylindrical object of uniform composition subjected to uniform, steady winds. He obtained erosion profiles with maxima 0.2–0.3 m above the ground, which correspond to the flux of kinetic energy due to saltating grains. These simulated profiles are in good agreement with observed erosional profiles due to sandblasting on lucite rods and on fenceposts in natural settings (Sharp 1964; Wilshire *et al.* 1981). Anderson (1986) concluded that cutting of the lower part of an obstacle is dominated by the peak winds. He concluded that small ventifacts (less than 0.2 m) are mostly eroded by the impacts of saltating sand grains whose paths are not much affected by the obstacle.

Anderson (1986) also calculated the erosional maxima due to the kinetic energy flux of suspended grains, and he concluded that these maxima would occur metres above ground level. However, his model of the vertical cylinder neglects both the boundary layer on the upwind side of the obstacle and the separation of streamlines in the lee. Relations of erosional markings on ventifacts to windflow systems observed in the field and laboratory settings lead us to conclude that the aerodynamic conditions neglected in his model are critical to the development of pressure gradients, which generate subsidiary currents and vorticity and thereby govern erosion by suspended particles. Instead, Anderson assumed that the delivery of kinetic energy to an obstacle, whether by grains travelling by saltation or in suspension, requires that the particles 'drop out' of the airflow, and that small particles, which are unable to diverge from the streamlines, are thus less likely than large particles to impact the surface. None the less, he invoked 'particles small enough to be significantly deflected by the air flow around the obstacle and to be spun in local vortices' to account for the erosion of stones or outcrops above the zone where saltation dominates. Thus, although his model does not incorporate the aerodynamic conditions in which vorticity operates, his conclusions appear to require them.

As pointed out by Greeley and Iversen (1985), for the formulation of ventifacts, the relative efficacy of sandblast versus erosion by suspended particles probably depends on the grain sizes of available abrasives. Obviously, ventifacts that are situated entirely within the saltation zone are subject to erosion by saltating sand, if available. Such ventifacts typically have slick surfaces that lack fine detail such as fluting on their windward faces. Slower winds carrying dust particles and operating over long periods of time can also account for the erosion of ventifacts and may be the dominant process in their development in stony, sand-poor regions.

Ventifact abrasion by sand

Ventifact abrasion by sand has been reported in many areas of the world, including coastal environments (King 1936), periglacial regions (Powers 1936; Tremblay 1961; Miotke 1982; McKenna-Neuman and Gilbert 1986; Nero 1988) and desert regions (Blake 1855; Blackwelder 1929; Maxson 1940; Sharp 1964; 1980; Sugden 1964; Selby 1977; Greeley and Iversen 1986; Smith 1984; Laity 1992; 1994; 1995). Laity (1994) summarised arguments for the role of sand in ventifact formation.

Sand is a much more effective agent of erosion than dust because the mass of material lost per particle impact is directly proportional to the kinetic energy of impacting grains (Greeley *et al.* 1984; Anderson 1986). Abrasion experiments show that the susceptibility of rocks to abrasion varies with particle diameter to approximately the third power (Greeley *et al.* 1982; 1984). Furthermore, experiments by Stewart *et al.* (1981) show a much greater decrease in abrasion by particles less than 90 μm in diameter than predicted by kinetic energy considerations alone. Clay particles present in entrained material are transferred to the rock surface cushioning it from subsequent abrasion by larger particles. In addition, an apparent kinetic energy threshold for impact fracture exists, suggesting that abrasion mechanisms change with kinetic energy. Anderson (1986) noted that for suspended grains particle deflection by airflow around the ventifact also leads to a reduction in the delivery of kinetic energy to the surface.

Ventifacts develop in a wide range of rock types. In California, ventifacts have been observed to form in basalt, granite, aplite, andesite, marble, dolomite and limestone (Figure 19.11, Laity 1995). Polishing and smoothing of rock surfaces occurs all lithologies, and in each case the abrasive agent is sand. Sand abrasion has been shown to be capable of developing small and delicately textured structures on ventifact surfaces, including fine lineations, flutes and helical forms. Research into the formation of active ventifacts is ongoing in the Little Cowhole Mountains, Mojave Desert, California (Laity 1995). Marble ventifacts form on a sandblasted ridge. Evidence of an active aeolian environment includes wind records from 1993 to 1995 which show speeds as high as 38 m s^{-1}, observations of an active reversing dune near the ridge summit, intermittent burial of ventifacts by sand, and abrasion of foam blocks, painted poles and ventifacts.

Ventifacts in the Little Cowhole Mountains, as well as elsewhere in California, commonly develop semi-planar faces, with the upper part of the abrasion face receding more rapidly than the lower part. This pattern of abrasion is similar to that recorded on lucite rods (Sharp 1964; 1980) and fence posts (Wilshire *et al.* 1981). The height of maximum erosion occurs where grain size, number and velocity combine to give the maximum impact energy. In rare cases, the abrasion maxima is observed to occur at heights of 1 to 1.5 m above the surface on large ventifacts located near hill crests or in

Figure 19.11 *(a) Large basalt ventifacts formed by sand abrasion on a ridge south of Owen's Lake, California. Northerly winds accelerating up the ridge have caused characteristic recession of the windward abrasion face which shows well-developed helical scoring. (b) Granitic boulder more than one and a half metres in height developed on a ridge crest in proximity to an active reversing dune in the Mojave Desert, California. Sand surrounds the boulder. Deep fluting and a receding angle characterizes the upper face of the boulder*

passes where topography accelerates windflow. A ramping effect caused by sand accumulating near the boulder base may be partly responsible for these unusually high zones of maximum abrasion.

Laity (1995) examined a 30-cm-high polished and finely grooved marble ventifact from the Little Cowhole Mountains with a scanning electron microscope in order to determine whether the increase in impact energy with height up the boulder face would be reflected in microscopic differences in the nature of abrasion. Megascopically, there was little difference between the top and base of the rock, with the surface appearing polished and finely grooved (Figure 19.12). However, microscopically there was considerable variation with height. At the top of the rock, cleavage fracture of crystalline material was widespread,

whereas at the base of the rock there was no evidence of cleavage, with the rocks appearing to have been rubbed and gouged away. Both marble ventifacts from the Little Cowhole Mountains and basalt ventifacts from Pisgah flow (Laity 1992) show that surfaces which appear smooth to the eye and touch can be very rough at the micro-scale and suggest that dust need not be invoked to explain polish or fine features.

In sandy areas, most ventifacts show significant erosion on their windward faces and no erosion on their lee surfaces (Blackwelder 1929; Sharp 1964; McKenna-Neuman and Gilbert 1986). Ventifacts throughout the Mojave Desert clearly show the anticipated relationship between maximum abrasion on the windward face and a facet that slopes away from the wind. Additionally, pitted surfaces were found

Figure 19.12 *Helical scores formed by sand abrasion in marble ventifacts in the Little Cowhole Mountains, California. Such features are not common in ventifacts, and appear to develop in areas of high wind intensity*

to be diagnostic of high-angle impact by sand grains on windward faces (Sharp 1949; Smith 1984; Laity 1987; 1992), and were not developed equally on all sides of rocks. Some fossil ventifacts on glacial moraines in the eastern Sierra Nevada exhibit various surficial alteration, including the accumulation of cosmogenic nuclides (e.g. ^{3}He, ^{10}Be, ^{26}Al, ^{36}Cl) and the growth of rock varnish (both of which can be used for absolute and relative dating), and the development of silica glaze and oxalate-rich crusts (Dorn 1995). These surficial alterations on the fossil ventifacts can be used to infer palaeoenvironmental conditions since the cessation or diminution of aeolian activity.

In summary, laboratory experiments have shown that dust is capable of changing the surface micromorphology of rocks, but that mass removal is almost immeasurable. For periglacial settings, Schlyter (1991; 1994) has argued that fossil ventifacts in Sweden were formed by suspended matter, i.e. snow or dust, whereas McKenna-Neuman and Gilbert (1986) have shown the significance of sand for active ventifact development in Arctic Canada. In arid regions, Laity (1994; 1995) has shown that ventifacts with polished surfaces and fine lineations occur where sand dominates the flux of particles, indicating that dust is not essential for the formation of small-scale features. Thus, the relative roles of sand and dust in abrasion remain a subject of lively discussion in ventifact research. Of particular value to this debate would be a long-term field study of active ventifacts in an arid region subject to dust alone. Nicking (personal communication, 1993) noted that ventifacts do not appear to be forming in areas of Africa which now have the highest aeolian dust concentrations of any place on Earth.

YARDANGS

Many shapes and erosional markings typical of ventifacts are duplicated on a much larger scale, on the wind-sculptured hills known as 'yardangs' (Figures 19.1 and 19.5). The classic form of yardangs described by Bosworth (1922) in siltstones of Peru is that of an inverted boat with a high prow and tapered profile that may include a keel (Figure 19.13). Ideal forms have streamlined, lemniscate shapes with length-to-width ratios of 3 : 1 or 4 : 1 (Grolier *et al.* 1980; Ward and Greeley 1984); this shape is best developed in fairly homogeneous materials and in regions where strong winds are highly directional, such as parts of coastal Peru (Figure 19.13), the Lut Desert of Iran (Figure 19.6), and many parts of the Western Desert of Egypt (Figure 19.1; also McCauley *et al.* 1977a; Whitney 1985). Yardangs are commonly carved in plateau outliers, and in their early development they may retain irregular shapes as well as ambiguous axial trends inherited from relict fluvial dissection patterns (Breed *et al.* 1984). Even such poorly shaped yardangs show many of the erosional markings typically developed on well-streamlined yardangs and on ventifacts. Variations occur in sculptural details influenced

Figure 19.13 *Classic yardangs carved in siltstone of the Pisco Formation of late Oligocene to late Miocene age in Ica Valley, coastal Peru. These yardangs are surrounded by a bare-rock, wind-denuded, sand-poor surface. Stair-step profiles are due to variations in resistance of horizontally bedded units in Pisco Formation. The dominant wind in this area is a southerly trade wind, reinforced by a diurnal sea breeze*

by rock type, local topography and time. Nonetheless, the relations of impact faces, lee zones, undercut areas, flutes and vortex pits on yardangs, as on ventifacts, are generally constant for a given locality.

Until recently, yardangs were defined as small features eroded only in soft materials of Quaternary age, such as the lake-bed deposits in northwestern China (Hedin 1903; 1905) and the lunette dunes at Rogers Lake, California (Blackwelder 1934; Ward and Greeley 1984). A survey of yardangs worldwide (McCauley *et al.* 1977a) described features metres to several kilometres in size, carved in materials ranging from semi-consolidated sediment to granites and other types of resistant rock. The largest yardangs on Earth ('cretes' of Mainguet 1968; 1972) are in a vast field around the Tibesti Plateau in the central Sahara (Figure 19.5). The Tibesti yardangs were the first wind-erosion features recognised on early images from space (Morrison and Chown 1965); they were generally described as grooved terrain (Verstappen and van Zuidam 1970; Worrall 1974), but Peel (1966; 1970) tentatively identified them as hard-rock yardangs.

In 1978, yardangs were discovered in dense crystalline limestone and in sandstone, shale and granite in Egypt's Western Desert (Grolier *et al.* 1980). This region is an ideal natural laboratory for studies of yardangs. Average annual rainfall is less than 1 mm (Henning and Flohn 1977), and, except around the springs and wells that support scattered oases, the region is a wasteland devoid of vegetation and running water. Thus, this desert lacks both the extensive vegetation cover and the localised but episodically heavy rainfall that interfere with wind activity and produce ambiguous, hybrid landforms in less arid regions. Only a few fairly uniform rock types are widely distributed in the Western Desert (mostly limestone, sandstone and shale, with isolated occurrences of gralute and lake-bed deposits). Each type has developed characteristic erosion shapes and textures that clearly illustrate the influences of rock structure and lithology on wind erosion. As recognised by Ball (1900) and subsequent workers in this region, the work of wind is everywhere apparent, in the aligned dissection patterns of bedrock surfaces (*kharafisch* in early descriptions), in abundant ventifacts strewn on

stony desert floors, in the parallel arrays of yardangs, and in southward-marching dune trains (Figures 19.1, 19.14 and 19.15). The relatively simple geological and topographical setting, the regionally constant strong winds (Wolfe and El-Baz 1979) and the hyper-arid climatic conditions (which have prevailed throughout most of the Quaternary) limit the variables that control wind erosion in most of the Western Desert. These variables are the effects of small-scale topography on local wind patterns; and the influence of rock type and structure on the erosional effectiveness of wind.

Most of the Western Desert consists of flat plains and plateau surfaces, where the effects of topography on the directional variability of air flow are minimal. In a few localities, demoiselles (hoodoos) of various shapes and sizes have formed adjacent to yardang fields (Figure 19.3). Such features in other deserts have generally been attributed to differential weathering, running water, or other processes (Bryan 1925; 1927; Grove 1960), but, in the Western Desert, demoiselles are carved in the same kinds of rock under the same conditions of hyper-aridity as nearby, obviously streamlined, wind-eroded forms. The peculiar shapes of some demoiselles may be attributed to local lithological variation and to structural control of wind erosion by joints and bedding planes, but many appear to have developed as a result of complex local wind patterns peculiar to a specific topographic setting.

Yardangs in the Western Desert are carved in nearly every type of cohesive material exposed to the wind; they range from less than a metre to a kilometre or more in length and typically are distributed in parallel arrays aligned with the prevailing north wind. 'Marble yardangs' tens to hundreds of metres long and a few metres to tens of metres high form one of the most extensive fields of wind-eroded bedrock features of Earth (Figures 19.1 and 19.15). They are distributed over an area of at least 35 000 km^2, on the Limestone Plateau north and east of the Kharga, Dakhla and Farafra depressions. The plateau surface has probably been exposed to subaerial weathering and erosion since at least Miocene time (Said 1962), under climatic conditions that varied from humid to hyper-arid. Wind erosion during arid climatic episodes has thus overprinted but not obliterated

a

b

Figure 19.14 *Yardangs in Western Desert, Egypt. (a) Metre-size 'mud lions' in Bahariya Depression; wind from north (right). (b) Kilometre-size yardangs in Nubia Sandstone and Dakhla Shale of Cretaceous age below scarp that bounds the Kharga Depression. These yardangs are plateau outliers that have been streamlined by the north wind (from left) and show few remaining traces of fluvial topography, unlike the same rock units exposed in scarp*

geomorphic characteristics, including relict karst features and traces of disintegrated, abandoned stream channels, that record climatic conditions quite different from the present hyper-aridity. The marble yardangs are carved in Thebes Limestone of early Eocene age, which is so fine grained and dense that it has been used to make statuary. This rock preserves a particularly detailed record of erosional patterns and textures produced by wind (Figures 19.4 and 19.16). Its yardangs commonly display highly burnished heads with splayed fluting as well as flutes and pits along their flanks and lee ends (Figure 19.15). Weathered and rilled surfaces that are relict from the earlier, wetter intervals are commonly intersected by clean, fluted

surfaces cut by wind. Detailed descriptions of the erosional shapes and markings on these yardangs were provided by Whitney (1983; 1985). Other yardangs in the Western Desert include sedimentary deposits of Quaternary age that have been eroded to metre- to decametre-size streamlined hillocks described by Bagnold (1939) as 'mud lions' (Figure 19.14); these features were the first yardangs recognised in Egypt, and they resemble the hills in China originally designated as yardangs by Hedin (1903; 1905). Much more massive yardangs, some a kilometre or more long and tens of metres high, are carved in the Nubia Sandstone and Dakhla Shale of Cretaceous age (Figure 19.14). Metre-size yardangs are carved in

a

b

Figure 19.15   *Ground views of 'marble yardangs' in field shown in Figure 19.1. (a) Sparse sand on plateau surface is shunted around bedrock yardangs to form aureoles and moats. Tyre tracks in middle ground for scale. (b) Yardang flanks show multiple, intersecting flute patterns produced by airflow in primary, negative and secondary directions. Large pit is probably inherited karst feature undergoing modification by wind erosion*

outcrops of El Tawila Granite of Precambrian age (Figure 19.17a).

Although abundant indirect evidence for the mechanisms of wind erosion has been gathered from field observations, comparison of yardang shapes and erosional markings with features observed or produced under controlled conditions is difficult, because experimental data on erosion of yardangs is very sparse. A field study of yardangs carved in semi-consolidated lunette dunes at Rogers Lake, California (Ward and Greeley 1984) included measurements of wind speeds obtained at four places in the yardang field: on a yardang crest and flank, on a trough floor, and at 3 m above the trough

floor. These authors concluded, however, that their results 'are only approximate, because multiple synchronous wind data were not obtained over the yardangs'. Wind-tunnel experiments with models of irregular topography and yardangs were conducted by McCauley *et al.* (1977b; 1979) and by Ward and Greeley (1984). These experiments used mixtures of sand, hominy grits, and moist coffee grounds to simulate the textures of semi-cohesive, granular sediments such as those in which the Rogers Lake yardangs are carved. In the first experiment, layered models showed rapid modification to streamlined forms under unidirectional winds. Models constructed of hard layers over

Figure 19.16 *Flutes are found on all surfaces of Thebes limestone yardangs. They are typically round floored, symmetrical in cross-section, and burnished (a) Flutes and vortex pits grade into one another, and flutes often grade into fingers (Whitney 1983). (b) Primary flutes, produced parallel to primary wind direction by positive or negative flow, commonly terminate in round or oval vortex pits beneath a steep-walled flute head. On yardangs shorter than about 1 m, the primary flutes typically are only 5 to 10 cm wide, but on larger yardangs, primary flutes are as wide as 60 cm and several metres long. Many primary flutes are crossed by minute, regularly spaced secondary flutes (transverse or oblique to primary wind direction) worn smooth by burnishing*

soft were modified by undercutting, collapse, and rapid deflation of debris from topographic troughs. Hard-over-soft models developed keels and long, concave, tapered prows, whereas soft-over-hard models developed blunt heads and broad, wide, convex backs. Individual grains were episodically plucked from the yardang surfaces by deflation. Abrasion was observed only in the troughs, where moving grains helped to dislodge particles from the surface. this experiment confirmed field observations that 'deflation as well as abrasion is critical to yardang formation' (McCauley *et al.* 1977b).

In a later experiment (Ward and Greeley 1984), a kidney-shaped sample of the synthetic material was subjected to a wind velocity of 10 m s^{-1}. Erosion in the wind tunnel proceeded through four stages: erosion of corners on the windward side of the object; erosion of the slope to windward; erosion of the lee corners and flanks; and erosion of the top surface. The latter two stages were accomplished by 'negative' flow (Whitney 1978) in a reverse direction to that of the primary wind (positive flow). The resulting wind-eroded form was streamlined with a width-to-length ratio of 1 : 4. The authors concluded that the streamlined shape of yardangs results primarily from abrasion in both the positive and negative flow directions, but that deflation of loose grains in zones of low pressure helps to shape these features and to maintain their form as streamlining progresses and turbulence is lessened, and reverse flow is thereby reduced.

Windflow in a confined tunnel cannot fully reproduce natural conditions, for the roof and

a

b

Figure 19.17   *Yardangs in the Western Desert, Egypt. (a) Metre-size hillocks carved in Precambrian basement rocks (El Tawila Granite). Although the rounded form of granite hillocks is almost certainly inherited from earlier, deep-weathering under moist tropical conditions, exhumed mounds are now being uniformly streamlined into prevailing wind (from left), and they have fluted, burnished surfaces typical of yardangs throughout this region. (b) Sequence of yardang shapes in Quaternary lake-bed deposits in Farafra Depression. If original shape of each hillock was an elongate block like that in centre, evolution of tapered form (right) indicates that more erosion has occurred along backs and sides of yardangs than at their high, blunt heads that are exposed to direct sandblasting from primary wind. Even where undercutting at head of yardang leads to breakage along vertical desiccation cracks, causing entire windward face to lose support and crumble, as in hillock to left, the windward end remains higher than the lee end. Meanwhile, wind erosion shortens and narrows lee end faster than windward end, maintaining tapered profile as yardang is reduced to smaller streamlined mound. Car tracks in right middle ground for scale*

sides of a wind tunnel constrain windflow under conditions unlike those in a natural yardang field. None the less, these laboratory simulations help demonstrate some of the aerodynamic erosional processes deduced by Whitney (1983) from observational evidence on yardangs in the Western Desert of Egypt and elsewhere. The basis for her reconstructions consists of the streamlined, tapered forms maintained by the yardangs, their convex–concave shape relations, flute and vortex pit patterns, and keels, as well as the patterns of associated aeolian deposits, including sand streamers, sand tails, ripples and aureoles (Figure 19.15).

The gross convex–concave shape relations of yardangs, together with their flute and pit patterns, suggest that erosion proceeds simultaneously on all surfaces by fine-particle abrasion and vorticity within a complex system of subsidiary airflows. Whitney (1978) used the term *interfacial flow* to differentiate this process from sandblasting. Particles loosened by weathering, mass-wasting, and vorticity are apparently lifted (deflated) from the rock surface into the interfacial flow in response to differential pressures, which are generated by the dividing and closing of the primary windflow streamlines around obstacles. This complex pattern of erosional processes occurs within

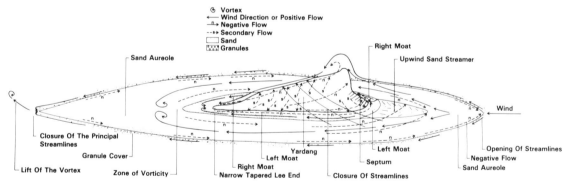

Figure 19.18 *Typical yardang in lake-bed deposits of Kharga Depression, Egypt. Bottom: schematic reconstruction of 'aerodynamic envelope' in which this form has developed, based on interpretation of its erosional markings (Whitney 1983). Erosion begins with impact of primary wind from the north and its division into positive flow streamlines that surround the obstacle and separate from boundary layer in lee zone, forming turbulent wake. This aerodynamic system creates pressure differentials that generate subsidiary currents, which flow over rock surface in positive, negative and secondary directions. Subsidiary currents carry suspended fine particles, eroded from or carried to the air–rock interface, and flow toward loci of low pressure; that is, to centres of vorticity, slope crests and junctures with accelerated flow lines. Currents impinge on surface irregularities and produce local vortices capable of bringing fine abrasive particles into contact with the surface. Turbulence in lee zone generates vorticity and negative flow that returns, in reverse direction, along flanks to windward. There, negative flow conflicts with oncoming primary wind, causing it to divide some distance upwind of the obstacle and to shunt loose grains away from its base, head and flanks. This loose material commonly collects in sand aureoles that surround the obstacle and converge to leeward. Deceleration of the primary wind due to conflict with negative flow may cause it to form a moat and to deposit sand streamer in front of yardang (Whitney 1983)*

the aerodynamic envelope that surrounds the yardang, as shown conceptually in Figure 19.18.

Closure of the positive flow streamlines in the lee of a yardang is a well-recognised aerodynamic phenomenon that produces turbulent flow, which in turn accelerates erosion by vorticity (Ward and Greeley 1984). Hedin (1903; 1905), who first described yardangs in lake-bed sediment in northwest China, said of the lee zone that an observer there is struck from every side by wind as though being in an eddy. Whitney (1983; 1985) attributes the typically rounded, fluted, burnished, and sometimes keeled back slopes of yardangs to the increased intensity of the eddy effect to leeward of the apex of the yardang, which accelerates positive flow along and down the top of the yardang. She attributed the formation of keels to strong currents of negative and secondary flow that rise along pressure gradients up the yardang flanks.

Some of the subsidiary flow lines follow bedding planes, joints and fractures, etching out various internal structures that intersect the surface of the yardang. Abrasion (by fine particles suspended in subsidiary flows and spun in local vortices), and deflation erode the sides and back of a yardang and reduce the rock mass substantially while the angle of the steep windward face is maintained by sandblasting and undercutting by the primary wind. Interfacial flow lines find each salient and depression — both large and minute — on rock surfaces and create reversals and centres of vorticity around each irregularity; that is, these flows form a hierarchy of aerodynamic systems around each large and small obstacle. Deflation and fine-particle abrasion eventually wear away the protuberances and create a form of least resistance to drag that approaches equilibrium with local wind conditions. As streamlining progresses, more material is removed from the lee area than from the head. Thus yardangs, like some ventifacts, may be eroded from the lee end forward.

In addition to abrasion and deflation, running water acts to influence yardang field development. Yardangs may be initiated along streamcourses that are enlarged and modified by wind. Gullies commonly develop on yardangs composed of relatively soft deposits, such as lake-bed silts and clays. In some cases, the gullies may be relics of earlier wetter climates and be associated with interyardang lacustrine deposits.

Halimov and Fezer (1989) analysed the morphologies of yardangs in the Qaidam Depression of central Asia and classified yardangs into eight forms. They concluded that intense abrasion at the prows of yardangs caused them to become shorter and smaller through time, such that they eventually developed into conical hills, mesas and pyramids. If yardangs are cut off from a source of abrasive sand their form slowly degrades. Brooks (1993) observed that yardangs on mesas separated from larger intact plateaus by scarp-parallel valleys were reduced by weathering to become low linear swells.

Conclusions

The recognition of large-scale wind-eroded forms carved in hard bedrock at heights above the limit commonly observed for sand in saltation has called attention to processes of wind erosion. According to some investigators, these erosional markings are out of the reach of saltating sand, have fragile, delicate textures, and are commonly oriented in directions other than that of the primary wind. Processes proposed for the erosion of these features suggest a greater reduction of rock mass in the lee of the yardang (or ventifact) than at its high blunt windward end. Field and laboratory studies demonstrate that deflation can effectively shape yardangs and ventifacts in granular material, but that deflation alone is difficult to invoke for erosion of hard rock surfaces. Deflation has long been recognised as responsible for excavating depressions on plains blanketed by sandy alluvial and aeolian sediment, but running water and other subaerial processes, rather than deflation, are now known to account for most reduction of large, rock-floored desert basins. Deflation has also largely been replaced by other processes as the agent responsible for the development of desert pavements. A third process, aerodynamic erosion by dust abrasion in air currents generated by pressure differentials and vorticity, has been suggested by field and laboratory experiments to be capable of eroding mineral specimens, various types of ventifact, and yardangs. Aerodynamic erosion is thought to account for the progressive reduction of the rock mass from the lee end forward, for multiple patterns of flutes and pits, for the development of keels and for other characteristics of shape that may not be explained solely by sandblasting and deflation.

According to other investigators, abrasion by sand causes greater mass reduction at the windward face of ventifacts, resulting in a characteristic semi-planar face, with the upper part of the face receding more rapidly than the lower part. In regions of unidirectional wind, the lee surface of the ventifact is uneroded. Smooth polish, flutes, pits and helical scores are observed to develop on ventifacts subject to sand abrasion alone. Fossil ventifacts develop characteristic weathering features, including staining, rock varnish and silica glaze and oxalate-rich crusts. Similarly, it appears that yardangs become progressively more degraded in form when cut off from a supply of abrasive sand.

References

Albritton, C.C., Brooks, J.E., Issawi, B. and Swedan, A. 1990. Origin of the Qattara Depression, Egypt. *Bulletin of the Geological Society of America*, 102: 952–960.

Anderson, R.S. 1986. Erosion profiles due to particles entrained by wind: application of an eolian sediment-transport model. *Bulletin of the Geological Society of America*, 97: 1270–1278.

Bagnold, R.A. 1939. An expedition to the Gilf Kebir and Uweinat, 1938. *Geographical Journal*, 93: 281–313.

Bagnold, R.A. 1941. *The physics of blown sand and desert dunes*. Chapman & Hall, London.

Ball, J. 1900. *Kharga Oasis, its topography and geology*. Survey Department, Cairo, Egypt.

Ball, J. 1927. Problems of the Libyan Desert. *Geographical Journal*, 70: 21–38; 105–128, 209–224.

Beadnell, H.J.L. 1909. Lake basins created by wind erosion. *Journal of Geology*, 3: 47–49.

Berkey, C.P. and Morris, F.K. 1927. *Geology of Mongolia*. American Museum of Natural History, New York, 2: 146–418.

Birot, P., Capot-Rey, R. and Druch, J. 1955. Récherches morphologiques dans le Sahara central. *Travaux Institut Recherches Sahariennes*, Algiers, 13: 13–74.

Blackwelder, E. 1929. Sandblast action in relation to the glaciers of the Sierra Nevada. *Journal of Geology*, 37: 256–260.

Blackwelder, E. 1930. Yardang and zastruga. *Science*, 72: 396–397.

Blackwelder, E. 1934. Yardangs. *Bulletin of the Geological Society of America*, 45: 159–166.

Blackwelder, E. 1935. The lowering of playas by deflation. *American Journal of Science*, 21: 140–144.

Blackwelder, E. 1954. Geomorphic processes in the desert. *Bulletin of the California Department of Natural Resources*, 170: 11–20.

Blake, W.P. 1855. On the grooving and polishing of hard rocks and minerals by sand. *American Journal of Science*, 20: 178–181.

Bosworth, T.O. 1922. *Geology and palaeontology of northwest Peru*. Macmillan, London: 269–311.

Breed, C.S. and Grow, T. 1979. Morphology and distribution of dunes in sand seas observed by remote sensing. *US Geological Survey, Professional Paper*, 1052: 253–302.

Breed, C.S., Fryberger, S.G., Andrews, S., McCauley, C., Lennartz, F., Gebel, D. and Horstman, K. 1979. Regional studies of sand seas using Landsat (ERTS) imagery. *US Geological Survey, Professional Paper*, 1052: 253–302.

Breed, C.S., McCauley, J.F. and Grolier, M J. 1984. *Multiprocess evolution of landforms in the Kharga region, Egypt — applications to Mars*. NASA Technical Memorandum, 86246: 225–227.

Brooks, I.A. 1983. Dakhleh Oasis — a geoarchaeological reconnaissance. *Journal of the Society for Study of Egyptian Antiquities*, 13: 167–187.

Brooks, I.A. 1986. *Quaternary geology and geomorphology of Dakhleh Oasis and environs, south-central Egypt: Reconnaissance findings*. Department of Geography, York University, Discussion Paper 32. Toronto, Canada.

Brooks, I.A. 1993. Geomorphology and Quaternary geology of the Dakhla Oasis region, Egypt. *Quaternary Science Reviews*, 12: 529–552.

Bryan, K. 1923. Wind erosion near Lee's Ferry, Arizona. *American Journal of Science*, 5(6): 291–306.

Bryan, K. 1925. Pedestal rocks in the arid Southwest. In *Contributions to the Geography of the United States, 1923–1924*. US Geological Survey, Bulletin, 760: 1–11.

Bryan, K. 1927. Pedestal rocks formed by differential erosion. In *Contributions to the Geography of the United States, 1926*. US Geological Survey, Bulletin, 790: 1–15.

Bryan, K. 1931. *Wind-worn stones or ventifacts — A discussion and bibliography*. National Research Council Circular 98. Washington, DC: 29–50.

Bryan, K. 1946. Cryopedology — The study of frozen ground and intensive frost-action with suggestions on nomenclature. *American Journal of Science*, 244: 622–642.

Cailleux, A. 1942. Les actions éoliennes periglaciaires en Europe. *Société Géologique de la France*, 192: 1–176.

Caton-Thompson, G. and Gardner, E.W. 1932. The prehistoric geography of Kharga Oasis. *Geographical Journal*, 80: 369–406.

Chepil, W.S. 1945. Dynamics of wind erosion. *Soil Science*, 60: 305–320, 397–341, 475–480.

Chepil, W.S. and Woodruff, N.P. 1963. The physics of wind erosion and its control. *Advances in Agronomy*, 15: 211–302.

Cooke, R.U. 1970. Stone pavements in deserts. *Annals of the Association of American Geographers*, 60: 560–577.

Dietrich, R.V. 1977. Impact abrasion of harder by softer materials. *Journal of Geology*, 85: 242–246.

Dorn, R.I. 1995. Alterations of ventifact surfaces at the glacier/desert interface. In V.P. Tchakerian (ed.), *Desert aeolian processes*. Chapman & Hall, London: 199–217.

Evans, J.W. 1911. Dreikanter. *Geological Magazine*, 8: 334–345.

Gilbert, G.K. 1895. Lake basins created by wind erosion. *Journal of Geology*, 3: 47–49.

Gile, L.H., Peterson, F.F. and Grossman, R.B. 1966. Morphological and genetic sequences of carbonate accumulation in desert soils. *Soil Science*, 101: 347–360.

Goudie, A.S. 1978. Dust storms and their geomorphological implications. *Journal of Arid Environments*, 1: 291–310.

Goudie, A.S. 1989. Wind erosion in deserts. *Proceedings of the Geologists Association*, 100: 89–92.

Goudie, A.S. and Wells, G.L. 1995. The nature, distribution and formation of pans in arid zones. *Earth Science Reviews*, 38: 1–69.

Greeley, R. and Iversen, J.D. 1985. *Wind as a geological process*. Cambridge University Press, Cambridge.

Greeley, R. and Iversen, J.D. 1986. Aeolian processes and features at Amboy Lava field, California. In F. El-Baz and M. Hassan (eds), *Physics of desertification*. Martinus Nijhoff, Dordrecht: 210–240.

Greeley, R., Leach, R.N., Williams, S.H., White, B.R., Pollack, J.B., Krinsley, D.H. and Marshall, J.R. 1982. Rate of wind abrasion on Mars. *Journal of Geophysical Research*, 87: 10009–10014.

Greeley, R., Williams, S.H., White, B.R., Pollack, J.B. and Marshall, J.R. 1984. Wind abrasion on Earth and Mars. In M.J. Woldenberg (ed.), *Models in geomorphology*. Allen & Unwin, Boston, 373–422.

Grolier, M.J., McCauley, J.F., Breed, C.S. and Embabi, N.S. 1980. Yardangs of the Western Desert. In F. El-Baz *et al.* (eds), Journey to the

Gilf Kebir and Uweinat, Southwest Egypt, 1978. *Geographical Journal*, 146: 86–87.

Grove, A.T. 1960. The geomorphology of the Tibesti Region. *Geographical Journal*, 126: 18–31.

Grove, A.T. 1969. Landforms and climatic change in the Kalahan and Ngamiland. *Geographical Journal*, 135: 191–212.

Halimov, M. and Fezer, F. 1989. Eight yardang types in central Asia. *Zeitschrift für Geomorphologie*, 33: 205–217.

Haynes, C.C. 1982. The Darb El-Arba'in desert: a product of Quaternary climate change. In F. El-Baz and T.A. Maxwell (eds), *Desert landforms of southwest Egypt: a basis for comparison with Mars*. NASA Contractor Report 3611, Washington, DC: 91–117.

Hedin, S. 1903. *Central Asia and Tibet*. Charles Scribner's Sons, New York.

Hedin, S. 1905. *Journey in Central Asia 1899–1902*. Lithographic Institute, General Staff Swedish Army, Stockholm, Sweden.

Henning. D. and Flohn, H. 1977. *Climate aridity index map*. United Nations Environment Programme, Nairobi.

Higgins, C.G. Jr 1956. Formation of small ventifacts. *Journal of Geology*, 64: 506–516.

Hobbs, W.H. 1917. The erosional degradational processes of deserts, with special reference to the origin of desert depressions. *Annals of the Association of American Geographers*, 7: 25–60.

Hume, W.F. 1925. *Geology of Egypt*. Government Press, Cairo.

Keyes, C.R. 1908. Rock-floor of intermont plains of the arid region. *Bulletin of the Geological Society of America*, 19: 63–92.

Keyes, C.R. 1909. Erosional origin of the Great Basin Range. *Journal of Geology*, 17: 31–37.

King, L.C. 1936. Wind-faceted stones from Marlborough, New Zealand. *Journal of Geology*, 44: 201–213.

Kuenen, P.H. 1928; Experiments on the formation of wind-worn pebbles. *Leidsche Geologische Medellinger*, 3: 17–38.

Kuenen, P.H. 1960. Experimental abrasion 4: Eolian action. *Journal of Geology*, 68: 427–449.

Laity, J.E. 1987. Topographic effects on ventifact development, Mojave Desert, California. *Physical Geography*, 8: 113–132.

Laity, J.E. 1992. Ventifact evidence for Holocene wind patterns in the east-central Mojave Desert. *Zeitschrift für Geomorphologie*, Supplementband, 84: 1–16.

Laity, J.E. 1994. Landforms of aeolian erosion. In A.D. Abrahams and A.J. Parsons (eds), *Geomorphology of desert environments*. Chapman & Hall, London: 506–535.

Laity, J.E. 1995. Wind abrasion and ventifact formation in California. In V.P. Tchakerian (ed.), *Desert aeolian processes.* Chapman & Hall, London: 295–321.

Lancaster, I.N. 1978. Composition and formation of southern Kalahari pan margin dunes. *Zeitschrift für Geomorphologie*, NF, 22: 148–169.

Mabbutt, J.A. 1977. *Desert landforms.* MIT Press, Cambridge, MA.

Mainguet, M. 1968. Le Bourkou — Aspects d'un modèle éolian. *Annales le Géographie*, 77: 296–322.

Mainguet, M. 1972. *Le modèle des grès.* Institute Geographie National, Paris.

Maxson, J.H. 1940. Fluting and faceting of rock fragments. *Journal of Geology*, 48: 717–751.

McCauley, J.F. 1973. Mariner 9 evidence for wind erosion in the equatorial and mid-latitude regions of Mars. *Journal of Geophysical Research*, 78: 4123–4137.

McCauley, J.F., Breed, C.S. and Grolier, M.J. 1977a. Yardangs. In D.O. Doehring (ed.), *Geomorphology in arid regions.* Annual Geomorphology Symposium, Binghamton, NY. Allen & Unwin, Boston: 233–269.

McCauley, J.F., Ward, A.W., Breed, C.S., Grolier, M.J. and Greeley, R. 1977b. *Experimental modeling of wind erosion forms.* Technical Memorandum X-3511: 150–152.

McCauley, J.F., Breed, C.S., El-Baz, F., Whitney, M.I., Grolier, M.J. and Ward, A.W. 1979. Pitted and fluted rocks in the Western Desert of Egypt — Viking comparisons. *Journal of Geophysical Research*, 84: 8222–8232.

McCauley, J.F., Breed, C.S. Grolier, M.J. and MacKinnon, D.A. 1981. The U.S. dust storm of February 1977. In T.J. Péwé (ed.), *Desert dust.* Geological Society of America, Special Paper, 186: 123–147.

McFadden, L.D., Wells, S.G., Dohrenwend, J.C. and Turrin, B.D. 1984. Cumulic soils formed in eolian parent materials on flows of the Cima Volcanic Field, Mohave Desert. *Geological Society of America Guidebook*, Annual Meeting, Reno, NV, 14: 134–149.

McFadden, L.D., Wells, S.G. and Jercinovich, M.J. 1987. Influences of eolian and pedogenic processes on the origin and evolution of desert pavements. *Geology*, 15: 504–508.

McKenna-Neumann, C. and Gilbert, R. 1986. Aeolian processes and landforms in glaciofluvial environments of south-eastern Baffin Island, N.W.T., Canada. In W.G. Nickling (ed.), *Aeolian geomorphology*, Allen & Unwin, Boston: 213–235.

Miotke, F. 1982. Formation and rate of formation of ventifacts in Victoria land. *Polar Geography and Geology*, 6: 93–113.

Morrison, A. and Chown, M.C. 1965. Photographs of the Western Sahara from the Mercury MA-4 satellite. *Photogrammetic Engineering*, 31: 350–362.

Nero, R.W. 1988. The ventifacts of the Athabasca sand dunes. *The Musk Ox*, 36: 44–50.

Passarge, S. 1904. *Die Kalahari.* Reimer, Berlin.

Peel, R.A. 1941. Denudational landforms of the Central Libyan Desert. *Journal of Geomorphology*, 4: 3–23.

Peel, R.A. 1960. Some aspects of desert geomorphology. *Geography*, 45: 241–262.

Peel, R.A. 1966. The landscape of aridity. *Transactions of the Institute of British Geographers*, 38: 1–23.

Peel, R.A. 1970. Landscape sculpture by wind. *21st International Geographical Congress Selected Papers*, 1: 99–104.

Péwé, T.J. (ed.) 1981. *Desert dust.* Geological Society of America, Special Paper, 186.

Potter, R.M. and Rossman, G.R. 1977. Desert varnish: the importance of clay minerals. *Science*, 196: 1446–1448.

Powers, W.E. 1936. The evidences of wind abrasion. *Journal of Geology*, 44: 214–219.

Price, W.A. 1968. Oriented lakes. In R.W. Fairbridge (ed.), *Encyclopedia of geomorphology.* Reinhold, New York: 784–796.

Said, R. 1962. *The geology of Egypt.* Elsevier, Amsterdam.

Schlyter, P. 1991. Recent and periglacial wind action in Scania and adjacent areas of S Sweden. *Zeitschrift für Geomorphologie*, Supplementband, 90: 143–153.

Schlyter, P. 1994. Paleo-periglacial ventifact formation by suspended silt or snow — site studies in south Sweden. *Geografiska Annaler*, 76A(3): 187–201.

Schoewe, W.H. 1932. Experiments on the formation of wind-faceted pebbles. *American Journal of Science*, 24: 111–134.

Selby, M.J. 1977. Palaeowind directions in the central Namib Desert, as indicated by ventifacts. *Madoqua*, 10: 195–198.

Sharp, R.P. 1949. Pleistocene ventifacts east of the Big Horn Mountains, Wyoming. *Journal of Geology*, 57: 175–195.

Sharp R.P. 1964. Wind-driven sand in Coachella Valley, California. *Bulletin of the Geological Society of America*, 75: 785–804.

Sharp, R.P. 1980. Wind-driven sand in Coachella Valley, California: further data. *Bulletin of the Geological Society of America*, 91: 724–730.

Simonett, D.S. 1968. Wind action. In R.W. Fairbridge (ed.), *Encyclopedia of geomorphology.* Reinhold, New York: 1233–1236.

Skidmore, E.L. and Woodruff, N.P. 1968. *Wind erosion forces in the United States and their use in predicting soil loss.* US Department of Agriculture Handbook, 346.

Small, R.J. 1978. *The study of landforms.* Cambridge University Press, London.

Smith, H.T.U. 1972. Playas and related phenomena

in the Saharan region. In C.C. Reeves, Jr (ed.), *Playa lake symposium proceedings.* Texas Technical University International Center for Arid and Semi-Arid Land Studies, 4, Lubbock: 63–87.

Smith, R.S.U. 1984. Eolian geomorphology of the Devils Playground, Kelso Dunes and Silurian Valley, California. In J. Lintz (ed.), *Western geological excursions*, Vol. 1, Geological Society of America 97th Annual Meeting Field Trip Guidebook, Reno, NV: 239–251.

Springer, M.E. 1958. Desert pavement and vesicular layer of some soils of the desert of the Lahonton Basin, Nevada. *Proceedings of the Soil Science Society of America*, 22: 63–66.

Squyres, C.H. and Bradley, W. 1964. Notes on the Western Desert of Egypt. In F.A. Reilly (ed.), *Guidebook to the geology and archaeology of Egypt.* Petroleum Exploration Society of Libya, 6: 99–105.

Stein, M.A. 1909. Explorations in Central Asia. *Geographical Journal*, 34: 5–36.

Stewart, G., Krinsley, D. and Marshall, J. 1981. An experimental study of the erosion of basalt, obsidian and quartz by fine sand, silt and clay. *NASA Technical Manual*, 84211: 214–215.

Stringfield, V.T., LaMoreaux, P.E. and LeGrand, H.E. 1974. Karst and paleohydrology of carbonate rock terranes in semiarid and arid regions with a comparison to humid karst of Alabama. *Bulletin of the Alabama Geological Survey*, 105, 106: 1–101.

Sugden, W. 1964. Origin of faceted pebbles in some recent desert sediments of southern Iraq. *Sedimentology*, 3: 65–74.

Suzuki, T. and Takahashi, K. 1981. An experimental study of wind abrasion. *Journal of Geology*, 89: 23–36.

Tolman, C.F. 1909. Erosion and deposition in the southern Arizona bolson region. *Journal of Geology*, 17: 136–163.

Tremblay, L.P. 1961. Wind striations in northern Alberta and Saskatchewan, Canada. *Bulletin of the Geological Society of America*, 72: 1561–1564.

Twidale, C.R. 1976. *Analysis of landforms.* Wiley, New York: 282–316.

Verstappen, H.T. and van Zuidam, R.A. 1970. Orbital photography and the geosciences — geomorphological example from the Central Sahara. *Geoforum*, 2: 33–47.

Wade, A. 1910. On the formation of 'dreikante' in desert regions. *Geological Magazine*, 7: 394–398.

Walther, J. 1892. A comparison of the deserts of North America with those of North Africa and northern India. *Science*, 19: 158.

Walther, J. 1912. *Das Gesetz der Wüstenbildung*, new edn. Von Quelle und Meyer, Leipzig.

Ward, A.W. and Greeley, R. 1984. Evolution of the yardangs at Rogers Lake, California. *Bulletin of the Geological Society of America*, 95: 829–837.

Wells, S.G., Dohrenwend, J.C., McFadden, L.D., Turrin, B.D. and Mahrer, K.D. 1985. Late Cenozoic landscape evolution of lava flow surfaces of the Cima Volcanic Field, Mojave Desert, California. *Bulletin of the Geological Society of America*, 96: 1518–1529.

Whitney, M.I. 1978. The role of vorticity in developing lineation by wind erosion. *Bulletin of the Geological Society of America*, 89: 1–18.

Whitney, M.I. 1979. Electron micrography of mineral surfaces subject to wind-blast erosion. *Bulletin of the Geological Society of America*, 90: 917–934.

Whitney, M.I. 1983. Eolian features shaped by aerodynamic and vorticity processes. In M.E. Brookfield and T.S. Ahlbrandt (eds), *Eolian sediments and processes.* Elsevier, Amsterdam: 223–245.

Whitney, M.I. 1985. Yardangs. *Journal of Geological Education*, 33: 93–96.

Whitney, M.I. and Brewer, H.B. 1968. Discoveries in aerodynamic erosion with wind tunnel experiments. *Michigan Academy of Science, Arts and Letters*, 53: 91–104.

Whitney, M.I. and Dietrich, R.V. 1973. Ventifact sculpture by windblown dust. *Bulletin of the Geological Society of America*, 84: 2561–2582.

Whitney, M.I. and Spletstoesser, J.F. 1982. Ventifacts and their formation: Darwin Mountains, Antarctica. In D. Yaalon (ed.), *The Proceedings of the International Conference of the International Society of Soil Science.* Catena, Supplement: 175–194.

Whitney, M.I., Brewer, H.B. and McNeel, W. Jr 1969. Wind tunnel and field data on erosion along pressure gradients. *Geological Society of America Abstracts for 1968*: 318–319.

Wilshire, H.G., Nakata, J.K. and Hallett, B. 1981. Field observations of the December 1977, wind storm, San Joaquin Valley, California. In T.J. Péwé (ed.), *Desert dust.* Geological Society of America, Special Paper 186: 233–251.

Wolfe, R.W. and El-Baz, F. 1979. *The wind regime of the Western Desert of Egypt.* NASA Technical Memorandum, 80339: 229–301.

Woodruff, N.P. and Siddoway, F.H. 1965. A wind erosion equation. *Soil Science*, 29: 602–608.

Worrall, G.A. 1974. Observations on some wind-formed features in the southern Sahara. *Zeitschrift für Geomorphologie*, NF, 18: 291–302.

Zingg, A.W. 1949. A study of the movement of surface wind. *Agricultural Engineering*, 30: 11–13, 19.

Zingg, A.W. 1953. Wind-tunnel studies of the movement of sedimentary material. *Proceedings Iowa State University, Institute of Hydraulics*, 5: 111–135.

Section 5
Geomorphology of the World's Arid Zones

20

Africa and Europe

Paul A. Shaw

Introduction

Africa, as the novelist Michael Crichton (1980) observed, is a trick of the Mercator projection. Despite its relatively small appearance on maps, this second largest of the continents contains 1.69 billion hectares of land classified as semi-arid or drier, 35% of the world's total, and includes 69% of the global land surface defined as hyper-arid (UNEP 1992). Two-thirds of Africa's surface is affected by aridity to some degree, and arid landforms, particularly aeolian forms, extend into currently humid areas.

Africa has two great desert regions. North of the equator the Sahara Desert covers 7 million km^2, an area equivalent to the USA with Alaska (Grove 1980), extending from the Atlas Mountains and Mediterranean coast to the savannas of the Sahel zone, and from the Atlantic to the Red Sea. It has provided the background for many of the classic studies of arid geomorphology (e.g. Bagnold 1941; Dresch 1982), yet the sheer scale of the desert and difficulty of access to it have prevented a full comprehension of its landscapes. Linguistic divisions, too, with strong contributions from French and German geomorphologists, have made its literature less accessible.

To the south of the equator the Kalahari, Namib and Karoo deserts occupy much of the western and central parts of the region which lies south of the Zambezi River, although dunes and sediments of the Mega Kalahari (Thomas 1984) extend northwards beyond the Congo River. The Namib and Kalahari deserts also provided an early impetus to research (e.g. Passarge 1904), but acquired geomorphological significance only in the post-colonial period, as access to certain areas improved, and remote sensing imagery became available (see, for example, Shaw 1993). Even today large areas of Africa's deserts lack basic topographic maps.

By contrast Europe is a relatively well-watered continent; 68% is classified as humid or cold, and there are no areas of extreme aridity (UNEP 1992). There are, however, extensive areas of semi-arid climate on the Mediterranean littoral, encompassing central and eastern Iberia, southern Italy, Sardinia and eastern Greece. Past aridity is also indicated by the loess accumulations of the past 2.4 Ma in western Europe and the Mediterranean, although depositional conditions seem to have been more humid than those experienced in Asia (see Chapter 21). The drier areas of Europe have a long history of human settlement, yet they have rarely been considered arid

Arid Zone Geomorphology: Process, Form and Change in Drylands, 2nd edition. Edited by David S. G. Thomas.

in the geomorphological sense. Within the past 20 years pressure on water resources, and the impending threat of desertification, have refocused attention on the Mediterranean rim, and there are now a number of geomorphological studies available, from western and southern European academics, on these regions (Embleton 1984).

The geological setting

Africa is a massive fragment of continental crust that has been moving northwards relative to the Earth's climatic belts since the breakup of Gondwanaland around 200 Ma ago. Stable by comparison with other continents, it is characterised by ancient erosion surfaces and sedimentary basins that give it flatness (Figure 20.1), identified by Mainguet (1983) as the primary characteristic of the greater Sahara. There is a strong Precambrian tectonic legacy throughout Africa. While Precambrian rocks occupy only 15% of the Saharan surface, for example, the distribution of sedimentary basins is influenced by structures developed during the Pan-African orogeny of 650–550 Ma (Williams 1993) and planed off by erosion cycles towards the end of the Precambrian (Mainguet 1983). Subsequently basin development in the Sahara has followed subsidence arising from downwarping or rifting (Burollet 1984) and the accumulation of great thicknesses of Phanerozoic sediments of different origins. The Ordovician glaciation affected approximately half of the Sahara region and left lineations and erratics (Beuf *et al.* 1971) which are still apparent in the landscape. It was followed by marine transgressions in the Silurian and the Mesozoic, interspersed by episodes of continental erosion and basin filling. These provided sediments, folded by Hercynian earth movements, now known as the Continentale Intercalaire in the western Sahara, and the Nubian Sandstone in the east (Grove 1980), which became source rocks for Cenozoic sedimentation.

In southern Africa Precambrian rocks again provide the basement upon which later sedimentation occurred, in this case the extensive Karoo Sequence (SACS 1980) recording transition from the Carboniferous (Dwyka) glaciation to the deposition of the terrestrial sandstones of the Ecca Formation, itself the source rock of the Kalahari sands (Smith 1984).

The fragmentation of Gondwanaland provided the impetus for the formation of marginal or pericratonic basins around the African coastline (Burollet 1984), and the flexuring of the continental crust which resulted in the formation of both the Great Escarpment of southern Africa, and the development of the intercratonic Kalahari–Cubango–Congo Basin (Thomas 1988a), in which the accumulation of Kalahari sediments commenced in the Cretaceous (Thomas and Shaw 1991a).

The Cenozoic in North Africa was marked by continued subaerial weathering and erosion, particularly in the southern Sahara, where silicate karst landscapes evolved (Busche and Erbe 1987). The Atlas region, folded in the early Cenozoic, was uplifted in the Miocene as the African plate moved northwards, while significant relief was created within the Saharan interior over hotter areas of the mantle (Burke and Wells 1989; Figure 20.2). Extensive areas of alkaline basalts were extruded in northeast Africa, to be upwarped and fractured by the Afro-Arabian rift system in the Pliocene. Volcanic activity also took place in the Tibesti region, while rifting and subsidence along the Great Rift extended southwards into southern Africa along the Gwembe Trough and Kalahari Rift (Shaw and Thomas 1993).

The collision of the African and European plates led not only to the formation of the Atlas Mountains and the Baetics in southern Iberia, but to the diminution of the Tethys Sea by the mid-Miocene, and to the Messinian salinity crisis of the late Miocene, in which marine conditions were replaced by an evaporitic basin (Hsü *et al.* 1977). Isolation from the Atlantic led to rapidly lowering base-levels, and downcutting of drainage at the continental margin, including an early course of the Nile, the Eonile, which carved a canyon 1300 km long and 2500 m deep (Said 1982), far greater than the dimensions of the Grand Canyon. Subsequent transgression of the Mediterranean area by the Atlantic in the Pliocene restored marine conditions, but made a major impact on the landscapes of the Mediterranean region (Hsü *et al.* 1977) through valley erosion, the intrusion of salt bodies and the activation of groundwater regimes.

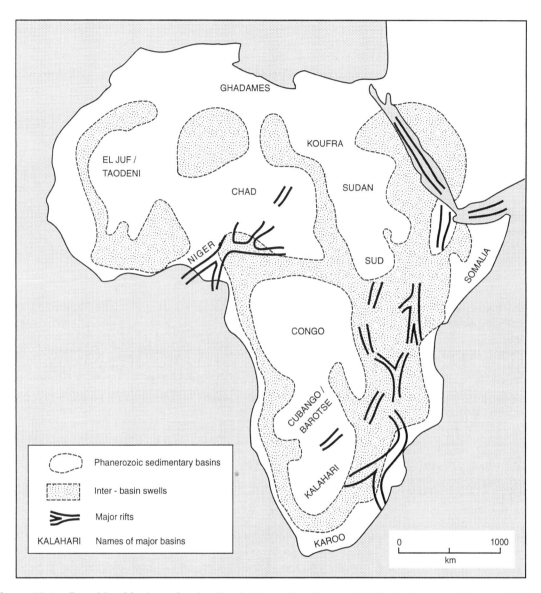

Figure 20.1 *Depositional basins and major rifts of Africa. After Holmes (1965), De Swardt and Bennet (1974), Burollet (1984) and Summerfield (1985)*

The establishment of a subaerial regime led to the initiation of drainage on the African continent, greatly affected by these subsequent tectonic and epeirogenic shifts. As a consequence the desert regions are characterised by extensive areas of endoreic drainage, alongside large exotic rivers such as the Nile, and areas of swamp, located on losimean fans (Stanistreet and McCarthy 1993) such as the Okavango Delta, the Nile Sudd, and the Niger Inland Delta, where derangement of the drainage has clearly taken place. In northern Africa McCauley *et al.* (1982; 1986) identified a major west-flowing drainage alignment from SIR-A imagery, draining from the Red Sea to the Atlantic via the Chad Basin and Niger Delta

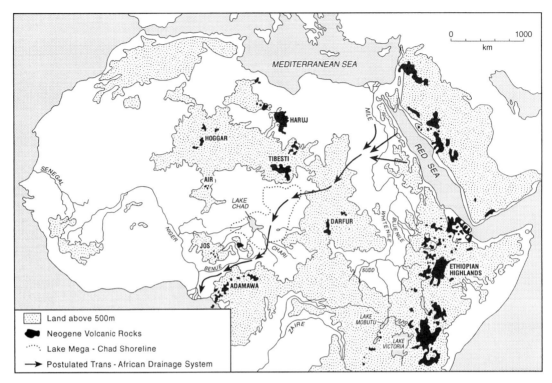

Figure 20.2   *North Africa — relief, present drainage and the alignment of the proposed Trans-Africa Drainage System. After McCauley et al. (1986) and Burke and Wells (1989)*

(Figure 20.2), although Burke and Wells (1989) suggest that some of the drainage traces belong to a proto-Nile, and not to the trans-Africa drainage system. The subsequent evolution of the Nile has also been complex, encompassing not only the erosion and infill into the proto-Mediterranean (Said 1982), but also changes in the headwater regions (Adamson and Williams 1980; Adamson *et al.* 1980). Drainage derangement from climatic causes continued into the Quaternary on major rivers such as the Niger and the Senegal (Talbot 1980).

In southern Africa short exoreic rivers with steep gradients developed on the Great Escarpment, while endoreic streams drained towards the Kalahari Basin. The Cubango–Okavango is the last remnant of the latter pattern, as downwarping of the Kalahari Rift has led to capture of the proto-Upper Zambezi and its tributaries by Indian Ocean drainage (Thomas and Shaw 1988; 1991; Figure 20.3).

The climatic setting

The principal causes of aridity are the location and dimension of the land mass within the subtropical zone, and the presence of cold oceanic currents on west coasts north and south of the equator. Rain-shadow effects are important locally, as for example, in the Tibesti and Hoggar mountains of the Sahara and in the Rift Valley of East Africa.

North and south of the equator there is a latitudinal arrangement of climatic zones, with a belt of summer rainfall, subject to interannual and intra-annual variation, towards the equator, and belts of winter rainfall, fed by the mid-latitude westerlies, in the Mediterranean and in the Southern Cape. Winter rainfall penetrates south of the Atlas Mountains in the northwest Sahara, and into the Red Sea Hills and northern Ethiopia to the east, whereas summer rainfall may affect the entire southern Sahara as far north as the latitude of Tibesti

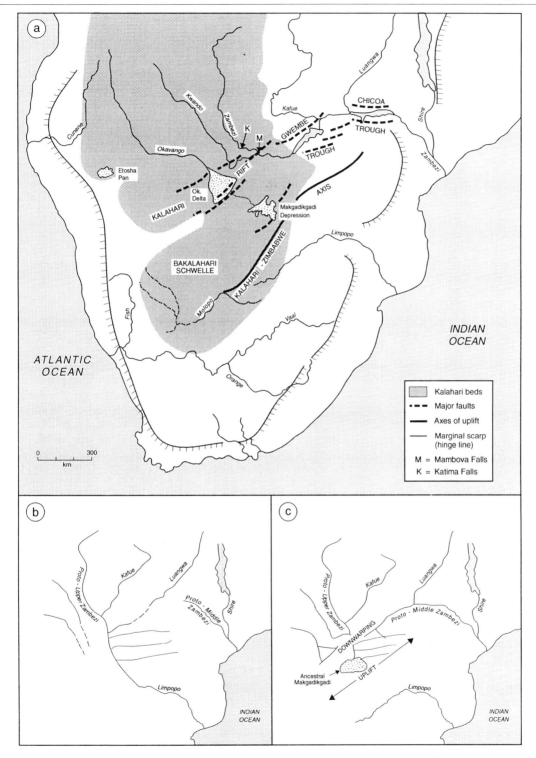

Figure 20.3 *Drainage development in southern Africa. (a) Modern drainage alignment; (b) the proto-Upper Zambezi as a tributary of the Limpopo after the division of Gondwanaland; (c) the capture of the Upper Zambezi in the early Pleistocene following uplift along the Kalahari–Zimbabwe axis. After Thomas and Shaw (1988)*

(Grove 1980). The dominance of dry, descending northeasterly airstreams throughout the year, and the continental position, lead to high maximum temperatures and great diurnal and annual temperature ranges in the region. Rainfall is below 100 mm per year over much of the Sahara, and evapotranspiration losses are among the highest in the world.

Southern Africa is narrower at 15–25° latitude, and the South Atlantic and Indian Ocean anticyclonic circulations mitigate the climate, leading to greater penetration of both Intertropical Convergence Zone (ITCZ) and cyclonic rainfall in the southern hemisphere, so that most areas receive upwards of 100 mm per year on a regular basis (Tyson 1986). There is a strong precipitation gradient across the subcontinent from west to east. However, variations in the Walker circulation occur frequently, and low-phase events may increase rainfall on the arid west coast, but lead to drought in southeastern Africa (Harrison 1984). The extreme aridity of the Namib coast is mitigated by sea fogs, occurring in excess of 100 days per year, which affect rock weathering tens of kilometres inland.

Climatic variation is considerable in both space and time. On the geological timescale the Sahara region became progressively drier from the Jurassic onwards, as it drifted through equatorial latitudes, though clockwise rotation led to desiccation earlier in the northwestern sector (Williams 1993). A seasonal climate was established by the late Eocene and true aridity, suggested by aeolian sand in ocean cores, by the Oligocene (Sarnthein 1978), although Sudano-Guinean woodland was the dominant vegetation cover at the time (Bonnefille 1983). By the late Tertiary the combined impacts of the Messinian salinity crisis, the uplift of the Tibetan Plateau, cooling of the ocean surface and progressive accumulation of the polar ice caps had firmly established arid conditions. During the Quaternary, fluctuations between arid and humid phases became a major feature of the climate, with wet conditions persisting in the Sahara as late as the mid-Holocene (see Williams 1993).

Aridity in southern Africa also has a long history. The Tsondab Sandstone Formation of the central and southern Namib represents the accumulation of a sand erg from the early Miocene (Ward *et al.* 1983; Wilkinson 1988), while initiation of upwelling of the Benguela Current can be traced back to the late Miocene from palynological evidence (Siesser 1980). Hyper-aridity and the accumulation of the Namib erg had commenced by the Pliocene (van Zinderen Bakker 1984). Moister episodes are indicated in the Quaternary record from increased fluvial activity (e.g. Rust and Vogel 1988), but there is little evidence that these were either prolonged or extensive. In the Kalahari Basin the stratigraphy has provided little information on Tertiary conditions, while the wetter episodes indicated by geomorphological evidence have proved difficult to correlate with those of northern Africa (Thomas and Shaw 1991b).

Landforms of the Sahara

The Sahara of the popular imagination is a flat, sand-covered expanse, yet there is considerable variety within it. Mountain ranges account for some 500 000 km^2 of the region (Gabriel 1991), with several peaks in the Tibesti Massif exceeding 3000 m. Although glacial landforms related to past cold periods are found only in the Atlas Mountains, cryonivational mass movement has been reported from the Hoggar (Rognon 1977) and Tibesti massifs (Messerli *et al.* 1980), and all high-altitude areas contribute floodwaters onto the surrounding plains; the Oued Tamanrasset, for example, flows 300–400 km every six to seven years from its source in the Hoggar (Rognon 1970).

Sedimentary strata are frequently preserved as plateaux, bounded by escarpments or cuestas, for example in southern Libya and eastern Niger (Busche and Erbe 1987). Both limestone and sandstone strata in many of these regions show evidence of karstification, such as the development of passages and dolines (Mainguet 1972), probably under the influence of spring activity prompted by higher ground water tables. In southern Libya karstification is associated with extensive silcrete and ferricrete duricrusts which are thought to have formed in the wetter climates of the early Miocene (Busche and Erbe 1987). Surface expressions of groundwater activity include the massive dry valleys in the Libyan Desert of Egypt, first reported by Peel (1941), and the development of scarp-foot depressions of the Arkiafera Plain, Tibesti (Hagedorn 1971), which have been attributed to intensified chemical weathering at the end of the Pliocene. In general scarp features, including landslides, appear to be inactive at present.

Low-relief areas are composed of gravel *regs*, rock *hamadas* and sand *ergs* (see Figure 17.4). These are linked by the erosion, transfer and deposition of sediment from source areas within the Sahara by both the regional clockwise wind circulation (Mainguet 1983), and by circulation patterns generated by upland areas (Oberlander 1994). Regs have been considered as both areas of residual coarse sand or gravel spreads from which deflation has taken place, as for example, in the *ténérés* of Libya (Capot-Rey 1970), or as hamadas which experience significant bedrock disintegration and deflation (Mainguet 1983). The evidence of wind scour is even more impressive in the dune-free hamada areas, where extensive yardang or *kalut* landscapes occur, as in northern Chad, and the Gilf Kebir region of Egypt (Chapter 19). Linear corrasion forms, termed *crest-couloir*, reach 350 km in length in the Tibesti region (Mainguet *et al.* 1980).

The Sahara contains numerous aeolian depocentres, of which 27 contain ergs in excess of 12 000 km², and the largest exceed 200 000 km² (Wilson 1973) (Figure 20.4 and Table 20.1). Sand depths are relatively deep, estimated at 20 m or more for three Algerian ergs (Wilson 1973). The dominant dune form, seen at the scale of satellite images, are *draa* or megadunes up to 300 m in height and in a variety of forms (Breed *et al.* 1979), including barchanoid, pyramidal (*rhourd*) and linear forms. Smaller-magnitude dune forms are superimposed upon the draa landscape. Despite their variety the erg morphology is generally consistent with intererg transfer of sediment, although definitions of fixed and active ergs (Wilson 1973) and ergs of sand loss or gain (Mainguet and Chemin 1983) are not universally accepted. In West Africa linear dunes have been traced from satellite imagery as far south as Kano, on the 1000 mm yr⁻¹ isohyet (Nichol 1991).

Important drainage landforms include major exotic rivers, such as the Nile (Said 1993), Niger and Senegal, dry and ephemeral valleys or *wadis*, and internal drainage basins. The latter may enclose playas (see Chapter 14),

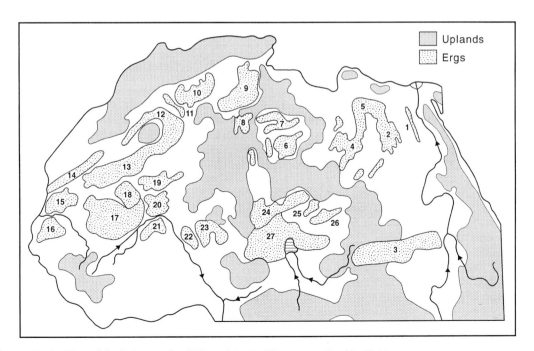

Figure 20.4 *Ergs of the Sahara, after Wilson (1973). The ergs are listed in Table 20.1*

Table 20.1 *Saharan ergs larger than 12 000 km², after Wilson (1973). The location of the ergs is shown in Figure 20.4*

No.	Name	Area (km²)	Status
1	Abu Moharik		Active
2	Great Sand Sea	105 000	Active
3	Sudanese Qoz	240 000	Fixed
4	Erg Rebiana	65 000	Active
5	Erg Calanscio	62 000	Active
6	Edeyen Murzuq	61 000	Active
7	Edeyen Ubari	61 000	Active
8	Issaouane-N-Irrararen	38 000	Active
9	Erg Oriental	192 000	Active
10	Erg Occidental	103 000	Active
11	Erg er Raoui	12 000	Active
12	Erg Iguidi	68 000	Active
13	Erg Chech-Adrar	319 000	Active
14	North Mauritanian Erg	85 000	Active
15	South Mauritanian Erg	65 000	Fixed
16	Trarza/Cayor Erg	57 000	Fixed
17	Ouarane/Aouker Erg	206 000	Active/fixed
18	El Mréyé	63 000	Active
19	Erg Tomboctou	66 000	Active
20	Erg Azouad	35 000	Fixed
21	Erg Gourma	43 000	Fixed
22	West Azouak	35 000	Fixed
23	East Azouak	34 000	Active/fixed
24	Erg Bilma-Ténéré	155 000	Active
25	Erg Foch	13 000	Active
26	Erg Djourab	45 000	Active
27	Erg Kanem	294 000	Fixed

such as the spring-fed *chotts* of Tunisia. Groundwater control of depressions is evident in many parts of the northern Sahara; the massive Qattara Depression of northern Egypt has been explained, among other hypotheses, by solution and collapse following the Messinian salinity crisis (Albritton *et al.* 1990). The largest of the internal basins is that of Lake Chad, a complex of lacustrine, deltaic and aeolian features (Grove and Warren 1968; Servant and Servant-Vildary 1980), which records not only changes in Quaternary environments, but is also related to the long-term evolution of the Saharan drainage (e.g. Faure *et al.* 1992).

In northeast Africa the upland landscape is dominated by the Blue and White Nile valleys, including large alluvial fans where the Ethiopian Highlands meet the Sudan Plain (Adamson and Williams 1980; Williams and Adamson 1980). South of the Nile headwaters small but significant rain-shadow deserts occur in the East African Rift system, an interesting juxtaposition of volcanic landforms and groundwater-fed salt lakes (Figure 20.5) such as Lakes Magadi (Eugster 1986; Damnati 1993; Williamson *et al.* 1993) and Natron (Roberts *et al.* 1993).

Landforms of the Namib

The Namib Desert stretches over 2000 km along the west coast of southern Africa, between the Olifants River in South Africa and the Carunjamba River in southern Angola (Lancaster 1989). The desert occupies a plain rising from the coast to the Great Escarpment

Figure 20.5 *The remnants of the bed of Lake Olorgesaille, a Quaternary playa truncated by faulting and drainage realignment in the Kenya Rift Valley*

120–200 km inland, and is traversed by a series of east–west-aligned rivers, of which only the Orange and Cunene have perennial flow (Figure 20.6).

A large part of the desert comprises gravel- or sand-covered surfaces which have been interpreted as late Cretaceous pediplains (Ollier 1977; Selby 1977), from which rise swarms of dolerite dykes (Van Zyl 1992) and massive inselberg complexes, such as Groot Spitskoppe (1728 m a.s.l.). The absence of deep-weathering features suggests that the inselbergs, including bornhardt forms, are controlled by structure (Selby 1982) and have evolved by mantle-controlled planation since the Mesozoic (Ollier 1978). Inselbergs near to the coast, such as Vogelfederberg (Figure 20.7), display honeycomb surfaces as a result of fog weathering (Olivier 1992).

Fluvial incision into the plains following Tertiary epeirogenic uplift has resulted in canyons up to 200 m deep with deeply dissected margins, a landscape known as *gramadulla* (Rust and Wieneke 1974). The presence of lacustrine deposits at high levels within the valleys, particularly those of the Kuiseb and Homeb, has suggested a complex evolutionary sequence with frequent drainage interruption (Marker 1977; Rust and Wieneke 1980; Ward 1982). Studies of flood deposits in the predominantly sand-river type channels shows a lack of synchroneity in flood events throughout the late Quaternary (Rust and Vogel 1988).

The Namib has five major dune fields (Figure 20.6), of which the largest, the Namib Sand Sea of 34 000 km^2 area, lies between Luderitz and the Kuiseb River, with a small outlier extending northwards towards Swakopmund. In the southern Namib the Obib Dunes occur on the sand-covered rock plain of the Sperrgebiet, while in the north dunefields occur on the Skeleton Coast (Lancaster 1982), in Angola around the mouth of the Cunene River (Bremner 1984), and in the Curoca-Bahia dos Tigres area of Angola (Torquato 1972). All of these dunefields have coastal rivers as the main source of sediment supply, of which the Orange is by far the most important (Lancaster and Ollier 1983), representing some 27% of the total sediment discharge from the southern African subcontinent (Dingle *et al.* 1983). Besler (1984) proposes the denudation of the Tsondab

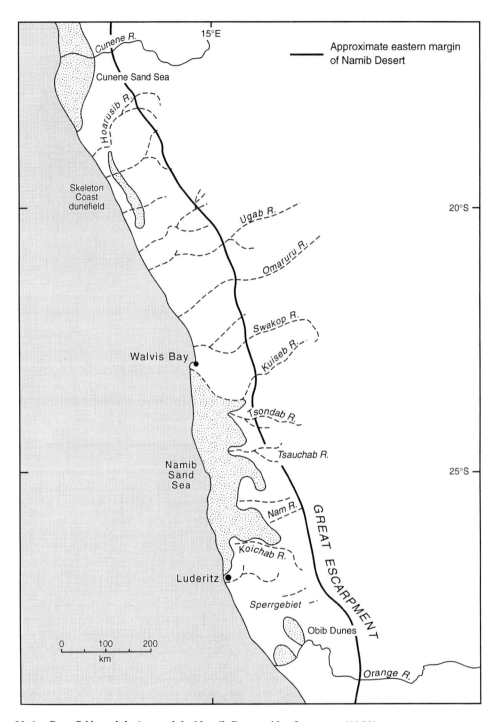

Figure 20.6   *Dunefields and drainage of the Namib Desert. After Lancaster (1989)*

Figure 20.7 *Vogelfederberg, an inselberg 40 km from the Namib coast, affected by weathering contributed to by fog moisture*

Sandstone as a possible origin for aeolian sediments, while eustatic sea-level change has also affected inputs from shelf sediments (Rogers 1977).

The dunes of the Namib Sand Sea have been described in detail by Lancaster (1989), who identifies a transition from crescentic and barchanoid dunes along the coast to linear dunes, covering 74% of the total dunefield, in the interior. The linear dunes, dominantly oriented parallel to the coast, become increasingly complex towards the eastern flank of the dunefield, and merge into areas of star dunes and star dune chains, with those of the Sossus Vlei field reaching 350 m in height. Lancaster (1989) estimates the rate of accumulation of the sand sea at 400 000 m^{3}yr^{-1}, giving an approximate age of 2–3 Ma, consistent with the climatic and geological history of the region.

Landforms of the Karoo

The Karoo occupies the western part of South Africa, between the Cape Fold Mountains and the Orange River, and is contiguous with both the Namib and Kalahari. Its extent varies according to whether it is defined by ecological characteristics (e.g. Klein 1980), topography (e.g. Wellington 1955), or by the distribution of Karoo System geology. Generally it is an extensive rock plain lying between 900–1200 m a.s.l., studded with inselbergs and hill ranges, and drained by seasonal rivers and endoreic drainage lines. It experiences rainfall between 150 and 400 m yr^{-1}.

The Karoo has received little scientific attention. Regional descriptions of the geomorphology have been provided by Dixey (1955) and Wellington (1955), with more detailed accounts of subregions from Van Rooyen and Burgher (1973) and McCarthy *et al.* (1985). Pans have received some attention (e.g. Le Roux 1978; Verster *et al.* 1992), as have inselberg slopes (Le Roux and Vrahimis 1987; 1990) and the linear dunes of the area adjacent to the Orange River (Le Roux 1990; 1992). Earth mounds, similar to the mima-mounds of North America, are widespread in the western Karoo, and have been attributed to burrowing organisms (Lovegrove and Siegfried 1986; 1989).

Landforms of the Kalahari

As with the Karoo, the definition of the areal extent of the Kalahari is dependent upon the

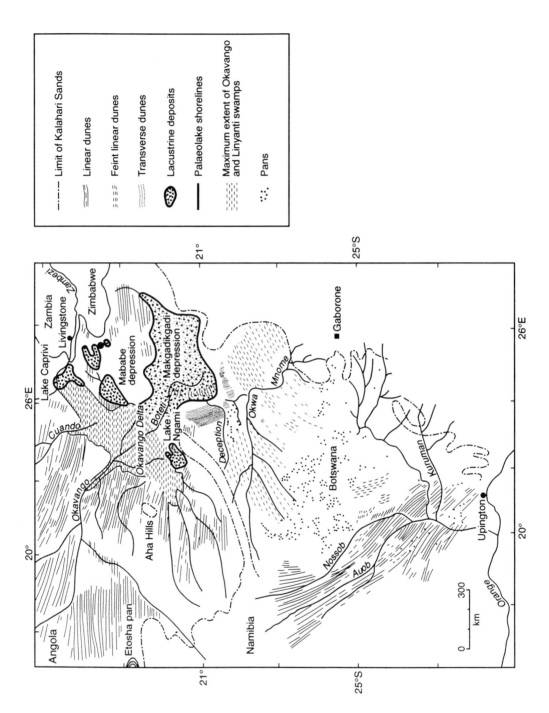

Figure 20.8   *The major dune fields of the Kalahari (after Thomas 1984)*

criteria used, from physiographic region (Wellington 1955), to ecozone (Werger 1978) to an area occupied by Kalahari Group sediments (e.g. Passarge 1904). On the largest scale the Mega Kalahari (Thomas 1984) refers to the extent of the Kalahari erg, which covers over 2.5 million km^2 from South Africa to Gabon and Congo north of the equator. Of this the drier parts occur south of the Zambezi River, with an arid core, the Kalahari 'Desert', in central and southern Botswana, where rainfall is as low as 150 $mm\,yr^{-1}$. In practice the Kalahari is an edaphic desert, with the absence of surface water arising from rainfall variability combined with high evaporation and infiltration rates, rather than from low rainfall *per se*.

The geomorphology of the Kalahari has been summarised by Thomas and Shaw (1991a). Essentially it is a gently undulating sand plain at approximately 1000 m a.s.l., interrupted occasionally by inselbergs or minor hill chains. The well-forested northern Kalahari is drained by tributaries of the Congo and Zambezi rivers, while the middle Kalahari has an endoreic network of ephemeral and dry channels draining towards the sump formed by the Makgadkgadi Basin, although the Okavango Delta, truncated by the terminal faults of the Kalahari Rift, does have some perennial outflow in that direction (Shaw and Thomas 1993). In the southern Kalahari the drainage lacks integration, being represented by hundreds of small pans, or the dry valley networks of the Molopo and Okwa-Mmone.

The distribution and characteristics of the pans and their associated lunette dunes have been described by Lancaster (1978a; 1978b; 1986a), while the morphology and function of the dry valleys have been considered by Shaw and De Vries (1988) and Nash *et al.* (1994a; 1994b). Both landforms are associated with extensive duricrust suites, and probably result from long-term groundwater activity, although aeolian deflation is certainly a modifying factor in the case of pans.

The pan and valley duricrusts in turn form part of a spectrum of silcretes, calcretes and intermediate hybrid forms which occur in a variety of geomorphological contexts, and in thicknesses of up to 100 m in the southern Kalahari (Shaw and De Vries 1988). The provenance of the calcretes has been examined by Goudie (1973) and Watts (1980), while Summerfield (1982; 1983) has identified silicification of Kalahari Sand, replacement of calcrete, pan and valley formation as the main origins of the silcrete. To this list of silcrete origins can be added the direct silicification of bedrock and spring tufas, contemporary biogenic activity in saline environments (Shaw *et al.* 1990), and fluvial and lacustrine origins (Thomas and Shaw 1991a). Although much of the duricrust is considered to be Tertiary in age, denoting long-term geomorphological stability (e.g. Goudie 1973), some types are late Quaternary or even contemporary in age.

In the middle and southern Kalahari the majority of all dunes are linear in form (Figure 20.8), with pan lunettes and small areas of parabolic dunes making up the remainder. Three dunefields have been identified on morphological and ecological criteria (Thomas 1984; Figure 20.8) forming a massive 'wheelround' which reflects the anticyclonic circulation of the subcontinent. The dunes of the more arid southern field have attracted morphological (Lancaster 1988; Thomas 1988b; Bullard *et al.* 1995), sedimentological (Lancaster 1986b; Thomas and Martin 1987) and process (Wiggs *et al.* 1994) studies, suggesting that morphometric and sedimentologic relationships between the dunes are complex, and that contemporary ideas of dune 'activity' in the context of vegetation cover may not be appropriate (Thomas and Shaw 1991b; see also Chapters 7 and 26). Although the northern and eastern dunefields are clearly relict features, the ages of dune formation are only now being established by direct dating.

Evidence of Quaternary environmental changes have come from the studies of the palaeolake basins of Etosha (Rust 1984) and the Okavango–Makgadikgadi (e.g. Shaw and Cooke 1986), and from scattered cave sites (e.g. Cooke 1984; Holmgren *et al.* 1995). Quaternary geomorphology in the Kalahari has been disadvantaged by the homogeneity of the Kalahari Group sediments, and the lack of datable materials.

Landforms of dryland Europe

Current dryness is best represented by the landscapes of Mediterranean Europe, an area

Figure 20.9   *Incision and drainage diversion of the Rio Feos, Almeria, Spain. The Aguas–Feos Gap in the middle distance is a response to the uplift and diapirism of the Sierra Cabrera*

strongly influenced by the Alpine orogeny and subsequent events. From Spain across to the Aegean fluvial erosion has been the dominant process, against a background of tectonism and eustatic sea-level change. Other processes and resultant landforms exist, but are poorly represented in the literature; for example, aerosols from African dust plumes are thought to have contributed to soil development (Prodi and Fea 1979; Rapp and Nihlén 1986), while salt weathering features and calcrete duricrusts, widely distributed in the region, are discussed only in local examples (e.g. Marques *et al.* 1990; Sancho *et al.* 1992). Other features, such as the coastal sabkhas (*salinas*) and inland playas (*lagunas*) of southern Spain, receive little attention.

The literature on fluvial geomorphology encompasses a range of themes. On a large scale, for example, basin evolution and drainage diversion, sometimes influenced by diapirism, has been described from southeast Spain, the driest part of Europe (Harvey and Wells 1987) (Figure 20.9) while on a smaller scale the impact of halokinetic processes of river terraces has been noted (Mather *et al.* 1991) and extreme flow events described (e.g. Harvey 1984). Southeast Spain, and the Methana area of Greece, have

also become important areas for the study of alluvial fans of both fluvial and debris flow types (Harvey 1990; 1992, see also Chapter 12).

Strong emphasis on fluvial process and form have led to studies of the spectacular badlands of Almeria (Scoging 1982; López-Bermúdez and Romero-Díaz 1989; Alexander *et al.* 1994), Greece (Harris and Vita-Finzi 1968) and Italy (Alexander 1982), including some discussion of historical anthropogenic impacts (e.g. Wise *et al.* 1982). Small-scale catchment studies and remote sensing have been utilised since 1991 by an international consortium of scientists under the the umbrella of the Mediterranean Desertification and Land Use Project (MEDALUS 1993), and their output, alongside recent books on the region (e.g. Woodward 1995), underline the emphasis placed on the role of running water.

References

Adamson, D.A. and Williams, F. 1980. Structural geology, tectonics and the control of drainage in the Nile Basin. In M.A.J. Williams and H. Faure (eds), *The Sahara and the Nile*. Balkema, Rotterdam: 225–252.

Adamson, D.A., Gasse, F., Street, F.A. and

Williams, M.A.J. 1980. Late Quaternary history of the Nile. *Nature*, 288: 50–55.

Albritton, C.C., Brooks, J.W., Issawi, B. and Swedan, A. 1990. Origin of the Qattara Depression, Egypt. *Bulletin of the Geological Society of America*, 102: 952–960.

Alexander, D. 1982. Difference between 'calanchi' and 'biancane' badlands in Italy. In R. Bryan and A. Yair (eds), *Badland geomorphology and piping*. GeoBooks, Norwich: 71–88.

Alexander, R.W., Harvey, A.M., Calvo, A., James, P.A. and Cerda, A. 1994. Natural stabilisation mechanisms on badland slopes: Tabernas, Almeria, Spain. In A. Millington and K. Pye (eds), *Environmental change in drylands: biogeographical and geomorphological perspectives*. Wiley, Chichester: 85–111.

Bagnold, R.A. 1941. *The physics of blown sand and desert dunes*. Methuen, London.

Besler, H. 1984. The development of the Namib Dune Field according to sedimentological and geomorphological evidence. In J.C. Vogel (ed.), *Late Cainozoic palaeoclimates of the southern hemisphere*. Balkema, Rotterdam: 445–453.

Beuf, S., Biju-Duval, B., de Charpal, O., Rognon, P., Gariel, O and Bennacef, A. 1971. *Les grès du Paléozoique inférieur au Sahara*. Publications de l'Institut Français du Pétrole, Paris. Editions Technip.

Bonnefille, R. 1983. Evidence for a cooler and drier climate in the Ethiopian uplands towards 2.4 my ago. *Nature*, 303: 487–491.

Breed, C.S., Fryberger, S.G., Andrews, S., McCauley, C., Lennartz, F., Gebel, D. and Horstman, K. 1979. *Regional studies of sand seas using Landsat (ERTS) imagery*. US Geological Survey, Professional Paper, 1052.

Bremner, J.M. 1984. The coastline of Namibia. *Geological Survey of South Africa/University of Cape Town Marine Geoscience Group Technical Report*, 15: 200–206.

Bullard, J.E., Thomas, D.S.G., Livingstone, I. and Wiggs, G.F.S. 1995. Analysis of linear sand dune morphological variability, southwestern Kalahari Desert. *Geomorphology*, 11: 189–203.

Burke, K. and Wells, G.L. 1989. Trans-African drainage system of the Sahara: was it the Nile? *Geology*, 17: 743–747.

Burollet, P.F. 1984. Intracratonic and pericratonic basins in Africa. *Sedimentary Geology*, 40: 1–11.

Busche, D. and Erbe, W. 1987. Silicate karst landforms of the southern Sahara (north-eastern Niger and southern Libya). *Zeitschrift für Geomorphologie*, NF, 64: 55–72.

Busche, D. and Hagedorn, H. 1980. Landform development in warm deserts — the central Saharan example, *Zeitschrift für Geomorphologie*, NF, 36: 123–139.

Capot-Rey, R. 1970. Remarques sur les ergs du Sahara. *Annales de Géographie*, 413: 2–19.

Cooke, H.J. 1984. The evidence from northern Botswana of climatic change. In J. Vogel (ed.), *Late Cenozoic palaeoclimates of the southern hemisphere*. Balkema, Rotterdam: 265–278.

Crichton, M. 1980. *Congo*. Alfred Knopf, USA.

Damanti, B. 1993. Sedimentology and geochemistry of lacustrine sequences of the Upper Pleistocene and Holocene in intertropical area (Lake Magadi and Green Lake): palaeoclimatic implications. *Journal of African Earth Sciences*, 16: 519–521.

De Swardt, A.M.J. and Bennet, G. 1974. Structural and physiographic evolution of Natal since the late Jurassic. *Transactions of the Geological Society of South Africa*, 77: 309–322.

Dingle, R.V., Siesser, W.G. and Newton, A.R. 1983. *Mesozoic and Tertiary geology of Southern Africa*. Balkema, Rotterdam.

Dixey, F. 1955. Some aspects of the geomorphology of central and southern Africa. *Transactions of the Geological Society of South Africa*, Supplement, 58.

Dresch, J. 1982. *Géographie des Régions Arides*. Presses Universitaires de France, Paris.

Embleton, C. 1984. *The geomorphology of Europe*. Macmillan, London.

Eugster, H.P. 1986. Lake Magadi, Kenya: a model for rift valley hydrochemistry and sedimentation? In L.E. Frostick (ed.), *Sedimentation in the African rifts*. Geological Society, Special Publication, 25, 177–189.

Faure, H., Breed, C.S. and McCauley, J.F. 1992. Paleodrainages of the eastern Sahara: the Nile problem and its relevance to the Chad Basin. *Journal of African Earth Sciences*, 14: 153–154.

Gabriel, B. 1991. Gebirgsregionen der Ostsahara. *Revue de Géographie Alpine*, 1: 101–116.

Goudie, A.S. 1973. *Duricrusts in tropical and subtropical landscapes*. Clarendon Press, Oxford.

Grove, A.T. 1980. Geomorphic evolution of the Sahara and the Nile. In M.A.J. Williams and H. Faure (eds), *The Sahara and the Nile*. Balkema, Rotterdam: 7–16.

Grove, A.T. and Warren, A. 1968. Quaternary landforms and climate on the south side of the Sahara. *Geographical Journal*, 134: 194–208.

Hagedorn, H. 1971. Untersuchungen über relieftypen arider Räume an Beispielen des Tibesti-Gebirge und seiner Umgebung. *Zeitschrift für Geomorphologie*, NF, 11: 251 pp.

Harris, D.R. and Vita-Finzi, C. 1968. Kokkinopilos — a Greek badland. *Geographical Journal*, 134: 537–545.

Harrison, M.S.J. 1984. A generalised classification of South African summer rain-bearing synoptic systems. *Journal of Climatology*, 4: 547–560.

Harvey, A.M. 1984. Geomorphic response to an extreme flood: a case from southeast Spain. *Earth Surface Processes and Landforms*, 9: 267–279.

Harvey, A.M. 1990. Factors affecting Quaternary alluvial fan development in southeast Spain. In A. Rachocki and M. Church (eds), *Alluvial fans: a field approach*. Wiley, Chichester: 247–269.

Harvey, A.M. 1992. Controls of sedimentary style on alluvial fans. In P. Billi, R. Hey, C. Thorne and P. Tacconi (eds), *Dynamics of gravel bed rivers*. Wiley, Chichester: 519–535.

Harvey, A.M. and Wells, S.G. 1987. Response of Quaternary fluvial systems to differential epeirogenic uplift: Aguas and Feos River systems, southeast Spain. *Geology*, 15: 689–693.

Holmes, A. 1965. *Principles of physical geology*, 2nd edn. Nelson, London.

Holmgren, K., Karlen, W. and Shaw, P.A. 1995. Palaeoclimatic significance of the stable isotopic composition and petrology of a late Pleistocene stalagmite from Botswana. *Quaternary Research*, 43: 320–328.

Hsü, K.J., Montadert, L., Bernoulii, D., Cita, M.B., Erickson, A., Garrison, R.E., Kidd, R.B., Mélières, F., Müller, C. and Wright, R. 1977. History of the Mediterranean salinity crisis. *Nature*, 267: 399–403.

Klein, R.G. 1980. Environmental and ecological implications of large mammals from Upper Pleistocene and Holocene sites in southern Africa. *Annals of the South African Museum*, 81: 223–283.

Lancaster, I.N. 1978a. The pans of the southern Kalahari, Botswana. *Geographical Journal*, 144: 80–98.

Lancaster, I.N. 1978b. Composition and formation of southern Kalahari pan margin dunes. *Zeitschrift für Geomorphologie*, NF, 22: 148–169.

Lancaster, N. 1982. Dunes of the Skeleton Coast, Namibia (South West Africa): geomorphology and grain size relationships. *Earth Surface Processes and Landforms*, 7: 575–587.

Lancaster, N. 1986a. Pans in the southwestern Kalahari: a preliminary report. *Palaeoecology of Africa*, 17: 59–67.

Lancaster, N. 1986b. Grain size characteristics of linear dunes in the south-western Kalahari. *Journal of Sedimentary Petrology*, 57: 573–574.

Lancaster, N. 1988. Development of linear dunes in the southwestern Kalahari. *Journal of Arid Environments*, 14: 233–244.

Lancaster, N. 1989. *The Namib Sand Sea*. Balkema, Rotterdam: 180 pp.

Lancaster, N. and Ollier, C.D. 1983. Sources of sand for the Namib Sand Sea. *Zeitschrift für Geomorphologie*, 45: 71–83.

Le Roux, J.S. 1978. The origin and distribution of pans in the Orange Free State. *South African Geographer*, 6: 167–176.

Le Roux, J.S. 1990. Linear dune trends and wind directions in the northern Cape Province, South Africa. *South African Geographer*, 17: 35–42.

Le Roux, J.S. 1992. Linear dunes near the Orange River: origin, mineralogy and climate. *South African Geographical Journal*, 74: 3–7.

Le Roux, J.S. and Vrahmis, S. 1987. The relationship between gradient and the size of rock fragments on debris slopes on dolerite-capped inselbergs in the southern Orange Free State. *South African Geographical Journal*, 69: 157–164.

Le Roux, J.S. and Vrahmis, S. 1990. The shape of rock particles and their distribution on debris slopes in a semi-arid environment. *South African Geographical Journal*, 72: 19–23.

López-Bermúdez, F. and Romero-Díaz, M.A. 1989. Piping erosion and badland development in southeast Spain. *Catena*, Supplement, 14: 59–73.

Lovegrove, B.G. and Siegfried, W.R. 1986. Distribution and formation of Mima-like earth mounds in the western Cape Province of South Africa. *South African Journal of Science*, 82: 432–437.

Lovegrove, B.G. and Siegfried, W.R. 1989. Spacing and origin of Mima-like earth mounds in the Cape Province of South Africa. *South African Journal of Science*, 85: 108–112.

Mainguet, M. 1972. *Le Modelé des Grès*. Institut Geographique National, Paris.

Mainguet, M. 1983. Tentative mega-morphological study of the Sahara. In R. Gardner and H. Scoging (eds), *Mega-geomorphology*. Clarendon Press, Oxford: 113–133.

Mainguet, M. and Chemin, M.C. 1983. Sand seas of the Sahara and Sahel: an explanation of their thickness and sand dune type by the sand budget principle. In M.E. Brookfield and T.S. Ahlbrandt (eds), *Eolian sediments and processes*. Elsevier, Amsterdam: 353–363.

Mainguet, M., Canon, L. and Chemin, M.C. 1980. Le Sahara: géomorphologie et paléogéomorphologie éoliennes. In M.A.J. Williams and H. Faure (eds), *The Sahara and the Nile*. Balkema, Rotterdam: 17–35.

Marker, M.E. 1977. Aspects of the geomorphology of the Kuiseb River, South West Africa. *Madoqua*, 10: 199–206.

Marques, M., Viera, M.C., Abreu, M.M., Prudencio, M.I. and Cabral, J.M. 1990. The caliche of Odivelas-Serpa area of Alentjo (Portugal): an approach to their palaeoenvironmental interpretation. *Chemical Geology*, 84: 176–178.

Mather, A.E., Harvey, A.M. and Brenchley, P.J. 1991. Halokinetic deformation of Quaternary terraces in the Sorbas Basin, southeast Spain. *Zeitschrift für Geomorphologie*, NF, 82: 87–97.

McCarthy, T.S., Moon, B.P. and Levin, M. 1985. Geomorphology of the western Bushmanland Plateau, Namaqualand, South Africa. *South African Geographical Journal*, 67: 160–178.

McCauley, J.F., Schaber, G.G., Breed, C.S., Grolier, M.J., Haynes, C.V., Issawi, B., Elachi, C. and Blom, R. 1982. Subsurface valleys and geoarchaeology of the eastern Sahara revealed by Shuttle radar. *Science*, 218: 1004–1020.

McCauley, J.F., Breed, C.S. and Schaber, G.G. 1986. The megageomorphology of the radar rivers of the eastern Sahara. In *The Second Spaceborne Imaging Radar Symposium, Pasadena, California*, NASA/JPL Publication No. 86-26: 25–35.

MEDALUS 1993. *Environment Programme — MEDALUS*. Executive Summary, Commission of the European Communities Directorate General for Science Research and Development.

Messerli, B., Winiger, M. and Rognon, P. 1980. The Saharan and East African uplands during the Quaternary. In M.A.J. Williams and H. Faure (eds), *The Sahara and the Nile*. Balkema, Rotterdam: 87–132.

Nash, D.J., Thomas, D.S.G. and Shaw, P.A. 1994a. Timescales, environmental change and dryland valley development. In A.C. Millington and K. Pye (eds), *Environmental change in drylands*. Wiley, Chichester: 25–41.

Nash, D.J., Shaw, P.A. and Thomas, D.S.G. 1994b. Duricrust development and valley evolution: process–landform links in the Kalahari. *Earth Surface Processes and Landforms*, 19: 299–317.

Nichol, J.E. 1991. The extent of desert dunes in northern Nigeria as shown by image enhancement. *Geographical Journal*, 157: 13–24.

Oberlander, T.M. 1994. Global deserts: a geomorphic comparison. In A.D. Abrahams and A.J. Parsons (eds), *Geomorphology of desert environments*. Chapman & Hall, London: 13–35.

Olivier, J. 1992. Some spatial and temporal aspects of fog in the Namib. *South African Geographer*, 19: 106–126.

Ollier, C.D. 1977. Outline geological and geomorphic history of the Central Namib Desert. *Madoqua*, 10: 207–212.

Ollier, C.D. 1978. Inselbergs of the Namib Desert: processes and history. *Zeitschrift für Geomorphologie*, 31: 161–176.

Passarge, F. 1904. *Die Kalahari*. Dietrich Riemer, Berlin.

Peel, R.G. 1941. Denudational landforms of the central Libyan Desert. *Journal of Geomorphology*, 4: 3–23.

Prodi, F. and Fea, G. 1979. A case for the transport and deposition of Saharan dust over the Italian peninsula and southern Europe. *Journal of Geophysical Research*, 84(C11): 6951–6960.

Rapp, A. and Nihlén, T. 1986. Dust storms and eolian deposits in North Africa and the Mediterranean. *Geoödynamik*, 7: 41–62.

Roberts, N., Taieb, M., Barker, P., Damanti, B., Icole, M. and Williamson, D. 1993. Timing of the Younger Dryas event in East Africa from lake-level changes. *Nature*, 366: 146–148.

Rogers, J. 1977. Sedimentation on the continental margins off the Orange River and the Namib Desert. *Geological Survey of South Africa/University of Cape Town Marine Geoscience Group Bulletin*, 7.

Rognon, P. 1970. Un massif montagneux en région tropicale aride: l'Atakor. *Annales de L'Universitie d'Abidjan*, Série G, vol. 2.

Rognon, P. 1977. *Formes périglacieres dans le massif de l'Atakor (Hoggar)*. Colloque, Université Strasbourg.

Rust, U. 1984. Geomorphic evidence of Quaternary environmental changes in Etosha, South West Africa/Namibia. In J. Vogel (ed.), *Late Cainozoic palaeoclimates of the southern hemisphere*. Balkema, Rotterdam: 279–286.

Rust, U. and Vogel, J.D. 1988. Late Quaternary environmental changes in the northern Namib Desert as evidenced by fluvial landforms. *Palaeoecology of Africa*, 19: 127–138.

Rust, U. and Wieneke, F. 1974. Studies on the gramadulla formation in the middle part of the Kuiseb River, South West Africa. *Madoqua*, series II, 3: 69–73.

Rust, U. and Wieneke, F. 1980. A reinvestigation of some aspects of the evolution of the Kuiseb River valley up-stream of Gobabeb, South West Africa. *Madoqua*, 12: 163–173.

SACS (South African Committee for Stratigraphy) 1980. *Stratigraphy of South Africa*, Part 1. Handbook of the Geological Survey of South Africa 8.

Said, R. 1982. The geological evolution of the River Nile in Egypt. *Zeitschrift für Geomorphologie*, NF, 26: 305–314.

Said, R. 1993. *The River Nile: geology, hydrology and utilisation*. Pergamon Press, Oxford.

Sancho, C., Melendez, A., Signes, M. and Bastida, J. 1992. Chemical and mineralogical characteristics of Pleistocene caliche deposits from the central Ebro Basin, NE Spain. *Clay Minerals*, 27: 293–308.

Sarnthein, M. 1978. Sand deserts during glacial maximum and climatic optimum. *Nature*, 272: 43–46.

Scoging, H. 1982. Spatial variations in infiltration, runoff and erosion on hillslopes in semi-arid Spain. In R. Bryan and A. Yair (eds), *Badland geomorphology and piping*. GeoBooks, Norwich: 89–112.

Selby, M.J. 1977. Bornhardts of the Namib Desert. *Zeitschrift für Geomorphologie*, 21: 1–13.

Selby, M.J. 1982. Form and origin of some bornhardts of the Namib Desert. *Zeitschrift für Geomorphologie*, 26: 1–15.

Servant, M. and Servant-Vildary, S. 1980. L'environnement Quaternaire du bassin du Tchad. In M.A.J. Williams and H. Faure (eds), *The Sahara and the Nile*. Balkema, Rotterdam: 133–162.

Shaw, P.A. 1993. Geomorphology in Botswana. In H.J. Walker and W.E. Grabau (eds), *The evolution of geomorphology*. Wiley, Chichester: 57–60.

Shaw, P.A. and Cooke, H.J. 1986. Geomorphic evidence for the late Quaternary palaeoclimates of the Middle Kalahari of northern Botswana. *Catena*, 13: 349–359.

Shaw, P.A. and De Vries, J.J. 1988. Duricrust, groundwater and valley development in the Kalahari of southeast Botswana. *Journal of Arid Environments*, 14: 245–254.

Shaw, P.A. and Thomas, D.S.G. 1993. Geomorphology, sedimentation and tectonics in the Kalahari Rift. *Israel Journal of Earth Science*, 41: 87–94.

Shaw, P.A., Cooke, H.J. and Perry, C.C. 1990. Microbialitic silcretes in highly alkaline environments: some observations from Sua Pan, Botswana. *South African Journal of Geology*, 93: 803–808.

Siesser, W.G. 1980. Late Miocene origin of the Benguela upwelling system off northern Namibia. *Science*, 208: 283–285.

Smith, R.A. 1984. *The lithostratigraphy of the Karoo Supergroup in Botswana*. Botswana Geological Survey Bulletin, 26.

Stanistreet, I.G. and McCarthy, T.S. 1993. The Okavango Fan and the classification of subaerial fan systems. *Sedimentary Geology*, 85: 115–133.

Summerfield, M.A. 1982. Distribution, nature and probable genesis of silcrete in arid and semi-arid southern Africa. In *Aridic soils and geomorphic processes*. Catena, Supplement, 1: 37–65.

Summerfield, M.A. 1983. Silcrete as a palaeoclimatic indicator: evidence from southern Africa. *Palaeogeography, Palaeoclimatology, Palaeoecology*, 41: 65–79.

Summerfield, M.A. 1985. Plate tectonics and landscape development on the African continent. In M. Morisawa and J.T. Hack (eds), *Tectonic geomorphology*. Binghampton Symposia in Geomorphology, International Series No. 15. Allen & Unwin, Boston: 27–51.

Talbot, M.R. 1980. Environmental responses to climatic change in the West African Sahel over the past 20,000 years. In M.A.J. Williams and H. Faure (eds), *The Sahara and the Nile*. Balkema, Rotterdam: 37–62.

Thomas, D.S.G. 1984. Ancient ergs of the former arid zones of Zimbabwe, Zambia and Angola. *Transactions of the Institute of British Geographers*, NS, 9: 75–88.

Thomas, D.S.G. 1988a. The nature and depositional setting of arid to semi-arid Kalahari sediments, southern Africa. *Journal of Arid Environments*, 14: 17–26.

Thomas, D.S.G. 1988b. Analysis of linear dune sediment-form relationships in the Kalahari Dune Desert. *Earth Surface Processes and Landforms*, 13: 545–553.

Thomas, D.S.G. and Martin, H.E. 1987. Grain-size characteristics of linear dunes in the southwestern Kalahari — discussion. *Journal of Sedimentary Petrology*, 57: 231–242.

Thomas, D.S.G. and Shaw, P.A. 1988. Late Cainozoic drainage evolution in the Zambezi Basin: geomorphological evidence from the Kalahari Rim. *Journal of African Earth Sciences*, 7: 611–618.

Thomas, D.S.G. and Shaw, P.A. 1991a. *The Kalahari environment*. Cambridge University Press, Cambridge.

Thomas, D.S.G. and Shaw, P.A. 1991b. 'Relict' desert dune systems: interpretations and problems. *Journal of Arid Environments*, 20: 1–14.

Torquato, J.R. 1972. Origin and evolution of the Mocamedes Desert (Angola). In *African geology: Quaternary rocks and geomorphology of Angola, Chad, Cote d'Ivoire, Nigeria and Sahara*. Department of Geography, University of Ibadan.

Tyson, P.D. 1986. *Climatic change and variability in Southern Africa*. Oxford University Press, Cape Town.

UNEP 1992. *World atlas of desertification*. Edward Arnold, London.

Van Rooyen, T.H. and Burger, R. Du T. 1973. Physiographic features of the central Orange River basin, with a note on pan formation. *South African Geographer*, 4: 218–227.

van Zinderen Bakker, E.M. 1980. Palynological evidence for late Cenozoic arid conditions along the Namibia coast from holes 532 and 530A, Leg 75, Deep Sea Drilling Project. In W.W. Hey (ed.), *Initial Reports of the Deep Sea Drilling Project*, Volume LXXV: 763–768.

van Zinderen Bakker, E.M. 1984. Aridity along the Namibian coast. *Palaeoecology of Africa*, 16: 149–152.

Van Zyl, J.A. 1992. The major landform regions of Namibia. *South African Geographer*, 19: 76–90.

Verster, E., van Deventer, P.W. and Ellis, F. 1992. Soils and associated materials of some pan floors and margins in southern Africa: a review. *South African Geographer*, 19: 35–47.

Ward, J.D. 1982. Aspects of a suite of Quaternary conglomeratic sediments in the Kuiseb Valley, Namibia. *Palaeoecology of Africa*, 15: 211–216.

Ward, J.D., Seely, M.K. and Lancaster, N. 1983. On the antiquity of the Namib. *South African Journal of Science*, 79: 175–183.

Watts, N.L. 1980. Quaternary pedogenic calcretes from the Kalahari (southern Africa): mineralogy, genesis and diagenesis. *Sedimentology*, 27: 661–686.

Wellington, J.H. 1955. *Southern Africa, a geographi-*

cal study, *Volume 1: Physical Geography*. Cambridge University Press, Cambridge.

Werger, M.J.A. 1978. Biogeographical division of southern Africa. In W. Werger (ed.), *Biogeography and ecology of Southern Africa*. Junk, The Hague: 145–170.

Wiggs, G.F.S., Livingstone, I., Thomas, D.S.G. and Bullard, J.E. 1994. The effect of vegetation removal on airflow structure and dune mobility in the southwest Kalahari. *Land Degradation and Rehabilitation*, 5: 53–71.

Wilkinson, M.J. 1988. The Tumas Sandstone Formation of the central Namib Desert; palaeoenvironmental implications. *Palaeoecology of Africa*, 19: 139–150.

Williams, M.A.J. 1993. Cenozoic climatic changes in deserts: a synthesis. In A.D. Abrahams and A.J. Parsons (eds), *Geomorphology of desert environments*. Chapman & Hall, London: 644–670.

Williams, M.A.J. and Adamson, D.A. 1980. Late Quaternary depositional history of the Blue and White Nile rivers in central Sudan. In M.A.J. Williams and H. Faure (eds), *The Sahara and the Nile*. Balkema, Rotterdam: 281–304.

Williamson, D., Taieb, M., Damanti, B., Icole, M. and Thouveny, N. 1993. Equatorial extension of the younger Dryas event: rock magnetic evidence from Lake Magadi (Kenya). *Global and Planetary Change*, 7: 235–242.

Wilson, I.G. 1973. Ergs. *Sedimentary Geology*, 10: 77–106.

Wise, S.M., Thornes, J.B. and Gilman, A. 1982. How old are the badlands? A case study from southeast Spain. In R. Bryan and A. Yair (eds), *Badland geomorphology and piping*. GeoBooks, Norwich: 259–278.

Woodward, J. (ed.) 1995. *Mediterranean Quaternary river environments*. Balkema, Rotterdam.

21
Asia

Edward Derbyshire and Andrew S. Goudie

Introduction

The drylands of central, east and south Asia extend across an area that represents a sizeable component of the Earth's arid regions (see Figure 1.1). They include sandy and stony ('gobi') deserts, sandy steppe, loessic steppe and scrub-woodland environments in a wide range of geomorphological and tectonic settings, from the Turpan Depression over 150 m below sea level to the desertic slopes of the Kunlun and Karakoram mountains and the eastern part of the Tibetan Plateau at altitudes of more than 5000 m (Figure 21.1). The Taklamakan and Junggar, the Karakum in Uzbekistan and the Thar in India are warm deserts, all having mean July temperatures of more than 24°C. In most of the Chinese deserts, in contrast, winter temperatures are mainly below −8°C with extensive areas below −12°C. Most of Asia's deserts lie at high altitudes, only the Thar, the Karakum and the small drylands of northeastern China lying below 1000 m above the sea. The great Plateau of Qinghai-Xizang (Tibet), with a surface around 5000 m above sea level, is a cold desert.

The geomorphological, sedimentary and ecological diversity of this extensive region is strongly influenced by a geological structure derived from its distinctive tectonic history. The region consists of a series of fold- and fault-bounded basins and mountain ranges, including the high-altitude plateau and range complex making up Tibet (Meyerhoff *et al.* 1991). The origins of this structural province lie in the collision of the Indian and Eurasian continental plates initiated between 54 and 49 Ma ago (Searle *et al.* 1987; Beck *et al.* 1995) and still continuing at a rate of between 4 and 5 cm a^{-1} (Patriat and Achache 1984). This continental–continental type of collision produced folding and thrusting of the Indian crust in the Himalaya and southern Tibet, and the homogeneous deformation of Tibet, involving shortening and doubling of the crustal thickness, and elevated the Himalayas and the Tibetan Plateau to an average altitude of between 4000 and 5000 m (Molnar 1988; Zhao *et al.* 1993). The buoyancy of the thickened crust was not sustained, however, and the Tibetan Plateau has spread under its own weight (e.g. England and Houseman 1988; Sandiford and Powell 1990; Zhou and Sandiford 1992; Turner *et al.* 1993). It is estimated that east–west extension on relatively young normal faults has averaged *c.* 1% Ma (Armijo *et al.* 1986). This process has added a number of 'pull apart' basins to those of compressional type (such as the Gonghe and

Arid Zone Geomorphology: Process, Form and Change in Drylands, 2nd edition. Edited by David S. G. Thomas.

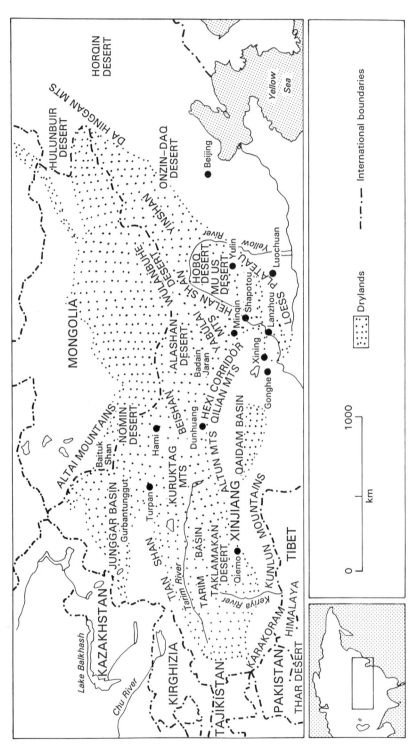

Figure 21.1   *Map of the deserts of central and eastern Asia*

Xining basins and the Hexi Corridor), and the large, relict marine basins underlain by oceanic lithosphere (including the Tarim, Junggar and North Caspian). Uplift of the Himalayas and Tibet, with the Tibetan Plateau perhaps reaching its present mean altitude as early as 14 Ma ago, is widely considered to have been critical to the development of the Asian monsoon (Molnar *et al.* 1993). Over much of the region, cold and dry winds from between north and west dominate under the influence of the Mongolian–Siberian high-pressure system. In summer, warm and moist oceanic air is drawn into the heart of the continent from the Indian Ocean and the South China Sea (Zhang and Lin 1992). Uplift of the Tibetan Plateau, the Himalayas and adjacent mountain ranges has also diversified the climates: the rise of the Himalaya has progressively impeded the ingress of moisture into central Asia. Thus, the distribution of drylands in Asia coincides closely with internal drainage regions. The considerable morphodynamic energy provided by the continuing tectonic evolution of Asia is expressed in high erosional potentials and very high rates of sediment production. Generation of sand and silt grains is favoured by direct (tectonic) crushing, glacier grinding, freeze–thaw action, salt weathering and cyclic hydration. These mater-

ials are reworked and exported from the region by some of the world's greatest rivers, including the Yellow, Yangtze, Indus, Ganges, Syr Darya and Amu Darya. The juxtaposition of high mountains and plateaux with frost action and glaciers, deep desert basins, thick mantles of wind-deposited silts (loess), and great rivers has produced a distinctive geomorphological assemblage and sedimentary accumulations containing very high-resolution records of climate change for the past 2 to 3 million years.

The drylands of China

Drylands occupy about 30% of the land area of China. Arid and semi-arid conditions in the northwest of the country appear to have been initiated in late Cretaceous–early Tertiary times (e.g. Zhao and Xing 1984), but estimates of the antiquity of the present pattern and environmental conditions of the Chinese drylands is contentious. The majority view in China is that aridification was closely linked to the uplift of the Tibetan Plateau, which had an altitude as low as 1000 m as late as the Pliocene. It is argued that violent uplift in the late Pliocene–early Quaternary established the essential characteristics of the deserts and older loess deposits of northern

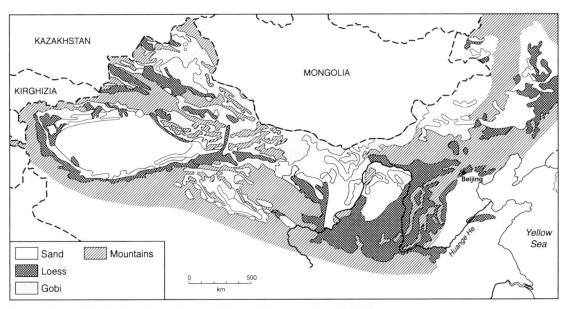

Figure 21.2 *Distribution of rock desert, gobi, sandy desert and loess in China.*

China, and that the Tibetan Plateau attained an altitude of 4000 m only as recently as the late Pleistoce–early Holocene (e.g. Li *et al.* 1979; Zhao and Xing 1984; Zhao 1986). In contrast, there is increasing support for the view that the Tibetan Plateau reached its present mean altitude 14–15 Ma ago, suggesting an element of circular reasoning in the earlier literature (cf. Searle 1995).

Southeastwards across western and central China, rocky and stony desert grades into sandy desert, then into sandy loess and, in turn, into thick loess with progression from arid, through semi-arid to subhumid climatic regimes (Figure 21.2). Much of the extensive gobi ('*gobi tan*': stony plain) in Mongolia and to the west of it (Dong *et al.* 1991) is essentially the product of huge alluvial fans deposited by violent floods at the foot of mountain ranges such as the Kunlun Shan, Qilian Shan and Tian Shan, all of which are seismically active and undergoing active uplift. Rapid morphoclimatic transitions characterise the slopes of such high ranges, arid plains and dune zones passing rapidly into desertic piedmonts and gorges which abut directly against the periglacial zone above (Figure 21.3).

The dune sands in the eastern deserts of China (eastern Inner Mongolia, Mu Us (Ordos),

Hobq, Wulanbuhe and Tengger: precipitation 200–400 mm a^{-1}) are generally stabilised or at least partly stabilised by vegetation under the present climate. However, the moist southeasterly monsoon rarely reaches the regions west of the Tengger Desert. Mean annual rainfall is generally less than 100 mm and mobile sand dunes are common. Although also very dry with only 100–200 mm annual rainfall, the Junggar (Gurbantunggut) Desert to the north of the Tian Shan is influenced by the westerlies and receives rainfall throughout the year. The aeolian sands of the cool, mid-latitude deserts of China are rather finer-grained than those of the subtropical deserts, a difference attributed by Yang (1991) to the greater exposure to frost of the central Asian deserts.

TAKLAMAKAN

The Taklamakan is the largest desert in China (Figure 21.4). It occupies an area of 337 600 km^2, mainly within the Tarim Basin but also extending into the intermontane basins of Turpan and Hami. It is warm temperate in type, with a mean annual precipitation of less than 90 mm throughout, values being as low as 10 mm in the centre. Alluvial fans and gravel aprons many tens of kilometres long are found over considerable

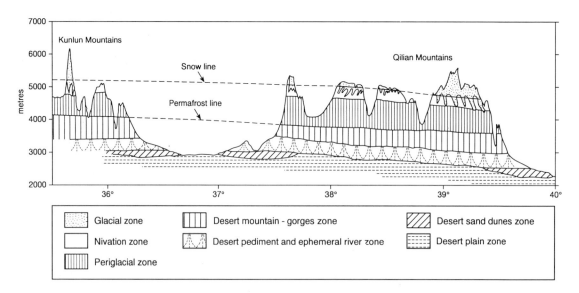

Figure 21.3 *Profile from Kunlun Pass to Yumen showing present-day morphoclimatic provinces across the eastern Qaidam Basin (after Wang and Derbyshire 1987)*

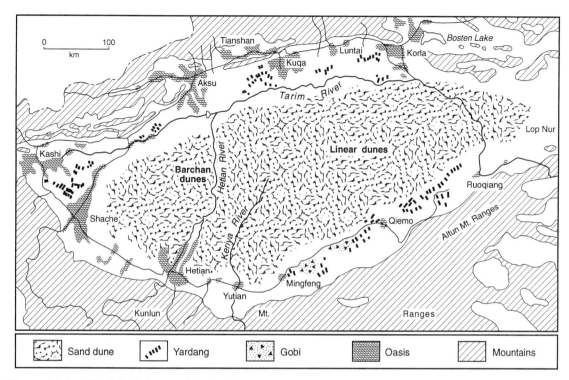

Figure 21.4 *Morphological types of the Taklamakan Desert (after Zhu 1984)*

lengths of the piedmont of the Tian Shan to the north, and particularly along the active front of the Kunlun Mountains to the south at altitudes of 1400–2000 m. The aeolian deposits show a broadly concentric sorting, the mountain slopes all around the basin being almost completely mantled by loess (up to 1 m thick) laid down during almost daily dust storms blowing out of the desert (Hövermann and Hövermann 1991). Below about 1400 m, the landscape of the Taklamakan is dominated by dune sands and gobi. The mineralogy of the sands suggest that all are derived from reworking of underlying alluvial and lacustrine deposits. Longitudinal, barchan and transverse dune types are all present, all three including complex varieties (Zhu and Lu 1991). Complex transverse and longitudinal forms are the commonest, the former generally being 5–15 km long and 80–200 m high, and the latter 10–20 km and 50–80 m, respectively. Barchan dunes occur mainly as single features and average 20 to 40 m in height. Horn-shaped dunes up to 100 m high also occur. Dune mobility is greatest in the southwest and southeast (*c.* 10 m a^{-1}) and the east

(5–10 m a^{-1}), and slowest (*c.* 1 m a^{-1}) on the large dunes in the central area of the desert (Zhu 1984). The oldest dune complex (>20 000 a BP) covers a large area of the eastern part of the basin. There is landform and sedimentary evidence indicating that fluvial and lacustrine environments were more extensive in the recent geological past. Many depressions contain ancient lacustrine deposits now undergoing strong deflation, with yardang development. At least two well-developed raised shorelines are widely evident. The Quaternary deposits underlying the upper Tarim River are over 400 m thick. Several fan-deltas, that associated with the Tarim River being up to 200 km wide, indicate much higher discharges from streams emerging from the Kunlun Mountains at several stages during the Quaternary. Two wide terraces, each with an aeolian mantle, 20 m and 10 m above the present channel, have been described from the Keriya River valley (Hövermann and Hövermann 1991). It has been estimated that more than one-quarter of the area of this desert shows evidence of general flooding up to 7500 years ago (Jäkel and Zhu 1991). There is both written and map evidence to suggest that the Keriya and Tarim rivers joined and crossed the whole extent of the Taklamakan about 2000 years ago, and again during the sixteenth and the early nineteenth centuries. Abandoned towns on the lower Keriya River have yielded remains with ^{14}C ages of between 2135 and 2684 BP (during the Western Han Dynasty, *c.* 150–250 BC) for a prosperous area which, in the words of the ancient chronicle, was part of the '… ancient Yumi Kingdom with 3340 households and a total population of 20 040 … and 3540 soldiers …' Maps of Xinjiang from the late Qing Dynasty (AD 1644–1911) show that, although by then only a seasonal stream, the Keriya still crossed the Taklamakan periodically (Chen 1991).

Vast areas of woodland (Tamarisk and Populus *spp.*) disappeared from the margins of the Tarim Basin in the Holocene as a result of depression of groundwater tables by human action (Tian 1991). Oases formerly existed more than 200 km north of the Kunlun front, but all modern oases are peripheral. Overrunning of oases by blown sand is well-documented in the Taklamakan, especially in the south. Records show that some important towns along the Old Silk Road (passing through Yutian) have been lost, the abandonment of Qiemo having been accelerated by its sacking during warfare in AD 636. Loulan (west of Lop Nor) had over 1600 inhabitants in the Han Dynasty (206 BC–AD 220), but was abandoned in AD 376 and is now covered by dunes and yardangs. Similarly, Niya (Jingyue) reached a population of 3000 but was abandoned in the eight-century and is now covered with barchans 3–5 m high.

The Turpan–Hami basin complex, some 500 km long from west to east, is separated from the Tarim Basin by the Kuruktag Mountains. The lowest part of the Turpan Basin, occupied by the Ayding salt lake, is 155 m below sea level: summer temperatures are the highest in China and mean annual precipitation is only 3.9 mm. The dominant surface cover throughout the Turpan–Hami depression is gobi.

HEXI CORRIDOR

The Hexi Corridor runs for about 1000 km WNW to ESE from the eastern end of the Tarim Basin to a point on the Yellow River about 125 km north of the city of Lanzhou. It is defined by the edge of the Mongolian Plateau to the north and the Qilian Mountains to the south. The part west of the town of Jaiyuguan (at the western end of the Great Wall) is a warm temperate desert with rainfall generally less than 50 mm a^{-1}. Large piedmont fans and residual gravels make up an extensive gobi landscape, with some well-developed yardangs on the finer-grained lithologies. Individual sand dunes locally reach heights of 100 m, e.g. at Dunhuang. Further east, mean altitudes rise from 1000 m to about 1500 m, mean temperatures declining and precipitation increasing. Some of the fans and associated alluvial deposits laid down by rivers from the Qilian Mountains to the south provide a surface dominated by gobi, but sand dunes are also present. The extensive piedmont fans filter the snow and ice meltwater streams from the Qilian slopes. The water emerges on the lower fan slopes to provide extensive oases of considerable historical and present-day importance. There are substantial subsurface water and oil resources.

Many of the ancient cities in the Hexi Corridor west of Jaiyuguan, sustained by meltwaters from the snow and glacier ice of the Qilian Mountains, have been overwhelmed by aeolian sediments because destruction of woodland on the lower mountain slopes raised the levels of stream-beds, made flow regimes more flashy and enhanced infiltration (Wang *et al.* 1991). Human settlement began in the lower Ruo Shui River, northeast of Jaiyuguan, in AD 102. Large-scale settlement and reclamation works were undertaken soon afterwards and further settlements added in the upper delta area between the ninth and early fourteenth centuries. However, this latter occupation was sustained only until AD 1359 and, by the middle of the Ming Dynasty (AD 1550), population greatly diminished as the important town of Heicheng was abandoned. This region was rendered a desert from about this period and the original Han oasis of Dunhuang is now threatened by dunes 50–150 m high, the modern oasis being 6 km from the original site.

INNER MONGOLIA

The mountain and plateau desert of Inner Mongolia extends northwards from the Hexi

Corridor, stretching from the Bei Shan in the west to the Helan Shan and the Yellow River in the east. This region contains the renowned 'Black Gobi' of the eastern Bei Shan, one of the most desolate arid landscapes in China. The eastern half of this region contains the mobile sand desert of Badain Jaran, over 40 000 km² in area. Its northern margins are made up of little-eroded, ancient pediments (Hövermann 1985). These give way southeastwards to extensive desert plains and dune fields (with dunes up to 300 m high) mantling a subdued mountain landscape, the strong winter northwesterlies transporting dune sand over passes along the crest of the Yabulai Mountains at altitudes between 1600 and 2200 m. Permanent interdune lakes, fed by groundwater, reach 1 km² in area. Southeastward of the dunefields, plains of sand and gobi cover an area of more than 1000 km². There are many relict lake beds now undergoing severe deflation to produce impressive yardang landscapes. Moister climatic conditions during which the dunes were at least partly fixed have been suggested on the basis of the remains of plant roots. U/Th and ¹⁴C dating indicates five such periods: 207 000 ± 10 000, 31 750 ± 485, 19 100 ± 770, 9435 ± 345 and 2070 ± 100 BP (Yang 1991). Two endoreic lakes investigated by Pachur *et al.* (1995), Gaxun Nur and Sogu Nur, provide complementary evidence. A shoreline 30 m above the present floor of Gaxun Nur provided algal mats dated by ¹⁴C at 29 400 ± 450 BP, and mollusc-bearing shoreline deposits 22 m above the Sogu Nur yielded a ¹⁴C age of 34 010 + 1540/−1290 BP. The slightly smaller Tengger Desert extends eastwards from the Great Wall at Minqin to the Helan Mountains, its southern limit being the Yellow River near Shapotou. Freshwater lakes appear to have been widespread in the Tengger Desert during the late Pleistocene, indicating a subhumid climate without any accompanying depression of temperatures to arctic values. The highest of six shorelines found around Baijian Hu (Hu = lake) has been ¹⁴C dated at 39 000 BP, with Ostracoda and geochemistry indicating that cool and humid conditions persisted until about 23 000 BP (Pachur *et al.* 1995).

Much of that part of the Inner Mongolian Plateau between the Yinshan–Da Hinggan Mountains and the international border is a desert steppe with precipitation less than 250 mm. Grassy plains alternate with dry steppe and surfaces of deflation. The Onzin–Daq and Horqin deserts lie on the western and eastern slopes, respectively, of the Da Hinggan Mountains, the Horqin Desert extending into the northeastern provinces of Liaoning and Heilongjiang. There are many dunefields but, where undisturbed by human action, most dunes are fixed by vegetation. The northernmost sandy land in China lies between 48 and 50°N within the great grasslands of Hulun Buir in the far northeast.

JUNGGAR

The Junggar Basin lies between the Tian Shan and Altai mountains in northernmost Xinjiang Province. Extending to 48°N, this temperate desert is significantly influenced by the Westerlies. Mean annual precipitation ranges from over 300 mm in the west to less than 50 mm in the east, and includes significant amounts of winter snow. With an area of about 175 000 km², the Junggar Basin is characterised by extensive gobi, particularly developed on the abundant peripheral piedmont fan gravels, and the mobile sand desert of Gurbantunggut in the centre of the basin and covering more than 70 000 km² (Figure 21.5). The belt of finer-grained sediments between the mobile sand and the northern Tian Shan piedmont fans is quite intensively settled and farmed. The western margins of the Junggar, along the Russian border, are transitional to desert steppe. East of the Baytik Mountains, on the Mongolian border, is the very dry Nomin Desert (mean annual precipitation 12.5 mm). This is a barren gobi landscape with well-developed rock varnished surfaces.

ORDOS

The Ordos Plateau, lying north of the Great Wall within the big bend of the Yellow River, has an altitudinal range between 1000 and 2000 m. Desert covers an area of about 85 000 km² (Figure 21.6). The northern sector, known as the Hobq Desert, contains extensive areas of mobile sands. To the south of it is the more semi-arid Mu Us Desert containing much

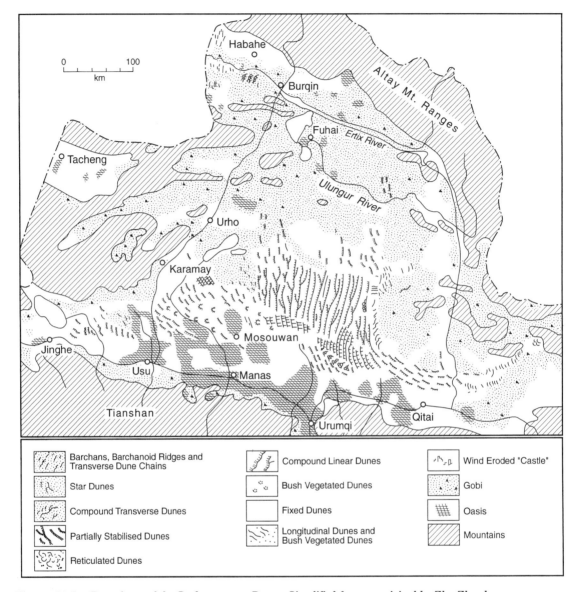

Figure 21.5   *Dune forms of the Gurbantunggut Desert. Simplified from an original by Zhu Zhenda*

fixed and semi-fixed dune sand as well as about 8000 km² of mobile dunes. Annual precipitation is less than 200 mm on the western margins, but rises to just over 400 mm in the east, and up to 80% of this may be concentrated in a period of only two months (July and August) in some years. Winters are dry and cold (January −13°C) under the influence of the Siberian high-pressure cell, and annual potential evapotranspiration is high (1000–3000 mm a⁻¹).

The Mu Us Sandy Land grades southeastwards into the sandy loess and then into the Huangtu Gaoyuan, or Loess Plateau of China. Where best developed in the central part of the Loess Plateau, this major sedimentary accumulation exceeds a thickness of 150 m and consists of alternating loess units and palaeosols indicating changes in climate between cold and dry of the glacial periods and warmer and moister periods coinciding with the interglacials. At

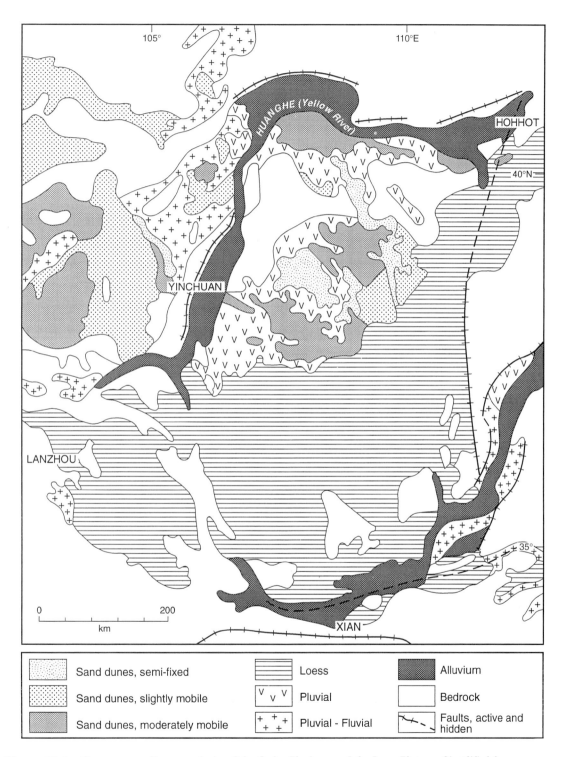

Figure 21.6 *Quaternary sedimentary facies of the Ordos Platform and the Loess Plateau. Simplified from 'Geomorphological Map of China and its Adjacent Area': (scale 1:4M) attached to Chen (1993)*

Luochuan, in northern Shaanxi Province, sandy loess and fine sand units indicate advances of the margins of the Mu Us which, by analogy with present conditions, may have totalled as much as 500 km. One of the more prolonged of such desert steppe periods in the Luochuan region (latitude 38°50'N) has been dated at around 77–87 ka BP (Liu *et al.* 1985). Seven units of aeolian sand, similar in their properties to the present dunes, indicate seven advances of the desert margin at Yulin, on the Mu Us–Loess Plateau margin, since about half a million years ago (T.S. Liu and Z.L. Ding, personal communication 1993).

The southern half of the Mu Us dryland and the Loess Plateau are important regions historically, strategically and economically. There is abundant evidence of environmental degradation arising from the human activity of the past 5000 years. The grassland vegetation in the Mu Us has attracted increasing numbers of both pastoralists and farmers especially over the past millennium and a half. The Chinese written records make it clear that, although there was no desert in the southern Mu Us during the fifth century AD ('everywhere is full of fresh water and streams are limpid') by the ninth century 'mighty sand dunes climbed over the city walls'. By the tenth century the substantial town of Tongwan had been buried by the advancing sand dunes. Comparable pressures continue to affect this region. The area classified as desert ('*shamo*': sandy wasteland) around the city of Yulin (northern Shaanxi Province) expanded by 1.7 times, and the desert margin encroached between 3 and 10 km, in the 20 years 1957–77, a period that witnessed a doubling of the population. Historical records suggest that the desert margin near Yulin advanced southwards between 30 and 40 km in the 150 year period 1830–1980, a rate approximately double that inferred from the Quaternary sedimentary record.

QINGHAI–XIZANG

The Qinghai-Xizang (Tibet) Plateau and its associated mountain ranges contain a number of cold and arid regions, a large part of the

Figure 21.7 *Shorelines of the diminishing Da Hai Lian Lake (2910 m), c. 40 km west of Gonghe town, Gonghe Basin, northeast Tibet*

northwestern plateau receiving less than 200 mm precipitation per year. The many basins between the Himalaya–Karakoram to the south and the Western Kunlun Mountains to the north are the sites of present-day or Pleistocene lakes that provide a record of relatively rapid changes from extremely dry to semi-arid or subhumid conditions (Gasse *et al.* 1991). At present, these basins are arid, their surfaces made up of gobi, extensive lacustrine plains and terraces, and denuded piedmont deposits. Similar landscapes are common in many other parts of the plateau. Although only semi-arid, for example, salt lakes with numerous raised shorelines, and complex barchan dune series with dune-dammed lakes, occur in the Gonghe Basin and elsewhere in northeastern Tibet (Figures 21.7 and 21.8).

The largest desert region within Tibet includes the Qaidam Basin and parts of its surrounding mountains including the Altun, western Qilian Shan and northern slopes of the central Kunlun Mountains. Annual rainfall totals are everywhere below 200 mm, and are as low as 10 mm in parts of the western Qaidam Basin. This is a cool desert: summer temperatures rarely exceed 18°C, and minima below −30°C are common in the long winters. The Qaidam Basin shows a broadly concentric zonation of landforms and surficial sediments: arid and semi-arid mountain gorges with patches of loess; pediments and alluvial fans with gobi, isolated sand dunes and thin loess; sand dune fields, including some of the highest linear dunes in the world (3200 m); and sand, silt and clay plains with salt lakes. Although the Quaternary deposits are known to be of the order of 2 km in thickness, their potential as a source of information on climatic change has yet to be fully realised (Wang *et al.* 1986; Derbyshire *et al.* 1991). Alternating episodes of freshwater, peaty, brackish and salt water conditions are known and have been crudely dated for the past 350 000 years at Dabusan Lake. As appears to be the case in the great loess record to the east, the sedimentary record

Figure 21.8 *Late Pleistocene and Holocene barchans with interdune ponds and sand-dammed lakes at 4200 m on the northeast part of the Tibetan Plateau. Mian Sha Ling (34°44' N, 98°08' E), east of the town of Madoi*

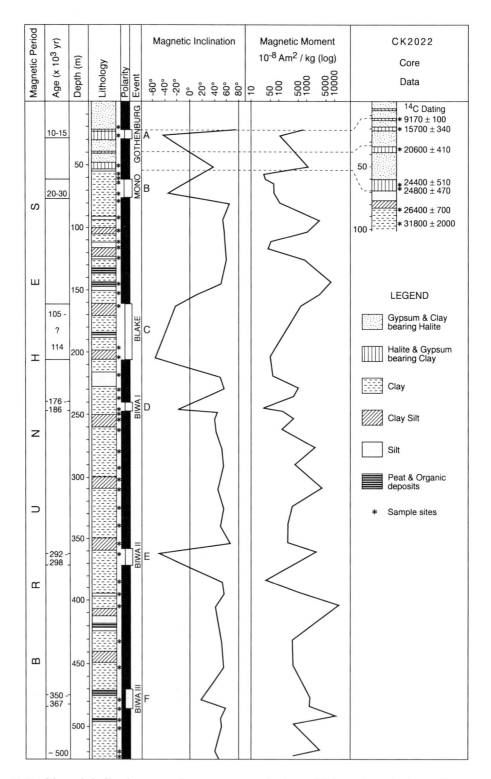

Figure 21.9 *Magnetic inclination, natural remanent magnetisation and lithostratigraphy for a sediment core from Dabusan Lake. The error on the inclination values is the 95% confidence interval (after Wang et al. 1986)*

here points to increasing desiccation since the Middle Pleistocene (Figure 21.9).

The deserts of India and Pakistan

The arid zone of the northwestern portion of the Indian subcontinent (Figure 21.10) stretches from the rocky mountains and basins of Baluchistan in the west to the ancient Aravalli Mountains in the east. The northern part in particular receives some winter rainfall from the westerlies while the summer southwest monsoon also produces some precipitation in the region so that only a small area has less then 100 mm of mean annual rainfall. This arid area is generally referred to as the Great Indian Desert or Thar (Singhvi and Kar 1992). It is, however, densely populated by humans, and in the Rajasthan portion of the desert the population density is 84 persons per square kilometre. Land degradation pressures are high. There is a striking contrast in relief between the arid foothills of the Karakorams in the north, the

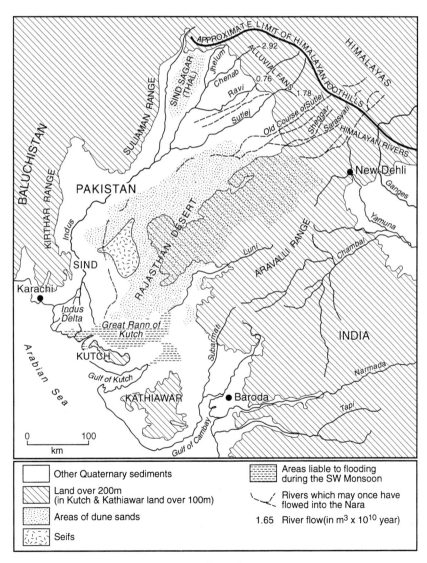

Figure 21.10 *The arid zone of the northwest of the Indian subcontinent*

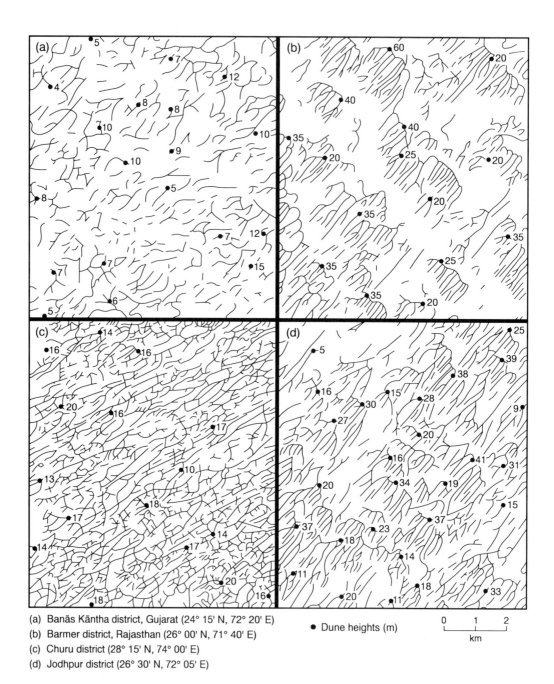

(a) Banãs Kãntha district, Gujarat (24° 15' N, 72° 20' E)
(b) Barmer district, Rajasthan (26° 00' N, 71° 40' E)
(c) Churu district (28° 15' N, 74° 00' E)
(d) Jodhpur district (26° 30' N, 72° 05' E)

● Dune heights (m)

Figure 21.11 *Characteristic types of Thar sand dunes. (a) Banas Kantha District (24°15'N, 72°20'E). (b) Barmer District (26°N, 71°40'E). (c) Churu District (28°15'N, 74°E). (d) Jodhpur District (26°30'N, 72°05'E). Derived from 1 : 50 000 Survey of India toposheets*

enormous alluvial plain and delta of the Indus River and the siliciclastic sabkhas of the Rann of Kutch in the south. The perennial Indus and its snow-fed tributaries dominate the area (Shroder 1993), and in the past these rivers have shown dramatic shifts in their courses (Wilhelmy 1969; Higgins *et al.* 1973).

Sand dunes are widespread in the Thar to the east of the Indus and also in the Thal Desert on the Sind Sagar doab between the Indus and the Chenab. The coastline of the Arabian Sea, the large expanses of alluvial plains, and the weathering of outcrops of sandstones and granites provide a plentiful source of sand which is then moulded predominantly by southwesterly winds. It is not, however, a high-energy aeolian environment. The dunes near the coast are highly calcareous and may be cemented to give aeolianites (calcarenites), locally called *miliolite*. Further inland they are dominantly quartzite, though even far inland the dunes contain foraminiferal tests (Goudie and Sperling 1977). The general pattern of the dunes has been mapped from Landsat imagery by Breed *et al.* (1979). This demonstrates that, uniquely for a large erg, clustered rake-like parabolics are a dominant dune form (Figure 21.11). In the south and east of the region many of the dunes are relicts and their palaeoclimatic significance has been analysed by Goudie *et al.* (1973) and by Wasson *et al.* (1983).

The arid zones of the Indian subcontinent are also zones of dust mobilisation and of loess deposition. The loess deposits are particularly extensive in Kashmir and on the Potwar Plateau near Islamabad (Rendell 1989), but loess also occurs in late Pleistocene river terrace deposits in Gujarat and in the Allahabad region of Uttar Pradesh (Williams and Clarke 1984). High aeolian dust loadings are present in late Pleistocene ocean core sections in the Indian Ocean (Kolla and Biscaye 1977), and indicate formerly more intense wind activity.

Indeed, ocean cores can be used to determine the probable Holocene and late Pleistocene climates of the region (see, for example, Sarkar *et al.* 1990). Duplessy (1982), working on the oxygen isotopic composition of planktonic foraminifera, found that the rain-bearing southwest summer monsoon was weaker at the Last Glacial Maximum than today. He also maintained that the Ganga and Brahmaputra rivers contributed much less water to the ocean at that time, since the salinity gradient in the Bay of Bengal was very much steeper than today. Ocean core evidence also suggests that the flux of terrigenous materials derived from the Indus was significantly lower during glacial periods than in the Holocene, indicating a reduced flow (Naidu 1991). This agrees with the climatic inferences gained from ancient dunes, miliolite deposits and lake stratigraphy (Allchin *et al.* 1977). Reviews of terrestrial stratigraphy for the Quaternary are provided by Chamyl and Mehr (1995) for Gujarat, and by Agrawal *et al.* (1985) for Kashmir.

Closed basins and their associated salt deposits are another important landform type in the desert. Some of these basins, called *dhands*, are deflational features blown out of the noses of parabolic dunes. Others may result from the blocking of drainage by sand migration (as in the case of the Sambhar Lake of Rajasthan), while others, such as the Jaisalmer and Pokaran Ranns may be related to faulting (Pandey and Chatterji 1970). It is unlikely on geochemical grounds, however, that the salt lakes are the remnants of the Tethys Sea (Ramesh *et al.* 1993). Some of the lakes have an interesting late Pleistocene and Holocene stratigraphy with alternations of aeolian, freshwater and hyper-saline sediments (see, for example, Singh *et al.* 1972; Wasson *et al.* 1984).

The best record of climatic change comes from the Didwana salt lake in Rajasthan (Singh *et al.* 1990). Hyper-arid conditions are evident from 18 000 to 13 000 years BP. Treeless steppe was prevalent, halite was deposited, and the power of the southwest monsoon was greatly reduced. Around about 13 000 years BP a change occurred to a shrub savanna grassland, and in the mid-Holocene, between *c.* 7500 and 6000 years BP, a rainfall maximum occurred when rainfall values may have been twice their present figure. Precipitation fell to present levels at or soon after 4200 BP. These fluctuations can be correlated with those known from other monsoonal environments (Gasse and Van Campo 1994).

The age of the Thar Desert is far from certain. Although many authors formerly argued that the desert is only of Holocene age, being the product of alleged post-glacial desiccation (see Meher-Homji 1973 for a discussion), more recent studies of the age of sand dunes and of the strati-

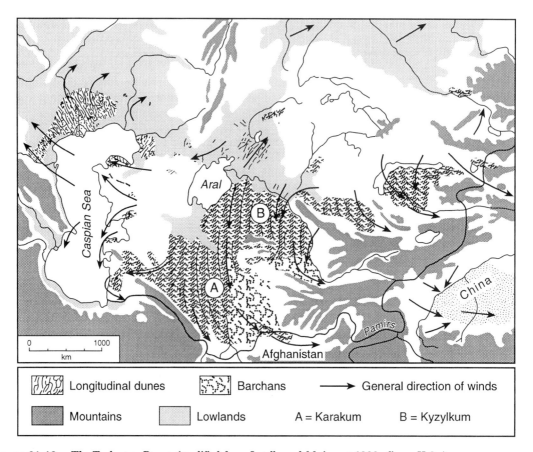

Figure 21.12   *The Turkestan Desert (modified from Letolle and Mainguet 1993, figure II.3a)*

graphy of lake basins, rules this out. The desert was more extensive in late Pleistocene times than it is today. The long-term climate history of the region is undoubtedly related intimately to the uplift history of the Karakorams, Himalaya and Tibetan Plateau in the Late Cenozoic.

The deserts of Central Asia

The Turkestan Desert (Figure 21.12), which lies between 36°N and 48°N and between 50°E and 83°E, is bounded on the west by the Caspian Sea, on the south by the mountains bordering Iran and Afghanistan, on the east by the mountains bordering Xinjiang, and on the north by the Kirghiz Steppe. Two great ergs are included: the Karakum ('black sands') in the south and the Kyzylkum ('red sands') to the west of Tashkent. Some background material

on the deserts of the former Soviet Union is given in Petrov (1976).

One of the most striking features of the area, and one which it shares with China, is the development of very thick (up to 200 m) and complex loess deposits dating back to the Pliocene. They are well-displayed in both the Tajik and Uzbek Republics, and rates of deposition were very high in late Pleistocene times (Lazarenko 1984). The nature of soils and pollen grains preserved in the loess profiles suggest a progressive trend towards greater aridity through the Quaternary, and this may be related to progressive uplift of the Ghissar and Tian Shan mountains (see Davis *et al.* 1980).

Closed basins, some of enormous size, are developed in the area. They contained greatly expanded water bodies in pluvial times, and the largest transgression occurred during early glacial phases (Figure 21.13). This was partly

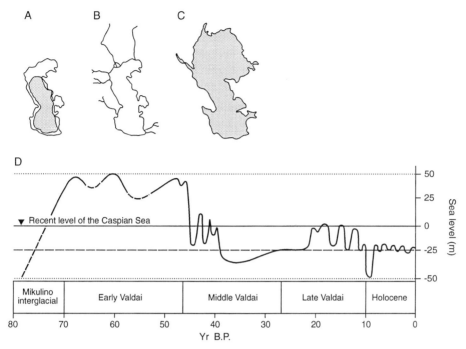

Figure 21.13 *The changing nature of the Caspian Sea. (A) The extent of the sea during the Mikulino interglacial, and (B) at the present day; (C) the greatly expanded sea during the Early Valdai glaciation; (D) the transgressions and regressions of the Caspian since the last interglacial (modified from Chepalyga 1984)*

because reduced temperatures caused less evaporative loss, partly because of inputs of glacial meltwater, and partly because the surface over large portions of their catchments was sealed by permafrost (Chelpalyga 1984). During the early Valdai glaciation the Aral and the Caspian were united, covered an area in excess of 1.1 million km², and extended some 1300 km up the Volga River from its present mouth. In recent years these two great inland seas have shown marked changes in their volume in response to rainfall fluctuations and anthropogenic impacts on their catchments. The Aral Sea is currently shrinking rapidly (Micklin 1992). Basins generated in part by aeolian deflation extend northwards from the true deserts of Kazakhstan and Uzbekistan into the steppes of West Siberia to the lee of the Urals. They appear to be comparable to those of the Argentinian Pampas, the High Plains of the USA or the interior of southern Africa (Goudie and Wells 1995).

The deserts of Afghanistan are intermediate between those of central Asia, the Middle East

and the Indian subcontinent. Much of the country is at a high altitude and is highly mountainous, but to the south and west of Kandahar and north and south of the Helmand River the relief is more gentle and there are, in the Registan Desert, some largely undescribed sand seas.

References

Agrawal, D.P., Kusumgar, S. and Krishnamurthy, R.V. 1985. *Climate and geology of Kashmir and Central Asia.* Today and Tomorrow, New Delhi.

Allchin, B.A., Goudie A.S. and Hegde, K.T.M. 1978. *The prehistory and palaeogeography of the Great Indian Desert.* Academic Press, London.

Armijo, R., Tapponnier, P., Mercier, J.L. and Han, T.L. 1986. Quaternary extension in southern Tibet: field observations and tectonic implications. *Journal of Geophysical Research,* 91: 13803–13872.

Beck, R.A., Burbank, D.W., Sercombe, W.J., Riley, G.W., Barndt, J.K., Berry, J.R., Afzal, J., Khan, A.M., Jurgen, H., Metje, J., Cheema, A., Shafique, N.A., Lawrence, R.D. and Khan, M.A. 1995. Stratigraphic evidence for an early collision between northwest India and Asia. *Nature*, 373: 55–58.

Breed, C.S., Fryberger, S.G., Andrews, S., McCauley, C., Lennartz, P., Gebel, D., Horstman, K. 1979. Regional studies of sand seas using Landsat (ERTS) imagery. *US Geological Survey, Professional Paper*, 1052: 305–397.

Chamyal, L. S. and Mehr, S.S. 1995. The Quaternary formations of Gujarat. *Memoirs Geological Society of India*, 32: 246–257.

Chen, H. 1991. The change of eco-environment and the rational utilization of water resources in the Keriya River valley. In D. Jäkel and Z. Zhu (eds), Reports on the 1986 Sino-German Kunlun Shan Taklimakan-Expedition. *Die Erde*, Erg. H 6: 133–147.

Chen, Z. (ed.) 1993. *An outline of China's geomorphology*. China Cartographic Publishing House, Beijing: 133 pp. and map supplement.

Chepalygya, A.L. 1984. Inland sea basins. In A. Velichko (ed.), *Late Quaternary environments of the Soviet Union*. Longman, Harlow: 229–247

Davis, R.S., Ranov, V.A. and Dodonov, A.E. 1980. Early man in Soviet Central Asia. *Scientific American*, 243: 91–102

Derbyshire, E. and Owen, L.A. (eds) 1989. Quaternary of the Karakoram and Himalaya. *Zeitschrift für Geomorphologie*, Supplementband, 76.

Derbyshire, E., Shi, Y., Li, J.J., Zheng, B.X., Li, S.J. and Wang, J.T. 1991. Quaternary glaciation of Tibet: the geological evidence. *Quaternary Science Reviews*, 10: 485–510.

Dong, G., Zhu, Z., Li, B. and Wang, G. 1991. On the origin and evolution of the modern Gobi-desert in North China. In T.S. Liu (ed.), *Quaternary geology and environment in China*. Science Press, Beijing: 34–40.

Duplessy, J.C. 1982. Glacial to interglacial contrasts in the Northern Indian Ocean. *Nature*, 295: 494–498.

England, P.C. and Houseman, G.A. 1988. The mechanics of the Tibetan Plateau. *Philosophical Transactions of the Royal Society of London*, 326: 301–320.

Gasse, F. and Van Campo, E.V. 1994. Abrupt postglacial climate events in west Asia and north Africa monsoon domains. *Earth and Planetary Science Letters*, 126: 435–456.

Gasse, F. and 13 co-authors 1991. A 13,000 year climate record from western Tibet. *Nature*, 353: 742–745.

Goudie, A.S. and Sperling, C.H.B. 1977. Long distance transport of foraminiferal tests by the wind in the Thar Desert, north-west India. *Journal of Sedimentary Petrology*, 47: 630–633.

Goudie, A.S. and Wells, G.L. 1995. The nature, distinction and formation of pans in arid zones *Earth Science Reviews*, 38: 1–69

Goudie, A.S., Allchin, B. and Hegde, K.T.M. 1973. The former extensions of the Great Indian Sand Desert. *Geographical Journal*, 139: 243–257.

Higgins, G.M., Mushtuq, A. and Brinkman, R. 1973. The Thal interfluve: geomorphology and depositional history. *Geologie en Mijnbouw*, 52: 147–155.

Hövermann, J. 1985. Das System der klimatischen Geomorphologie auf landschaftskundlicher Grundlage. *Zeitschrift für Geomorphologie*, Supplementband, 56: 143–153.

Hövermann, J. and Hövermann, E. 1991. Pleistocene and Holocene geomorphological features between the Kunlum Mountains and the Taklimakan desert. In D. Jäkel and Z. Zhu (eds), Reports on the 1986 Sino-German Kunlun Shan Taklimakan-Expedition. *Die Erde*, Erg. H 6: 35–50.

Jäkel, D. and Zhu, Z. 1991. Reports on the 1986 Sino-German Kunlun Shan Taklimakan-Expedition. *Die Erde*, Erg. H 6: 200 pp. and map volume.

Kolla, V. and Biscaye, P.C. 1977. Distribution and origin of quartz in the sediments of the Indian Ocean. *Journal of Sedimentary Petrology*, 47: 642–649.

Lazarenko, A.A. 1984. The loess of Central Asia. In A. Velichko (ed.), *Late Quaternary environments of the Soviet Union*. Longman, Harlow: 125–131.

Letolle, R. and Mainguet, M. 1993. *Aral*. Springer-Verlag, Paris.

Li Jijun, Wen Shixuan, Zhong Qingong, Wong Fubao, Zheng Benxing and Li Bingyuan 1979. A discussion on the period, amplitude and type of uplift of the Qinghai Xizang Plateau. *Scientia Sinica*, 22: 1314–1328.

Liu, T.S., An, Z.S., Yuan, B.Y. and Han, J.M. 1985. The loess paleosol sequence in China and climatic history. *Episodes*, 8: 21–28.

Meher-Homji, V.M. 1973. Is the Sind-Rajasthan Desert the result of a recent climatic change? *Geoforum*, 15: 47–57.

Meyerhoff, A.A., Kamen-Kaye, M., Chen, C. and Taner, I. 1991. *China — stratigraphy, palaeogeography and tectonics*. Kluwer, Dordrecht: 188 pp.

Micklin, P.P. 1992. The Aral crisis: introduction to the special issue. *Post-Soviet Geography*, 33: 269–282.

Molnar, P. 1988. A review on the geophysical constraints on the deep structure of the Tibetan Plateau, the Himalaya and the Karakoram, and their tectonic implications. *Philosophical Transactions of the Royal Society of London*, Series A, 326: 33–88.

Molnar, P., England, P. and Martinod, J. 1993. Mantle dynamics, uplift of the Tibetan Plateau, and the Indian monsoon. *Reviews of Geophysics*, 31: 357–396.

Naidu, P.D. 1991. Glacial to interglacial contrasts in

the calcium carbonate content and influence of Indus discharge in two eastern Arabian cores. *Palaeogeography, Palaeoclimatology, Palaeoecology*, 86: 255–63.

Pachur, H.-J., Wünnemann, B. and Zhang, H. 1995. Lake evolution in the Tengger desert, Northwestern China, during the last 140,000 years. *Quaternary Research*, 44: 171–180.

Pandey, S. and Chatterji, P.C. 1970. Genesis of 'Mitha Ranns', 'Kharia Rann', and 'Kanodurala Ranns' in the Great Indian Desert. *Annals of Arid Zone*, 9: 175–180.

Patriat, P. and Achache, J. 1984. India–Eurasia collision chronology and its implications for crustal shortening and driving mechanisms of plates. *Nature*, 311: 615–621.

Petrov, M.P. 1976. *Deserts of the world*. Wiley, New York.

Ramesh, R., Jani, R.A. and Bhushan, R. 1993. Stable isotope evidence for the origin of salt lakes in the Thar Desert. *Journal of Arid Environments*, 25: 117–123.

Rendell, H.M. 1989. Loess deposition during the late Pleistocene in northern Pakistan. *Zeitschrift für Geomorphologie*, Supplementband, 76: 247–255.

Sandiford, M. and Powell, R. 1990. Some isostatic and thermal consequences of the vertical strain geometry in convergent orogens. *Earth and Planetary Science Letters*, 98: 154–165.

Sarkar, A., Ramesh, R., Bhattacharya, S. K. and Rajagopalam, A.G. 1990. Oxygen isotope evidence for a stronger winter monsoon during the last glaciation. *Nature*, 343: 549–551.

Searle, M.P. 1995. The rise and fall of Tibet. *Nature*, 373: 17–18.

Searle, M.P., Windley, B.F., Coward, M.P., Cooper, D.J.W., Rex, A.J., Rex, D., Tingdong, L., Xuchang, X., Jan, M.Q., Thakur, V. and Kumar, S. 1987. The closing of Tethys and the tectonics of the Himalayas. *Bulletin of the Geological Society of America*, 98: 678–701.

Shroder, J. F. (ed.), 1993. *Himalaya to the sea*. Routledge, London.

Singh, G., Joshi, R.D. and Singh, A.P. 1972. Stratigraphic and radiocarbon evidence for the age and development of three salt lake deposits in Rajasthan, India. *Quaternary Research*, 2: 496–501.

Singh, G., Wasson, R.J. and Agrawal, D.P. 1990. Vegetational and seasonal climatic changes since the last full glacial in the Thar Desert, northwest India. *Review of Palaeobotany and Palynology*, 64: 351–358.

Singhvi, A.K. and Kar, A. (eds) 1992. *Thar Desert in Rajasthan*. Geological Society of India, Bangalore.

Tian, Y. 1991. Tokai on the delta at the lower reach of the Keriya River — a natural vegetation complex reflecting ecological degradation. In D. Jäkel and Z. Zhu (eds), Reports on the 1986 Sino-German Kunlun Shan Taklimakan-Expedition. *Die Erde*, Erg. H 6: 99–112.

Turner, S., Hawkesworth, C., Jiaqi, L., Rogers, N., Kelley, S. and van Calsteren, P. 1993. Timing of Tibetan uplift constrained by analysis of volcanic rocks. *Nature*, 364: 50–54.

Wang, J.T. and Derbyshire, E. 1987. Climatic geomorphology of the northeastern part of the Qinghai-Xizang Plateau, People's Republic of China. *Geographical Journal*, 153: 59–71.

Wang, J.T., Derbyshire, E. and Shaw, J. 1986. Preliminary magnetostratigraphy of Dabusan Lake, Qaidam Basin, China. *Physics of the Earth and Planetary Interiors*, 44: 41–46.

Wang, J.T., Derbyshire, E., Meng, X.M. and Ma, J.H. 1991. Natural hazards and geological processes: an introduction to the history of natural hazards in Gansu Province, China. In T.S. Liu (ed.), *Quaternary geology and environment in China*. Science Press, Beijing: 285–296.

Wasson, R.J., Rajaguru, S.N., Misra, V.N., Agrawal, D.P., Dhir, R.P., Singhvi, A.K. and Kameswara Rao, K. 1983. Geomorphology, late Quaternary stratigraphy and palaeoclimatology of the Thar dune field. *Zeitschrift für Geomorphologie*, Supplementband, 45: 117–151.

Wasson, R.J., Smith, G.I. and Agrawal, D.P. 1984. Late Quaternary sediments, minerals and inferred geochemical history of Didwana Lake, Thar Desert, India. *Palaeogeography, Palaeoclimatology, Palaeoecology*, 46: 345–372.

Wilhelmy, H. 1969. Das Urstromtal am Ostrand der Indusbene und der Sarasvati-Problem. *Zeitschrift fur Geomorphologie*, Supplementband, 8: 76–93.

Williams, M.A.J. and Clark, M.F. 1984. Late Quaternary environments in north central India. *Nature*, 308: 633–635.

Yang, X. 1991. Geomorphologische Untersuchungen in Trockenräumen NW-Chinas unter besonderer Berücksichtigung von Badanjilin und Takelamagan. *Göttinger Geographische Abhandlungen*, 96: 124 pp.

Zhang, J. and Lin, Z. 1992. *Climate of China*. Wiley, New York, 376 pp.

Zhao, S. 1986. *Physical geography of China*. Science Press, Beijing, and Wiley, New York, 209 pp.

Zhao, S. and Xing, J. 1984. Origin and development of the shamo (sandy deserts) and the gobi (stony deserts) of China. In R.O. White (ed.), *The evolution of the East Asian environment*, 1: 230–251.

Zhao, W., Nelson, K.D. and Project INDEPTH Team 1993. Deep seismic reflection evidence for continental underthrusting beneath southern Tibet. *Nature*, 366: 557–559.

Zhou, S. and Sandiford, M. 1992. On the stability of

isostatically compensated mountain belts. *Journal of Geophysical Research*, 97: 14207–14223.

Zhu, Z. 1984. Aeolian landforms in the Takalamakan Desert. In F. El-Baz (ed.), *Deserts and arid lands*. Martinus Nijhoff, The Hague: 133–143.

Zhu, Z. and Lu, J. 1991. A study on the formation and development of aeolian landforms and the trend of environmental changes in the lower reaches of the Keriya River, Central Taklimakan Desert. In D. Jäkel and Z. Zhu (eds), Reports on the 1986 Sino-German Kunlun Shan Taklimakan-Expedition. *Die Erde,* Erg. H 6: 89–97.

Zhu, Z., Liu, S., Wu, Z. and Di, X. 1986. *Deserts of China*. Academia Sinica, Lanzhou.

22

Middle East and Arabia

Robert J. Allison

Introduction

The term 'Middle East' appears to have originated during the 1850s (Davison 1960; Koppes, 1976), to identify a group of countries centred on the Persian Gulf. The term is now most commonly used to encompass the countries at the eastern end of the Mediterranean Sea and the nations of the Arabian Peninsula which are bounded by the Red Sea, Persian Gulf, Arabian Sea and Indian Ocean (Figure 22.1). Subregional definitions are largely utilitarian but the area as a whole can be conveniently divided into two major groups of countries, based primarily on physical characteristics and boundaries. The first comprises nations and states which either border or lie close to the eastern margin of the Mediterranean. They include Israel, Jordan, Lebanon and Syria. The second group of countries comprises Saudi Arabia, Yemen, the People's Democratic Republic of Yemen, the Sultanate of Oman and the smaller states of Bahrain, Kuwait, Qatar and the United Arab Emirates. There are also the nations of Iraq and Iran. The latter country marks the northeastern margin of the Persian Gulf and the desert areas in Iran abut those of central Asia such as the Thar of Pakistan and India. The surrounding seas and oceans are therefore physical features which both delimit the region as a

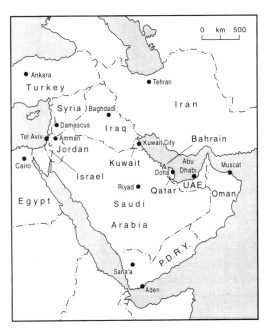

Figure 22.1 *Countries of the Middle East and Arabian Peninsula*

whole and determine the internal subdivisions. Occasionally Turkey is referred to as a Middle East country but the degree and nature of aridity is considerably less than in the core Middle

Arid Zone Geomorphology: Process, Form and Change in Drylands, 2nd edition. Edited by David S. G. Thomas.
© 1997 John Wiley & Sons Ltd.

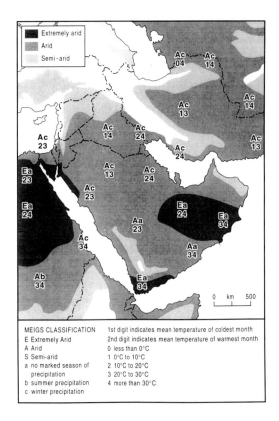

MEIGS CLASSIFICATION | 1st digit indicates mean temperature of coldest month
E Extremely Arid | 2nd digit indicates mean temperature of warmest month
A Arid | 0 less than 0°C
S Semi-arid | 1 0°C to 10°C
a no marked season of | 2 10°C to 20°C
precipitation | 3 20°C to 30°C
b summer precipitation | 4 more than 30°C
c winter precipitation |

Figure 22.2 *The distribution of semi-arid, arid and extremely arid areas in the Middle East and Arabian Peninsula*

Eastern countries, comprising dry-subhumid and semi-arid conditions (UNEP 1992). Moving through the Middle East from north to south there is increasing aridity (Figure 22.2), giving rise to ever greater tracts of desert, until states such as Saudi Arabia comprise almost entirely extremely arid environments.

Physical background

Sitting aside the Tropic of Cancer, the deserts of the Middle East and Arabian Peninsula (Figure 22.3) extend from the Mediterranean seaboard to the shores of the Sultanate of Oman and Yemen. The region includes many of the major desert zones of the world. Almost all of the Middle East and Arabian Peninsula falls within the arid zone although there are notable spatial variations in climate, topography, land-

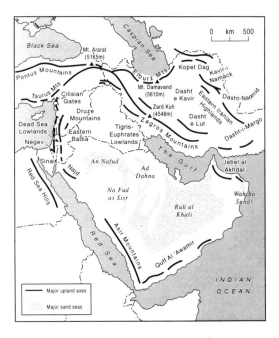

Figure 22.3 *The main deserts of the Middle East and Arabian Peninsula*

scape, soils and vegetation. The five parameters combine to produce distinctive geomorphological identities. There are also temporal changes in climate which become increasingly significant to the north as a marked seasonality begins to emerge. Winter precipitation (Table 22.1) and ground surface runoff are two examples of variables which change markedly across the region.

Geologically the Middle East and Arabian Peninsula is a complex region, reflecting its origin as a remnant of the Gondwanaland continental land mass (Figure 22.4). Three major and a number of smaller lithospheric plates can be identified. Of greatest importance are the African, Eurasian and Arabian plates, whose boundaries are clearly identifiable as the Azores–Gibraltar ridge and its extension across north Africa, the Red Sea Rift and the Alpine zone of Iran. Tectonic movement associated with the plates and activity across their margins is often recorded. The oldest rocks are Precambrian and Palaeozoic, exposed on the stable landmasses of Africa and the Arabian Peninsula. From south to north the rocks become younger and are predominantly sedimentary, although there are occasional igneous intrusions. It is the

Table 22.1 *Precipitation statistics for selected locations in the Middle East and Arabian Peninsula*

Country	Location	Max. (mm)	Min. (mm)	Mean (mm)
Egypt	Port Said	129.6	13.0	63
	Cairo	63.4	1.5	22
	Asŷut	25.0	0.0	5
Turkey	Izmir	1116.5	331.2	695
	Ankara	500.8	247.5	362
	Konya	500.5	143.7	316
Israel	Jerusalem	957.7	273.1	529
	Eilat	96.9	5.5	27
Jordan	Amman	476.5	128.3	273
Iraq	Mosul	585.2	208.2	340
	Baghdad	336.0	72.3	151
	Ar Rutbah	269.0	46.9	121
Bahrain		169.4	10.1	76
South Yemen		93.0	7.6	39

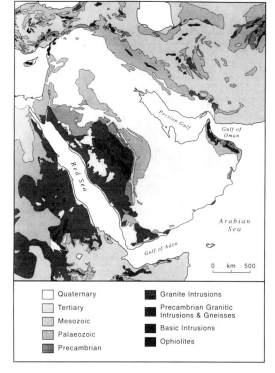

Figure 22.4 *Geology of the Middle East and Arabian Peninsula*

sedimentary materials for which the region is best known geologically, with calcareous deposits forming the reservoirs for oil reserves on the one hand and zones of groundwater storage on the other. Tertiary sediments of sand, clay, marl and limestone are focused on the low-lying regions. A stratigraphic characteristic throughout much of the Middle East and Arabian Peninsula is the presence of intrusive volcanic rocks, basalt being a good example. Volcanic peaks often radiate basalt lava flows and volcanic ash falls in every direction. One of the most famous is Mount Ararat. In other places the lava has extruded in linear flows, along major fissures for example, rather than from single isolated vents. Good examples are the Jebel Druze in Syria and associated basalt fields in the Eastern Badia of Jordan between Syria and Saudi Arabia. The basalt lava fields are often young geologically. Those in Jordan are of Tertiary and Quaternary origin for example (Quennell 1959; Bender 1974; Al-Homoud *et al.* 1995).

The Middle East and Arabian Peninsula is a transition zone between equatorial and mid-latitude climates. In summer temperatures are high almost everywhere (Table 22.2). During winter months there are more marked contrasts in temperature across the region, with a gradual decline towards the north. Precipitation totals relate to two factors. The first is sources of moisture, which include the Mediterranean, Black and Caspian seas. Generally, precipitation

Table 22.2   *Summer temperature statistics for selected locations in the Middle East and Arabian Peninsula*

Country	Location	Absolute Max. (°C)	Absolute Min. (°C)	Average Max (°C)	Mean daily range (°C)
Egypt	Quseir	40	23	34	8
Bahrain	Bahrain	44	24	37	8
Kuwait	Kuwait	48	26	40	10
Oman	Muscat	45	25	36	6
Saudi	Jeddah	42	21	38	12
Saudi	Riyadh	45	19	42	17
Iraq	Baghdad	50	17	43	19
Iran	Tehran	43	15	37	15

Note: Temperatures are July recordings at established weather stations.

Figure 22.5   *The Oman Mountains*

totals decline away from the Mediterranean Sea. The second factor is altitude. In addition, there are occasional local influences such as the effects of mid-latitude cyclonic disturbances. There are also orographic controls over small areas. Just as locations such as the Jabal Akhdar in the Sultanate of Oman ameliorates temperature, so the upland massif promotes orographic rain, lush vegetation and a generally hospitable climate. Of all the countries in the Middle East and Arabian Peninsula, it is only Lebanon which does not include areas of extreme aridity because mean annual rainfall totals do not fall below 200 mm yr^{-1}. One of the variables which influences contemporary geomorphological processes in desert environments is the intensity of individual rainstorms. Rainfall intensity data are scarce for countries of the Middle East and Arabian Peninsula but some analysis of magnitude and frequency has been made for data from Tehrän. High-magnitude/low-frequency storms, which are likely to have a major influence as an agent of landscape change, do occur relatively frequently when compared with other arid zone regions, a factor of some importance when considering the characteristics of landforming processes.

Although the Middle East and Arabian Peninsula has the largest single area of sand dunes of any desert region (McKee 1979), as might be expected in such a large area the physical nature of the deserts and their component landforms vary considerably. There are many other forms of desert landscape ranging from flat, stone-strewn plains to the rugged interiors of the Oman Mountains (Figure 22.5) and Druze Mountains in Syria. There are also, across some of the flattest landscapes on Earth, seasonal lakes, salt pans and playas of particularly large proportions (Figure 22.6). The wadis which flow into them are in places deeply incised, indicating high discharges in the past and a considerable capacity to transport sediment.

An important point to consider is the extent to which the present desert ground surface conditions in the Middle East and Arabian Peninsula reflect current geomorphological processes on the one hand and past climates on the other (Roberts 1983). There is evidence to suggest, in the raised channels on the flanks of the Wahiba Sands in Oman (Figure 22.7) or in the heavily incised wadis of the Eastern Badia of Jordan, for example, that facets of the landscape today reflect past climates where surface

Figure 22.6 *A playa in the Eastern Badia, Jordan, known locally as Qaa*

Figure 22.7   *Raised channels of the Wadi Andam drainage system, Sultanate of Oman*

runoff regimes were much greater and wadi flow more continuous. Pluvial periods are thought to have occurred in parts of the Middle East (Figure 22.8) in association with an expansion of ice sheets to the north in Europe during the last glacial. Similarly, in some of the highlands, river terrace sequences and gravel deposits are indicative of wetter climatic regimes and greater surface runoff volumes. Archaeological and palaeo-geomorphological evidence suggests that the region began to enter its present phase of aridity some 7000–5000 BP (Allison 1988a; Gardner 1988). Further indicators of change in the geomorphological environment are found in erosion surfaces which are present throughout the region. In western Saudi Arabia, for example, there is a well-developed pedeplain, the Najd, on Precambrian rocks with isolated inselbergs rising

above it. Parts of the Najd may be a Precambrian surface which was buried, evidenced by the remnants of covering Tertiary lava flows for example, and subsequently exhumed. Three further erosion surfaces at levels of 1650, 1200 and 900 m can also be found extending across large parts of the Sinai and into central Arabia.

One characteristic of the Arabian Peninsula is its mineral wealth, particularly the oil. The Arabian Peninsula contains an estimated 65% of the world hydrocarbon reserve. Most of the oil fields lie along the east of the Peninsula due to the geological structure and presence of Mesozoic sedimentary rocks. Twenty-eight of the 33 major oilfields on Earth lie within the Arabian Peninsula, each with an estimated reserve in excess of 5 billion barrels. Saudi Arabia is the largest producer, with proven reserves in excess of 258 million barrels.

A

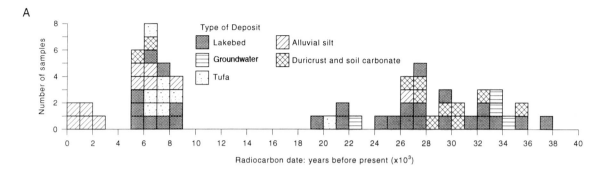

B

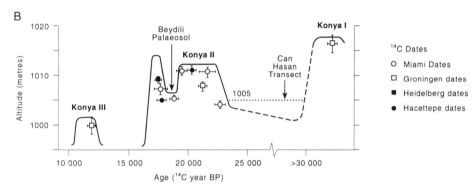

Figure 22.8   *Recent environmental history and climate change for the Middle East and Arabian Peninsula*

Despite the massive hydrocarbon resource, a characteristic of the desert regions of the Arabian Peninsula countries in recent years has been attempts to find alternative natural resources in an attempt to diversify the economy and provide better financial stability. Many of the deserts are now yielding quantities of metalliferous minerals such as zinc, lead, iron ore, copper, gold and chromite (Abu-Ajamieh *et al.* 1988). Agriculture is increasingly practised in some areas but only under irrigation and there is a question about the most sensible use of limited groundwater reserves when other forms of income are more readily available, allowing the preservation of water for more pressing human needs. There are two major groundwater zones in the Arabian Peninsula. One is the sedimentary rocks which reach thicknesses in excess of 500 m and range from Cambrian to Quaternary in age. They result in 11 major aquifers, some of which are linked, holding water between 16 000 and 35 000 years old. Other groundwater reserves occur in wadi and other alluvial sediments close to the ground surface, although the total extent of the water resource is more limited and the age of the water is significantly less than that found at depth.

The eastern Mediterranean

The countries bordering the eastern Mediterranean display a distinct seasonality of climate and, as a consequence, there is considerable spatial and temporal variation in aridity. The Mediterranean Sea itself has an ameliorating effect throughout much of the year along the coastal fringe. In the interior, cool, wet, winter conditions result in zones where rain-fed agriculture is possible in the late winter and spring but in the summer a complete absence of surface water results in desert conditions. Seasonality of climate decreases from north to south. When combined with topographic controls, such as the orographic effects of the Druze Mountains on the Syrian Desert and Jordan Badia in comparison to the flatter topography of the Negev in

Israel, there are notable spatial patterns with greater aridity in the south.

SYRIAN DESERT AND JORDAN BADIA

The desert zone in southeastern Syria and Jordan falls within an arid climatological zone where rainfall is erratic and seldom exceeds 200 mm annually. There is a seasonal difference in temperature, with winter minimum and summer maximum values of $-5°C$ and $46°C$, respectively. Geologically the area is dominated by sedimentary rocks, the thickness of which increases in a northeast direction, where progressively younger sediments are exposed (Bender 1974). Much of the Eastern Badia in Jordan, most notably the footslopes of the Druze Mountains in Syria and land extending south through Jordan to Saudi Arabia, is covered by extensive basalt plains. They have their origin in late Tertiary and early Quaternary volcanic intrusions and lava flows. The total area covered by basalt exceeds 11 000 km^2 and forms part of the major North Arabian Volcanic Province, which stretches from the southern rim of the Damascus Basin in Syria, across Jordan and into Saudi Arabia. Sediments flanking the basalt plateau are Late Cretaceous and Eocene in age.

Geomorphologically the Syrian Desert and Jordan Badia are dominated by extensive tracts of undulating, low hills, with local elevation differences of 25 to 30 m between high points and low points across the landscape. There is some topographic variation within the basalt region, depending on the age of a particular lava flow, but almost everywhere topography is characterised by gentle concavo-convex slopes. Broadly speaking, where basalt is present the ground surface topography of the older flows is more subdued, with well-developed wadi networks and a fine colluvial rill pattern. More recent flows display greater topographic irregularity, a lower degree of connectivity to wadi channel networks and many depressions which are filled with fine sediment. The majority of the ground surface is covered by typical desert pavement although on the footslopes of the Druze, rocks and boulders have been cleared from the land for agricultural purposes. The physical characteristics of the ground surface

rock cover is controlled by geology. Where sedimentary rocks are present the pavement or reg comprises a combination of small rock fragments and caliche deposits. On the basalt plateau the size of clasts is controlled by rock physical properties and their control on rock weathering and disintegration. In some locations large boulders are predominant, with slope processes being controlled by the degree of boulder burial and relative proportions of covered and exposed ground. In other places the basalt weathers to much smaller rock fragments which cover the entire ground surface. When basalt ground surface cover is disturbed, the fine-grained, dark-brown to orange coloured sediments which are present beneath the reg are subject to deflation and erosion by water during wet seasons.

The other noticeable geomorphological features in the region are fine-grained, water lain, contemporary sedimentary deposits. They vary in size and character, but four main types can be identified, based on their size, shape and drainage characteristics. First, there are extensive deposits which are fed by wadis but do not have any outward drainage. They are usually saline and display many of the characteristics of playa deposits. Second, there are deposits which are both fed and drained by wadis. They are seldom saline or are of low salinity. There are linear features which follow the line of wadis, usually occurring at locations where the ground surface is seasonally covered by wadi flow and resulting in the precipitation of slack-water deposits. Finally, there are smaller features which occur in depressions. Some both receive and deliver water. Others act as sediment traps due to local drainage characteristics. The wadi systems associated with the pans and playas are extensive and often well-developed, with regional flow directions being from north to south. The Wadi Rajil, for example, has its headwaters in Syria, crosses Jordan and extends south into Saudi Arabia.

THE NEGEV

The deserts of the Negev and Sinai act as links between the arid zones of north Africa, the eastern Mediterranean and Arabia. Indeed the term Negev comes originally from the Hebrew language and literally translates as *dryness*. Much of the region has no typical annual average rainfall. In some years there may be an absolute drought, while in others there may be up to 150 mm of precipitation. There is a decrease in annual precipitation from north to south which is related to the distance from the centre of the winter depressions of the eastern Mediterranean. The Negev is triangular in shape with the northern margin running from a point near Gaza on the Mediterranean coast to the edge of the Dead Sea and the southern tip centred on Eilat. To the west the Negev abuts the Sinai Desert and to the east the region merges with the Badia of Jordan. In total the desert covers approximately 12 500 km². It is predominantly a rocky desert and lacks the massive sand seas typical of the Sahara to the west and Arabian Peninsula to the southeast. In the north the Negev is formed of a high, broken plateau, cut by wadis which in may places flow in deep ravines. Rocks are generally sedimentary with varying parallel layers of hard and soft limestone. To the south the Negev merges into the Sinai Peninsula and sandstones dominate the stratigraphy although near Eilat sedimentary rocks give way to dark, magmatic materials which also form part of the mountainous highland massif of Sinai.

Geomorphologically it is possible to identify four different zones within the Negev (Figure 22.9). In the north is an area of undulating, rolling plains which vary in elevation between 200 and 450 m. The area is dissected by wadis and towards the Mediterranean coast is a region of sand dunes, dominated by linear dunes and vegetated to varying degrees. The linear dunes of the Negev, and adjacent parts of Sinai, are significant for investigations by Tsoar (1978; 1983) into the nature of the dynamics of linear dunes and the role vegetation plays in their surface morphology (see Chapter 17).

Much of the ground surface, particularly to the south of Beer Sheva, has a cover of loess which reaches a thickness of several metres in places. The fine-grained, windblown material appears to have its origins in the Sinai to the southwest and beyond. Much of the material has been reworked since its original deposition either by wadis or by overland flow processes on hillslopes. Consequently there are large alluvial fan sequences in places and thicker

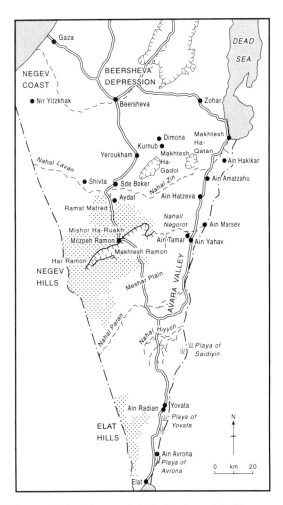

Figure 22.9 *Geomorphological regions of the Negev*

deposits in depressions and at topographic lows where adjacent hillslopes have been denuded of loess. Second, there is a central highland area, covering around 2000 km². Elevations vary between 450 and 1000 m above sea level, with a series of parallel ridges and valleys which trend in a northeast to southwest direction. The western margin of the region drops towards the Mediterranean and wadis accordingly flow in the same direction. The eastern fringe is marked by a series of steep cliffs and narrow gorges which border the Avara valley. Much of the central highland area is covered by desert pavement with the precise nature of the reg depending on the underlying geology. The third subdivision of the Negev is to the south, a relatively small area at around 70 km², but identifiable due to igneous rocks and mountainous topography. Wherever the surface is flat, hamadas have developed, while in other locations wadi incision has resulted in the development of a canyon-like landscape. Finally there is the Arava valley which extends from the Dead Sea to the Red Sea Gulf of Aqaba. The valley is around 100 km in length and varies between 8 and 20 km in width. The northern part of the valley is below sea level and in many places, particularly depressions and topographic lows, the level of ground salinity is high and sabkhas form (Yaalon 1962). The largest is at the southern end of the Dead Sea, while smaller areas are found at Yovata and Elat.

One particular aspect of the geomorphology of the Negev is the importance of water as an erosive agent in arid environments (Shick 1980). Much of the recent research in the desert is focused on the dynamics of fluvial processes, sediment transport and wadi network evolution (see, for example, Yair 1990; 1994; Yair and Enzel 1987). There is evidence of both overland flow processes and rill, gully and wadi erosion processes throughout the Negev. Floods are occasionally severe and deep canyons and ravines may form, particularly where tectonic factors are favourable. High erosion rates are partly a consequence of surface properties and there are locations where the infiltration rate can drop to zero in as little as 10 minutes (Yair and Lavee 1985). In addition there are the intensity characteristics of individual rainstorm events, with high-magnitude/low-frequency storms being common and around 77% of the total annual rainfall occurring in storms with an intensity between 5 and 10 mm per hour (Kutiel 1978).

The Arabian Peninsula

It is the Arabian Peninsula perhaps more than any other part of the world with which deserts are associated and indeed arid, barren tracts of land encompass virtually the whole of the region, save for one or two pockets, such as mountainous areas which receive greater precipitation due to orographic effects. Despite the picture of aridity that the Arabian environment generates, there is a growing body of evidence

that testifies to significant climatic and environmental changes affecting the geomorphic and sedimentary evolution of the region in the Quaternary period (e.g. Edgell 1989; Sanlaville 1992). In southeastern Arabia, for example, more humid climates leading to fan development have been identified at 142–108, 42–45 and 30–21 K year BP (Sanlaville 1992; Glennie 1995), with more humid conditions also recorded for the Holocene climate optimum.

Bordered by the Red Sea, Arabian Sea, Persian Gulf and Euphrates River, the Arabian Peninsula covers approximately 2.3 million km². The western margin follows the geologically controlled Great Rift Valley, which diverges at the northern end of the Red Sea. One limb sweeps north between Israel and Jordan, dropping down to the lowest point on the land surface of the Earth at the Dead Sea at −396 m and finally petering out in southern Syria. Some of the exposed rocks date back 4600 million years.

East of the Rift, the Arabian Peninsula sweeps into rock- and boulder-strewn desert plains, gradually passing into great sand seas. Relief is low, with altitude in all but a few places seldom rising above 500 m, a consequence of the regional setting upon one of the main tectonic world plates. There is great contrast between the flat, gravel-strewn fans and plains such as the Quff Al Awamir of the People's Democratic Republic of Yemen and the ophiolitic upland massif of Jabal Akhdar which reaches an altitude of 3018 m in the Sultanate of Oman. Large dunes extend over 795 000 km² of the Arabian Peninsula (Wilson 1973), with the main dunefields being the Rub' al Khali of Saudi Arabia and Wahiba Sands of the Sultanate of Oman in the south and the An Nafud, Ad Dahna and numerous smaller sand seas in the north. Geologically many of the dunefields are underlain by basement crystalline rocks which are often capped by lithified sediments.

The degree of soil profile development varies across the Arabian Peninsula. The most basic subdivision is one of poorly developed horizonation in the northern part of the region and no soil development in the south, due to the dominance of mobile sand. The best developed soils are found in preferentially favourable zones such as along the course of large wadis, on alluvial plains and along the course of rivers. Small areas of soil occasionally develop in interdune areas and in topographic lows due to the preferential movement of fines which improve the potential for horizon development and profile evolution. In places where vegetation grows, either naturally fringing oases or where agriculture is practised, the soils are usually light brown to yellow in colour. There may be some red iron staining, particularly where there has been some discoloration of sediment during previously wetter climates and subsequent remobilisation of minerals such as iron oxide.

RUB' AL KHALI

The Rub' al Khali, or Empty Quarter as it is sometimes called, is the sandiest of all deserts and also one of the driest, with a total absence of rivers or perennial springs of any size. Sitting in a structural basin and covering 560 000 km^2, the erg extends from the United Arab Emirates westward over a distance of some 1500 km to the foothills of the Yemen Mountains (Breed *et al.* 1979). In places the dunes are as much as 300 m high with a mixture of barchan, compound, linear and star dunes. Crescentric dunes dominate the northeastern sector of the desert, with star dunes along the eastern and southern margins and linear dunes in the west. The largest linear dunes have interdune corridors between 2 and 2.1 km wide. The origin of the sediments and the age of the Rub' al Khali remain subjects of debate although there is some suggestion that the largest dunes are Pleistocene in age, providing a link between the evolution of this erg and the age of the larger dunes in the Wahiba Sands in the Sultanate of Oman. Lower glacial-age sea levels, resulting, for example, in the drying up of the Arabian Gulf at the height of the last glacial, may have provided a source for dune sediments (Glennie 1995).

Largely in the northern Saudi sector of the Arabian Peninsula, on the margin of the Rub' al Khali, are the sand seas of the An Nafud, Ad Dahna, Nafud ath Thuwayrat, Nafud as Sirr and a number of small, pocket-dune areas. The largest, the An Nafud, covers 72 000 km^2, with dunes reaching heights of 100 m. The individ-

ual dunefields form an arc that extends southeast from An Nafud to the northern margin of the Rub' al Khali (Holm 1960) with a variety of different dune shapes and sizes. Throughout the Arabian Peninsula the movement of sediment by aeolian processes is controlled by the Sahara High of the African continent in winter and by a large low-pressure cell on the Indian subcontinent in the summer. The consequence in the winter is a clockwise flow of air resulting in a westward flow of wind in the northern Arabian Peninsula and a south to southwestward flow over the Rub' al Khali. In the summer the northwesterly 'Shamal' wind blows over much of the peninsula.

WAHIBA SANDS

To the east of the Rub' al Khali, in the Sultanate of Oman, lies the 16 000 km^2 erg of the Wahiba Sands (Figure 22.3), occupying the eastern part of the interior basin. Here is the most extensively examined sand sea of the Middle East and Arabian Peninsula in recent years, with much work being completed on both contemporary and palaeo-deposits in an attempt to elucidate patterns of sand sea evolution (Goudie *et al.* 1987; Dutton 1988).The sand sea is bounded to the north by the mountainous Al Hajar al Sharqi, while the Wadi Batha which drains from the northern mountains cuts into the Wahiba along its northern margin. Flanking the eastern margin of the sands is a series of large sabkhas. The western margin is bounded by a sequence of channels of the Wadi Andam and the southern boundary is the coastline of the Gulf of Oman.

The Wahiba, with maximum dimensions of 200 km from north to south and 100 km from east to west, can be divided into three main dune areas (Figure 22.10). First, there is an extensive tract of land in the north, running up to the foothills of the Oman Mountains comprising north- to south-trending large, linear dunes. Dune dimensions are large, with crest separation distances often exceeding 1.5 km and individual forms often exceeding 100 m in altitude between the top of the dune and interdune areas. The region has been described as 'High Wahiba' (Jones *et al.* 1988) due to the dune dimensions. Second, there is a zone of

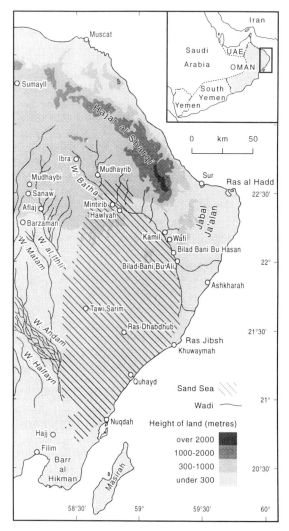

Figure 22.10 *Geomorphology of the Wahiba Sands and surrounding area. Dune area is cross-hatched*

transverse dunes close to the coastline in the south. The sediments comprising these dunes appear to be more recent in origin than the sands of High Wahiba (Allison 1988a), of marine origin and much more active in terms of their present-day mobility and process dynamics. Finally, there are peripheral dune areas. In some places, such as in the southeastern corner of the Wahiba Sands, the dunes are complex in form and highly mobile. At other points around the margin of the sands the dunes are interspersed with areas of desert pavement, outcrops of bedrock and aeolianite,

the product of a previously active sand desert, now cemented by natural precipitates to form a silica-rich sedimentary sequence which underlies large parts of the currently active dune belt (Gardner 1988). The variations in sediments and dune morphology suggest that there have been a number of periods of sediment accumulation under the influence of different dominant winds and varying degrees of aridity over the last 20 000 years (Allison 1988b; see Figure 17.2).

Across those parts of the Wahiba devoid of sand, the ground surface usually comprises a gravel or stone reg (see Figure 6.12 for example). The reg surfaces may be residual or transported. The residual deposits are a consequence of overland flow and deflation processes removing finer particles. Transported regs on the other hand are a consequence of sediment movement and deposition by water moving across the landscape. Sediment is laid down as alluvial fans and wadi deposits, many of which fringe or run out from mountainous, upland areas. The gravel plains have been cut and incised in places by wadis, an example being the Wadi Andam and Wadi Batha which run down the western and eastern sides of the Wahiba Sands. Areas of aeolian sand and flat, stone-strewn surfaces are occasionally broken by highlands. Altitude influences temperature patterns and promotes orographic rain in otherwise precipitation deficient zones, frequently giving rise to lush vegetation. An example is the Jebel Akhdar in Oman and the southern extension of the Asir region in Yemen, where climate and soil resources in an area with an altitude of more than 3700 m have allowed the local population to develop terrace agriculture.

Iran

Beyond the Persian or Arabian Gulf towards central Asia lies Iran, the eastern outpost of the Arabian Peninsula. Much of the country comprises a continental plateau interior and, as might be expected, temperatures are extreme, sometimes exceeding 50°C in summer and dropping below freezing in the middle of winter. Add to this strong local winds which sweep across the plateau surface and a mean annual rainfall which seldom exceeds 200 mm yr^{-1}, and the consequence is a harsh, hostile environment. Coupled with a lack of surface water is a dearth of groundwater. Potable water is in short supply and not only are reserves lacking for agriculture but there are occasions when water for human consumption has to be limited.

Iran is generally a country of low relief but there are two significant mountain zones. One, the Elburz Mountains, flanks the southern margin of the Caspian Sea, with altitudes exceeding 5000 m and separating the coastal lowlands which drop below sea level in places from the sabkha-dominated interior of the Dasht-e Kavir. Second, there are the Zargos Mountains which trend northwest to southeast, with topographic highs of 4328 m at Oshtorah Kur and 4547 m at Zard Kuh. The Zargos fall away to the west into an area of coastal lowland, swamp and marsh environments at the head of the Persian Gulf around Abadan in Iran, Al Basrah in Iraq and Kuwait.

The desert zone of Iran comprises several major units (Figure 22.11), differentiated either on the basis of surface conditions, or by being separated by mountain areas, for in reality the central areas of the country are dominated by dryland conditions. Major desert areas are the Dasht-e Kavir, a salt desert, and the Dasht-e Lut, a sandy desert. The Dasht-e Kavir, or Great Kavir, is a largely uninhabited area, characterised by almost flat pans and playas with a high natural salt content (Beaumont

1968). The name Kavir-i-Namak is also used as an alternative name for part of this area. Sodium and potassium salts frequently crystallise at the ground surface to form a deceptive crust which masks underlying soft, waterlogged sediment and networks of subsurface drainage channels. The region extends over more than 500 km from east to west. The Dasht-e Lut in the southeast is for the most part covered by loose sand and desert pavement. Various other discrete desert basins lie between these two major desert areas and the Zagros Mountains, such as the deserts around Ardekan, Kashan, Nain and Yazd. These variously consists of playas, some possess marked strandlines, and dunefields (e.g. Huber 1955; Krinsley 1970), and where their extent is punctuated by isolated mountains and hills then loess and climbing and falling dune features may exist, such as shown in Figure 26.11.

South of the Dasht-e Lut lies the Jazmurian Desert or Hamum-e Jaz Murian, which possesses a central playa lake and is notable for its extensive areas of nebkha dunes. Finally, the Dasht-i-Naomid and the Dasht-i-Margo traverse the international boundaries between Iran and Afghanistan. In most places the landscape is characterised by sequences of desert mountains, low rounded hills, alluvial fans around the edge of topographic highs and intervening plains which have been cut by intermittent or ephemeral watercourses. Ground surface materials are often coarse-textured and soils are poorly developed, with a high calcareous component throughout the profile. Fans and other alluvial deposits may have some sign of soil profile evolution on them but they are always weak and widely scattered.

The eastern margin of the Middle East and Arabian Peninsula region ends at the international boundaries between Iran, Afghanistan and Pakistan. Aridity gradually falls towards the Indus River, which separates the deserts of Eastern Arabia from the Thar, which straddles the border between Pakistan and India.

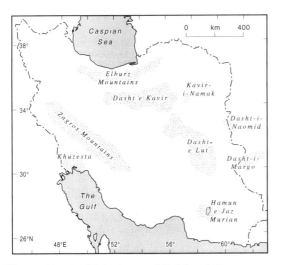

Figure 22.11 *Main desert subdivisions in Iran*

References

Abu-Ajamieh, M.M., Bender, F.K., Eicher, R.N., El-Kaysi, K.K., Nimri, F., Qudah, B.H. and Sheyyab, K.H. 1988. *Natural resources in Jordan*. Natural Resources Authority, Amman: 224 pp.

Ailmen, M.H., Doudoux-Fenet, D., Ferrere, J. and Palau-Caddoux, M. 1957. *Sables quaternaires du Sahara nord-occidental (Saoura-Ougarta)*. Algeria Service da la Carte Geólogique Bulletin No.15: 207 pp.

Al-Homoud, A.S., Allison, R.J., Sunna, B.F. and White, K. 1995. Geology, geomorphology, hydrology, groundwater and physical resources of the desertified Badia environment in Jordan. *GeoJournal*, 37: 51–67.

Allison, R.J. 1988a. Sediment types and sources in the Wahiba Sands. *Journal of Oman Studies*, Special Report No. 3: 161–168.

Allison, R.J. 1988b. Sand dune dynamics and sand sea evolution. *Landscape Research*, 13: 6–11.

Bender, F. 1974. *Geology of Jordan*. Gebrüder Bornträger, Berlin: 196 pp.

Beaumont, P. 1968. Salt weathering on the margin of the Great Kavir, Iran. *Bulletin of the Geological Society of America*, 79: 1683–1684.

Breed, C.S., Fryberger, S.G., Andrews, S., McCauley, C., Lennartz, F., Gebel, D. and Horstman, K. 1979. Regional studies of sand seas using Landsat (ERTS) imagery. In E.D. McKee (ed.), *A study of global sand seas*. US Geological Survey Professional Paper, 1052: 305–397.

Davison, R.H. 1960. Where is the Middle East? *Foreign Affairs*, 38: 665–675.

Dutton, R.W. 1988. Introduction, overview and conclusions. The scientific results of the Royal Geographical Society's Oman Wahiba Sands Project 1985–1987. *Journal of Oman Studies*, Special Report No. 3: 1–17.

Edgell, G. 1989. Evolution of the Rub al Khali desert. *Earth Science*, 3: 109–126.

Evenari, M., Shanan, L. and Naphtali, T. 1971. *The Negev. The challenge of a desert*. Harvard University Press, Cambridge, MA.

Fairbridge, R.W. 1970. An ice age in the Sahara. *Geotimes*, 5: 18–20.

Fontes, J.C. and Gasse, F. 1989. On the ages of humid Holocene and late Pleistocene phases in north Africa — remarks on 'Late Quaternary climatic reconstruction for the Maghreb (North Africa)' by P. Rognon. *Palaeogeography, Palaeoclimatology, Palaeoecology*, 70: 393–398.

Gardner, R.A.M. 1988. Aeolianites and marine deposits of the Wahiba Sands: character and palaeoenvironments. *Journal of Oman Studies*, Special Report No. 3: 75–94.

Glennie, K.W. 1995. *Dunes and inland sabkhas of the*

Emirates. Fieldtrip guidebook. United Arab Emirates University, Al Ain: 26 pp.

Goudie, A.S., Warren, A., Jones, D.K.C. and Cooke, R.U. 1987. The character and possible origins of the aeolian sediments of the Wahiba Sand Sea, Oman. *Geographical Journal*, 153: 231–256.

Hillel, D. 1982. *Negev. Land, water and life in a desert environment*. Praeger, New York.

Holm, D.A. 1960. Desert geomorphology in the Arabian Peninsula. *Science*, 132: 1369–1379.

Huber, H. 1955. *Geological report on the Ardakan-Kalut area*. National Iranian Oil Company, Tehran 144 pp.

Jones, D.K.C., Cooke, R.U. and Warren, A. 1988. A terrain classification of the Wahiba Sands of Oman. *Journal of Oman Studies*, Special Report, 3: 19–32.

Krinsley, D.B. 1970. *A geomorphological and palaeoclimatological study of the playas of Iran*. US Geological Survey, Final Scientific Report, CP 70–800.

Koppes, C.R. 1976. Captain Mahan, General Gordon and the origin of the term Middle East. *Middle Eastern Studies*, 12: 95–98.

Kutiel, H. 1978. The distribution of rain intensities in Israel. Unpublished MSc thesis, Hebrew University of Jerusalem.

McKee, E.D. (ed.) 1979. *A study of global sand seas*. US Geological Survey, Professional Paper, 1052: 429 pp.

Quennell, A.M. 1959. *Handbook of the geology of Jordan*. Government of the Hashemite Kingdom of Jordan, Amman: 82 pp.

Roberts, N. 1983. Age, palaeoenvironments, and climatic significance of late Pleistocene Konya Lake, Turkey. *Quaternary Research*, 19: 154–171.

Samuel, R. 1973. *The Negev and Sinai*. Wiedenfield & Nicholson, London.

Sanlaville, P. 1992. Changement climatiques dans la Peninsule Arabique durant le Pilestocene superieur et L'Holocene. *Paleorient*, 18: 5–26.

Shick, A. 1980. A tentative sediment budget for an extremely arid watershed in the southern Negev. In D.O. Doehring (ed.), *Geomorphology in arid regions*. Allen & Unwin, London: 139–163.

Street, A. and Gasse, F. 1981. Recent developments in research into the Quaternary climatic history of the Sahara. See Allah (1981), 7–28.

Tsoar, H. 1978. *The dynamics of longitudinal dunes*. Final technical report, European Research Office, US Army.

Tsoar, H. 1983. Dynamic processes acting on a longitudinal (seif) dune. *Sedimentology*, 30: 567–578.

UNEP 1992. *World atlas of desertification*. Edward Arnold, London.

Wilson, I.G. 1973. Ergs. *Sedimentary Geology*, 10: 77–106.

Yaalon, D.H. 1962. On the origin and accumulation of salts in groundwater and soils of Israel. *Bulletin of the Research Council of Israel*, 11G: 378–393.

Yair, A. 1990. Runoff generation in a sandy area. The Nizzana sands, Western Negev, Israel. *Earth Surface Processes and Landforms*, 15: 597–609.

Yair, A. 1994. The ambiguous impact of climate change at a desert fringe: northern Negev, Israel. In A.C. Millington and K. Pye (eds), *Environmental change in drylands*. Wiley, Chichester: 198–227.

Yair, A. and Enzel, Y. 1987. The relationship between annual rainfall and sediment yield in arid and semi-arid areas. The case of the northern Negev. *Catena*, Supplement, 10: 121–135.

Yair, A. and Lavee, H. 1985. Runoff generation in arid and semi-arid zones. In M.G. Anderson and T.P. Burt (eds), *Hydrological forecasting*. Wiley, Chichester: 183–220.

23

North America

Vatche P. Tchakerian

Introduction

The North American drylands include extensive areas of western Canada and the USA that experience semi-arid to arid conditions, including large tracts of the Prairies and Great Plains. The more specific desert areas within this general dryland domain constitute some of the most geomorphologically and biogeographically diverse deserts in the world. The Basin and Range Province and the Colorado Plateau provide the physiographic framework for North American deserts. In the Basin and Range Province, the Sonoran, the Chihuahuan, the Great Basin and the Mojave Desert are the main arid zones (Figure 23.1). The Basin and Range Province is characterised by block-faulted, subparallel, north–south-trending mountain ranges separated by extensive basins, with the mountain ranges getting progressively lower in mean elevation from north to south. In contrast, the Colorado Plateau exhibits horizontal to gently dipping sedimentary strata, with extensive mesa and scarp terrain, sandstone canyons, and an intricately dissected fluvial

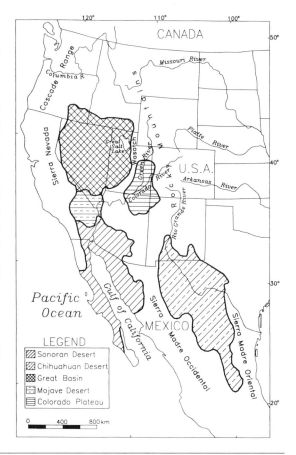

Figure 23.1 *The North American arid zones (after Cooke* et al. *1993)*

Arid Zone Geomorphology: Process, Form and Change in Drylands, 2nd edition. Edited by David S. G. Thomas.

landscape, the latter largely the work of the Colorado and Green rivers.

The Basin and Range Province

THE SONORAN DESERT

The Sonoran Desert, centred primarily on the Mexican states of Sonora and Baja California and southern Arizona, is about 300 000 km² (Crosswhite and Crosswhite 1982), and occupies 19% of the North American arid zone. It contains such physiographically diverse regions as the Salton Trough (including the Colorado Desert in California), which is part of the spreading ridge system of the Gulf of California, Baja California (including the unique Viscaino Desert), and the mountains and alluvial plains of southern Arizona and northwestern Sonora. The Sonoran Desert is predominantly under the influence of the subtropical high with peak precipitation during the summer 'monsoon' months. The coast of Baja California is also influenced by the cool, California Current. The eastern boundary of the desert includes the Sierra Madre Occidental, which receive signifi-

cant amounts of precipitation from orographic lifting of Pacific air masses during the winter months. One of the two hyper-arid regions in North America is centred in the Salton Trough and the delta of the Colorado River and includes the Gran Desierto del Altar of Sonora, with average precipitation less than 100 mm yr^{-1} (Table 23.1). Drainage systems are integrated within the Salt and Gila river systems in Arizona, the Rio Yaqui, Sonora and Magdalena in Sonora, and the Colorado River and the Gulf of California.

Block-faulted mountain ranges, separated by extensive alluvial plains and pediments, cover much of the Sonoran Desert. Average elevations for the ranges are between 500 and 1500 m (Figure 23.2), with isolated peaks reaching 2500 m. The mountain ranges are lower than their counterparts in the Great Basin and the Mojave Desert. Most of the tectonic activity and faulting ended by the early Pliocene (Morrison 1991). The ensuing period of tectonic quiescence led to the weathering of the mountain fronts, and the establishment of wide basins with thick alluvial deposits. Pediments with sharp piedmont angles (40 to 50% of the intermontane basins are occupied by

Table 23.1 *Climate of selected North American arid zone cities*

Arid zone	Elevation (m)	Mean July temperature (°C)	Mean January temperature (°C)	Mean annual precipitation (mm)
Sonoran Desert				
Tucson, Arizona	785	30.1	10.2	285
Guaymas, Sonora	10	31.1	18.2	233
Chihuahuan Desert				
El Paso, Texas	1194	27.4	6.6	201
Torreon, Coahuila	1137	28.1	15.2	230
Great Basin				
Elko, Nevada	1547	20.9	−5.2	231
Salt Lake City, Utah	1286	25.5	−2.2	375
Mojave Desert				
Las Vegas, Nevada	660	32.1	6.9	105
Barstow, California	707	28.6	8.2	145
Colorado Plateau				
Winslow, Arizona	1487	27.1	−0.3	183
Moab, Utah	1225	27.8	−1.1	227

Source: Universidad Nacional Autonoma de Mexico (1970); National Oceanic and Atmospheric Adminstration (1993).

Figure 23.2 *Western front of the Ajo Mountains, Arizona, showing mountain elevations typical of the Sonoran Desert. Note the typical Sonoran vegetation of giant* saguaro *and* palo verde *on the pediments*

pediments), boulder-strewn hillslopes and Pleistocene fluvial terraces (such as along the Rio Magdalena in Sonora), are also prominent. Hence, some of the oldest stable geomorphic surfaces in the Basin and Range Province are located in the Sonoran Desert.

Notable contributions to arid zone geomorphology in the Sonoran Desert include works by Kirk Bryan (1925) on boulder-mantled hillslopes and pediments, Mammerickx (1964) and Melton (1965) on pediments, Norris and Norris (1961) on the Algodones Dunes in the Colorado Desert, and Sharp (1964) on aeolian erosion. One of the most interesting works has been the ongoing field investigations on desert hillslopes of Abrahams, Parsons, and colleagues at the Walnut Gulch Experimental Watershed site, in southern Arizona (e.g. Abrahams *et al.* 1988). Using rainfall simulation, they discovered a complex relation between overland flow, slope and sediment yield, with a critical threshold value at 12°. Sediment yield values decreased both below and above this value. Since most pediment slopes average near this threshold value (Mabbutt 1977), this could represent some sort of an upper limit for pediment slopes (Cooke *et al.* 1993). The significance of the 12° threshold (if any) needs to be further investigated (especially with different lithologies). Overland flow and rainsplash seem to be more significant in transporting clasts on slopes, rather than hillslope gradient (Abrahams *et al.* 1988).

A spectacular site for geomorphological research is the Pinacate shield volcanic complex in northwestern Sonora, Mexico. The landscape has its origin in the opening of the Gulf of California and the subsequent volcanism associated with such spreading centres. Numerous volcanoes, cinder cones, lava flows, maars (explosion craters) and other extrusive igneous features are present (Figure 23.3). Some of the eruptions are as recent as 2000 years ago (Lynch 1982). However, the volcanic geomorphology of the region remains yet to be studied. The westernmost section of the Pinacate volcanic field merges with the sand dunes of the Gran Desierto del Altar (Figure 23.4), North America's only significant erg, the latter with an area of 5700 km², and significant aeolian accumulations, including crescentic and star dunes (Lancaster 1995). Stabilised and active dunes with different orientations, suggest significant fluctuations in source areas owing to

Figure 23.3   *Elegante Crater, a classic maar in the Pinacate Volcanic Preserve, northwestern Sonoran Desert, Mexico. Sierra Pinacate, a complex shield volcano, can be seen in the background*

Figure 23.4   *The western edge of the Gran Desierto del Altar erg, northwestern Sonoran Desert, Mexico. Sierra Blanca in the background. The dark area between Sierra Blanca and the edge of the sand dunes is one of the youngest lava flows in the Pinacate Volcanic Preserve. (Photo courtesy of R. & P. Boyer, CEDO)*

eustatic, climatic and tectonic changes of the region and to changes in the position of the Colorado River delta (Lancaster 1995).

Well-developed coastal dune systems are located in western Sonora and Baja California. Examples include a coastal dune complex, with foredunes and landward parabolic dunes and vegetated linear dunes, between Puerto Peñasco and Guaymas in Sonora, the dunes in the San Quintin Basin in northwestern Baja California (Orme and Tchakerian 1986) and a complex transverse and barchanoid field near Guerrero Negro (Figure 23.5) in central Baja California (Inman *et al.* 1966). Additional aeolian studies include works by Norris and Norris (1961) on the Algodones Dunes, Norris (1966) and Haff and Presti (1995) on the barchan dunes of the Salton Sea (Figure 23.6), and Péwé (1981), on desert dust.

THE CHIHUAHUAN DESERT

The Chihuahuan Desert is the largest and least studied of the North American deserts with an area of 450 000 km², a mean elevation of 1400 m (Medellin-Leal 1982), and occupying 32% of the North American arid zone. It is centred in the state of that name in Mexico with smaller sections in Texas and New Mexico (Figure 23.1). It is located between the ranges of the Sierra Madre Occidental and the Sierra Madre Oriental. The causes for aridity include the rain-shadow effects of the Sierra Madre ranges, the subtropical high and continentality. Most of the precipitation falls during the summer months, with a peak in September as a result of the occasional intrusion of cyclonic storms from the Caribbean (Table 23.1). Most drainage is endoreic, with the exception of the Rio Grande River, the only major exotic river that crosses the Chihuahuan Desert.

Geomorphological studies have been rather few in the Chihuahuan Desert. The work of McKee (1966) remains the seminal study on the gypsum dunes at White Sands National Monument. In West Texas, P.B. King's (1948) work on the geology and geomorphology of the Guadalupe Mountains remains a good introduction for this section of the Chihuahuan Desert. A comprehensive review by Medellin-Leal (1982) still offers the best introduction to the physical geography of this much neglected arid zone.

Figure 23.5 *Transverse dunes with slip-face projections, Guerrero Negro, Baja California, Mexico. (Photo courtesy of K. Mulligan)*

Figure 23.6   *Barchan dunes west of the Salton Sea, Imperial Valley, California. (Photo courtesy of K. Mulligan)*

Late Cenozoic deformation and subsequent weathering, erosion and fluvial activity, combined with Quaternary volcanism associated with the formation of the Basin and Range Province, provide the main geologic and geomorphic framework. The presence of many bolsons (closed intermontane basins with endoreic drainage) is one of the most unique geomorphic features of this arid zone. The largest, Bolson de Mapimi, occupies an area of about 12 000 km². Many of the bolsons contain ephemeral lakes and salt lagoons and are favourable sites for the accumulation of thick alluvial deposits and gypsum dunes. Examples include the Tularosa Basin in New Mexico and the Salt Basin in West Texas.

The Tularosa Basin, part of the Rio Grande rift zone (the latter consisting of a series of *en echelon* grabens), is a classic bolson with extensive alluvial and playa deposits. The weathering and transport of gypsum from the surrounding mountains to the playa in the centre of the graben (Lake Lucero), has led to the formation of barchan, transverse and parabolic gypsum dunes at White Sands National Monument. The same geologic and geomorphic forces also formed the equally distinct aeolian, alluvial and lacustrine environments in the nearby Salt Basin of West Texas.

Here, parabolic and transverse gypsum dunes (Figure 23.7) are superimposed over older, late Pleistocene, stabilised quartz dunes, indicating an episodic history of aeolian and lacustrine activity (Wilkins and Tchakerian 1994).

THE GREAT BASIN

The Great Basin, with an area of 409 000 km², comprises 28% of the arid zone in North America. It is bounded on the east by the Rocky Mountains, and by the Sierra Nevada and the Cascade Range on the west (Figure 23.1). Owing to its high latitude and a mean elevation of about 1500 m, it is the coolest of the North American deserts, with great annual ranges of temperature and precipitation (Table 23.1). At Death Valley, California, the elevation of the basin floor (see Figure 14.1) is at −86 m (lowest point in the western hemisphere), while Telescope Peak in the Panamint Range about 25 km to the west, towers over 3300 m above the basin (Figure 23.8). The Great Basin is a continental interior desert with rain-shadow conditions owing to its position with respect to the nearby ranges. Precipitation (including

Figure 23.7 *Gypsum dunes, Salt Basin, West Texas. Guadalupe Mountains in the background*

Figure 23.8 *Death Valley, California. Panamint Mountains, with snow-capped Telescope Peak (3368 m) in the background. The playa surface (mostly below sea level) of Pleistocene Lake Manly is in the foreground. Note the massive alluvial fans and bajadas*

snow) occurs in the winter months and is associated with frontal storms from the Pacific Ocean. In the winter, cold temperatures also result from the intrusion of continental polar air masses from Canada (Logan 1968). Over 150 north–south trending, block-faulted mountain ranges with their adjacent alluvial basins have been identified (Dohrenwend 1987). Weathering and subsequent erosion, and the removal of debris by running water, wind and slope processes, led to the infilling of the basins with thick alluvial deposits and the formation of numerous playa lakes. The sediments in the lakes contain a plethora of palaeoenvironmental information (Smith and Street-Perrott 1983).

The most notable geomorphic study is Gilbert's (1890) monograph on Lake Bonneville. By analysing such relict features as barrier bars, Gilbert was able to deduce shoreline processes, including wave dynamics. He explained the rise and fall of Lake Bonneville with respect to climatic oscillations, and, based on the elevation of raised shorelines, established the nature of isostatic forces and crustal deformation (Baker and Pyne 1978). Russell's studies on Mono Lake (1889) and Lake Lahontan (1885) are also seminal. Russell produced the first detailed geomorphic and stratigraphic analysis of the Lahontan basin and identified the various lacustrine clays, muds and evaporites. Lake Lahontan contains the most complete stratigraphic record of palaeoclimates in the North American arid zone (Morrison 1991). Subsequent works in the Great Basin, led to the establishment of a detailed stratigraphy, geomorphology and chronology of lake-level fluctuations and climate change (see, for example, Morrison 1964; 1991; Benson *et al.* 1990; Currey 1990; Thompson *et al.* 1993). Sharp's (1940) comprehensive study of the Ruby Mountains in Nevada is also significant.

About 100 of the basins in the Great Basin contained pluvial lakes during late Quaternary time (Benson *et al.* 1990; Sack 1994). The largest, Lake Bonneville (see Figure 26.10), with an area greater than 50 000 km^2 and a maximum depth of 330 m, was centred in Utah, with the present Great Salt Lake only a small remnant of the original lake (Figure 23.9). Another pluvial lake, Lake Lahontan, with an area of 23 000 km^2 and a maximum depth of over 275 m, was centred in northwestern Nevada (Figure 23.9). Searles

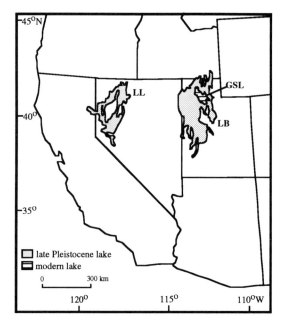

Figure 23.9   *Major Pleistocene lakes in the Great Basin. LL = Lake Lahontan, LB = Lake Bonneville, GSL = Great Salt Lake (after Sack 1994)*

Lake (Figure 23.10) sediments have also been used for palaeoclimatic reconstruction (Smith and Street-Perrott 1983).

Geomorphic evidence from the Great Basin indicates that lake levels were highest between 20 and 12 ka (Smith and Street-Perrott 1983; Benson *et al.* 1990; Thompson *et al.* 1993). Between 12 and 10 ka, lake levels experienced fluctuations reflecting the complex nature of climate–lake-level dynamics and subbasin geomorphology. Beginning around 9 ka, lake levels in the Great Basin dropped significantly with more arid conditions and warmer than present temperatures. Most of the lakes in the Great Basin contracted or dried completely during the mid-Holocene as observed from lake sediments, pollen ratios and packrat midden assemblages (Thompson *et al.* 1993),although regional variations in moisture, evaporation rates and summer enhanced 'monsoonal' incursions of precipitation, enabled some lakes to maintain high water elevations (Spaulding and Graumlich 1986). This mid-Holocene enhanced aridity and warmer than present temperatures was prevalent throughout much of

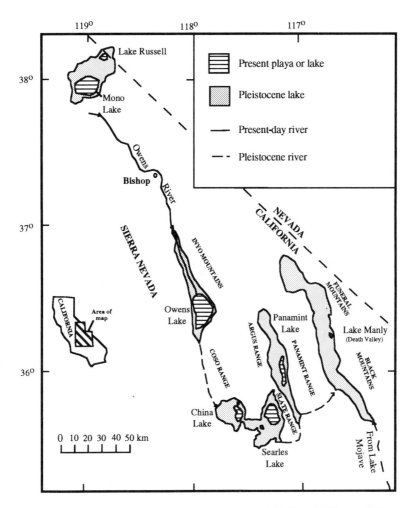

Figure 23.10 *Pleistocene lakes in the Owens River system and Death Valley, California (after Benson* et al. *1990)*

the southwestern deserts of North America (Antevs 1955; Van Devender *et al.* 1987; Spaulding 1991; Tchakerian 1994).

Death Valley (the other hyper-arid region in North America), a classic graben between two block-faulted mountain systems, the Black and Funeral mountains on the east and the Panamint Range on the west, has been the focus of great attention, because of the continuous downdropping of the basin, hyper-aridity, and the landforms. The floor of Death Valley is occupied by one of the world's largest playas, covering more than 520 km^2 and crusted by a variety of evaporites (see Figure 26.9), the latter's distribution (horizontally and

vertically) controlled primarily by differences in solubility, groundwater fluctuations and basin geomorphology (Hunt 1975). During the late Pleistocene, a much larger pluvial lake, Lake Manly (Blackwelder 1933; Hunt 1975), occupied the basin and served as the ultimate sink for a series of interconnected pluvial lakes and stream systems, including the Owens and Mojave river systems (Figure 23.10). However, it is the alluvial fans and the massive bajadas developed in the lower piedmonts of the Panamint Range (Figures 23.8 and 23.11), that have received intense scrutiny (Blackwelder 1931; Denny 1965; Hunt and Mabey 1966; Hooke 1968; 1972; Hunt 1975; Bull 1977;

Figure 23.11 *SPOT satellite image of southern Death Valley, with the Panamint Range and its spectacular alluvial fan and bajada systems on the west, and the Black Mountains, with their smaller fan systems, on the east. Differences reflect the tectonic setting of the basin. (Photo courtesy of R. Dorn)*

Dorn 1988; see Chapter 12). The alluvial fans on the Panamint Range are long, segmented and entrenched (Figure 23.12). Continuing tectonic activity, changes in climate (fluctuations in discharge), the hyper-aridity and the sparse vegetation cover, have all been instrumental in the formation and preservation of these impressive fan complexes. Dorn (1988) identified three semi-arid to arid cycles of alluvial fan development (from about 650 ka to 13 ka) in southern Death Valley, based on rock varnish dating (both cation-ratio and AMS radiocarbon dates) and microchemical analysis.

THE MOJAVE DESERT

With an area of about 140 000 km², the Mojave Desert is the smallest of the North American deserts and occupies 10% of the arid zone (Figure 23.1). The San Andreas Fault and the Transverse Ranges form the western boundary, and the Garlock Fault, the northern. In the south and the north, it merges with the Sonoran Desert and the Great Basin, respectively. The Mojave Desert is considered a type of 'transitional' desert between the more classic Basin and Range section to the north and the Sonoran Desert to the south. Geomorphologically, its landform assemblages and drainage systems, have more in common with the Great Basin than the Sonoran Desert. The Mojave Desert is predominantly a rain-shadow desert, although continentality seems to also contribute to its aridity (although to a lesser degree than in the Great Basin). Because of its more southerly location, winters are more moderate than the Great Basin, with snow and freezing temperatures restricted to higher elevations. Most of the precipitation falls during the winter months and averages less than 150 mm (Table 23.1).

Figure 23.12 *Uppermost section of Hanaupah Canyon Fan, Panamint Range, Death Valley, California, showing fanhead entrenchment and abandoned alluvial fan surfaces. (Photo courtesy of R. Dorn)*

The drainage pattern in the Mojave Desert is largely endoreic. North–south trending, fault-block mountains and basins with playas, characterise the Mojave Desert. However, in most of the Mojave Desert (especially in the western sections), extensional tectonism ended during the Miocene, followed by a long period of weathering and erosion, basin sedimentation and pedimentation (Figure 23.13), with climate as the controlling factor in landscape evolution (Dohrenwend *et al.* 1991). Some tectonic activity is still evident in the south-central section, perhaps as a result of its proximity to the active San Andreas Fault (Ford *et al.* 1990; Dohrenwend *et al.* 1991).

Significant contributions to the geomorphology of the Mojave Desert include the many studies of Elliot Blackwelder on desert playas, yardangs, weathering, Pleistocene glaciation and pluvial lakes (1931; 1934; 1954), and Robert Sharp on granite landforms, pediments and aeolian geomorphology (1957; 1964; 1966). The microbial origin of rock varnish was first investigated and 'fine-tuned' in the Mojave Desert and the Great Basin (Dorn and Oberlander 1982; Dorn *et al.* 1989; 1992). Other stimulating geomorphological research include studies on

pediments (Cooke 1970; Oberlander 1974; Dohrenwend 1994) and the role of wind (past and present) in the Mojave Desert (Smith 1967).

In the Mojave Desert, the late Pleistocene to Holocene climatic transition and its influence on geomorphic systems has been analysed by various investigators using geomorphic, sedimentological, rock varnish and soil-stratigraphic analysis (McFadden *et al.* 1986; Dohrenwend 1987; Dorn *et al.* 1987; Wells *et al.* 1987). These studies and others indicate, late Pleistocene alluvial fan surfaces with relict sheetflood bedforms and subdued bar and slew microtopography, strongly varnished and well-developed desert pavements and, palaeosols with argillic B-horizons and calcrete. On the other hand, Holocene fan surfaces have very well-developed bar and swale topography, with larger, less weathered and thinly varnished clasts and poorly developed, thin soils. These studies also suggest a late Pleistocene environment, with thicker soils and denser vegetation cover on desert hillslopes, limiting the erosive capability of overland flow, with clasts transported and deposited in close proximity to source regions. In addition, most clasts are finer and exhibit no dramatic reduction in particle

Figure 23.13   *Pediment surface, Granite Mountains, Mojave Desert, California*

size with increasing distance from source areas (Mayer *et al.* 1984). In contrast, Holocene hillslopes have coarser clasts, with particle size decreasing rapidly with increasing distance from source areas, indicating a more erosive overland flow, owing largely to the reduction in vegetative cover and to thinner soils (Dohrenwend 1987; Wells *et al.* 1987; Dohrenwend *et al.* 1991). Most desert soils are polygenetic, having formed during arid, semi-arid and subhumid conditions (McFadden *et al.* 1986; Dohrenwend 1987).

With the exception of the sand dunes in the Gran Desierto del Altar in the Sonoran Desert, the Mojave Desert is the only arid zone with significant aeolian landforms, keeping in mind that sand dunes form only a minor component of the total surficial deposits in the North American arid zone (Tchakerian 1994; see Chapter 1). The aeolian landforms include ventifacts, yardangs, lunettes, sand dunes (both fossil and active) and sand ramps (Smith 1982).

One of the largest and best studied dunefields in the Mojave Desert is the Kelso Dunes (Figure 23.14). Dune formation reflects local topographic controls, changes in the frequency, magnitude and direction of sand-transporting

winds and the fluctuations in sand supply and vegetation cover (Sharp 1966; Smith 1967; Tchakerian 1989; Lancaster 1993; 1994). Both linear and crescentic dunes are represented with some older dunes superimposed on younger deposits reflecting changes in the wind regime and sediment supply. Luminescence ages for dune sediments at the Kelso Dunes indicate extensive reworking of the sands owing to the fact that most ages are between 50 and 4000 years (Lancaster 1993). However, luminescence dates from nearby sand sheets and sand ramps indicate two major periods of late Pleistocene aeolian deposition between >35 to 25 ka and 15 to 10 ka (Rendell *et al.* 1994).

Most of the aeolian sediments in the Mojave Desert are found as sand sheets, climbing and falling dunes and sand ramps. The latter are especially noteworthy because of their unique geomorphic setting. Sand ramps (Chapter 17) contain mostly aeolian sands deposited as sand sheets against the flank of mountain fronts (Figure 23.15), with significant fluvial and mass-movement units (Tchakerian 1991; Lancaster and Tchakerian 1995). They occur astride well-defined local and regional sand transport corridors (Zimbelman *et al.* 1995),

Figure 23.14 *Kelso Dunes, Mojave Desert, California. View west to the Kelso and Bristol mountains and the Devil's Playground sands. Note the active linear dune ridge flanked by degraded, vegetation-stabilised crescentic dunes*

Figure 23.15 *Sand ramp on the western slopes of the Sheephole Mountains, southern Mojave Desert, California. Multiple aeolian depositional episodes with fluvial and mass-movement units comprise this 40 m exposed section*

and contain multiple aeolian depositional units separated by palaeosols with well-preserved Bk-horizons and calcified root nodules. They are a major source of palaeoenvironmental information and provide a record of the response of aeolian processes to climatic perturbations. Most are relict and formed during periods of higher sediment supply and reduced vegetation cover, as a result of the gradual desiccation of desert lakes, especially during the late Pleistocene (Tchakerian 1991; Lancaster and Tchakerian 1995).

The Colorado Plateau

Located where Utah, Colorado, New Mexico and Arizona converge, the Colorado Plateau covers an area of about 375 000 km². However, the main semi-arid zone covers an area of about 170 000 km² (excluding the more montane zones such as the Grand Canyon region and the high plateaus of Utah) and constitutes 11% of the North American arid zone (Figure 23.1). This core area is centred in southeastern Utah (Canyon Lands Section), and characterised by horizontal to gently folded sedimentary strata, dissected by the Colorado and Green rivers (Figure 23.16), with spectacular cliffed escarpments and sandstone canyons, and a southern region (Navajo Section) around Monument Valley in Arizona, with mesa and butte topography and isolated volcanic plugs (Graf *et al.* 1987).

The average elevation of the Colorado Plateau is about 1600 m and average precipitation values between 100 and 400 mm (Table 23.1). Most of the precipitation falls during the summer from convective thunderstorms with snow common during the winter. The aridity is controlled by the continentality and the rain-shadow effects of the ranges surrounding the Colorado Plateau. The current geomorphic landscape is the result of the compression and uplift of the region during the early Tertiary, deformation associated with extensional tectonics during the formation of the Basin and Range Province to the west from the late Oligocene to the middle Miocene, and the Pliocene incision and evolution of the Colorado River system (Graf *et al.* 1987; Patton *et al.* 1991). The various climatic oscillations, from semi-tropical to arid, have left their imprint

on this complex landscape. If not for these processes, the Colorado Plateau would resemble a Saharo-Arabian type platform and erg desert (Oberlander 1994).

The Colorado Plateau is the birthplace of process fluvial geomorphology with classic works by Marvine, Powell, Gilbert, Dutton, Gregory and Hunt (Graf 1988). Noteworthy are the pioneering studies of John Wesley Powell (1875), in fluvial geomorphology, including the concepts of superposition, antecedence and base-level, among others, and Dutton's (1881; 1882) work on base-level, the Grand Canyon, and early explanations of alluvial fans. The most brilliant and influential was G.K. Gilbert. Gilbert's (1877) seminal report on the laccoliths of the Henry Mountains in southeastern Utah contains the earliest concepts in process geomorphology, including his 'laws' of declivities, later known as dynamic equilibrium. Gilbert also used thermodynamics to explain the various geomorphic processes and landforms. Gregory's (1917; 1938) contributions to fluvial and aeolian geomorphology and weathering, deserve much wider recognition.

Fluvial geomorphology and canyon formation and erosion have dominated the geomorphic studies in this arid zone. Graf (1982; 1983; 1985) studied the spatial variation of erosion and sedimentation in streams and their relation to changing stream power, as well as the effects of dams on vegetation and channel morphology along the Colorado River and its tributaries. Hereford (1984) used ground-based photographs and repeat photography to study the alluvial history of streams. Studies by Oberlander (1977) on segmented cliffs in sandstones (see Chapter 8), Laity and Malin (1985) on sapping and the development of amphitheatre-shaped valley networks (see Chapter 15), and Schmidt (1989) on scarp recession and talus flatirons, are especially consequential. Comprehensive reviews of desert slope development on bedrock are by Howard and Selby (1994) and in Chapter 8. Studies in aeolian geomorphology in the Navajo Section of the Colorado Plateau include works by Hack (1941) on dune types and Breed *et al.*'s (1984) exhaustive review of the aeolian erosional and depositional landforms, including the stabilised linear dune systems in northeastern Arizona, the largest of its kind in North America.

Figure 23.16 *Dead Horse Point, Colorado Plateau, Utah. Incision by the Green River in Palaeozoic sedimentary strata*

Badlands

Particularly within the Colorado Plateau, discussed above, and in the Canadian Province of Alberta, significant areas of badland areas occur. Within the North American arid zone, as in many other settings, their presence is more likely to be principally the result of lithological controls rather than those due specifically to climatic, tectonic or physiographic influences. Most North American badlands are found within the shales and associated lithologies and regoliths of Jurassic and Cretaceous age rocks. Important badland locations include southwestern Colorado, southeastern Utah, and northern Arizona, especially within the Mancos Shales and the Chinle and Morrison Formations, South Dakota, within the Brule Formation, Alberta, Canada, in Dinosaur Badlands Provincial Park, and southern California, in Anza-Borrego Desert State Park and surrounding areas. Significant contributions from North America to research on badlands include the Canada-based research of Bryan *et al.* (1978) and Campbell (1974; 1982), while the earlier

studies of Schumm (1956; 1962) conducted at Perth Amboy, outside the arid zone in New Jersey, and in South Dakota, are still considered benchmark works on badland processes. Badland geomorphology in an arid context is fully covered in Chapter 13.

Conclusions

In this relatively small region of the Earth's drylands, the North American arid zone affords a unique geomorphological setting for the study of desert processes and landforms. Tectonic movements, changes in base-level, climatic perturbations, changes in vegetation cover and density, and the unique geologic structure, have provided the theatre over which geomorphic processes have acted with great lucidity. This fascinating landscape also enabled the pioneering geographers and geologists of the nineteenth century (such as Powell and Gilbert) to initiate some of the first steps towards deciphering the complexity of Earth surface processes in drylands, and establish some of the earliest concepts in process geomorphology.

References

Abrahams, A.D., Luk, S.-H. and Parsons, A.J. 1988. Threshold relations for the transport of sediment by overland flow on desert hillslopes. *Earth Surface Processes and Landforms*, 13: 407–419.

Antevs, E.A. 1955. Geologic–climatic dating in the West. *American Antiquity*, 20: 317–335.

Baker, V.R. and Pyne, S. 1978. G.K. Gilbert and modern geomorphology. *American Journal of Science*, 278: 97–123.

Benson, L. V., Currey, D.R., Dorn, R.I., Lajoie, K.R., Oviatt, C.G., Robinson, S.W., Smith, G.I. and Stine, S. 1990. Chronology of expansion and contraction of four Great Basin systems during the past 35,000 years. *Palaeogeography, Palaeoclimatology, Palaeoecology*, 78: 241–286.

Blackwelder, E. 1931. Desert plains. *Journal of Geology*, 39: 133–140.

Blackwelder, E. 1933. Lake Manly. An extinct lake in Death Valley. *Geographical Review*, 23: 464–471.

Blackwelder, E. 1934. Yardangs. *Bulletin of the Geological Society of America*, 24: 159–166.

Blackwelder, E. 1954. Pleistocene lakes and drainage in the Mojave Region, southern California. *California Division of Mines Bulletin*, 170V: 35–40.

Breed, C.S., McCauley, J.F., Breed, W.J., McCauley, C.K. and Cotera, A. 1984. Eolian (wind-formed) landscapes. In T.L. Smiley, J.D. Nations, T.L. Pewe and J.P. Schafer (eds), *Landscapes of Arizona*. University of Arizona Press, Tucson: 359–413.

Bryan, K. 1925. *The Papago County*. US Geological Survey, Water Supply Paper, 499.

Bryan, R.B., Yair, A. and Hodges, W.K. 1978. Factors controlling the initiation of runoff and piping in Dinosaur Provincial Park badlands, Alberta, Canada. *Zeitschrift für Geomorphologie*, Supplementband, 34: 48–62.

Bull, W.B. 1977. The alluvial-fan environment. *Progress in Physical Geography*, 1: 222–270.

Campbell, I.A. 1974. Measurement of erosion on badland surfaces. *Zeitschrift für Geomorphologie*, Supplementband, 21: 122–137.

Campbell, I.A. 1982. Surface morphology and rates of change during a ten-year period in the Alberta badlands. In R.B. Bryan and A. Yair (eds), *Badland geomorphology and piping*. GeoBooks, Norwich: 221–236.

Cooke, R.U. 1970. Morphometric analysis of pediments and associated landforms in the western Mojave Desert. *American Journal of Science*, 269: 26–38.

Cooke, R.U., Warren, A. and Goudie, A. 1993. *Desert geomorphology*. UCL Press, London.

Crosswhite, F.S. and Crosswhite, C.D. 1982. The Sonoran Desert. In G.L. Bender (ed.), *Reference handbook on the deserts of North America*. Greenwood Press, Westport, CT: 163–295.

Currey, D.R. 1990. Quaternary paleolakes in the evolution of semidesert basins, with special emphasis on Lake Bonneville and the Great Basin, U.S.A. *Palaeogeography, Palaeoclimatology, Palaeoecology*, 76: 189–214.

Denny, C.S. 1965. *Alluvial fans in the Death Valley region, California and Nevada*. US Geological Survey, Professional Paper, 466.

Dohrenwend, J.C. 1987. Basin and Range. In W.L. Graf (ed.), *Geomorphic systems of North America*. Geological Society of America Centennial Special Volume 2, Boulder, CO: 303–342.

Dohrenwend, J.C. 1994. Pediments in arid environments. In A.D. Abrahams and A.J. Parsons (eds), *Geomorphology of desert environments*. Chapman & Hall, London: 321–353.

Dohrenwend, J.C., Bull, W.B., McFadden, L.D., Smith, G.I., Smith, R.S.U. and Wells, S.G. 1991. Quaternary geology of the Basin and Range Province in California. In R.B. Morrison (ed.), *Quaternary nonglacial geology: conterminous U.S.* Geological Society of America, The Geology of North America, Volume K-2, Boulder, CO: 321–352.

Dorn, R.I. 1988. A rock varnish interpretation of alluvial-fan development in Death Valley, California. *National Geographic Research and Exploration*, 4: 56–73.

Dorn, R.I. and Oberlander, T.M. 1982. Rock varnish. *Progress in Physical Geography*, 6: 317–367.

Dorn, R.I., Tanner, D., Turrin, B.D. and Dohrenwend, J.C. 1987. Cation-ratio dating of Quaternary materials in the east-central Mojave Desert, California. *Physical Geography*, 8: 72–81.

Dorn, R.I., Jull, A.J.T., Donahue, D.J., Linick, T.W. and Toolin, L.J. 1989. Accelerator mass spectroscopy radiocarbon dating of rock varnish. *Bulletin of the Geological Society of America*, 101: 1363–1372.

Dorn, R.I., Clarkson, P.B., Nobbs, M.F., Loendorf, L.L. and Whitley, D.S. 1992. New approach to the radiocarbon dating of rock varnish, with examples from drylands. *Annals of the Association of American Geographers*, 82: 136–151.

Dutton, C.E. 1881. The physical geology of the Grand Canyon district. *US Geological Survey, Second Annual Report*: 47–166.

Dutton, C.E. 1882. *Tertiary history of the Grand Canyon region*. US Geological Survey, Monograph, 2.

Ford, J.P., Dokka, R.K., Crippin, R.E. and Blom, R.G. 1990. Faults in the Mojave Desert, California, as revealed on enhanced Landsat images. *Science*, 248: 1000–1003.

Gilbert, G.K. 1877. *Report on the geology of the Henry Mountains*. Government Printing Office, Washington, DC.

Gilbert, G.K. 1890. *Lake Bonneville*. US Geological Survey, Monograph, 1.

Graf, W.L. 1982. Spatial variation of fluvial processes in semiarid lands. In C.E. Thorn (ed.), *Space and time in geomorphology*. Allen & Unwin, London: 193–217.

Graf, W.L. 1983. Downstream changes in stream power in the Henry Mountains, Utah. *Annals of the Association of American Geographers*, 73: 373–387.

Graf, W.L. 1985. *The Colorado River: instability and basin management*. Association of American Geographers: Washington, DC.

Graf, W.L. 1988. *Fluvial processes in dryland rivers*. Springer-Verlag, Berlin.

Graf, W.L., Hereford, R., Laity, J.E. and Young, R.A. 1987. Colorado Plateau. In W.L. Graf (ed.), *Geomorphic systems of North America*. Centennial Special Volume 2, Geological Society of America, Boulder, CO: 259–302.

Gregory, H.E. 1917. *Geology of the Navajo Country*. US Geological Survey, Professional Paper, 93.

Gregory, H.E. 1938. *The San Juan Country: A geographic and geologic reconnaissance of southeastern Utah*. US Geological Survey, Professional Paper, 188.

Hack, J.T. 1941. Dunes of the western Navajo Country. *Geographical Review*, 31: 240–263.

Haff, P.K. and Presti, D.E. 1995. Barchan dunes of the Salton Sea region, California. In V.P. Tchakerian (ed.), *Desert aeolian processes*. Chapman & Hall, London: 153–177.

Hereford, R. 1984. Climate and ephemeral-stream processes: twentieth-century geomorphology and alluvial stratigraphy of the Little Colorado River, Arizona. *Bulletin of the Geological Society of America*, 95: 654–668.

Hooke, R. LeB. 1968. Steady-state relationships on arid-region alluvial fans in closed basins. *American Journal of Science*, 266: 609–629.

Hooke, R. LeB. 1972. Geomorphic evidence for late-Wisconsin and Holocene tectonic deformation, Death Valley, California. *Bulletin of the Geological Society of America*, 83: 2073–2098.

Howard, A.D. and Selby, M.J. 1994. Rock slopes. In A.D. Abrahams and A.J. Parsons (eds), *Geomorphology of desert environments*. Chapman & Hall, London: 123–172.

Hunt, C.B. 1975. *Death Valley: geology, ecology and archaeology*. University of California Press, Berkeley.

Hunt, C.B. 1975. and Mabey, D.R. 1966. *Stratigraphy and structure, Death Valley, California*. US Geological Survey, Professional Paper, 494-A.

Inman, D.L., Ewing, G.C. and Corliss, L.B. 1966. Coastal sand dunes of Guerrero Negro, Baja California, Mexico. *Bulletin of the Geological Society of America*, 77: 787–802.

King, P.B. 1948. Geology of the southern Guadalupe Mountains, Texas. US Geological Survey, Professional Paper, 215.

Laity, J.E. and Malin, M.C. 1985. Sapping processes and the development of theater-headed valley networks on the Colorado Plateau. *Bulletin of the Geological Society of America*, 96: 203–217.

Lancaster, N. 1993. Kelso Dunes. *National Geographic Research and Exploration*, 9: 44–59.

Lancaster, N. 1994. Controls on aeolian activity: some new perspectives from the Kelso Dunes, Mojave Desert, California. *Journal of Arid Environments*, 27: 113–125.

Lancaster, N. 1995. Origin of the Gran Desierto Sand Sea, Sonora, Mexico: evidence from dune morphology and sedimentology. In V.P. Tchakerian (ed.), *Desert aeolian processes*. Chapman & Hall, London: 11–35.

Lancaster, N. and Tchakerian, V.P. 1995. Geomorphology and sediments of sand ramps in the Mojave Desert. *Geomorphology* (in press).

Logan, R.F. 1968. Causes, climates and distribution of deserts. In G.W. Brown (ed.), *Desert biology*, Vol. I. Academic Press, New York: 21–50.

Lynch, D.J. 1982. Pinacate's largest lava flow. *Noticias del Cedo*, 4: 1–13.

Mabbutt, J.A. 1977. *Desert landforms*. MIT Press, Cambridge, MA.

Mammerickx, J. 1964. Quantitative observations on pediments in the Mojave and Sonoran Deserts (southwestern United States). *American Journal of Science*, 262: 417–435.

Mayer, L., Gerson, R. and Bull, W.B. 1984. Alluvial gravel production and deposition; useful indicator of Quaternary climatic changes in deserts. In A.P. Schick (ed.), *Channel processes: water, sediment, catchment controls*. Catena, Supplement, 5: 137–151.

McFadden, J.C., Wells, S.G. and Dohrenwend, J.C. 1986. Cumulic soils formed in eolian parent materials on flows of the Cima volcanic field, Mojave Desert, California. *Catena*, 13: 361–389.

McKee, E.D. 1966. Structures of dunes at White Sands National Monument, New Mexico. *Sedimentology*, 7: 1–69.

Medellin-Leal, F. 1982. The Chihuahuan Desert. In G.L. Bender (ed.), *Reference handbook on the deserts of North America*. Greenwood Press, Westport, CT: 163–295.

Melton, M.A. 1965. Debris-covered hillslopes of the southern Arizona desert — consideration of their stability and sediment contribution. *Journal of Geology*, 73: 715–729.

Morrison, R.B. 1964. Lake Lahontan: geology of southern Carson Desert, Nevada. *US Geological Survey, Professional Paper*, 401.

Morrison, R.B. 1991. Quaternary stratigraphic, hydrologic, and climatic history of the Great Basin, with emphasis on Lakes Lahontan, Bonneville, and Tecopa. In R.B. Morrison (ed.), *Quaternary nonglacial geology: conterminous U.S.* Geological Society of America, The Geology of North America, Volume K-2, Boulder, CO: 283–320.

National Oceanic and Atmospheric Administration 1993. *Local climatological data: annual summaries for 1993.* National Climatic Center, Ashville, NC.

Norris, R.M. 1966. Barchan dunes of Imperial Valley, California. *Journal of Geology*, 74: 292–306.

Norris, R.M. and Norris, K.S. 1961. Algodones dunes of southeastern California. *Bulletin of the Geological Society of America*, 72: 605–620.

Oberlander, T.M. 1974. Landscape inheritance and the pediment problem in the Mojave Desert of southern California. *American Journal of Science*, 274: 849–875.

Oberlander, T.M. 1977. Origin of segmented cliffs in massive sandstones of southeastern Utah. In D.O. Doehring (ed.), *Geomorphology in arid regions.* Allen & Unwin, Boston: 79–114.

Oberlander, T.M. 1994. Global deserts: a geomorphic comparison. In A.D. Abrahams and A.J. Parsons (eds), *Geomorphology of desert environments.* Chapman & Hall, London: 13–35.

Orme, A.R. and Tchakerian, V.P. 1986. Quaternary dunes of the Pacific coast of the Californias. In W.G. Nickling (ed.), *Aeolian geomorphology.* Allen & Unwin, Boston: 149–175.

Patton, P.C., Biggar, N., Condit, C.D., Gillam, M.L., Love, D.W., Machette, M.N., Mayer, L., Morrison, R.B. and Rosholt, J.N. 1991. Quaternary geology of the Colorado Plateau. In R.B. Morrison (ed.), *Quaternary nonglacial geology: conterminous U.S.* Geology of North America, Volume K-2, Geological Society of America, Boulder, CO: 373–406.

Péwé, T.L. (ed.) 1981. *Desert dust.* Geological Society of America, Special Paper, 186.

Powell, J.W. 1875. *Exploration of the Colorado River of the west and its tributaries, 1869–1872.* US Government Printing Office, Washington, DC.

Rendell, H.M., Lancaster, N. and Tchakerian, V.P. 1994. Luminescence dating of late Quaternary aeolian deposits at Dale Lake and Cronese Mountains, Mojave Desert, California. *Quaternary Science Reviews*, 13: 417–422.

Russell, I.C. 1885. *Geological history of Lake Lahontan: A Quaternary lake of northwestern Nevada.* US Geological Survey Monograph, 11.

Russell, I.C. 1889. Quaternary history of Mono Valley, California. US Geological Survey Eighth Annual Report, 1886–1887. Government Printing Office, Washington, DC: 261–394.

Sack, D. 1994. Geomorphic evidence of climate change from desert-basin paleolakes. In A.D. Abrahams and A.J. Parsons (eds), *Geomorphology of desert environments.* Chapman & Hall, London: 616–630.

Schmidt, K.H. 1989. The significance of scarp retreat for Cenozoic landform evolution on the Colorado Plateau, USA. *Earth Surface Processes and Landforms*, 14: 93–105.

Schumm, S.A. 1956. Evolution of drainage systems and slopes at Perth Amboy, New Jersey. *Bulletin of the Geological Society of America*, 67: 597–646.

Schumm, S.A. 1962. Erosion on miniature pediments in Badlands National Monument, South Dakota. *Bulletin of the Geological Society of America*, 73: 719–724.

Sharp, R.P. 1940. Geomorphology of the Ruby–East Humboldt Range, Nevada. *Bulletin of the Geological Society of America*, 51: 337–372.

Sharp, R.P. 1957. Geomorphology of the Cima Dome, Mojave Desert, California. *Bulletin of the Geological Society of America*, 68: 273–290.

Sharp, R.P. 1964. Wind-driven sand in the Coachella Valley, California. *Bulletin of the Geological Society of America*, 75: 785–804.

Sharp, R.P. 1966. Kelso Dunes, Mojave Desert, California. *Bulletin of the Geological Society of America*, 77: 1045–1074.

Smith, G.I. and Street-Perrott, F.A. 1983. Pluvial lakes of the western United States. In S.C. Porter (ed.), *Late Quaternary environments of the United States I, The Late Pleistocene.* University of Minnesota Press, Minneapolis: 190–211.

Smith, H.T.U. 1967. *Past versus present wind action in the Mojave Desert region, California.* US Air Force Cambridge Research Labs Publication, AFLCRL-67-0683.

Smith R.S.U. 1982. Sand dunes in North American deserts. In G.L. Bender (ed.), *Reference handbook on the deserts of North America.* Greenwood Press, Westport, CT: 481–526.

Spaulding, W.G. 1991. A middle Holocene vegetation record from the Mojave Desert of North America and its palaeoclimatic significance. *Quaternary Research*, 35: 427–437.

Spaulding, W.G. and Graumlich, L.J. 1986. The last pluvial climate episodes in the deserts of southwestern North America. *Nature*, 320: 441–444.

Tchakerian, V.P. 1989. Late Quaternary aeolian geomorphology, east-central Mojave Desert, California. Unpublished PhD dissertation, University of California, Los Angeles.

Tchakerian, V.P. 1991. Late Quaternary aeolian geomorphology of the Dale Lake Sand Sheet, southern Mojave Desert, California. *Physical Geography*, 12: 347–369.

Tchakerian, V.P. 1994. Palaeoclimatic interpretations from desert dunes and sediments. In A.D. Abrahams and A.J. Parsons (eds), *Geomorphology of desert environments*. Chapman & Hall, London: 631–643.

Thompson, R.S., Whitlock, C., Bartlein, P.J., Harriosn, S.P. and Spaulding, W.G. 1993. Climatic changes in the western United States since 18,000 yr B.P. In H.E. Wright, Jr., J.E. Kutzbach, T. Webb III, W.F. Ruddiman, F.A. Street-Perrott and P.J. Bartlein (eds), *Global climates since the last glacial maximum*. University of Minnesota Press, Minneapolis: 468–513.

Universidad Nacional Autonoma de Mexico 1970. *Carta de climas*. Instituto de Geographia, Mexico City.

Van Devender, T.R., Thompson, R.S. and Betancourt, J.B. 1987. Vegetation history of the deserts of southwestern North America: the nature and timing of the late Wisconsin-Holocene transition. In W.F. Ruddiman and H.E. Wright, Jr (eds), *North America and adjacent oceans during the last deglaciation*. Geology of North America, Volume K3, Geological Society of America, Boulder, CO: 323–352.

Wells, S.G., McFadden, L.D. and Dohrenwend, J.C. 1987. Influence of late Quaternary climatic changes on geomorphic and pedogenic processes on a desert piedmont, eastern Mojave Desert, California. *Quaternary Research*, 27: 130–146.

Wilkins, D.E. and Tchakerian, V.P. 1994. Palaeoclimatic implications from aeolian quartz dunes, Salt Basin Graben, Hudspeth County, Texas. *Association of American Geographers, Abstracts with Programs*, 90: 407–408.

Zimbelman, J.R., Williams, S.H. and Tchakerian, V.P. 1995. Sand transport paths in the Mojave Desert, southwestern United States. In V.P. Tchakerian (ed.), *Desert aeolian processes*. Chapman & Hall, London: 101–129.

24

South America

I.A. 'Lillan' Berger

Introduction

The deserts of South America stretch in an almost continuous belt along the Andes from the hyper-arid, subtropical Pacific coast, diagonally across the high-altitude desert of the Altiplano-Puna, to the arid east coast in the temperate latitudes of Patagonia. The Andes also dominate tectonically, but the setting ranges from an active subduction margin along the west coast, to the uplifted plateau of the Altiplano-Puna, through to the continental craton and passive Atlantic margin. Accordingly the deserts are dominated by a basin and range structure. Additional semi-arid zones occur in the northeast of Brazil and along the Caribbean coast of Venezuela and Colombia.

In each desert the relative importance of the major factors which cause aridity — dynamic anticyclonic subsidence, orographical influences, continentality and cold offshore currents — differs (Thompson 1975; Cooke *et al.* 1993), and in addition the characteristics of the arid climates vary according to latitude and altitude (UNESCO 1979; Figure 24.1). Evidence from aeolian palaeoforms indicates that during glacial maxima of the Quaternary, arid conditions extended over large areas east of the Andes, from the Caribbean Sea to the River Plate estuary, where present-day vegetation includes savanna woodland, open savanna, chaco woodland and pampa grassland (Figure 24.2). On the other hand it seems clear that the Atacama Desert has been predominantly hyper-arid since the middle Miocene, whereas palaeolake shorelines indicate 'pluvial' episodes in the Bolivian and Chilean Altiplano, and in Patagonia (Clapperton 1993).

The most important controls on, and geomorphological features of, the desert regions along the Andean belt will be considered, and the salt deposits of the Atacama and Altiplano-Puna deserts examined in more detail. The literature on the desert geomorphology of the Monte and Patagonian deserts is not extensive. Several regional studies of geomorphological aspects of the Atacama Desert exist (Paskoff 1970; Mortimer 1973; 1980; Stoertz and Ericksen 1974; Mortimer and Saric 1975; Ericksen 1981; Oberlander 1994), but despite a recent upsurge in interest, much also remains to be investigated west of the Andes.

West of the Andes

The great trade wind desert of South America, the Peru–Chile desert, is considered to be

Arid Zone Geomorphology: Process, Form and Change in Drylands, 2nd edition. Edited by David S. G. Thomas.

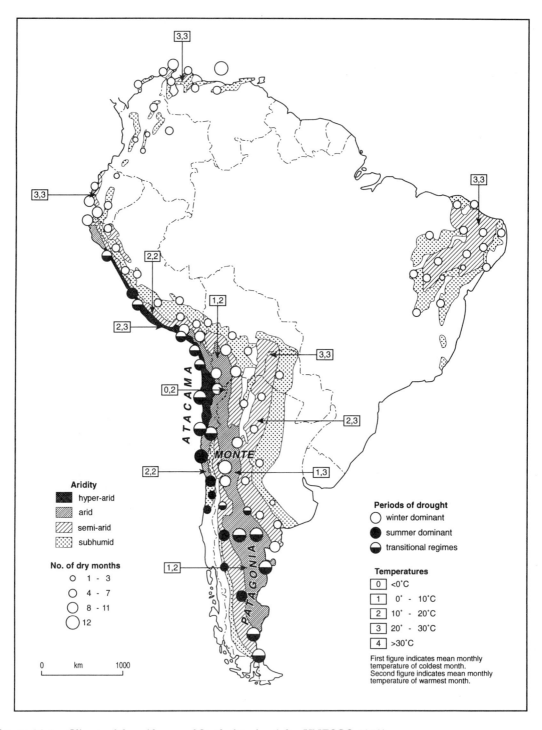

Figure 24.1 *Climate of the arid zones of South America (after UNESCO 1979)*

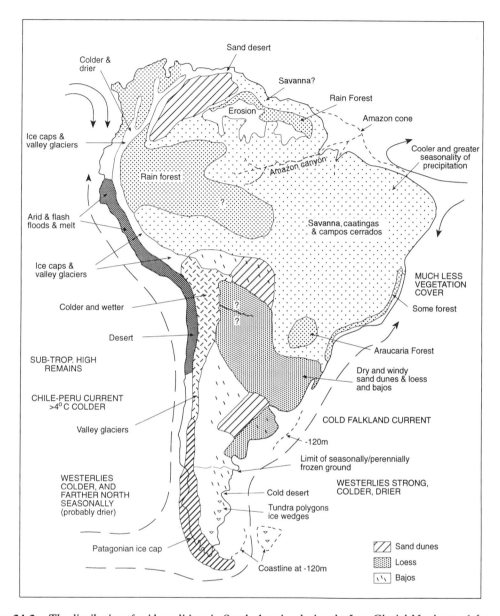

Figure 24.2 *The distribution of arid conditions in South America during the Late Glacial Maximum (after Clapperton 1993)*

among the driest in the world. It extends across the equator from southern Ecuador to stretch some 3700 km along the coast and western slopes of the Central Andes of Peru and northern Chile, and into the high-altitude (3800 m a.s.l.) intermontane plateau of the Altiplano-Puna, reaching into southern Bolivia and northwestern Argentina (Figure 24.3).

Tectonic uplift and the eastward migration of the Andean volcanic arc, controlled by subduction of the Nasca oceanic plate beneath the South American continent, have created one of the greatest relative reliefs on Earth, from Andean peaks up to 7000 m a.s.l. to the offshore Peru–Chile trench some 7600 m b.s.l. over a distance of less than 300 km (Galli-Olivier 1969;

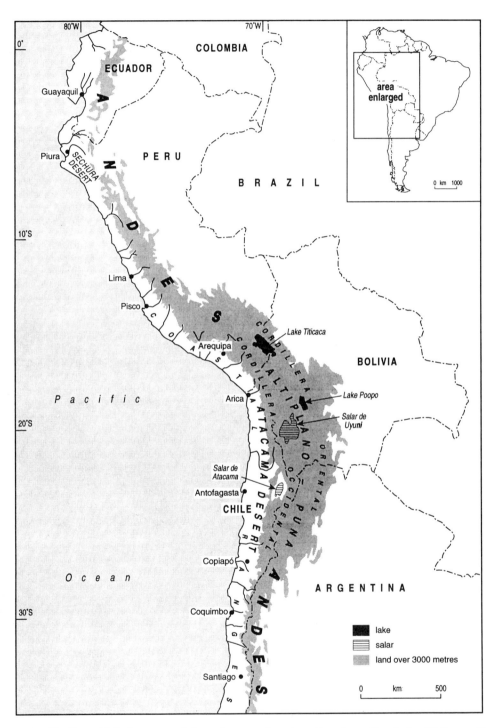

Figure 24.3   *The Peru–Chile desert extends from the coast of southern Ecuador to Coquimbo, Chile, reaching inland to include the high-altitude desert of the Altiplano-Puna*

Palacios 1993). Northern Chile and southern Peru are dominated by a longitudinal basin and range topography (Coastal Ranges, Central Valley, Pre-Cordillera, High Andes), which originated from block tectonism and the beginning of major Andean uplift in the late Oligocene (Lahsen 1982; Naranjo and Paskoff 1985), whereas further north the Andes rise close to the Pacific shoreline until the coastal plain widens in the Sechura Desert, northernmost Peru (De Vries 1988). The Altiplano, a large (500 000 km^2), closed basin between the Cordilleras Occidental and Oriental of the High Andes, is composed of many individual, endoreic basins formed by Tertiary and Quaternary tectonic and volcanic activity (Gardeweg and Ramírez 1985; De Silva 1989). Contrary to early belief, uplift in the southern Central Andes appears to have been limited since the last major episode in the late Miocene (Clark *et al.* 1967; Mortimer 1973; Tosdal *et al.* 1984; Kött *et al.* 1995; Vandervoort *et al.* 1995), and geochronologies of coastal terrace sequences indicate differential but low typical rates of uplift (0.1–0.2 m ka^{-1}) during the Quaternary (Goy *et al.* 1992; Hsu 1992; Leonard and Wehmiller 1992; Marcharé and Ortlieb 1992).

Aridity is induced primarily by subtropical atmospheric subsidence, reinforced by upwelling of the north-flowing, cold Humboldt Current and by the Andean rain shadow (Thompson 1975). The South Pacific Anticyclone is further stabilised by the flexure of the coastline, but localised, torrential precipitation may occur at intervals of two to ten years during El Niño–ENSO anomalies (Enfield 1989). The region has been predominantly arid since at least the late Eocene (Marty *et al.* 1988), and the onset of current hyper-aridity has been attributed to an increase in upwelling and/or intensity of the Humboldt Current in the middle Miocene, due to the development of the Antarctic ice cap, and the elevation of the Andes to a least 2–3000 m a.s.l. to create an effective rain shadow (Alpers and Brimhall 1988; Clark *et al.* 1990). As a core desert, the coast and western slopes of the Andes appear to have been dominantly arid throughout the Quaternary (Ortlieb and Macharé 1989; Clapperton 1993), with glaciation restricted to only the highest mountains in the arid, southern Altiplano (Hollingworth and Guest 1967;

Ramírez 1988). Lake shorelines and sediments (Clapperton 1993; Grosjean 1994; Grosjean *et al.* 1995; Servant *et al.* 1995; Valerogarces *et al.* 1996), soil development (Messerli *et al.* 1993; Veit 1993), and archaeological sites (Lynch 1986; 1990) do, however, indicate an increase in precipitation in the Altiplano and Chilean Pre-Cordillera during late Quaternary glacial intervals and periods of the Holocene. The persistence of El Niño events throughout at least the Holocene is suggested by flood deposits in northern Peru (Wells 1990), but at present evidence for earlier and 'mega' El Niño events is incomplete (De Vries 1987; Ortlieb and Macharé 1993; Meggers 1994).

The considerable latitudinal extent of the Peru–Chile desert, the increase in elevation eastwards, and the basin and range structure lead to varied regional and local conditions of aridity. The most intense aridity occurs in the Atacama Desert of northern Chile (<10 mm rainfall per year), where the Andean snowline lies at some 6000 m a.s.l. (Abele 1992). An increase in precipitation and lowering of the snowline occurs southwards into the semi-arid Norte Chico, linked with winter incursions of the polar front (Miller 1976), and also northwards due to more intense El Niño events and Andean overspill of summer precipitation from the Amazon, the so-called 'Bolivian Winter' of the Altiplano (Johnson 1976). Anomalously high relative humidity (>70%) and cloudiness, and low mean annual temperatures (16–22°C) characterise the coastal desert between 8°S and 30°S due to the influence of the winter advection fog; the *garúa* of Peru supports a variety of coastal *loma* vegetations (Rauh 1984), while the *camanchaca* of the Atacama Desert occurs at a greater altitude, is less moist, and considerably restricted by the Coastal Ranges, although its range and moisture content increase southwards (Cereceda and Schemenauer 1991). Inland, relative humidity, cloudiness, and mean annual temperature decrease with altitude, while the mean seasonal and diurnal temperature range increases. Regional winds along the coast are mostly southeasterly to southwesterly, but they become variable inland with a strong diurnal mountain-valley component, and frequently reach high velocities in the Altiplano.

The Peru–Chile desert is primarily a rocky desert. Ventifacted bedrock and clasts are common, and well-developed, mobile barchan fields occur in southern Peru (Hastenrath 1967; 1987: see Chapter 17 for discussion of their migration), but true ergs are absent. Tectonic controls and endogenic landforms, such as fault scarps, thrust sheets and diverted drainage, are ubiquitous (see, for example, Cooke and Mortimer 1971; Paskoff 1978; Jolley *et al.* 1990), a function of active seismicity and volcanicity, and low denudation rates due to long-term aridity and relatively little Andean uplift since the late Miocene. Regionally this is reflected in the widespread preservation of a series of broadly correlated pediplains dating from episodic uplift during the Miocene (Tosdal *et al.* 1984), and evidence abounds for continued neotectonic activity, such as segmented alluvial fans along the Atacama Fault Zone (Figure 24.4) and deformation of marine terraces along the Coastal Scarp (Armijo and Thiele 1990). Exogenic processes become relatively more important in the south (Paskoff 1970; Mortimer 1973; Veit 1993) and north (Waylen and Caviedes 1986; Wells 1987) with the increase in precipitation (see above) and the regular occurrence of

exoreic rivers crossing the desert from the Andes (Mortimer 1980; Abele 1992). In the driest part of the Atacama Desert, however, traversed only by the Río Loa, the overriding importance of endogenic processes is unique (Mortimer and Saric 1972; 1975). Consequently, the barren landscape of this region of endoreic basins, widespread salars, and predominantly fossil groundwater (Messerli *et al.* 1993), has been described in Chapter 8 as 'erosionally paralysed' (Figure 24.5). Rock varnish on the debris-covered desert surface is mostly of the alkaline, iron-rich type (Oberlander 1982; Jones 1991; Dorn *et al.* 1992), and unusual geomorphological features include seismically induced slope creep (Oberlander 1994), well-spaced desert pavements (Cooke 1970; Berger 1993; Oberlander 1994), and an abundance of salts.

SALTS IN THE ATACAMA AND ALTIPLANO-PUNA DESERTS

The salt assemblages that have accumulated in the Atacama and Puna deserts since the middle

Figure 24.4 *Fault scarp and fissures across a segmented alluvial fan straddling the Atacama Fault Zone east of the Coastal Ranges, near Antofagasta*

Figure 24.5 *Creep of surface clasts appears to be the dominant process on these salt-encrusted slopes, Atacama Desert*

Miocene are exceptional in quantity and nature. Salt-encrusted playas (salars) are widespread (Figure 24.6), much of the topography is mantled by a saline crust and soil — often with deposits at depth in both alluvium and bedrock — and considerable outcrops of tectonically deformed, Neogene continental evaporites are present in the Pre-Andean basins and the Puna (Figure 24.7). An unusual, regional paucity of carbonate minerals, and abundance of sulphates and chlorides has been observed (Stoertz and Ericksen 1974; Vila 1975; Alonso 1991; Risacher and Fritz 1991), and local concentrations of nitrates, borates, iodates, chromates, potassium and lithium are in some cases of unique character, containing rare minerals such as perchlorate (Ericksen 1981).

The nature and distribution of salt deposits depends on salt sources, and processes of distribution and selective concentration (Cooke *et al.* 1993). A primary source of salts in the Atacama and Puna deserts is provided by the ocean, volcanic activity and the breakdown of rock, and a secondary, cyclic input derives from past salt accumulations preserved in geological deposits. Variations in environment and relief, from the High Andes in the east to the Pacific Coast in the west, lead to regional differences in salt input and distribution processes (Figure 24.8). Most marked is the regional, climatic asymmetry noted by Mueller (1968), whereby precipitation in the Andes, a major source area of volcanic salts, flows westwards as groundwater and occasional surface runoff into the hyper-arid Central Valley, an area of salt deposition and restricted further mobility due to the rarity of rainfall events. Additional, local sources of salt in the Central Valley are the widespread salars, the salt-rich coastal fog (camanchaca), which may penetrate some 50–100 km inland, and the further dispersion of salts by both local and regional winds.

A range of theories have been invoked to explain the select distribution and the mineral assemblages associated with the nitrate deposits along the eastern margin of the Coastal Ranges, a debate reviewed by Ericksen (1981; 1983) but as yet unresolved. More recently, sources

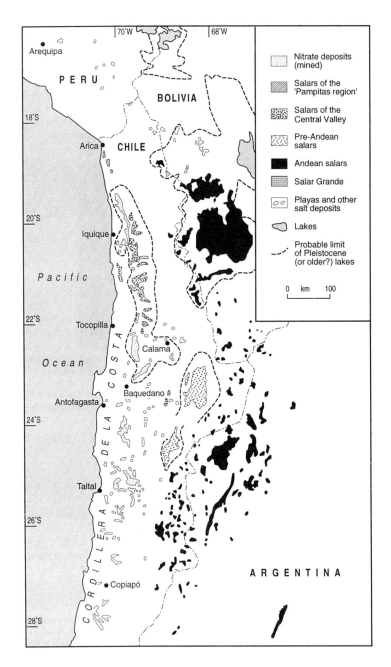

Figure 24.6 *The distribution of salars, saline lakes and salt deposits in the Atacama and Altiplano-Puna deserts (after Chong 1984)*

from Andean groundwaters and from the atmosphere (coastal fog, rainfall, and aeolian salts) have been emphasised, and deposit concentration and segregation attributed to multiple episodes of leaching, water table fluctuations, and capillary migration, according to salt solubility and deliquescence properties (Van Moort 1985; Searl and Rankin 1993).

Volcanism, specifically the weathering of Andean volcanic rocks and thermal activity, is the major source of a variety of relatively rare salts (such as borate, potassium and lithium) that are regionally widespread in salt deposits and groundwater, and present at high local concentrations in many basins of the Pre-Cordillera and Puna (Vila 1976; Heathcote,

Figure 24.7 *A view eastwards of the Valle de la Luna in the tectonically deformed evaporitic deposit of the Cordillera de la Sal, Salar de Atacama*

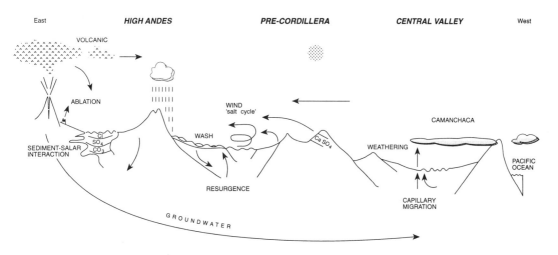

Figure 24.8 *A schematic illustration of salt input and distribution processes as they may occur along a cross-section of northern Chile — many may operate across the region*

1989). In fact, the large borate reserves of Neogene evaporites in the Puna have been linked, speculatively, to an increase in the rate of supply of geothermal solutions as a result of some dynamic event in magmatic arc history at c. 7 Ma ago (Alonso *et al.* 1991). The large sodium chloride component in salars and groundwaters of the region has been related to the resolution of Neogene continental evaporites, and the repeated resolution and precipitation of ancient salar crusts (Alpers and Whittemore 1990; Risacher 1992). The halite may have an original, volcanogenic source (Alonso *et al.* 1991), but alternatively may derive from the resolution of Cretaceous or earlier evaporites (Dingman 1962; Risacher 1992).

The regional paucity of carbonate minerals and widespread abundance of sulphates has received little specific attention, but a recent study of the geochemical evolution of inflowing waters to the Lipez salars, in the southern Bolivian Puna, suggests that the two may be related (Risacher and Fritz 1991). Contact with excess sulphur, widely dispersed in salar catchments by wind erosion of nearby crater and fumarole deposits, results in the oxidation of sulphur to sulphate and the neutralisation of carbonate, and thus the development of near-neutral brines where alkaline soda brines might otherwise be expected. Sulphates, especially calcium sulphate, are certainly prominent at and near the ground surface throughout the region, as has been noted by several authors (Stoertz and Ericksen 1974; Berger 1993; Searl and Rankin 1993; Oberlander 1994). This reflects both an abundance of calcium sulphate sources (Ericksen 1981) — from present and

past volcanism (including weathering of supergene copper deposits (Alpers and Whittemore 1990)), Neogene evaporites, precipitation from Amazon air masses (Risacher and Fritz 1991), the camanchaca (Schemenauer and Cereceda 1992), and salar deposits — and widespread aeolian redistribution (Berger and Cook 1997), while prevalence at and near the ground surface is favoured by a relatively low solubility.

Besides reflecting a series of past and present geomorphological processes within a context of largely continuous, long-term aridity, the salt deposits of the Atacama and Puna play an important geomorphological role themselves in this arid desert, most notably in salt tectonics (Wilkes and Görler 1988; Goudie 1989), salt-crust development (Stoertz and Ericksen 1974; Vila 1975; Chong 1984; 1988; Oberlander 1994), salt heave and slope equiplanation (Oberlander 1994), and salt weathering (Mortensen 1933; Wright and Urzúa 1963; Grenier 1968; Berger 1993). Many of the processes involved are moisture-dependent, and our understanding of their mode and rate of operation is severely limited by an inadequate

Figure 24.9 *Lichen are destroying rock varnish on an alluvial fan in the Central Valley which lies within reach of the camanchaca*

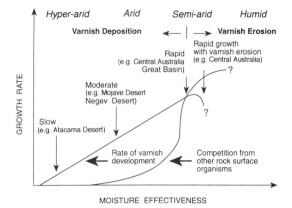

Figure 24.10 *A theoretical model of rock varnish growth in which slow varnishing characterises the hyper-arid Atacama Desert and competition from rock surface organisms may completely inhibit varnish development when the environment becomes too moist as in parts of Australia (after Dorn and Dragovich 1990)*

knowledge of the details of both regional and local moisture availability. Snowfall, for instance, appears to be underestimated in precipitation values for the Andes (Messerli *et al.* 1993), although rapid loss may result from sublimation. Furthermore, observations of rock varnish destruction by lichen, in a part of the hyper-arid Central Valley reached by the camanchaca (Figure 24.9), suggest that the latter may be a source of effective moisture, and therefore that the model of rock varnish development presented by Dorn and Dragovich (1990; Figure 24.10) is very much scale dependent (Berger 1993).

East of the Andes

MONTE

The Monte Desert lies in the subtropical, continental interior of Argentina, in the lee of the Central Andes (Figure 24.11). It is dominated by the basin and range structure of the Sierras Pampeanas, a series of north-trending basement blocks upfaulted in a tectonic foreland environment in association with the uplift of the Altiplano-Puna immediately to the north; a modern analogue to the basement uplifts of the Laramide province of the southwest United States (Jordan and Allmendinger 1986).

Regional topography decreases southwards, and to the east abuts dramatically with the undeformed lowlands of the semi-arid western Chaco and the Argentinian Pampa stretching to the Atlantic.

Pacific air masses are effectively blocked by the high altitude of the Andean Cordillera at this latitude, and aridity is enforced by the Sierras Pampeanas, which obstruct moisture-laden, subtropical air masses from the northeast. A narrow belt of high, orographic precipitation (>1000 mm a^{-1}) and subtropical jungle to the east of the easternmost ranges thus contrasts with the xeric basins of typical *monte* vegetation to the west, where precipitation falls to below 200 mm a^{-1}. Geological and geomorphological evidence indicates that major uplift of the Sierras Pampeanas occurred between 4.0 and 3.4 Ma ago, culminating after 2.9 Ma ago, and that arid conditions have been largely prevalent since then (Strecker *et al.* 1989).

Precipitation is associated with a thermal low-pressure cell over the Gran Chaco in the summer, falling as heavy showers concentrated in a small number of days, sporadic and highly variable from year to year. This coincides with high temperature maxima ($>40°C$) due to intense heating of the continental interior, but daily and annual temperature ranges are considerable with temperature minima falling below zero, and relative humidity is characteristically low (40–60%). Local relief frequently modifies the dominant southeasterly trade winds, and occasionally the hot and dry, katabatic, *zonada* descends from the Andes with concomitant dust storms (Prohaska 1976). South of Mendoza a zone of transition begins to the winter precipitation and westerly wind regime of arid Patagonia, and within the Monte great variations in local and microclimatic conditions occur as a result of the extreme topographical contrasts (Mares *et al.* 1984).

The Sierras Pampeanas consist of Precambrian to Lower Palaeozoic metamorphic and igneous rocks that have been uplifted asymmetrically along high-angle reverse faults, commonly exposing an Upper Palaeozoic erosion surface. The intervening basins have been infilled with Tertiary and Quaternary sediments, and are characterised by a piedmont zone of pediments and alluvial fans with frequent fault scarps, grading into alluvial plains,

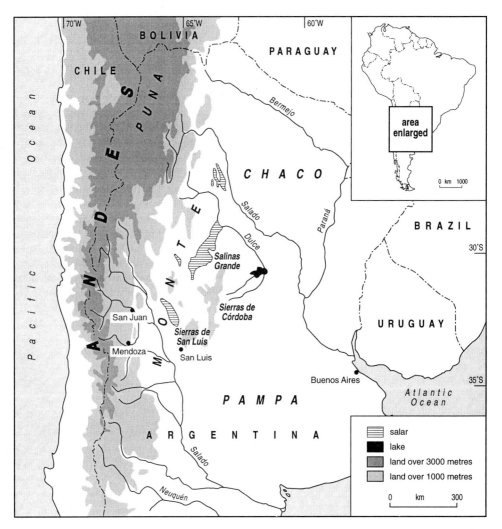

Figure 24.11   *The Monte Desert lies in the lee of the Andes, within the basin and range structure of the Sierras Pampeanas*

playas and salt pans. Although drainage is for the most part endoreic and characterised by ephemeral channels, groundwater and some rivers rising in the Andes to the west are supplied by snow and glaciermelt, and give rise to local *bañados* (swamps) where overspill occurs in years of high discharge. Large pediment surfaces are prominent in the region and have attracted considerable attention (see, for example, Stingl *et al.* 1983; Garleff *et al.* 1983), but local studies have also emphasised the variety of landforms present, which include badlands, arroyos, active and stabilised dune-fields, deflation basins, and a variety of salt

deposits (Cordini 1948; Polanski 1963; Suvires 1989).

The relative impact of neotectonics and Quaternary changes in climate on landform evolution in the Monte Desert are frequently difficult to distinguish. Nevertheless, as evidence for Plio-Quaternary neotectonic activity is all-pervasive (see, for example, Costa 1989), and it appears that glaciations at these latitudes were restricted to the high peaks of the Andes while the basins to the east remained arid, several studies (e.g. Fernández 1984) have attributed the major control on landform development to neotectonic activity. Notably,

Strecker *et al.* (1989) have dated the formation of five pediment levels of the eastern Santa María valley to the period between 2.5 and 0.3 Ma BP, relating the deformation and subsequent dissection of the three older pediments to uplift pulses and base-level changes at 2.5 Ma, 1.2 Ma, and 1.2–0.6 Ma BP. Thus it is apparent that strong tectonic movements occurred until at least the mid- to late Quaternary, and local earthquake records suggest that seismicity continues.

Given largely continuous Quaternary aridity, the concurrence in the Monte Desert of relatively small systems of active dunes and larger palaeoforms may be due to variations in sediment availability, thus an increase in glaciofluvial sediment entering the region from the Andes during glacial periods. Indeed, several studies have identified a relationship between local, fluvial sediments and the formation of both active dunes, such as the Cafayate dunefield (Salta Province), and stable palaeoforms (Cordini 1948; Polanski 1963; Cortelezzi *et al.* 1984). On a larger scale, a variety of palaeoaeolian forms (loess, sand sheets, dunefields, and deflation basins) to the east of the Monte, in the presently subhumid to humid Chaco and Pampa (Iriondo 1989), have been related to increased sedimentation on alluvial fans, floodplains, and an emerged continental shelf during the Last Glacial Maximum, and to local reworking and redeposition during dry periods of the late Holocene (Iriondo 1993; Zárate and Blasi 1993). Fine, lacustrine sediments, exposed in the Puna upon desiccation at the end of glacial intervals, may have acted as an additional source for the Pampean loess deposits, transported eastwards as great dust plumes by the subtropical jet stream and subsequently redeposited in the Pampa (Clapperton 1993). The formation of calcretes in the Pampa has also been attributed to an aeolian input during Quaternary periods of increased aridity, originating from Andean volcanoes to the west (Tricart 1989).

PATAGONIA

The Patagonian Desert, the only east coast desert to exist in temperate latitudes, remains largely uninvestigated (Soriano 1983; Vogt and Del Valle 1994). The desert stretches between latitudes 39°S and 50°S and is fringed by semi-arid steppe along the Andean foothills and in the Tierra del Fuego to the south (Figure 24.12). It is essentially a stony desert, dominated by low-lying plateaux sloping eastwards from the Andes to the Atlantic, which are incised by wide, flat-floored valleys, and characterised by numerous endoreic basins. A degree of differential neotectonic uplift is suggested by Holocene (Codignotto *et al.* 1992) and Plio-Quaternary coastal terraces up to 186 m a.s.l. (Clapperton 1993).

The plateaux are predominantly formed by sheet basalts and coarse gravels, lying, often unconformably, over a horst and graben structure of Mesozoic–Cenozoic volcanic and sedimentary rocks and Precambrian basement (North and South Patagonian Massifs), and provide the context for contemporary arid processes and features. The widespread nature of the basalt plateaux and the areas of mongenetic cinder cones, lava flows and explosion craters reflect the significant impact of Tertiary–Quaternary back-arc volcanism on the region. Similarly, the extensive plains of the Patagonian Gravel Formation testify to the importance of 'glacio'-fluvial and 'pluvio'-fluvial processes in transporting enormous volumes of sediment into the desert during Cenozoic glaciations. These influences on the Patagonian Desert are exemplified by the occurrence of a remarkable Mio-Pliocene glacial sequence of interbedded basalt and till, which has been preserved in the long term by the limited tectonic uplift and predominant aridity of the region (Clapperton 1993).

Aridity is primarily due to the effect of the Andes, blocking rain-bearing westerly air masses and giving rise to the persistent, adiabatic, dry winds that characterise Patagonia for much of the year. This is reinforced by the narrow configuration of the landmass, precluding east coast precipitation from the low-pressure system centred offshore in the Atlantic, and by the cold Falkland Current, which inhibits any significant precipitation from depressions associated with the northerly migration of the Polar Front during winter (Thompson 1975). General aridity may date from the elevation of the Andes to a height allowing glaciation in the late Miocene, but

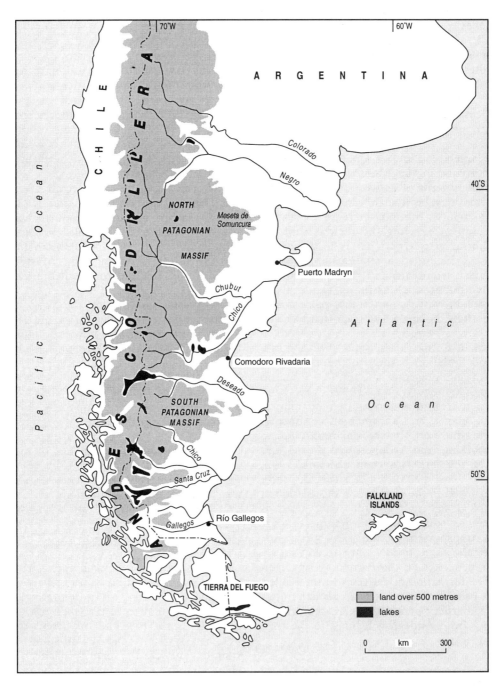

Figure 24.12   *South of the lower Rio Colorado, the tablelands of Patagonia gradually slope eastwards from the Andes*

although the desert tablelands remained beyond the glacial limit, proximity to the Patagonian ice cap and evidence from Quaternary–Holocene palaeolake shorelines suggest important fluctuations in the nature of Patagonian aridity (Clapperton 1993).

The region is classified as a cool winter, mild summer desert with irregular precipitation (UNESCO 1979). Precipitation increases from a low annual mean of *c.* 150 mm on the Atlantic coast to some 400 mm a^{-1} in the steppes at the foot of the Andes, where there is a winter maximum, attributed to a 'spillover' of precipitation from the Andean crest (Clapperton 1993). Temperatures decrease southwards from an annual mean of 14°C in the north to 5°C in Tierra del Fuego (Soriano 1983), but no significant change occurs inland, as the increase in elevation is counteracted by the cooling, oceanic effect on the coast (Meigs 1966). The mean position of the monthly 0°C isotherm to the south of the continent precludes low-altitude permafrost and limits the extent of

seasonally frozen ground (Clapperton 1993). One of the most dominant features of the climate is the constancy and strength (mean daily value of *c.* 7 m s^{-1}) of the dry westerly wind (Prohaska 1976).

Much of the desert surface is protected by low, sparse xerophitic shrubs, interspersed with stone pavements and surface crusts, and underlain by a vesicular horizon (Figueira 1984; Rostagno *et al.* 1991; Bouza *et al.* 1993). Pavement clasts may be varnished and ventifacted (Fidalgo and Riggi 1965), and at depth, fracturing of clasts within the Patagonian Gravel Formation has been attributed to gypsum crystallisation (Cortelezzi and Kilmurray 1965). The widespread occurrence of calcretes, closely associated with fossil ice-wedge pseudomorphs, has been attributed to seasonal and long-term changes in groundwater (Galloway 1985), but recent evidence suggests an aeolian origin from the extensive, carbonate-rich continental platforms that emerged during Pleistocene glacial maxima (Vogt 1992; Vogt

Figure 24.13   *'Erosion-tongues' in Patagonia (after Soriano and Movia 1986)*

and Del Valle 1994). A subject of early debate, it appears that the large endoreic depressions of Patagonia generally have a tectonic or glacial origin, and that the numerous, small enclosed basins have been formed by differential erosion (primarily deflation) of regolith, fluvial and glaciofluvial deposits, possibly during glacial intervals (Clapperton 1993). The capacity for aeolian erosion under present conditions is demonstrated by the large areas affected by 'erosion-tongues', composite aeolian features which may advance at rates of >500 m a^{-1} (Gargantini and Movia 1989), and one of the desertification processes linked to the introduction of sheep last century (Soriano and Movia 1986; Figure 24.13). Apart from the importance of gully erosion (Coronato and Del Valle 1993), fluvial erosion has received little attention at a scale other than the great fans forming the northern Patagonian Gravel Formation, but a clear interrelationship is evident between current fluvial and aeolian processes (Soriano 1983).

Conclusion

Studies of South American arid geomorphology reflect a strong European influence, the approach of individual French and German scientists, and more recently the Anglo-American approach, adapted to the major controls and features of the individual deserts. Thus, an appreciation of the fundamental control exerted by neotectonics in the Monte Desert led Polanski (1963) to develop an 'active' morphostructural approach, based on landform genesis and formative processes within a geological context, and to exert considerable influence as the 'father of Argentinian geomorphology' (González-Díaz 1993). Of wider impact, and based on his work in the Atacama Desert, Mortensen (1930; 1933) contributed the concept of 'morphological convergence' to climatic geomorphology, and was the first to demonstrate theoretically the effectiveness of salt hydration in rock weathering (Araya 1993). Recent technological developments in dating and the use of satellite imagery suggest that future work on the frequently extreme conditions of the South American deserts will not only continue to contribute to an understanding of desert processes and landform evolution specific to this region, but also to arid zone geomorphology as a whole.

References

Abele, G. 1992. Landforms and climate on the western slope of the Andes. *Zeitschrift für Geomorphologie*, Supplementband, 84: 1–11.

Alonso, R.N., Jordan, T.E., Tabbutt, K.T. and Vandervoort, D.S. 1991. Giant evaporite belts in the Neogene central Andes. *Geology*, 19: 401–404.

Alpers, C.N. and Brimhall, G.H. 1988. Middle Miocene climatic change in the Atacama Desert, northern Chile: evidence from supergene mineralization at La Escondida. *Bulletin of the Geological Society of America*, 100: 1640–1656.

Alpers, C.N. and Whittemore, D.O. 1990. Hydrogeochemistry and stable isotopes of ground and surface waters from two adjacent closed basins, Atacama Desert, northern Chile. *Applied Geochemistry*, 5: 719–734.

Araya, J.F. 1993. Geomorphology in Chile. In H.J. Walker and W.E. Grabau (eds), *The evolution of geomorphology*. Wiley, Chichester: 83–92.

Armijo, R. and Thiele, R. 1990. Active faulting in northern Chile: ramp stacking and lateral decoupling along a subduction plate boundary? *Earth and Planetary Science Letters*, 98: 40–61.

Berger, I.A. 1993. Salts and surface weathering features on alluvial fans in northern Chile. Unpublished PhD thesis, University of London.

Berger, I.A. and Cook, R.U. 1977. The origin and distribution of salts on alluvial fans in the Atacama Desert, northern Chile. *Earth Surface Processes and Landforms* (in press).

Bouza, P., Del Valle, H.F. and Imbellone, P.A. 1993. Micromorphological, physical, and chemical characteristics of soil crust types of the central Patagonia Region, Argentina. *Arid Soil Research and Rehabilitation*, 7: 355–368.

Cereceda, P. and Schemenauer, R.S. 1991. The occurrence of fog in Chile. *Journal of Applied Meteorology*, 30: 1097–1105.

Chong, G. 1984. Die Salare in Nordchile — Geologie, Struktur und Geochemie. *Geotektonische Forschungen*, 67: 1–146.

Chong, G. 1988. The Cenozoic saline deposits of the Chilean Andes between 18°00' and 27°00' South latitude. *Lecture Notes in Earth Sciences*, 17: 137–151.

Clapperton, C.M. 1993. *Quaternary geology and geomorphology of South America*. Elsevier, Amsterdam.

Clark, A.H., Mayer, A.E.S., Mortimer, C., Sillitoe,

R.H., Cooke, R.U. and Snelling, N.J. 1967. Implications of the isotopic ages of ignimbrite flows, southern Atacama Desert, Chile. *Nature*, 215: 723–724.

Clark, A.H., Tosdal, R.M., Farrar, E. and Plazolles, A. 1990. Geomorphologic environment and age of supergene enrichment of the Cuajone, Quellaveco, and Toquepala porphyry copper deposits, southeastern Peru. *Economic Geology*, 85: 1604–1628.

Codignotto, J.O., Kokot, R.R. and Marcomini, S.C. 1992. Neotectonism and sea-level changes in the coastal zone of Argentina. *Journal of Coastal Research*, 8: 125–133.

Cooke, R.U. 1970. Stone pavements in deserts. *Annals of the Association of American Geographers*, 60: 560–577.

Cooke, R.U. and Mortimer, C. 1971. Geomorphological evidence of faulting in the southern Atacama Desert, Chile. *Revue de Géomorphologie Dynamique*, 20: 71–78.

Cooke, R.U., Warren, A. and Goudie, A.S. 1993. *Desert geomorphology*. UCL Press, London.

Cordini, I.R. 1948. Cuerpos salinos de la Argentina. *Revista Asociación Geológica Argentina*, 3: 145–200.

Coronato, F.R. and Del Valle, H.F. 1993. Methodological comparisons in the estimate of fluvial erosion in an arid closed basin of northeastern Patagonia. *Journal of Arid Environments*, 24: 231–239.

Cortelezzi, C.R. and Kilmurray, J.O. 1965. Surface properties and epigenetic fractures of gravels from Patagonia, Argentina. *Journal of Sedimentary Petrology*, 35: 976–980.

Cortelezzi, C.R., Pavlicevic, R.E. and Rivelli, F.R. 1984. Estudio sedimentológico de las arenas de las dunas de Cafayate, Provincia de Salta, República Argentina. *Geociências*, 3: 47–65.

Costa, C.H. 1989. Remote sensing applied to neotectonics: case studies in San Luis Province (Argentina). *Zeitschrift für Geologische Wissenschaften*, 17: 25–36.

De Silva, S.L. 1989. Altiplano–Puna volcanic complex of the central Andes. *Geology*, 17: 1102–1106.

De Vries, T.J. 1987. A review of geological evidence for ancient El Niño activity in Peru. *Journal of Geophysical Research*, 92 (C13): 14471–14479.

De Vries, T.J. 1988. The geology of late Cenozoic marine terraces (tablazos) in northwestern Peru. *Journal of South American Earth Sciences*, 1: 121–136.

Dingman, R.J. 1962. Tertiary salt domes near San Pedro de Atacama, Chile. *US Geological Survey, Professional Paper*, 450-D: D92–94.

Dorn, R.I. and Dragovich, D. 1990. Interpretation of rock varnish in Australia: case studies from the arid zone. *Australian Geographer*, 21: 18–32.

Dorn, R.I., Clarkson, P.B., Nobbs, M.F., Loendorf, L.L. and Whitley, D.S. 1992. New approach to the radiocarbon dating of rock varnish, with examples from drylands. *Annals of the Association of American Geographers*, 82: 136–151.

Enfield, D.B. 1989. El Niño, past and present. *Reviews of Geophysics*, 27: 159–187.

Ericksen, G.E. 1981. Geology and origin of Chilean nitrate deposits. *US Geological Survey, Professional Paper*, 1188: 1–37.

Ericksen, G.E. 1983. The Chilean nitrate deposits. *American Scientist*, 71: 366–374.

Fernández, B.L. 1984. Stratigraphy of the Quaternary piedmont deposits of the Río de Las Tunas Valley, Mendoza, Argentina. *Quaternary of South America and Antarctic Peninsula*, 2: 31–40.

Fidalgo, F. and Riggi, J.C. 1965. Los Rodados Patagonicos en la Meseta del Guenguel y alrededores (Santa Cruz). *Revista de la Asociación Geológica Argentina*, 20: 273–325.

Figueira, H.L. 1984. Horizonte vesicular: morfología y génesis en un aridisol del Norte de la Patagonia. *Ciencia del Suelo*, 2: 121–129.

Galli-Olivier, C. 1969. A primary control of sedimentation in the Peru–Chile trench. *Bulletin of the Geological Society of America*, 80: 1849–1852.

Galloway, R.W. 1985. Fossil ice wedges in Patagonia and their palaeoclimatic significance. *Zeitschrift für Geomorphologie*, NF, 29: 389–396.

Gardeweg, M. and Ramírez, C.F. 1985. *Hoja Río Zapaleri. II Región de Antofagasta*. Carta Geológica de Chile No. 66, Servicio Nacional de Geología y Minería, Santiago.

Gargantini, C.I. and Movia, C.I. 1989. Monitoring aeolian erosion by digital analysis in remote sensing. *Photo Interprétation*, 89: 31–35.

Garleff, K., Stingl, H. and Lambert K.-H. 1983. Fussflächen- und Terrassentreppen im Einzugsbereich des oberen Río Neuquén, Argentinien. *Zeitschrift für Geomorphologie*, Supplementband, 48: 247–259.

González-Díaz, E.F. 1993. Geomorphology in Argentina. In H.J. Walker and W.E. Grabau (eds), *The evolution of geomorphology*. Wiley, Chichester: 19–28.

Goudie, A.S. 1989. Salt tectonics and geomorphology. *Progress in Physical Geography*, 13: 597–605.

Goy, J. L., Macharé, J., Ortlieb, L. and Zazo, C. 1992. Quaternary shorelines in southern Peru: a record of global sea-level fluctuations and tectonic uplift in Chala Bay. *Quaternary International*, 15/16: 99–112.

Grenier, M.P. 1968. Observasions sur les taffonis du désert Chilien. *Association de Géographes Français, Bulletin*, 365: 193–211.

Grosjean, M. 1994. Palaeohydrology of the Laguna Lejía (north Chilean Altiplano) and climatic implications for late-glacial times. Palaeogeography, Palaeoclimatology, Palaeoecology, 109: 89–100.

Grosjean, M., Gegh, M.A., Messerli, B. and Schotterer, U. 1995. Late-glacial and early Molocene lake sediments, ground-water formation and climate in the Atacama Altiplano 22–24° S. *Journal of Palaeolimnology*, 14: 241–252.

Hastenrath, S. 1967. The barchans of the Arequipa region, southern Peru. *Zeitschrift für Geomorphologie*, NF, 11: 300–331.

Hastenrath, S. 1987. The barchan dunes of southern Peru revisited. *Zeitschrift für Geomorphologie*, NF, 31: 167–178.

Heathcote, J.A. 1989. Brine geochemistry of the Salar de Atacama, Chile. In The 6th International Symposium on Water–Rock Interaction, Malvern, 3–8 August 1989 (unpublished).

Hollingworth, S.E. and Guest, J.E. 1967. Pleistocene glaciation in the Atacama Desert, northern Chile. *Journal of Glaciology*, 6: 749–751.

Hsu, J.T. 1992. Quaternary uplift of the Peruvian coast related to the subduction of the Nazca Ridge: 13.5 to 15.6 degrees south latitude. *Quaternary International*, 15/16: 87–97.

Iriondo, M. 1989. A Late Holocene dry period in the Argentine plains. *Quaternary of South America and Antarctic Peninsula*, 7: 197–218.

Iriondo, M. 1993. Geomorphology and late Quaternary of the Chaco (South America). *Geomorphology*, 7: 289–303.

Johnson, A.M. 1976. The climate of Peru, Bolivia and Ecuador. In W. Schwerdtfeger (ed.), *Climates of Central and South America*. World Survey of Climatology, Vol. 12. Elsevier, Amsterdam: 147–218.

Jolley, E.J., Turner, P., Williams, G.D., Hartley, A.J. and Flint, S. 1990. Sedimentological response of an alluvial system to Neogene thrust tectonics, Atacama Desert, northern Chile. *Journal of the Geological Society of London*, 147: 769–784.

Jones, C.E. 1991. Characteristics and origin of rock varnish from the hyper-arid coastal deserts of northern Peru. *Quaternary Research*, 35: 116–129.

Jordan, T.E. and Allmendinger, R.W. 1986. The Sierras Pampeanas of Argentina: a modern analogue of Rocky Mountain foreland deformation. *American Journal of Science*, 286: 737–764.

Kött, A., Gaupp, R. and Wörner, G. 1995. Miocene to recent history of the western Altiplano in northern Chile revealed by lacustrine sediments of the Lauca Basin (18°5′–18°40′ S/69°30′–69°05′ W). *Geologische Rundschau*, 84: 770–780.

Lahsen, A. 1982. Evolución tectónica, solevantamiento y actividad volcánica de los Andes del Norte de Chile durante el Cenozoico Superior. *Congreso Geológico Chileno*, 3: B1–27.

Leonard, E.M. and Wehmiller, J.F. 1992. Low uplift rates and terrace reoccupation inferred from mollusc aminostratigraphy, Coquimbo Bay area, Chile. *Quaternary Research*, 38: 246–259.

Lynch, T.F. 1986. Climate change and human settlement around the late-glacial Laguna de Punta Negra, northern Chile: the preliminary results. *Geoarchaeology*, 1: 145–162.

Lynch, T.F. 1990. Quaternary climate, environment, and the human occupation of the south-central Andes. *Geoarchaeology*, 5: 199–228.

Macharé, J. and Ortlieb, L. 1992. Plio-Quaternary vertical motions and the subduction of the Nazca Ridge, central coast of Peru. *Tectonophysics*, 205: 97–108.

Mares, M.A., Morello, J. and Goldstein, G. 1984. The Monte Desert and other subtropical semi-arid biomes of Argentina, with comments on their relation to North American arid areas. In M. Evenari, I. Meir-Noy and D.W. Goodall (eds), *Hot deserts and arid shrublands*, Vol. 1A. Elsevier, Amsterdam: 203–237.

Marty, R., Dunbar, R., Martin, J.B. and Baker, P. 1988. Late Eocene diatomite from the Peruvian coastal desert, coastal upwelling in the eastern Pacific, and Pacific circulation before the terminal Eocene event. *Geology*, 16: 818–822.

Meggers, B.J. 1994. Archaeological evidence for the impact of mega-Niño events on Amazonia during the past two millennia. *Climatic Change*, 28: 321–338.

Meigs, P. 1966. *Geography of coastal deserts*. Arid Zone Research 28. UNESCO, Paris.

Messerli, B., Grosjean, M., Bonani, G., Buergi, A., Geyh, M.A., Graf, K., Ramseyer, K., Romero, H., Schotterer, U., Schreier, H. and Vuille, M. 1993. Climate change and natural resource dynamics of the Atacama Altiplano during the last 18,000 years: a preliminary synthesis. *Mountain Research and Development*, 13: 117–127.

Miller, A. 1976. The climate of Chile. In W. Schwerdtfeger (ed.), *Climates of Central and South America*, World Survey of Climatology, Vol. 12. Elsevier, Amsterdam: 113–129.

Mortensen, H. 1930. Einige Oberflächenformen in Chile und auf Spitsbergen in Rahmen einer vergleizehenden Morphologie der Klimazonen. *Petermann's Geographische Mitteilungen*, 209: 147–156.

Mortensen, H. 1933. Die 'Salzsprengung' und ihre Bedeutung für die Regionalklimatische Gliederung der Wüsten. *Petermann's Geographische Mitteilungen*, 79: 130–135.

Mortimer, C. 1973. The Cenozoic history of the southern Atacama Desert, Chile. *Journal of the Geological Society of London*, 129: 505–526.

Mortimer, C. 1980. Drainage evolution in the Atacama Desert of northernmost Chile. *Revista Geológica de Chile*, 11: 3–28.

Mortimer, C. and Saric, N. 1972. Landform evolution in the coastal region of Tarapacá Province, Chile. *Revue de Géomorphologie Dynamique*, 21: 162–170.

Mortimer, C. and Saric, N. 1975. Cenozoic studies in northernmost Chile. *Geologische Rundschau*, 64: 395–420.

Mueller, G. 1968. Genetic histories of nitrate deposits from Antarctica and Chile. *Nature*, 219: 1131–1134.

Naranjo, J.A. and Paskoff, R. 1985. Evolución Cenozoica del piedmonte andino en la Pampa del Tamarugal, Norte Grande de Chile (18°–21°S). *Congreso Geológico Chileno*, 4: 149–165.

Oberlander, T.M. 1982. Interpretation of rock varnish from the Atacama Desert. *Association of American Geographers, Program Abstracts*: 311.

Oberlander, T.M. 1994. Global deserts: a geomorphic comparison. In A.D. Abrahams and A.J. Parsons (eds), *Geomorphology of desert environments*. Chapman & Hall, London: 13–35.

Ortlieb, L. and Macharé, J. 1989. Evolución climática al final del Cuaternario en las regiones costeras del norte Peruano: breve reseña. *Institut Français et Andines, Bulletin*, 18: 143–160.

Ortlieb, L. and Macharé, J. 1993. Former El Niño events: records from western South America. *Global and Planetary Change*, 7: 181–202.

Palacios, C.M., Townley, B.C., Lahsen, A.A. and Egana, A.M. 1993. Geological development and mineralization in the Atacama segment of the South American Andes, northern Chile (26°15'–27°25'S). *Geologische Rundschau*, 82: 652–662.

Paskoff, R.P. 1970. *Recherches géomorphologiques dans le Chili semi-aride*. Biscaye Frères, Bordeaux.

Paskoff, R.P. 1978. Sur l'évolution géomorphologique du grand escarpement côtier du désert Chilien. *Géographie Physique et Quaternaire*, 32: 351–360.

Polanski, J. 1963. Estratigrafía, neotectónica y geomorfología del Pleistoceno pedemontano entre los ríos Diamante y Mendoza (Provincia de Mendoza). *Revista Asociación Geológica Argentina*, 17: 127–349.

Prohaska, F. 1976. The climate of Argentina, Paraguay and Uruguay. In W. Schwerdtfeger (ed.), *Climates of Central and South America*, World Survey of Climatology, Vol. 12. Elsevier, Amsterdam: 13–112.

Ramírez, C.F. 1988. Evidencias de glaciación en el macizo de los volcanes Púlar y Pajonales, Región de Antofagasta. *Congreso Geológico Chileno*, 5: D143–157.

Rauh, W. 1984. The Peruvian–Chilean Deserts. In M. Evenari, I. Meir-Noy and D.W. Goodall (eds), *Hot deserts and arid shrublands*, Vol. A. Elsevier, Amsterdam: 239–267.

Risacher, F. 1992. Géochimie des lacs salés et croûtes de sel de l'altiplano bolivien. *Sciences Géologiques, Bulletin*, 45: 135–214.

Risacher, F. and Fritz, B. 1991. Geochemistry of Bolivian salars, Lipez, southern Altiplano: origin of solutes and brine evolution. *Geochimica et Cosmochimica Acta*, 55: 687–705.

Rostagno, C.M., del Valle, H.F. and Videla, L. 1991. The influence of shrubs on some chemical and physical properties of an aridic soil in northeastern Patagonia, Argentina. *Journal of Arid Environments*, 20: 179–188.

Schemenauer, R.S. and Cereceda, P. 1992. The quality of fog water collected for domestic and agricultural use in Chile. *Journal of Applied Meteorology*, 31: 275–290.

Searl, A. and Rankin, S. 1993. A preliminary petrographic study of the Chilean nitrates. *Geological Magazine*, 130: 319–333.

Servant, M., Fournier, M., Argollo, J., Servant-Vildary, S., Sylvestre, F., Wirrmann, D. and Ybert J.P. 1995. The last glacial/interglacial transition in the South Tropical Andes (Bolivia) based on comparisons of locustrine and glacial fluctuations. *Comptes Rendus de l'Academie des Sciences Series II*, 320: 729–736.

Soriano, A. 1983. Deserts and semi-deserts of Patagonia. In N.E. West (ed.), *Temperate deserts and semi-deserts*. Elsevier, Amsterdam: 423–460.

Soriano, A. and Movia, C.P. 1986. Erosión y desertización en la Patagonia. *Interciencia*, 11: 77–83.

Stingl, H., Garleff, K. and Brunotte, E. 1983. Pedimenttypen im westlichen Argentinien. *Zeitschrift für Geomorphologie*, Supplementband, 48: 213–224.

Stoertz, G.E. and Ericksen, G.E. 1974. *Geology of salars in northern Chile*. US Geological Survey, Professional Paper, 811.

Strecker, M.R., Cerveny, P., Bloom, A.L. and Malizia, D. 1989. Late Cenozoic tectonism and landscape development in the foreland of the Andes: northern Sierra Pampeanas (26°–28°S), Argentina. *Tectonics*, 8: 517–534.

Suvires, G. 1989. Quaternary landforms in southeastern San Juan Province, Argentina. *Quaternary of South America and Antarctic Peninsula*, 7: 93–117.

Thompson, R.D. 1975. *The climatology of the arid world*. University of Reading, Department of Geography, Geographical Papers, 35.

Tosdal, R.M., Clark, A.H. and Farrar, E. 1984. Cenozoic polyphase landscape and tectonic evolution of the Cordillera Occidental, southernmost Peru. *Bulletin of the Geological Society of America*, 95: 1318–1332.

Tricart, J.L.F. 1989. Roles du volcanisme explosif et du vent dans la formation des précipites calcaires quaternaires ('tosca') de la Pampa Argentine.

Association Française pour l'Etude du Quaternaire, Bulletin, 37: 45–54.

UNESCO 1979. *Map of the world distribution of arid regions*. MAB Technical Note 7. UNESCO, Paris.

Valerogarces, B.L., Grosjean, M., Schwalb, A., Geyh, M., Messerli, B. and Kelts, K. 1996. Limnogeology of Laguna Miscanti – evidence for mid to late Holocene moisture changes in the Atacama Altiplano (northern Chile) *Journal of Palaeolimnology*, 16: 1–21.

Van Moort, J.C. 1985. Natural enrichment processes of nitrate, sulphate, chloride, iodate, borate, perchlorate and chromate in the caliches of northern Chile. *Congreso Chileno*, 4: 3/674–702.

Vandervoort, D.S., Jordan, T.E., Zeitler, P.K. and Alonso, R.N. 1995. Chronology of internal drainage development and uplift, southern Puna plateau, Argentine central Andes. *Geology*, 23: 145–148.

Veit, H. 1993. Upper Quaternary landscape and climate evolution in the Norte Chico: an overview. *Mountain Research and Development*, 13: 138–144.

Vila, T.G. 1975. Geología de los depósitos salinos andinos, Provincia de Antofagasta, Chile. *Revista Geológica de Chile*, 2: 41–55.

Vila, T.G. 1976. Modelo de distribución y origen de algunos elementos en salmueras de depósitos salinos andinos, Norte de Chile. *Congreso Geológico Chileno*, 1: E65–82.

Vogt, T. 1992. Western Anti-Atlas (Morocco) and Central Patagonia (Argentina) calcretes: the calcium carbonate origin. *Zeitschrift für Geomorphologie*, Supplementband, 84: 115–127.

Vogt, T. and Del Valle, H.F. 1994. Calcretes and cryogenic structures in the area of Puerto Madryn (Chubut, Patagonia, Argentina). *Geografiska Annaler*, 76A: 57–75.

Waylen, P.R. and Caviedes, C.N. 1986. El Niño and annual floods on the north Peruvian littoral. *Journal of Hydrology*, 89: 141–156.

Wells, L.E. 1987. An alluvial record of El Niño events from northern coastal Peru. *Journal of Geophysical Research*, 92(C13): 14463–14470.

Wells, L.E. 1990. Holocene history of the El Niño phenomenon as recorded in flood sediments of northern coastal Peru. *Geology*, 18: 1134–1137.

Wilkes, E. and Görler, K. 1988. Sedimentary and structural evolution of the Cordillera de la Sal, II Región, Chile. *Congreso Geológico Chileno*, 5: A173–188.

Wright, A.C.S. and Urzúa, H. 1963. Meteorización en la región costera del desierto del Norte de Chile. *Communicaciones y Resumenes de Trabajos, Conferencia Latino-Americana para el Estudio de las Regiones Andinas, Buenos Aires, República Argentina*: 27–28.

Zárate, M. and Blasi, A. 1993. Late Pleistocene–Holocene aeolian deposits of the southern Buenos Aires Province, Argentina: a preliminary model. *Quaternary International*, 17: 15–20.

25
Australia

Jacky Croke

Introduction

Drylands occupy almost three-quarters of Australia earning it the title of the driest continent on Earth after Antarctica. Characteristic features of Australia's low-latitude, low-relief deserts include extensive duricrust plains, a continental-scale whorl of longitudinal dunes, drainage-aligned systems of playas and clay pans, and a network of disorganised internal drainage patterns. The physiographic and geomorphological diversity of the Australian arid region is clearly illustrated in Mabbutt's (1969) six-fold classification of desert types (Figure 25.1). Many features, such as the abundance of relict weathering profiles, unusual inverted relief features, arcs of groundwater-charged mound springs, and the nature and extent of precious opal deposits, are unique to the arid Australian landscapes. The Australian arid zone displays a high degree of coincidence with rock type and structure. The majority of the arid zone lies within two of the continent's three major tectonic divisions, namely the Western Platform and the Central Basin. The Western Platform consists of a basement of Archaean granite, gneiss and greenstone, and is exposed in updomed shield areas (Mabbutt 1969). It is extensively pre-served in the Yilgarn Shield in the southern half of Western Australia. The Arunta Shield of central Australia, likewise, consists of Archaean schists, granites, gneiss and resistant metaquartzites. The portion of the Central Basin within the arid zone consists mainly of the Great Artesian Basin and part of the Murray Basin in the south. The Great Artesian Basin is an area of marine Mesozoic claystone and sandstones with a partial cover of Tertiary sands. Here the duricrusted remnants form tablelands, mesa cappings and gibber mantles which cover vast low-relief plains. The central portion of the basin contains wide alluvial plains, dunefields and playas.

Precipitation/evapotranspiration (P/Etp) ratios in the Australian arid zone range between 0.20 and 0.03. Australia displays some of the most marked spatial and temporal variation in annual rainfall in the world (Finlayson and McMahon 1988), and precipitation throughout the arid zone is particularly variable. Average annual rainfall ranges between 200 and 400 mm but no part of the Australian arid zone averages less than 100 mm of rain a year. Much of the Australian desert receives rainfall from the tropical monsoon which crosses the continent in the summer months, December–March, and descends south into the desert region. Periods of enhanced

Arid Zone Geomorphology: Process, Form and Change in Drylands, 2nd edition. Edited by David S. G. Thomas.
© 1997 John Wiley & Sons Ltd.

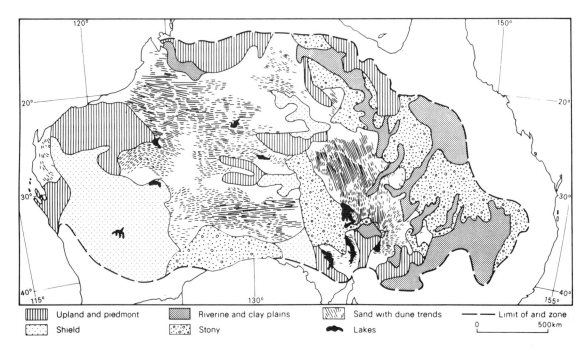

Figure 25.1   *Mabbutt's (1969) classification of desert types in Australia. Sandy deserts dominate the continent covering approximately 40% of the Australian land surface*

rainfall are believed to be linked to variations in the Southern Oscillation Index (SOI) and occurrences of El Niño (Allan 1985). Areas along the southern margin of the desert are influenced by the local westerly circulation which produces a zone of unreliable, winter rainfall dominance.

In addition to the obvious effect of continentality, the most dominant factor responsible for the aridity of the Australian deserts is the continent's latitudinal position, astride the tropics in a zone of semi-permanent high-pressure cells (Williams 1984). Australia's arid features reflect prolonged periods of stability and weathering over a range of timescales. Many stable landforms represent relict features formed under different climatic conditions from those prevailing today, in particular, those of the more humid Tertiary climates. For example, widespread palaeodrainage systems in arid Western Australia, preserved in an extensive, relict duricrusted landscape, formed during the late Cretaceous to early Tertiary (Van de Graaff *et al.* 1977). It is the evolution of aridity over time in Australia that accounts for many of its unique and intriguing landforms and processes.

Australian aridity: a process through time

The origins and patterns of aridity in Australia can be viewed on two timescales. The earliest signs of aridity are associated with events in early Cenozoic time (see reviews by Bowler 1982; 1986a; Williams 1984). However, the full development of arid environments, as indicated by saline evaporitic lake facies and dune successions in southern Australia, occurred in response to the rhythmic oscillations between wet and dry episodes that have characterised the Quaternary (see reviews by Kershaw 1981; 1985; Bowler and Wasson 1984; Bowler 1986a; Chappell 1991; Nanson *et al.* 1992a). The onset of Pleistocene aridity and dune building in central Australia has been interpreted from aeolian sequences at Lake Amadeus in the Arunta Shield area dated to approximately 700 ka (Chen and Barton 1991).

The original development of aridity in Australia was traditionally explained through continental drift. This explanation assumed that the climate and position of the arid belts remained stable while the continent's position changed. However, Bowler (1982) indicated that the relatively slow movement of continental plates was subordinate to major changes in global circulation. Evidence suggests that intensified subtropical high-pressure belts migrated from south to north, overtaking the continent on its slower northward drift. Taking account of proximity to coastlines and general continental configuration, Bowler (1982) argues that the earliest severely arid conditions are likely to have been encountered in the Nullarbor region about 6 million years ago, thence extending inland as the high-pressure cells moved further towards the equator and intensified. In southeastern Australia, Miocene humid environments with associated high lake levels culminated in drying at about 5 Ma, close to the Miocene–Pliocene boundary (Bowler 1976; 1982). A drier climate with seasonally dry lakes prevailed during most of the Pliocene. Reactivation of the lake systems occurred about 2.5 Ma (Bowler 1986a), heralding the beginning of Quaternary climatic oscillations. The zone of high-pressure cells appear to have taken up their present position by about 1 million years ago. Central Australia would already have been dry, rather as it is today, and southern Australia was dominated by winter rain and westerly circulations. As Bowler (1982) points out, however, it was not until the Quaternary that the system entered one of its most dramatic phases.

Quaternary oscillations in climate were extreme, varying between periods of massive lake expansion to periods of extensive dune building and saline playa development. Raised shoreline features around the southern margin of Lake Eyre in the central arid zone stand +10 m above sea level, compared with present playa levels of −15 m a.s.l., and have been dated to the last interglacial period (Callen 1986). These features, together with deep-water lacustrine facies in stratigraphic sequences, indicate a major lake expansion phase during this last interglacial period (Figure 25.2) (Nanson *et al.* 1992a; Croke *et al.* 1995; Magee *et al.* 1995). During these more humid and wetter phases of the last 400 ka the arid belt may have disappeared completely, especially if contractions of the semi-arid belts on northern and southern desert margins prove to have been synchronous. At the time of peak aridity (18 to 20 ka) the

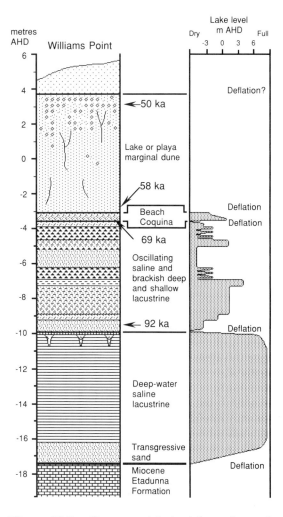

Figure 25.2. *Quaternary lake-level fluctuations and episodes of aeolian activity at Madigan Gulf in the southeastern corner of Lake Eyre (Magee et al. 1995). Lake Eyre is Australia's largest saline playa, and forms the terminus for the drainage systems of the Channel Country and northwestern rivers in arid central Australia*

Australian desert expanded on its southern margin with the development of longitudinal dunes as far south as Kangaroo Island (Bowler 1978a; Sprigg 1979) and even to the northeastern tip of Tasmania, which was joined to the mainland of Australia at that time, during maximum sea-level lowering (Bowden 1978).

While it has generally been accepted that much of the central part of the Australian arid zone was settled as late as 10 000–12 000 BP, thermoluminescence dates from a rock shelter to the west of the Macdonnell Ranges suggests human occupation had occurred by 22 000 years ago (Smith 1987). It is possible that human disruption of the vegetation ecosystems may have contributed to changes in microclimate during the late Quaternary, thereby affecting the arid zone (Singh and Geissler 1985; Miller and Magee 1992).

Desert geomorphology

Mabbutt's (1969; 1984) reviews of arid landforms provide good general summaries of Australia's desert features. Regional studies of desert landforms, soils and vegetation are presented in Wells and Callen (1986) and Tyler *et al.* (1990).

DUNEFIELDS

It has been suggested that the Australian Continental Dunefield represents over 38% of the world's aeolian landscapes, and covers approximately 40% of Australia's land surface (Wasson *et al.* 1988). The dunefield is made up of seven individual but interconnected desert regions (Figure 25.3). Most dunes are of simple longitudinal form and can be up to 300 km in length, with an average interdune spacing of 160–200 m and a height of 10–35 m. The dunes are oriented in a continental-scale anticlockwise whorl (Jennings 1968; Brookfield 1970). Dune orientations are not exactly replicated by modern wind directions but there is considerable similarity (Wasson *et al.* 1988).

Many aspects of the dunefields, such as their origin, asymmetric form, colour, sediment provenance and age are the subject of ongoing research and debate. In terms of their origin, Twidale (1972) argued that the linear (longitudinal) dunes developed and extended downwind from debris mounds located on the northern sides of playas and alluvial plains, particularly in the eastern and southern sectors of the Simpson Desert. Helicoidal secondary wind currents generated by the prevailing SSE winds are believed to have maintained the dune pattern (Bagnold 1953; Mabbutt and Sullivan 1968; Folk 1971).

The Strezlecki and Simpson desert dunes show marked asymmetry with steeper eastern slopes (Wopfner and Twidale 1967; Mabbutt *et al.* 1969; Folk 1971; Twidale 1981) causing Rubin (1990) to argue that they are migrating eastward (see also Chapter 17). He suggests rates of a few hundred metres eastward over the last 100 ka or so. Mabbutt *et al.* (1969) suggested three mechanisms that may have caused these asymmetric dune slopes in the Simpson Desert; two of these imply a cross-crest sand movement to the east. For the Simpson and Strezlecki deserts, however, Brookfield (1970) showed a clear bias in sand-transporting winds from the east across the dune orientation. Nanson *et al.* (1995) believe that asymmetry of longitudinal dunes is an indication that recent wind directions have not exactly paralleled the dune trend but have favoured a vector across the dune, toward the steeper slope. The latter suggest that as yet there are no chronological data to propose that the longitudinal dunes of the Simpson Desert have migrated in either direction more than 100 m or so from their Pleistocene core.

Dune colours are generally pale close to the major drainage systems (Diamantina River and Cooper Creek) and become red in the northern Simpson and southern Strezlecki deserts, away from the major watercourses (Wasson 1983b). Some studies have suggested progressive ageing and reddening of sand grains downwind from sand sources (Wopfner and Twidale 1967). Wasson (1983a; 1983b) has argued that different coloured dunes result from different sand provenances and are unrelated to dune age. Nanson *et al.* (1992b) also suggest that colour shows little correlation with dune age. The dune sand comes from diverse local sources including unconsolidated lacustrine and fluvial sediments as well as weathered country rock, and there is no evidence of long-distance sand transportation (Wasson 1983b).

Most recently, Pell and Chivas (1995) used variations of surface features of detrital quartz grains from the Australian deserts to identify provenance areas and the means of transport into the deserts. The predominance of chemically produced surface features and the poorly rounded and irregular nature of most grains indicate that the majority of sand in the Australian Continental Dunefield is currently stabilised and has not been transported long distances during its sedimentary history. They

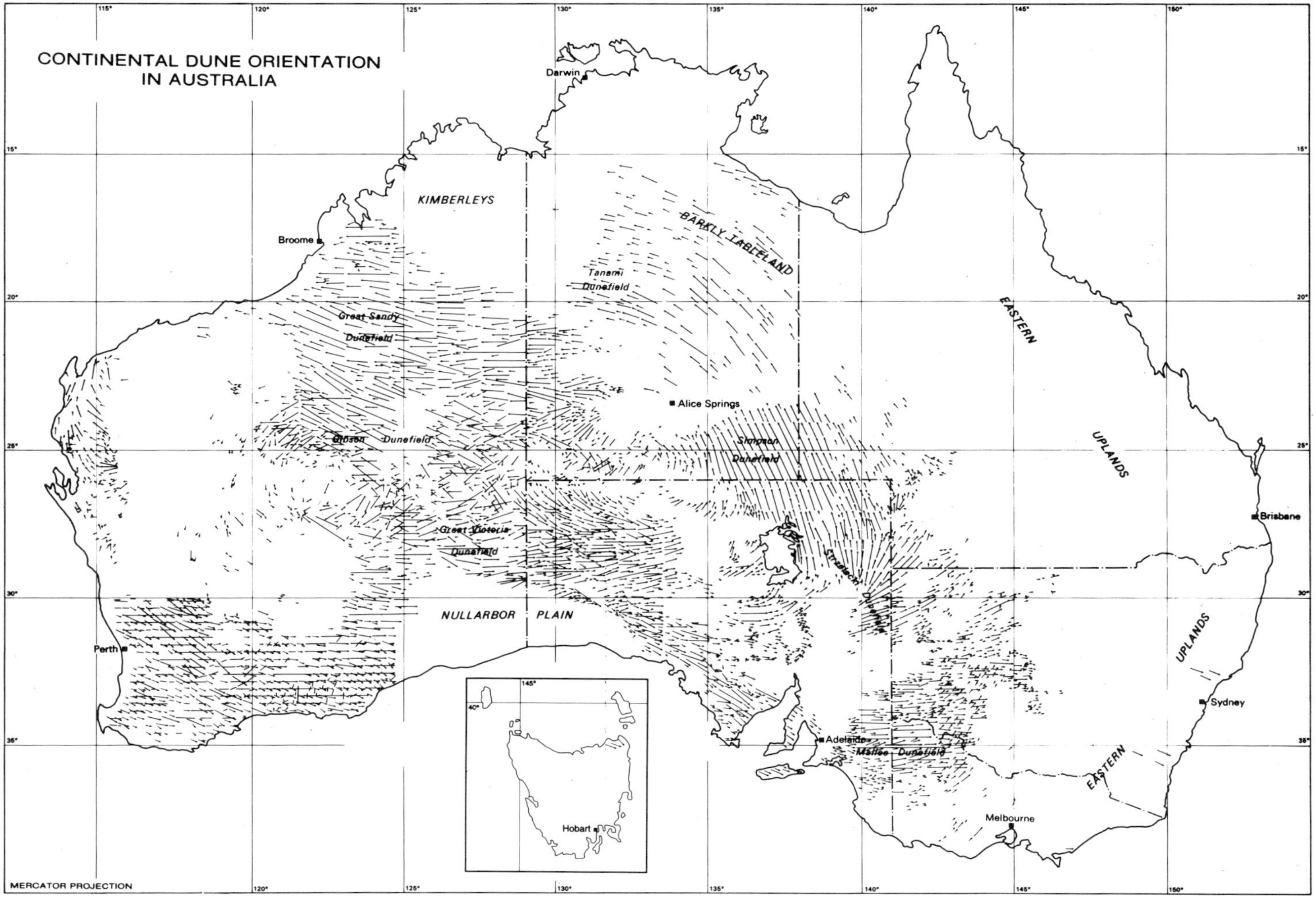

Figure 25.3. *The continental-scale whorl of longitudinal dunes in Australia. The Australian Continental Dunefield is dominated by parallel linear dunes and consists of seven individual, but interconnected desert regions*

conclude that most sand was probably washed into sedimentary basins and then reworked more recently by wind into dunes. Wind is seen only as the agent to shape dunes from existing sediment sources and is not the principal mechanism for long-distant transport.

Wopfner and Twidale (1988) suggest that most of the dunes in the Simpson Desert are 'intrinsically Holocene'. Wells and Callen (1986) and Williams (1982) give evidence of earlier episodes of dune development in the Lake Eyre and Lake torrens regions, respectively. Callen and Nanson (1992) used thermoluminescence dating (see Chapter 27) to indicate that dunes in the basin range from 125 ka through to the present, with the most intense aeolian activity at or about the Last Glacial Maximum (Wasson *et al.* 1988). Nanson *et al.* (1995) dated an aeolian sequence at the extreme edge of the Simpson Desert which yielded an age of 274 ka, suggesting that the dunefields have survived many subsequent shifts in wind pattern during past glacial cycles. Ash and Wasson (1983) showed that over much of the Australian desert dunefield the vegetation is insufficient fully to inhibit major sand movement (see Chapters 7, 16 and 17). The limiting factor appears to be wind, suggesting that the modern environment is far less windy than the glacial environment. A minimum increase in windiness of 20–30% would mobilise most of the dunes in Australia (Bowler and Wasson 1984).

LAKES (PLAYAS) AND LUNETTES

Lacustrine systems in the Australian arid and semi-arid zones are ephemeral or dry due to relatively small catchments, low relief and high evaporation relative to inflow. Even Lake Eyre, as an example of an extremely large catchment $(1.3 \times 10^6 \text{ km}^2)$ with a catchment-to-lake amplification ratio of about 135, lies in the highest evaporation regime on the continent and rarely operates as a surface-water system. The sedimentology and geomorphology of arid and semi-arid lakes are dominantly controlled by groundwater processes. This was first recognised by Jutson (1934) in an early study of Western Australian salt lakes, but the dominance of saline-groundwater-controlled processes in arid zone

lake systems has only recently been more thoroughly investigated and validated (Magee 1991; Magee *et al.* 1995).

Many important aspects of Australia's salt lakes have been studied as part of the SLEADS (Salt Lakes Evaporites and Aeolian Deposits) multidisciplinary projects which investigated various aspects of selected playa/salt-lake environments. Results from these core-projects have been published in three special issues of *Palaeogeography, Palaeoclimatology and Palaeoecology* (1984 Vol. 54; 1988 Vol. 85; 1991 Vol. 113). The history, chemistry and biota of Australian salt lakes have also been reviewed by De Deckker (1983; 1988).

Bowler (1986b) examined the spatial variability and hydrologic evolution of Australian arid lacustrine systems and classified them according to a scheme which relates catchment area/lake area ratios to climatic and runoff functions (see Chapter 14). Bowler's classification recognised a hydrologic continuum from wet to dry which operates both spatially, as climate varies across the continent, and temporally as climate altered during the glacial/interglacial cycles of the Quaternary. Full freshwater conditions, where wind-induced wave erosion and sediment transport occur, produce smooth shorelines and quartz-rich beaches and foredunes on their downwind margins. As hydrological budgets become negative, the influence of groundwater and salinity begin to dominate. Sand-sized sediment aggregates and salts are transported by the dominant wind to form clay-rich dunes or lunettes on the downwind shoreline, and finer dust material is spread further down-basin (Bowler 1973; 1983; Chen *et al.* 1991; Magee 1991). At the extremely dry end of the spectrum surface-water cover is rare and groundwater processes dominate producing systems with highly irregular outlines, termed boinkas by Macumber (1980).

Inherited from the wetter climatic phases of the Quaternary, many lakes in the semi-arid zone that have large catchments preserve evidence of enhanced surface-water conditions in the form of lacustrine sedimentary sequences and shoreline features expanded well beyond the present basin. Examples in southern Australia are the Willandra Lakes (Bowler 1986b; Magee 1991) and Lake Tyrrell (Bowler and

Teller 1986) and, on the northern desert margin Lakes Woods and Gregory (Bowler 1990). Only rarely do lakes in the arid core have sufficiently large catchments to show evidence of prior surface-water-dominated phases. Examples are Lake Eyre (Croke *et al.* 1995; Magee *et al.* 1995) and Lake Frome (Bowler *et al.* 1986).

RIVERS, FLOODPLAINS AND ALLUVIAL FANS

The characteristics of Australian arid zone drainage systems have been reviewed by Mabbutt (1969; 1984). However, three inland river systems have been studied in greater detail; the large anastomosing rivers of the Channel Country, the river systems of the 'stony desert' to the northwest of Lake Eyre, and those draining the Central Highlands of the Macdonnell and Musgrave ranges near Alice Springs in the heart of the arid zone.

The rivers of the Channel Country represent a unique assemblage of fluvial forms and sedimentary facies which are discussed in greater detail in Chapter 10. The rivers of the 'stony desert' along the northwest margin of Lake Eyre are shorter, of higher energy, and have a coarser sediment load than the large alluvial systems of the Channel Country. Channel form is only clearly recognisable in downstream reaches where they are deeply incised into older Tertiary sediments. The form and processes of these rivers have been strongly influenced by changing base-levels as Lake Eyre responded to major infillings and deflationary episodes during the Quaternary (Croke *et al.* 1996). Sedimentary facies from these rivers reflect the transition from humid Tertiary climates where channel systems were largely braided to the seasonally dominated meandering sequences of the middle Pleistocene, and, finally, to the ephemeral and aeolian dominated systems of the Holocene (Croke *et al.* 1995).

Fluvial systems draining the highlands of the central desert region preserve a good record of large-scale hydrological events (Williams 1970; 1971), providing data for the reconstruction of Holocene palaeodischarges using preserved slack-water deposits (Baker *et al.* 1983; Pickup *et al.* 1988; Patton *et al.* 1993). Pickup (1991)

summarises many of the characteristic features of event frequency and landscape stability on the floodplain systems of the central arid region around Alice Springs. Although Pickup uses the term 'floodplain' he suggests that these landforms contain a mixture of fan and floodplain features and, in many cases, behave more like low-angle alluvial fans than floodplains. Bourke (1994) describes processes of floodplain erosion and stripping in response to large scale-flood events in the Todd River of the same region.

In the southern desert margins, work has concentrated on the meandering and highly sinuous river systems of the Riverine Plain of southeastern Australia (Schumm 1968; Bowler 1986a; Page *et al.* 1991; Page 1994). Much of the Riverine Plain was deposited by an extensive distributary network of ancient sand-bedded streams and large meandering channels which contained bankfull discharges up to five times greater than the present channels (Schumm 1968; Page *et al.* 1991). Page *et al.* (1991) have dated these alluvial deposits using thermoluminescence and provide a record of fluvial, aeolian and lacustrine deposition on the Riverine Plain during the last 100 ka.

Arid alluvial fans are typically not well developed in the stable tectonic settings of shield and platform deserts as found in Australia (Mabbutt 1977). Studies in the inland arid zone are restricted to alluvial fans found along the faulted western front of the Flinders Ranges (Williams 1973) and to piedmont angles in the desert margins of South Australia (Twidale 1967). Some work has also been carried out on the sedimentation history of alluvial fans in the semi-arid desert margins of western New South Wales (Wasson 1979).

DURICRUSTS

Australian deserts preserve excellent examples of vast duricrusted (see Chapter 6) plains composed predominantly of silcrete and ferricrete. The term 'ferricrete' is now used in preference to the older, more ambiguous 'laterite' term (Ollier and Galloway 1990). Their spatial distribution in Australia is outlined in Figure 25.4 and by Twidale and Hutton (1986). A more complete treatise of silcretes in Australia is presented by Langford-Smith

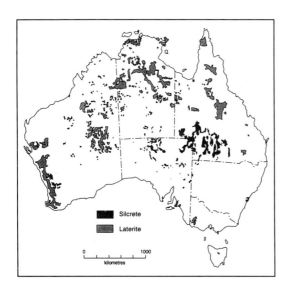

Figure 25.4. *The spatial distribution of silcrete and ferricrete on the Australian continent (Twidale and Hutton 1986)*

(1978). Silcretes are believed to have first formed in the late Jurassic, on stable low-relief landscapes (Langford-Smith 1978; Twidale 1983). Two major subsequent phases of development in Australian silcretes have been recognised, the last occurring in the late Pleistocene (Langford-Smith 1978; Ollier 1978; Ambrose and Flint 1981; Callen 1983; Twidale 1983). It is now generally accepted that these surfaces are relicts of a wetter climate (Ambrose and Flint 1981; Callen 1983). Ancient silicified shorelines of a large Miocene lake around Lake Eyre in central Australia suggests a complex mode of silica formation which occurred as silica-rich groundwaters met saline lake waters (Ambrose and Flint 1981). Angular silcrete fragments known as 'gibbers' are very common on the surfaces of the 'stony deserts'. The abundant supply of readily soluble silica throughout these duricrusted landscapes has contributed significantly to the nature and extent of precious opal deposits in the Australian arid zone (Darragh *et al.* 1966).

Ferricretes are best preserved along the margins of the continent (Twidale and Hutton 1986). Unlike silcretes, these are believed to be the products of warm, seasonal environments (Twidale 1983). Erosive incision into these duricrusted landscapes have produced a net-

work of second- and third-order landforms, such as inverted drainage features (Jessup 1961; Ollier 1988) and steep vertical 'breakaways' (Blume and Barth 1979).

References

Allan, R.J. 1985. *The Australian summer monsoon, teleconnections and flooding in the Lake Eyre Basin.* South Australian Geographer, Paper 2. Royal Geographical Society of Australasia (South Australia Branch), Adelaide, 47 pp.

Ambrose, G.J. and Flint, R.B. 1981. A regressive Pliocene lake system and silicified strandlines in northern South Australia: implications for regional stratigraphy and silcrete genesis. *Journal of the Geological Society of South Australia*, 28: 81–94.

Ash, J.E. and Wasson, R.J. 1983. Vegetation and sand mobility in the Australian desert dunefield. *Zeitschrift für Geomorphologie*, NF, Supplementband, 45: 7–225.

Bagnold, R.A. 1953. The surface movement of blown sand in relation to meteorology. In *Desert research*. Proceedings of an International Symposium, Jerusalem, Research Council of Israel Publication 2: 83–93.

Baker, V.R., Kochel, R.C., Patton, P.C. and Pickup, G. 1983. Palaeohydrologic analysis of Holocene flood slack-water sediments. *International Association of Sedimentologists, Special Publication*, 6: 229–239.

Blume, H.P. and Barth, H.K. 1979. Laterische Krustenstufen in Australien. *Zeitschrift für Geomorphologie*, NF, 31: 443–448.

Bourke, M.C. 1994. Cyclical construction and destruction of flood dominated floodplains in semiarid Australia. In Proceedings of the Canberra Symposium, December 1994. International Association of Hydrological Sciences Publication, No. 24: 113–123.

Bowden, A.R. 1978. *Geomorphic perspective on shallow groundwater potential coastal northeastern Tasmania.* Australian Water Resources Council, Technical paper, 36. Australian Government Services, Canberra.

Bowler, J.M., 1973. Clay dunes: their occurrence, formation and environmental significance. *Earth Science Reviews*, 9: 315–338.

Bowler, J.M. 1976. Aridity in Australia: age, origins and expressions in aeolian landforms and sediments. *Earth Science Reviews*, 12: 279–310.

Bowler, J.M. 1978a. Glacial age aeolian events in high and low latitudes: a southern hemisphere perspective. In: E.M. Van Zinderen Bakker (ed.), *Antarctic glacial history and world palaeoenvironments.* Balkena, Rotterdam.

Bowler, J.M. 1978b. Quaternary climate and tectonics in the evolution of the Riverine Plain, southeastern Australia. In J.L. Davies and M.A.J. Williams (eds), *Landform evolution in Australasia*. Australian National University Press, Canberra: 70–112.

Bowler, J.M. 1982. Aridity in the late Tertiary and Quaternary of Australia. In W.R. Barker and P.J.M. Greenslade (eds), *Evolution of the flora and fauna of arid Australia*. Peacock Publications/ Australian Systematic Botany Society /Anzas, Adelaide: 35–45.

Bowler, J.M. 1983. Lunettes as indices of hydrologic change: a review of Australian evidence. *Proceedings of the Royal Society of Victoria* 95: 147–168.

Bowler, J.M. 1986a. Quaternary landform evolution. In D. Jeans (ed.), *A geography of Australia*. University of Sydney Press, 117–147.

Bowler, J.M. 1986b. Spatial variability and hydrologic evolution of Australian lake basins: analogue for Pleistocene hydrologic change and evaporite formation. *Palaeogeography, Palaeoclimatology, Palaeoecology* 54: 21–41.

Bowler, J.M. 1990. Lake Gregory — geomorphology and palaeohydrology. In S.A. Halse (ed.), *The natural features of Lake Gregory: a preliminary review*. Occasional Paper 2/90. Department of Conservation and Land Management, Perth, Western Australia.

Bowler, J.M. and Teller, J.T. 1986. Quaternary evaporites and hydrological changes, Lake Tyrell, northwest Victoria. *Australian Journal of Earth Sciences*, 33: 43–63.

Bowler, J.M. and Wasson, R.J. 1984. Glacial age environments of inland Australia. In J.C. Vogel (ed.), *Late Cainozoic palaeoclimates of the southern hemisphere*. Balkema, Rotterdam: 183–208.

Bowler, J.M., Huang, Q., Chen, K., Head, M.J. and Yuan, B. 1986. Radiocarbon dating of playa-lake hydrologic changes: examples from northwestern China and central Australia. *Palaeogeography, Palaeoclimatology, Palaeoecology*, 54: 241–260.

Brookfield, M. 1970. Dune trends and wind regime in Central Australia. *Zeitschrift für Geomorphologie*, Supplementband, 10: 121–153.

Callen, R.A. 1983. Late Tertiary 'grey billy' and the age and origin of surficial silicifications (silcrete) in South Australia. *Journal of the Geological Society of South Australia*, 30: 393–410.

Callen, R.A. 1986. Early and Middle Cainozoic sediments. In R.T. Wells and R.M. Callen, (eds), *The Lake Eyre Basin — Cainozoic sediments, fossil vertebrates and plants, landforms, silcretes and climatic implications*. Australasian Sedimentologists Group Field Guide Series, Geological Society of South Australia, 4: 21–32.

Callen, R.A. and Nanson, G.C. 1992. Discussion: formation and age of dunes in the Lake Eyre depocentres. *Geologische Rundschau*, 81: 589–593.

Chappell, J.C. 1991. Late Quaternary environmental changes in eastern and central Australia, and their climatic interpretation. *Quaternary Science Reviews*, 10: 377–390.

Chen, X.Y. and Barton, C.E. 1991. Onset of aridity and dune building in central Australia: sedimentology and magnetostratigraphic evidence from Lake Amadeus, central Australia. *Palaeogeography, Palaeoclimatology, Palaeoecology*, 84: 55–73.

Chen, X.Y., Bowler, J.M. and Magee, J.M. 1991. Aeolian landscapes in central Australia: gypsiferous and quartz dune environments from Lake Amadeus. *Sedimentology*, 38: 519–538.

Croke, J.C., Magee, J.W. and Price, D.M. 1995. Stratigraphy and sedimentology of the Neales delta system, west Lake Eyre, central Australia: from Palaeocene to Holocene. *Sedimentary Geology* (in press).

Croke, J.C., Magee, J.W. and Price, D.M. 1996. Major episodes of Quaternary activity in the lower Neales, west Lake Eyre, central Australia. *Palaeogeography, Palaeoclimatology, Palaeoecology* (in press).

Darragh, P.J. Gaskin, A.J. Terrell, B.C. and Sanders, J.V. 1966. Origin of precious opal. *Nature*, 209: 13–16.

De Deckker, P. 1983. Australian salt lakes: their history, chemistry and biota — a review. *Hydrbiologia*, 105: 231–244.

De Deckker, P. 1988. Biological and sedimentary facies of Australian salt lakes. *Palaeogeography, Palaeoclimatology, Palaeoecology*, 62: 237–270.

Finlayson, B.L. and McMahon, T. 1988. Australia v the world: a comparative analysis of streamflow characteristics. In R.F. Warner (ed.), *Fluvial geomorphology of Australia*. Academic Press, Sydney: 17–40.

Folk, R.L. 1971. Longitudinal dunes of the northwestern edge of the Simpson desert, Northern territory, Australia: geomorphology and grain size relationships. *Sedimentology*, 16: 5–54.

Jennings, J.N. 1968. A revised map of the desert dunes of Australia. *Australian Geographer*, 10: 408–409.

Jessup, R.W. 1961. A Tertiary–Quaternary pedological chronology for the southeastern portion of the Australian arid zone. *Journal of Soil Science*, 12: 199–213.

Jutson, J.T. 1934. The physiography (geomorphology) of Western Australia. *Bulletin of the Geological Survey of Western Australia*, 95.

Kershaw, A.P. 1981. Quaternary vegetation and environments. In A. Keast (ed.), *Ecological biogeography of Australia*. Junk, The Hague: 83–101.

Kershaw, A.P. 1985. An extended Late Quaternary vegetation record from north eastern Queensland and its implications for the seasonal tropics of Australia. *Proceedings of the Ecological Society of Australia*, 13: 179–189.

Krieg, G.W., Callen, R.A., Gravestock, D.I. and Gatehouse, C.G. 1990. Geology. In M.J. Tyler, C.R. Twydale, M. Davies and C.B. Wells (eds), *Natural history of the North East Deserts*. Royal Society of South Australia, Adelaide: 1–26.

Langford-Smith, T. 1978. *Silcrete in Australia*. University of New South Wales Press, Armidale.

Langford-Smith, T. 1982. The geomorphic history of the Australian deserts. *Striae*, 17: 4–19.

Mabbutt, J.A. 1969. Landforms of arid Australia. In R.O. Slayter and R. Perry (eds), *Arid lands of Australia*. Australian National University Press, Canberra: 11–32.

Mabbutt, J.A. 1977. *Desert landforms, Vol. 2: An introduction to systematic geomorphology*. Australian National University Press, Canberra.

Mabbutt, J.A. 1984. Landforms of the Australian deserts. In F. El-Baz (ed.), *Deserts and arid lands*. Martinus Nijhoff, The Hague.

Mabbutt, J.A. and Sullivan, M.E. 1968. The formation of longitudinal dunes — evidence from the Simpson Desert. *Australian Geographer*, 10: 483–487.

Mabbutt, J.A., Wooding, R.A. and Jennings, J.N. 1969. The asymmetry of Australian desert sand ridges. *Australian Journal of Science*, 32: 159–160.

Macumber, P.G. 1980. The influence of groundwater discharge on the Mallee landscape. In R.R. Storrier and M.E. Stannard (eds), *Aeolian landscapes in the semi-arid zone of southeastern Australia*. Australian Society of Soil Science, Riverine Branch: 67–84.

Magee, J.M. 1991. Late Quaternary lacustrine, groundwater, aeolian and pedogenic gypsum in the Prungle Lakes, south eastern Australia. *Palaeogeography, Palaeoecology, Palaeoclimatology*, 84: 3–42.

Magee, J.M., Bowler, J.M., Miller, G.H. and Williams, D.L.G. 1995. Stratigraphy, sedimentology, chronology and palaeohydrology of Quaternary lacustrine deposits at Madigan Gulf, Lake Eyre, South Australia. *Palaeogeography, Palaeoecology, Palaeoclimatology*, 113: 3–42.

Miller, G.H. and Magee, J.M. 1992. Drought in the Australian outback: anthropogenic impact on regional climate. *Abstracts, American Geophysical Union, Fall Meeting*: 104.

Nanson, G.C., Price, D.M., Short, S.A., Page, K.J. and Nott, J.F. 1991. Major episodes of climatic change in Australia over the last 300,000 years. In R. Gillespie (ed.), *Quaternary dating workshop 1990*. Department of Biogeography and Geomorphology, RSpacS, Australian National University, Canberra: 45–50.

Nanson, G.C., Price, D.M. and Short, S.A. 1992a. Wetting and drying of Australia over the past 300 Ka. *Geology*, 30: 791–794.

Nanson, G.C., Chen, X.Y. and Price, D.M. 1992b. Lateral migration, thermoluminescence chronology and colour variation of longitudinal dunes near Birdsville in the Simpson Desert, central Australia. *Earth Surface Processes and Landforms*, 17: 807–819.

Nanson, G.C., Chen, X.Y. and Price, D.M. 1995. Aeolian and fluvial evidence of changing climate and wind patterns during the past 100 ka in the western Simpson Desert, Australia. *Palaeogeography, Palaeoecology, Palaeoclimatology*, 113: 87–102.

Ollier, C.D. 1988. The regolith in Australia. *Earth Science Reviews*, 25: 355–62.

Ollier, C.D. and Galloway, R.W. 1990. The laterite profile, ferricrete and unconformity. *Catena*, 17: 97–109.

Page, K.J. 1994. Late Quaternary stratigraphy and chronology of the Riverine Plain, southeast Australia. Unpublished PhD thesis, University of Wollongong, NSW, Australia: 228 pp.

Page, K.J., Nanson, G.C. and Price, D.M. 1991. Thermoluminescence chronology of Late Quaternary deposits on the Riverine Plain of south eastern Australia. *Australian Geographer*, 22: 14–23.

Patton, P.C. Pickup. G. and Price, D.M. 1993. Holocene palaeofloods of the Ross River, central Australia. *Quaternary Research*, 40: 201–212.

Pell, S.D. and Chivas, A.R. 1995. Surface features of sand grains from the Australian Continental Dunefield. *Palaeogeography, Palaeoecology, Palaeoclimatology*, 113: 119–132.

Pickup, G., 1991. Event frequency and landscape stability on the floodplain systems of arid central Australia. *Quaternary Science Reviews* 10: 463–473.

Pickup, G., Baker, V.R. and Allan, G. 1988. History, palaeochannels and palaeofloods of the Finke River, central Australia. In R.F. Warner (ed.), *Essays in Australian fluvial geomorphology*. Academic Press, Sydney.

Rubin, D.M. 1990. Lateral migration of linear dunes in the Strezlecki Desert, Australia. *Earth Surface Processes and Landforms*, 15: 1–14.

Schumm, S.A. 1968. *River adjustment to altered hydrologic regimen — Murrumbidgee River and palaeochannels, Australia*. US Geological Survey, Professional Paper, 598.

Singh, G. and Geissler, E.A. 1985. Late Cainozoic history of vegetation, fire, lake levels and climate at Lake George, New South Wales, Australia. *Philosophical Transactions of the Royal Society of London*, 311B: 379–447.

Sprigg, R.C. 1979. Standard and submerged sea-beach systems of southeast South Australia and the aeolian desert cycle. *Sedimentary Geology*, 22: 53–96.

Smith, M.A. 1987. Pleistocene occupation in arid Central Australia. *Nature*, 328: 710–711.

Twidale, C.R. 1967. Origin of the piedmont angle as evidenced in South Australia. *Journal of Geology*, 75: 393–411.

Twidale, C.R. 1972. Evolution of sand dunes in the Simpson Desert, central Australia. *Transactions of the Institute of British Geographers*, 56: 77–109.

Twidale, C.R. 1981. Age and origin of longitudinal dunes in the Simpson and other sand ridge deserts. *Die Erde*, 112: 231–247.

Twidale, C.R. 1983. Australian laterites and silcretes: ages and significance. *Revue Geologie Dynamic et Geographie Physique*, 24: 28.

Twidale, C.R. and Hutton, J.T. 1986. Silcrete as a climatic indicator: discussion. *Palaeogeography, Palaeoclimatology, Palaeoecology*, 52: 351–360.

Tyler, M.J., Twidale, C.R., Davies, M. and Wells, C.B. 1990. *Natural history of the North East Deserts*. Royal Society of South Australia.

Van der Graaff, W.J.E., Crowe, R.W.A., Bunting, J.A. and Jackson, M.J. 1977. Relict early Cenozoic drainage in arid western Australia. *Zeitschrift für Geomorphologie*, NF, 21: 379–400.

Wasson, R.J. 1979. Sedimentation history of the Mundi Mundi alluvial fans, western NSW. *Sedimentary Geology*, 22: 21–51.

Wasson, R.J. 1983a. The Cainozoic history of the Strzelecki and Simpson dune fields (Australia) and the origin of the desert dunes. *Zeitschrift für Geomorphologie*, Supplementband, 45: 85–115.

Wasson, R.J. 1983b. Dune sediment types, sand colour, sediment provenance and hydrology in the Simpson–Strzelecki Dunefield, Australia. In M.E. Brookfield and T.S. Ahlbrandt (eds), *Eolian sediments and processes*. Elsevier, Amsterdam: 165–195.

Wasson, R.J. 1986. Geomorphology and Quaternary history of the Australian Continental Dunefields. *Geographical Review of Japan*, 59B: 55–67.

Wasson, R.J., Fitchett, K., Mackey, B. and Hyde, R. 1988. Large-scale patterns of dune type, spacing and orientation in the Australian Continental Dunefield. *Australian Geographer*, 19: 89–104.

Wells, R.T. and Callen, R.A. 1986. *The Lake Eyre Basin — Cainozoic sediments, fossil vertebrates and plants, landforms, silcretes and climatic implications*. Australasian Sedimentologists Group, Field Guide Series No. 4: 176 pp.

Williams, D.L.G. 1982. Multiple episodes of Pleistocene dune building at the head of Spencer Gulf, South Australia. *Search*, 13: 88–90.

Williams, G.E. 1970. The central Australian stream floods of February–March 1967. *Journal of Hydrology*, 11: 185–200.

Williams, G.E. 1971. Flood deposits of the sand-bed ephemeral streams of central Australia. *Sedimentology*, 17: 1–40.

Williams, M.A.J. 1984. Cenozoic evolution of arid Australia. In H.G. Cogger and E.E. Cameron (eds), *Arid Australia*. Australian Museum, Sydney: 59–79.

Wopfner, H. and Twidale, C.R. 1967. Geomorphological history of the Lake Eyre Basin. In J.N. Jennings and J.A. Mabbutt (eds), *Landform studies from Australia and New Guinea*. Cambridge University Press, Cambridge: 119–143.

Wopfner, H. and Twidale, C.R. 1988. Formation and age of desert dunes in the Lake Eyre depocentres in central Australia. *Geologische Rundschau*, 77: 815–834.

Section 6
Extensions and Change in the Arid Realm

26

Reconstructing ancient arid environments

David S. G. Thomas

Introduction

Morphological and sedimentary evidence indicate that the world's deserts have experienced significant episodic expansions and contractions during the Quaternary period. The geological record further indicates that the positions of deserts have changed in response to major global tectonic and climatic developments throughout Earth history. As ancient evaporate and aeolian sediments can be important sources of hydrocarbons, the elucidation of the whereabouts and nature of ancient deserts is of paramount economic importance.

The sporadic existence of deserts has been identified as far back as the Proterozoic (Glennie 1987; Figure 26.1). Various criteria may be used to credit an ancient sediment with an arid zone origin, including mineralogy, grain-size distribution, bedding structure and the micromorphology of individual particles (Table 26.1). The utility of these criteria is dependent upon the application of results from studies of today's arid zone phenomena; the so-called *uniformitarian* or *modern analogue* approach.

Studies of ancient desert sediments have led, for example, to the identification of different dune types from bedding structures in the rock record (see, for example, Ahlbrandt and Fry-

berger 1982), to the recognition of complicated depositional systems in ancient sandstones (Crabaugh and Kocurek 1993) and, perhaps most significantly, to the recognition that ancient desert systems were as variable and complex as those of today. However, the environmental significance of pre-Devonian aeolian sediments is likely to be rather different from that of modern sand seas. This is because prior to the colonisation of land surfaces by plants, wind activity may have been a more potent geomorphological agent, even in relatively humid environments (Wilson 1973; Glennie 1987).

For the Quaternary period, the sedimentological record of arid zone expansions and contractions is often complemented by morphological evidence, in the form of features such as degraded or fossilised sand dunes, inactive rock varnished, alluvial fans, and palaeolake shorelines. In recent decades, the identification of these landforms, often poorly defined or masked by vegetation on the ground, has been greatly enhanced by the availability of high-quality remotely sensed data.

While morphological evidence is confined to the land surface (though 'drowned' features have been identified in some coastal locations: Fairbridge 1964; Sarnthein 1972; Jennings

Arid Zone Geomorphology: Process, Form and Change in Drylands, 2nd edition. Edited by David S. G. Thomas.

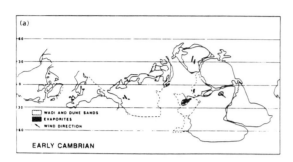

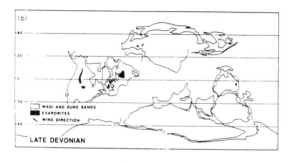

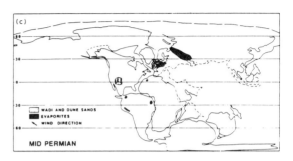

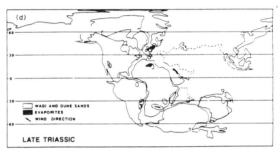

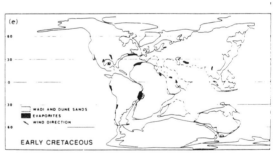

Figure 26.1   *Arid zones from the geological record (after Glennie 1987)*

1975), sedimentary data are more wide-ranging. The great advances in Quaternary science which have been forthcoming from the analysis of ocean sedimentary cores (see, for example, discussions in Bowen 1979; Imbrie and Imbrie 1979) have also yielded valuable information on fluvial and aeolian inputs (see Chapter 18) into the oceans from arid areas.

Though beyond the scope of this volume, biological evidence, especially palynological studies (see, for example, Livingstone 1975; Flenley 1979; Servant *et al.* 1993) including the investigation of floral and faunal distributions (Haffer 1979), has proved to be a valuable source of information on the expansion and contractions of drier Quaternary environments. The following sections examine the identification and interpretation of palaeolandform and

sedimentary evidence of Quaternary desert fluctuations; the chapter is concluded by a discussion of the climatic factors which may have caused perturbations in the extent of the global arid zone during the late Quaternary period.

Arid zone expansions: morphological evidence

The palaeoenvironmental interpretation of morphological evidence of desert expansions is dependent upon the correct identification of a landform, evidence that it is indeed 'relict' or 'fossil' and the subsequent application of environmental analogues from studies of equivalent active modern forms. As the precise

Table 26.1 *Criteria for the identification of arid zone sediments*

Aeolian deposits
(a) Deposits may vary considerably in thickness
(b) Laminae dips range from 0 to 34° (repose-angle for sand) unless affected by post-depositional Earth movements. Angles may be reduced by post-depositional compression
(c) Laminae bedding identifiable with structures in modern dunes (e.g. Ahlbrandt and Fryberger 1982)
(d) Grain sizes range from coarse silt to coarse sand (*c.* 60–2 000 μm, majority in 125–300 μm range
(e) Low silt and clay content (25%, but see Table 26.3)
(f) Large particles often rounded, smaller are subrounded to subangular (see also Thomas 1987a)
(g) May be cemented (aeolianite: Gardner 1983) by haematite or calcium carbonate
(h) May be reddened (Gardner and Pye 1981)
(i) Distinctive surface micromorphology when viewed using a scanning electron microscope (e.g. Krinsley and Doornkamp 1973)

Water-lain deposits
(a) Commonly calcite cemented. Locally cemented by gypsum or anhydrite
(b) Conglomerates may be common
(c) Sand fraction may be absent — removed by deflation
(d) Mudflow conglomerates present
(e) Sharp upward decrease in grain size. Indicates rapid water-level fall
(f) Clay pellets, pebbles or flakes common. Due to effects of salt efflorescence (e.g. Bowler 1986)
(g) Mud cracks common
(h) Often interbedded with aeolian deposits

Source: Modified after Glennie (1987).

environmental and climatic prerequisites for the formation of some landforms which have been used in palaeoenvironmental reconstructions are unclear, the conclusions drawn remain somewhat speculative. Nonetheless, morphological evidence has undoubtedly contributed to a better understanding of former desert extensions; indeed Fairbridge (1970, p. 99) suggested that 'inherited landforms' are the most diagnostic evidence of past climates.

Table 26.2 lists landforms which have been used to indicate former extensions of arid and semi-arid conditions. In each case, landforms are found out of equilibrium with present geomorphic processes and/or eroded and degraded, and have been interpreted as being inherited from a past climatic regime.

RIVERS

One of the most debatable and complex lines of morphological evidence of former aridity comes from fluvial systems. As Baker (1983) has noted, the state of Pleistocene fluvial palaeohydrology shows both promise and problems, which is particularly true of arid regions. Morphological and sedimentological evidence are strongly coupled, with channel-fills and terrace sequences providing the information of past flow conditions. However, as channel flow responds to a range of controls including, for example, climatic and tectonic influences, it can prove difficult to determine palaeoclimatic signals in fluvial sediments and forms (e.g. Frostick and Reid 1989; Reid 1994).

Arid conditions may favour channel aggradation. Reduced rainfall, often accompanied by a marked seasonality, means diminished stream power and the clogging up of channels with relatively coarse sediment, derived from slopes, because of reduced vegetation cover. Conversely, wetter conditions favour incision and terrace development. The terraces and sediments of the Nile have been interpreted in such a way by Adamson *et al.* (1982), and the evidence used to indicate the late Pleistocene extension of arid conditions. In Amazonia, Andean headwaters also aggraded during the late Pleistocene arid phase (Baker 1978), but lowland rivers, including the Amazon itself, incised their courses because sea levels were lower (Tricart 1975; 1984).

Table 26.2   *Examples of morphological evidence of arid and semi-arid zone extensions*

Landform	Indicative of	Location	References
Slopes			
Pediments in humid environments	Former extensions of arid slope systems	Brazil	Bigarella and de Andrade (1965)
Colluvial mantles	Former devegetation due to drier climates	Swaziland	Watson *et al.* (1985)
Dissected alluvial fans	Former high sediment–water ratios due to aridity	Southwestern USA	Lustig (1965)
Alluvial fan construction	Reduction of vegetation in source areas due to aridity	Death Valley S Australia	Hunt and Mabey (1966) Williams (1973)
Rivers			
Clogged drainage	Former river blockage by dunes	Niger	Talbot (1980)
Incised channels	Devegetation and lower sea levels	Amazonia	Tricart (1975; 1984)
Aggraded rivers	Lower and more seasonal discharge	Nile	Adamson *et al.* (1982)
Dunes			
Vegetated linear dunes	Former drier climates and past circulation patterns	Australia Southern Sahara Kalahari	Bowler *et al.* (1976) Grove (1958) Thomas (1984)
Drowned barchans	Former aridity	Botswana	Cooke (1984)
Vegetated parabolic dunes	Drier conditions in past	Colorado	Muhs (1985)
Gullied dunes	Former aridity	Sudan	Talbot and Williams (1978)
Lithified dunes	Former aridity	India	Sperling and Goudie (1975)

Studies of modern arid zone rivers (see Chapters 10 and 11) in fact show that the interpretation of fluvial evidence of former arid and semi-arid zone extensions is far from clear-cut (see Goldberg 1984). Because the flow regimes of streams in dry areas are rarely constant, channel forms and processes are often not in equilibrium (Graf 1983; Hereford 1984). The transport of coarse sediment is very irregular (Schick *et al.* 1987), while the major determinant of channel morphology may well be the size of infrequent, large floods (Hereford 1986; Nanson 1986). These can cause channels to be flushed out and incised, floodplains to be stripped, or channels to be clogged by the arrival of large quantities of sediment from neighbouring slopes. There can also be peculiarities in downstream flow, imparted by the shifting of sediment from one part of the network to another. Overall, the observations of Graf (1983) and Hereford (1984; 1986) suggest that clear-cut relationships between form, process and rainfall are not readily identifiable in arid streams, making the interpretation of recent forms difficult and perhaps individualistic in different drainage basins. The picture is further complicated by the role of base-level changes (Tricart 1975) which result from the sea-level movements that are an important facet of major global climatic and environmental changes.

Modern studies have also shown that channel and slope systems can be poorly linked in dry regions (see, for example, Lekach and Schick 1982). Overall, rivers may in fact be relatively insensitive to all but major changes in climate (Reid 1994), while the sensitivity to climatic fluctuations may be a reflection of the range of flows and floods experienced in an individual climatic regime.

SLOPES

Under semi-arid and arid conditions, when vegetation cover is sparse, sediment transported may be more active on slopes than in channels. This may favour the accumulation of thick, lower-slope colluvium aprons. Such deposits have been identified in southeastern Africa and interpreted as evidence of widespread late Pleistocene semi-arid conditions (Price Williams *et al.* 1982; Watson *et al.* 1985).

Alluvial fans may also act as buffers between arid zone slope and fluvial systems, with their sediments providing a valuable record of environmental changes in source areas (see Chapter 12). The proposition that alluvial fans aggrade more favourably under conditions of reduced vegetation cover has been made by authors such as Dorn (1988) and Blair *et al.* (1990). However, the palaeoenvironmental interpretation of fan and slope deposits may be as contentious as that of channel features and sediments (Dorn 1994a), although systematic analysis of sediment changes and inferred flow variations may enhance interpretations (e.g. Harvey and Wells 1994).

During arid phases slopes near dry lakes may receive gypsum deflated from lake-floor evaporite deposits. In Tunisia, these have contributed to slope-mantling pedogenic gypsum crusts (Watson 1988). Variations in crust micromorphology reflect climatic conditions at the time of development and therefore yield palaeoenvironmental information.

SAND DUNES

Inactive or degraded sand dunes have been widely used as indicators of palaeodeserts, with their widespread identification being facilitated by the examination of satellite imagery and aerial photography. In linear dunefields distinct vegetation zonation aids identification (Figure 26.2). Compared with interdune areas, the dune crests may be relatively sparsely vegetated, as in much of Australia (Madigan 1936) or densely covered by trees because the soft sands favour the development of deep root systems, as in the northern Kalahari (Thomas 1984). In areas where smaller dune types occur and their morphological patterns are less

obvious, identification is still feasible using imagery because quartz-rich sand has a high reflectivity (Muhs 1985).

The extent of ancient dunefields is now well established in Africa (Figure 26.3), India, Australia, North America and to a lesser extent in South America (see Figure 17.1). Reviews of the fixed sand seas of the southern hemisphere have been provided by Thomas and Goudie (1984), for Australia by Wasson (1986) and for North America by Wells (1983).

A range of criteria have been used to designate fixed or relict dune status (Figure 26.3). In the Thar Desert (India), Makgadikgadi Basin (Botswana) and Chad Basin (Nigeria), dunes have been flooded and overlain by lacustrine deposits (Grove 1958; Singh 1971, Cooke 1984). Other dunes have been subjected to eluviation (Smith 1963), pedogenesis and calcification, resulting from an increase in available moisture and reduction in aeolian activity (Allchin *et al.* 1978).

The presence of dune surface vegetation is perhaps the most widely used designator of relict status, but needs careful interpretation as an indicator of inactivity. This is especially so for linear dunes which are in any case relatively immobile forms (Thomas and Shaw 1991b), and which may widely support a element of plant cover even under arid conditions (Tsoar and Møller 1985). While there can be little dispute that fixed dunes covered by dense woodland are inactive, it is not so clear in the case of sparsely or partially vegetated dunes. In present arid environments, there is not a distinct threshold between aeolian activity and inactivity, but rather different levels of activity (Livingstone and Thomas 1993) and a gradual decline in sand movement in the direction of increasing vegetation density (Ash and Wasson 1983; Wiggs *et al.* 1995). Depending on wind velocities, significant sand transport may occur where up to 35% of the ground surface is plant-covered — a situation which occurs in several dunefields which have been designated fixed or inactive, for example, the southwestern Kalahari and parts of the Australian and Indian sand seas.

From a palaeoenvironmental perspective, it may be important, but difficult, to distinguish between phases of dune emplacement and phases of dune surface activity. In parts of the Simpson Desert, Australia, Nanson *et al.*

Figure 26.2   *Landsat image (5 August 1973) of part of southwestern Zambia, southeastern Angola and the Caprivi Strip, Namibia. Vegetation contrasts between the crests of fixed dune ridges and interdune areas allow the identification of ancient dune systems in the area covered by the left half of the image*

(1992a) have used luminescence dating to demonstrate that partially vegetated linear dunes are largely Holocene in age but with dune cores as old as 80 ka. In some areas, the presence of palaeosols within dune sediments may indicate the occurrence of multiple phases of aeolian activity separated by wetter hiatuses.

Present-day dune inactivity is not necessarily a function of increased effective precipitation but may be due to a decline in windiness (Ash and Wasson 1983; Wasson 1984; Bullard *et al.* 1995). For example, as Williams (1985, p. 226) has noted:

In the case of the source bordering dunes of Australia, Sudan and presumably, of Zaire and Amazonia, the three prerequisites for dune formation seem to be a seasonally abundant supply of river sands, strong unidirectional seasonal winds, and a less dense tree cover than today. Such palaeo dunes do not connote extreme aridity.

In an attempt to take account of the possible changes in moisture balance and wind strengths to explain dune inactivity and the level of climatic changes that have occurred since dunes were active, Wasson (1984), Lancaster (1988) and Talbot (1994) have all proposed the used of dune mobility indices (Figure 26.4).

Relict dunes systems can be used to reconstruct palaeocirculation patterns as well as the former extent of aridity (Lancaster 1981; Wells 1983; Wasson 1984; Thomas and Shaw 1991b). The orientation of the relict dunes is compared with the resultant direction of modern sand-moving winds, calculated from wind data (see Chapter 17 for method). Results have frequently suggested that circulation patterns differed from those of today when dune development occurred. Because dunes form in a wide variety of wind regimes today, often with considerable seasonal or diurnal directional variability, the conclusions which have been drawn are frequently speculative.

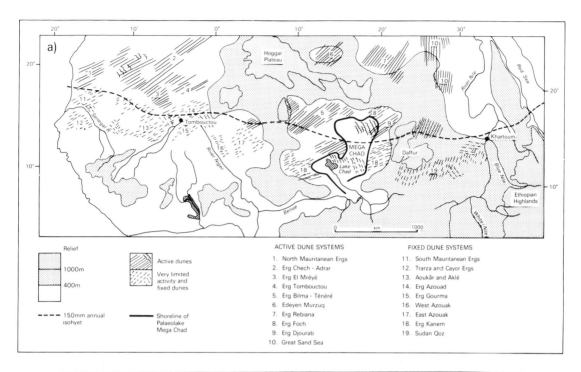

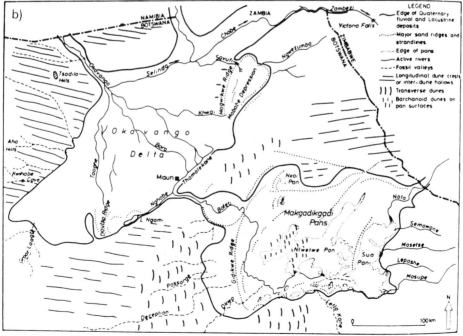

Figure 26.3 *(a) The distribution of active and fixed dune systems in the southern Sahara and Sahel zone, north Africa. Note that the degree of activity and stability is not considered. Drawn from data in Grove (1958), Grove and Warren (1968) and other sources. (b) Fixed dunes and lacustrine features in northern Botswana. After Cooke (1984)*

Table 26.3   *Some possible differences between active and ancient aeolian sands*

Active sands	Ancient sands
95% of particles are sand-sized	Higher component of 'fines' due to subsequent inputs or weathering
Bimodal or unimodal size distribution	May be altered by post-depositional inputs/losses
Larger particles more rounded	Rounding may be reduced by post-depositional chemical weathering
Bedding structures are often present	Structure destroyed by burrowing animals and plant root growth
Low carbonate content	Carbonate contents increased through organic inputs

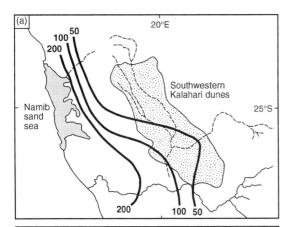

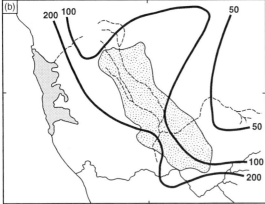

Figure 26.4   *Suggested spatial variations in the distribution of dune mobility index M values between today (a) and the Last Glacial Maximum (b), in the southwestern Kalahari, according to Lancaster (1988). M = W/(P/PE) where W is the % time wind exceeds the threshold velocity for sand transport and P/PE is the ratio of annual precipitation to annual potential evapotranspiration. Index values are interpreted in terms of M > 200 = dunefield fully active; 100–200 = dune plinths and interdunes vegetated; 50–100 = dune crests only active and <50 = dunes inactive*

Arid zone expansions: sedimentological evidence

The value of sedimentological evidence of palaeodeserts extends back beyond the Quaternary period to earlier times for which morphological evidence is absent. 'Comparative sedimentology' relies on the same uniformitarian principles which are applied in the interpretation of relict landforms, and suffers the same problems. Particularly important are the effects of post-depositional modification (diagenesis) which can alter or mask important diagnostic characteristics.

Sedimentary evidence of Quaternary arid zone expansions comes from terrestrial and marine sources. While land-based evidence can be derived from the deposits of any of the geomorphological regimes discussed in previous chapters, aeolian and evaporite deposits are perhaps most important.

AEOLIAN DEPOSITS

In the case of geologically recent desert expansions, aeolian sands are frequently found in conjunction with relict dune forms (see the previous section). When landform associations are absent, aeolian deposits can be identified by a range of sedimentary characteristics (Table 26.1), though these may be altered significantly after deposition (Table 26.3). At Didwana in the Thar Desert, India, for example, a 20-m-high section in aeolian sands exposed in a canal cutting (Figure 26.5) indicates that multiple periods of aridity have occurred since 160 000 BP, the basal luminescence date (Dhir *et al.* 1992).

Interpreting the environmental significance of sand with aeolian attributes is not necessarily

Figure 26.5 *Major sequence of aeolian sediments spanning 160 000 years (from the foot of the trenched section in the foreground to the top of the untrenched section in the background) at Didwana in the Thar Desert, India*

simple. In the Kalahari, the sands found not only in conjunction with relict aeolian land-forms but also with those of lacustrine and fluvial origins have many of the attributes of aeolian deposits (Thomas 1987c). This is because sedimentary characteristics can be inherited from preceding phases of deposition and reworking, or even from parent sediments. The same applies to reddened dune sands, which have sometimes been assumed to indicate aeolian stability and humidity (see Gardner and Pye 1981). Therefore aeolian attributes in a sediment do not necessarily indicate that aeolian processes were the last mechanism of deposition.

Compared with aeolian sands, silt (loess) deposits are poorly represented on the terrestrial margins of subtropical deserts (Tsoar and Pye 1987). This contrasts markedly with the situation in the mid-latitudes, where loess deposits are an important component of the Quaternary record (Smally and Vita-Finzi 1968; Catt 1977), derived through the deflation of glacial outwash and till deposits (Catt 1978). The extensive loess deposits of eastern Asia and South America (Figure 17.1) may, however, be partially derived from more extensive Quaternary arid zones (Goudie 1983a). Loess deposits and intervening palaeosols in north-central China provide a record of climate fluctuations spanning 2.6 Ma (Rutter *et al.* 1991; Li and Zhou 1993).

The lack of loess deposits on the fringes of most desert regions is probably due to a combination of factors, of which two are perhaps most significant. First, desert-derived silt is frequently transported far beyond immediate desert margins (Middleton 1987); consequently, it is well represented in the oceanic record (see Chapter 18 and below). Second, when silt-falls occur in desert regions, distinct loess deposits may not result because the material is incorporated into pre-existing sediments such as stoney *regs* (Amit and Gerson 1986; Yair 1987).

ROCK VARNISH

In some drylands, especially in the American southwest, valuable palaeoenvironmental information has been gained from rock varnish studies (see also Chapter 6 and Dorn 1990). The potential for rock varnish palaeoenvironmental contributions is considerable: varnish is widespread in deserts, can possess a long record, and may yield information on environmental parameters not available from other sources (Dorn 1994). However, only varnishes in subaerial situations that represent a continuous deposition are suitable for investigation, while assumptions have to be made about rates of deposition and it is not possible to date individual varnish layers.

The manganese component of rock varnish, as represented in the Fe : Mn ratio, is a particularly useful palaeoenvironmental tool, representing changes in alkalinity (Dorn and Oberlander 1982). Variations in varnish alkalinity are in turn likely to represent changes in regional vegetation and where appropriate palaeolake extent, and therefore rainfall (Jones 1991; Dorn 1994b; Figure 26.6).

It is also suggested that the micromorphology of rock varnish may yield information on past periods of aeolian activity (Dorn 1986). When aeolian dust is abundant, the assimilation of particles into the varnish results in a structure of parallel microplatelets, but, with less dust in the atmosphere, accumulation occurs around distinct nuclei (Dorn and Oberlander 1982). The superimposition of different micromorphologies can therefore result from changes in aeolian activity through time which can be dated by the cation-ratio method (Whalley 1983). For example, in the southwestern United States rock varnish surfaces record periods of dustiness alternating with less dusty episodes, the latter coinciding with high lake-level stages (Dorn 1986).

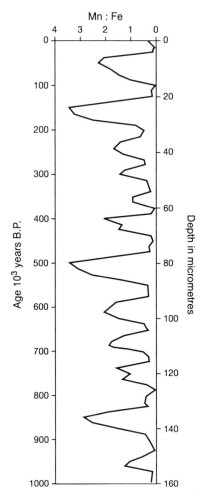

Figure 26.6 *Microlamination of Mn : Fe ratio variations from a rock varnish on the Cima volcanic field lava flow, Mojave Desert. The age scale is based on a K/Ar date from the lava on which the varnish has formed and an assumption of a linear rate of varnish formation with the top representing the present day (after Turrin et al. 1985)*

The stable isotope content of organic material incorporated in rock varnish is a further palaeoenvironmental data source. Dorn and De Niro (1985) found a significant correlation between ^{13}C content of varnish and moisture in the environment, with Dorn *et al.* (1987) using ^{13}C changes in varnish layers to deduce palaeoclimatic impacts on alluvial fan development.

EVAPORITES

Evaporite deposits are an important component of many arid and semi-arid pan or playa
environments (see Chapter 14) and the coastal
sabkhas of some arid regions (Glennie 1987).
Lowenstein and Hardie (1985) identified a
three-stage cycle of ephemeral-playa sedimentation which results in the modification of salt
crystal structure through dissolution and
redisposition and the alternation of mud and
salt layers. The sequences they identify in
modern playa sediments (Figure 26.7) are also
identifiable in the sedimentary record and
assist in the environmental interpretation of
playa deposits preserved in the Quaternary
record (see, for example, Hunt *et al.* 1966;
Smith *et al.* 1983).

The precipitation of different salts in playa
environments can result from variations in
evaporation rates and surface-water conditions
(Chapter 14). Although the minerals within
playa salt deposits are somewhat dependent
upon the chemistry of inflowing waters
(Lowenstein and Hardie 1985), differences in
evaporite deposits can yield information about
the degree of salinity and hence evaporation
rates during past periods of playa sedimentation
(Ullman and McLoed 1986). Bromide and
chloride concentrations can be particularly
valuable in this respect (Allison and Barnes
1985; Ullman 1985). Oxygen isotope ratios
from lacustrine sediments may also yield useful
information though applications may be more
restricted than from ocean sediments (Stuiver
1970). Lowenstein *et al.* (1994) have analysed
the fluid inclusions in halite from a 15 m lake
sediment core extracted from the Qaidam
Basin, western China for major elements and
stable isotopes to produce a uranium series
chronology of aridity and atmospheric temperature spanning 50 000 years.

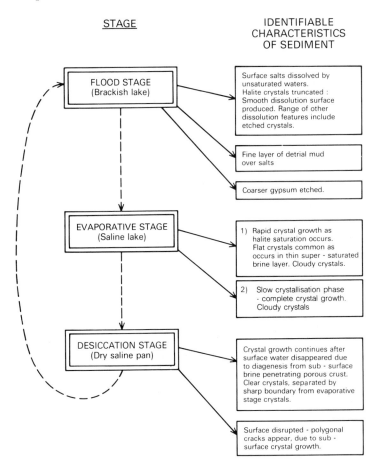

Figure 26.7 *Sequence of playa sediment evolution identified by Lowenstein and Hardie (1985)*

MARINE SEDIMENTS AND PALAEOARIDITY

Analyses of terrestially derived sediments in ocean cores have contributed greatly to interpreting the timing, intensity and extent of Quaternary arid zone extensions. Much land-derived detrius reaches the oceans through fluvial and aeolian pathways. Down-core changes in the abundance of inputs from these sources provide information on weathering and erosion processes and consequently the hydrological regimes on shore, together with an indication of the intensity and direction of wind systems (see Bowles 1975; Pokras and Mix 1985; Rea *et al.* 1985). Analyses of quartz and clay materials from onshore sources have in some cases been supplemented by studies of land-derived pollen (see Parmenter and Folger 1974; Melia 1984).

Terrestrial inputs are added to the relatively constant supply of oceanic organic carbonates (Bradley 1985). Therefore, core sections with a relative abundance of carbonates indicate limited terrestrial inputs, with the inverse relationship applying when carbonates are poorly represented. Analysis of particle size (Schroeder 1985), mineralogy (Windom 1975; Hashimi and Nair 1986), quartz grain coloration (Diester-Haass 1976) and particle surface microrelief (Krinsley 1978) can be used to determine the transportational pathway and hence palaeoenvironmental significance of the terrigenous components of ocean-core sediments.

Aeolian dust inputs are represented by a high proportion of quartz silt, which may have diagnostic red-staining compared with paler fluvial silts (Diester-Haass 1976). An increased abundance of aeolian silts has been attributed to four main factors (Street 1981): enhanced wind strengths; changes in wind directions; a thickening of the dust-transporting layers of the atmosphere; and enhanced aridity in the source area. Analysis of the accumulation rates and grain-size characteristics of aeolian dust in ocean sediments can yield information about the source area and the wind systems responsible for transportation (Rea *et al.* 1985).

Enhanced inputs of aeolian dust, and in some cases dune sand, to Atlantic Ocean cores off west Africa (Sarnthein and Diester-Haass 1977;

Kolla *et al.* 1979), downwind from the Australian deserts (Thiede 1979) and in the Arabian Sea (Kolla and Biscaye 1977) occurred during Quaternary cold periods. The relationship holds for at least the last 600 000 years (Emiliani 1966; Bowles 1975), while dust is detectable almost back to 2 million years BP in some core sediments (Parmenter and Folger 1974).

Comparison of the spatial distribution of Holocene and late Quaternary aeolian deposits in the east Atlantic cores (Figure 26.8) led

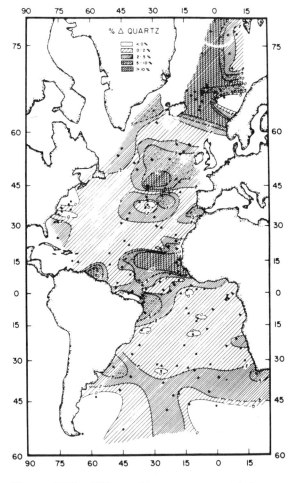

Figure 26.8 *Difference in quartz percentage in late-glacial and Holocene sections of Atlantic sediment cores. Enhanced late-glacial concentrations in the northern hemisphere tropical zone are interpreted as the product of dust inputs deflated from an expanded Sahara and/or due to stronger circulation systems. After Kolla et al. (1979)*

Diester-Haass (1976) to propose a southward shift and Kolla *et al.* (1979) the southward expansion of the Sahara during the Last Glacial Maximum. The identification of freshwater diatoms in later Quaternary Atlantic dust as far south as 20°N, interpreted as deflated material from dry lake floors by Parmenter and Folger (1974), also supports the view of Saharan extension at this time.

As well as indicating that some deserts were more extensive during the Last Glacial Maximum, the widespread occurrence of aeolian dust in ocean cores also suggests that conditions were windier (see Bowler 1978; Sarnthein and Koopman 1980; Sarnthein *et al.* 1981). The size of aeolian dust particles may not however be a good indication of the distance of travel as once supposed, for Schroeder (1985) has shown that the size of mixed-nuclei dust particles responds to water-table fluctuations in the source area, not travel distance. The ratio of high- to low-density dust particles, used by Diester-Haass (1976), may be a better indicator of the transportation ability of the wind and hence atmosphere circulation strengths at the time of deposition.

The analysis of the dust content of northwest Pacific Ocean cores (Rea and Leinen 1988) supports other evidence which shows that Asian continental aridity was reduced at 18 000 BP and more intensified between 9000 and 6000 BP. However, despite changes in the degree and extent of aridity in the source areas, and the northward spread of the zone of strong dust-transporting westerly winds between 11 000 and 6000 BP, the flux of maximum dust supply was always located between 38°N and 40°N (Rea and Leinen 1988). This would seem to imply that the spatial links between dust source (and hence continental aridity), atmospheric transport and deposition in the oceans are rather complex.

Fluvial inputs to ocean core sediments also provide an estimation of the extent of tropical palaeoaridity. Sediments in the offshore Amazon fan, dated to the last glacial period, contain a high proportion of poorly weathered feldspathic sand and little kaolinitic clay (Damuth and Fairbridge 1970; see also Singer 1984 for a review of the palaeoenvironmental significance of clay minerals). This has been interpreted as an indication of less chemical weathering in source areas due to the spread of aridity to the Brazilian Shield (Damuth and Fairbridge 1970) or Amazon headwaters (Milliman and Summerhayes 1975). A similar interpretation has been applied to late Pleistocene sediments from the Indian continental shelf (Hashimi and Nair 1986).

Evidence of arid zone contractions

Various geomorphological sources have yielded information evidencing past periods of greater humidity in present-day arid areas (Table 26.4). Studies of closed-basin lake-level fluctuations from shoreline, sedimentary and organic evidence, with limnological studies including the analysis of sediment chemistry, fauna including gastropod and mollusc remains, microfauna including diatoms and ostracods and microflora including pollen, have become significant indicators of environmental changes in many areas of the tropics and subtropics (see reviews by Street and Grove 1979; Street-Perrott and Roberts 1983; Street-Perrott *et al.* 1985; Currey 1990).

Closed basins are important because their levels adjust to changes in inputs, whereas open lake-basin levels are more likely to remain stable, as input increases can be balanced by changes in overflow. When the palaeolake basins occur within today's arid zone, they provide important evidence of more vigorous hydrological regimes in the past and therefore probably periods of desert contraction (see, for example, Servant and Servant-Vidary 1980; Roberts and Wright 1993; Metcalfe *et al.* 1994).

In the southwestern United States, presently hyper-arid to semi-arid, evidence for more humid conditions during the last Pleistocene is found in the tectonically controlled 'Basin and Range' physiographic province (Smith and Street-Perrott 1983). Shoreline and sedimentary features show that many basins, including Death Valley (Figure 26.9), were once occupied by lakes, some of which were linked together to form huge palaeolakes such as Bonneville (51 700 km^2 in area) and Lahontan (22 442 km^2) (Broecker and Kaufmann 1965; Flint 1971; Goudie 1983a) – see also Chapter 23.

Thick accumulations of playa deposits, interdigitated with palaeosols and in some cases

Table 26.4 *Examples of morphological evidence of former more humid conditions in drylands*

Landforms	Indicative	Location	References
Rivers			
Gravel-bed channels remnants left in bas-relief by erosion of surrounding soft sediments	Wetter climates/greater channel flow	Oman	Maizels (1987)
Underfit modern channels	Greater flow	Argentina	Baker (1978)
Dry valley networks	Higher rainfall and channel flow or higher groundwater tables and spring sapping	Kalahari, Botswana	Nash *et al.* (1994)
Valleys crossed by dunes	Wet–arid climate changes	SE Sahara	Pachur and Kropelin (1987)
Lakes			
Strandlines	Higher former lake levels	Kalahari	Thomas and Shaw (1991a)
Slopes and hillsides			
Fossil screes	Colder, possibly moister conditions	Libya	McBurney and Hey (1955)
Cave and sinter development	Greater local rainfall	Kalahari	Cooke (1975)
Alluvial fan aggradation	Increased effective precipitation	California	Harvey and Wells (1994)
Deep-weathering profiles	Wetter climates	Niger	Williams *et al.* (1987)

aeolian sediments, indicate multiple periods of Quaternary high lakes (Smith 1968; Flint 1971; Smith *et al.* 1983). The extensiveness and complexity of sedimentary and morphological features has made reconstruction of the Pleistocene record complex, well illustrated in the case of Lake Bonneville (Figure 26.10).

Palaeolake Mega Chad in west Africa (Grove and Warren 1968) and Lake PalaeoMakgadikgadi in Botswana (Grey and Cooke 1977) were very extensive (350 000 km^2 and over 60 000 km^2, respectively) and relatively shallow at the maximum extents. Their limits are defined by beach strandlines which occur as part of palaeolandform suites that include fixed dune systems (Figures 26.2 and 26.3), and a number of shorelines at different altitudes, representing palaeolakes of various antiquities and sizes (Shaw and Cooke 1986; see Figure 26.3).

In Australia, groundwater and surface-water fluctuations have been identified as important controls of lake-basin development (Bowler 1986). The presence of lacustrine features and

fringing transverse or lunette dunes in lake basins is not necessarily indicative of distinct periods of positive and negative water balance, for seasonal climatic regimes can result in the contemporaneous development of such landforms (Bowler 1976; 1978). In the Kalahari, periods of permanent water bodies in similar 'pan' lake basins have been deduced from the presence of datable lacustrine stromatolites (Lancaster 1979).

In north Africa and Australia, many lake basins were dry during the Last Glacial Maximum, supporting the oceanic evidence of desert expansion and aridity at that time (Rognon and Williams 1977). In contrast, the lake basins in the southwestern United States and northern and middle Kalahari basins preserve evidence which indicates that enhanced moisture availability characterised these areas during much of the late Pleistocene, and early to mid-Holocene were lacustrine phases in north Africa and Australia, though there are regional differences in the timings of lake rises and falls (Street 1981).

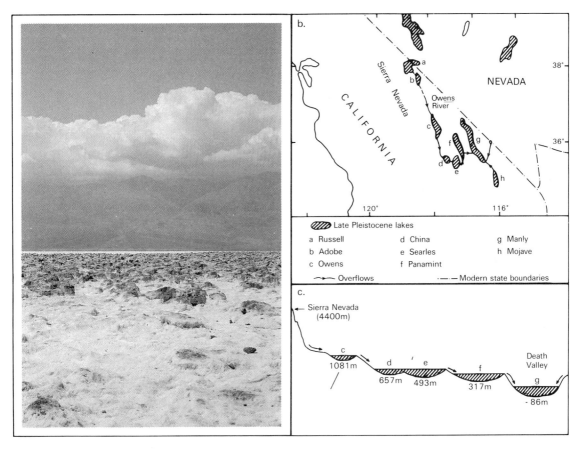

Figure 26.9 *(a) Death Valley, California, has a mean annual precipitation of only 41 mm. Note evaporite deposits in the foreground and alluvial fan in background. (b) and (c) During the late Pleistocene the valley was occupied by Lake Manly, which at its greatest was 183 m deep and 1600 km in area. This was one of a chain of lakes, linked by overspill channels, supplied by moisture from the Sierra Nevada by the Owens River. (b) and (c) respectively based on diagrams in Flint (1971) and Smith and Street-Perrott (1983)*

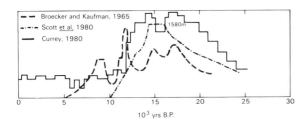

Figure 26.10 *Three interpretations of the late Quaternary history of Lake Bonneville, southwestern USA. Higher curves represent higher lake levels, for example peak of curve of Scott* et al. *(1980) is the 'Bonneville level' at 1580 m above sea level. Differences arise through problems with interpretation of radiocarbon dates, the use of different dating techniques, disturbance of shorelines by isostatic adjustment, and correlating data from distant locations. Other interpretations also exist, for example, Broeker and Orr (1958) and Morrison and Frye (1965)*

Table 26.5 Examples of estimates of changes in climatic parameters in the southwestern USA, based on lake-basin studies

Location	Mean annual temperature (°C)	Summer temperature (°C)	Annual precipitation (mm)	Annual evaporation (mm)	Evaporation (%)	References
Lake Lahontan, Nevada	−5	−5	+840	−410	34	Antevs (1952)
Lake Lahontan, Nevada	−3	−	+270	−180	30	Broecker and Orr (1958)
Lake Lahontan, Nevada	−6.5	−9	+230	−380	16	Mifflin and Wheat (1979)
Lake Estancia, New Mexico	−8	−8	+180	−250	34	Leopold (1951)
Lake Estancia, New Mexico	−10.5	−1	−40	−510	42	Brackenridge (1978)
Lake Estancia, New Mexico	−	−7	−46	−330	45	Galloway (1970)
Spring Valley, Nevada	−8	−8	+210	−480	30	Snyder and Langbein (1962)
Spring Valley, Nevada			0		43	Brackenridge (1978)

Although periods of high lakes are often described as pluvials, it is often difficult to determine whether increased rainfall or reduced evaporation because of lower temperatures was in fact responsible (Smith and Street-Perrott 1983; Bradley 1985; see Table 26.5). Various hydrological-balance models have been produced to try to resolve these issues (see Brackenridge 1978; Street 1979; Kutzbach 1980; Sack 1994). In some cases, palynological data have assisted in determining the relative importance of changes in precipitation and evaporation (see Wickens 1975; Livingstone 1980). The recent sedimentological, morphological and chemical classification of modern Australia lake basins, which occur along a distinct hydrological gradient (see Chapter 14), may also assist in a clearer interpretation of the Quaternary record.

In the Makgadikgadi Basin, Botswana, budgetary studies have shown that the highest palaeolake stage cannot be accounted for in climatological terms alone (Grove 1969; Shaw 1985). This has been confirmed by a study which has identified links between the palaeolake and increased fluvial inputs from distant sources (Shaw and Thomas 1988). Where fluvial inputs have clearly made substantial contributions to high lake stages, but the palaeolake tributaries are now dry, as in the case of Lake Bonneville, it is extremely difficult to quantify their inputs in budgetary studies (Smith and 1983).

In many deserts, relict fluvial features have been used to suggest more humid regimes in the past. As discussed earlier, reconstructions of fluvial regimes in desert regions present many problems. Where old drainage lines are found, choked today with aeolian sands, the evidence for former greater humidity would appear to be more clear-cut (see Boocock and Van Straten 1962; Mulcahy and Betenay 1972; Rognon 1987). However, recent investigations in the eastern Kalahari have indicated that some dry valley systems have evolved through the influence of higher groundwater tables and spring sapping, with the magnitude of local humidity increases necessary to account for their development being less than once supposed (Shaw and De Vries 1988; Nash *et al.* 1994; see Chapter 15).

Caves can provide valuable evidence of palaeohumidity. For example, Cooke (1975) and Cooke and Verhagen (1977) investigated and dated periods of cave construction and sinter development in dolomite hills in the northwestern Kalahari, providing unequivocal evidence of phases of late Pleistocene and Holocene humidity. Stable isotope analyses from speleothems in dryland caves yield direct and indirect information on temperature and precipitation changes (Holmgren *et al.* 1995). Periods of aridity are represented by cave infilling by aeolian sands. In the arid central Sahara, cave paintings, picturing animals, fishermen and boats, are believed to indicate human habitation and higher rainfall during the mid-Holocene (Monod 1963; Bradley 1985).

Dating arid zone fluctuations

Establishing a chronology of arid zone expansions and contractions during the Quaternary period is subject to the application of relative and absolute dating techniques (see Chapter 27). In this respect, the advances in isotopic dating methods of the last few decades (see Goudie 1983a; Bradley 1985; and other Quaternary texts for discussions of these methods) have contributed to the overall picture of dynamic tropical environments, to the recognition that some deserts are extremely old (Street 1981) and to a clearer picture of the geomorphic responses of current desert and peridesert areas to climatic changes. Recent developments and applications of some dating methods such as the luminescence family of techniques (e.g. Singhvi *et al.* 1982) and the application of radiometric techniques to new data sources, such as rock varnish (e.g. Dorn 1986), have had major implications for the development of chronologies of climate and environmental changes in drylands (Figure 26.11). This in part helps to overcome the problems in a dryland context associated with, and relative paucity of materials available for, more generally used methods such as radiocarbon dating (e.g. Deacon *et al.* 1984; Williams 1985; Thomas and Shaw 1991b; Nanson *et al.* 1992a; 1992b).

A broad framework for arid zone expansion has been achieved by the placement of aeolian sediments in the long timescales of oceanic cores. Together, the palaeomagnetic and iso-

Figure 26.11   *Interpretation of this recently discovered 25-m-high gully-cut section in a falling dune near Ardekan, central Iran, which reveals phases of aeolian deposition separated by palaeosol development, should be greatly enhanced by the application of luminescence dating to samples from the aeolian units*

topic evidence which they preserve has led to the establishment of the 'master chronology' of Cenozoic temperature and ice volume changes. Within this, major layers of aeolian sediments have been correlated to periods of global cooling extending back 38 million years (Sarnthein and Diester-Haass 1977; Sarnthein 1978). This has contributed directly to the replacement of the 'glacial equals pluvial' hypothesis with one which equates glaciation with tropical aridity (Street 1981; Goudie 1983b; Williams 1985). Like its predecessor, this general framework is not without its problems (see Shaw and Cooke 1986).

Characteristics and causes of late Quaternary arid zone extensions

The growth of the paradigm which equates high-latitude glaciations with low-latitude aridity (Williams 1975; Sarnthein 1978; Williams *et al.* 1993) makes it appropriate to examine the distribution of expanded deserts, their nature, and climatic mechanisms responsible for their development. Because the evidence is clearest, this will be done with

respect to the period of the last glacial (*c.* 80 000 BP to *c.* 12 000 BP with maximum ice extent at *c.* 24–20 000 BP), though ocean-core evidence suggests that such a reconstruction is equally appropriate for earlier Pleistocene glaciations.

CLIMATIC CONSIDERATIONS

The thermal depression accompanying global ice expansion had a number of profound effects on atmospheric circulation and temperature regimes. Among the most important of these (cf. Nicholson and Flohn 1980) were the steepening of meridional (pole–equator) temperature gradients, an equatorward displacement of climatic belts and baraclinic zones, and the reduction of interhemispheric temperature contrasts.

The southward displacement of the northern hemisphere baraclinic zone and the associated upper westerly Jet Stream caused an increase in moisture-bearing weather systems reaching parts of the southwestern United States, the Mediterranean region and north Africa for at least part of the late-glacial period. This is borne witness to by the oceanic and lake-level records (Diester Haass 1976; Smith and Street-Perott 1983; Street-Perrott and Roberts 1983), though pollen analyses show that conditions were drier in some parts of, or for some of the time in, the Mediterranean region (Rossignol-Strick and Duzer 1980).

The southward shift of the belt of northern hemisphere subtropical high-pressure cells was accompanied by their expansion and intensification, because of the enhanced Hadley cell circulation caused by steeper temperature gradients (Flohn and Nicholson 1980). This led to the southward extension of the Sahara arid zone, indicated by low lake levels (Street and Grove 1979; Servant and Servant-Vildray 1980), fluvial studies (Adamson *et al.* 1980), the activation of aeolian processes and dune systems in the Sahel belt (see Talbot 1980) — though little of this evidence is directly dated — and the dust layers in ocean cores (Sarnthein 1978; Kolla *et al.* 1979). The palynological record indicates that even beyond the expanded arid zone, conditions were drier and cooler in parts of tropical Africa (e.g. Livingstone 1975; Kadomura and Hori 1990).

At the Last Glacial Maximum, tropical land areas were about 2–6°C cooler than today (mean annual values; but see Street 1981, for consideration of limitations of such estimates). The steeper thermal gradient led to higher wind speeds (see Petit *et al.* 1981); therefore the late Pleistocene low-latitude deserts were probably windier as well as colder than their modern counterparts. Over the oceans, this could have increased evaporation, counteracting the effects of reduced sea-surface temperatures on atmospheric moisture levels. The stronger trade winds caused by steeper pressure gradients also led to enhanced upwelling of cold ocean waters in subtropical locations, further enhancing the aridity of coastal deserts such as the Namib and Atacama (Williams *et al.* 1993).

Several factors, however, seem to mitigate against this. First, the stronger Hadley cell circulation resulted in a greater upwelling of cold water in equatorial oceans (Hays *et al.* 1976; Molina-Cruz 1977; Prell *et al.* 1980), so that tropical waters were up to 8°C cooler than today in the main areas of upwelling. Second, the equatorward compression of the atmospheric circulation, and the expansion of subtropical highs, resulted in the latter becoming even more persistent features than they are today (see Brookfield 1970; Kolla and Biscaye 1977). The penetration of northern hemisphere southwest trade winds into west Africa and northwestern India would consequently have been severely restricted (Nicholson and Flohn 1980), contributing to desert expansion in both areas. Third, lower glacial-age sea levels increased the continentality of areas with broad continental shelves, further limiting the penetration of precipitation. In some areas, for example northeastern Australia, this was a major contributory factor in the spread of aridity (see Chappell 1978).

THE DISTRIBUTION OF LATE-GLACIAL TROPICAL ARIDITY

In north Africa and Australia, late Quaternary environmental changes, inducing late-glacial aridity, are both well-documented and well-dated (see, for example, Rognon and Williams 1977; Street-Perrott and Roberts 1983; Bowler and Wasson 1984). In north Africa, the expansion of aridity south of the Sahara occurred from about 22 000 BP to 12 000 BP (Butzer 1972; Rognon and Williams 1977), and was followed by an episode of greater humidity and high lakes (Street and Grove 1979). Pokras and Mix (1985) have therefore proposed a climatic model which equates north African tropical aridity with phases of ice growth, and maximum humidity with deglaciation. The Younger Dryas cooling event at 11–10 000 BP, recognised in the climate records of the mid-latitudes, has now also been identified as causing a period of enhanced low-latitude aridity, as evidenced from East African lake-level records (Roberts *et al.* 1993).

The Australian last glacial was a two-staged affair. Many areas, including the western and southern interior, saw full but oscillating lakes from 50 000 BP, and marked aridity between 25 000 BP and 16 000 BP, through to 12 000 BP in the arid core (Bowler and Wasson 1984). Luminescence dates indicate, however, that aeolian activity has occurred in the Australian interior for much of the last 80 000 years (Nanson *et al.* 1992a). In India sedimentary evidence shows that the arid period, centred on Rajasthan, commenced at about 25 000 BP and persisted to as late as 1000 BP in some areas (Wasson *et al.* 1983). Dune building in the Thar Desert did not commence until 14 000 BP, with the re-establishment of the southwestern monsoon (Chawla *et al.* 1991; Dhir *et al.* 1992). Lacustrine evidence, however, suggests that the late Holocene desiccation of northwest India was separated from late Pleistocene aridity by a period of more equitable climates (Singh *et al.* 1972).

Several authorities have interpreted the evidence from north Africa and Australia (Williams 1975; 1985; Rognon and Williams 1977), or from oceanic cores (Sarnthein 1978), to infer that tropical aridity was the norm during the late glacial. For example, Sarnthein (1978) has written that 'Today about 10 percent of the land area between 30°N and 30°S is covered by active sand deserts ... [These were] much more widespread 18 000 years ago ... [characterising] almost 50 percent of the land area between 30° and 30°S, forming two vast belts'.

Climatic modelling of the late-glacial circulation (Nicholson and Flohn 1980) has also assumed a mirroring of conditions in the tropics north and south of the equator (Figure 26.12), based primarily on the assumption that this would be achieved through the reduction of interhemisphere temperature contrasts at times of maximum ice expansion.

Although some evidence is contradictory, the overall climatic signal derived from proxy data from the southern African interior would appear to contrast significantly with that from north of the equator (Figure 26.13). In the northern Kalahari, the period 40 000–20 000 BP saw fluctuating conditions with humidity evidenced by high lake levels (Cooke 1980; Helgren and Brooks 1983) and aridity by aeolian sands incorporated between datable cave deposits (Cooke 1975). Luminescence dates from the extensive Kalahari dune systems will greatly enhance the interpretation of the southern African chronology.

Although drier conditions prevailed locally for short periods, including in the Makgadikgadi Basin at 19 000 BP (Shaw and Cooke 1986), a significant body of evidence from cave, lacustrine and fluvial deposits indicate that the late-glacial saw enhanced humidity at times during the Late Glacial in the Kalahari (Butzer *et al.* 1978; Lancaster 1979; Shaw and Cooke 1986; Shaw and Thomas 1988), in adjacent areas (Butzer *et al.* 1973; Kent and

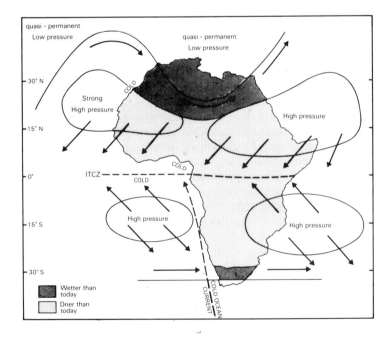

Figure 26.12 *Nicholson and Flohn's (1980) model of late-glacial circulation over Africa. Dark shading: areas wetter than today; light shading: drier than today*

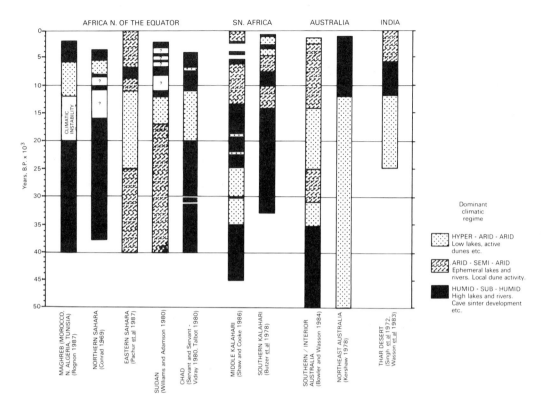

Figure 26.13 *Simplified chronologies of late Quaternary expansions and contractions of tropical and subtropical arid zones. Transitions may often have been less abrupt than indicated, while changes of less than 1000 years' duration are not shown. All chronologies constructed primarily from geomorphological/sedimentological data, except northeast Australia, which is solely palynological in origin*

Gribnitz 1985) and even in parts of Namibia (Vogel and Visser 1981; Rust *et al.* 1984).

The analysis of palaeotemperature data by Harrison *et al.* (1984) has indicated that there were considerable departures in the timings and extent of changes in the meridional temperature gradient in the northern and southern hemispheres. Their calculations suggest that, during the late-glacial, climatic belts in both hemispheres could have been displaced southwards. This lends support to the views of Tricart (1956), Newell (1973) and Lancaster (1979), who suggested that, over Africa, the meteorological equator (the intertropical convergence zone) may have been displaced south of the equator during the last glacial. If this were the case, it would provide a climatic mechanism to explain the late-glacial contraction of the southern African arid zone and the expansion of its northern counterpart.

Conclusion

There are many issues which remain unresolved in the reconstruction of the extent of past arid zones. Some of these concern the evaluation of morphological and sedimentological evidence. Bowler and Wasson (1984) illustrated how unusual aspects of the relict arid zone landscapes of Australia could be better understood when explanations incorporated an awareness of the environmental conditions preceding a climatic change. This concept of 'geomorphological inheritance' (Bowler and Wasson 1984) also illustrates that the direction of an environmental change influences the geomorphological response. Knox (1972), for example, demonstrated how a change from aridity to humidity would be represented differently in the geomorphological record when compared with one in the opposite direction.

Other studies have shown that it can be exceedingly difficult to determine the relative contributions of temperature (via evaporation) and precipitation changes to the spread of aridity (Williams 1985). But perhaps of overwhelming importance is to recognise the dangers of accepting the replacement of a climatic model for the Quaternary which equated glacial periods with pluvials by an equally simple one which links the growth of ice in high latitudes to the spread of aridity in the tropics.

References

Adamson, D.A., Gasse, F., Street, F.A. and Williams, M.A.J. 1980. Late Quarternary history of the Nile. *Nature*, 287: 50–55.

Adamson, D.A., Gillespie, R. and Williams, M.A.J. 1982. Palaeogeography of the Gezira and of the lower Blue and White Nile valleys. In M.A.J. Williams and D.A. Adamson (eds), *A land between two Niles*. Balkema, Rotterdam: 165–219.

Ahlbrandt, T.S. and Fryberger, S.G. 1982. Introduction to eolian deposits. In R.A. Scholle and D. Spearing (eds), *Sandstone depositional environments*. Memoirs of the American Association of Petroleum Geologists, 31: 11–47.

Allchin, B., Goudie, A.S. and Hegde, K.T.M. 1978. *The prehistory and palaeogeography of the great Indian desert*. Academic Press, London.

Allison, G.B. and Barnes, C.J. 1985. Estimation of evaporation from the normally 'dry' Lake Frome in South Australia. *Journal of Hydrology*, 78: 229–242.

Amit, R. and Gerson, R. 1986. The evolution of Holcene reg (gravelly) soils in deserts — an example from the Dead Sea region. *Catena*, 13: 59–79.

Antevs, E.A. 1952. Cenozoic climates of the Great Basin. *Geologische Rundschau*, 40: 94–108.

Ash, J.E. and Wasson, R.J. 1983. Vegetation and sand mobility in the Australian desert dunefield. *Zeitschrift für Geomorphologie*, Supplementband, 45: 7–25.

Baker, V.R. 1978. Adjustment of fluvial systems to climate and source terrain in tropical and subtropical environments. In A.D. Miall (ed.), *Fluvial sedimentology*. Canadian Society of Petroleum Geologists, Memoir, 5: 211–230.

Baker, V.R. 1983. Late-Pleistocene fluvial systems. In H.E. Wright, Jr (ed.), *Late Quarternary environments of the United States, Vol. 1. The Late Pleistocene*. University of Minnesota Press, Minneapolis: 115–129.

Bigarella, T.J. and de Andrade, G.O. 1965. Contribution to the study of the Brazilian Quaternary. *Geological Society of America, Special Publication*, 84: 433–451.

Blair, T.C., Clark, J.S. and Wells, S.C. 1990. Quaternary continental stratigraphy, landscape evolution and application to archaeology: Jarilla piedmont and Tularosa graben floor, White Sand Missile Range, New Mexico. *Bulletin of the Geological Society of America*, 102: 749–759.

Boocock, C. and Van Straten, O.J. 1962. Notes on the geology and hydrogeology of the central Kalahari Region, Bechuanaland Protectorate. *Transactions of the Geological Society of South Africa*, 65: 125–171.

Bowen, D.Q. 1979. Geographical perspective on the Quaternary. *Progress in Physical Geography*, 3: 167–186.

Bowles, F.A. 1975. Paleoclimatic significance of quartz/illite variations in cores from the eastern equatorial North Atlantic. *Quaternary Research*, 5: 225–235.

Bowler, J.M. 1976. Aridity in Australia: age, origins and expression in aeolian landforms and sediments. *Earth Science Reviews*, 12: 279–310.

Bowler, J.M. 1978. Glacial age aeolian events at high and low latitudes: a southern hemisphere perspective. In E.M. van Zinderen Bakker (ed.), *Antarctic glacial history and world palaeoenvironments*. Balkema, Rotterdam: 149–172.

Bowler, J.M. 1982. Aridity in the Late Tertiary and Quaternary of Australia. In W.R. Barker and P.J.M. Greenslade (eds), *Evolution of the flora and fauna of arid Australia*. Peacock, Australia: 35–46.

Bowler, J.M. 1986. Spatial variability and hydrologic evolution of Australian lake basins: analogue for Pleistocene hydrologic change and evaporite formation. *Palaeogeography, Palaeoclimatology, Palaeoecology*, 54: 21–41.

Bowler, J.M. and Wasson, R.J. 1984. Glacial age environments of inland Australia. In J.C. Vogel (ed.), *Late Cainozoic palaeoclimates of the southern hemisphere*. Balkema, Rotterdam: 183–208.

Bowler, J.M., Hope, G.S., Jennings, J.N., Singh, G. and Walker, O. 1976. Late Quaternary climates of Australia and New Guinea. *Quaternary Research*, 6: 359–394.

Brackenrige, G.R. 1978. Evidence for a cold, dry full-glacial climate in the American southwest. *Quaternary Research*, 9: 22–40.

Bradley, R.S. 1985. *Quaternary palaeoclimatology*. Allen & Unwin, Boston.

Broecker, W.S. and Kaufmann, A. 1965. Radiocarbon chronology of Lake Lahontan and Lake Bonneville II, Great Basin. *Bulletin of the Geological Society of America*, 76: 537–566.

Broecker, W.S. and Orr, P.C. 1958. Radiocarbon chronology of Lake Lahontan and Lake Bonneville. *Bulletin of the Geological Society of America*, 70: 1009–1032.

Brookfield, M. 1970. Dune trends and wind regime in central Australia. *Zeitschrift für Geomorphologie*, Supplementband, 10: 121–158.

Bullard, J.E., Thomas, D.S.G., Livingstone, I. and Wiggs, G. 1995. Wind energy variation in the SW Kalahari desert and implications for linear dunefield activity. *Earth Surface Processes and Landforms*, 21: 263–278

Butzer, K.W. 1972. *Environment and archaeology — an ecological approach to prehistory.* Methuen, London.

Butzer, K.W., Fock, G.J., Stuckenrath, R. and Zilch, A. 1973. Paleohydrology of late Pleistocene Lake Alexanderfontein, Kimberley, Southern Africa. *Nature*, 243: 328–330.

Butzer, K.W., Stuckenrath, R., Bruzewicz, A.J. and Helgren, D.M. 1978. Late Cenozoic palaeoclimates of the Gaap Escarpment, Kalahari margin, South Africa. *Quaternary Research*, 10: 310–339.

Catt, J.A. 1977. Loess and coversands. In F.W. Shotton (ed.), *British Quaternary studies — recent advances.* Oxford University Press, Oxford: 221–229.

Catt, J.A. 1978. The contribution of loess to soils in lowland Britain. In S. Limbrey and J. G. Evans (eds), *The effects of man in the landscape: the lowland zone.* CBA Research Report 21: 12–20.

Chappell, J. 1978. Theories of Upper Quaternary ice ages. In A.B. Pittock *et al.* (eds), *Climatic change and variability: a southern perspective.* Cambridge University Press, Cambridge: 211–225.

Chawla, S., Dhir, R.P. and Singhvi, A.K. 1991. Thermoluminescence chronology of sand profiles in the Thar Desert and their implications. *Quaternary Science Reviews*, 1: 25–32.

Cole, K.L. 1986. The lower Colorado River Valley: a Pleistocene desert. *Quaternary Research*, 25: 395–400.

Conrad, G. 1969. *L'evolution continental posthercynienne du Sahara algerien.* CNRS, Paris, 10.

Cooke, H.J. 1975. The palaeoclimatic significance of caves and adjacent landforms in western Ngamiland, Botswana. *Geographical Journal*, 141: 430–444.

Cooke, H.J. 1980. Landform evolution in the context of climatic change and neo-tectonism in the middle Kalahari of north central Botswana. *Transactions of the Institute of British Geographers*, NS, 5: 80–99.

Cooke, H.J. 1984. The evidence from northern Botswana of late Quaternary climatic change. In J.C. Vogel (ed.), *Late Cainozoic palaeoclimates of the southern hemisphere.* Balkema, Rotterdam: 265–278.

Cooke, H.J. and Verhagen, B.T. 1977. The dating of cave development — an example from Botswana. In *Proceedings of the 7th International Speleological Congress, Sheffield.*

Crabaugh, M. and Kocurek, G. 1993. Entrada Sandstone: an example of a wet aeolian system. In K. Pye (ed.), *The dynamic and environmental context of aeolian sedimentary systems.* Geological Society of London, Special Publication, 72: 103–126.

Currey, D.R. 1980. Coastal geomorphology of Great Salt Lake and vicinity. In J.W. Gwynn (ed.), *Great Salt Lake: a scientific, historic and economic overview.* Utah Geological and Mineral Survey Bulletin, 116: 69–82.

Currey, D.R. 1990. Quaternary palaeolakes in the evolution of semi-desert basins, with special emphasis on Lake Bonneville and the Great Basin, USA. 76: 189–214.

Damuth, J.E. and Fairbridge, R.W. 1970. Arkosic sands of the last glacial stage in the tropical Atlantic off Brazil. *Bulletin of the Geological Society of America*, 81: 189–206.

Deacon, J., Lancaster, N. and Scott, L. 1984. Evidence for late Quaternary climatic change in southern Africa. Summary of the proceedings of the SASQUA Workshop held in Johannesburg, September 1983. In J.C. Vogel (ed.), *Late Cainozoic palaeoclimate of the southern hemisphere.* Balkema, Rotterdam: 391–404.

Dhir, R.P., Kar, A., Wadhawan, S.K., Rajaguru, S.N., Misra, V.N., Singhvi, A.K. and Sharma, S.B. 1992. *Thar Deserts in Rajasthan: land, man and environment.* Geological Society of India, Bangalore.

Diester-Haass, L. 1976. Late Quaternary climatic variations in northwest Africa deduced from east Atlantic sediment cores. *Quaternary Research*, 6: 299–314.

Dorn, R.I. 1986. Rock varnish as an indicator of aeolian environmental change. In W.G. Nickling (ed.), *Aeolian geomorphology.* The Binghamton Symposia in Geomorphology: International Series, 17. Allen & Unwin, Boston: 291–307.

Dorn, R.I. 1988. A rock varnish interpretation of alluvial fan development in Death Valley, California. *National Geographic Research*, 4: 56–73.

Dorn, R.I. 1990. Quaternary alkalinity fluctuations recorded in rock varnish microlamination on western USA volcanics. *Palaeogeography, Palaeoclimatology, Palaeoecology*, 76: 291–310.

Dorn, R.I. 1994a. The role of climate change in alluvial fan development. In A.D. Abraham and A.J. Parson (eds), *Geomorphology of desert environments.* Chapman & Hall, London: 593–615.

Dorn, R.I. 1994b. Rock vanish as evidence of climatic change. In A.D. Abrahams and A.J.

Parsons (eds), *Geomorphology of desert environments*. Chapman & Hall, London: 539–552.

Dorn, R.I. and De Niro, M.J. 1985. Stable cation isotope ratios on rock varnish organic matter: a new palaeoenvironmental indicator. *Science*, 227: 1472–1474.

Dorn, R.I. and Oberlander, T.M. 1982. Rock varnish. *Progress in Physical Geography*, 6: 317–367.

Dorn, R.I., DeNiro, M.J. and Ajse, H. 1987. Isotopic evidence for climatic influence on alluvial fan development in Death Valley, California. *Geology* 15: 108–110.

Emiliani, C. 1966. Palaeotemperature analysis of Caribbean cores P6304-8 and P6304-9 and a generalised temperature curve of the past 425,000 years. *Journal of Geology*, 74: 109–126.

Fairbridge, R.W. 1964. African ice-age aridity. In A.E.M. Nairn (ed.), *Problems in palaeoclimatology*. Wiley Interscience, London: 356–363.

Fairbridge, R.W. 1970. World paleoclimatology of the Quaternary. *Revue de Géographie Physique et Géologie Dynamique*, 12: 97–104.

Flenley, J.R. 1979. The late Quaternary vegetation history of the equatorial mountains. *Progress in Physical Geography*, 3: 488–509.

Flink, R.F. 1971. *Glacial and Quaternary geology*. Wiley Interscience, New York.

Flohn, H. and Nicholson, S. 1980. Climatic fluctuations in the arid belt of the 'Old World' since the last glacial maximum: possible causes and future implications. *Palaeoecology of Africa*, 12: 3–21.

Frostick, L.E. and Reid, I. 1989. Climate versus tectonic control on far sequences: lessons from the Dead Sea. *Journal of the Geological Society of London*, 146: 527–538.

Galloway, R.W. 1970. The full-glacial climate in the southwestern United States. *Annals of the Association of American Geographers*, 60: 245–256.

Gardener, R. and Pye, K. 1981. Nature, origin and palaeoenvironmental significance of red coastal and desert dune sands. *Progress in Physical Geography*, 5: 514–534.

Gardner, R.A.M. 1983. Aeolianite. In A.S. Goudie and K. Pye (eds), *Chemical sediments and geomorphology*. Academic Press, London: 265–300.

Glennie, K.W. 1970. *Desert sedimentary environments*. Elsevier, Amsterdam.

Glennie, K.W. 1987. Desert sedimentary environments, present and past — a summary. *Sedimentary Geology*, 50: 135–166.

Goldberg, P. 1984. Late Quaternary history of Qadesh Barnea, northeastern Sinai. *Zeitschrift für Geomorphologie*, NF, 28: 193–218.

Goudie, A.S. 1983a. *Environmental change*, 2nd edn. Oxford University Press, Oxford.

Goudie, A.S. 1983b. The arid earth. In R.A.M.

Gardner and H. Scoging (eds), *Mega geomorphology*. Oxford University Press, Oxford: 152–171.

Graf, W.L. 1983. Flood related channel change in an arid region. *Earth Surface Processes and Landforms*, 8: 125–140.

Grey, D.R.C. and Cooke, H.J. 1977. Some problems in the Quaternary evolution of the landforms of northern Botswana. *Catena*, 4: 123–133.

Grove, A.T. 1958. The ancient ergs of Hausaland, and similar formations on the south side of the Sahar. *Geographical Journal*, 124: 528–533.

Grove, A.T. 1969. Landforms and climatic change in the Kalahari and Ngamiland. *Geographical Journal*, 135: 191–212.

Grove, A.T. and Warren, A. 1968. Quaternary landforms and climate on the south side of the Sahara. *Geographical Journal*, 134: 194–208.

Haffer, J. 1969. Speciation in Amazonian forest birds. *Science*, 165: 131–137.

Harrison, S.P., Metcalfe, S.E., Street-Perrott, F.A., Pittock, A.B., Roberts, C.N. and Salinger, M.J. 1984. A climatic model of the last glacial/intergalcial transition based on palaeotemperature and palaeohydrological evidence. In J.C. Vogel (ed.), *Late Cainozoic palaeoclimates of the southern hemisphere*. Balkema, Rotterdam: 21–34.

Harvey, A.M. and Wells, S.G. 1994. Late Pleistocene and Holocene changes in hillslope sediment supply to alluvial fan systems, Zzyzx California. In A.C. Millington and K. Pye (eds), *Environmental change in drylands: biogeographical and geomorphological perspectives*. Wiley, Chichester: 67–84.

Hashimi, N.H. and Nair, R.B. 1986. Climatic aridity over India 11,000 years ago: evidence from feldspar distribution in shelf sediments. *Palaeogeography, Palaeoecology, Palaeoclimatology*, 53: 309–319.

Hays, J.D. Imbrie, J. and Shackleton, N.J. 1976. Variations in the earth's orbit: pacemaker of the ice ages. *Science*, 194: 1121–1132.

Helgren, D.M. and Brooks, A. 1983. Geoarchaeology of Gi, a Middle Stone Age site in the northwest Kalahari. *Journal of Archaeological Science*, 10: 181–187.

Hereford, R. 1984. Climate and ephemeral stream processes: twentieth century geomorphology and alluvial stratigraphy of the Little Colorado River, Arizona. *Bulletin of the Geological Society of America*, 95: 654–668.

Hereford, R. 1986. Modern alluvial history of the Paria River Drainage Basin, Southern Utah. *Quaternary Research*, 25: 293–311.

Holmgren, K., Carless, W. and Shaw, P.A. 1995. Paleoclimatic significance of the stable isotope composition and petrology of a late Pleistocene stalagmite from Botswana. *Quaternary Research*, 43: 320–328.

Hunt, C.B. and Mabey, D.R. 1966. *Stratigraphy and structure of Death Valley, California.* US Geological Survey, Professional Paper, 494A.

Hunt, C.B., Robinson, T.W., Bowles, W.A. and Washburn, A.L. 1966. *Hydrologic Basin, Death Valley, California.* US Geological Survey, Professional Paper, 494B.

Imbrie, J. and Imbrie, K.P. 1979. *Ice ages: solving the mystery.* Macmillan, London.

Jennings, J.N. 1975. Desert dunes and estuarine fill in the Fitzroy estuary (North West Australia). *Catena,* 2: 215–262.

Jones, C.E. 1991. Characteristics and origin of rock varnish from the hyper arid coastal deserts of northern Peru. *Quaternary Research,* 33: 116–129.

Kadomura, H. and Hori, N. 1990. Environmental implications of slope deposits and humid tropical Africa: evidence from southern Cameroon and western Kenya. *Geographical Reports of Tokyo Metropolitan University,* 25: 213–236.

Kent, L.E. and Gribnitz, K.H. 1985. Freshwater shell deposits in the northwestern Cape Province: further evidence for a widespread wet phase during the late Pleistocene in southern Africa. *South African Journal of Science,* 81: 361–370.

Kershaw, A.P. 1978. Record of last interglacial–glacial cycle from northeastern Queensland. *Nature,* 272: 159–161.

Knox, J.C. 1972. Valley alluviation in southwestern Wisconsin. *Annals of the Association of American Geographers,* 62: 401–410.

Kolla, V. and Biscaye, P.E. 1977. Distribution and origin of quartz in the sediments of the Indian Ocean. *Journal of Sedimentary Petrology,* 47: 642–649.

Kolla, V., Biscaye, P.E. and Hanley, A.F. 1979. Distribution of quartz in late Quaternary Atlantic sediments in relation to climate. *Quaternary Research,* 11: 261–267.

Krinsley, D.H. 1978. The present state and future prospects of environmental discrimination by scanning electron microscopy. In W.B. Whalley (ed.), *Scanning electron microscopy in the study of sediments.* University of East Anglia, Norwich: 169–179.

Krinsley, D.H. and Doornkamp, J. 1973. *Atlas of quartz sand surface textures.* Cambridge University Press, Cambridge.

Kutzbach, J.E. 1980. Estimate of past climate at paleolake Chad, North Africa, based on a hydrological and energy-balance model. *Quaternary Research,* 14: 210–223.

Lancaster, N. 1979. Evidence for a widespread late Pleistocene humid period in the Kalahari. *Nature,* 279: 145–146.

Lancaster, N. 1981. Palaeoenvironmental implications of fixed dune systems in southern Africa. *Palaeogeography, Palaeoclimatology, Palaeoecology,* 33: 327–346.

Lancaster, N. 1988. Development of linear dunes in the southwestern Kalahari. *Journal of Arid Environments,* 14: 233–244.

Lekach, J. and Schick, A.P. 1982. Suspended sediment in desert floods in small catchments. *Israel Journal of Earth Sciences,* 31: 144–156.

Leopold, L.B. 1951. Pleistocene climate in New Mexico. *American Journal of Science,* 249: 152–168.

Li, P.-Y. and Zhou, L.-P. 1993. Occurrence and palaeoenvironmental implications of the Late Pleistocene loess along the eastern coast of the Bohai Sea China. In K. Pye (ed.), *The dynamics and environmental context of aeolian sedimentary systems.* Geological Society of London, Special Publication, 72: 293–309.

Livingstone, D.A. 1975. Late Quaternary climatic change in Africa. *Annual Review of Ecology and Systematics,* 6: 249–280.

Livingstone, D.A. 1980. Environmental changes in the Nile headwaters. In M.A.J. Williams and H. Faure (eds), *The Sahara and the Nile.* Balkema, Rotterdam: 339–359.

Livingstone, I. and Thomas, D.S.G. 1993. Modes of linear dune activity and their palaeoenvironmental significance: a evaluation with reference to southern African examples. In K. Pye (ed.), *The dynamics and environmental context of aeolian sedimentary systems.* Geological Society of London, Special Publication, 72: 91–101.

Lowestein, T.K. and Hardie, L.A. 1985. Criteria for the recognition of salt-pan evaporites. *Sedimentology,* 32: 627–644.

Lowenstein, T.K., Spencer, R.J., Wembo, Y., Casas, E., Pengxi, Z., Baozhen, Z., Haibo, F. and Krouse, H.R. 1994. Major-element and stable isotope geochemistry of fluvial inclusions in halite, Qaidam basin, western China: implication for late Pleistocene/Holocene brine evolution and paleoclimates. In M.R. Rosen (ed.), *Paleoclimate and basin evolution of playa systems.* Geological Society of America, Special Paper, 289: 19–32.

Lustig, L.K. 1965. *Clastic sedimentation in Deep Springs Valley, California.* US Geological Survey, Professional Paper, 352F: 131–192.

Madigan, C.T. 1936. The Australian sand ridge deserts. *Geographical Review,* 26: 205–227.

Maizels, J.K. 1987. Plio-Pleistocene raised channel systems of the western Sharqiya (Wahiba) Oman. In L.E. Frostick and I. Reid (eds), *Desert sediments: ancient and modern.* Geological Society of London, Special Publication, 35: 31–50.

McBurney, C.B.M. and R.W. Hey 1955. *Prehistory and Pleistocene geology in Cyrenian Libya.* Cambridge University Press, Cambridge.

Melia, M.B. 1984. The distribution and relationship between palynomorphs in aerosols and deep sea sediments off the coast of northwest Africa. *Marine Geology*, 58: 345–372.

Metcalfe, S.E., Street-Perrott, F.A., O'Hara, S.L., Hales, P.E. and Periott, R.A. 1994. The palaeolimnological record of environmental change: examples from the arid frontier of Mesoamerica. In A.C. Millington and K. Pye (eds), *Environmental change in drylands: biogeographical and geomorphological perspective*. Wiley, Chichester: 131–145.

Middleton, N.J. 1987. *Desertification and wind erosion in the western Sahel: the example of Mauritania*. School of Geography, Oxford Research Paper Series, 40.

Mifflin, M.D. and Wheat, M.M. 1979. Pluvial lakes and estimated pluvial climates of Nevada. *Nevada Bureau of Mines and Geology Bulletin*, 94.

Milliman, J.D. and Summerhayes, C.P. 1975. Quaternary sedimentation on the Amazon continental margin: a model. *Bulletin of the Geological Society of America*, 86: 610–614.

Molina-Cruz, A. 1977. The relation of the southern trade winds to upwelling processes during the last 75,000 years. *Quaternary Research*, 8: 324–338.

Monod, Th. 1963. The late Tertiary and Pleistocene in the Sahara. In F.C. Howell and F. Bouliere (eds), *African ecology and human evolution*. Viking Publications in Anthropology 36. Wenner-Gren, New York: 117–229.

Morrison, R.B. and Frye, J.C. 1965. Correlation of the Middle and Late Quaternary successions of the Lake Lahontan, Lake Bonneville, Rocky Mountain (Wasatch Range), southern Great Plains and eastern Midwest area. *Nevada Bureau of Mines Report*, 9: 1–45.

Muhs, D.R. 1985. Age and paleoclimatic significance of Holocene dune sand in northeastern Colorado. *Annals of the Association of American Geographers*, 75: 566–582.

Muhs, D.R. and Moat, P.B., 1993. The potential response of aeolian sands to greenhouse warming and precipitation reduction on the Great Plains of the USA. *Journal of Arid Environments* 25: 351–362.

Mulcahy, M.J. and Bettenay, E. 1972. Soil and landscape studies in Western Australia. 1. The major drainage division. *Journal of the Geological Society of Australia*, 18: 349–357.

Nash, D.J., Thomas, D.S.G. and Shaw, P.A. 1994. Timescale, environmental change and dryland valley development. In A.C. Millington and K. Pye (eds), *Environmental change in drylands: biogeographical and geomorphological perspectives*. Wiley, Chichester: 25–41.

Nanson, G.C. 1986. Episode of vertical accretion and catastrophic stripping: a model of disequilibrium flood plain development. *Bulletin of the Geological Society of America*, 97: 1467–1475.

Nanson, G.C., Chen, X.Y. and Price, D.M. 1992a. Lateral migration, thermoluminescence chronology and colour variation of longitudinal dunes near Birdsville in the Simpson Desert Central Australia. *Earth Surface Processes and Landforms*, 17: 807–820.

Nanson, G.C., Price, D.M., Young, R.W. and Rust, B.R. 1992b. Stratigraphy, sedimentology and late Quaternary chronology of the Channel Country of western Queensland. In R.F. Warner (ed.), *Fluvial geomorphology of Australia*. Academic Press, Sydney: 151–175.

Newell, R.E. 1973. Climate and the Galapagos Islands. *Nature*: 245: 91–92.

Nicholson, S.E. and Flohn, H. 1980. African environmental and climatic changes and the general atmospheric circulation in Late Pleistocene and Holocene. *Climatic Change*, 2: 313–348.

Pachur, H.-J. and Kropelin, S. 1987. Wadi Howar: palaeoclimatic evidence from an extinct river system in the southeastern Sahara. *Science*, 237: 298–300.

Pachur, H.-J., Roper, H.P., Kropelin, S. and Groschin, M. 1987. Late Quaternary hydrography of the Eastern Sahara. *Berliner Geowissenschaften Abhandlungen (A)*, 5: 331-8 74.

Parmenter, C. and Folger, D.W. 1974. Eolian biogenic detritus in deep sea sediments: a possible index of equatorial Ice Age aridity. *Science*, 185: 695–698.

Petit, J.R., Briat, M. and Royer, A. 1981. Ice age aerosol content from East Antarctic ice core samples and past wind strength. *Nature*, 293: 391–394.

Pokras, E.M. and Mix, A.C. 1985. Eolian evidence for spatial variability of late Quaternary climates in tropical Africa. *Quaternary Research*, 24: 137–149.

Prell, W.L., Hutson, W.H., Williams, D.F., Be, A.W.H., Geitzenauer, K. and Molfino, B. 1980. Surface circulation of the Indian Ocean during the last Glacial Maximum, approximately 18,000 BP. *Quaternary Research*, 14: 309–336.

Price Williams, D., Watson, A. and Goudie, A.S. 1982. Quaternary colluvial stratigraphy, archaeological sequences and palaeoenvironment in Swaziland. *Geographical Journal*, 138: 50–68.

Pye, K. 1982. Thermoluminescence dating of sand dunes. *Nature*, 229: 376.

Rea, D.K. and Leinen, M. 1988. Asian aridity and the zonal westerlies: late Pleistocene and Holocene record of eolian deposition in the northwest Pacific. *Palaeogeography, Palaeoclimatology, Palaeoecology*, 66: 1–8.

Rea, D.K., Leinen, M. and Janececk, T.R. 1985. Geologic approach to the long-term history of atmospheric circulation. *Science*, 227: 721–725.

Reid, I. 1994. River landforms and sediments: evidence of climatic changes. In A.D. Abrahams and A.J. Parsons (eds), *Geomorphology of desert environments*. Chapman & Hall, London: 571–592.

Roberts, N. and Wright, H.E., 1993. Vegetational, lake level and climate history of the Near East and Southwest Asia. In Wright, H.E., Kutcback, J.E., Webb, T., Ruddmans, W.F., Street-Perrott, F.A. and Barllein. P.J. (eds), *Global climates since the last glacial maximum*. University of Minnesota Press, Minneapolis: 199–217.

Roberts, N., Taneb, M., Barker, P., Damnati, B., Toole, M. and Willamson, D. 1993. Timing of the younger Dryas event in East Africa from lake level change. *Nature*, 366: 146–148

Rognon, P. 1987. Late Quaternary climatic reconstruction for the Maghreb (north Africa). *Palaeogeography, Palaeoclimatology, Palaeoecology*, 58: 11–34.

Rognon, P. and Williams, M.A.J. 1977. Late Quaternary climatic changes in Australia and north Africa: a preliminary interpretation. *Palaeogeography, Palaeoclimatology, Palaeoecology*, 21: 285–327.

Rossignol-Strick, M. and Duzer, D. 1980. Late Quaternary West African climate inferred from palynology of Atlantic deep sea cores. *Palaeoecology of Africa*, 12: 227–228.

Rust, U. 1984. Geomorphic evidence of Quaternary environmental changes in Etosha, Southwest Africa/Namibia. In J.C. Vogel (ed.), *Late Cainozoic palaeoclimates of the southern hemisphere*. Balkema, Rotterdam: 279–286.

Rust, U., Schmidt, H.H. and Dietz, K.R. 1984. Palaeoenvironments of the present day arid southwestern Africa, 30,000–5,000 BP. Results and problems. *Palaeoecology of Africa*, 16: 109–148.

Rutter, N., Ding, Z.L. and Liu T. 1991. Comparison of isotope stage 1-61 with the Baoji-type pedostratigraphic section of north-central Cluva. *Canadian Journal of Earth Sciences*, 28: 985–990.

Sack, D, 1994. Geomorphic evidence of climate change from desert-basin palaeolakes. In A.D. Abraham and A.J. Parsons (eds), *Geomorphology of desert environments*. Chapman & Hall, London: 616–630.

Sarnthein, M. 1972. Sediments and history of the post-glacial transgression in the Persian Gulf and north-west Gulf of Oman. *Marine Geology*, 12: 245–266.

Sarnthein, M. 1978. Sand deserts during Glacial Maximum and climatic optimum. *Nature*, 272: 43–46.

Sarnthein, M. and Diester-Haass, L. 1977. Eolian-sand turbidites. *Journal of Sedimentary Petrology*, 47: 868–890.

Sarnthein, M. and Koopman, B. 1980. Late Quaternary deep-sea record of northwest African dust supply and wind circulation. *Palaeoecology of Africa*, 12: 238–253.

Sarnthein, M., Tetzlaff, G., Koopmann, B., Walter, K. and Pflaumann, U. 1981. Glacial and interglacial wind regimes over the eastern sub-tropical Atlantic and northwest Africa. *Nature*, 293: 193–196.

Schick, A.P., Lekach, J. and Hassan, M.A. 1987. Vertical exchange of coarse bedload in desert streams. In L. Frostrick and I. Reid (eds), *Desert sediments: ancient and modern*. Geological Society of London, Special Publication, 35. Blackwell, Oxford: 7–16.

Schroeder, J.H. 1985. Eolian dust in the coastal desert of the Sudan: aggregate cemented by evaporites. *Journal of African Earth Sciences*, 3: 370–380.

Scott, W.E., McCoy, W.D., Shroba, R.R. and Miller, R.D. 1980. New interpretations of Late Quaternary history of Lake Bonneville, western United States. *Abstracts and Programme, American Quaternary Association, Sixth Biennial Meeting, University of Maine*: 168–169.

Servant, M. and Servant-Vidary, S. 1980. L'environment quaternaire du basin du Tchad. In M.A.J. Williams and H. Faure (eds), *The Sahara and the Nile*. Balkema, Rotterdam: 133–162.

Servant, M., Maley, J., Turcq, B., Absy, M.L., Brenac, P., Fournier, M. and Ledru, M.P. 1993. Tropical forest changes during the late Quaternary in African and Southern American lowlands. *Global and Planetary Change*, 7: 25–40.

Shaw, P.A. 1985. Late Quaternary landforms and environmental change in northeast Botswana: the evidence of Lake Ngami and the Mababe Depression. *Transactions of the Institute of British Geographers*, NS, 10: 333–346.

Shaw, P.A. and Cooke, H.J. 1986. Geomorphic evidence for the Late Quaternary palaeoclimate of the Middle Kalahari of Northern Botswana. *Catena*, 13: 349–359.

Shaw, P.A. and De Vries, J.J. 1988. Duricrusts, ground water and valley development in the Kalahari of southeastern Botswana. *Journal of Arid Environments*, 14: 245–254.

Shaw, P.A. and Thomas, D.S.G. 1988. Lake Caprivi — a Late Quaternary link between the Zambezi and Middle Kalahari drainage systems. *Zeitschrift für Geomorphologie*, NF, 32: 329–337.

Singer, A. 1984. The paleoclimatic interpretation of clay minerals in sediments — a review. *Earth Science Review*, 21: 251–293.

Singh, G. 1971. The Indus valley culture seen in context of post-glacial climate and ecological studies in northwest India. *Archaeology and Anthropology in Oceania*, 6: 177–189.

Singh, G., Joshi, R.D. and Singh, A.B. 1972. Stratigraphic and radiocarbon evidence for the age and development of three salt lake deposits in Rajasthan, India. *Quaternary Research*, 2: 496–505.

Singhvi, A.K., Sharma, Y.P. and Agrawal, D.P. 1982. Thermoluminescence dating of sand dunes in Rajasthan, India. *Nature*, 295: 313–315.

Smalley, I.J. and Vita-Finzi, C. 1968. The formation of fine particles in sandy deserts and the nature of 'desert' loess. *Journal of Sedimentary Petrology*, 38: 766–774.

Smith, G.I. 1968. Late Quaternary geologic and climatic history of Searles Lake, southeastern California. In R.B. Morrison and H.E. Wright, Jr (eds), *Means of correlating Quaternary successions*. Volume 8, Proceedings of 7th Congress of the International Association of Quaternary Research. University of Utah Press, Salt Lake City: 293–310.

Smith, G.I. and Street-Perrott, F.A. 1983. Pluvial lakes of the western United States. In S.C. Porter (ed.), *Late Quaternary environments in the United States. Volume 1: The Late Pleistocene*. University of Minnesota Press, Minneapolis: 190–214.

Smith, G.I., Barczak, V.J., Moulton, G.F. and Liddicoat, T.C. 1983. *Core KM-3, on surface-to-bedrock record of Late Cainozoic sedimentation in Searles Valley, California*. US Geological Survey, Professional Paper, 1256.

Smith, H.T.U. 1963. *Eolian geomorphology, wind direction and climatic change in North Africa*. US Air Force, Cambridge Research Laboratories, Report AF19 (628).

Snyder, C.T. and Langbein, W.B. 1962. The Pleistocene Lake in Spring Valley, Nevada, and its climatic implications. *Journal of Geophysical Research*, 67: 2385–2394.

Sombroek, A.G., Mbuvi, J.P. and Okwaro, H.W. 1976. *Soils of the semi-arid savanna zone of northeastern Kenya*. Kenya Soil Survey, Paper M2.

Sperling, C.H.B. and Goudie, A.S. 1975. The miliolite of western India: a discussion of the aeolian and marine hypotheses. *Sedimentary Geology*, 13: 71–75.

Street, F.A. 1979. Late Quaternary precipitation estimates for the Ziway-Shala Basin, southern Ethiopia. *Palaeoecology of Africa*, 11: 135–143.

Street, F.A. 1981. Tropical palaeoenvironments. *Progress in Physical Geography*, 5: 157–185.

Street, F.A. and Grove, A.T. 1979. Global maps of lake-level fluctuations since 20,000 yr BP. *Quaternary Research*, 12: 83–118.

Street-Perrott, F.A. and Roberts, N. 1983. Fluctuations in closed-basin lakes as an indicator of past atmospheric circulation patterns. In F.A. Street-Perrott, M. Beran and R. Ratcliffe (eds), *Variations in the global water budget*. Reidel, Dordrecht: 331–345.

Street-Perrott, F.A., Roberts, N. and Metcalfe, S. 1985. Geomorphic implications of Late Quaternary hydrological and climatic change in the northern hemisphere tropics. In I. Douglas and T. Spencer (eds), *Environmental change and tropical geomorphology*. Allen & Unwin, London: 165–183.

Stuiver, M. 1970. Oxygen and carbon isotope ratios of freshwater carbonates as climatic indicators. *Journal of Geophysical Research*, 75: 5247–5257.

Talbot, M.R. 1980. Environmental responses to climatic change in the West African Sahel over the past 20,000 years. In M.A.J. Williams and H. Faure (eds), *The Sahara and the Nile*. Balkema, Rotterdam: 37–62.

Talbot, M.R. 1994. Late Pleistocene rainfall and dure building in the Sahel. *Paleoecology of Africa*, 16: 203–214.

Talbot, M.R. and Williams, M.A.J. 1978. Erosion of fixed sand dunes in the Sahel, central Niger. *Earth Surface Processes*, 3: 107–113.

Tchakerian, V.P. 1991. Late Quaternary aeolian geomorphology of the Dale Lake sand sheet, southern Mojaire Desert California. *Physical Geography*, 12: 347–369.

Thiede, J. 1979. Wind regime over the Late Quaternary southwest Pacific Ocean. *Geology*, 7: 259–262.

Thomas, D.S.G. 1984. Ancient ergs of the former arid zones of Zimbabwe, Zambia and Angola. *Transactions of the Institute of British Geographers*, NS, 9: 75–88.

Thomas, D.S.G. 1987a. The roundness of aeolian quartz sand grains. *Sedimentary Geology*, 52: 149–153.

Thomas, D.S.G. 1987b. *Research strategies and methods for Quaternary science: the case of southern Africa*. School of Geography, Oxford, Research Paper Series 39.

Thomas, D.S.G. 1987c. Discrimination of depositional environments, using sedimentary characteristics, in the Mega Kalahari, central southern Africa. In L. Frostock and I. Reid (eds), *Desert sediments: ancient and modern*. Geological Society of London, Special Publication, 35. Blackwell, Oxford: 293–306.

Thomas, D.S.G. and Goudie, A.S. 1984. Ancient ergs of the southern hemisphere. In J.C. Vogel (ed.), *Late Cainozoic palaeoclimates of the southern hemisphere*. Balkema, Rotterdam: 407–418.

Thomas, D.S.G. and Shaw, P.A. 1991a. *The Kalatain environment*. Cambridge University Press, Cambridge.

Thomas, D.S.G. and Shaw, P. 1991b. Relict dune systems: interpretations and problem. *Journal of Arid Environments*, 20: 1–114.

Tricart, J. 1956. Tentative de correlation des périodes pluviales africaines et des périodes glaciaires. *Comptes Rendus sommaires des Séances de la Société Géologique de France*, 9–10: 164–167.

Tricart, J. 1975. Influence des oscillations climatiques recentes sur le modèle en Amazonie orientale (Région de Santarem) d'après les images de radar latéral. *Zeitschrift für Geomorphologie*, NF, 19: 140–163.

Tricart, J. 1984. Evidence of Upper Pleistocene dry climates in northern South America. In I. Douglas and T. Spencer (eds), *Environmental change and tropical geomorphology*. Allen & Unwin, London: 197–217.

Tsoar, H. and Møller, J.T. 1986. The role of vegetation in the formation of linear dunes. In W.G. Nickling (ed.), *Aeolian geomorphology*. Allen & Unwin, Boston: 75–95.

Tsoar, H. and Pye, K. 1987. Dust transport and the question of desert loess formation. *Sedimentology*, 34: 139–153.

Turrin, B.D. Dohrenwend, J.C., Drake, R.E. and Curtiss, G.H. 1985. K–Ar ages from the Cima volcanic field, eastern Mojave Desert, CA. *Isochron West*, 44: 9–16.

Ullman, W.J. 1985. Evaporation rates from a salt pan: estimates from chemical profiles in near surface groundwaters. *Journal of Hydrology*, 79: 365–373.

Ullman, W.J. and McLoed, L.C. 1986. The Late Quaternary salinity record of Lake Frome, South Australia: evidence from Na^+ in stratigraphically preserved gypsum. *Palaeogeography, Palaeoclimatology, Palaeoecology*, 54: 153–169.

Vogel, J.C. and Visser, E. 1981. Pretoria radiocarbon dates II. *Radiocarbon*, 23: 43–60.

Wasson, R.J. 1984. Late Quaternary palaeoenvironments in the desert dunefields of Australia. In J.C. Vogel (ed.), *Late Cainozoic palaeoclimates of the southern hemisphere*. Balkema, Rotterdam: 419–432.

Wasson, R.J. 1986. Geomorphology and Quaternary history of the Australian continental dunefields. *Geographical Review of Japan*, 59B: 55–67.

Wasson, R.J., Rajaguru, S.N., Misra, V.N., Agrawal, D.P., Ohir, R.P., Singhvi, A.K. and Rao, K.K. 1983. Geomorphology, Late Quaternary stratigraphy and palaeoclimatology of the Thar Desert. *Zeitschrift für Geomorphologie*, Supplementband, 45: 117–152.

Watson, A. 1988. Desert gypsum crusts as paleoenvironmental indicators: a micropetrographic study of crusts from southern Tunisia and the central Namib Desert. *Journal of Arid Environments*, 15: 9–12.

Watson, A., Price Williams, D. and Goudie, A.S. 1985. The palaeoenvironmental interpretation of colluvial sediments and palaeosols in Southern Africa. *Palaeogeography, Palaeoclimatology, Palaeoecology*, 45: 255–249.

Wells, G.L. 1983. Late-glacial circulation over central north America revealed by aeolian features. In F.A. Street-Perrott, M. Beran and R. Ratcliffe (eds), *Variations in the global water budget*. Reidel, Dordrecht: 317–330.

Wells, P.V. 1976. Macrofossil analysis of wood rat (*Neotoma*) middens as a key to the Chihuahun Desert. *Science*, 153: 970–973.

Whalley, W.B. 1983. Desert varnish. In A.S. Goudie and K. Pye (eds), *Chemical sediments and geomorphology*. Academic Press, London: 197–226.

Wickens, G.E. 1975. Changes in the climate and vegetation of the Sudan since 20,000 BP. *Boissiera*, 24: 43–65.

Wiggs, G.F.S., Livingstone, I., Thomas, D.S.G. and Bullard J.E. 1995. Dune mobility and vegetation cover in the southwest Kalahari Desert. *Earth Surface Processes and Landforms*, 20(6): 515–529.

Williams, G.E. 1973. Late Quaternary piedmont sedimentation, soil formation and palaeoclimates in arid South Australia. *Zeitschrift für Geomorphologie*, NF, 17: 102–125.

Williams, M.A.J. 1975. Late Pleistocene tropical aridity synchronous in both hemispheres? *Nature*, 253: 617–618.

Williams, M.A.J. 1985. Pleistocene aridity in tropical Africa, Australia and Asia. In I. Douglas and T. Spencer (eds), *Environmental change and tropical geomorphology*. Allen & Unwin, London: 219–233.

Williams, M.A.J. and Adamson, D.A. 1980. Late Quaternary depositional history of the Blue and White Nile rivers in central Sudan. In M.A.J. Williams and H. Faure (eds), *The Sahara and the Nile*. Balkema, Rotterdam: 281–304.

Williams, M.A.J., Abell, P.I. and Sparks, B.W. 1987. Quaternary landforms, sediments, depositional environments and gastropod isotope ratios at Adrar Bous, Ténéré Desert of Niger. In Frostick, L. and Reid, I. (eds), *Desert sediments ancient and modern*. Geographical Society of London, Special Publication, 35: 105–125.

Williams, M.A.J., Dunkerley, D.L., DeDeckter, P., Kershaw, A.P. and Stokes, T. 1993. *Quaternary environments*. Edward Arnold, London.

Wilson, I.G. 1973. Ergs. *Sedimentary Geology*, 10: 77–106.

Windom, H.L. 1975. Eolian contributions to marine sediments. *Journal of Sedimentary Petrology*, 45: 520–529.

Wintle, A.G. and Huntley, D.J. 1982. Thermoluminescence dating of sediments. *Quaternary Science Reviews*, 1: 31–54.

Yair, A. 1987. Environmental effects of loess penetration into the northern Negev Desert. *Journal of Arid Environments*, 13: 9–24.

27

Dating of desert sequences

Stephen Stokes

Perspectives on arid zone geochronology

Previous chapters clearly document the diversity of both form and process in arid zone environments. This diversity is expressed in a range of variables including composition, genesis, extent and lateral association. It follows therefore that the accurate and comprehensive dating of arid zone landscapes requires the application of a range of methods capable of isolating time-dependent signals from widely differing materials. The actual development of detailed chronological data in many arid zone environmental settings has however at best been limited, and at worst entirely lacking (Thomas and Shaw 1991).

A number of factors typical of arid zone environments have posed particular problems for geochronologists. These include:

1. Short-term, episodic accumulation of material is commonplace (e.g. sand dunes, alluvial fans). This is in contrast to the (more) continuously accumulating nature of many other environmental archives (e.g. lacustrine, deep-sea and ice-core sequences).
2. Arid zone environments typically neither produce nor preserve large quantities of biomass. This has presented a serious limitation to the widespread application of conventional radiocarbon dating methods, which have proved to be highly effective in other environmental settings. In addition, many arid zone landscape components are considerably older than the maximum possible time range of the radiocarbon dating method (*c.* 40 ka).
3. Weathering and vadose diagenesis result in the generation of a spectrum of potentially datable precipitated deposits, but percolating groundwaters rich in solutes whose deposition can contaminate underlying datable sediment fractions (e.g. Chen and Polach 1986).
4. Arid zone soils are typically thin and poorly developed (see Chapter 5).
5. Many arid zone landscape components may be exposed at or near the surface for extended periods of time. This may invalidate methods which assume rapid emplacement of a sediment within an effectively 'closed' environmental archive (see, for example, Roper 1993).

In attempting to understand the chronology and development of arid and semi-arid landscapes, it is necessary to consider the evidence of both constructional or depositional and where possible erosional or degradational features (Dorn and Phillips 1991). The recognition of evidence indicating that the influence of abrupt global climatic events is preserved in some arid zone depositional settings (e.g.

Arid Zone Geomorphology: Process, Form and Change in Drylands, 2nd edition. Edited by David S. G. Thomas.
© 1997 John Wiley & Sons Ltd.

Table 27.1   *Some examples of landscape elements and events datable in arid zone environmental settings*

Landscape element	Nature of datable event	Applicable dating methods	Case example
Inorganic components			
Mountains, low-angle bedrock surface, alluvial fans	Erosion and surface exposure	Cation-ratio, AMS ^{14}C (varnish OM), cosmogenic isotopes	Bull (1991); Dorn (1983; 1988); Dorn et al. (1987; 1989; 1992 a, b)
River plains and dry watercourses	Erosion and deposition	^{14}C, U-series	Brookes (1989); Toomey et al. (1993)
Badlands and gully fill	Erosion and deposition	^{14}C, luminescence	Botha et al. (1994)
Closed lake basins, playas and desert flats	Lacustrine deposition, wetting/drying	^{14}C, luminescence, U-series, cosmogenic isotopes	Street and Grove (1976); Brookes (1989); Herczeg and Chapman (1991); Miller et al. (1991); Metcalfe (1995)
Sand seas	Deposition	^{14}C, luminescence	Readhead (1988); Wells et al. (1990); Madole (1994; 1995); Stokes and Gaylord (1993); Forman et al. (1995)
Recent volcanic deposits	Eruption	Fission track, U-series, K/Ar	Walther et al. (1991)
Biogenic components and chemical precipitates			
Plant remains	Biological productivity	^{14}C	Haynes (1987)
Ostrich eggshells	Biological productivity	AAR, ^{14}C, TIMS U-series	Miller et al. (1991)
Land snail shell	Biological productivity	AAR, ^{14}C	Goodfriend (1987)
Body fossils	Biological productivity	^{14}C, U-series, ESR	Goodfriend and Stipp (1983)
Halite	Evaporite deposition	U-series, cosmogenic isotopes (^{36}Cl)	Herczeg and Chapman (1991); Phillips et al. (1993)
Travertine	Spring seepage	^{14}C, U-series	Chen and Polach (1986)
Calcrete	Carbonate deposition	^{14}C, U-series	Herczeg and Chapman (1991)
Gypsum	Evaporite deposition	U-series	Hass et al. (1986)
Soil components (humates, carbonates, etc.)	Pedogenesis	^{14}C	

AMS–accelerator mass spectrometry, ^{14}C–radiocarbon dating method, U-series–uranium-series dating method, AAR–amino acid racemisation, TIMS–thermal ionisation mass spectrometry.

Street-Perrott and Perrott 1990; Roberts *et al.* 1993) further emphasises the punctuated nature of some arid zone environmental components, and hence the importance of generating moderate- to high-resolution chronologies of arid sedimentary environments as an archive of such change.

Early researchers relied heavily on the use of relative methods of age assessment including simple stratigraphic superposition, morphology, weathering and soil development. Examples of such approaches are provided in Yaalon (1971), Jennings (1975), Birkeland (1984) and Chapter 26. In addition to the use of relative methods of age evaluation, the timing of changes in deserts has also been estimated by the application of conceptual models connecting dune reddening or degree of pedogenesis with relative age (e.g. Muhs 1985; Jorgensen 1992) or general phases of high-latitude glacial expansion with low-latitude aridity and desert expansion (e.g. Sarnthein 1978). While such approaches have provided useful insights into the nature of changes in deserts, they reveal little about the nature and rapidity of change, and the exact timing of changes has remained for the most part locked within the world's major desert basins.

Recently, the relative approach has become eclipsed by expansion in the number of dating methods capable of establishing either direct numerical or calibrated ages for both surficial and buried contexts (Table 27.1). It is these methods that are the focus of this chapter. The specific focus is on the more significant methods and case examples of their application in arid zone settings. Many such methods remain developmental, but their widespread application in the future promises to revolutionise our understanding and interpretations of changing dryland environments.

Over recent, relatively short (century-scale) timescales, historical accounts, photogrammetric and direct offset and erosion measurements may be used to directly gauge rates of processes and changes in desert systems (e.g. Haynes 1989; Maxwell and Haynes 1989; Gaylord and Stetler 1994; Jacobberger-Jellison 1994; Muhs and Holliday 1995). As the main focus of this chapter is on the use of numerical and calibrated-age dating methods over Quaternary timescales, discussion of these

contemporary–historical approaches is not provided. Additionally, in many arid and semi-arid basins landscape components may be chronologically evaluated in relation to associated volcanic deposits which are datable by fission-track, K–Ar and other methods (e.g. Walther *et al.* 1991). As these deposits are not specifically derived from desert sediments they are for the most part excluded from this discussion. Readers are however referred to Aitken (1990) and Geyh and Schleicher (1990) where full accounts may be found.

Classification of dating methodologies

Table 27.2 summarises the two-phase approach adopted in this chapter to classifying the dating methods applicable in arid environments. Methods are subdivided first into those applicable to either buried (stratigraphic) or surficial deposits. Within these subdivisions methods are grouped into one of four categories based on the recommendations of Colman *et al.* (1987) that such classifications should be organised on the level of quantitative information provided and the degree of confidence contained in the age estimate.

Numerical-age methods are those that produce results on an absolute timescale.

Calibrated-age methods measure systematic changes resulting from processes dependent on environmental variables such as climate and lithology; they require the use of independent chronological control to 'calibrate' the observed trends into an absolute timescale.

Relative-age methods provide an age sequence, and in some cases a measure of the magnitude of age difference between members of a sequence. Like calibrated-age methods, relative approaches usually involve a systematic change in some measured parameter. There are numerous examples of such approaches in arid zone research. Goudie *et al.* (1993) used the repose angle of degraded linear dunes to generate a relative chronology of dune emplacement in northwestern Australia. Haynes (1982) and Haynes *et al.* (1993) developed a detailed pedostratigraphic scheme for discriminating the relative ages of sand sheets and associated lacustrine and archaeological deposits in the southeastern Sahara.

Table 27.2 *A classification of methods capable of providing age-related information in arid zone settings*

Numerical ages	Calibrated ages	Relative ages	Correlated ages
Dating of buried sedimentary and organic materials			
^{14}C dating of plant (including algal) and animal remains	Obsician hydration	Morphology	Archaeological associations
^{14}C dating of soil humates and carbonates	Amino acid racemisation/epimerisation (land snails, molluscs, ostrich eggshells)	Stratigraphic superposition	Palaeological associations
^{14}C dating of calcrete		Degree of dune reddening	Oxygen isotopes in lake sediments
U-series dating of precipitated carbonates (tufa, calcrete, lacustrine marl)	Accumulation of soil carbonate	Degree of induration	Palaeolomagnetic variations in lake sediments
U-series dating of organic carbonates (land snails, ostrich eggshells)	Weathering of minerals (e.g. hornblende)	Individual soil properties (e.g. Fe-oxides, clay accumulation)	Stable isotopes in soils and sediments
Uranium-trend dating of soil carbonates		Dune repose angle	
Luminescence dating of detrital sediments (quartz, feldspar, etc.)			
Surface exposure dating methods			
Accumulation of cosmogenic radionu-cleides (^{36}Cl, ^{26}Al, ^{10}Be, ^{14}C, ^{129}I, ^{41}Ca)	Ratio of mobile to immobile cations in rock varnish	Morphology	Archaeological associations
Accumulation of cosmogenic stable nuclides	Weathering rinds	Relative position	Fluctuations in stable isotopes and Mn:Fe ratios in rock varnish
^{14}C dating of organic matter in rock coatings and carbonate rinds	Weathering of minerals (e.g. hornblende)	Percentage cover of varnish on rocks	Palaeomagnetic variations

Correlated-age methods produce ages only by demonstrating equivalence to independently dated deposits or events. A useful example of the application of such methods is that of the high-resolution correlative chronology for the extensive loess and intercalated palaeosol sequences of China and elsewhere based on magnetostratigraphy (Heller and Evans 1995). The magnetic susceptibility and other sedimentary parameters (e.g. grain size, [10]Be content) of loess sequences have been used as a means of correlation to oceanic oxygen isotope records, which are independently dated via radiocarbon, K/Ar, and spectral analysis methods. It is commonplace in current geochronological research to supplement where possible relative or correlated-age dating strategies where used with numerical and calibrated-age dating techniques.

The remainder of this chapter outlines key principles and recent developments applicable to drylands. Reviews of other methods may be found in Cullingford *et al.* (1980), Aitken (1990), Geyh and Schleicher (1990), Dorn and Phillips (1991) and Beck (1994).

Conventional and AMS radiocarbon dating

The application of radiocarbon dating methods extends to all sectors of geoscientific and archaeological research on the late Quaternary (Aitken 1990; Williams *et al.* 1993). Radiocarbon dating has in particular been critical in the establishment of chronologies over approximately the last 30 ka. In arid zone settings, radiocarbon dating has been instrumental in developing a detailed understanding of the nature and timing of environmental change within a variety of environmental archives; few of the advances which have occurred in our understanding of the later Quaternary evolution of arid zone sequences would have been possible in the absence of radiocarbon dates. For example, radiocarbon analysis of biogenic and inorganic carbon-bearing components has demonstrated the occurrence of extensive lacustrine and fluvial systems throughout the Sahara and Kalahari deserts both prior and subsequent to the Last Glacial Maximum (LGM) (e.g. Petit-Maire and Riser 1981;

Haynes 1987; Pachur and Kropelin 1987; Brookes 1989; Pachur and Hoelzman 1991); it has fuelled the paradigm of globally expanded deserts during the LGM (Sarnthein 1978), while at the same time demonstrating the apparently ambiguous pluvial lake systems of the southwestern USA which attained their maximum extent at the same time (Street-Perrott and Harrison 1985); it has assisted workers in their description and definition of Holocene climatic optima and in assessing the local environmental responses to the optima (see, for example, Haynes 1982; Ritchie 1994; Ritchie *et al.* 1985); and it has been critical in the correlation of a wide range of aeolian, fluvial, lacustrine, pedogenic and other units (e.g. Street and Grove 1976; Haas *et al.* 1986; Roberts *et al.* 1993; Toomey *et al.* 1993; Madole 1995). Additionally, given its position as a central chronological tool, the radiocarbon method has importantly represented a long-standing base-level comparator against which many of the more developmental methods have been tested.

BASIC PRINCIPLES

A radioactive isotope of carbon (^{14}C) is continuously formed in the atmosphere from the interaction of cosmic-ray neutrons with a stable isotope of nitrogen (^{14}N). The ^{14}C content of the atmosphere remains at a low, near-equilibrium concentration from where it is oxidised into carbon dioxide. Variations in the atmospheric concentration of ^{14}C are related to solar and other modulations of the cosmic-ray flux. Interactions in the atmosphere–Earth systems result in the fixation of carbon into a variety of biogenic and inorganic forms from where the radioactive decay of ^{14}C takes place. The radioactive decay of ^{14}C occurs, via emission of a beta (β) particle, as an exponentially declining trend with a half-life of 5730 years. Given assumptions regarding the initial ^{14}C activity of a sample, measurement of the ratio of stable (^{12}C) to radioactive (^{14}C) isotopes allows the determination of the time which has elapsed since biological or inorganic fixation. The age range of the method spans from a few centuries up to approximately 40 ka.

The actual measurement procedure comprises either the counting of radioactive (β-particle) decay events (commonly referred to as conventional radiocarbon dating), or direct counting of abundances of stable and unstable carbon atoms by accelerator mass spectrometric (AMS) methods (Elmore and Phillips 1987; Linick *et al.* 1989; Hedges 1991). A major disadvantage of conventional radiocarbon dating is that only about 1% of the ^{14}C atoms in a sample will emit a β-particle in about 80 years — only a very small portion of the total sample carbon content is therefore measured. As all the carbon atoms present in a sample are counted in the AMS method, the required sample size is less by several orders of magnitude. Not only does the AMS methods' small sample size requirements (minimum size *c.* 100 µg; Hedges 1991) provide access to a wider range of sample types than conventional methods (e.g. individual seed grains, included organic debris from rock varnish layers), but it also allows determinations to be made on separate chemical components from within a sample — some of which may be more reliable for dating than others (see below). Ultimately both methods are limited by the half-life of ^{14}C.

MATERIALS DATABLE BY THE
RADIOCARBON METHOD

A variety of materials are datable by the radiocarbon method, many of which are in fact relatively low in carbon content (i.e. <0.1%; Haas *et al.* 1986). While typically scarce, charcoal, wood and macrofossil fragments (including insects, chronomids and other aquatic organisms, bones, molluscs and snails) constitute the sample of preference for dating (Pachur and Kropelin 1987; Gillespie *et al.* 1991). The advantage of macrofossils lies in their ability to be biologically, and in some cases specifically, identified. It is noted, however, that the robust nature of such materials renders them susceptible to redeposition, and consumption of carbon-bearing materials by some organisms may result in anomalous age determinations (see below). Goudie *et al.* (1993), for example, found that AMS radiocarbon ages generated on particulate charcoal lenses underestimated the true age of linear

dunes in northwestern Australia by over 10 ka when the data were compared with independent estimates based on morphological and luminescence dating criteria. Bone may also be problematic as it is prone to accumulation of post-depositional carbon in many environmental settings although appropriate pre-treatment procedures may result in accurate age determinations (Stafford *et al.* 1987; 1988). Additional datable animal remains include faecal pellets, ivory, horn, hair and egg shells (e.g. Haynes 1982; Brookes 1989; Ellis and Van Devender 1990; Van Devender *et al.* 1994). Obviously the lower abundance of organic lifeforms in dryland relative to more humid areas may lead to a reduced likelihood of datable material being located in arid zone sediments.

Application of radiocarbon methods to soil and sediment components has been an area of considerable investigation and debate (e.g. Geyh *et al.* 1983). This in particular relates to the open system nature of soils and the corresponding high likelihood of introduction of old or young carbon-bearing compounds and solutions via percolating groundwaters or modern root penetration. While dating undifferentiated soils or sediments may therefore result in erroneous ages, the analysis of rigorously pre-treated samples appears to produce accurate sediment age estimates (Fowler *et al.* 1986; Gillespie *et al.* 1992). Minimum ages of periods of soil formation also appear to be obtainable provided care is taken in sampling and pre-treatment. This was demonstrated for a series of Holocene soil and marsh sediments from North Texas by Haas *et al.* (1986). Of the chemically extractable fractions obtainable from soils and sediments, water-insoluble lipids and organic residues (humins) are of preference as they are unlikely to be strongly influenced by post-depositional processes. The more soluble humates (humic and fulvic acids) are found in some cases to be highly mobile in sediment systems (Head *et al.* 1989), and may therefore produce incorrect sediment ages (Hedges 1991). In specific cases, however, soil and sediment dates based on humates appear capable of producing acceptable ages (e.g. Hass *et al.* 1986; Haynes 1987; Gillespie *et al.* 1992; Haynes *et al.* 1993; Madole 1994).

Application of radiocarbon methods to soil carbonates, and discrete calcrete and tufa deposits, has generally resulted in erroneous radiocarbon ages (Chen and Polach 1986). This has been attributed to the incorporation of old carbonate in percolating groundwaters.

AMS RADIOCARBON DATING OF ORGANIC MATERIALS IN ROCK VARNISH

An interesting and highly significant application of the AMS radiocarbon dating method is in the analysis of small quantities of organic material included within rock varnish layers (see, for example, Dorn *et al.* 1987; 1989; 1992a and Chapter 6). Dorn *et al.* (1992a) established that detrital fragments of organic material are neither complexed into the oxides which make up the bulk of the varnish or widely dispersed throughout the rock varnish matrix. Instead, the organic matter is found to be concentrated either at the varnish surface (i.e. contemporary), under the varnish at the rock interface, or in voids in the weathering rind that are subsequently entombed by the overlying varnish. Some examples of SEM micrographs of organic fragments found within varnish are displayed in Figure 27.1. Correspondingly, the events dated by the entrapment of the organic material into rock varnish have been interpreted as representing minimum ages of surface formation; the organic matter being emplaced by lichens, cyanobacteria or fungal growth within surface hollows and voids exposed on rock surfaces. Comparisons with independent dating methods indicate that the method is capable of producing accurate age estimates over approximately the past 25 ka (Figure 27.2).

This relatively new method represents a significant methodological advancement for workers attempting to constrain the ages of exposed surfaces in drylands which may be used independently, or in tandem with cation-ratio or other dating methods. Examples of the use of the method in arid zone settings include studies on pluvial lake highstands in the western United States which have been more precisely constrained to the period spanning 13–14 ka (Dorn *et al.* 1990). Also, Dorn *et al.* (1992a) have developed a chronologically constrained model of late Pleistocene through latest Holocene alluvial fan activity in Peru based on AMS dating of rock varnish. In Australia,

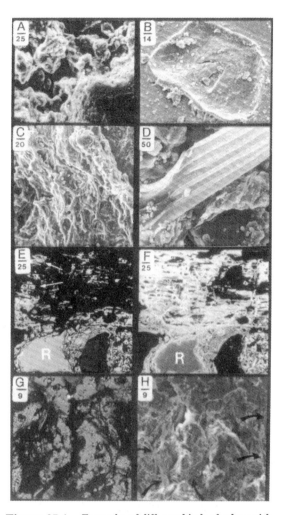

Figure 27.1 *Examples of different kinds of subvarnish organic matter, found in this case associated with petroglyphs (data and photos courtesy R.I. Dorn). (A) Unidentified mounds of fungal proportions; (B) organic matter scraped from under varnish; (C) fungal filaments (?desiccated lichen) identified in a layer underneath a layer of amorphous silica; (D) probable plant (?grass) fibre found within varnish; (E and F) BSE and SE images of carbonised woody tissue found under varnish layer; (G and H) BSE and SE image of organic filaments within a weathering rind under a varnish layer. For further details of samples see Nobbs and Dorn (1993)*

Dragovich (1986) was able to demonstrate that the main phase of varnish development on rock outcrops bearing aboriginal engravings occurred since at least 10 ka.

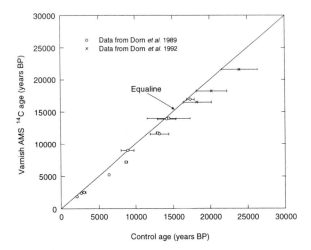

Figure 27.2   *A comparison of ^{14}C AMS dates of subvarnish organic matter and independent estimates of sample ages (data from Dorn* et al. *1989; 1992)*

FRACTIONATION CORRECTION, CALIBRATION PROCEDURES AND POTENTIAL CAUSES OF NON-SYSTEMATIC ERRORS IN RADIOCARBON DATES

While the basic assumptions of the radiocarbon method are sound, a number of complicating factors must be considered in order to guarantee the generation of accurate absolute dates. Isotopic fractionation resulting from metabolic processes causes different kinds of samples to exhibit slightly different initial ^{14}C activities to that of the atmospheric carbon dioxide reservoir from where they were derived. As the fractionation of the stable carbon isotopes ($^{13}C/^{12}C$) is proportional to that of the ratio $^{14}C/^{12}C$, it is possible to use the ratio of the more abundant and more easily measured stable isotopes to estimate, and correct for, the degree of fractionation within any given dating sample (Geyh and Schleicher 1990).

A further complication to radiocarbon dating lies in the need for calibration of radiocarbon dates to determine absolute ages. This is necessary because the assumed constant ^{14}C production rate (and therefore also the atmospheric ratio of $^{14}C/^{12}C$) is only approximately true due to subtle changes in the cosmic-ray flux. Calibration is achieved by comparison to absolute chronologies generated from tree-rings, varves, and from uranium-series dates on corals (Bard *et al.* 1990; Stuvier *et al.* 1993).

In addition to correction for fractionation and calibration it is necessary to consider non-systematic factors which may influence radiocarbon evaluations for a given suite of samples. These include:

1. Contamination from intrusion of younger material (e.g. roots, humic acids, bacteria), or the introduction of older materials (such as fossil carbonates as found in many closed-basin lake systems (Riggs 1983)). Burrowing of animals may also alter the stratigraphic levels of certain carbon-bearing sediment components. As the actual content of ^{14}C reduces exponentially with time, the effect of a given degree of contamination is highly dependent upon the actual context age (Aitken 1990).

2. The 'hard-water effect' whereby ecological and hydrological systems fed by groundwater containing dissolved fossil carbonates produce organic materials (e.g. sapropels, shells) which are effectively depleted in ^{14}C, and correspondingly produce radiocarbon ages which may be in excess by a millennia or more (e.g. Haynes 1987; Goodfriend and Stipp 1983). While this may be problematic in many circumstances, Haynes (1987) was able to use the age offset between sapropels, carbonates and charcoal dates to develop an index of evaporation for early Holocene palaeolakes in the southwestern Sahara.

3. At the young age range of the method, burning of ^{14}C-free fossil fuels and the generation of ^{14}C by atmospheric testing of nuclear devices has resulted in competing effects which limit the generation of radiocarbon dates over the last few centuries (Geyh and Schleicher 1990).

While difficulties in isolating sufficient datable materials free of the complicating factors outlined above are considerable, the radiocarbon method continues to be the central technique for generating numerical age estimates in arid zone research. An additional novel dating application utilising radiocarbon and applicable in arid zones relates to the build-up of hypogene ^{14}C within exposed rock surfaces. This is discussed in a later section.

Luminescence dating methods

The direct dating of terrestrial aeolian deposits has until recently been regarded as problematic (e.g. Thomas and Shaw 1991). Luminescence methods are now overcoming this, providing moderate precision age estimates for aeolian sediments and other dryland deposits. Importantly, these methods appear, in ideal circumstances, to be capable of providing direct and systematic numerical-age chronologies of deposition over timescales which may span the last 800 ka or more (see, for example, Berger *et al.* 1992; Huntley *et al.* 1993a; 1993b; 1994; Berger 1994). While there has been recognition that thermoluminescence methods have been particularly useful in the dating of aeolian sediments for some time (e.g. Singhvi *et al.* 1982; 1986; Pye and Johnson 1987), developments in the field over the past decade have improved key aspects of the dating methodology and prompted rapid expansion and diversification of the field. Recent reviews may be found in Wintle and Huntley (1982), Forman (1989), Wintle (1990; 1993) and Aitken (1992; 1994).

PRINCIPLES OF LUMINESCENCE DATING

Luminescence dating methods provide a means of directly determining the age of mineral constituents of a sediment context, and are based on estimating the time elapsed since a sediment was last exposed to daylight. They are radiation damage (or sometimes termed 'trapped charge') dating methods involving quantifying the accumulation of a radiation-related signal. Initial developments in the field, resulting in thermoluminescence (TL) dating,

focused on the use of heat to stimulate a time-dependent signal from mineral grains which had been 'zeroed' by exposure to heat during formation or usage (e.g. pottery, burnt flint). With the realisation that exposure to daylight was also capable of 'zeroing' previously accumulated radiation damage, the method was applied to sediments (for explanation see Wintle and Huntley 1982); the date so established being the time elapsed since a sediment was last exposed to daylight (a depositional age). The process is shown schematically in Figure 27.3.

A significant drawback to TL methods was the difficulty in establishing the completeness of the resetting (bleaching) process at deposition. Indeed, irrespective of the duration of light exposure a residual or 'hard-to-bleach' TL component will remain (Figure 27.3). Although there are procedures to overcome this issue, a significant advantage of the related and more recently developed optical dating methods is the effectiveness of signal resetting at deposition. In fact, complete signal resetting may be achieved in minutes of light exposure (Figure 27.4). Procedural details and further information

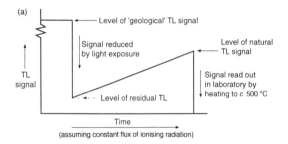

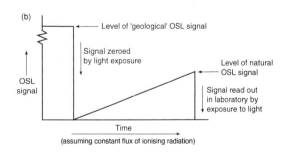

Figure 27.3 *Schematic representation of the accumulation of luminescence in minerals (A) TL, (B) OSL. Note: the diagrams differ only in the presence of a residual signal at deposition for TL, and the mode of signal read-out in the laboratory*

a

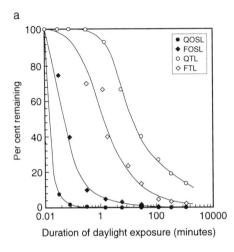

b

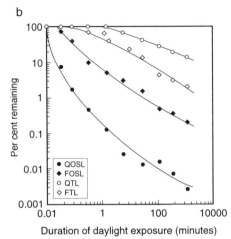

Duration of daylight exposure (minutes) Duration of daylight exposure (minutes)

Figure 27.4 *Examples of TL and OSL bleaching curves for sand-sized quartz and feldspar grains. QOSL — quartz optically stimulated luminescence; FOSL — feldspar optically stimulated luminescence, QTL — quartz thermoluminescence; FTL — feldspar thermoluminescence (data from Godfrey-Smith et al. 1988). Data plotted on both linear (a) and logarithmic (b) vertical axis scales*

relating to optical dating can be found in Smith *et al.*; (1990), Aitken (1992; 1994), and references therein.

In TL dating thermal energy is supplied by an oven in which a sample is heated in a nitrogen or argon atmosphere to temperatures in the range 450–550°C (Figure 27.5). The ensuing progressive emptying of traps of successively higher temperatures results in the generation of the typical 'glow curve' form. Although exhibiting a considerable degree of sample-to-sample variability, minerals exhibit diagnostic glow curve peaks which emit luminescence at specific wavelengths (Aitken 1992).

Optical dating involves the generation of a signal from a sample by optical stimulation (optically stimulated luminescence [OSL] being produced). Convention now is to further classify the form of signal according to the wavelength of optical stimulation (for example green light [GLSL] and infra-red [IRSL] stimulated luminescence are frequently described in the literature).

Unlike TL, the eviction of charge from traps is observed only as a rapidly depleting signal when expressed as a function of light exposure (Figure 27.5). Minerals are variously sensitive to differing wavelengths of light; quartz, for example, is sensitivity to green (c. $\lambda = 500$ nm) and shorter wavelengths; feldspars exhibit a wide range of sensitivities to visible, red and near-IR photons (Huntley *et al.* 1988; 1989; 1991).

THE DERIVATION OF A LUMINESCENCE AGE

The age of a sample is derived from the general age equation:

$$\text{age} = \frac{\text{palaeodose}}{\text{dose rate}} \quad (27.1)$$

where the palaeodose (*P*) (sometimes referred to as equivalent dose [*ED*] or accumulated

Figure 27.5 *Equipment and examples of output generated in luminescence dating. (a) Thermoluminescence apparatus, (b) optically stimulated luminescence apparatus. In (a) note measurements are undertaken in a nitrogen (or argon) atmosphere, and that the read-out process is destructive and a further heating results only in the generation of black-body radiation which is not related to the charge trapping process (natural = as found in the field). Addition of further radiation in the laboratory results both in an increase in the intensity of TL peaks at high (c. >300°C) and the population of geologically unstable traps at lower temperatures. In (b) note that optical excitation of a sample results in a rapidly depleting OSL signal*

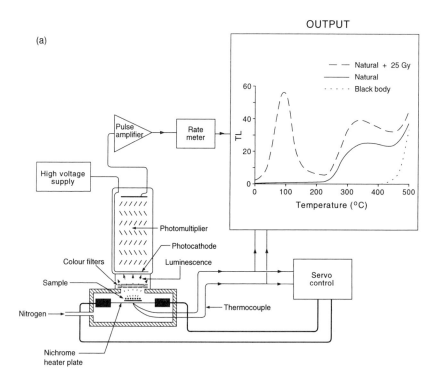

(a)

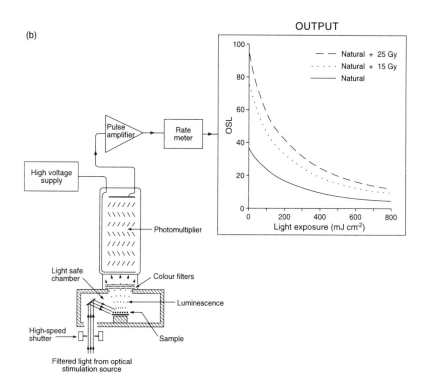

(b)

dose [*AD*]), is the accumulated radiation damage (measured in grays [Gy]); and the dose rate is the rate at which the sample absorbs energy from its immediate proximity; this comprises alpha, beta, and gamma radiation from ^{40}K, ^{238}U, ^{235}U and ^{232}Th and their protégé products, together with a typically small cosmic ray contribution (Aitken 1985).

There are a variety of methods which have been employed for evaluation of the palaeodose. Most utilise measurements of the response of single or multiple aliquots of refined samples of quartz, feldspar or mineral mixtures to laboratory radiation to quantify the radiation sensitivity of the mineral used, and all require some form of thermal pre-treatment to remove unstable luminescence components from those which are stable over geological time periods. Palaeodoses are typically calculated either for progressive temperature (TL) or exposure time (OSL)

intervals as a means of testing the stability and reproducibility of the estimate. Dose-rate evaluation may be achieved by a wide variety of field- or laboratory-based spectroscopic, nuclear and chemical methods (Aitken 1985).

The maximum age range of TL and OSL methods is controlled both by the quantity of radiation damage which a given mineral species can accommodate, and the rate at which the damage accumulates. At typical environmental dose rates quartz saturates at between 100 and 150 ka, although under certain circumstances ages of over 700 ka may be obtainable (Huntley *et al.* 1993a; 1993b). Feldspars demonstrate a considerably greater capacity to accumulate charge, and correspondingly exhibit a maximum age range up to 800 ka, though many feldspars exhibit anomalous effects which may result in age underestimation. Fine-grained mineralic mixtures from loess have also been dated up to 800 ka (Berger *et al.* 1992).

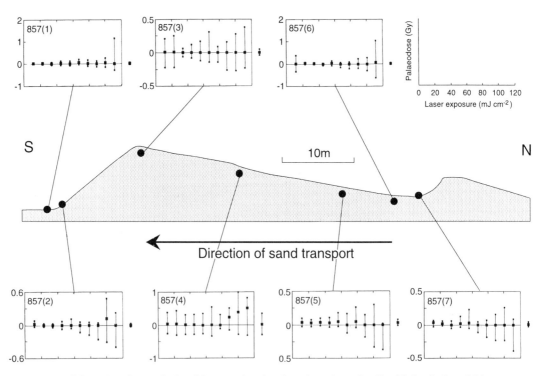

Figure 27.6   *OSL palaeodoses calculated for a modern barchan dune from the Abu Muharik dunefield, approximately 10 km southeast of Kharga, Egypt. In each case palaeodoses have been calculated for a range of laser exposure intervals, as well as being calculated for the total laser exposure (final estimate plotted to the right hand side of graphs). In all cases palaeodoses are observed to be within one sigma error of zero. For further details see Stokes (1994)*

The minimum age range of the methods is determined by both the completeness of resetting of previously accumulated luminescence at deposition (including the degree to which the solar spectrum is filtered in subaqueous depositional environments), the sample's sensitivity to ionising radiation, and subtle charge reorganisation effects which may occur during deposition, or which are the result of laboratory pre-treatment procedures. Studies by Huntley (1985) and Berger (1990) indicate that a 'realistic' minimum age range for the TL methods in well-bleached depositional environments is of the order of 1 to 2 ka. The rapidity of resetting of the OSL signals of both quartz and feldspar, however, means that minimum ages are considerably lower, even less than a few hundred years (Stokes 1992a; 1994). In ideal situations, such as those associated with the exposure of sand grains

during migration of a barchan dune, an effectively zero palaeodose can be reproducibly demonstrated (Figure 27.6).

SOME RECENT APPLICATIONS OF LUMINESCENCE METHODS

Though luminescence methods are still essentially experimental, comparisons with other methods have produced highly encouraging results and indicate that luminescence methods are capable of producing accurate chronologies over at least the past 150 ka and possibly longer (see, for example, Forman and Maat 1990; Stokes 1992a; 1993; Stokes and Gaylord 1993; Wendorf *et al.* 1994 and Figure 27.7).

Table 27.3 gives examples of recent applications of luminescence dating in arid zone settings. Luminescence methods have specific utility for

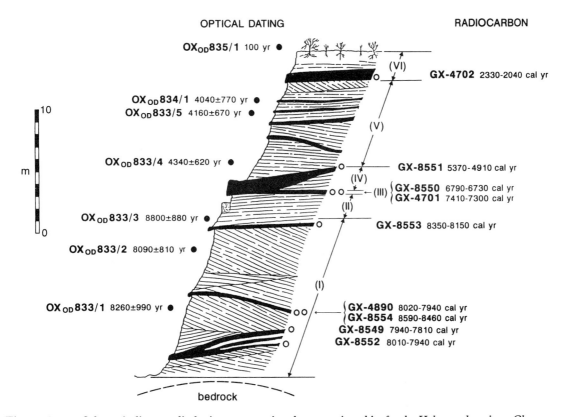

Figure 27.7 *Schematic diagram displaying comparative chronostratigraphies for the Holocene deposits at Clear Creek Ferris Dunefield, Wyoming, USA. Black lenses indicate interdune deposits. Radiocarbon dates are based on wood, charcoal and bone fragments (see Stokes and Gaylord 1993)*

Table 27.3 *A selection of recent studies which have employed luminescence dating methods in the construction of chronologies for selected dryland depositional environments*

Depositional environment	Dosimeter/ luminescence technique	Context age range (ka)	References
Aeolian			
Warm desert	TL, q	0–18	Singhvi *et al.* (1982)
		0–40	Chawla *et al.* (1992)
		40–60	Chen *et al.* (1990)
	TL/OSL, q	2–28	Rendell *et al.* (1994)
	OSL, q	8–20	Goudie *et al.* (1993)
	OSL, q	0–6	Stokes and Breed (1993)
	IRSL, kf	0–6	Edwards (1993)
Cold desert	TL, kf	9–11	Lundqvist and Mejdahl (1987)
	TL, q, kf	10–12	Dijkmans *et al.* (1988)
		11–18	Kolstrup *et al.* (1990)
	OSL, q	0–8	Stokes and Gaylord (1993)
		11–13	Stokes (1991)
Loess	TL, fg	0–90	Frechen (1991)
		0–130	Pye and Johnson (1987)
		0–140	Zoeller *et al.* (1994)
		10–>170	Parks and Rendell (1992)
		5–300	Singhvi *et al.* (1987)
		18–800	Berger *et al.* (1992)
Lunettes	TL, q	0–60	Chen *et al.* (1990)
	TL, q	0–140	Buch and Zoeller (1992)
	OSL, q	0–5	Stokes (1992)
Fluvial			
Meandering channel	TL, q	0–100	Page *et al.* (1991)
Plunge pools	TL, q	5–22	Nott and Price (1994)
Alluvial fan	IRSL, kf	6–8	Clarke (1994)
Soil/colluvium	TL, fg	0–60	Berger and Mahaney (1990)
	IRSL, fg, mg	0–4	Aitken and Xie (1992)
	IRSL, kf	0–36	Botha *et al.* (1994)

Key: q–quartz, kf–potassium feldspar, fg–4–11 mm polymineral grains, mg–20–35 mm polymineral grains, gl–glass.

dating aeolian and other sediments over time ranges well beyond those of the radiocarbon methods. Dating of these aeolian sequences has allowed reassessments of the timing of aridity, especially the identification of more recent phases of dune reactivation (Chalwa *et al.* 1992; Clarke, 1994; Rendell *et al.* 1994) and the determination of bedform accumulation rates (Stokes and Breed 1993). Dating of alluvial channel and fan sequences has in addition provided insights into periods of pluvial activity.

Botha *et al.* (1994) have demonstrated that IRSL techniques may be capable of dating colluvial sequences. These depositional environments are not expected to involve extended periods of daylight exposure during deposition, and although the application of TL methods was problematic, generally good agreement was observed between IRSL dates and radiocarbon dates on soil organic matter and carbonate nodules for colluvium up to 36 ka in age in southern Africa.

Uranium-series dating of precipitated carbonates, sulphates and halites

BASIC PRINCIPLES OF URANIUM-SERIES DATING METHODS

Uranium-series dating methods are numerical-age techniques applicable to a variety of materials over the time range from 10^4 to 10^6 years. In arid areas, the uranium-series dating techniques have been applied to a range of inorganic (e.g. secondary carbonate and salt) and biogenic (e.g. enamel, bone, shell) deposits (e.g. Kaufman and Broecker 1965; Hillaire-Marcel *et al.* 1986; Lao and Benson 1988; Herczeg and Chapman 1991; Wendorf *et al.* 1994).

Uranium in surface sediments is typically bound in silicate and oxide minerals. Over time (a few million years) it will approach secular equilibrium with its decay chain protégé products. Weathering of uranium-bearing minerals produces water-soluble complexes that may become separated from their less soluble protégé products. When this mobile uranium precipitates (inorganically or biogenically) as a trace constituent of surficial minerals it develops a new succession of protégé isotopes.

There are two uranium decay series, each commencing with a different parent uranium isotope (Figure 27.8). The relatively low abundance of ^{235}U complicates its accurate measurement and therefore uranium-series dating usually concentrates on the protégé products of ^{238}U. Methods employing ^{230}Th (the protégé isotope of ^{234}U) are most frequently used due to its convenient half-life ($T_{1/2}$) of 75 ka, and corresponding age range of approximately 350 ka. By measuring the activity ratio of protégé isotopes to parents (e.g. $^{230}Th/^{234}U$) and knowing the half-lives of the daughter products, it is possible to determine the time which has elapsed since precipitation. Figure 27.9 demonstrates the process of increase of ^{230}Th within a system over time and the corresponding ratio between ^{230}Th and ^{234}U. With the basic principles outlined above, two assumptions are required to allow the generation of accurate uranium-series dates:

1. A sample must have behaved as a closed system since the time of formation/deposition (i.e. there has been neither loss nor gain of any isotopes except by radioactive decay).

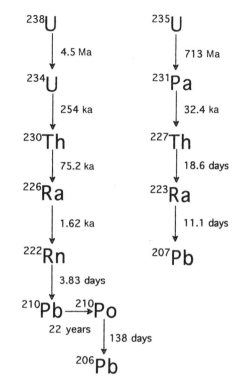

Figure 27.8   *Decay scheme of the longer-lived uranium isotopes (adapted from Schwarcz, 1989)*

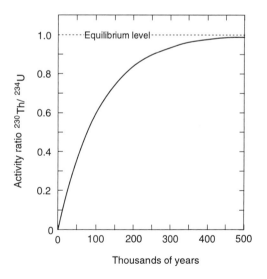

Figure 27.9   *Build-up of ^{230}Th with time illustrated via plotting its activity ratio with ^{234}U. Note the 75.4 thousand year half-life of ^{230}Th corresponds to an activity ratio of 0.5*

622 *Arid zone geomorphology*

2. At the time of formation, the activity ratios of the isotopes being used to determine the age must be either zero or some determinable level.

The conventional means of activity measurement is by alpha spectroscopy (the particles emitted during the decay of key isotopes being measured). More recently, ratios of uranium isotopes have been directly counted by thermal ionisation mass spectrometric techniques (TIMS) (Edwards *et al.* 1986). As in the development of AMS dating in radiocarbon, TIMS has resulted in improvements in precision of analyses and a corresponding increase in the age range of the method (details are given below), reductions in minimum sample size requirements, and the ability to measure entirely new types of precipitated terrestrial deposits (see next section).

In reality, the assumptions listed above are rarely entirely fulfilled in environmental systems. In particular, it is commonplace for 'detrital' thorium to be incorporated at the time of formation via dust (a particular issue in drylands) or incorporation of detrital limestone. Closed system behaviour is complicated in many situations by the porous nature of many pedogenic and lacustrine deposits, and recrystallisation of aragonite into calcite may cause further reorganisation of the isotopic state of a deposit (Szabo and Rosholt 1982). A series of methods have been developed to check, and in some cases to correct, for the occurrence of detrital contaminants and open system behaviour.

Other details of uranium-series dating can be found in Ivanovitch and Harmon (1982), Edwards *et al.* (1986) and Schwarcz (1989).

APPLICATIONS OF URANIUM-SERIES DATING IN ARID ZONE ENVIRONMENTS

Table 27.4 exemplifies the application of uranium-series dating in drylands. The dating of inorganically precipitated carbonates has shown a great deal of promise for arid zone chronologies. Open system behaviour, particularly relating to the dating of pedogenic carbonates and porous tufa deposits, and the presence of detrital contamination, however, always introduces a degree of uncertainty to the reliability of the dates (see, for example, Herczeg and Chapman 1991). Exciting results have been generated by Szabo *et al.* (1989; 1995) who, based on uranium-series dating of surface calcretes, root casts and carbonate-cemented sediments have identified wet phases in the now hyper-arid eastern Sahara at approximately 45, 140 and 210 ka. Stratigraphically sensible Th/U ages have also recently been described, based on pedogenic carbonate, from alluvial fan sequences by Bull (1991).

Phillips *et al.* (1993) have recently employed $^{230}Th/^{234}U$ dating of halite deposits to demonstrate temporal changes in the rate and mode of lacustrine (halite- versus gypsum-dominated) sedimentation in closed lake basins and playas in western China. They used this data to infer that wind intensity may be a more significant variable than aridity in controlling enhanced regional dust and loess deposition.

Ostrich eggshell fragments have proven to be an ideal substrate for TIMS $^{230}Th/^{234}U$ dating. The eggshells incorporate uranium within the organic matrix of the shells and subsequently remain as a closed system. Ostrich eggshells are distinctive in their considerable mass and robustness, which coupled with the wide geographical range of the genera gives them a high likelihood of preservation in arid and semi-arid zone depositional records. The utility of this approach is shown by Schwarcz and Morawska (1993) who constrained the timing of last interglacial and earlier lacustrine episodes southwestern Egypt. Analyses of lacustrine marl and carbonate-cemented sands from the same deposits were somewhat more problematic for age evaluation due to influences from detrital contamination and open system behaviour. Similar complicating effects were described from the analysis of tooth enamel from the same site (McKinney 1993).

A recent addition to the family of uranium-series dating methods is the uranium-trend technique which allows the dating of detrital sediments over a time range of 4 to 900 ka (Muhs *et al.* 1989). In this approach carbonate from soils and sediments are considered as an open system. The concentrations of ^{238}U, ^{234}U, ^{232}Th and ^{230}Th are measured to determine an isochron. This method has not been widely applied to arid zone sediments but the principle has considerable potential (Bull 1991).

Table 27.4 *Summary descriptive classification of selected terrestrial inorganic and biogenic precipitates potentially datable using uranium-series methods (adapted from Szabo and Rosholt 1982)*

Substrate	Description/comment	Review/case example
Inorganic carbonates		
Travertine	Mostly finely crystallised carbonates precipitating from groundwater (may be hard and dense or porous and soft)	Livnat and Kronfeld (1985)
Caliche or pedogenic carbonate	Secondary accumulations of cementing carbonates in the host material within the zones of soil development	Szabo *et al.* (1989) Herczeg and Chapman (1991)
Caliche-rind	Hard and dense carbonate coating on pebbles in the zones of weathering and soil development	Bull (1991)
Calcrete	Massive surficial conglomerates of cemented rock fragments and minerals where soil morphology is not present	Schwarcz and Morawska (1993)
Lacustrine marls	Authigenic carbonates precipitating in lakes	Lao and Benson (1988)
Tufa	$CaCO_3$ deposits that display fossilised vegetable mats indicating that they originate from spring discharges	Szabo *et al.* (1989)
Inorganic sulphates and chlorides		
Gypsum	Evaporitic crusts and cements produced following desiccation of ground or surface waters enriched in Ca^{2+} and SO_4^{2-}	Herczeg and Chapman (1991)
Halite	Incorporation of uranium during evaporite formation	Phillips *et al.* (1993)
Biogenic deposits		
Bone, enamel, antler	Rapid uptake of post-mortem uranium, results complicated by moisture and high porosity	McKinney (1993)
Eggshell	Mass spectrographic analysis allows examination of small (<1 g samples). Closed system behaviour is observed following initial uptake	Schwarcz and Morawska (1993)
Calcareous exoskeletal material (molluscs, gastropods, etc.)	Post-mortem uptake of uranium. Many examples of complications due to open system behaviour and recrystallisation of carbonate during burial	Schwarcz and Morawska (1993)
Rock varnish	Uranium is incorporated from external sources during varnish formation. The varnish then forms a closed system	Knasus and Ku (1980)

Surface exposure and evaporite precipitation ages determined from cosmogenic isotopes

In addition to ^{14}C, a number of other radiogenic and stable isotopes may be created by the interaction of extraterrestrial cosmic rays and terrestrial atoms; these are known as *cosmogenic isotopes*. Their formation may be atmospheric (of meteoric origin), as is the case for ^{14}C, or result from interactions with terrestrial rocks in the upper few metres of earth regolith (of hypogene origin).

The advent of accelerator mass spectroscopy (AMS) has allowed accurate analysis of cosmogenic isotope concentrations, which are typically low (at best a few atoms per gram per year) in whole-rock samples and individual minerals and precipitates. The development of experimental numerical-age dating methods which may quantify the timing of precipitation of chloride evaporitic minerals (e.g. halite) and the exposure ages of rock surfaces has ensued. Table 27.5 gives the cosmogenic isotopes that have potential for dating applications. As the isotopes exhibit a wide range of half-lives, they provide a potential for establishing ages over a considerable range of time; from a few thousand to in excess of 5 million years.

PRINCIPLES AND EXAMPLES OF DATING SURFACE EXPOSURES USING COSMOGENIC ISOTOPES

Figure 27.10 shows the basic principles of the production, redistribution and build-up or decay of cosmogenic isotopes in surface or near-surface environments. Central to dating via cosmogenic isotopes is the assumption of a constant flux of cosmic rays and their secondary products for any given area, and therefore a constant (and known) rate of cosmogenic isotope production at source through time. *In situ* (hypogene) accumulation of cosmogenic isotopes has been described for exposed country rock, alluvial fan surfaces, flood and shoreline deposits, and lava flows (see, for example, Klein *et al.* 1986; Cerling 1990; Jull *et al.* 1992; Laughlin *et al.* 1994). Much of the atmospheric cosmogenic ^{36}Cl production is transported by the hydrological system to the oceans, while some remains trapped in evaporitic deposits associated with closed basins (Bentley *et al.* 1986).

Cosmogenic isotopes are separated from specific target minerals or bulk rock samples and are typically measured by mass spectrometry. Other aspects of the dating of evaporites and geomorphic surfaces are, however, somewhat different; the differences relating to whether the isotopes accumulate *in situ* or are introduced subsequent to their formation, and whether the isotopic system is one of gradual accumulation of a stable isotopic end product, gradual decay of an unstable isotope, or from the attainment of a steady-state equilibrium between rates of formation and decay. Three key types of cosmogenic isotope systems are described below (see also Figure 27.10).

Cosmogenic isotopes in precipitated minerals

The cosmogenic isotope of chlorine (^{36}Cl) has previously been used to date halite and other

Table 27.5 *Some cosmogenic isotopes used for surface exposure dating (adapted from Geyh and Schleicher 1990 and Kurz and Brook 1994)*

Cosmogenic isotope	Half-life (years)	Stable isotope	Production rate (atoms $g^1 a^{-1}$)	Age range
^{3}He	stable	–	160	1 ka–*c.* 3 Ma
^{10}Be	1.5×10^6	^{9}Be	6	3 ka–4 Ma
^{26}Al	7.16×10^5	^{27}Al	37	5 ka–2 Ma
^{36}Cl	3.08×10^5	$^{35,37}Cl$	8	5 ka–1 Ma
^{21}Ne	stable	–	45	7 ka–10 Ma(?)
^{14}C	5730	^{14}N	20	1 ka–18 ka
^{41}Ca	103×10^3	$^{40,42}Ca$	?	to 300 ka

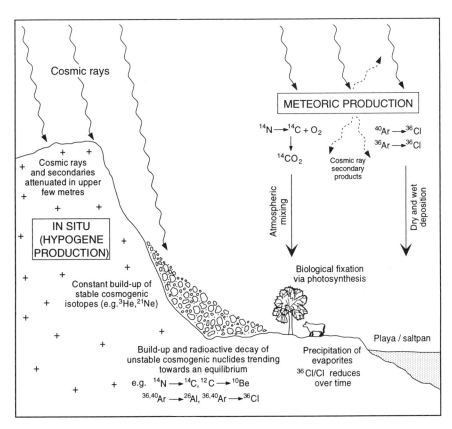

Figure 27.10 *Schematic diagram showing the various production sites and formation mechanisms of cosmogenic isotopes in exposed rock surfaces and precipitated evaporitic minerals (see text for details). Note: atmospheric production and pathway into biosphere is also indicated*

saline lake deposits (e.g. Bentley *et al.* 1986; Phillips *et al.* 1993). Cosmogenic chlorine has been employed in dating applications due to its relative abundance in evaporite deposits, its appropriate half-life (3.01×10^5 years), and its low chemical activity and hydrophilic nature, resulting in its relatively inert behaviour and lack of widespread mixing in most surface geochemical environments (Bentley *et al.* 1986). Accumulation of unstable cosmogenic chlorine (^{36}Cl) in halite and other evaporitic minerals can be considered as similar to the biogenic and inorganic accumulation of meteorically produced ^{14}C in plant and animal tissue and the various forms of inorganic carbonate. The quantities which must be known or measured are therefore the initial abundance of ^{36}Cl (collectively derived from atmospheric, epigene and hypogene sources); the current abundance of

^{36}Cl (usually expressed relative to stable chlorine); and the radioactive half-life of ^{36}Cl. The age of an evaporite may thus be estimated from the standard equation for radioactive decay:

$$N = N_0 e^{-\lambda t} \qquad (27.2)$$

where N_0 is the initial concentration of ^{36}Cl at the time of precipitation, t is time, and λ is the decay constant of ^{36}Cl.

The time range of dating precipitates is controlled by the half-life of ^{36}Cl and the ability to measure ^{36}Cl activity at low levels. Ages as young as 5 ka and as old as 1 Ma are theoretically determinable using the ^{36}Cl method (Table 27.5). Limiting factors include difficulties in precisely estimating the ^{36}Cl production rate at deposition, and the possible introduction of post-depositional groundwater-derived or hypogene ^{36}Cl.

Build-up of stable cosmogenic isotopes in geomorphic surfaces

Hypogene accumulation of cosmogenic isotopes provides a means of directly dating the timing of surface exposure. A family of developmental numerical-age techniques have accordingly been developed (for review see Kurz and Brook 1994; and Zreda and Phillips 1994). Assumptions central to the methods include:

1. The rock contains little or none of the cosmogenic isotope of interest at time of exposure ($t = 0$).
2. The rocks have remained in a stable position since deposition.
3. The surfaces of rocks have not undergone any significant mass loss due to erosion.
4. The rocks have not at any time been shielded from cosmic rays (e.g. by mobile sand dunes).
5. The rocks have behaved as a closed system (i.e. no loss or contamination).

Stable cosmogenic isotopes are expected to accumulate at a constant rate, controlled by the isotope production rate (P, measured in atoms $g^{-1} a^{-1}$) which varies as a function of both latitude and longitude. Given an estimate of P for an area, the exposure age of a surface may be estimated from the equation:

$$N = Pt \qquad (27.3)$$

where N is the measured number of cosmogenic isotopes (atoms g^{-1}) and t is time (a). Limiting factors in the application of stable cosmogenic isotopes include the effects of soil cover and erosion, isotope loss via diffusion (which may be accelerated in areas of accelerated chemical weathering), and, in the case of ${}^3He$, the need to correct for any magmatic ${}^3He$ present at deposition. The absolute upper age limit of the method is controlled by diffusion of isotopes out of minerals and remains debated. Retention rates vary according to the host mineral, but at best limit the method to the last 2 Ma, and possibly much less, in the case of ${}^3He$, and to a maximum of 10 Ma in the case of ${}^{21}Ne$ (Kurz and Brook 1994). ${}^3He$ ages as young as 3 ka have been reported and show excellent agreement with radiocarbon estimates on associated plant remains (Laughlin *et al.* 1994).

Due to its high production rate within olivine and clinopyroxene, ${}^3He$ has been the most intensively studied stable cosmogenic isotope. Typically landscape components are dated in relation to ages of basic volcanic rocks which host olivine, clinopyroxene, and other minerals where the cosmogenic isotopes accumulate in abundance (see, for example, Cerling 1990).

Build-up of unstable cosmogenic isotopes in geomorphic surfaces

In the case of unstable isotopes the rate of accumulation in rock surfaces decreases with time until the production rate equals the radioactive decay rate. In the case of terrestrial rock surfaces, assuming that no cosmogenic isotopes are present at the time of initial exposure at or near (<2 m depth) the surface, the quantity of a given cosmogenic nuclide present is simply calculated by subtracting the loss by radioactive decay (which is zero for stable nuclides) from the production rate. This may be expressed mathematically by the equation:

$$dN/dt = P - N\lambda \qquad (27.4)$$

where N is the isotope concentration (atoms g^{-1}), t is time (a), P is the production rate (atoms $g^{-1} a^{-1}$) and λ is the decay constant of the cosmogenic nuclide (a^{-1}). This equation

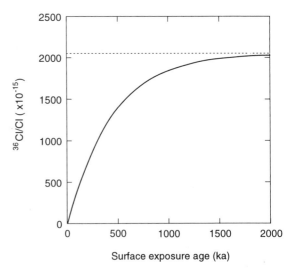

Figure 27.11 *Build-up of cosmogenic ${}^{36}Cl$ on the surface in rocks (assuming no erosion). The build-up curve is steepest at early times and flattens gradually to reach a steady state (from Zreda and Phillips 1994)*

may be solved to give:

$$N = P/\lambda \ (1 - e^{-\lambda t}) \qquad (27.5)$$

The time range of dating for a given cosmogenic isotope in a surficially exposed rock is therefore controlled by the time which elapses before the production rate reaches a steady-state equilibrium with the decay rate. Hypogene ^{36}Cl for example, with a half-life of 3.08×10^5 years, rapidly approaches a production–decay equilibrium at around 2 Ma, limiting its realistic datable age range to 1 Ma (Figure 27.11). The use of hypogene ^{14}C for dating surface exposures results in a considerably reduced age range (up to *c.* 30 ka).

Other limiting factors on surface exposure age evaluation via unstable cosmogenic isotopes are the same as those for stable cosmogenic isotopes.

In addition to measurements made on single isotopes, the use of paired unstable cosmogenic nuclides (e.g. $^{26}Al/^{10}Be$) may allow the estimation of long-term accretion and erosion rates, and residence times of sedimentary particles in depositional systems (see, for example, Klein *et al.* 1986; Nishiizumi *et al.* 1993). Clear demonstration of the accuracy and reliability of the techniques is, however, yet to be presented.

Dating rock varnish coatings using the cation-ratio method

Biogenic rock varnish coatings or patinas formed on surface boulders, rock engravings, petroglyphs and surface artefacts provide an opportunity to obtain chronological control on the timing of exposure of underlying surfaces.

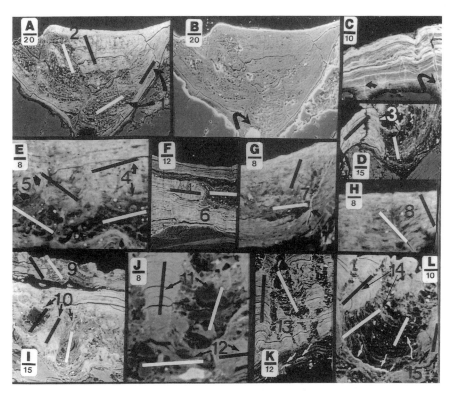

Figure 27.12 *SEM images (BSE except B) of rock varnish profiles and associated partial chemical analyses (WDS) of both porous and layered (less porous) regions (white and black superimposed lines in figure, respectively). A and B demonstrate the compositional nature of the BSE response versus the topographic nature of the SE response. Images A–E and G–I are of varnish layers on petroglyphs, F — desert pavement, J and L — till, K — basalt flow. Data from Dorn and Krinsley (1991) (courtesy of R.I. Dorn)*

Table 27.6   *Methods of estimating the age of rock varnish formation (after Dorn 1994)*

Method	Theory	Precision level*	Comments/case examples
Appearance	Varnish darkens over time	Relative	Controlled by factors other than time
Thickness	As varnish gets older, it grows vertically	Relative	Also controlled by microenvironment
Cover of black surface varnish	Varnish grows laterally away from nucleation centres	Relative	Derbyshire et al. (1984)
Other bottom varnish growth	As age increases, undersides of clasts are coated with Fe-clay rock varnish (Mn-poor)	Relative	Derbyshire et al. (1984)
Trace element trends	Assumes varnish derived from underlying rock. Trace element profiles with depth reflect time	Relative	Bard (1979)
Metal scavenging	Zn, Cu, Ni and other metals increase over time as they are scavenged by Mn–Fe-oxides	Relative	Dorn et al. (1992a)
Palaeomagnetism	Magnetic field aligned when Fe-oxides precipitate	Correlative	Clayton et al (1990)
Tephra-chronology	Glass fragments from known volcanic eruptions might be identifiable in rock varnish	Correlative	Harrington (1988)
Varnish geochemical layering	Sequences of chemical (e.g. Mn:Fe, Pb, $\delta^{13}C$) and textural changes correlated from site to site	Correlative	Dorn (1988; 1992; 1994)
Stratigraphy	Dating material on or under varnish constrains varnish age	Correlative and numerical	Dragovich (1986)
Cation-ratio	Mobile cations are leached faster than immobile cations (K+Ca)/Ti	Calibrated	Dorn (1989); Dorn et al. (1990)
KAr dating	As varnish clays accumulate they may undergo a diagenesis that refixes K; or date K in Mn-oxides	Numerical	Dorn (1989); Vasconcelos et al. (1992)
Uranium-series	Uranium precipitates with Mn-oxides and then decays	Numerical	Knauss and Ku (1980)
Radiocarbon	Accreting varnish encapsulates underlying organic matter	Numerical	Dorn et al. (1992b)

Relative, correlative, calibrated and *numerical* are Quaternary dating terms recommended by Colman *et al.* (1987), see text for details.

Such surface varnishes are common in arid regions (Chapters 6, 23 and 26) and may exhibit a detailed microstratigraphy when examined in thin-section (Figure 27.12). As the ages thus established relate to the timing of colonisation of exposed substrates they represent minimum ages of surface formation. A wide variety of approaches have been adopted for the generation of chronological information from rock varnish. A summary of these methods is presented in Table 27.6.

The two most widely used approaches to establish age control for rock varnishes are the above-mentioned AMS radiocarbon dating of subvarnish organic matter and the relative-age method of cation-ratio dating. The method has almost exclusively been developed by R.I. Dorn in his analyses of environmental and archaeological settings in the arid southwestern USA, Peru and Australia (see, for example, Dorn and Oberlander 1981; Dorn 1983; 1989; Dorn *et al.* 1990). While the cation-ratio method requires calibration to other techniques to derive absolute age estimates, the relatively low costs (compared with AMS radiocarbon or cosmogenic methods), its high speed of analysis, and ability to generate relative or absolute chronologies in varnishes which do not contain subvarnish organic matter mean that the method has considerable utility despite its numerous limitations.

This method is based on the observation that the ratio of certain cations ([potassium and calcium]/titanium) in varnish decreases with age (Figure 27.13). It is generally accepted that this is the result of preferential leaching of the more mobile potassium and calcium (Dorn 1983). The minimum age range of the method relates to the time required for initiation of a visible rock varnish. Visible varnishes form over periods of the order of a few thousand years, while incipient varnish development may be observed microscopically over as little as a hundred years (Bull 1991). Cation-ratios are calculated either by mechanical removal of large (mm³–cm³) samples of varnish from an underlying substrate and analysis via proton-induced X-ray emission spectroscopy (PIXE), electron microprobe, or inductively coupled plasma spectroscopic (ICP) methods, or by *in situ* SEM analysis. Relative-age determinations are generated for a given area by comparing cation-ratios from a range of samples. Calibrated ages may be determined by comparison to cation-leaching curves generated from varnish samples of known age (typically calibrated by AMS radiocarbon or K/Ar methods).

The reliability and accuracy of the cation-ratio method have been debated (e.g. Bierman and Gillespie 1991; Bierman *et al.* 1991). It would however, appear to be capable of producing

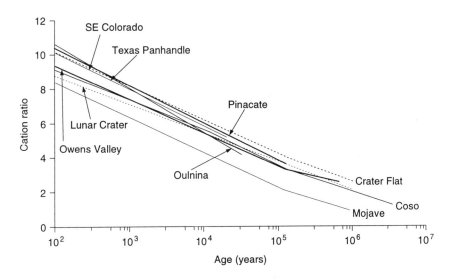

Figure 27.13 *Cation-leaching curves constructed by PIXE analyses of rock varnishes removed from surfaces of known age. Details provided in Dorn (1994)*

useful relative-age data where care has been taken to ensure the avoidance of rock varnish samples influenced by the numerous negatory environmental factors which may cause non-time-dependent changes in the varnish composition. These include a range of ecological, pedogenic and topographic factors which are discussed in detail in Dorn (1994). Published blind tests of the cation-ratio method have produced good agreement with independent chronological controls in some cases (e.g. Harrington and Whitney 1987; Pineda *et al.* 1988; 1990).

While the bulk of the published cation-ratio studies have focused on archaeological contexts or dates on deglaciated or young volcanic terranes (e.g. Dorn *et al.* 1987; 1992b), the method has been successfully used to constrain the chronology of colluvium and debris-flow levee development in Death Valley, California (Dorn 1994) and alluvial-related sequences in the southwestern United States (Bull 1991).

Other methods

AMINO ACID RACEMISATION/ EPIMERISATION STUDIES

Amino acid racemisation/epimerisation dating are relative-age methods involving measurement of the extent to which certain types of amino acids within protein residues have transformed from one of two chemically identical forms to the other. Materials which may preserve such protein residues within sediment bodies include bones and other body components (e.g. mummies), mollusc shells and eggshells. At formation, only L-form amino acids are present. Over time, and in part controlled by temperature and other factors (hydrolysis, pH), some of the L-form acids are converted to a D-form until an equilibrium is reached. *Racemisation* and *epimerisation* reactions differ in that racemisation involves only amino acids with a single chiral carbon atom, whereas epimerisation involves amino acids with two chiral carbon atoms (Aitken 1990). The use of a range of amino acids which undergo racemisation or epimerisation over a range of timescales makes it possible to apply the method over timescales ranging from a few years to hundreds of thousands of years.

Where it is possible precisely to establish burial temperature and other parameters for a given context it is theoretically possible directly to calculate a numerical age. This is rarely the case and it is more common to either generate a calibrated-age chronology based on comparison to radiocarbon or uranium-series dates (e.g. Miller *et al.* 1991), or to assume regional uniformity of palaeotemperatures and determine *aminozones* (zones of uniform racemisation/epimerisation ratio) and use these as a basis for stratigraphic correlation.

Terrestrial mollusc shells appear to be an excellent material for amino acid dating over Holocene timescales in arid zone environments due to their non-porous structure and their likelihood of experiencing only terrestrial conditions (compared with aquatic or marine molluscs which would be expected to have been exposed to both submerged and terrestrial conditions). Goodfriend (1987) demonstrated this potential by comparing epimerisation ratios of the land snail *Trochoiea seetzeni* against a radiocarbon-based chronology. Goodfriend obtained correlations between epimerisation ratios and ^{14}C dates as high as 0.95. Importantly, the application of such an approach allows the development of a calibrated-age chronostratigraphy at a considerably lesser cost than the widespread use of only radiocarbon analyses (which may in themselves be problematic for such sample contexts). Use of molluscs over longer (i.e. Pleistocene) time periods may in some cases be complicated by diagenetic effects. Epimerisation of isoleucine from within ostrich eggshells has, however, provided an additional relative- or calibrated-age method for constraining the age of late Pleistocene arid zone sequences up to 200 ka in age (e.g. Miller *et al.* 1991).

OBSIDIAN HYDRATION DATING

Ages for buried or surface contexts may also in some instances be evaluated by an analysis of the hydration rims of archaeological artefacts produced from obsidian. This method is based on the time- and temperature-dependent hydration of freshly chipped obsidian which occurs progressively in a non-linear fashion as water is diffused into the glass matrix. Hydration rims grow according to the equation

$$x^2 = kt \qquad (27.6)$$

where x is the rim thickness (mm), k is the diffusion constant, and t is time (a). k is dependent on temperature, chemical composition of the obsidian, and relative humidity. Uncertainties regarding the relative levels and stability of temperature and humidity during the period since flaking of the artefact render numerical ages to be problematic; the method has, however, been used widely as a relative-age technique in the arid southwestern United States (see, for example, Beck and Jones 1994).

ELECTRON SPIN RESONANCE

Electron spin resonance (ESR) dating is a numerical-age method closely related to the luminescence methods described earlier. In ESR dating, age evaluation is achieved by determining the population of trapped unpaired electrons as measured by a high-frequency magnetic field (for details see Aitken 1990). While applications of ESR dating in arid zone settings are not widespread, the method has potential for dating a variety of biogenic (e.g. molluscs, tooth enamel, bone) and inorganic (e.g. travertine, silcrete, gypsum) crystalline deposits over a time range spanning a few thousand years through to over 2 Ma (Schwarcz 1994). The full potential of the method in arid zone environments, particularly for secondary carbonate deposits, is yet to be fully explored.

Conclusions

This extensive yet non-exhaustive overview of numerical- and relative-age dating methods applicable to (primarily later Quaternary) arid zone environments clearly demonstrates the rapid expansion and advancement which has taken place in geochronological methods over recent years. With regard to chronological control, the status of arid zone surficial and depositional (buried) landscape components is poised to improve dramatically. While AMS dating of biogenic, sediment and soil components has provided a great deal of new geochronological data, its is clearly necessary to look towards other methods to establish temporal frameworks for older sequences, for surface exposure ages, and for the majority of depositional sequences which are devoid of materials datable via the radiocarbon methods. In this regard, cosmogenic isotopes, uranium series and luminescence dating methods in particular hold a great deal of promise for future research. These approaches may be complemented by the many other applicable calibrated, correlative and relative-age methods (Table 27.2).

None of the methods described above are universally applicable in either spatial or temporal dimensions, and, additionally, all poses limitations which may restrict the precision of results, or entirely prohibit the application of such methods in certain environmental situations. To extract chronological information accurately from arid zone environments, and in doing so establishing the timing and nature of change of both depositional sequences and surface exposures, a multi-dating strategy incorporating a spectrum of methods would appear to be the logical model for future research. This strategy results not only in cross-checking of generated chronologies in specific applications, but also contributes to the ongoing testing and developments of the dating methods themselves. This approach is already generating highly important data, as has been demonstrated in a number of recent studies (e.g. Stokes and Gaylord 1993; Wendorf *et al.* 1994; Zreda and Phillips 1994; Forman *et al.* 1995; Madole 1995).

References

Aitken, M. 1985. *Thermoluminescence dating.* Academic Press, London: 359 pp.

Aitken, M.J. 1990. *Science-based dating in archaeology.* Longmans Archaeology Series: 274 pp.

Aitken, M.J. 1992. Optical dating. *Quaternary Science Reviews*, 11: 127–131.

Aitken, M.J. 1994. Optical dating: A non-specialist review. *Quaternary Geochronology*, 13: 503–508.

Aitken, M.J. and Xie, J. 1992. Optical dating using infrared diodes: young samples. *Quaternary Science Reviews*, 11: 147–152.

Bard, J.C. 1979. The development of a Palination dating technique for Great Basin petrographs utilising neutron activation and X-ray fluorescence analysis. Unpublished PhD dissertation, University of California, Berkeley.

Bard, E., Hamelin, B., Fairbanks, R.G. and Zindler, A. 1990. Calibration of the ^{14}C timescale over the past 30 000 years using mass spectrometric U–Th ages from Barbados corals. *Nature*, 345: 405–410.

Beck, C. 1994. (ed.). *Dating in exposed and surface contexts*. University of New Mexico Press, Albuquerque: 239 pp.

Beck, C. and Jones, G.T. 1994. *Dating surface assemblages using obsidian hydration*. In Beck, C. (ed). *Dating in exposed and surface contexts*. University of New Mexico Press, Albuquerque: 47–76.

Bentley, H.W., Phillips, F.M. and Davis, S.N. 1986. Chlorine-36 in the terrestrial environment. In P. Fritz and J.C. Fontes (eds), *Handbook of environmental isotope geochemistry*. Elsevier, Amsterdam: 427–478.

Berger, G.W. 1990. Effectiveness of natural zeroing of the thermoluminescence in sediments. *Journal of Geophysical Research*, 95 (B): 12375–12397.

Berger, G.W. 1994. Thermoluminescence dating of sediments older than *c.* 100 ka. *Quaternary Geochronology*, 13 (5–7): 447–456

Berger, G.W. and Mahaney, W.C. 1990. Test of thermoluminescence dating of buried soils from Mt Kenya, Kenya. *Sedimentary Geology*, 66: 45–56.

Berger, G.W., Pillans, B.J. and Palmer, A.S. 1992. Dating loess up to 800 ka by thermoluminescence. *Geology*, 20: 403–406.

Bierman, P. and Gillespie, A. 1991. Accuracy of rock-varnish chemical analyses: implications for cation-ratio dating. *Geology*, 19: 196–199.

Bierman, P., Gillespie, A. and Kuehner, S. 1991. Precision of rock-varnish chemical analyses and cation-ratio ages. *Geology*, 19: 135–138.

Birkeland, P.W. 1984. *Soils and geomorphology*. Oxford University Press, New York: 372 pp.

Botha, G.A., Wintle, A.G. and Vogel, J.C. 1994. Episodic late Quaternary palaeogully erosion in northern KwaZulu-Natal, South Africa. *Catena*, 23: 327–340.

Brookes, I.A. 1989. Early Holocene basinal sediments of the Dakhleh Oasis region, south central Egypt. *Quaternary Research*, 32: 139–152.

Buch, M.W. and Zoeller, L. 1992. Pedostratigraphy and thermoluminescence chronology of the western margin- (lunette-) dunes of Etosha Pan, Northern Namibia. *Wurzb. Geogr. Arb.*, 84: 361–384.

Bull, W.B. 1991. *Geomorphic responses to climate change*. Oxford University Press, New York: 326 pp.

Cerling, T.E. 1990. Dating geomorphologic surfaces using cosmogenic ^{3}He. *Quaternary Research*, 33: 148–156.

Chawla, S., Dhir, R.P. and Singhvi, A.K. 1992. Thermoluminescence chronology of sand profiles in the Thar Desert and their implications. *Quaternary Science Reviews*, 11: 25–32.

Chen, X.Y., Prescott, J.R. and Hutton, J.T. 1990. Thermoluminescence dating on gypseous dunes of Lake Armadeus, central Australia. *Australian Journal of Earth Sciences*, 37: 93–101.

Chen, Y. and Polach, H. 1986. Validity of ^{14}C ages of carbonates in sediments. *Radiocarbon*, 28 (2A): 464–472.

Colman, S.M., Pierce, K.L. and Birkeland, P.W. 1987. Suggested terminology for Quaternary dating methods. *Quaternary Research*, 28: 314–319.

Clarke, M.L. 1994. Infra-red stimulated luminescence ages from aeolian sand and alluvial fan deposits from the eastern Mohave Desert, California. *Quaternary Geochronology*, 13: 533–538.

Clayton, J.A., Verosub, K.L. and Harrington, C.D. 1990. Magnetic techniques applied to the study of rock varnish. *Geophysical Research Letters*, 17: 787–790.

Cullingford, R.A., Davidson, D.A. and Lewin, J. (eds) 1980. *Timescales in geomorphology*. Wiley, Chichester: 360 pp.

Derbyshire, E., Jijun, L., Perrot, F.A., Shuying, X. and Waters, R.S. 1984. *Quaternary glacial history of the Hunza Valley, Karakoram Mountains, Pakistan*. In K.J. Miller (ed), *The International Karakoram Project*, vol. 2. Cambridge University Press: 456–495.

Dijkmans, J.W.A., Wintle, A.G. and Mejdahl, V. 1988. Some thermoluminescence properties and dating of eolian sands from the Netherlands. *Quaternary Science Reviews*, 7 (3/4): 349–356.

Dorn, R.I. 1983. Cation-ratio dating: A new rock varnish age-determination technique. *Quaternary Research*, 20: 49–73.

Dorn, R.I. 1988. A rock varnish interpretation of alluvial fan development in Death Valley, California. *National Geographic Research*, 4: 56–73.

Dorn, R.I. 1989. Cation-ratio dating of rock varnish: A geographic perspective. *Progress in Physical Geography*, 13: 559–596.

Dorn, R.I. 1992. Palaeoenvironmental signals in rock varnish on petroglyphs. *American Indian Rock Art*, 18: 1–15.

Dorn, R.I. 1994. Surface exposure dating with rock varnish. In C. Beck (ed.), *Dating in exposed and surface contexts*. University of New Mexico Press, Albuquerque: 77–114.

Dorn, R.I. and Krinsley, D.H. 1991. Cation leaching sites in rock varnish. *Geology*, 19: 1077–1080.

Dorn, R.I. and Oberlander, T.M. 1981. Microbial origin of desert varnish. *Science*, 213: 1245–1247.

Dorn, R.I. and Phillips, F.M. 1991. Surface exposure dating: review and critical evaluation. *Physical Geography*, 12 (4): 303–333.

Dorn, R.I., Turrin, B.D., Jull, A.J.T., Linick, T.W. and Donahue, D.J. 1987. Radiocarbon and cation-ratio ages for rock varnish on Tioga and Tahoe morainal boulders of Pine Creek, Eastern Sierra Nevadas, California, and paleoclimatic implications. *Quaternary Research*, 28: 38–49.

Dorn, R.I., Jull, A.J.T., Donahue, D.J., Linick, T.W. and Toolin, L.J. 1989. Accelerator mass spectrometry radiocarbon dating of rock varnish. *Bulletin of the Geological Society of America*, 101: 1363–1372.

Dorn, R.I., Cahill, T.A., Eldred, R.A., Gill, T.E., Kusko, B.H., Bach, A.J. and Elliot-Fist, D.L. 1990. Dating rock varnishes by the cation-ratio method with PIXE, ICP, and the electron microprobe. *International Journal of PIXE*, 157–195.

Dorn, R.I., Clarkson, P.B., Nobbs, M.F., Loendorf, L.L. and Whitley, D.S. 1992a. Radiocarbon dating inclusions of organic matter in rock varnish, with examples from drylands. *Annals of the Association of American Geographers*, 82: 136–151.

Dorn, R.I., Jull, A.J.T., Donahue, D.J., Linick, T.W., Toolin, L.J., Moore, R.B., Rubin, M., Gill, T.E. and Cahill, T.A. 1992b. Rock varnish on Hualalai and Mauna Kea volcanoes, Hawaii. *Pacific Science*, 46: 11–34.

Dragovich, D. 1986. Minimum age of some desert varnish near Broken Hill, New South Wales. *Search*, 17: 149–151.

Edwards, L.R., Chen, J.H. and Wasserberg, G.J. 1986. ^{238}U-^{234}U-^{230}Th-^{232}Th systematics and the precise measurement of time over the past 500 000 years. *Earth and Planetary Science Letters*, 81: 175–192.

Edwards, S.R. 1993. Luminescence dating of sand from the Kelso Dunes, California. In K. Pye, (ed.), *The dynamics and environmental context of aeolian sedimentary systems*. Geological Society of London, Special Publication, 72: 59–68.

Ellis, S.A. and Van Devender, T.R. 1990. Fossil insect evidence for Late Quaternary climate change in the Big Bend region, Chihuahuan Desert, Texas. *Quaternary Research*, 34: 249–261.

Elmore, D. and Phillips, F.M. 1987. Accelerator mass spectrometry for measurement of long-lived radioisotopes. *Science*, 236: 543–550.

Forman, S.L. 1989. Applications and limitations of thermoluminescence to date Quaternary sediments. *Quaternary International*, 1: 47–59.

Forman, S.L. and Maat, P. 1990. Stratigraphic evidence for Late Quaternary dune activity near Hudson on the Piedmont of Northern Colorado. *Geology*, 18: 745–748.

Forman, S.L., Oglesby, R., Markgraf, V. and Stafford, T. 1995. Paleoclimatic significance of Late Quaternary eolian deposition on the Piedmont and High Plains, Central United States. *Global and Planetary Change*, 11: 35–55.

Fowler, A.J., Gillespie, R. and Hedges, R.E.M. 1986. Radiocarbon dating of sediments. *Radiocarbon*, 28(2A): 441–450.

Frechen, M. 1992. Systematic thermoluminescence dating of two loess profiles from the Middle Rhine Area (F.R.G.). *Quaternary Science Reviews*, 11: 93–101.

Gaylord, D.R. and Stetler, L.D. 1994. Aeolian-climatic thresholds and sand dunes at the Hanford site, south-central Washington, USA. *Journal of Arid Environments*, 28(2): 95–116.

Geyh, M.A. and Schleicher, H. 1990. *Absolute age determination*. Springer-Verlag, Berlin: 503 pp.

Geyh, M.A., Roeschmann, G., Wijmstra, T.A. and Middeldorp, A.A. 1983. The unreliability of ^{14}C dates obtained from buried sandy podzols. *Radiocarbon*, 25: 409–416.

Gillespie, R., Magee, J.W., Luly, J.G., Dlugokencky, E., Sparks, R.J. and Wallace, G. 1991. AMS radiocarbon dating in the study of arid environments: Examples from Lake Eyre, South Australia. *Palaeogeography, Palaeoclimatology, Palaeoecology*, 84: 333–338.

Gillespie, R., Prosser, I.P., Dlugokencky, E., Sparks, R.J., Wallace, G. and Chappell, J.M.A. 1992. AMS dating of alluvial sediments on the southern tablelands of New South Wales, Australia. *Radiocarbon*, 34(1): 29–36.

Godfrey-Smith, D.I., Huntley, D.J. and Chen, W.-H. 1988. Optical dating studies of quartz and feldspar sediment extracts. *Quaternary Science Reviews*, 7: 373–380.

Goodfriend, G.A. 1987. Evaluation of amino-acid racemization/epimerisation dating using radiocarbon-dated fossil land snails. *Geology*, 15: 698–700.

Goodfriend, G.A. and Stipp, J.J. 1983. Limestone and the problem of radiocarbon dating of land snail shell carbonate. *Geology*, 11: 575–577.

Goudie, A.S., Stokes, S., Livingstone, I., Bailiff, I.K. and Allison, R.J. 1993. Post-depositional modification of the linear sand ridges of the West Kimberley area of Northwest Australia. *Geographical Journal*, 159 (3): 306–317.

Haas, H., Holliday, V. and Stuckenrath, R. 1986. Dating of Holocene stratigraphy with soluble and insoluble organic fractions at the Lubbock Lake Archaeological Site, Texas: an ideal case study. *Radiocarbon*, 28: 473–485.

Harrington, C.D. 1988. Recognition of components of volcanic ash in rock varnishes and the dating of volcanic ejects plumes. *Geological Society of America, Abstracts with Programs*, 20: 167.

Harrington, C.D. and Whitney, J.W. 1987. Scanning electron microscope method for rock varnish dating. *Geology*, 15: 967–970.

Haynes, C.V. Jr 1982. Great sand sea and Selima sand sheet, eastern Sahara: geochronology of desertification. *Science*, 217: 629–633.

Haynes, C.V. Jr 1987. Holocene migration rates of the Sudano-Sahelian wetting front, ArbaŌin Desert, Eastern Sahara. In A.E. Close (ed)., *Prehistory of arid North Africa. Essays in honor of Fred Wendorf.* Southern Methodist University Press, Dallas: pp 67–84.

Haynes, C.V. Jr 1989. Bagnold's barchan: a 57-yr record of dune movement in the Eastern Sahara and implications for dune origin and palaeoclimate since Neolithic times. *Quaternary Research,* 32: 153–167.

Haynes, C.V. Jr, Maxwell, T.A. and Johnson, D.L. 1993. Stratigraphy, geochronology, and origin of the Selima sand sheet, eastern Sahara, Egypt and Sudan. In U. Thorweihe and H. Schandelmeier (eds), *Geoscientfic research in northeast Africa.* Balkema, Rotterdam: 621–626.

Head, M.J., Zhou, W. and Zhou, M. 1989. Evaluation of ^{14}C ages of organic fractions of paleosols from loess-paleosol sequences near Xian, China. *Radiocarbon,* 31: 680–694.

Hedges, R.E.M. 1991. AMS dating: present status and potential applications. *Quaternary Proceedings,* 1: 5–10.

Heller, F. and Evans, M.E. 1995. Loess magnetism. *Reviews in Geophysics,* 33: 211–240.

Herczeg, A.L. and Chapman, A. 1991. Uranium-series dating of lake and dune deposits in southeastern Australia: a reconnaissance. *Palaeogeography, Palaeoclimatology, Palaeoecology,* 84: 285–298.

Hillaire-Marcel, C., Carro, O. and Casanova, J. 1986. ^{14}C and Th/U dating of Pleistocene and Holocene stromatolites from East African paleolakes. *Quaternary Research,* 25: 312–329.

Huntley, D.J. 1985. On the zeroing of the thermoluminescence of sediments. *Physics and Chemistry of Minerals,* 12: 122–127.

Huntley, D.J., Godfrey-Smith, D.I. and Thewalt, M.L.W. 1985. Optical dating of sediment. *Nature,* 313: 105–107.

Huntley, D.J., Godfrey-Smith, D.I., Thewalt, M.L.W. and Berger, G.W. 1988. Thermoluminescence spectra of some minerals relevant to thermoluminescence dating. *Journal of Luminescence,* 39: 123–136.

Huntley, D.J., McMullen, W.G., Godfrey-Smith, D.I. and Thewalt, M.L.W. 1989. Time-dependent recombination spectra arising from optical ejection of trapped charges in feldspars. *Journal of Luminescence,* 44: 41–46.

Huntley, D.J., Godfrey-Smith, D.I. and Haskell, E.H. 1991. Light-induced emission spectra from some quartz and feldspars. *Nuclear Tracks and Radiation Measurements,* 18: 127–131.

Huntley, D.J., Hutton, J.T. and Prescott, J.R. 1993a. The stranded beach-dune sequence of south-east South Australia: a test. *Quaternary Science Reviews,* 13: 201–207.

Huntley, D.J., Hutton, J.T. and Prescott, J.R. 1993b. Optical dating using inclusions within quartz. *Geology,* 21: 1087–1090.

Huntley, D.J., Hutton, J.T. and Prescott, J.R. 1994. Further thermoluminescence dates from the dune sequence in the southeast of South Australia. *Quaternary Science Reviews,* 13: 201–207.

Ivanovitch, M. and Harmon, R.S. (eds) 1982. *Uranium series disequilibrium: applications to environmental problems.* Clarendon Press, Oxford: 571 pp.

Jacobberger-Jellison, P.A. 1994. Detection of post-drought environmental conditions in the Tombouctou region. *International Journal of Remote Sensing,* 15(16): 3183–3197.

Jennings, J.N. 1975. Desert dunes and estuarine fill in the Fitzroy Estuary (NW Australia). *Catena,* 2: 215–262.

Jorgensen, D.W. 1992. Use of soils to differentiate dune age and to document spatial variations in eolian activity, northeast Colorado, U.S.A. *Journal of Arid Environments,* 23: 19–34.

Jull, A.J.T., Wilson, A.E., Burr, G.S., Toolin, L.J. and Donahue, D.J. 1992. Measurement of cosmogenic ^{14}C produced by spallation in high-altitude rocks. *Radiocarbon,* 34 (3): 737–744.

Kaufman, A. and Broecker, W.S. 1965. Comparison of ^{230}Th and ^{14}C ages for carbonate minerals from lakes Lahontan and Bonneville. *Journal of Geophysical Research,* 70: 4039–4054.

Klein, J., Gigengack, R., Middleton, R., Sharma, P., Underwood, J.R. and Weeks, R.A. 1986. Revealing histories of exposure using in situ produced ^{26}Al and ^{10}Be in Libyan desert glass. *Radiocarbon,* 28(2A): 547–555.

Kolstrup, E., Grun, R., Mejdahl, V., Packman, S.C. and Wintle, A.G. 1990. Stratigraphy and thermoluminescence dating of Late Glacial cover sand in Denmark. *Journal of Quaternary Science,* 5: 207–224.

Knauss, K.G. and Ku, T.L. 1980. Desert varnish: Potential for age determination via uranium series isotopes. *Journal of Geology,* 88: 95–100.

Knuepfer, P.L.K. 1988. Estimating ages of late Quaternary stream terraces from analysis of weathering rinds and soils. *Bulletin of the Geological Society of America,* 100: 1224–1236.

Ku, T.-L., Bull, W.B., Freeman, S.T. and Knauss, K.G. 1979. ^{230}Th-^{234}U dating of pedogenic carbonates in gravelly desert soils of Vidal Valley, southeastern California. *Bulletin of the Geological Society of America,* 90: 1063–1073.

Kurz, M.D. and Brook, E.J. 1994. Surface exposure dating with cosmogenic nuclides. In C. Beck (ed.), *Dating in exposed and suface contexts.* University of New Mexico Press, Alberquerque: 139–160.

Lao, Y. and Benson, L. 1988. Uranium-series age estimates and paleoclimatic significance of Pleistocene tufas from the Lahontan Basin, California and Nevada. *Quaternary Research*, 30: 165–176.

Laughlin, A.W., Poths, J., Healey, H.A., Reneau, S. and Woldegabriel, G. 1994. Dating Quaternary basalts using the cosmogenic ^{3}He and ^{14}C methods with implications for excess ^{40}Ar. *Geology*, 22: 135–138

Linick, T.W., Damon, P.E., Donahue, D.J. and Jull, A.J.T. 1989. Accelerator mass spectrometry: the new revolution in radiocarbon dating. *Quaternary International*, 1: 1–6.

Livnat, A. and Kronfeld, J. 1985. Paleoclimatic implications of U-series dates for lake sediments and travertines in the Arava rift valley, Israel. *Quaternary Research*, 24: 164–172.

Loendorf, L.L. 1991. Cation-ratio varnish dating and petroglyph chronology in Southeastern Colorado. *Antiquity*, 65: 246–255.

Lundqvist, J. and Mejdahl, V. 1987. Thermoluminescence dating of eolian sediments in central Sweden. GFF, 109: 147–158.

Madole, R.F. 1994. Stratigraphic evidence of desertification in the West-Central Great Plains within the past 1000 years. *Geology*, 22: 483–486.

Madole, R.F. 1995. Spatial and temporal patterns of late Quaternary eolian deposition, eastern Colorado, U.S.A. *Quaternary Science Reviews*, 14: 155–177.

Maxwell, T.A. and Haynes, C.V. Jr 1989. Large-scale, low-angle bedforms (chevrons) in the Selima sand sheet. *Science*, 243: 1179–1182.

McKinney, C.R. 1993. A stratigraphic test of uranium-series dating of tooth enamel. In F. Wendorf, R. Schild and A.E. Close (eds), *Egypt during the last interglacial*. Plenum Press, New York: 218–223.

Metcalfe, S.E. 1995. Holocene environmental change in the Zacapu Basin, Mexico: A diatom-based record. *The Holocene*, 5 (2): 196–208.

Miller, G.H., Wendorf, F., Ernst, R., Schild, R., Close, A.E., Friedman, I., Schwarcz, H.P. 1991. Dating lacustrine episodes in the eastern Sahara by epimerization of isoleucine in ostrich eggshells. *Palaeogeography, Palaeoclimatology, Palaeoecology*, 84: 175–189.

Muhs, D.R. 1985. Age and paleoclimatic significance of Holocene sand dunes in Northeastern Colorado. *Annals of the Association of American Geographers*, 75: 566–582.

Muhs, D.R. and Holliday, V.T. 1995. Evidence of active dune sand on the Great Plains in the 19th century from accounts of early explorers. *Quaternary Research*, 43: 198–208.

Muhs, D.R., Rosholt, J.N. and Bush, C.A. 1989. Uranium trend dating method: principles and application for southern California marine terrace deposits. *Quaternary International*, 1: 19–34.

Nishiizumi, K., Kohl, C.P., Arnold, J.R., Dorn, R.I., Klein, J., Fink, D., Middleton, R. and Lal, D. 1993. Role of in situ cosmogenic nuclides ^{10}Be and ^{26}Al in the study of diverse geomorphic processes. *Earth Surface Processes and Landforms*, 18: 407–425.

Nobbs, M.F. and Dorn, R.I. 1993. New surface exposure ages for petroglyphs from the Olary Province, South Australia. *Archaeology in Oceania*, 28: 18–39.

Nott, J. and Price, D. 1994. Plunge pools and palaeoprecipitation. *Geology*, 22: 1047–1050.

Pachur, H.-J. and Hoelzmann, P. 1991. Paleoclimatic implications of late Quaternary lacustrine sediments in western Nubia, Sudan. *Quaternary Research*, 36: 257–276.

Pachur, H.-J. and Kropelin, S. 1987. Wadi Howar: palaeoclimatic evidence from an extinct river system in the southeastern Sahara. *Science*, 237: 297–300.

Page, K.J., Nanson, G.C. and Price, D.M. 1991. Thermoluminescence chronology of late Quaternary deposition on the riverine plain of southeastern Australia. *Australian Geographer*, 22 (1): 14–23.

Parish, R. 1994. *The influence of feldspar weathering on luminescence signals and the implications for luminescence dating of sediments*. In D.A. Robinson and R.B.G. Williams (eds), *Rock weathering and landform evolution*. Wiley, Chichester: 243–258.

Parks, D.A. and Rendell, H.M. 1992. Thermoluminescence dating and geochemistry of loessic deposits in southeast England. *Journal of Quaternary Science*, 7: 99–107.

Petit-Maire, N. and Riser, J. 1981. Holocene lake deposits and palaeoenvironments in central Sahara, northeastern Mali. *Palaeogeography, Palaeoclimatology, Palaeoecology*, 35: 45–61.

Phillips, F.M., Zreda, M.G., Ku, T.-L., Luo, S., Huang, Q., Elmore, D., Kubik, P.W. and Sharma, P. 1993. ^{230}Th/^{234}U and ^{36}Cl dating of evaporite deposits from the western Qaidam Basin, China: Implications for glacial-period dust export from Central Asia. *Bulletin of the Geological Society of America*, 105: 1606–1616.

Pineda, C.A., Jacobson, L. and Peisach, M. 1988. Ion beam analysis for the determination of cation-ratios as a means of dating Southern African rock varnishes. *Nuclear Instruments and Methods in Physics Research*, B35: 463–466.

Pineda, C.A., Peisach, M., Jacobson, L. and Sampson, C.G. 1990. Cation-ratio differences in rock patina on hornfels and calcedony using thick target PIXE. *Nuclear Instruments and Methods in Physics Research*, B49: 332–335.

Pye, K. and Johnson, R. 1987. Stratigraphy, geochemistry and thermoluminescence ages of lower Mississippi Valley loess. *Earth Surface Processes and Landforms*, 13: 103–124.

Readhead, M.L. 1988. Thermoluminescence dating study of quartz in aeolian sediments from south-eastern Austalia. *Quaternary Science Reviews*, 7: 257–264.

Rendell, H.M. and Townsend, P.D. 1988. Thermoluminescence dating of a 10m loess profile in Pakistan. *Quaternary Science Reviews*, 7: 251–255.

Rendell, H.M., Calderon, T., Perez-Gonzalez, A., Gallardo, J., Millan, A. and Townsend, P.D. 1994. Thermoluminescence and optically stimulated luminescence dating of Spanish dunes. *Quaternary Geochronology*, 13: 429–432.

Riggs, A.C. 1983. Major carbon-14 deficiency in modern snail shells from southern Nevada Springs. *Science*, 224: 59–61.

Ritchie, J.C. 1994. Holocene pollen spectra from Oyo, northwestern Sudan: Problems of interpretation in a hyperarid environment. *The Holocene*, 4(1): 9–15.

Ritchie, J.C., Eyles, C.H. and Haynes, C.V. 1985. Sediment and pollen evidence for an early to mid-Holocene humid period in the eastern Sahara. *Nature*, 314: 352–355.

Roberts, N., Taleb, M., Barker, P., Damnati, B., Icole, M. and Williamson, D. 1993. Timing of the Younger Dryas event in East Africa from lake-level changes. *Nature*, 366: 146–148.

Roper, H.P. 1993. Calcretes in the western desert of Egypt. In U. Thorweihe and H. Schandelmeier (eds), *Geoscientfic research in Northeast Africa*. Balkema, Rotterdam: 635–639.

Sarnthein, M. 1978. Sand deserts during glacial maximum and climatic optimum. *Nature*, 272: 43–46.

Schwarcz, H.P. 1989. Uranium series dating of Quaternary deposits. *Quaternary International*, 1: 7–17.

Schwarcz, H.P. 1994. Current challenges in ESR dating. *Quaternary Geochronology*, 13: 501–605.

Schwarcz, H.P. and Morawska, L. 1993. *Uranium series dating of carbonate from Bir Tafawi and Bir Sahara East*. In F. Wendorf, R. Schild and A.E. Close (eds), *Egypt during the last interglacial*. Plenum Press, New York: 205–217.

Singhvi, A.K., Sharma, Y.P. and Agrawal, D.P. 1982. Thermoluminescence dating of sand dunes in Rajesthan, India. *Nature*, 295: 313–315.

Singhvi, A.K., Deraniyangala, S.U. and Sengupta, D. 1986. Thermoluminescence dating of Quaternary red-sand beds: a case study of coastal dunes in Sri Lanka. *Earth and Planetary Science Letters*, 80: 139–144.

Singhvi, A.K., Bronger, A., Pant, R.K. and Sauer, W. 1987. Thermoluminescence dating and its implications for the chronostratigraphy of loess–palaeosol sequences in the Kashmir Valley (India). *Chemical Geology*, 65: 45–56.

Smith, B.W., Rhodes, E.J., Stokes, S. and Spooner, N.A. 1990. Optical dating of quartz. *Radiation Protection Dosimetry*, 34 (1/4): 75–78.

Stafford, T.W., Jull, A.J.T., Brendel, K., Duhamel, R.C. and Donahue, D. 1987. Study of bone radiocarbon dating accuracy at the University of Arizona NSF accelerator facility for radioisotope analysis. *Radiocarbon*, 29(1): 24–44.

Stafford, T.W., Brendel, K. and Duhamel, R.C. 1988. Radiocarbon ^{13}C and ^{15}N analysis of fossil bone: removal of humates with XAD-2 resin. *Geochimica et Cosmochimica Acta*, 52: 2257–2267.

Stokes, S. 1991. Quartz-based optical dating of Weichselian coversands from the eastern Netherlands. *Geologie en Mijnbouw*, 70: 327–337.

Stokes, S. 1992a. Optical dating of young (modern) sediments using quartz: Results from a selection of depositional environments. *Quaternary Science Reviews*, 11: 153–159.

Stokes, S. 1992b. Optical dating of independently-dated Late Quaternary eolian deposits from the Southern High Plains. *Current Research in the Pleistocene*, 9: 125–129.

Stokes, S. 1993. Optical dating of sediment samples from Bir Tafawi and Bir Sahara East: an initial report. In F. Wendorf, R. Schild and A.E. Close (eds), *Egypt during the last interglacial*. Plenum Press, New York: 227–229.

Stokes, S. 1994. Optical dating of selected late quaternary aeolian sediments from the southwestern United States. Unpublished DPhil thesis, Oxford University.

Stokes, S. and Breed, C.S. 1993. A chronostratigraphic re-evaluation of the Tusayan Dunes, Moenkopi Plateau and Southern Ward Terrace, Northeastern Arizona. In K. Pye (ed.), *The dynamics and environmental context of aeolian sedimentary systems*. Geological Society of London, Special Publication, 72: 75–90.

Stokes, S. and Gaylord, D.R. 1993. Optical dating of Holocene dune sands in the Ferris Dune Field, Wyoming. *Quaternary Research*, 39: 274–281.

Stokes, S. and Rhodes, E.J. 1989. Limiting factors in the optical dating of quartz from young sediments. In *Long and short range limits in luminescence dating*. RLAHA Occasional Publication No. 9: 105–110.

Street, F.A. and Grove, A.T. 1976. Environmental and climatic implications of Late Quaternary lake-level fluctuations in Africa. *Nature*, 261: 385–390.

Street-Perrott, F.A. and Harrison, S.P. 1985. Lake levels and climate reconstruction. In A.D. Hecht (ed.), *Palaeoclimate analysis and modelling*. Wiley, New York: 291–340.

Street-Perrott, F.A. and Perrott, R.A. 1990. Abrupt climatic fluctuations in the tropics: the influence of Atlantic Ocean circulation. *Nature*, 343: 607–612.

Stuvier, M., Long, A., Kra, R.S. and Devine, J.M. (eds), 1993. *Calibration 1993*. Radiocarbon, 35: 244 pp.

Szabo, B.J. and Rosholt, J.N. 1982. Surficial Continental Sediments. In M. Ivanovitch and R.S. Harmon (eds), *Uranium series disequilibrium: applications to environmental problems*. Clarendon Press, Oxford: 246–268.

Szabo, B.J., McHugh, W.P., Schaber, G.G., Haynes, C.V. and Breed, C.S. 1989. Uranium-series dated authigenic carbonates and Acheulian sites in southern Egypt. *Science*, 243: 1053–1056.

Szabo, B.J., Haynes, C.V. and Maxwell, T.A. 1995. Ages of Quaternary pluvial episodes determined by uranium-series and radiocarbon dating of lacustrine deposits of Eastern Sahara. *Palaeogeography, Palaeoclimatology, Palaeoecology*, 113: 227–242.

Thomas, D.S.G. and Shaw, P.A. 1991. 'Relict' desert dune systems: interpretations and problems. *Journal of Arid Environments*, 20: 1–14.

Toomey, R.S., Blum, M.D. and Valastro, S. Jr 1993. Late Quaternary climates and environments of the Edwards Plateau, Texas. *Global and Planetary Change*, 7: 299–320.

Van Devender, T.R., Burgess, T.L., Piper, J.C. and Turner, R.M. 1994. Paleoclimatic implications of Holocene plant remains from the Sierra Bacha, Sonora, Mexico. *Quaternary Research*, 41: 99–108.

Vasconcelos, P.M., Becker, T.A., Renne, P.R. and Brimball, G.H. 1992. Age and duration of weathering by ^{40}K-^{40}Ar and $^{40}Ar/^{39}Ar$ analysis of potassium–manganese oxides. *Science*, 258: 451–455.

Walther, R.C., Manega, P.C., Hay, R. L., Drake, R.E. and Curtis, G.H. 1991. Laser-fusion $^{40}Ar/^{39}Ar$ dating of Bed I, Olduvai Gorge, Tanzania. *Nature*, 354: 145–149.

Wells, S.G., McFadden, L.D. and Schultz, J.D. 1990. Eolian landscape evolution and soil formation in the Chaco Dune Field, Southern Colorado Plateau, New Mexico. *Geomorphology*, 3: 517–546.

Wendorf, F., Schild, R., Close, A.E., Schwarcz, H.P., Miller, G.H., Grun, R., Bluszcz, A., Stokes, S., Morawska, L., Huxtable, J., Lundberg, J., Hill., C.L. and McKinney, C. 1994. A chronology for the Middle and Late Pleistocene wet episodes in the Eastern Sahara. In O. Bar Yosef and R.S. Kra (eds), *Late Quaternary chronology and palaeoclimates of the Eastern Mediterranean*. *Radiocarbon*: 147–168.

Williams, M.A.J., Dunkerley, D.L., DeDecker, P., Kershaw, A.P. and Stokes, T.J. 1993. *Quaternary environments*. Routledge, Chapman & Hall: 330 pp.

Wintle, A.G. 1990. A review of current research on TL dating of loess. *Quaternary Science Reviews*, 9: 385–397.

Wintle, A.G. 1993. Luminescence dating of aeolian sands: an overview. In K. Pye (ed.), *The dynamics and environmental context of aeolian sedimentary systems*. Geological Society of London, Special Publication, 72: 49–58.

Wintle, A.G. and Huntley, D.J. 1982. Thermoluminescence dating of sediments. *Quaternary Science Reviews*, 1: 31–53.

Wintle, A.G., Lancaster, N.A. and Edwards, S.R. 1994. Infrared stimulated luminescence (IRSL) dating of Late-Holocene aeolian sands in the Mojave Desert, California, USA. *The Holocene*, 4 (1): 74–78.

Yaalon, D.H. 1971. *Soil forming processes in time and space*. In D.H. Yaalon (ed.), *Paleopedology*. Israel University Press, Jerusalem: 29–39.

Zoeller, L., Oches, E.A. and McCoy, W.D. 1994. Towards a revised chronostratigraphy of loess in Austria with respect to key sections in the Czech Republic and Hungary. *Quaternary Geochronology*, 13: 465–472.

Zreda, M.G. and Phillips, F.M. 1994. Surface exposure dating by chlorine-36 accumulation. In Beck, C. (ed.), *Dating in exposed and surface contexts*. University of New Mexico Press: 161–184.

28

Human impacts on dryland geomorphic processes

Sarah L. O'Hara

Introduction

The world's drylands have a long history of human occupation and exploitation, although until the twentieth century these often inhospitable regions were sparsely populated (Figure 28.1). The advent of better health care, increased levels of hygiene and improved water resource provision has resulted in a rapid increase in the number of people living in these areas and at present dryland environments are the source of livelihood for a growing percentage of the world's population (UNEP 1992; Table 28.1).

An increasing number of investigations are addressing the problem of both long- and short-term human impacts on dryland environments. Palaeoenvironmental investigations, for example, indicate that many of the important dryland societies that emerged during prehistory had an ability to manipulate natural resources resulting in widespread and often detrimental environmental change (Jacobsen and Adams 1958; Mainguet 1991; van Andel 1991; O'Hara *et al.* 1993a). It is therefore inevitable that recent dramatic increases in population have placed a considerable strain on already scarce resources; land has been lost to urbanisation, there have been significant changes in land use as the need

to provide food and fuel becomes more pressing, and the exploitation of available water resources has been such that few rivers, lakes and aquifers remain untouched. Indeed the direct and indirect consequences of human actions in dryland environments are numerous and few facets of these landscapes remain unchanged. In this chapter some of the different ways in which humans have influenced geomorphic processes in dryland environments will be highlighted, through an examination of four areas of geomorphology: weathering, soils, aeolian and fluvial processes.

Human influences on weathering processes

The processes of weathering in dryland environments have long been of interest to geomorphologists (see Chapters 3 and 4). Although the importance of salt as a weathering agent has only recently been recognised, it is now realised that it is a dominant contributor to weathering in drylands (Doornkamp and Ibrahim 1990) causing the rapid disintegration of material to silt-sized sediment (Goudie 1984). There is a growing body of evidence as to the deleterious impacts of salt weathering, although these are

Arid Zone Geomorphology: Process, Form and Change in Drylands, 2nd edition. Edited by David S. G. Thomas.
© 1997 John Wiley & Sons Ltd.

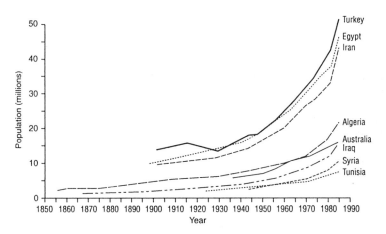

Figure 28.1   *Population growth in selected arid countries since the mid-nineteenth century (after Beaumont 1989)*

Table 28.1   *Arid and semi-arid nations of the world. Population data from the Population Refrence Bureau (1985)*

Percentage of nation arid or semi-arid	Country	Population 1985 (millions)
100	Bahrain, Djibouti, Egypt, Kuwait, Mauritania, Oman, Qatar, Saudi Arabia, Somalia, South Yemen, UAE	75.4
75–99	Afghanistan, Algeria, Australia, Botswana, Cape Verde, Chad, Iran, Iraq, Israel, Jordan, Kenya, Libya, Mali, Morocco, Namibia, Niger, North Yemen, Pakistan, Senegal, Sudan, Syria, Tunisia, Upper Volta	350.0
50–74	Argentina, Ethiopia, Mongolia, South Africa, Turkey	153.1
25–49	Angola, Bolivia, Chile, China, India, Mexico, Tanzania, Togo, USA	2173.6
<25	Benin, Brazil, Canada, Central African Republic, Ecuador, Ghana, Lebanon, Lesotho, Madagascar, Mozambique, Nigeria, Paraguay, Peru, Sri Lanka, USSR, Venezuela, Zambia, Zimbabwe	663.1
	Total	3415.2

mainly in respect to urban locations and focus on its effects on built structures such as houses, roads and pillars (Cooke *et al.* 1982; Doornkamp and Ibrahim 1990). Goudie (1977), for example, noted that increased salinity levels associated with irrigation had resulted in the recently excavated archaeological remains at Mohenjo-Davo in Pakistan being severely degraded.

While many dryland regions such as coastal margins and areas of internal drainage are naturally saline, human activities have served to increase salinity levels and, while no quantitat-

ive figures are available, it is likely that many dryland areas will experience increased rates of salt weathering. It is likely therefore that salt weathering processes will continue to be enhanced by human activities and will ultimately influence a variety of processes including rock disintegration, soil formation and erosion.

Human impacts on dryland soils

Soils in dryland environments are more closely linked with geomorphic processes than they are in more humid regions, partly due to their greater exposure, but also as a result of the nature of many subsurface soil horizons (Cooke *et al.* 1993). Thus soils are highly vulnerable to change, in particular as a result of human activity, be it deliberate or inadvertent.

SALINISATION

Salinisation is a major problem in many dryland environments. Where it is a direct result of anthropogenic activity it is referred to as secondary salinisation. The main way by which humans have increased salinity levels is via hydrological changes resulting in increased groundwater levels. Where groundwater levels rise close to the surface, salts can be drawn into the upper parts of the soil horizon where they accumulate. The clearance of natural vegetation and its replacement by crops and pasture can result in such a process occurring; in southwestern Australia, for example, the removal of native evergreen forest by European settlers has resulted in an estimated 200 000 hectares of land which originally had low saline levels experiencing problems of salinisation (Rhoades 1990; Goudie 1993) and being rendered useless for agricultural purposes (Mackay 1990). A somewhat different salinisation process has been reported from the American southwest, where the planting of salt cedars (*Tamarix* spp.) which tap deep saline groundwaters and extrude salts from their leaves has increased surface salinity levels (Farvar and Milton 1972).

Problems of secondary salinisation, however, are mostly associated with irrigation systems and it is estimated that over 50% of the world's 400 million hectares of irrigated land suffers severe salinisation (Rhoades 1990). Particularly important are the over-application of water to the land which, when coupled with inadequate drainage, can result in waterlogging. The high levels of evaporation experienced in drylands draw salts in to the upper part of the soil column where they accumulate, often forming a salt efflorescence (Figure 28.2). The effects on soil properties have been outlined by Mabbutt (1986) who noted that increased cracking and puffing of salt-affected surfaces significantly altered infiltration and runoff. In localities where salt weathering leads to the dispersal of

clays, a surface crust can form, further reducing infiltration and increasing the likelihood of runoff and erosion.

A similar scenario has been observed in those areas where salt accumulation forms an impervious subsoil layer which can result in the upper layers of the soil becoming waterlogged. Furthermore, the production of fine materials by salt weathering increases the amount of material available for deflation by wind and water and can lead to the development of salt scalds (Mabbutt 1986).

ANTHROPOGENIC IMPACTS ON SOIL EROSION IN DRYLANDS

Soil erosion can be defined as the physical detachment, entrainment and redeposition of soil particles. It is a natural process which is a measure of the erosivity of the physical agent and the erodibility of the soil itself. The natural balance between these two factors has been significantly altered by human activities, and in many parts of the world, including dryland environments, there is clear evidence of anthropogenic-accelerated erosion. One of the main factors leading to increased erosion is the removal of natural vegetation, primarily for agricultural purposes, but also to provide grazing lands, to meet fuel demands, and for construction purposes.

Anthropogenic-accelerated erosion is not a recent phenomenon. Plato, writing in the fourth century BC, commented on the disappearance of forests and denudation of cattle pasture. Other evidence of long-term human-induced erosion has been provided by a number of palaeoenvironmental investigations (van Andel *et al.* 1990). In the Mediterranean, for example, deforestation by the Greeks and Romans was on such a massive scale that in many areas the topsoil has been completely washed away. Likewise, palaeolimnological investigations in central Mexico indicate that there was a significant increase in erosion following the introduction of sedentary agriculture some 3500 years ago (O'Hara *et al.* 1993a; 1994). While it has been argued that changes in climate over the Holocene will have resulted in increased erosion, the findings from the central Mexican investigation indicate that human-induced erosion rates were far more

Figure 28.2 *Salt efflorescence associated with traditional irrigation systems, central Tunisia*

significant than those associated with even major changes in climate (O'Hara *et al.* 1993b). Similarly, investigations in Australia indicate that erosion rates increased significantly following the arrival of Europeans and that sediment yields were some 50–90 times greater in the post-European period that during the Holocene as a whole (Pickard 1994). Clearly humans have had a significant impact on all aspects of the erosion cycle and as such human impacts on soil erosion are now viewed to be one of the most serious problems facing dryland regions at present.

WIND EROSION IN DRYLAND ENVIRONMENTS

Wind erosion is a particular problem where moisture deficits, relatively sparse vegetation cover and thin soils combine to create highly erosive conditions on highly erodible surfaces. Wind erosion was identified in the recent UNEP world desertification survey (UNEP 1992) as affecting up to 39% of the susceptible drylands (dryland areas susceptible to human degradation, which is 40% of world land area). A particularly important component of dryland wind erosion is the entrainment, transport and off-site deposition of fine material or dust (Pye 1987; see Chapter 18 this volume).

Aeolian dust (suspended particles generally smaller than 50 μm) is commonplace and dust storms are a well-known natural hazard in dryland regions (see Chapter 18). Livingstone and Warren (1996) note that on a global scale the quantities of dust in motion are of the same order as the quantities of sediment carried by rivers. The major source of this dust is from the aeolian deflation of sediments and soils (Pye 1987) from the semi-arid areas of the world

(Coudé-Gaussen 1984). Of particular importance are those areas where a combination of wind and water action occurs such as floodplains, alluvial fans, wadis, salt pans and former lake beds (see Table 18.3). Such areas are particularly attractive to human occupation and as a result have experienced considerable disturbance-enhancing deflation activity.

Variations in the intensity and frequency of dust-raising activity occur and can be largely attributed to global climatic change, although anthropogenic factors are playing an increasingly significant role in the production of atmospheric dust (see Chapter 18). The pattern of anthropogenically accelerated aeolian activity closely reflects that of desertification risk (Figure 28.3), the problem being more severe on desert margins where pressures on land and climatic uncertainties are greatest (Cooke and Doornkamp 1990; Chapter 18).

Particularly susceptible to disturbance are those areas where pioneer settlers have attempted to exploit plains areas such as in the United States and the former Soviet Union. Investigations into the role of specific agricultural practices in dust generation (Ley and McTainsh 1994) have noted that increased frequency of ploughing, use of mechanised systems and the actual strategy of planting may significantly disrupt surface cohesion and increase the potential for aeolian dust entrainment. Livestock production may also lead to local devegetation and enhanced susceptibility to wind erosion, although Perkins and Thomas (1993) suggested that this may be insignificant under borehole-centred livestock management systems, due to impacts being spatially confined.

One of the best examples of anthropogenically enhanced dust generation is the 1930s Dust Bowl of the United States (see, for example, Worster 1979) which affected most of the Great Plains, in particular Kansas, southeast Colorado, the Oklahoma Panhandle, the northern part of the Texas Panhandle and northeast New Mexico (Figure 28.4). Agricultural mismanagement and the application

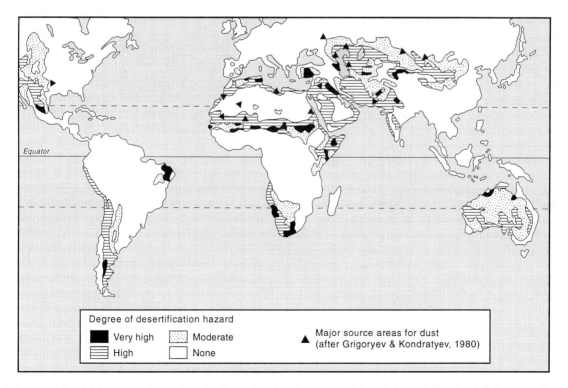

Figure 28.3 *Major areas of desertification hazard and major sources of dust (after UN 1977 and Grigoryev and Kondratyev 1980). Note that desertification hazard in UN (1977) simply equates to the distribution of drylands*

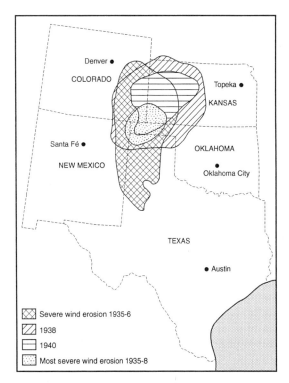

Figure 28.4 *The 'Dust Bowl' region of the United States in the 1930s (after Worster 1979)*

of inappropriate techniques by pioneering farmers who ploughed up the grassy plains for wheat cultivation are blamed for the tragedy of the 'Dirty Thirties', creating deleterious surface conditions that were exploited by the wind during a drought period. The drive of economic and social forces have been highlighted as a major factor in bringing this disaster about, with a culture set on dominating and exploiting the land and its natural resources (Worster 1979). The removal of natural vegetation with little heed for climatic and pedological suitability resulted in a dramatic increase in large-scale wind erosion during the drought years of the 1930s. During the great wind storm of March 1935, for example, the aeolian loads for central Kansas are calculated to have been in excess of 35×10^6 kg km^{-2} (Skinner and Porter 1987). By 1937 the US Soil Conservation Service estimated that 43% of a 25 000 km^2 area in the heart of the Dust Bowl had been seriously damaged by wind erosion.

Despite the devastating social and economic impacts of the Dust Bowl there has been continued dust storm activity throughout the area particularly during periods of drought. In the 1970s, for example, economic and political measures resulted in vast tracts of marginal land being put into dryland wheat cultivation, while the introduction of the Federal Wheat Disaster Assistance Program to compensate farmers for losses due to wind erosion, means that farmers are more likely to plough up marginal lands. In addition the removal of wind-breaks put in place after the 1930s Dust Bowl has allowed new centre-pivot irrigation technology to be accommodated. The resultant increase in dust storm activity clearly highlights the need for prudent land management; a single dust storm in late February 1977 affected an area of nearly 400 000 km^2 (Purvis 1977), while during the period between November 1989 and February 1990 the US Soil Conservation Service reported that two million hectares of land had been damaged by wind (Johnson and Lewis 1994).

Dried lake beds are a major source of dust (see Chapters 14 and 18), and deflation from these surfaces has been exacerbated by human factors. Problem areas are often associated with drainage or diversion of waters from lake systems, as has happened in Mexico City, Owens Valley and the Aral Sea (Goudie 1994; see Chapter 18, this volume). Owens Lake was a relatively small water body (110 km^2) that was located in the southern part of the Owens Valley, California. Early European settlers found it an attractive location and it rapidly became a small but important area for agriculture with the development of irrigation farming.

In the early part of the twentieth century, however, the waters of this area were being secretly purchased by the Los Angeles Water Company (Reisner 1986) who diverted them to Los Angeles to provided water for domestic and industrial use. By 1927 the lake was all but dry leaving the unconsolidated lake sediments exposed to the elements. Entrainment of these fine sediments by strong winter and spring winds has created an environmental problem of immense proportions and sediment derived from this tiny area accounts for 1% of the total dust production in the United States every year. Dust levels during one storm in February

1989, for example, had a concentration of 1861 $\mu g\,m^{-3}$, which is 37 times the health standard of the State of California (Knudson 1991).

Similarly, water diversion schemes developed by the government of the former Soviet Union have had profound and in some cases devastating environmental impacts. In recent years much attention has focused on the Aral Sea which lies in the countries or states of Kazakhstan and Uzbekistan. In 1960 the lake, the fourth largest inland water body in the world, covered an area of 60 000 km^2. Fed by the Amu Darya and Syr Darya rivers the lake was the focal point for a large population and provided an important economic base. The diversion of water to major irrigation schemes such as those along the Karakum canal, resulted in a substantial reduction in water discharge to the Aral Sea which is now over 14 m lower than its 1960 level, resulting in a 40% decrease in surface area and a 60% decrease in volume (Kotlyakov 1991). At the present rate of decline the sea will disappear in 30 years (Lewis 1992). The exposure of over 20 000 km^2 of its former floor has significantly increased dust storm potential and at present an estimated 43 million tonnes of dust are blown from the area each year, which is equivalent to 520 kg of sand and salt per hectare of land (Lewis 1992).

THE IMPACTS OF WATER EROSION

Until recently, soil erosion by water received little attention as it was considered to play only a minor role in the degradation of dryland environments (Perez-Trejo 1992). Recent investigations, however, have highlighted the importance of this process particularly in the context of increased human activity in these environments with an estimated 9% of the worlds susceptible drylands being affected by human-induced water erosion (Thomas and Middleton 1994). A survey of suspended sediment loads of rivers indicates that parts of China, India, western USA, central Russia and the Mediterranean are particularly prone to erosion by water (Walling and Webb 1983; Table 28.2). The clearance of natural vegetation and increased land-use pressure on many drylands has increased the threat of anthropogenic-accelerated water erosion. Regions bordering the Mediterranean Sea, for example, now experience some of the highest erosion rates in the world (Woodward 1995; Table 28.2).

Attempting to establish the relative contribution of natural and anthropogenic agents to water erosion has proved difficult, although estimates have recently been provided by Dedkov and Mozzherin (1992, Figure 28.5), who have compiled a global data base of sediment yield data

Table 28.2 *Sediment loads and yields for selected dryland rivers (data from Walling and Webb 1983; McNeill 1992; Milliman and Syvitski 1992)*

River	Country	Area (10^6 km^2)	Load (10^6 t yr^{-1})	Yield (t km^{-2} yr^{-1})
Semani	Albania	0.0052	22.0	4200
Lamone	Italy	0.0052	12.00	2400
Isser	Algeria	0.0036	6.10	1700
Agrioum	Algeria	0.00064	33.92	5300
Tlata	Morocco	0.00018	39.60	22000
Kasseb	Tunisia	0.0001	0.51	5070
Dali	China	0.0001	2.60	25600
Dali	China	0.0038	6.19	16300
Huang Ho	China	0.7500	1600	2130
Perkeria	Kenya	0.0013	25.4	19520
Orange	S Africa	1.0006	150	150
Nile	Egypt	3.000	111	37
Murray–Darling	Australia	1.000	32	32

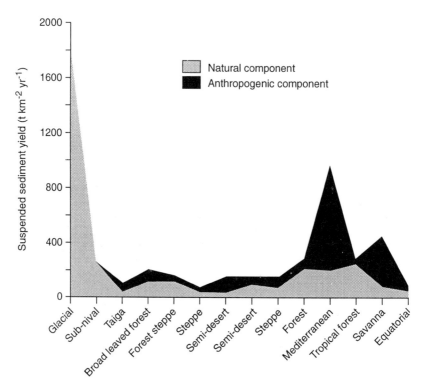

Figure 28.5 *The relative importance of natural and anthropogenic contributions to the suspended sediment yields of mountain river basins in different climate regions (after Dedkov and Mozzheim 1992)*

from 1872 mountain streams. The degree of human disturbance within each basin was based on a simple index of the amount of forested and cultivated land within the catchment. Based on these data they concluded that anthropogenic accelerated erosion was significant in many areas, but no more so that the Mediterranean where it accounted for over 70% of erosion by water (Figure 28.5).

These finding are certainly supported by more site-specific investigations. In many dryland countries the rapid increase in population since the late nineteenth century has significantly increased land-use pressure resulting in the loss of vegetation cover and widespread gullying and degradation of agricultural land (Figure 28.6), with intense grazing around settlements becoming extremely problematic (Wellens and Millington 1990). In Albania, 80% of the land is threatened by water erosion due to the combination of unfavourable natural conditions and poor land management practices (Zachar 1982).

The governments of a number of Mediterranean countries have implemented various policies aimed at increasing the use of machinery and decreasing the number of people employed in the agricultural sector. Morgan (1995) noted that countries that had adopted such strategies were experiencing accelerated rates of erosion by water.

While increased land-use pressure has caused problems of erosion, so has abandonment of agricultural land. A significant increase in erosion has been reported in the Haraz Mountains of the Yemen, where abandonment of the terraced slopes as a result of successive drought episodes and migration to neighbouring oil-rich states has resulted in the collapse of the terrace system. Vogel (1990) reported that erosion rates in this area are now as high as 1 to 3 cm yr^{-1}. Increased soil erosion as a result of terrace abandonment has also been reported in other dryland contexts (e.g. Millington 1989; Tiffen *et al.* 1994).

Figure 28.6 *Deeply eroded landscapes in Swaziland as a result of increased agricultural activity since the 1800s (photograph courtesy of D.S.G. Thomas)*

Human impacts on dune processes

In many dryland areas there is sufficient rain to support vegetation which can result in sand dunes being stabilised. This is particularly true of linear dune systems, such as those of the southwest Kalahari (Thomas and Shaw 1991; Livingstone and Thomas 1994; Wiggs *et al.* 1995; Chapter 7). Stabilised dune systems have proved attractive for human exploitation as they can often support a richer vegetation than other terrain types (Pye and Tsoar 1990) and although dune soils are frequently skeletal they are more easily cultivated and are less prone to salinisation than heavier soils in low-lying areas.

Increased exploitation of dune systems can degrade vegetation cover, eliminating the protection afforded to the sandy surface and as such increase the erodibility of the system. Furthermore, it can change the nature of the wind structure at the dune/air interface, thereby increasing its erosivity. The situation is further exacerbated as one of the effects of vegetation loss is to increase moisture loss further enhancing sediment transport. Thus any change in vegetation cover as a result of

anthropogenic factors such as grazing, agriculture and burning will have the net effect of destabilising the system. An investigation of the effects of fire on dune systems in the southwest Kalahari, for example, reported a three-fold increase in dune activity in the immediate post-burn period (Wiggs *et al.* 1994).

An excellent example of the impact of humans on dune formation and activity is provided by Tsoar and Møller (1986) who used a series of aerial photographs, dating from the 1940s onwards, to document changing land-use pressure on an area of linear dunes in the Sinai and Negev deserts. Periods of intense grazing over this period resulted in rolling linear ridges changing to linear braided and sharp-crested linear dunes. The development of the latter dune type was considered important as they indicate that they are affected by the two main wind directions and not the strongest winds, as is the case with vegetated dune systems (see Chapter 17 for further discussion).

Aerial photographs were also used by Khalaf (1989) to determine the impact that humans have had on dune formation in Kuwait. The country, which experienced a 10-fold increase in population between 1950 and 1985, is now heavy reliant on resources found within its desert environment which has lead to widespread degradation. A comparison of satellite photography from 1977 (MSS Landsat) and 1986 (SPOT) indicated that there had been a 35 km extension in the southern limit of mobile sand sheets. Furthermore, extensive areas of the 'rugged-vegetation sand-sheet ecosystem' had been degraded, resulting in soil loss and increased deflation of the sand sheets causing the morphology to change from domed sand accumulations to rippled sand sheets and granule ripples.

There are numerous examples of the impacts of increased land-use pressure in dryland environments. In Rajasthan, northwest India, the rapid increase in population since the 1940s has placed considerable strain on resources. Particularly important has been the loss of land for grazing with more and more livestock being grazed on marginal lands which have become susceptible to

wind erosion (Kumar and Bhandari 1993), and this has turned formerly productive agricultural lands into areas susceptible to aeolian sand transport and soil degradation (Johnson and Lewis 1994).

Dryland environments are susceptible to even relatively low levels of interference and the increased use of animals and vehicles for transport is leading to considerable erosion in some areas and has been commented on by a number of researchers (Wilshire 1980; Marston 1986; Khalaf 1989). The responses include accelerated deflation due to disruption of vegetation cover, and armour layers of coarse material or surface crusts as well as changes in microtopography. Sheridan (1979) noted that in north Africa tracks made by General Patton's tanks in the 1940s were still clearly visible some 35 years later.

Modifications of dune systems can, however, be even more dramatic. On the southern margins of the Karakum desert in Turkmenistan,

Figure 28.7 *Linear dunes in southern Turkmenistan that have been levelled using heavy machinery in order to increase the amount of land available to cultivation. Agriculture is only possible with irrigation, and large tracts of land are rapidly becoming severely salinised*

for example, linear dunes are being levelled to increase the area of land available for agriculture (Figure 28.7). The process generally involves the use of heavy machinery and the reclaimed land is used to grow a variety of crops under irrigation. Unfortunately these systems have been poorly managed and land is being rapidly abandoned, creating problems with windblown sand.

Equally destructive is the increasing level of quarrying for sands and gravels for industrial and construction purposes. In Kuwait one of the main commercial activities in the desert is the quarrying and processing of material for use in urban development. Clearance of vegetation and the selective removal of coarse material has resulted in an increase in fines available for deflation (Khalaf 1989).

Increased human use of drylands has not only increased the problem with wind blowing sands, but has also meant that settlements have been located where sands are naturally mobile, affecting agriculture, transport networks and industrial installations as well as the health and well-being of the population. A variety of techniques have been adopted to control blowing sands (Watson 1990), which range from placing barriers and fences in the direction of moving sands (Figure 28.8) to treatment of the sand surface with a range of chemicals. In central Asia and Saudi Arabia asphalt has been used to stabilise sands that are adjacent to roadways (Petrov 1983; Watson 1990). Other techniques have also included increasing the grain size of the surface material causing a protective lag to form on the surface hence preventing deflation. Although such measures are still on a relatively small scale, the control of windblown sands is becoming more common and could have a significant impact on aeolian processes in areas where settlements are found.

Human impacts on dryland river systems

The importance of water in shaping the geography of human settlements, not only in terms of their location, but their political, economical and social evolution, is well-documented (e.g. Wittfogel 1958; Butzer 1976; Cosgrove and Petts 1990). In dryland nations an ability to manipulate and exploit scarce water resources has been especially important in the development of these regions. Initially such modifications were to increase agricultural potential by irrigating land away from natural watercourses, and ancient irrigation systems can be found

Figure 28.8 *Vegetation fences used to stabilise dunes in central Tunisia*

throughout the arid realm, many of which continue to flourish. As already mentioned there has been a substantial increase in the amount of land under irrigation since the 1940s, and it is not surprising, therefore, that the most widespread association of human activity with the hydrological balance, relief, slope and stream network has been through the operation of irrigation systems (Smith 1969).

Humans can have both a direct and indirect influence on dryland rivers. One of the most important indirect influences comes from changing land use within a drainage catchment. The removal of vegetation, for example, can have a profound impact on catchment hydrology as it plays a pivotal role in determining the timing and amount of runoff and influences both sediment load and discharge (Graf 1988).

Vegetation is also important in areas immediately adjacent to the river channel itself, influencing both processes operating in the channel and offering protection against bank erosion. Graf (1978) noted that major throughflow channels in the Colorado river system experienced a 27% reduction in width following the establishment of tamarisk along the banks of the river in the 1930s. Equally as much, the removal of riparian vegetation can increase the rate of erosion, firstly by removing the root systems but also by temporarily increasing the amount of water flowing through the system (Graf 1985).

The impacts of grazing on catchment processes have also been the subject of much investigation. The development of extensive arroyo systems in the American southwest in the latter parts of the 1800s has been attributed, in part, to the introduction of cattle by Europeans (for extensive reviews see Cooke and Reeves 1976 and Graf 1983), although there remains considerable debate as to the exact nature of its effects. Lusby *et al.* (1971) undertook a long-term study on the effects of grazing using a series of paired-catchment experiments. They reported that grazed plots experienced an increase in sediment yield of about 50% while runoff was 43% greater than that on a similar, adjacent ungrazed area.

There is little published work on the impacts of agriculture on fluvial processes in dryland environments (Graf 1988), but the expansion of irrigation agriculture and the emplacement of artificial drainage networks will undoubtedly exert some influence on fluvial systems. One example is

provided by Graf (1988) who describes the effects of a small-scale irrigation scheme on Queens Creek, southern Arizona, noting that channelisation of the river and construction of levees prevented the normally dispersed water flow from flowing across fields intended for agriculture, thus changing the drainage pattern of the system.

THE IMPACT OF URBANISATION ON RIVERS

The dramatic increase in the number and size of settlements within drylands linked to population growth (Figure 28.1) will undoubtedly have had a profound influence on fluvial processes. Much of the evidence available to date has been drawn from experiences in more humid environments (e.g. Wolman 1967), although it is likely that certain process will be similar. During the period of construction when natural vegetation is removed and the surface is disrupted by the use of heavy machinery, increased erosion may occur; this is most likely to be a result of wind erosion with water erosion being restricted to rare rainfall events.

Once completed urbanised areas encompass large areas of impervious surfaces which are markedly different from the often highly porous material that characterises many dryland regions and this can result in a significant increase in runoff generation with subsequent increase in channel flow during rainfall events (Leopold 1968, Figure 28.9). Changes in the channel network are also prevalent with many smaller streams being obliterated, although the overall system is extended by the incorporation of drains, channels and conduits into the drainage network. Graf (1977) found that in a 7.7 km² drainage basin with 25% of the area urbanised, the drainage density increased by 50%. The total length increased dramatically as did the number of links in the network. The impact of such changes on river systems is to increase the volume of runoff, decrease the lag time between precipitation and runoff and to increase the flashy nature of the flow.

INTENTIONAL MODIFICATIONS OF DRYLAND RIVERS

Rivers can be modified in a number of ways to suit the perceived requirements of humans. The

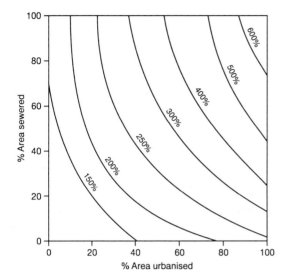

Figure 28.9 *The impact of urbanisation on the mean annual flood from a drainage basin 2.6 km² in extent (after Leopold 1968)*

diversion of water from natural courses for the purpose of irrigation, for example, goes back some 7000 years (Heathcote 1983). The earliest known dam to be built was the Sadd el Kafara (Dam of the Pagans) constructed *c*. 4800 years ago near modern Cairo. Elsewhere in north Africa there are remains of dams built during the Roman period and used to store storm waters in ephemeral wadi systems. Since the late 1800s there has been a dramatic increase in the scale and number of projects which have a direct impact on fluvial systems and today few major dryland rivers remain untouched. The impacts of such modifications are numerous and have been the subject of considerable interest from a wide range of disciplines. The importance of drylands rivers and the significant level to which many of these systems have been modified has prompted a number of in-depth studies of these systems. Oft-quoted, for example, is the Colorado River, one of the most regulated and legislated rivers in the world (e.g. Graf 1985; Beaumont 1989; Petts 1994).

From a geomorphological perspective the effect of damming a river and creating a reservoir is to raise local base-level so that it is contiguous with the surface of the impounded water. Fluctuations in the level of water in the reservoir will result in continual base-level

change and may cause instability in the area immediately upstream. The creation of a still standing water body causes deposition of material previously transported by the river with accumulation of sediment both within the reservoir and in the immediate upstream region (Borland 1971; Figure 28.10), although upstream influences seem to be limited to only a few kilometres (Leopold and Bull 1979).

The effect of increased sedimentation is to reduce channel capacities thereby increasing the likelihood of flooding, causing channel instability as reduced channel gradients promote lateral migration (Schumm 1977). As coarse material is deposited first there is a tendency for braiding and the formation of unstable banks. The effect of the creation of a water body can also change slope stability by increasing porewater pressure; subsequent fluctuation in the level of the reservoir can thus trigger landslips (Graf 1985). Where the original gradient of the river is low the upstream impacts of a dam can be quite significant, as in the case of the Elephant Butte Reservoir situated on the Rio Grande River, New Mexico. In this instance the reservoir had a considerable impact which extended several kilometres upstream, making it necessary to construct levees to prevent flooding of the surrounding agricultural land.

The downstream impacts from a dam are more complex and less well-understood. Several

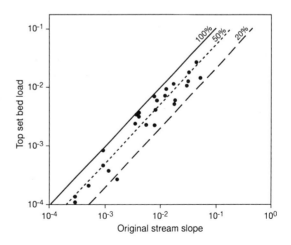

Figure 28.10 *The effect of dam construction on upstream channel gradients in arid and semi-arid western United States. The lines indicate percentage reduction in gradient (after Borland 1971)*

important studies, however, have considered the implications for dryland rivers (see, for example, American Society of Civil Engineers 1978). Dams influence downstream geomorphic processes because they impose radical changes on flow regime as well as acting as a sediment trap. The flashy, intermittent flow that often characterises dryland rivers is frequently altered with peak flow reduced and low flow increased resulting in a more consistent flow as in the Colorado (Graf 1985) and the Murray (Jacobs 1990) rivers.

As a reservoir acts as a huge sediment trap there is a reduction in the amount of sediment being carried in the river immediately downstream of the dam. This was observed on the Glen Canyon River where the amount of sediment transported by the river dropped from 1500 parts per million prior to construction of the dam to a mere 7 p.p.m. afterwards (Graf 1988). More spectacular was the impact of the Aswan Dam, which reduced the amount of sediment carried by the Nile River to 8% of its former load (Goudie 1994). The loss of sediment to coastal regions was highlighted by Kassas (1972) who noted that the Nile Delta was being starved of essential building material. It is estimated that by the year 2100 the combination of sea-level rise and a cessation in the build-up of the delta will result in the sea extending 20 to 30 km inland in the area between Alexander and Port Said (Milliman *et al.* 1989).

Although the Nile has lost much of its sediment load it still retains the potential to transport sediment. The river erodes material immediately downstream of the dam which has resulted in a dramatic reduction in its channel gradient (Figure 28.11). The distance of this effect does vary depending on the nature of the river and the size and style of the dam although it can be quite considerable. The Sariyar Dam in Turkey, for example, influences downstream processes for over 300 km (Simon and Senturk 1977).

Erosion can lead to the preferential removal of fines resulting in a build-up of coarse material changing the nature of the river by armouring (Little and Mayer 1972). The extent of armouring can vary but is usually only a few kilometres, although Williams and Wolman (1984) reported effects over 100 km below the Hoover Dam on the Colorado River. The

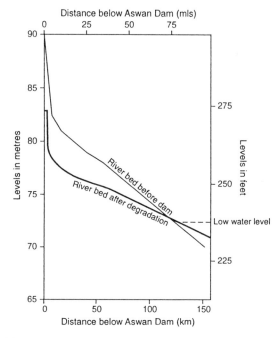

Figure 28.11   *The reduction in the gradient of the Nile River following the construction of the Aswan Dam (after Shalash 1974)*

combination of changing flow regimes and reduced sediment load can further alter channel forms in the area immediately downstream of a dam. Below large structures flows can be sufficiently high to remove even coarse material, but there is relatively little sediment deposited during low-flow events and thus many river features such as mid and channel bars are removed and not replaced.

Many major rivers have a series of regulatory features and these often interrelate with one another. The Murray River of southern Australia has been modified by humans since Europeans first arrived on the Australian continent over 200 years ago. Rising in the Eastern Highlands the river traverses its way 2500 km across the continent before flowing out into the Southern Ocean. In the early stages of European colonisation the Murray and its tributaries served as a major transport route and in wetter periods over 5000 km of the river system were navigable (Jacobs 1990). Flow in the river basin was influenced not only by climatic conditions in the mountains but also by strength and position of the monsoon, resulting

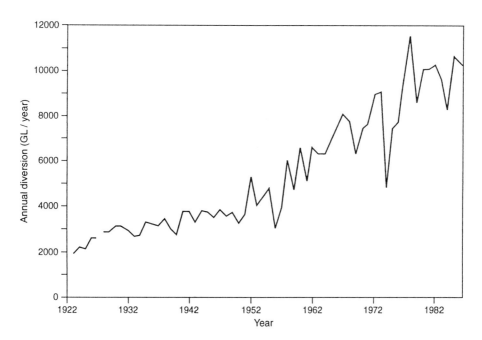

Figure 28.12   *Annual diversions in the Murray–Darling Basin from the 1920s (after Close 1990)*

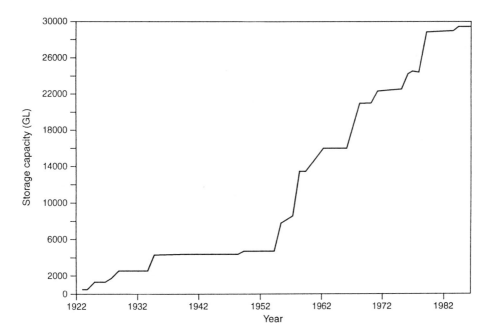

Figure 28.13   *The increase in the total water storage capacity in the Murray–Darling Basin since the 1920s (after Close 1990)*

in flow conditions fluctuating markedly from year to year. Failure of the monsoon could lead to the river drying up in places as was seen early in 1995.

A system of locks and weirs was installed to improve navigation on the river by increasing the water level upstream. The siting of these instalments was such that the pooling effect of one weir extended upstream to the next. While the weirs do not actually affect the flow of the river they have increased the water level and also maintained flow in many previously intermittent smaller tributaries. Increases in channel width have also been noted causing a reduction in flow velocity.

The situation has been further exacerbated by the diversion of water from the Murray–Darling catchment which has increased spectacularly since Australians began diverting water in the 1870s (Figure 28.12). At present of the order of 10 000 to 11 000 gigalitres (1 gigalitre = 1 000 000 000 litres) are taken from the system, much of which is used to irrigate over 700 000 hectares of agricultural land. Further impacts on the flow regime have resulted from the construction of over 90 major storage facilities in the catchment (Figure 28.13) which have influenced the river in three ways. First by redistributing flow within the season; second by reducing flood events and increasing flows in drought; and third by ensuring sufficient water supply for periods of peak demand it has increased the amount of water extracted from the system.

The impact of diversions and dams have been considerable, especially at the mouth of the river where it enters the Southern Ocean. Under natural conditions there was almost always some flow out of the mouth, but current flows are now only a third of previous levels and periods of little flow are a common occurrence. In 1981 the mouth of the river silted up completely and a new channel was only incised when the river subsequently flooded (Close 1990).

The impacts of anthropogenic-accelerated global warming

In recent years there has been growing concern as to the impacts that global warming will have on dryland environments (Goudie 1989; 1994;

Nash and Gleick 1991; Muhs and Maat 1993). A shift towards even drier conditions in some regions has been predicted by a number of General Circulation Models, and will have significant implications on geomorphic processes that operate in such environments. Muhs and Maat (1993), for example, concluded that the predicted decrease in effective precipitation (precipitation minus evaporation) for the Great Plains area of the United States would reduce vegetation cover to such an extent that large areas of now stable and semi-stable dune and sand fields would be reactivated (Figure 28.14). Findings from an extensive study of desertification in the Mediterranean have indicated that the predicted increase in temperature over the next 50 years, together with an estimated decrease in rainfall of between 11 and 20%, will result in large areas of the southern Mediterranean being more susceptible to natural erosion processes over the coming decades (MEDALUS 1993).

Changes in the hydrological cycle are also to be expected, although the complex interrelationships between climate, vegetation and hydrology make it difficult to determine the exact nature of such changes. Revelle and Waggoner (1983) suggest that a 2°C increase in temperature together with a 10% decrease in precipitation will result in a 76% decrease in the water supply of the Rio Grande River and a 40% decrease in the upper reaches of the Colorado River. However, as Goudie (1990) points out, this reduction may be partly or wholly offset by increased runoff brought about by a reduction in evapotranspiration due to greater carbon dioxide levels. Whether one would outweigh the other is difficult to determine but clearly there would be changes in the timing and scale of runoff process in these and other dryland drainage basins. Clearly, fluctuations in the climate have enormous bearing on processes that operate within dryland environments and some indication of the scale and extent of these impacts may be drawn from the numerous investigations that have considered the impact of late Quaternary climatic change on geomorphic processes (Knox 1972; McDowell 1983; O'Hara and Campbell 1993). It is unlikely, however, that the impacts of future climate change will be significant when compared with those that are a direct result of human actions (Goudie 1994).

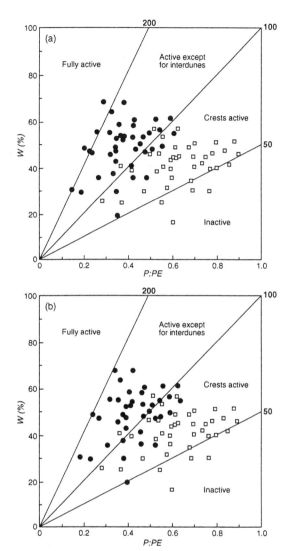

Figure 28.14 *Percentage of time that wind is above the threshold velocity for sand (W) vs. ratio of mean annual precipitation to evaporation (P:PE) under present climate conditions and a greenhouse climate in the great plains, USA, using (a) the GISS GCM with a 20% increase in W and (b) the GFDL GCM with a 20% increase in activity (after Muhs and Maat 1993)*

Conclusion

Humans are having an ever increasing influence on geomorphic processes that operate in dryland environments. Despite increased concern as to the nature and extent of these impacts, there is still very little information available and

our understanding of cause and effect is complicated by the fact that it is difficult to disentangle those changes bought about due to natural and anthropogenic climate change and human activities. It is evident, however, that many aspects of drylands geomorphology are being changed by human activities, be they direct or indirect causes.

References

American Society of Civil Engineers 1978. *Environmental effects of large dams*. ASCE, New York.

Beaumont, P. 1989. *Drylands: environmental management and development*. Routledge, London.

Borland, W.M. 1971. Reservoir sedimentation. In H.W. Shen (ed.), *River Mechanics*. Shen, Fort Collins: 29/1–29/38.

Butzer, K.W. 1976. *Early hydraulic civilizations. A study in cultural ecology*. University of Chicago Press.

Close, A. 1990. The impact of man on the natural flow regime. In N. Mackay and D. Eastburn (eds), *The Murray*. Murray–Darling Basin Commission, Australia.

Cooke, R.U. and Doornkamp, J.C. 1990. *Geomorphology and environmental management*. Clarendon Press, Oxford.

Cooke, R.U. and Reeves, R.W. 1976. *Arroyos and environmental change in the American South-West*. Clarendon Press, Oxford.

Cooke, R.U., Brunsden, D., Doornkamp, J.C. and Jones, D.K.C. 1982. *Urban geomorphology in drylands*. Oxford University Press, Oxford.

Cooke, R.U., Warren, A. and Goudie, A.S. 1993. *Desert geomorphology*. UCL Press, London.

Cosgrove, D. and Petts, G. 1990. *Water engineering and landscape: water control and landscape transformation in the modern period*. Belhaven Press, London.

Coudé-Gaussen, G. 1984 Le cycle des poussieres eoliennes desertiques actuelles et la sedimentation des loess peridesertiques quaternaires. *ELF Aquitaine*, Bulletin 8: 167–182.

Dedkov, A.P. and Mozzheim, V.I. 1992. Erosion and sediment yields in mountain regions of the world. In D.E. Walling, T.R. Davies and B. Hasholt (eds), *Erosion, flows and environment in mountain regions*. Proceedings of the Chengdo Symposium, July, 1992. International Association of Hydrological Sciences, Publication, 209: 29–36.

Doornkamp. J.C. and Ibrahim, H.A.M. 1990. Salt weathering. *Progress in Physical Geography*, x: 335–348.

Farvar, M.T. and Milton, J.P. (eds) 1972. *The careless technology: ecology and international development*. Natural History Press, New York.

Goudie, A.S. 1977. Sodium sulphate weathering and the disintegration of Mehenjo-Daro, Pakistan. *Earth Surface Processes*, 2: 75–86.

Goudie, A.S. 1984. Salt efflorescence and salt weathering in the Jurza Valley, Karakoram Mountains, Pakistan. In K.J. Miller (ed.), *The International Karakoram Project* (2 vols). Cambridge University Press, Cambridge: 607–615.

Goudie, A.S. 1989. The global geomorphological future. *Zeitschrift für Geomorphologie*, Supplementband, 79: 51–62.

Goudie, A.S. 1993. *The human impact on the environment*, 4th edn. Blackwell, Oxford.

Goudie, A.S. 1994. Deserts in a warmer world. In A.C. Millington and K. Pye (eds), *Environmental change in drylands: biogeographical and geomorphological perspectives*. Wiley, Chichester: 1–24.

Graf, W.L. 1975. The impact of urbanization on fluvial geomorphology. *Water Resources Research*, 11: 690–692.

Graf, W.L. 1977. Network characteristics of suburbanizing streams. *Water Resources Research*, 13: 459–463.

Graf, W.L. 1978. Fluvial adjustments to the spread of tamarisk in the Colorado Plateau region. *Bulletin of the Geological Society of America*, 89: 1491–1501.

Graf, W.L. 1983. The arrays problem. Palaeohydrology and palaeohydraulics in the short term. In K.J. Gregory (ed.) *Background to palaeohydrology*. Wiley, Chichester.

Graf, W.L. 1985. *The Colorado River: instability and basin management*. Association of American Geographers, Washington, DC.

Graf, W.L. 1988. *Fluvial processes and dryland rivers*. Springer-Verlag, Berlin.

Grigoryev, A.A. and Kondratyev, K.J. 1980. Atmospheric dust observed from space. *WMO Bulletin*, 3–9.

Heathcote, R.L. 1983. *The arid lands. Their uses and abuses*. Longman, Harlow.

Jacobs, T. 1990. Regulation and management of the River Murray. In N. Mackay and D. Eastburn (eds), *The Murray*. Murray–Darling Basin Commission, Australia.

Jacobsen, T. and Adams, R.M. 1958. Salt and silt in ancient Mesopotamian agriculture. *Science*, 128: 1251–1258.

Johnson, D.L. and Lewis, L.A. 1994. *Land degradation: creation and destruction*. Blackwell, Oxford.

Kassas, M. 1972. Impact of river control schemes on the shorelines of the Nile Delta. In M.T. Farvar and J.P. Milton (eds), *The careless technology: ecology and international development*. Natural History Press, New York: 179–188.

Khalaf, F.I. 1989. Desertification and aeolian processes in the Kuwait Desert. *Journal of Arid Environments*, 16: 125–145.

Knox, J.C. 1972. Valley alluviation in southwestern Wisconsin. *Annual of the Association of American Geographers*, 62: 401–410.

Knox, J.C. 1983. Response of river systems to Holocene climate. In H.E. Write (ed.), *Late Quaternary environments of the United States*. University of Minnesota Press, Minneapolis: 26–41.

Knudson, T. 1991. Mountain lake turns to desert. *Sacramento Bee*, 10 June 1991.

Kotlyakov, V.M. 1991. The Aral Sea basin. *Environment*, 33: 36–38.

Kumar, M. and Bhandari, M.M. 1993. Human use of sand dune ecosystems in the semi-arid zone of the Rajastan Desert, India. *Land Degradation and Rehabilitation*, 4: 21–36.

Leopold, L.B. 1968. *Hydrology for urban land planning: a guidebook on the hydrological effects of urban land use*. US Geological Survey, Professional Paper, 689-D.

Leopold, L.B. and Bull, W.B. 1979. Base level aggradation and grade. *Proceedings of the American Philosophical Society*, 123: 168–202.

Lewis, R.A. (ed.) 1992. *Geographic perspective on Soviet Central Asia*. Routledge, New York.

Ley, J. and McTainsh, G. 1994. Physical characteristics of wind blown sediments from field studies in south eastern Australia. In *Response of eolian processes to climate change*. Desert Research Institute, Zzyzx, California, Occasional Paper 2.

Little, W.C. and Mayer, P.G. 1972. *The role of sediment gradation on channel armouring*. Georgia Institute of Technology, School of Engineering. Unnumbered Publications.

Livingstone, I. and Thomas, D.S.G. 1994. Models of dune activity and their palaeoenvironmental significance: an evaluation with reference to southern African examples. In K. Pye (ed.), *The dynamics and context of aeolian sedimentary systems*. Geological Society of London, Special Publication, 112: 91–101.

Livingstone, I. and Warren, A. 1996. *Aeolian geomorphology: an introduction*. Longman, London.

Lusby, G.C., Reid, V.H. and Knipe, O.D. 1971. *Effects of grazing on the hydrology and biology of the Badger Wash Basin in Western Colorado, 1953–1966*. US Geological Survey, Water Supply Paper, 1532-D.

Mabbutt, J.A. 1986. Desertification in Australia. In *Arid land development and the combat against desertification: an integrated approach*. UNEP, Moscow: 101–112.

Mackay, N. 1990. Understanding the Murray. In N. Mackay and D. Eastburn (eds), *The Murray*. Murray–Darling Basin Commission, Australia.

Mainguet, M. 1991. *Desertification: natural background and human mismanagement*. Springer-Verlag, Berlin.

Marston, R.A. 1986. Manoeuvre-caused wind erosion impacts, South Central New Mexico. In

W.G. Nickling (ed.), *Aeolian geomorphology*. Allen & Unwin, Boston: 273–290.

McDowell, P.F. 1983. Evidence of stream response to Holocene climatic change in a small Wisconsin watershed. *Quaternary Research*, 19: 199–216.

McNeill, J.R. 1992. *The mountains of the Mediterranean world: An environmental history*. CUP, Cambridge.

MEDALUS 1993 *Mediterranean desertification and land use*. Executive summary abstracted from the report of the first phase of the MEDALUS project. Commission of the European Committee Director General XII for Science, Research and Development.

Milliman, J.D. and Syvitski, J.P.M. 1992. Geomorphic and tectronic controls of sediment dicharge to the ocean: the importance of small mountain rivers. *Journal of Geology*, 100: 525–544.

Milliman, J.D., Broadis, J.M. and Gable, F. 1989. Environmental and economic implications and rising sea level and subsiding deltas. The Nile and Bengal examples. *Ambio*, 18: 340–345.

Millington, A.C. 1989. African soil erosion: nature undone and the limits of technology. *Land Degradation and Rehabilitation*, 1: 279–290.

Morgan, K.P.C. 1995. *Soil erosion and conservation*. Longman, London.

Muhs, D.R. and Maat, P.B. 1993. The potential response of eolian sands to greenhouse warming and precipitation reduction on the Great Plains of the USA. *Journal of Arid Environments*, 25: 351–361.

Nash, L.L. and Gleick, P.H. 1991. Sensitivity of streamflow in the Colorado Basin to climatic changes. *Journal of Hydrology*, 125: 221–241.

O'Hara, S.L. and Campbell, I.A. 1993. Holocene geomorphology and stratigraphy of the Lower Falcon Valley, Dinosaur Provincial Park, Alberta, Canada. *Canadian Journal of Earth Sciences*, 30: 1846–1852.

O'Hara, S.L., Street-Perrott, F.A. and Burt, T.P. 1993a. Accelerated soil erosion around a Mexican highland lake caused by prehispanic agriculture. *Nature*, 363: 48–51.

O'Hara, S.L., Street-Perrott, F.A. and Burt, T.P. 1993b. Soil erosion and climate: Reply to Vita-Finzi. *Nature*, 364: 197.

O'Hara, S.L., Metcalfe, S.E. and Street-Perrott, F.A. 1994. On the arid margin: the relationship between climate, humans and the environment. A review of evidence from the highlands of central Mexico. *Chemosphere*, 29: 965–981.

Perez-Trejo, F. 1992. Desertification and land degradation in the European Mediterranean. Prepared by F. Perez-Trejo for the European Commission EPOCH programme. Director General XII. Contract no. EPOC-0036-GB.

Perkins, J.S. and Thomas, D.S.G. 1993. Environmental responses and sensitivity to permanent cattle ranching, semi-arid western central Botswana. In D.S.G. Thomas and R.J. Allison (eds), *Landscape sensitivity*. Wiley, Chichester: 273–286.

Petrov. V.E. 1983. Mobile sand fixation in the arid zone of the RFSFR. *Problems of Desert Environments*, 5: 67–70.

Petts, G. 1994. Large-scale river regulations. In N. Roberts (ed.), *The changing global environment*. Blackwell, Oxford: 262–284.

Pickard, J. 1994. Post-European changes in creeks of semi- and rangelands, 'Polpah Station', New South Wales. In A.C. Millington and K. Pye (eds), *Environmental change in drylands: biogeographical and geomorphological perspectives*. Wiley, Chichester: 271–284.

Population Reference Bureau 1985. *World population data sheet 1985*. Washington D.C.

Purvis, J.C. 1977. *Satellite photos help in dust episode in South Carolina*. Information note 77/8. US National Weather Service. Natural Environment Satellite Service, Satellite Applications.

Pye, K. 1987. *Aeolian dust and dust deposits*. Academic Press, London.

Pye, K. and Tsoar, H. 1990. *Aeolian sands and sand dunes*. Unwin Hyman, London.

Reisner, M. 1986. *Cadillac Desert: The American West and its disappearing water*. Viking Penguin, London.

Revelle, R.R. and Waggoner, P.E. 1983. Effect of carbon dioxide-induced climatic change on water supplies in the western United States. In *Carbon Dioxide Assessment Committee, changing climate*. National Academy Press, Washington, DC: 419–432.

Rhoades, J.D. 1990. Soil salinity — causes and controls. In A.S. Goudie (ed.), *Techniques for desert reclamation*. Wiley, Chichester: 109–134.

Schumm, S.A. 1977. *The fluvial system*. Wiley, New York.

Simons, D.B. and Senturk, F. 1977. *Sediment transport technology*. Water Resources Publications, Littleton.

Shalash, S. 1974. Facts about degradation. Unpublished report, Department of Hydrology, Egyption Ministry of Irrigation, Cairo.

Sheridan, D. 1979. *Off-road vehicles on public land*. US Council on Environmental Quality, Washington, DC.

Skinner, B.J. and Porter, S.C. 1987. *Physical geology*. Wiley, New York.

Smith, C.T. 1969. The drainage basin as an historical basis for human activity. In R.J. Corley (ed.), *Water, earth and man*. Methnen, London: 101–110.

Thomas, D.S.G. and Middleton, N.J. 1994. *Desertification: exploding the myth*. Wiley, Chichester.

Thomas, D.S.G. and Shaw, P.A. 1991. Relict desert dune systems: interpretations and problems. *Journal of Arid Environments*, 20: 1–14.

Tiffen, M., Mortimore, M. and Gichuki, F. 1994. *More people, less erosion. Environmental recovery in Kenya*. Wiley, Chichester.

Tsoar, H. and Møller, J.T. 1986. The role of vegetation in the formation of linear sand dunes. In W.G. Nickling (ed.), *Aeolian geomorphology*. Allen & Unwin, Boston: 75–95.

UN 1977. *World map of desertification at a scale of 1:25 000 000*. UNCOD, Nairobi.

UNEP 1992. *World atlas of desertification*. Edward Arnold, Sevenoaks.

van Andel, T.H., Zangger, E. and Demtrack, A. 1990. Landuse and soil erosion in Prehistoric Greece. *Journal of Field Archaeology*, 17: 379–396.

Vogel, H. 1990. Deterioration of a mountain agro-ecosystem in the Third World due to emigration of rural labour. In B. Messerli and H. Hurni (eds), *African mountains and highlands: problems and perspectives*. African Mountain Association: 389–406.

Walling, D.E. and Webb, B.W. 1983. Patterns of sediment yield. In K.J. Gregory (ed.), *Background to palaeohydrology*. Wiley, Chichester: 69–100.

Watson, A. 1990. The control of blowing sand and mobile desert dunes. In A.S. Goudie (ed.), *Techniques for desert reclamation*. Wiley, Chichester: 35–85.

Wellens, J. and Millington, A.C. 1990. Desertification. In A.M. Mannion and S.R. Bowlby (eds), *Environmental issues in the 1990s*. Wiley, Chichester: 245–261.

Wiggs, G.F.S., Livingstone, I., Thomas, D.S.G. and Bullard, J.E. 1994. Effects of vegetation removal on airflow patterns and dune dynamics in the southwest Kalahari Desert. *Land Degradation and Rehabilitation*, 5: 13–24.

Wiggs, G.F.S., Thomas, D.S.G., Bullard, J.E. and Livingstone, I. 1995. Dune mobility and vegetation cover in the southwest Kalahari Desert. *Earth Surface Processes and Landforms*, 20: 515–529.

Williams, G.P. and Wolman, M.G. 1984. *Downstream effects of dams on alluvial rivers*. US Geological Survey, Professional Paper, 1286.

Wilshire, H.G. 1980. Human causes of accelerated wind erosion in California's deserts. In D.R. Coates and J.D. Vitek (eds), *Thresholds in geomorphology*. Allen & Unwin, London: 415–433.

Wittfogel, K.A. 1958. *Oriental despotism: a comparative study of total power*. C.T. Yale, New Haven.

Wolman, M.G. 1967. A cycle of sedimentation and erosion in urban river channels. *Geografiska Annaler*, 49A: 385–395.

Woodward, J.C. 1995. Patterns of erosion and suspended yield in Mediterranean river basins. In I.D.L. Foster, A.W. Gurnell and B.W. Webb (eds), *Sediment and water quality in river catchments*. Wiley, Chichester: 365–389.

Worster, D. 1979. *Dust bowl*. Oxford University Press, New York.

Zachar, D. 1982. *Soil erosion*. Developments in Soil Science 10, Elsevier, Amsterdam.

29

Extraterrestrial arid surface processes

Gordon L. Wells and James R. Zimbelman

Introduction

The development of arid zone geomorphology has progressed over the past century from ground-level observations of a limited number of locations through the field traverse mapping of sizeable arid areas to a broad-scale overview of global desert regions assisted by aerial photography and satellite imagery. Extraterrestrial investigations of planetary surfaces and landforms have proceeded in exactly the opposite manner. Telescopic observations first provided global overviews of distant planets. The early spacecraft missions of the 1960s relayed hemispheric and regional views, while later missions returned more detailed images comparable to satellite images of the Earth. To date, detailed views seen from the surfaces of other terrestrial planets are restricted to two locations on Mars and four landing sites on Venus. As a result, the approach and scope of our inquiry into arid processes are necessarily different when studying extraterrestrial surfaces. In this review, we attempt to build a picture of the arid surface of Mars starting with a discussion of possible microscale and mesoscale processes and later examining the creation of individual arid landforms and terrains. We conclude with an overview of recent discoveries regarding the

surface of Venus and attempt to forecast the course of future planetary missions of interest to arid zone scientists.

Spacecraft exploration of Mars began with three Mariner flyby missions during the 1960s. The initial impression was of a surface much like that of the Moon. This judgement changed following the Mariner 9 mission in 1971–72, which relayed more detailed images showing a surface with many landforms common to Earth. Beginning in the summer of 1976, views from the surface were transmitted by two landing craft. On 20 July 1976, Viking 1 landed on the vast volcanic plain of Chryse Planitia (22.5°N, 47.8°W). Several weeks later, Viking 2 touched down upon Utopia Planitia (48.0°N, 225.6°W), another northern hemisphere plain. During the time of Viking lander data collection, two Viking orbiters imaged the surface from an altitude of about 1500 km.

Though Mars is a small planet, its land-surface area is nearly equal to that of the Earth. Another close similarity is daylength. Because the Martian day is slightly longer than Earth's, events at the Viking landing sites were measured in *Sols*, or Martian solar days, following each landing. Mars also has distinct seasons, but due to the eccentricity of its orbit, the seasons are of unequal length. The

Arid Zone Geomorphology: Process, Form and Change in Drylands, 2nd edition. Edited by David S. G. Thomas.

Table 29.1   *Physical characteristics of Earth, Mars and Venus*

	Earth	Mars	Venus
Planetary mass (kg)	5.98×10^{24}	6.44×10^{23}	4.87×10^{24}
Planetary radius (km)	6369	3394	6052
Land surface area (km^2)	1.47×10^6	1.45×0^6	4.61×10^6
Acceleration due to gravity (m s^{-1})	9.81	3.72	8.85
Composition of atmosphere	N_2 (0.78) O_2 (0.21) H_2O (<0.03) A (0.01)	CO_2 (0.95) N_2 (0.027) A (0.016)	CO_2 (0.97) N_2 (0.03) $SO_2 + S_2$ (<0.001)
Mean surface pressure (mb)	1016	7.5	~90 000
Surface temperature (°C)	−53–57	−128--−28	374–465
Rotation period (hours and minutes)	23 h 56 m	24 h 37 m	5832 h (retrograde)
Revolution period (yr)	1.000	1.881	0.615
Orbit eccentricity	0.017 Range: 0.01–0.05	0.093 Range: 0.009–0.14	0.007
Obliquity* (degrees)	23.5 Range: 21.8–24.4	25.2 Range: ~12–38 (0.2–51.4)*	~179

Source: Hartmann (1972); Leovy (1979); Carr (1981).
*Obliquity history values calculated by Bills (1990).

convention for reference to Martian seasons is the aerocentric longitude (L_s), where in the northern hemisphere the vernal equinox occurs at $L_s = 0°$, summer solstice at $L_s = 90°$, autumnal equinox at $L_s = 180°$ and winter solstice at $L_s = 270°$.

The basic physical characteristics of Earth, Mars and Venus (Table 29.1) permit a wide range of surface environments. One of the major goals in the development of geomorphological theory is the creation of explanations for landscape processes based upon general principles and not merely the results of descriptive speculation. The surfaces of Mars and Venus offer laboratories for testing the assumptions of terrestrial theories under conditions where such variables as gravity, atmospheric pressure and surface temperature are different.

Rock disintegration processes on Mars

The primordial mechanism for rock disintegration on planetary surfaces was brecciation caused by impacting meteorites. Following the first several hundred million years of solar system history, a decline in the rate of meteorite impacts reduced the quantity of newly fragmented materials. On Earth, Mars and Venus, the availability of volatiles allowed the chemical breakdown of impact breccia (regolith), and on Earth the plate tectonic regime produced a chemically differentiated crust suited to chemical alteration by a variety of processes. The chemical weathering processes occurring on Mars are in several ways unlike those on Earth, because the Martian lithosphere does not undergo plate motion, subduction and crustal recycling, and its regolith is composed primarily of mafic to ultramafic basalts.

PRESENCE OF WATER

Attempts to estimate the total amount of outgassed water during Martian history have arrived at figures representing an equivalent global surface layer of liquid water ranging in depth from 100 to 500 m (Carr 1981; 1986; Squyres 1984; Donahue 1995). The entire inventory of observable water in the modern atmosphere and ice caps represents a surface layer less than 10 m in depth (Rossbacher and Judson 1981). Analyses performed on soil samples collected at the Viking landing sites confirm that at least some of the water has become chemically bound with surficial sediments. The materials at the landing sites show high concentrations of iron (as anticipated from the rusty red surface coloration of the Martian landscape), sulphur, calcium and magnesium (Baird *et al.* 1976). The relative abundances suggest source rocks of mafic composition, probably basalts. The Viking results led Baird *et al.* (1976) to conclude that the fine-grained materials on Mars are predominantly iron-rich clays with additional components of magnesium sulphate (10%), carbonate (5%) and iron oxides (5%). The production of smectite clays on Mars may follow several pathways (Carr 1981), including hydrothermal alteration in the volcanic regions (Griffith and Arvidson 1994), groundwater alteration in subsurface aquifers (Burns and Fisher 1990; Burns 1994) and, perhaps, ultraviolet-catalysed reactions with a monolayer of water coating fine-grained basaltic sediments (Huguenin 1976; Stephens *et al.* 1994).

FROST WEATHERING AND DISAGGREGATION PROCESSES

The frost point of water on Mars is $-75°C$, and all regions poleward of $40°$ latitude receive at least some seasonal H_2O frost accumulation usually mixed with CO_2 frost in a clathrate structure. At both Viking landing sites, water vapour was saturated in the atmosphere, and lander measurements appear to confirm a seasonal exchanged between surface materials and the atmosphere (Moore *et al.* 1987; Moore and Jakosky 1989). Laboratory experiments by Huguenin (1982), which simulated the Martian frost cycle, produced chemical weathering of silicates (olivine) as a result of the incorporation of hydrogen ions from the H_2O frost. In addition to chemical weathering, frost may serve mechanically to disaggregate fine-grained surface materials in the high-latitude regions. The precipitation of frost occurs when atmospheric dust particles, which serve as nucleation sites for equatorial water vapour, arrive over the winter hemisphere high latitudes. The dust–H_2O ice particles may also condense CO_2 ice at its frost point of $-125°C$. In regions near the poles, the dust–H_2O–CO_2 ice particles coat

the surface as a seasonal extension of the polar cap. If these condensates occupy pore spaces among the fine surficial sediments and condense additional trace amounts of surface water vapour, frost-wedging may liberate fine particles that would then be available for aeolian transport.

SALT WEATHERING

An unexpected result of the Viking chemical analyses of Martian surface samples was the discovery of a high concentration of sulphate ($MgSO_4$ and $NaSO_4$), carbonate ($CaCO_3$) and chloride (NaCl) salts in the Martian soil. The amount of sulphur was found to be enriched in samples taken from thin layers of duricrust atop aeolian drift deposits and in pebble-sized soil aggregates, when their composition was compared to that of unconsolidated fine materials. The excess sulphur content strongly indicates that salt weathering is an active process on Mars (Clark *et al.* 1976; Clark and Van Hart 1981), and that thin layers of iron-rich clay particles are cemented by sulphate and carbonate minerals (Toulmin *et al.* 1977). Diffusion of salts to the surface by water vapour passing through the substrate could lead to the formation of widespread duricrust layers over periods of 10^5–10^6 years (Jakosky and Christensen 1986) The most dramatic 'geomorphological event' to occur during observations at the Viking 1 landing site was the slope failure and slumping of duricrust material covering a drift deposit near the base of a large boulder (Figure 29.1). The drift deposits in the vicinity of the Viking 1 lander give the appearance of being stripped by deflation. If this interpretation is correct, duricrust formation has proceeded with sufficient rapidity to encase aeolian drift materials as the deposits are in the process of being deflated and removed from the area. In areas of stable or accumulating aeolian deposits, salt weathering may lead to the production of considerably thicker duricrusts.

Salt weathering may attack competent Martian rocks along several possible routes (Malin 1974) even in the absence of liquid water. As occurs in terrestrial environments discussed in Chapter 3, hydration by water vapour of salts trapped in surface rock pores may physically

Figure 29.1 *Two views of the Viking 1 landing site taken on (A) VL-1 Sol 25 and (B) Sol 239 record a slump of duricrust material upon a drift deposit. [(A) VL1 #11A143; (B) #11C162]*

destroy rocks, as may the diurnal thermal expansion and contraction of salt crystals trapped inside pore spaces of rocks having a different expansion coefficient.

AEOLIAN ABRASION

Apparently wind-faceted and pitted rocks are plentiful near the Viking landing sites (Figure 29.2), and wind-tunnel experiments (Greeley *et al.* 1984) simulating Martian conditions ·have confirmed high rates of aeolian abrasion. In fact, the simulated rates are enormous, if basalt or quartz grains are used as impactors. Abrasion rates in the range 0.0003–0.002 cm yr^{-1} have been calculated by Greeley *et al.* (1984) and result from the high initial velocities of Martian particles in saltation and their high terminal velocities under conditions of low atmospheric drag. Paradoxically, the Martian surface near the landing sites is covered with large and probably very ancient rocks (Mutch and Jones 1978; Sharp and Malin 1984), while orbital imagery reveals concentrations of small- to medium-sized craters near the Viking 1 landing site, which indicate a surface age of more than 3 billion years (Arvidson *et al.* 1979). At the simulated rate of aeolian abrasion, all rocky landforms on the scale of 100 m would be abraded to near base-level in less than 50 million years.

At least three factors may explain the discrepancy between the rate of aeolian abrasion modelled by wind-tunnel experiments and the degree of surface modification actually observed. Broad regions of Mars presently serve as sediment sinks for aeolian materials, while other areas show signs of the removal of formerly extensive aeolian mantles. Perhaps in some regions (e.g. around the South Pole) a blanket of aeolian deposits protected the ancient surface from abrasion (Plaut *et al.* 1988). A second possibility follows from the character of saltation on Mars. Both landing sites display rocky surfaces. In the Martian atmosphere, typical saltation path lengths span several metres and require sandy surfaces over such distances for a saltation cloud to develop fully. The distribution of rocks could effectively dampen the saltation flux and reduce the rate of abrasion (Williams and Greeley 1986). Finally,

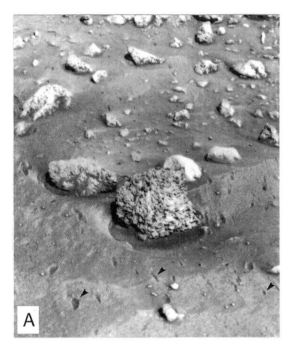

Figure 29.2 *Two views under different illumination conditions show wind-faceted and pitted rocks at the Viking 1 landing site. Arrows point to small impacts penetrating duricrust by clasts dislodged during the firing of the lander descent rocket. [(A) VL1 #11B144; (B) #11A078]*

most saltating particles on Mars probably do not consist of intact mineral grains. Energetic collisions during saltation would lead to short lifetimes for basalt fragments and other 'kamikaze' grains similar to terrestrial quartz sands (Sagan *et al.* 1977). Clay aggregates dislodged from duricrusts provide the most reasonable candidates for saltation (Greeley 1979), and aggregates formed as a sublimation residue of dust-nucleated ice may supply saltable materials in the polar regions (Saunders and Blewett 1987; Thomas and Weitz 1989; Herkenhoff and Murray 1990a). Impact abrasion by these low-density particles would be much less effective than with intact mineral grains.

Transport by suspension, saltation and creep on Mars

As on Earth (see Chapter 16), the primary modes of aeolian transport on Mars are suspension, saltation and creep. However, the boundary conditions governing Martian aeolian processes result in transport events having different magnitudes, frequencies and durations to those of comparable events on Earth (Greeley and Iversen 1985).

DUST DEVILS

During summer months, in the lower boundary layer of the clear, cold, thin Martian atmosphere, the large vertical temperature gradient above the relatively warm surface may support intense free convection. Under these conditions during periods of near calm, thermals will rise into the deep Martian convective layer. Gentle surface winds crossing topographic irregularities may create sufficient basal shear to induce vortices lifting dust into the rising thermal. The resulting dust devils have been observed on high-resolution Viking orbiter images of Arcadia Planitia and Phlegre Montes between 30°N and 45°N (Thomas and Gierasch 1985). Of the 99 dust vortices detected, most dust funnels achieved heights of 1.0–2.5 km. Estimated dust mass suspended aloft in the typical Martian dust devil is 3000 kg (Thomas and Gierasch 1985).

Although never directly imaged, circumstantial evidence points to the existence of much larger Martian aeolian vortices. Dark linear filaments approximately 0.5 km wide with lengths of 2–75 km lie along approximately parallel tracks on the plains of the southern hemisphere between 30°S and 60°S (Grant and Schultz 1987). The filaments display short gaps and make their appearance seasonally from midsummer until early autumn with slight changes in orientation detected from year to year. These dark filaments may result from surface dust removal and concentration of sand-sized sediments by tornadic vortices sweeping across the southern hemisphere mid-latitudes. Confirmation of such storms occurring in a southern hemisphere 'tornado alley' awaits their detection by future Mars-orbiting spacecraft.

BAROCLINIC AND KATABATIC DUST STORMS

Global weather systems may produce local dust storms whenever the poleward temperature gradient is sufficiently large to generate intense zonal circulation across the mid-latitudes in the form of baroclinic waves. Numerical simulations of Martian general circulation indicate that surface wind velocities in excess of the saltation threshold should accompany the passage of strong baroclinic waves (Mass and Sagan 1976; Pollock *et al.* 1981; Barnes *et al.* 1993; Greeley *et al.* 1993), and examples of dust-transporting extratropical cyclones have been observed in a few instances (Hunt and James 1979; Kahn *et al.* 1992). On the surface, meteorological instruments aboard the Viking landers recorded peak gust velocities of $20–26 \text{ m s}^{-1}$ during the passage of dust storms (Hess *et al.* 1977; Moore *et al.* 1987).

Most observations of local dust storms document a second type of circulation that operates on a regional scale. These dust storms are produced by katabatic outflow from receding frost outliers of the polar caps and diurnal winds descending areas of high relief (Lee *et al.* 1982; Magalhaes and Gierasch 1982; Kahn *et al.* 1992). The great majority of dust storms

take place from late spring until after perihelion passage, $L_s = 250°$, near the time of the southern hemisphere summer solstice (Peterfreund 1985), when the latitudinal temperature gradient close to the retreating frost outlier reaches its most extreme value, as does surface heating near the subsolar point. The frequent occurrence of point-source dust plumes rising from areas recently covered by the seasonal frost cap leads to speculation about the role of frost-wedging in the production of fines for transport by suspension after water ice has sublimed (Figure 29.3). During the modern climatic epoch on Mars, the combination of fine particles released by frost-wedging and strong katabatic surface winds sweeping along the subliming frost cap has created a significant dust source in the southern hemisphere (Christensen 1988). Based upon Viking Infrared Thermal Mapper (IRTM) observations, Viking camera images and terrestrial telescopic sightings, Peterfreund (1985) located 156 dust storm events, 79% of which originated in the southern hemisphere.

MARTIAN GLOBAL DUST STORMS

When the Mariner 9 spacecraft approached Mars in November 1971, the entire surface was obscured by a global dust storm. Such enormous dust storms follow an evolutionary sequence governed by the thermal properties of the Martian atmosphere and global circulation. Of the large dust storms, which have been observed to grow to global proportions, all originated south of the equator near the time of southern hemisphere perihelion, $L_s = 250° \pm 60°$ (Briggs *et al.* 1979; Kahn *et al.* 1992).

Once entrained in the Martian atmosphere, dust absorbs incoming radiation, which is then reradiated as sensible heat. Differential heating of dusty and clear air parcels promotes convective turbulence leading to high-velocity surface winds along the dust storm margin which, in turn, serve to raise additional dust. On a larger scale, regional dust storms form an atmospheric thermal blanket that interacts with clear-air regions of lower heat capacity to produce strong, thermally driven tidal winds (Leovy and Zurek 1979). These winds will be particularly effective in raising dust from areas along the

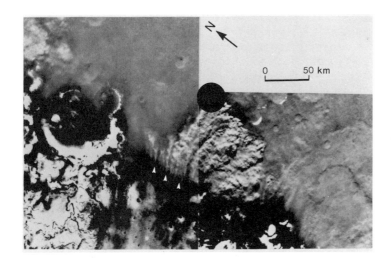

Figure 29.3 *Point-source dust plumes (arrows) rise from an area recently covered by the seasonal H_2O frost cap in the Aonia Terra region (68°S, 69°W) of the Martian southern hemisphere. [VO #248B55, #248B56]*

sunset terminator, where a region of dust warmed by radiative heating adjoins a rapidly cooling region of clear air. A regional dust storm driven by convective turbulence and thermal tides may propagate to achieve global dimensions.

During the late spring and summer, global circulation on Mars is dominated by a cross-equatorial Hadley cell (Haberle *et al.* 1993). The rising limb of the cell is located near the subsolar point in the summer hemisphere, and the descending limb over the winter hemisphere mid-latitudes (Toon *et al.* 1980; Pollock *et al.* 1981; Magalhaes 1987). The presence of a dusty atmosphere in the southern hemisphere near the time of northern winter solstice greatly amplifies the Hadley circulation (Haberle *et al.* 1993). Once pervaded with dust, the global atmosphere becomes stabilised because only the upper portion is effectively heated (Haberle 1986a; 1986b). Convective turbulence is smothered near the surface by an almost isothermal environmental lapse rate. Over a period of several weeks, dust settles out of the atmosphere. The northern polar region, with its condensing frost cap, acts as a primary sediment sink where the dust from global storms serves in the condensation process to nucleate H_2O and CO_2 ice (Toon *et al.* 1980; Barnes 1990).

Global dust storms constitute the primary geomorphological process occurring on Mars during the present climatic epoch. Laboratory measurements show that small amounts of deposited dust (less than 10^{-3} g cm^{-2}) can result in reflectance values comparable to optically thick dust layers, even when deposited upon a dark surface (Wells *et al.* 1984). Viking lander cameras at both landing sites recorded the deposition of a thin (less than 1 mm thick), red dust layer following the two global dust storms of 1977 (Guinness *et al.* 1982). The surface brightening was most noticeable in the areas surrounding the Viking 2 landing site on Utopia Planitia (Figure 29.4). As viewed from the Viking orbiters, the northern hemisphere surfaces of Syrtis Major and Acidalia Planitia brightened measurably immediately after the 1977 global storms, then gradually darkened again over the following months (Christensen 1988). By contrast, Hellas Planitia, a southern hemisphere basin and site of frequent dust outbreaks, appears to be a sediment source currently undergoing net erosion by aeolian processes (Moore and Edgett 1993).

BRIGHT AND DARK WIND-STREAKS

Extending from many topographic highpoints, crater rims in particular, are light or dark albedo streaks in the equatorial and mid-latitude portion of Mars between 40°N and 40°S (Figure 29.5). Both types of streak are intimately connected to the occurrence of global dust storms. Bright streaks are interpreted to be deposits of fine material which have been concentrated preferentially in protected

Figure 29.4   *The brightening of the surface around the Viking 2 landing site following dust deposition by the 1977 global storms is apparent in two surface views made under similar illumination conditions on (A) VL-2 Sol 30 prior to the dust storms and (B) after the dustfall. [(A) VL2 #22A252; (B) #22G085]*

wind-shadow zones during dustfalls from global storms (Veverka *et al.* 1981; Greeley *et al.* 1992b). Light-coloured streaks are typically wide, exhibit a feather-like structure downwind and reach lengths of 5–20 km. The dark streaks result from deflation in the lee of topographic obstructions by paired horizontal vortices shed from obstacles that accelerate local airflow (Greeley *et al.* 1974). The dark erosional streaks that expose the underlying surface are narrow, sharply defined and attain lengths of 5–30 km (Greeley and Iversen 1985). Whereas the bright streaks are relatively stable features exhibiting only minor changes over the course of several Martian years, the dark streaks largely vanish after deposition of a thin layer of surface dust from a single global dust storm and begin

to reform in the period immediately after the dustfall (Thomas and Veverka 1979).

Local differences in bright and dark wind-streak directions and the creation of bright depositional streaks rather than dark erosional streaks have been attributed to differing meteorological conditions in a hypothesis developed by Veverka *et al.* (1981). When radiative heating of dust promotes static stability during the dustfall stge of a global storm, airflow in the boundary layer becomes more laminar, leading to concentrated dust deposition in long wind-shadow zones in the lee of obstacles. Sediments in bright wind-streaks are deposited at this time. After the atmosphere clears, the effect of flow-blocking by terrain diminishes, and vertical motions again

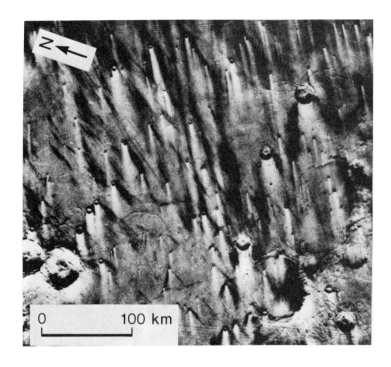

Figure 29.5 *Both bright and dark albedo streaks extend downwind from crater rims in the Syrtis Major region (4°N, 297°W). [VO #496A45]*

predominate in the boundary layer, allowing vortices to scour the recent dustfall layer to create dark wind-streaks.

The global distribution of bright wind-streak directions serves to map the general circulation of Mars at the time of streak formation. Since most bright streaks are formed during global dust storms, the trans-equatorial Hadley circulation of the southern hemisphere late spring and summer is traced by these surface 'weather vanes' (Thomas and Veverka 1979; Lee *et al.* 1982; Greeley *et al.* 1992b; 1993).

MARTIAN AEOLIAN MANTLE

The transport of aeolian sediment across large regions on Mars leads to areas of net sediment accumulation and removal (Thomas 1982; Christensen 1986b; Kahn *et al.* 1992). Regions of relatively high albedo (0.25–0.4) are interpreted to contain more dust than regions of lower albedo (0.1–0.2) (McCord and Westphal 1971; Pleskot and Miner 1981). The subdued surface morphology within regions of higher reflectance is consistent with the presence of a uniformly distributed mantling material (Zimbelman and Kieffer 1979). Our knowledge

of the absolute depth of aeolian deposits in an area is limited by the spatial resolution of spacecraft cameras. However, some aspects of the physical properties of Martian surface materials can be inferred from data transmitted by the IRTM sensors on the Viking orbiters.

The equatorial and mid-latitude regions display areas of relatively low thermal inertia interspersed with areas of relatively high thermal inertia (Figure 29.6A). The distribution of particle sizes inferred from thermal inertia values divides Mars into longitudinal sectors (Figure 29.6B). Global circulation patterns influenced by large-scale topography may determine the regional concentrations of fine- and coarse-grained surface materials. It appears significant that the areas of lower thermal inertia are closely correlated with the areas of higher albedo (Kieffer *et al.* 1977).

The absolute thickness of surface materials distributed across the Martian surface can be broadly constrained by remote sensing measurements. In high thermal inertia regions, the fine-particle thickness may be as much as tens of centimetres. Earth-based radar measurements of low thermal inertia areas show a large amount of scattering by surface roughness (Harmon *et al.* 1982). This result suggests an upper limit to the

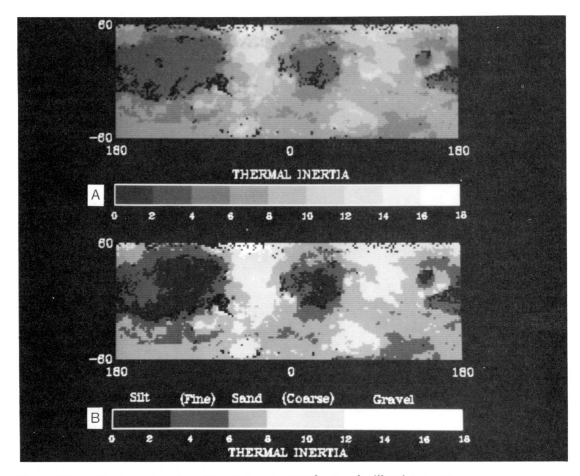

Figure 29.6   *Martian surface thermal inertia in units of $10^{-3}\,cal\,cm^{-3}\,s^{-1/2}\,K^{-1}$ derived from temperatures measured by the Viking IRTM. (A) Data separated by thermal inertia units (0 = lowest; 18 = highest). (B) Data separated according to average particle size for an idealised uniform particle surface (Kieffer et al. 1973)*

depth of fine materials of 1–2 m (Christensen 1986a; 1986b). Moreover, in low thermal inertia areas, fine materials are not sufficiently thick to mask the thermal effects of more competent materials, such as large blocks or boulders, which are observed in varying abundances over most of the Martian surface (Christensen 1986c). These observations indicate that potentially mobile fine sediments are dispersed around the entire planet.

TRANSPORT BY SALTATION AND CREEP

Experiments performed using the Mars Surface Wind Tunnel (MARSWIT) and Venus Wind Tunnel (VWT) at the NASA Ames Research Center have clarified differences in particle saltation that take place on Earth, Mars and Venus (Figure 29.7). On Mars, the most easily entrained sediment grains have a diameter of 0.115 mm (Iverson and White 1982). Thermal inertia measurements of aeolian sediments held within crater dunefields indicate the presence of considerably coarser grains with diameters of 0.4–0.6 mm (Edgett and Christensen 1991). The lower atmospheric pressure results in threshold friction velocities roughly one order of magnitude higher than terrestrial velocities for lifting particles of similar size. On Mars, a threshold friction speed of 2.4 m s^{-1} is required to entrain a 0.1 mm particle with the density of basalt (2.5 g cm^{-3}). With the arrival of the first

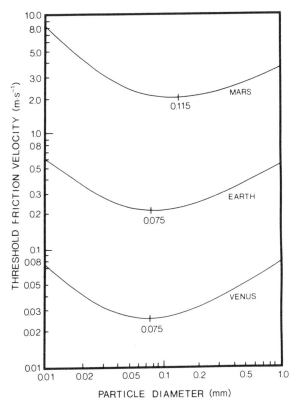

Figure 29.7 *Threshold friction velocities for entraining aeolian sediments on Mars, Earth and Venus as determined by wind-tunnel experiments (modified after Iversen and White 1982)*

global dust storm of 1977, wind velocities measured 1.6 m above the surface at the Viking 1 landing site reached 17.7 m s^{-1} with instantaneous gusts of more than 25 m s^{-1} (Ryan and Henry 1979). These velocities are close to the required threshold for saltation, when extrapolated to the surface. In a typical terrestrial sand desert, significant saltation may occur during about 10% of the days in a year. Saltation events occur much less frequently on Mars, and probably happen during only 0.01–0.1% of the days in a Martian year. Despite its relative rarity, saltation on Mars should be much more vigorous than on Earth because initial particle velocities are much higher, and atmospheric drag is minimal, leading to high terminal velocities.

Surface creep, or the rolling of grains in traction across the surface, may play a substantially smaller part in aeolian transport on Mars,

given the boundary conditions for saltation to be initiated and the anticipated nature of saltation clouds. The very long path lengths of saltating particles cause fewer impacts per unit area under wind conditions just above the saltation threshold. On Mars, surface creep will be inhibited by the relatively small number of neighbouring impacts setting stationary particles rolling across the surface.

Depositional bedforms on Mars

The appearance of Martian dunes, dunefields and ergs suggests a strong connection to bedform sand transport processes in terrestrial deserts. Yet there exist intriguing differences between the two planets in the global distribution of sand dunes and relative abundance of particular dune types.

BARCHANOID AND TRANSVERSE DUNES

Nearly all dunes on Mars are either barchanoid or transverse bedforms (Figure 29.8). Individual barchans appear in planform to be almost identical to their terrestrial counterparts and occur as simple bedforms and as compound megadunes, where smaller, secondary dunes rest atop the major bedform (Breed *et al.* 1979; Tsoar *et al.* 1979). In dunefields, barchans have collided to form barchanoid ridge dunes, and areas with an abundant sediment supply contain transverse ridge dunes. With the exception of isolated dunefields within crater basins (Lancaster and Greeley 1987; Edgett and Christensen 1991), most regions of Martian dunes are situated in the high latitudes between 70°N and 85°N (Ward *et al.* 1985). Whether this asymmetry reflects characteristics of Martian global circulation, latitudinal differences in available sources of sediment or a combination of these factors remains an open question.

A more fundamental question is the composition of Martian dunes. The petrological evolution of Mars has limited the amount of siliceous rock for breakdown into quartz grains, and highly energetic particle collisions over the course of geological time would have long ago smoothed the surface and reduced the saltating population of intact mineral grains to dust-sized

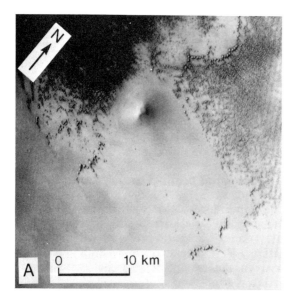

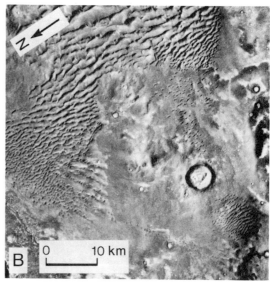

Figure 29.8   *Martian barchan and transverse dunes. (A) Barchans of the North Polar erg (77°N, 97°W). (B) Transverse bedforms in the ancient cratered terrain of the southern hemisphere (47°S, 340°W). [(A) VO #519B27; (B) #575B60]*

fragments. Perhaps the saltating sediments of Mars need not be long-lived to construct bedforms. On Earth, aggregates of clay occasionally are transported to create small dunes with slip faces, even though the entire dune structure becomes cemented after minimal rainfall (Bowler 1973). Dunes composed of silt–clay aggregates are suggested to occur on Mars (Greeley 1979). As a possible terrestrial analogue, the gypsum dunes of White Sands, New Mexico, are barchanoid and transverse bedforms (McKee 1966) constructed by strongly cohering grains. Breed *et al.* (1979) noted close morphological similarities between Martian barchanoid ridges in the northern polar region and those at White Sands in terms of scale width, length and spacing. They also detected groove-like features on Martian dunes comparable to the wind-eroded surfaces of partially cemented gypsum dunes at White Sands.

Many dunes in the northern polar region are covered by frost in the winter. Frost-wedging, chemical weathering and deflation of their surfaces and interdunes may dislodge aggregates from duricrust layers, liberating particles for saltation. So long as Martian saltation events are in equilibrium with aggregate production, the dunes will remain intact. If aggregate production

diminishes, the dunes may become etched and grooved by wind erosion.

SCARCITY OF LINEAR DUNES

Unlike terrestrial sand deserts, Mars lacks large concentrations of linear dunes. Painstaking investigation has uncovered only a few examples of linear dunes and other bedforms resulting from bi- and multidirectional sediment-moving winds. The least ambiguous linear bedforms lie within a 70-km-diameter crater in the southern hemisphere (59°S, 343°W), where 4–5 km dune ridges extend from a field of barchanoid dunes (Edgett and Blumberg 1994). In this case, local topographic obstructions appear to influence the direction of winds crossing the crater floor. Limited numbers of Martian star dunes in the Mare Tyrrhenum region are similarly situated in an area subject to local topographic disruptions of regional airflow (Edgett and Blumberg 1994).

Several reasons may explain the scarcity of linear bedforms on Mars. In their discussion of the northern polar region, two investigations (Tsoar *et al.* 1979; Lee and Thomas 1995) point to barchans with elongated wings as examples of seif dunes. Tsoar *et al.* (1979) suggest that the dunes are in an early stage of bedform evolution. This idea fails to accord with the high degree of bedform organisation displayed by regional megadune patterns, which requires a minimum of several thousand years to develop on Earth and probably much longer on Mars. In an alternative hypothesis, Breed *et al.* (1979) regard the dunes of the northern polar region and other Martian dunefields to be geomorphologically mature formations, where most dunes now lie in topographically restricted areas, such as crater basins. Saltable sediments no longer pass across great distances between sedimentary basins. An analogy can be made to the mass-transport flux of aeolian sediments in the Saharan sand basins (Mainguet and Chemin 1985). In the Sahara, linear dunes are the primary mode of bedform transport from the upwind source regions of a basin and across interbasin sills into topographically confined sediment sinks.

A likely explanation for the scarcity of linear dunes is that Martian saltation events take place under a limited range of meteorological conditions. The climatological setting for these low-frequency events may include only the most energetic phases of Hadley circulation over the equatorial regions, zonal circulation crossing the mid-latitudes, katabatic outflow from the polar caps and downslope winds from high terrain. Each of these systems creates relatively unidirectional airstream trajectories that would lead to the production of barchanoid and transverse bedforms. An atmospheric General Circulation Model of Mars, which incorporates several boundary conditions for surface shear stress reflecting different saltation threshholds, generated theoretical values for the resultant drift potential divided by the total drift potential (RDP/DP ratios) for global regions (Lee and Thomas 1995). The calculated RDP/DP ratios commonly exceed 0.9 and point to unidirectional sediment movement under a unimodal wind regime in most regions of Mars.

ERGS

The North Polar sand sea (Figure 29.9) forms a dark circumpolar collar that reaches its maximum extent between 77–83°N and 110–220°W (Tsoar *et al.* 1979). With an area of approximately 680 000 km^2, the North Polar erg is larger than any individual terrestrial sand sea containing mobile dunes and holds an estimated 1158 km^3 of dune sediment (Lancaster and Greeley 1990).

Two principle circulation patterns are responsible for shaping the dunes of the North Polar erg. Low-pressure fronts tracking along the northernmost latitude of zonal circulation produce dune-forming westerly and southwesterly winds below 80°N. Easterly and northeasterly sand-moving winds have created the dunes above 80°N between 120°W and 240°W in the region nearest to the perennial polar ice cap (Tsoar *et al.* 1979). The airflow regime is the consequence of katabatic winds in the cyclonic circumpolar vortex which follow an outflow pattern traced by frost streaks into dunefields surrounding the perennial ice cap (Howard 1980). The concentration of dark polar dunes into a narrow latitudinal band may be aided by strong surface winds induced along the sharp regional temperature gradient and albedo boundary in a manner similar to the creation of a terrestrial sea breeze (Thomas and Gierasch 1995). The location of dunefields downwind of the polar cap suggests that polar layered deposits are the most likely source of dune sediment (Thomas 1982; Thomas and Weitz 1989; Lancaster and Greeley 1990).

In contrast to Tsoar *et al.* (1979), Breed *et al.* (1979) argued that the North Polar erg is ancient, citing the absence of sand-passing longitudinal dunes, the high degree of organisation in dune patterns and signs of wind erosion upon some dune surfaces. Another feature that signifies an older age for the erg is the occurrence of several very large barchanoid ridges in the upwind areas of dunefields. These 'framing dunes' are now displaced from neighboring dune ridges located downwind by a distance several times the mean wavelength separating other ridges in the dunefields. Such a circumstance can arise only after many reconstitution cycles (the period required for a bedform to move its own length). In an earlier stage of bedform development, more rapidly migrating upwind dunes collided with a barchanoid ridge causing its volume and height to increase and rate of advance to slow. The adjacent

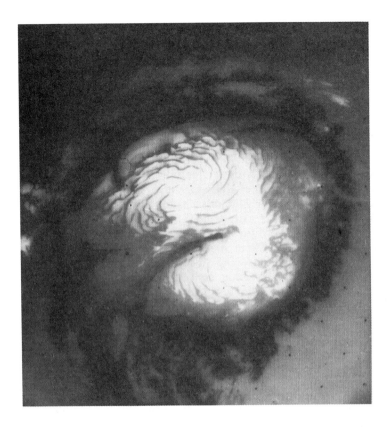

Figure 29.9   *The dark dune collar of the North Polar erg surrounds the perennial H₂O ice cap of the North Pole. [Mariner 9 #4297-46]*

downwind ridges of the dunefield then began to move away from the larger barchanoid ridge. The existence of windward framing dunes separated at some distance from other ridges together with the likelihood that reconstitution cycles for Martian megadunes are considerably longer than the *c.* 1000-year periods for equivalent bedforms on Earth belie the creation of the North Polar erg on the *c.* 10^4–10^6-year timescale of the present climatic period.

Aeolian erosion on Mars

In the vicinity of the Viking landing sites, pitted rocks with wind-faceted forms have been created by aeolian abrasion (Figure 29.2). Though abrasion has modified many smaller rocks at the landing sites, the presence of boulder fields of rather pristine appearance and probable age of more than a billion years demonstrates that, in regions where lag materials protect the surface, wind erosion occurs very slowly. On a broader scale, the stripping of aeolian mantles and other friable deposits in several regions confirms that aeolian erosion proceeds more rapidly where surface units are composed of fine-grained, loosely consolidated materials (Figure 29.10).

AERODYNAMICALLY SCULPTURED LANDFORMS

Yardangs are most plentiful on the equatorial plains overlain by layered deposits. Mature terrestrial yardangs have a width-to-length ratio of about 1:4 (Ward and Greeley 1984). By contrast, in many cases Martian yardangs appear to be extremely elongated, a factor that Ward (1979) ascribes to more intensive abrasion within the corridors separating Martian yardangs than that which occurs on Earth. The paucity of detailed field research and laboratory experiments dealing with the evolution of terrestrial yardangs limits an attempt to understand Martian yardang elongation, but the spacing of joints and fracture patterns within the layered deposits and granulometric characteristics of

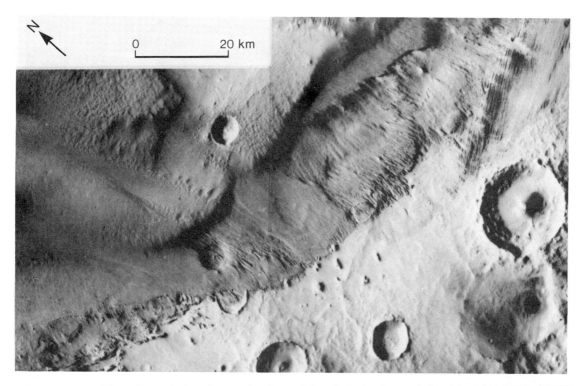

Figure 29.10 *The aeolian stripping of a mantle of layered deposits in the Amazonis Planitia region (11°S, 177°W) exhumes an ancient cratered surface. [VO #438S01, #438S03]*

the sediments may also contribute to the development of elongated yardangs.

Though some yardangs occur in the mid-latitude and polar regions of Mars, the majority are located on the equatorial plains. These low-latitude yardang concentrations resemble terrestrial wind-sculptured landforms developed upon ignimbrite sheets in the central Andes (Figure 29.11). The proximity of the Amazonis Planitia layered deposits to the Olympus Mons and Tharsis Montes volcanic massifs has led some investigators to conclude that the layered deposits were emplaced by pyroclastic flows (Scott and Tanaka 1982; Ward *et al.* 1985). Other features suggest they are layers of an aeolian mantle deposited in a sedimentary basin to the west of Tharsis Montes. For instance, several of the layers are capped by a resistant unit approximately 5–10 m thick. Terrestrial ignimbrites form welded horizons only near the base of flow surges and have easily eroded surfaces. An indurated horizon within the uppermost portion of a layer in the Martian deposits appears out of

character with ignimbrite formation, but is in keeping with the anticipated weathering processes of a loess-like aeolian mantle.

DEFLATION BASINS

The most Earth-like type of wind-eroded basin on Mars takes the form of a cusp-shaped depression in which the apex of the cusp points upwind along the axis of deflation (Figure 29.16). The blunted, downwind portion of cuspate basins can be more than 100 m deep, while the apex grades to the level of the surrounding surface. On Earth, shallower basins but with similar shapes in planform have been created upon sandy surfaces along the margins of ancient lake-beds in the western United States, where loosely consolidated sediments favour lateral deflation in the direction of the formative wind (Goudie and Wells 1995).

The most unusual Martian deflation basins are distinctly crescentic pits developed upon the

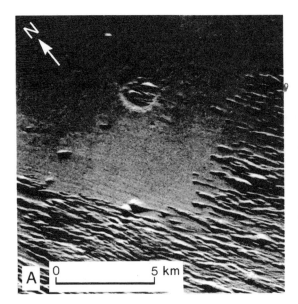
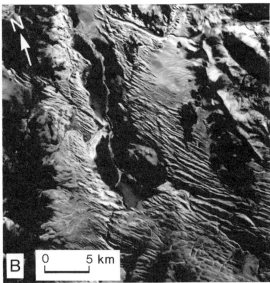

Figure 29.11   *Martian yardangs (A) of the Amazonis Planitia region (1°N, 171°W) closely resemble terrestrial yardangs (B) formed upon the Cerro Galan ignimbrite of northwestern Argentina (25.5°S, 66.8°W). [(A) VO #7728A64; (B) Landsat TM 50787-13503]*

equatorial layered deposits. These small basins have classically aerodynamic shapes similar to an inverted mould cast from a barchan dune. All occur at the windward base of incipient yardangs, where the most intense abrasion occurs upon yardangs formed on Earth (Ward and Greeley 1984). No basins of similar morphology have been reported from investigations of terrestrial yardang localities.

Slope processes on Mars

Gravity serves to modify slopes on all planetary bodies. On most planetary surfaces, this involves the slumping of debris, including the rolling of individual rocks detached from upper slopes by local meteorite impacts or seismic tremors. Martian mass movements extend from the gradual slope retreat and talus creation of dry mass-wasting to the sudden and catastrophic emplacement of giant debris avalanches.

EXPOSURE OF SUBSURFACE VOLATILES

Of all the unresolved issues of consequence to the formation of landforms on Mars, the one

with the greatest implication for possible slope processes is the abundance of H_2O ice in the regolith near the surface. Studies of Martian impact crater morphology have generally taken the presence of craters surrounded by unusually large ejecta-flow deposits in the latitudinal bands along 40°N and 45°S to be evidence of H_2O near the surface (Mouginis-Mark 1979; 1987; Squyres 1979; Barlow 1994). Areas of fretted terrain in the northern hemisphere may present the best case for volatile-controlled surface modification. These regions are composed of isolated mesas with steep valley wall profiles separated by level surfaces and are found between 30°N and 50°N and in isolated locations at equivalent latitudes of the southern hemisphere (Carr 1986). Under current Martian conditions, H_2O ground ice is unstable between 40°N and 40°S and will sublime to the atmosphere from near-surface regolith (Farmer and Doms 1979). The occurrence of fretted terrain only in the areas above equatorial latitudes has prompted several investigators to suggest that ground ice processes may mobilise slope materials in the mid-latitude regions (Carr 1981; 1986; Rossbacher and Judson 1981; Lucchitta 1984).

The most obvious means for destabilising slopes in the fretted terrain is the exposure of subsurface volatiles contained behind valley walls either suddenly by tectonic activity or the shock of a meteorite impact or more gradually through the undercutting of scarps by deflation. Once exposed to the surface, the volatiles will sublime and destabilise the resulting slope, which may then collapse to form a debris flow or slowly disintegrate to build a debris apron.

TERRAIN SOFTENING

Squyres and Carr (1986) have identified a global-scale degradation of surface relief at latitudes poleward of 30°. The terrain softening is interpreted to represent a pervasive expression of surface debris movement in response to downslope creep caused by interstitial ice. The three landforms most often associated with terrain softening are convex lobate debris aprons (Squyres 1978), lineated valley fill consisting of multiple ridges parallel to valley walls (Squyres 1979) and concentric crater-fill indicated by circular ridges and grooves within craters (Squyres 1979; Squyres and Carr 1986). In each case, the volatile content is inferred from the ridge-and-groove topography taken to indicate the flow of surface material in a given direction (Squyres 1989).

The distinctive morphology of each type of terrain feature associated with icy regolith has been illustrated primarily with Viking orbiter images of moderate resolution. The available images of greater detail supply clues for an alternative to the volatile-related origin of some terrain-softening features (Zimbelman 1987). The northern mid-latitude regions of Mars show evidence of mantling by materials that uniformly buried surface landforms and were subsequently partially or completely exhumed (Williams and Zimbelman 1988; Moore 1990). High spatial resolution images of deposits both within and around various craters exhibit layering (Zimbelman *et al.* 1988). These characteristics support an aeolian alternative for the deposition and removal of the ubiquitous mantling materials. The distinctive morphology of certain terrain-softening features may be explained as surface expressions of multi-layered aeolian deposits subjected to differential weathering and erosion. At equivalent latitudes in the southern hemisphere basin of Hellas

Planitia, a widespread aeolian mantle blankets much of the region (Tanaka and Leonard 1994).

MASS WASTING

Within the equatorial canyons, numerous talus chutes form a spur-and-gully morphology entrenched into canyon walls (Lucchitta 1978a). The shapes and spacings of the talus chutes change from irregular crenations dissecting the resistant caprock of canyon walls to parallel chutes uniformly incised into mesas of layered units located within the canyons (Figure 29.12). The distinctive change in erosional styles within the same region may reflect differing mechanical properties of materials composing wall rocks and layered units. Mass-wasting processes causing lateral slope retreat along escarpments may explain the fretted terrain landscapes in the northern hemisphere mid-latitudes and similar areas of the southern hemisphere highlands (Squyres 1979).

DEBRIS AVALANCHES

At the larger scale, gravity driven processes on Mars have resulted in the creation of colossal landslide debris avalanches triggered by catastrophic slope failure. The most conspicuous avalanche deposits occur in the deep, broad-floored Valles Marineris canyon system between 50°W and 90°W (Figure 29.13A). In this region several dozen debris avalanche deposits ranging in size between 40 and 7000 km² have been discovered (Lucchitta 1978b; 1979; 1987; McEwen 1989).

The landslides in Valles Marineris were triggered along canyon walls with heights of 2–7 km and slopes of about 30°. The uppermost sections of the canyon walls are interpreted to be resistant sills of lava, while the underlying materials may be composed of regolith (Lucchitta 1979). The proximal areas of many debris avalanches display giant backward-rotated slide blocks of the resistant units underlying the plateau surface. Hummocky topography often appears downflow from the slide blocks and probably represents coherent megablocks transported far from the collapse

Figure 29.12   *Mass wasting of two different lithologies in the Candor Chasma region (6°S, 71°W) produces irregular crenations (A) along canyon walls and parallel chutes (B) incised into layered deposits with the canyon. [Viking Orbiter mosaic]*

amphitheatre. The distal reaches of most Martian landslides are characterised by longitudinal ridges and grooves radiating outward tens of kilometres from the avalanche source (Figure 29.13B). In several instances flow levees have formed along the lateral margins of an avalanche, but are usually absent along the distal margin.

One possible mechanism for the emplacement of debris avalanches in Valles Marineris has been suggested to be high-velocity mudflows resulting from the sudden outbreak of liquid water from an aquifer held behind an ice

lens rising high in the canyon wall (Lucchitta 1979; 1987). The predominance of longitudinal ridge and groove terrain, the extremely low apparent coefficient of friction implied by long avalanche travel distances and the estimated high velocities of emplacement led Lucchitta to the conclusion that Martian avalanches must have been lubricated by water. Recent work on terrestrial volcanic debris avalanches challenges this interpretation.

In a study of the major volcanic debris avalanches in the central Andes, Francis and Wells (1988) discovered a suite of components

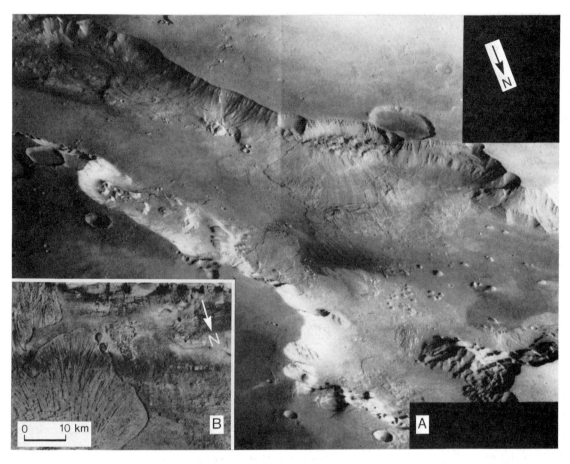

Figure 29.13   *The deposits of giant landslides are found on the floor of Valles Marineris.*
(A) An oblique view across Ganges Chasma (9°S, 45°W) shows landslides with coherent slide blocks alongside
collapse amphitheatres and debris streams radiating far from the avalanche source areas. (B) A detailed view
illustrates the longitudinal ridges and grooves in the distal portion of a debris avalanche deposit in Coprates Chasma
(11°S, 67°W). [(A) Viking Orbiter mosaic; (B) VO #080A01]

(rotated slide blocks, hummocky topography, longitudinal ridges and grooves, and marginal levees) reflecting elements of Martian landslides. While Lucchitta (1979; 1987) presumes that the occurrence of longitudinal ridge and groove terrain is rare for terrestrial avalanches, the deposits emplaced from the Lastarria, Llullaillaco, Aucanquilcha, San Pedro and Irruputunca volcanoes on the Andean Altiplano are dominated by such radiating linear features (Naranjo and Francis 1987; Francis and Wells 1988; de Silva and Francis 1991). The well-studied Socompa volcano debris avalanche deposit in northeastern Chile possesses each of the Martian landslide components (Francis and Self 1987;

Francis 1993), yet lacks any evidence for water-mobilised emplacement in the form of debris-flow–mudflow transitions, runoff channels or ice escape structures. The Andean landslide deposits with ridge and groove terrain appear to have been formed as dry debris avalanches deposited by enormous sturzstroms (Hsü 1975; Melosh 1979; Francis and Wells 1988; Campbell 1989). Calculations of the yield strength of materials involved in Martian landslides are similar to values obtained for dry debris avalanches on Earth (McEwan 1989). From these lines of evidence, it may be concluded that Martian landslides flowed across the surface in a similar manner without the aid of a lubricating fluid.

Climatic change on Mars

The most ancient terrains on Mars show ample evidence of water flowing across the surface during the first several hundred million years of the planet's geological history (Pieri 1980; Baker 1982). More controversial interpretations of the Martian landscape have offered clues for the existence of ancient oceans (Baker *et al.* 1991), lacustrine deposits (Parker *et al.* 1989; Scott *et al.* 1992; Williams and Zimbelman 1994) and glacial features (Lucchitta 1982; Kargel and Strom 1992), though such contentions remain highly speculative. During the present climatic epoch and probably for the past 3 billion or more years, liquid water has not existed on the Martian surface in equilibrium with the atmosphere. Despite the absence of liquid water and an Earth-like hydrological cycle, some regions of Martian landforms indicate a response to climatic fluctuations occurring on a timescale between 10 000 and 1 million years.

MARTIAN MILANKOVITCH CYCLES

On Earth, slight changes in the angle of the spin axis to the orbital plane (obliquity), the axial tilt direction (precession) which controls the seasonal timing of the closest approach to the Sun, and the shape of the orbit around the Sun (eccentricity) produce cyclical differences in the intensity of incoming radiation (Table 29.2). Mars also displays cyclical variations in its orbital geometry (Figure 29.14). The changes are much greater for Mars because its obliquity cycle encompasses a much greater range of values. The most important characteristic of the Martian Milankovitch cycles is their combined effect upon the amplitude of insolation change. Terrestrial ice ages and deglaciations appear to be determined by high-latitude insolation

variations of about 8% (Kutzbach 1987). Martian changes of high-latitude insolation approach 50% during a million-year cycle (Murray *et al.* 1973), and obliquity forced insolation changes may have been even greater over the past 10 million years (Bills 1990).

EVIDENCE FOR CHANGING CIRCULATION PATTERNS

The polar regions contain layered deposits almost unblemished by impact craters, which suggests a surface age of only a few million years. Each layer extends uniformly over a vast area and is approximately 30 m thick. High- and low-albedo units alternate to form a sequence of layered deposits exposed along spiral troughs radiating outward from the poles (Figure 29.15). At the highest latitudes of the northern hemisphere, a perennial H_2O ice cap rests upon the polar layered deposits (Kieffer *et al.* 1976), whereas a perennial CO_2 frost cap covers the South Pole (Kieffer 1979). The albedo of the north polar ice cap is only about 0.4, an indication of substantial aeolian sediment mixed with the ice (Kieffer *et al.* 1977).

A model proposed to account for deposition of polar layered deposits is regulated by the climatic change created by Milankovitch cycles (Toon *et al.* 1980). At times of low obliquity (12–20°), atmospheric CO_2 condenses in the polar regions and leads to a decrease in atmospheric pressure to about 1 mb, thus halting the movement of windblown sediments. During periods of low obliquity, largely sediment-free H_2O ice and CO_2 ice are deposited at the poles. When Martian obliquity reaches high values (30–38°), CO_2 ice sublimes from the poles and regolith to raise atmospheric pressure to perhaps 30 mb (Toon *et al.* 1980; François *et al.* 1990). The increased temperature gradient near

Table 29.2 *Milankovitch cycles for Mars and Earth (years)*

	Mars	Earth
Obliquity	1.2×10^5 and 1.3×10^6	9.5×10^4
Eccentricity	9.5×10^4 and 2.0×10^6	4.1×10^4
Precession	7.2×10^4 and 1.7×10^5	2.3×10^4 and 1.9×10^4

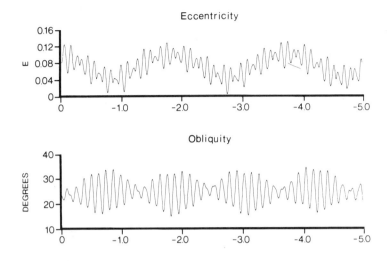

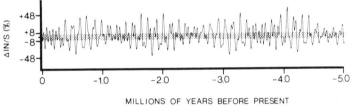

Figure 29.14 *Cyclical changes of Martian orbital eccentricity and obliquity cause periodic fluctuations in the amount of incident radiation. The shaded portion of the northern hemisphere insolation curve represents the range of variation in the Earth's insolation during the past 5 million years*

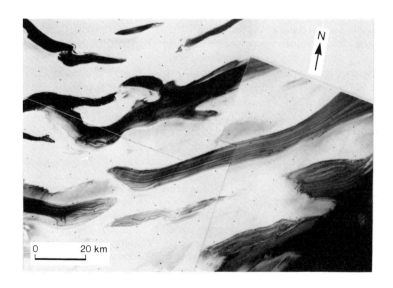

Figure 29.15 *Alternating bright and dark units compose the terrain of layered deposits in the north polar region (82°N, 269°W). [VO #0065B67-69]*

the polar caps and higher atmospheric pressure promote frequent global dust storms and saltation events during summer in both hemispheres. During periods of high obliquity, condensates of dusty ice accumulate on the winter pole. The colour and albedo characteristics of polar layered deposits are consistent with a mixture of dusty ice and saltated sediment (Herkenhoff and Murray 1990b). Following sublimation, a dark layer of aeolian sediment will insolate the underlying ice to produce alternating bright and dark bands deposited under high- and low-obliquity conditions. A 120 000-year cycle controls layer deposition, according to this hypothesis (Toon *et al.* 1980).

Layered deposits without volatiles are found at equatorial latitudes. In the region of Amazonis Planitia (7°N, 141°W), the deposits are 3.5 km in thickness with individual layers of about 100 m. Two sets of yardangs with directions nearly perpendicular to each other have formed upon the layered deposits (Figure 29.16). The older, northward-trending yardangs are partially buried by the uppermost layer. Deflation by easterly winds created the younger yardangs and exhumed the older set.

A model to explain the sequence of layer deposition and yardang formation by winds of different directions invokes climatic change governed by the Milankovitch cycles (Zimbel-man and Wells 1987). During periods of low obliquity, no aeolian sedimentation occurs, and an indurated horizon may form in the upper portion of the surface layer. At a time of high obliquity (or intermediate obliquity, high eccentricity and perihelion during northern hemisphere summer), cross-equatorial airflow returning to the northern hemisphere etched the set of yardangs aligned north–south (Figure 29.17A). During a phase of intermediate obliquity and low to intermediate eccentricity, convergent circulation along the equator produces easterly winds that transport sediments from the Tharsis Montes region and beyond for deposition in Amazonis Planitia (Figure 29.17B). Possible southern sources for the aeolian mantle burying the north–south yardang set are excluded by the apparent lack of sediments within that region during the preceding interval when yardangs were etched by southerly winds. Following the decline of available sediments from eastern source regions, the uppermost layer was excavated by easterly winds to expose the older yardang set and create a new set of east–west yardangs (Figure 29.17C). Though the Amazonis Planitia layered deposits comprise one of the youngest surfaces on Mars, it is unlikely that the episodic layer deposition and yardang formation represent the consequences of a single obliquity cycle.

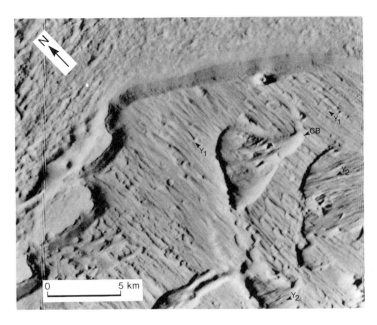

Figure 29.16   *Two sets of yardangs were formed by winds of different directions in the Amazonis Planitia region (7°N, 141°W). The older yardangs (Y_1) were buried by a deposit 100 m in thickness and later exposed by easterly winds that formed the younger yardangs (Y_2) and a cuspate deflation basin (CB). [VO #471S18]*

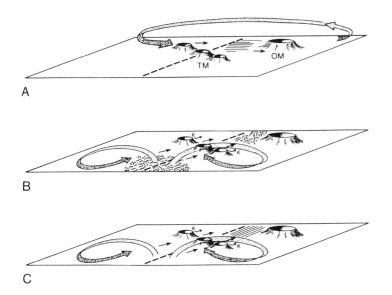

Figure 29.17 *A model to account for episodic layer deposition and yardang formation in the Amazonis Planitia region to the south of Olympus Mons (OM). (A) The older yardang set is eroded by circulation linked to a period of high obliquity (or intermediate obliquity, high eccentricity and northern hemisphere perihelion). (B) During a period of intermediate obliquity, convergent circulation transport sediments from the east of Tharsis Montes (TM) with additional material carried by katabatic winds (K) from the volcanic massif. (C) The younger yardangs form after the eastern source of sediments diminishes*

Arid surface processes on Venus

With a surface temperature hotter than the melting point of lead, Venus offers a near-ultimate case for the action of arid surface processes. While seemingly exotic in many aspects, the forces shaping the Venusian land-scape can produce results which are surprisingly terrestrial in their final forms, making Venus more a twin of our own planet than many would expect based upon general physical principles (Table 29.1). Information gained from the Magellan radar mapping mission of the early 1990s established that, like Earth, most of the surface of Venus is geologically young with an age of less than 500 million years (Head 1994; Strom *et al.* 1994). Unlike Earth, however, with its oceanic basins, mobile conti-nental plates and continually recycled crust, Venus experienced a global volcanic flooding event which covered the entire planet with mafic silicate basalts approximately 300 million years ago (Kargel 1994). Physiographically, 65% of the Venusian surface consists of rolling plains, 27% lies in basins below the mean surface datum (the 6051 km planetary radius), and 8% forms highlands with altitudes greater than 2 km (Sharpton and Head 1986). Most of Venus is covered by exposed bedrock with only about one-quarter of the surface overlain by unconsolidated sediments (Greeley 1994).

The abundance of carbon dioxide in the atmosphere, coupled with trace amounts of sulphur dioxide, sulphuric acid and water vapour, produces an extreme greenhouse effect, warming the surface by an additional 227°C (Cattermole 1994). High surface temperatures are held constant because the obliquity of the Venusian rotation axis creates no seasonal changes of insolation between hemispheres (Sagan 1976), and heat retention by the dense atmosphere allows only a 1–3°C difference between day and night-time temperatures (Zolotov and Volkov 1992). Differences in surface temperature and pressure stem solely from changes in altitude and vary from 465°C and 96 bars at the lowest elevations to 374°C and 41 bars at peak elevations in the highlands (Greeley 1994). The high level of atmospheric carbon dioxide is probably buffered by the mineral reaction between calcite and quartz producing wallastonite and carbon dioxide, a reaction favoured by the range of surface tem-peratures and pressures on Venus (Urey 1952; McGill *et al.* 1983). Chemically reactive gases (CO, CO_2, COS, SO_2, H_2O, H_2S, HCl, HF) in the lower Venusian atmosphere raise the possibility of surface weathering by gas–solid interactions (Florensky *et al.* 1976; Nozette and Lewis 1982; Zolotov and Volkov 1992). Although mixing ratios of corrosive gases in the lower atmosphere are poorly known, chemical analyses performed using X-ray florescence at three landing sites (Venera 13, 14; Vega 2) suggest an enrichment of sulphur in surface materials as a result of chemical weathering processes (Basilevsky *et al.* 1992). Another sign of weathering is the near-infrared brightness of soil and surface rocks at the Venera 13 landing site, which may be attributable to the formation of ferric oxides in an oxidising environment (Pieters *et al.* 1986).

Altitude-dependent chemical weathering processes may explain an unusual feature detected by the Magellan radar imaging system. At elevations approximately 2.8 km above the mean planetary radius, a sharp boundary occurs between lower regions having low radar reflection coefficients and highland areas with high reflection coefficients (Arvidson *et al.* 1991; 1992; Tyler *et al.* 1991). Contacts between radar-bright and radar-dark surfaces can be traced for hundreds of kilometres with little change in elevation, leading to the conclusion that a global chemical phase change controlled by atmospheric temperature and pressure occurs at a specific altitude above the surface. Such a phase change would alter the dielectric constant of exposed rock surfaces by producing a different kind of weathering crust containing magnetite, haematite, ilmenite or other minerals that remain unstable at lower elevations (Arvidson *et al.* 1991; Zolotov and Volkov 1992). At still higher elevations, surfaces again become radar-dark in areas where highland summit rocks may be coated with pyrite-bearing minerals stable only at the highest elevations on Venus (Arvidson *et al.* 1992). An intriguing consequence of these chemical equilibrium changes is the possibility that metal halides liberated in vapour state in the hot (*c.* 460°C) lowlands of Venus are transported by atmospheric circulation to higher, cooler (*c.* 380°C) elevations, where they precipitate to create a uniform surface layer (Brackett *et al.* 1994). Crystal growth in rock pore spaces by condensing metal halides may mechanically destroy highland rocks and may explain the increased radar roughness characteristics noted for highland terrains.

Despite ample evidence for active chemical weathering processes on Venus, most fine granular material on the surface appears to originate from ejecta blankets deposited during impact crater formation (Garvin 1990) with,

perhaps, minor additional contributions from volcanic eruptions of basaltic cinders and ashes (Basilevsky *et al.* 1992). In contrast to Earth and Mars, the sandblasting of surface rocks does not play a significant role in particle formation. Aeolian abrasion of coherent rocks is orders of magnitude less efficient on Venus than on Earth and Mars because low surface wind velocities fail to transmit sufficient kinetic energy to airborne mineral grains (McGill *et al.* 1983). As documented in a series of panoramic views relayed by the Venera 13 lander showing the sequential removal of fine materials from the lander support ring, aeolian transport of sediments by saltation and suspension occurs on Venus (Basilevsky *et al.* 1992). Wind-tunnel experiments under Venusian surface conditions confirm that saltation involving a 0.06 mm grain will take place at a threshold friction speed of 0.17 ms^{-1} (Iversen and White 1982; White 1986), which may be extrapolated to a wind velocity of approximately 1 ms^{-1} measured at a level 1 m above the surface (Figure 29.7). Saltation path lengths are short and trace low-angle trajectories on Venus, and should create ripple-scale bedforms having higher amplitudes and shorter wavelengths than terrestrial sand ripples (Greeley and Iversen 1985). An unusual characteristic of basaltic materials placed in saltation under the high temperatures and pressures of the Venusian surface is the growth of thin weathering veneers created by debris comminuted from basaltic grains onto impacted rock surfaces (Greeley *et al.* 1987).

In a manner similar to comparable features on Mars, wind-streaks on Venus are closely associated with impact craters, volcanic domes and other topographic barriers that disrupt and accelerate airflow (Figure 29.18). Radar-bright streaks typically extend 10–20 km in the lee of obstructions and represent zones swept free of aeolian sediments (Greeley *et al.* 1992a); thus, the radar-bright streaks of Venus correspond to the visibly dark crater streaks seen on Mars (Figure 29.5). Other wind-streaks are radar-dark and represent the concentration of a surface layer of unconsolidated aeolian sediments at least 10–70 cm deep (Greeley *et al.* 1992a). The global pattern of wind-streak directions can be used to construct a map of prevailing surface winds related to the global Hadley circulation (Schubert 1983; Greeley *et al.*

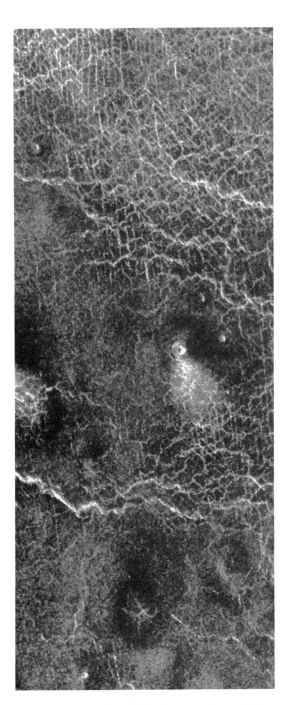

Figure 29.18   *Radar-bright wind-streaks extend in fan-shaped patterns in the lee of small volcanic cones in the Guinevere Planitia region of Venus (22.3°N, 332.1°E). [Magellan F-MIDR #20N334]*

1994). On Venus, the general circulation is symmetrical about the equator and consists of rising limbs of Hadley cells along the equator, poleward winds aloft, downflow over the polar regions and equatorward surface winds. A global map of wind-streak directions beautifully reflects the strongly meridional character of Venusian surface winds (Greeley *et al.* 1994).

Evidence for aeolian deposition on Venus includes landscape features created by both dust- and sand-sized particles. Dark horseshoe-shaped haloes associated with young impact craters are thought to be the consequence of pre-impact shock waves which swept dust from the surface to altitudes greater than 10 km, where the dust clouds interacted with easterly winds of the atmospheric super-rotation (Arvidson *et al.* 1992; Greeley *et al.* 1992a). The parabolic haloes of radar-dark sediments extend hundreds of kilometres to the west of several of the largest and youngest craters in a pattern which suggests aeolian dispersal by easterly winds. Given the low rate of large impact crater formation during the past billion years, the presence of parabolic haloes adjacent to approximately 10% of all large craters leads to the conclusion that dust deposits may persist as distinct units for tens of millions of years on the Venusian surface (Arvidson *et al.* 1992).

Imaging radar systems do not facilitate detection of aeolian bedforms because dry, granular sediments are poor radar reflectors. Despite this inherent difficulty in the use of Magellan radar imagery, two small dunefields have been located on Venus. The Algaonice dunefield (25°N, 340°E) of transverse ridge bedforms covers 1290 km² and lies near the radar-bright ejecta deposits of Algaonice Crater (Greeley *et al.* 1992a). The Fortuna–Meshkenet dunefield (67.7°N, 90.5°E) is also situated in a heavily cratered region and contains 17 120 km² of transverse dunes. Bedform lengths of the transverse dunes range from 0.5 to 10 km with average widths of 200 m and crest-to-crest spacings of approximately 500 m (Greeley *et al.* 1992a). The apparent scarcity of dunes on Venus may be related to the limited supply of saltable sediments available from crater ejecta blankets. Alternatively, there may remain many undiscovered dunes on Venus which are simply too small to be detected by the Magellan imaging radar system.

Another kind of aeolian deposit may be more ubiquitous than any other discussed in this section. It is possible that the bedded slab-like rocks seen in surface views of the Venera 10, 13 and 14 landing sites are indurated aeolian sediments (Florensky *et al.* 1977; McGill *et al.* 1983). The low bulk density (*c.* 2 g cm^{-3}) of rock samples examined at the sites is consistent with the expected properties of an extraterrestrial eolianite (McGill *et al.* 1983).

Although wind erosion is orders of magnitude less effective on Venus than on Earth and Mars, yardang-like features are present in the region of Aphrodite Terra (9°N, 60.5°E) and have been ascribed to the aeolian stripping of a friable ejecta deposit derived from Meade Crater (Greeley *et al.* 1992a). The parallel ridges and grooves attain lengths of 25 km with widths of 0.5 km and are aligned in the same direction as wind-streaks in the region.

Mass-wasting processes are apparent in surface views of the Venusian landscape and from radar imagery of highland regions. At the Venera 9 landing site, the panoramic scene displayed detached rocks and slab-like boulders descending a 20° slope (Florensky *et al.* 1977). Steep slopes in mountainous regions serve as sources for slump blocks and large landslides observed on radar images (Arvidson *et al.* 1991; Malin 1992). Debris avalanche deposits resembling those emplaced by giant terrestrial and Martian landslides occur in the region of Maxwell Montes (Vorder-Bruegge 1994).

Careful inspection of the landforms discovered to date on Venus has yet to reveal any evidence for climatic change. Since Venus does not experience significant cyclical variations in its orbital geometry, a possible mechanism for global climatic change is not readily apparent. It would seem that Venus has suffered a monotonously torrid environment for at least the last several hundred million years.

Future planetary research

After reading this review of the current state of knowledge concerning arid surface processes on Mars and Venus, one may be interested in the prospects for resolving some areas of debate with information returned from future planetary missions. Several missions are planned for the remainder of the century.

The Mars Global Surveyor (MGS) spacecraft is scheduled to provide the initial recovery following the loss of the Mars Observer spacecraft in 1993. Three of the MGS instruments will be of particular importance to investigations of arid surface processes on Mars. The Mars Observer Camera (MOC) will provide global synoptic images of clouds and large regions of the surface at low spatial resolution (7.5 km per pixel) and many very high resolution (1.4 m per pixel) views of selected sites around the planet (Malin *et al.* 1992). The Thermal Emission Spectrometer (TES) will gather emission spectra of the surface and atmosphere in the 6 to 50 μm range at a spatial resolution of 3 km (Christensen *et al.* 1992). The Mars Observer Laser Altimeter (MOLA) will obtain altimetry measurements with 1.5 m vertical resolution over 160 m swaths, which will generate surface elevations and gradients over a global grid of 0.2° × 0.2° cells (Zuber *et al.* 1992). The Mars Global Surveyor is scheduled to be placed into a mapping orbit around Mars in 1998. Images and measurements by the MOC, TES and MOLA will permit new numerical tests to be applied to many of the hypotheses discussed in this chapter regarding surfaces processes, landform development, the existence of volatiles and the climatic history of Mars.

Another NASA spacecraft should arrive on Mars before the MGS mission. Mars Pathfinder is scheduled to place an instrumented lander on the Martian surface in July of 1997. The lander's colour imaging system will have the capability to generate stereoscopic views and to image wind socks at three discrete heights above the Martian surface, providing the first extraterrestrial measurements of boundary layer structure. A small solar-charged, battery-powered, 'rover' vehicle will carry an alpha proton X-ray spectrometer to be placed in direct contact with rocks identified using the lander's colour imaging system (Golombek and Spear 1994). Mars Pathfinder is designed primarily as an engineering test-bed, but important scientific results should be obtained during its mission.

Several other advanced spacecraft missions are currently under study by NASA, the European Space Agency and the space agencies of Japan and Russia. If funded and launched, most of the missions will make use of miniaturised science and engineering components that reduce the required size and mass of a spacecraft while enhancing the potential scientific return from their data.

The ultimate goal of various spacecraft missions is to prepare the way for human exploration. Only then can some of the issues raised in this review reach a satisfactory resolution. Afterwards we can look forward to the kind of stimulating debate concerning the arid surfaces of Mars and Venus that followed unexpected discoveries during the initial exploration of remote regions of the Earth.

References

Arvidson, R., Guiness, E. and Lee, S. 1979. Differential aeolian redistribution rates on Mars. *Nature*, 278: 533–535.

Arvidson, R.E., Baker, V.R., Elachi, C., Saunders, R.S. and Wood, J.A. 1991. Magellan: initial analysis of Venus surface modification. *Science*, 252: 270–275.

Arvidson, R.E., Greeley, R., Malin, M.C., Saunders, R.S., Izenberg, N., Plaut, J.J., Stofan, E.R. and Shepard, M.K. 1992. Surface modification of Venus inferred from Magellan observations of plains. *Journal of Geophysical Research*, 97: 13303–13317.

Baird, A.K., Toulmin, P., Clark, B.C., Rose, H.J. Jr, Keil, K., Christian, R.P. and Gooding, J.L. 1976. Mineralogic and petrologic implications of Viking geochemical results from Mars: interim report. *Science*, 194: 1288–1293.

Baker, V.R. 1982. *The channels of Mars*. University of Texas Press, Austin.

Baker, V.R., Strom, R.G., Gulick, V.C., Kargel, J.S., Komatsu, G. and Kale, V.S. 1991. Ancient oceans, ice sheets and the hydrological cycle on Mars. *Nature*, 352: 589–594.

Barlow, N.G. 1994. Impact craters as indicators of subsurface H_2O on Mars. *Lunar and Planetary Science Conference*, 25: 59-60 (abstract).

Barnes, J.R. 1990. Transport of dust to high northern latitudes in a Martian polar warming. *Journal of Geophysical Research*, 95: 1381–1400.

Barnes, J.R., Pollack, J.B., Haberle, R.M., Leovy, C.B., Zurek, R.W., Lee, H. and Schaeffer, J. 1993. Mars atmospheric dynamics as simulated by the NASA Ames general circulation model, 2. Transient baroclinic eddies. *Journal of Geophysical Research*, 98: 3125–3148.

Basilevsky, A.T., Nikolayeva, O.V. and Kuzmin, R.O. 1992. Resurfacing. In V.L. Barsukov, A.T. Basilevsky, V.P. Volkov and V.N. Zharkov (eds), *Venus geology, geochemistry, and geophysics*. University of Arizona Press, Tucson: 153–160.

Bills, B.G. 1990. The rigid body obliquity history of Mars. *Journal of Geophysical Research*, 95: 14137–14153.

Bowler, J.M. 1973. Clay dunes: their occurrence, formation and environmental significance. *Earth Science Reviews*, 9: 315–378.

Brackett, R.A., Fegley, B. Jr and Arvidson, R.A. 1994. Vapor transport, weathering, and the highlands of Venus. *Lunar and Planetary Science Conference*, 25: 157–158 (abstract).

Breed, C.S., Grolier, M.J. and McCauley, J.F. 1979. Morphology and distribution of common 'sand' dunes on Mars: comparison with the Earth. *Journal of Geophysical Research*, 84: 8183–8204.

Briggs, G.A., Baum, W.A. and Barnes, J. 1979. Viking orbiter imaging observations of dust in the martian atmosphere. *Journal of Geophysical Research*, 84: 2795–2820.

Burns, R.G. 1994. Schwertmannite on Mars: deposition of this ferric oxyhydroxysulfate mineral in acidic saline meltwaters. *Lunar and Planetary Science Conference*, 25: 203–204 (abstract).

Burns, R.G. and Fisher, D.S. 1990. Iron–sulfur mineralogy of Mars: magmatic evolution and chemical weathering products. *Journal of Geophysical Research*, 95: 14415–14421.

Campbell, C.S. 1989. Self-lubrication for long runout landslides. *Journal of Geology*, 97: 653–665.

Carr, M.H. 1981. *The surface of Mars*. Yale University Press, New Haven, CT.

Carr, M.H. 1986. Mars: a water-rich planet? *Icarus*, 68: 187–216.

Cattermole, P. 1994. *Venus*. Johns Hopkins University Press, Baltimore.

Christensen, P.R. 1986a. Seasonal variability of Martian surface albedos: implications for dust deposition and removal. *Lunar and Planetary Science Conference*, 17: 121–122 (abstract).

Christensen, P.R. 1986b. Regional dust deposits on Mars: physical properties, age, and history. *Journal of Geophysical Research*, 91: 3534–3536.

Christensen, P.R. 1986c. The spatial distribution of rocks on Mars. *Icarus*, 68: 217–238.

Christensen, P.R. 1988. Global albedo variations on Mars: implications for active aeolian transport, deposition, and erosion. *Journal of Geophysical Research*, 93: 7611–7624.

Christensen, P.R., Anderson, D.L., Chase, S.C., Clark, R.N., Kieffer, H.H., Malin, M.C., Pearl, J.C., Carpenter, J., Bandiera, N., Brown, F.G. and Silverman, S. 1992. Thermal Emission

Spectrometer Experiment: Mars Observer Mission. *Journal of Geophysical Research*, 97: 7719–7734.

Clark, B.C. and Van Hart, D.C. 1981. The salts of Mars. *Icarus*, 45: 370–378.

Clark, B.C., Baird, A.K., Rose, H.J. Jr, Toulmin, P., Keil, K., Castro, A.J., Kelliher, W.C., Rowe, C.D. and Evans, P.H. 1976. Inorganic analyses of Martian surface samples at the Viking landing sites. *Science*, 194: 1283–1288.

de Silva, S.L. and Francis, P.W. 1991. *Volcanoes of the central Andes*. Springer-Verlag, New York.

Donahue, T.M. 1995. Evolution of water reservoirs on Mars from D/H ratios in the atmosphere and crust. *Nature*, 374: 432–434.

Edgett, K.S. and Blumberg, D.G. 1994. Star and linear dunes on Mars. *Icarus*, 112: 448–464.

Edgett, K.S. and Christensen, P.R. 1991. The particle size of martian aeolian dunes. *Journal of Geophysical Research*, 96: 22765–22776.

Farmer, C.B. and Doms, P.E. 1979. Global seasonal variation of water vapor on Mars and the implications for permafrost. *Journal of Geophysical Research*, 84: 2881–2888.

Florensky, C.P., Ronca, L.B, Basilevsky, A.T., Burba, G.A., Nikolaeva, O.V., Pronin, A.A., Trakhtman, A.M., Volkov, V.P. and Zazetsky, V.V. 1976. The surface of Venus as revealed by Soviet Venera 9 and 10. *Bulletin of the Geological Society of America*, 88: 1537–1545.

Francis, P.W. 1993. *Volcanoes: a planetary perspective*. Clarendon Press, Oxford.

Francis, P.W. and Self, S. 1987. Collapsing volcanoes. *Scientific American*, 256: 90–97.

Francis, P.W. and Wells, G.L. 1988. Landsat Thematic Mapper observations of debris avalanche deposits in the Central Andes. *Bulletin of Volcanology*, 50: 601–621.

François, L.M., Walker, J.C.G. and Kuhn, W.R. 1990. A numerical simulation of climatic changes during the obliquity cycle on Mars. *Journal of Geophysical Research*, 95: 14761–14778.

Garvin, J.B. 1990. The global budget of impact-derived sediment on Venus. *Earth, Moon, Planets*, 50/51: 175–190.

Golombek, M.P. and Spear, A.J. 1994. Mars Pathfinder science investigations and objectives. *Proceedings of the Congress of the International Astronomical Federation*, 45: 1 (abstract).

Goudie, A.S. and Wells, G.L. 1995. The nature, distribution and formation of pans in arid zones. *Earth Science Reviews*, 38: 1–69.

Grant, J.A. and Schultz, P.H. 1987. Possible tornado-like tracks on Mars. *Science*, 237: 883–885.

Greeley, R. 1979. Silt–clay aggregates on Mars. *Journal of Geophysical Research*, 84: 6248–6254.

Greeley, R. 1994. *Planetary landscapes*. Chapman & Hall, London.

Greeley, R. and Iversen, J.D. 1985. *Wind as a geological process on Earth, Mars, Venus and Titan*. Cambridge University Press, Cambridge.

Greeley, R., Iversen, J.D., Pollack, J.B., Udovich, N. and White, B. 1974. Wind tunnel simulations of light and dark streaks on Mars. *Science*, 183: 847–849.

Greeley, R., Williams, S., White, B.R., Pollack, J., Marshall, J. and Krinsley, D. 1984. *Abrasion by aeolian particles: Earth and Mars*. NASA Contractor Report 3788. Washington, DC.

Greeley, R., Marshall, J.R. and Pollack, J.B. 1987. Physical and chemical modification of the surface of Venus by windblown particles. *Nature*, 327: 313–315.

Greeley, R., Arvidson, R.E., Elachi, C., Geringer, M.A., Plaut, J.J., Saunders, R.S., Schubert, G., Stofan, E.R., Thouvenot, E.J.P., Wall, S.D. and Weitz, C.M. 1992a. Aeolian features on Venus: Preliminary Magellan results. *Journal of Geophysical Research*, 97: 13319–13345.

Greeley, R., Lancaster, N., Lee, S. and Thomas, P. 1992b. Martian aeolian processes, sediments, and features. In H.H. Kieffer, B.M. Jakosky, C.W. Snyder and M.S. Matthews (eds), *Mars*. University of Arizona Press, Tucson: 730–766.

Greeley, R., Skypeck, A. and Pollack, J.B. 1993. Martian aeolian features and deposits: comparisons with general circulation model results. *Journal of Geophysical Research*, 98: 3183–3196.

Greeley, R., Schubert, G., Limonadi, D., Bender, K.C., Newman, W.I., Thomas, P.E., Weitz, C.M. and Wall, S.D. 1994. Wind streaks on Venus: clues to atmospheric circulation. *Science*, 263: 358–361.

Griffith, L.L. and Arvidson, R.E. 1994. Mars — It's what's inside that counts. *Lunar and Planetary Science Conference*, 25: 481–482 (abstract).

Guiness, E.A., Leff, C.E. and Arvidson, R.E. 1982. Two years of surface changes seen at the Viking landing sites. *Journal of Geophysical Research*, 87: 10051–100058.

Haberle, R.M. 1986a. The climate of Mars. *Scientific American*, 254 (5): 54–62.

Haberle, R.M. 1986b. Interannual variability of global dust storms on Mars. *Science*, 234: 459–461.

Haberle, R.M., Pollack, J.B., Barnes, J.R., Zurek, R.W., Leovy, C.B., Murphy, J.R., Lee, H. and Schaeffer, J. 1993. Mars atmospheric dynamics as simulated by the NASA Ames general circulation model, 1. The zonal-mean circulation. *Journal of Geophysical Research*, 98: 3093–3123.

Harmon, J.K., Campbell, D.B. and Ostro, S.J. 1982. Dual-polarization radar observations of Mars: Tharsis and environs. *Icarus*, 52: 171–187.

Hartmann, W.K. 1972. *Moons and planets*. Bogden & Quidly, Tarrytown, NY.

Head, J.W. III. 1994. Venus after the flood. *Nature*, 372: 729–730.

Herkenhoff, K.E. and Murray, B.C. 1990a. High-resolution topography and albedo of the south polar layered deposits on Mars. *Journal of Geophysical Research*, 95: 14511–14529.

Herkenhoff, K.E. and Murray, B.C. 1990b. Color and albedo of the south polar layered deposits on Mars. *Journal of Geophysical Research*, 95: 1343–1358.

Hess, S.L., Henry, R.M., Leovy, C.B., Ryan, J.A. and Tillman, J.E. 1977. Meteorological results from the surface of Mars: Viking 1 and 2. *Journal of Geophysical Research*, 82: 4559–4574.

Howard, A.D. 1980. *Effect of wind on scarp evolution on the Martian poles*. NASA Technical Memo, TM 82385: 333–335.

Hsü, K.J. 1975. Catastrophic debris streams (sturzstroms) generated by rockfalls. *Bulletin of the Geological Society of America*, 86: 129–140.

Huguenin, R.L. 1976. Mars: chemical weathering as a massive volatile sink. *Icarus*, 28: 203–212.

Huguenin, R.L. 1982. Chemical weathering and the Viking biology experiments on Mars. *Journal of Geophysical Research*, 87: 10069–10082.

Hunt, G.E. and James, P.B. 1979. Martian extratropical cyclones. *Nature*, 278: 531–532.

Iversen, J.D. and White, B.R. 1982. Saltation threshold on Earth, Mars and Venus. *Sedimentology*, 29: 111–119.

Jakosky, B.M. and Christensen, P.R. 1986. Global duricrust on Mars: analysis of remote-sensing data. *Journal of Geophysical Research*, 91: 3547–3559.

Kahn, R.A., Martin, T.Z. and Zurek, R.W. 1992. The Martian dust cycle. In H.H. Kieffer, B.M. Jakosky, C.W. Snyder and M.S. Matthews (eds), *Mars*. University of Arizona Press, Tucson: 1017–1053.

Kargel, J.S. 1994. An alluvial depositional analog for some volcanic plains on Venus. *Lunar and Planetary Science Conference*, 25: 667–668 (abstract).

Kargel, J.S. and Strom, R.G. 1992. Ancient glaciation on Mars. *Geology*, 20: 3–7.

Kieffer, H.H. 1979. Mars south polar spring and summer temperatures: a residual CO_2 frost. *Journal of Geophysical Research*, 84: 8263–8288.

Kieffer, H.H., Chase, S.C., Miner, E.D., Munch, G. and Neugebauer, G. 1973. Preliminary report on infrared radiometric measurements from the Mariner 9 spacecraft. *Journal of Geophysical Research*, 78: 4291–4312.

Kieffer, H.H., Chase, S.C., Martin, T.Z., Miner, E.D. and Palluconi, F.D. 1976. Martian north pole summer temperatures: Dirty water ice. *Science*, 194: 1341–1344.

Kieffer, H.H., Martin, T.Z., Peterfreund, A.R., Jakosky, B.M., Miner, E.D. and Palluconi, F.D. 1977. Thermal and albedo mapping of Mars during the Viking primary mission. *Journal of Geophysical Research*, 82: 4249–4291.

Kutzbach, J.E. 1987. Model simulations of the climatic patterns during the deglaciation of North America. In W.F. Ruddiman and H.E. Wright, Jr (eds), *North American and adjacent oceans during the last deglaciation*. The geology of North America, Vol. K-3. Geological Society of America, Boulder, CO: 425–446.

Lancaster, N. and Greeley, R. 1987. Mars: morphology of southern hemisphere intracrater dunefields. *Reports of Planetary Geology and Geophysics Program — 1986*. NASA Technical Memo, TM 89810: 264–265.

Lancaster, N. and Greeley, R. 1990. Sediment volume in the north polar sand seas of Mars. *Journal of Geophysical Research*, 95: 10921–10927.

Lee, S.W. and Thomas, P.C. 1995. Longitudinal dunes on Mars: relation to current wind regimes. *Journal of Geophysical Research*, 100: 5381–5395.

Lee, S.W., Thomas, P.C. and Veverka, J. 1982. Wind streaks in Tharsis and Elysium: implications for sediment transport by slope winds. *Journal of Geophysical Research*, 87: 10025–10041.

Leovy, C.B. 1979. Martian meteorology. *Annual Review of Astronomy and Astrophysics*, 17: 387–413.

Leovy, C.B. and Zurek, R.W. 1979. Thermal tides and martian dust storms: direct evidence for coupling. *Journal of Geophysical Research*, 84: 2956–2968.

Lucchitta, B.K. 1978a. Morphology of chasma walls, Mars. *Journal of Research of the US Geological Survey*, 6: 651–662.

Lucchitta, B.K. 1978b. A large landslide on Mars. *Bulletin of the Geological Society of America*, 89: 1601–1609.

Lucchitta, B.K. 1979. Landslides in Valles Marineris, Mars. *Journal of Geophysical Research*, 84: 8097–8113.

Lucchitta, B.K. 1982. Ice sculpture in the Martian outflow channels. *Journal of Geophysical Research*, 87: 9951–9973.

Lucchitta, B.K. 1984. Ice and debris in the fretted terrain, Mars. *Journal of Geophysical Research*, 89B: 409–414.

Lucchitta, B.K. 1987. Valles Marineris: wet debris flows and ground ice. *Icarus*, 72: 411–429.

Magalhaes, J.A. 1987. The martian Hadley circulation: Comparison of 'viscous' model predictions to observations. *Icarus*, 70: 442–468.

Magalhaes, J. and Gierasch, P. 1982. A model of martian slope winds: implications for eolian transport. *Journal of Geophysical Research*, 87: 9975–9984.

Mainguet, M. and Chemin, H.-C. 1985. Sand seas of the Sahara and Sahel: an explanation of their thickness and sand dune type by the sand budget principle. In M.E. Brookfield and T.S. Ahlbrandt (eds), *Eolian sediments and processes.* Elsevier, Amsterdam: 343–352.

Malin, M.C. 1974. Salt weathering on Mars. *Journal of Geophysical Research,* 79: 3888–3894.

Malin, M.C. 1992. Mass movements on Venus: preliminary results from Magellan cycle 1 observations. *Journal of Geophysical Research,* 97: 16337–16353.

Malin, M.C., Danielson, G.E., Ingersoll, A.P., Masursky, H., Veverka, J., Ravine, M.A. and Soulanille, T.A. 1992. Mars Observer Camera. *Journal of Geophysical Research,* 97: 7699–7718.

Mass, C. and Sagan, C. 1976. A numerical circulation model with topography for the martian southern hemisphere. *Journal of Atmospheric Sciences,* 33: 1418–1430.

McCord, T.B. and Westphal, J.A. 1971. Mars: narrow-band photometry, from 0.3 to 2.5 microns, of surface regions during the 1969 apparition. *Astrophysical Journal,* 168: 141–153.

McEwen, A.S. 1989. Mobility of large rock avalanches: evidence from Valles Marineris, Mars. *Geology,* 17: 1111–1114.

McGill, G.E., Warner, J.L., Malin, M.C., Arvidson, R.E., Eliason, E., Nozette, S. and Reasenberg, R.D. 1983. Topography, surface properties, and tectonic evolution. In D.M. Hunten, L. Colin, T.M. Donahue and V.I. Moroz (eds), *Venus.* University of Arizona Press, Tucson: 69–130.

McKee, E.D. 1966. Structures of dunes at White Sands National Monument, New Mexico. *Sedimentology,* 7: 1–69.

Melosh, H.J. 1979. Acoustic fluidization: a new geologic process? *Journal of Geophysical Research,* 84: 7513–7520.

Moore, H.J. and Jakosky, B.M. 1989. Viking landing sites, remote sensing observations, and physical properties of martian surface materials. *Icarus,* 81: 164–184.

Moore, H.J., Hutton, R.E., Clow, G.D. and Spitzer, C.R. 1987. *Physical properties of the surface materials at the Viking landing sites on Mars.* US Geological Survey, Professional Paper, 1389.

Moore, J.M. 1990. Nature of the mantling deposit in the heavily cratered terrain of northeastern Arabia, Mars. *Journal of Geophysical Research,* 95: 14279–14289.

Moore, J.M. and Edgett, K.S. 1993. Hellas Planitia, Mars: site of net dust erosion and implications for the nature of basin floor deposits. *Geophysical Research Letters,* 20: 1599–1602.

Mouginis-Mark, P. 1979. Martian fluidized crater morphology: variations with crater size, latitude, altitude, and target material. *Journal of Geophysical Research,* 84: 8011–8022.

Mouginis-Mark, P.J. 1987. Water or ice in the martian regolith?: clues from rampart craters seen at very high resolution. *Icarus,* 71: 268–286.

Murray, B.C., Ward, W.R. and Yeung, S.C. 1973. Periodic insolation variations on Mars. *Science,* 180: 638–640.

Mutch, T.A. and Jones, K.L. 1978. *The martian landscape.* NASA Publication 425.

Naranjo, J.A. and Francis, P. 1987. High velocity debris avalanche at Lastarria volcano in the north Chilean Andes. *Bulletin of Volcanology,* 49: 509–514.

Nozette, S. and Lewis, J.S. 1982. Venus: chemical weathering of igneous rocks and buffering of atmospheric composition. *Science,* 216: 181–183.

Parker, T.J., Saunders, R.S. and Schneeberger, D.M. 1989. Transitional morphology in west Deuteronilus Mensae, Mars: implications for modification of the lowland/upland boundary. *Icarus,* 82: 111–145.

Peterfreund, A.R. 1985. Local dust storms and global opacity on Mars as detected by the Viking IRTM. In S. Lee (ed.), *Workshop on dust on Mars.* LPI Technical Report 85-02. Lunar and Planetary Institute, Houston: 10–14.

Pieri, D.C. 1980. Martian valleys: Morphology, distribution, age, and origin. *Science,* 210: 895–897.

Pieters, C.M., Head, J.W., Patterson, W., Pratt, S., Garvin, J., Barsukov, V.L., Basilevsky, A.T., Khodakovsky, I.L., Selivanov, A.S., Panfilov, A.S., Gektin, Yu.M. and Narayeva, Y.M. 1986. The color of the surface of Venus. *Science,* 234: 1379–1383.

Plaut, J.J., Kahn, R., Guinness, E.A. and Arvidson, R.E. 1988. Accumulation of sedimentary debris in the south polar region of Mars and implication for climate history. *Icarus,* 75: 357–377.

Pleskot, L.K. and Miner, E.D. 1981. Time variability of martian bolometric albedo. *Icarus,* 45: 427–441.

Pollock, J.B., Leovy, C.B., Greiman, P.W. and Mintz, Y. 1981. A martian general circulation experiment with large topography. *Journal of Atmospheric Sciences,* 38: 3–29.

Robertson, D.F. 1987. U.S. Mars Observer seeks global picture. *Astronomy,* 15 (11): 33–37.

Rossbacher, L.A. and Judson, S. 1981. Ground ice on Mars: inventory, distribution and resulting landforms. *Icarus,* 45: 39–59.

Ryan, J.A. and Henry, R.M. 1979. Mars atmospheric phenomena during major dust storms, as measured at surface. *Journal of Geophysical Research,* 84: 2821–2829.

Sagan, C. 1976. Erosion and the rocks of Venus. *Nature,* 261: 31.

Sagan, C., Pieri, D. and Fox, P. 1977. Particle

motion on Mars inferred from the Viking lander cameras. *Journal of Geophysical Research*, 82: 4430–4438.

Saunders, R.S. and Blewett, D.T. 1987. Mars north polar dunes: possible formation from low-density sediment aggregates. *Astonomicheskii Vestnik*, 21: 181–188.

Schubert, G. 1983. General circulation and the dynamical state of the Venus atmosphere. In D.M. Hunten, L. Colin, T.M. Donahue and V.I. Moroz (eds), *Venus*. University of Arizona Press, Tucson: 681–765.

Scott, D.H. and Tanaka, K.L. 1982. Ignimbrites of Amazonis Planitia region of Mars. *Journal of Geophysical Research*, 87: 1179–1190.

Scott, D.H., Chapman, M.G., Rice, J.W. and Dohm, J.M. 1992. New evidence of lacustrine basins on Mars: Amazonis and Utopia Planitia. *Lunar and Planetary Science Proceedings*, 22: 53–62.

Sharp, R.P. and Malin, M.C. 1984. Surface geology from Viking landers on Mars: a second look. *Bulletin of the Geological Society of America*, 95: 1398–1412.

Sharpton, V.L. and Head, J.W. III 1986. A comparison for the regional slope characteristics of Venus and Earth: implications for geologic processes on Venus. *Journal of Geophysical Research*, 91: 7545–7554.

Squyres, S.W. 1978. Martian fretted terrain: flow of erosional debris. *Icarus*, 34: 600–613.

Squyres, S.W. 1979. The distribution of lobate debris aprons and similar flows on Mars. *Journal of Geophysical Research*, 84: 8087–8096.

Squyres, S.W. 1984. The history of water on Mars. *Annual Review of Earth and Planetary Sciences*, 12: 83–106.

Squyres, S.W. 1989. Urey Prize lecture: water on Mars. *Icarus*, 79: 229–288.

Squyres, S.W. and Carr, M.H. 1986. Geomorphic evidence for the distribution of ground ice on Mars. *Science*, 231: 249–252.

Stephens, S.K., Stevenson, D.J., Keyser, L.F. and Rossman, G.R. 1994. Carbonate formation on Mars: Implications of recent experiments. *Lunar and Planetary Science Conference*, 25: 1343–1344 (abstract).

Strom, R.G., Schaber, G.G. and Dawson, D.D. 1994. The global resurfacing history of Venus. *Lunar and Planetary Science Conference*, 25: 1353–1354 (abstract).

Tanaka, K.L. and Leonard, G.J. 1994. Eolian history of the Hellas region of Mars. *Lunar and Planetary Science Conference*, 25: 1379–1380 (abstract).

Thomas, P. 1982. Present wind velocity on Mars: relation to large latitudinally zoned sediment deposits. *Journal of Geophysical Research*, 87: 9999–10008.

Thomas, P. and Gierasch, P.J. 1985. Dust devils on Mars. *Science*, 231: 175–177.

Thomas, P. and Gierasch, P.J. 1995. Polar margin dunes and winds on Mars. *Journal of Geophysical Research*, 100: 5397–5406.

Thomas, P. and Veverka, J. 1979. Seasonal and secular variations of wind streaks on Mars: an analysis of Mariner 9 and Viking data. *Journal of Geophysical Research*, 84: 8131–8146.

Thomas, P.C. and Weitz, C. 1989. Sand dune materials and polar layered deposits on Mars. *Icarus*, 81: 185–215.

Toon, O.B., Pollock, J.B., Ward, W., Burns, J.A. and Bilski, K. 1980. The astronomical theory of climatic change on Mars. *Icarus*, 44: 552–607.

Toulmin, P. III, Baird, A.K., Clark, B.C., Keil, K., Rose, H.J. Jr, Christian, R.P., Evans, P.H. and Kelliher, W.C. 1977. Geochemical and mineralogical interpretation of the Viking inorganic chemical results. *Journal of Geophysical Research*, 82: 4625–4634.

Tsoar, H., Greeley, R. and Peterfreund, A.R. 1979. Mars: the north polar sand sea and related wind patterns. *Journal of Geophysical Research*, 84: 8167–8180.

Tyler, G.L., Ford, P.G., Campbell, D.B., Elachi, C., Pettengill, G.H. and Simpson, R.A. 1991. Magellan: electrical and physical properties of Venus' surface. *Science*, 252: 265–270.

Urey, H.G. 1952. *The planets, their origin and development*. Yale University Press, New Haven.

Veverka, J., Gierasch, P. and Thomas, P. 1981. Wind streaks on Mars: meteorological control of occurrence and mode of formation. *Icarus*, 45: 154–166.

Vorder Bruegge, R.W. 1994. Depositional units in western Maxwell Montes: implications for mountain building processes on Venus. *Lunar and Planetary Science Conference*, 25: 1447–1448 (abstract).

Ward, A.W. 1979. Yardangs on Mars: evidence of recent wind erosion. *Journal of Geophysical Research*, 84: 8147–8166.

Ward, A.W. and Greeley, R. 1984. Evolution of the yardangs at Rogers Lake, California. *Bulletin of the Geological Society of America*, 95: 829–837.

Ward, A.W., Doyle, K.B., Helm, P.J., Weisman, M.K. and Witbeck, N.E. 1985. Global map of eolian features on Mars. *Journal of Geophysical Research*, 90: 2038–2056.

Wells, E.N., Veverka, J. and Thomas, P. 1984. Mars: experimental study of albedo changes caused by dust fallout. *Icarus*, 58: 331–338.

White, B.R. 1986. Particle transport by atmospheric winds on Venus: an experimental wind tunnel study. In W.G. Nickling (ed.), *Aeolian geomorphology*. The Binghampton Symposium in Geomorphology, International Series, 17. Allen & Unwin, Boston: 57–73.

Williams, S.H. and Greeley, R. 1986. Wind erosion on Mars: impairment by poor saltation cloud development. *Lunar and Planetary Science Conference*, 17: 952–953 (abstract).

Williams, S.H. and Zimbelman, J.R. 1988. Aeolian gradation on Mars: widespread and ancient. *Lunar and Planetary Science Conference*, 19: 1281–1282 (abstract).

Williams, S.H. and Zimbelman, J.R. 1994. 'White Rock': an eroded Martian lacustrine deposit(?). *Geology*, 22: 107–110.

Zimbelman, J.R. 1987. Spatial resolutions and the geologic interpretation of Martian morphology: Implications for subsurface volatiles. *Icarus*, 71: 257–267.

Zimbelman, J.R. and Kieffer, H.H. 1979. Thermal mapping of the northern equatorial and temperate latitudes of Mars. *Journal of Geophysical Research*, 84: 8239–8251.

Zimbelman, J.R. and Wells, G.L. 1987. Geomorphic evidence for climatic change at equatorial latitudes on Mars. *Geological Society of America 100th Annual Meeting*, 19: 905 (abstract).

Zimbelman, J.R., Clifford, S.M. and Williams, S.H. 1988. Terrain softening revisited: photogeological considerations. *Lunar and Planetary Science Conference*, 19: 1321–1322 (abstract).

Zolotov, M.Yu. and Volkov, V.P. 1992. Chemical processes on the planetary surface. In V.L. Barsukov, A.T. Basilevsky, V.P. Volkov and V.N. Zharkov (eds), *Venus geology, geochemistry, and geophysics*. University of Arizona Press, Tucson: 177–199.

Zuber, M.T., Smith, D.E., Solomon, S.C., Muhleman, D.O., Head, J.W., Garvin, J.B., Abshire, J.B. and Bufton, J.L. 1992. The Mars Observer laser altimeter investigation. *Journal of Geophysical Research*, 97: 7781–7797.

30

Arid zone geomorphology: perspectives and challenges

David S. G. Thomas

There is more to be understood about these land-forms than those of any other major environment.

(Warren 1987)

A century ago ... Geomorphology was a science filled with wonder and excitement.

(Baker 1985)

The preceding chapters have provided a wide-ranging assessment of the state of geomorphological knowledge of drylands and the geomorphological characteristics of the world's main arid areas. It is clear that a full understanding of the nature of dryland geomorphology not only requires an explanation of the landforms present and the processes that have shaped them, but at any specific location necessitates recognition of temporal changes in the operation of processes. In the longer perspective this requires understanding of changes in controlling parameters, particularly climatic changes, and, more recently, recognition of the influences that human activities can have on environmental processes.

The need for a composite approach has not always been recognised. Approaches to the study of arid geomorphology have varied, not only through the last century or so, but to some extent according to the features being investigated. The intent of this concluding piece is to highlight and examine the more important trends and themes that have influenced the study of arid zone geomorphology and to provide consideration of its future directions.

Paradigms of wind and of water and the question of representativeness

The inaccessibility, remoteness and inhospitable nature of many desert areas were undoubtedly factors which contributed to the limited data base upon which early theories concerning arid zone geomorphology were based. The descriptions and reports of arid landforms that appeared in the late nineteenth and early twentieth centuries were frequently selective, as geomorphologists and visitors to desert environments, more used to temperate and humid regions, showed a tendency to concentrate on exciting, spectacular and bizarre features, as noted by Goudie in Chapter 3, at the expense of the more common and mundane. One consequence of the unrepresentativeness of these pioneering studies and the lack of an overview of arid environments *in toto*, or at least of a characteristic sample, was the rapidity with which ideas changed (Goudie 1985). Of particular note in this respect is that during the last hundred years there have been significant shifts in the relative importance attached to the process domains of wind and water.

The late nineteenth and early twentieth centuries witnessed a propensity towards exaggerating the impact of wind as a geomorphological agent in drylands (Cooke and Warren 1973; Tchakerian 1995; Breed *et al.*, Chapter 19 this volume), though in fact 'extravagant aeolation' never attained the currency in scientific circles with which it is sometimes credited today. The importance of water in contributing to the shaping of many arid landscapes, through both the effects of more humid palaeoclimates and low-frequency rainfall events under contemporary regimes, has, of course, been widely recognised since, as represented in the third section of this book. Following the supposed heyday of the aeolationists, there was a substantial growth in the acceptance of runoff, even if operating very infrequently, as the most important agent of landform development in arid environments. But, as Peel (1960; 1966) observed, this view was in itself unrepresentative, growing to a large degree from the preponderance of published work based on studies conducted in the southwest United States, which 'is for the most part, however, not very typical of the world's dry lands' (Peel 1966).

The viewing of wind and water as competing paradigms in the study of arid geomorphology has largely passed, to be replaced by considerations of their respective importance in time and space. To a large extent this is due to the expansion of both the range of data sources and quality of information available in the last few decades to geomorphologists working in drylands. A century ago, sources of information were restricted to a limited number of ground-level observations. Today, we have the benefit of spatially extensive overviews gained from aerial photography and satellite imagery. These have allowed the distribution and relative importance of terrain and landform types to be determined both within and between the Earth's arid lands. Additionally, as adequately demonstrated in this volume, remotely sensed data have contributed to the recognition of the former extensions and contractions of arid zones (Chapter 26), to a greater understanding

of the relationships between dune forms, distributions and regional wind regimes (Chapter 17), to the monitoring of desert-derived dust movement within the atmosphere (Chapter 18) and to the comparison of Earth and planetary arid systems (Chapter 29).

Approaches to explanations

The information gained from remotely sensed sources, while of paramount importance in redressing some of the inadequacies of early arid zone geomorphology, is not appropriate for the resolution of issues concerning the way in which processes operate or the development of specific landforms. Early explanations of landform development frequently preceded data collection and were often based on nothing more than limited description or surmise. Some of these explanations even attained the totally unjustified status of quasi-laws. For example, the explanation of linear dune development under the influence of lower-atmosphere counter-rotating vortices, very tentatively put forward by Bagnold (1953), was, for over two decades, widely and usually unquestionably adopted as *the* mechanism of linear dune development (it still is in some texts, such as Chorley *et al.* 1984). Despite Bagnold's cautionary note that his main reason for suggesting helical vortices as a casual mechanism in linear dune development was 'to stimulate the collection ... of much more data on the ground wind', the detailed studies which he sought to inspire were not forthcoming until the work of Tsoar (1978; 1983; see Chapter 17 and Thomas 1988a).

There is growing evidence, represented in many chapters in this volume, to suggest that detailed monitoring, measurement and observation are playing a more prominent role in the quest for the explanation of geomorphological phenomena in drylands. Process studies in the field, for example to understand the operation of aeolian processes and refine the theoretical foundations outlined by R. Bagnold (Tchakerian 1995, see Chapter 16); laboratory experimentation coupled with high-precision field monitoring to ensure correct parameterisation, for example in the quest for understanding rock weathering in drylands (Chapters 3 and 4); and the application of modelling techniques (Chapter 9) indicate how the careful selection and application of a variety of techniques is leading to major advances in our understanding of dryland geomorphology.

Challenges to explanation

The spread and depth of the information gained from rigorous investigations are as yet far from even across the spectrum of arid zone landforms and deposits. This is apparent within the third section of this volume, which deals with the work of water. For example, the body of knowledge concerning the morphology and development of alluvial fans and badlands is impressive. These features, neither of which are confined to arid environments, sometimes attain considerable spatial significance at the regional scale, but are limited in aerial importance when the Earth's arid lands are considered as a whole (see Table 1.5, for example). Contrastingly, as Reid and Frostick (Chapter 11) observe, the data base for the explanation of arid channel forms and processes is as yet rather restricted, though growing. One important reason for this is the practicality of monitoring flow events that may be widely dispersed in both time and space, which may cause a tendency for investigations to be conducted in semi-arid areas, where the probabilities of rainfall events are higher than in more arid locations.

One solution to the problem of investigating ephemeral and episodic fluvial events has been to make inferences about processes from their ensuing deposits (see, for example, Chapter 11). An alternative is to employ conceptual or mathematical modelling procedures, though the scarcity of suitable arid land empirical data to plug into such models can handicap their calibration, development and usefulness. Recourse to adapting data or models from more humid situations has often proved equally inappropriate and deficient, while the variability of surface conditions in drylands, at a range of spatial scales, hinders the application of simple models and the development of pandemic explanations (see Knighton and Nanson, Chapter 10) and argues, as Baird notes in Chapter 9, for more complex approaches.

Additional problems in attempting to model the development of some arid landforms and in

the quest for general explanations are inherent in the nature of drylands. First, as discussed by several authors contributing to this volume, the spottiness of rainfall events makes the spatial prediction of runoff difficult. Second, some features, especially bedrock slope forms found in the driest environments (see Chapter 8), appear either to be evolving so slowly, or to be features inherited from previous climates, so that it is extremely difficult to relate their form to current processes. Third, and perhaps most importantly, surface and material characteristics probably exert a greater influence on dryland geomorphological phenomena than in more humid environments, as identified in the chapters by Oberlander and Campbell (Chapters 8 and 13).

This not only affects the operation of processes such as aeolian entrainment and abrasion, and runoff generation and sediment mobilisation by water but influences their efficacy in shaping landforms. Perhaps more than in any other major environment, therefore, an understanding of arid zone landform development necessitates much more than a knowledge of process mechanics. It is insufficient, however, simply to take account of surface materials and bedrock lithologies. Equally important is to consider the nature and often neglected, oversimplified, and unrecognised variations of desert vegetation, which can play a significant geomorphic role at the interface between atmospheric processes and the ground surface (Chapter 7). Encouragingly, empirical and modelled studies of the biogeomorphology of arid environments are now appearing (Ash and Wasson 1983; Thornes 1985; Thomas 1988b).

Further advances, commonly in the area of process studies, are being made through the use of laboratory simulation. Laboratory work was employed in early investigations of desert weathering processes (Tarr 1915), though these have since received many criticisms (see Chapters 3 and 4 in this volume), largely founded in the poor reproduction of real environment conditions achieved in the pioneer simulations. This can in part be traced to the limited information available on desert temperature and moisture regimes, while the use of inappropriate rock samples cannot be ignored either. Weathering simulations are now both more sophisticated and better grounded in empirical data.

Table 30.1 *Predicted changes in dryland climates based on three high-resolution models of a doubling of atmospheric CO_2 content*

	Europe	North Africa	Southern Africa	Asia	Middle East and Arabia	North America	South America	Australia
Areas included	Drylands bordering the Mediterranean	Sahara and surrounding areas	Kalahari, Karoo and Namib	From Iran to China, including Thar	Arabia, and eastern Mediterranean	N Mexico, western USA and Great Plains	Atacama, Peru Chaco, Patagonia	
Summer temp. change (°C)	+4–6	+4–6.	+2–4	+6	+4–6	+4–6 in USA +2–4 in Mexico	+2–4, +2–6 in Atacama, Patagonia	+2–4
Winter temp. change (°C)	+2–6	+2–4, +4–6 in Sahel	+2–4, +4–6 in Karoo	+4–8 in east, +4–6 in west	+2–4	+4–6 USA +2–4 Mexico	+2–4	+2–4
Summer precip. change	Decrease	Decrease	Decrease	Highly varied, increase in Thar	Decrease	Highly varied	Slight decrease possible increase in Peru	Variable, mainly increased
Winter precip. change	Varied decrease	Decrease	Slight decrease	Slight decrease	Decrease	Highly varied	Slight decrease	Increase in east, decrease in west
Soil moisture change	Significant decrease	Decrease	Significant decrease	Slight increase (winter), general decrease (summer), increase in Thar	Decrease	Highly varied	General decrease	General increase in east, decrease in west

Note: The term 'varied' in the table indicates that the three models predict different directions of the relevant climatic change. Based on information in Houghton *et al.* (1990) and Williams and Balling (1995).

The laboratory simulation of processes of lower-atmosphere sediment movement, which are often difficult to observe and monitor in the field, has provided a major stimulus to advance in the fields of aeolian entrainment, deposition and erosion, as indicated within the fourth section of this book. The driving force behind some of these small-scale simulations has, paradoxically, been the space research programmes (Greeley and Iversen 1985; see also Chapter 29) which have provided the large-scale views of arid zones on Earth and other planets. However, one of the biggest challenges facing an integrated geomorphological understanding of the dynamics of drylands is marrying the outcomes of reductionist process studies with the improved picture of the spatial complexity of arid landscapes gained from larger-scale studies.

Arid zone geomorphology in a more crowded, warmer world

The challenge to understand and explain is complicated and enhanced by two further factors that together suggest that the geomorphology of the world's drylands will experience significant changes in future decades (e.g. Goudie 1994). First, geomorphic processes, and the temporal and spatial patterning of their operation, are increasingly being altered by humans (Chapter 28). Though human agency in drylands is by no means new in the twentieth century, it is certainly greater and more diverse than in the past. Countering both negative impacts, such as the multifarious components of desertification, and misunderstandings of the nature and extent of these actions (Thomas and Middleton 1994), requires better explanations of the behaviour and dynamics of natural systems and the processes that operate in them. Separating natural and anthropogenic components of processes, and background and enhanced rates of their operation, represent major challenges to geomorphologists working in drylands, where marked variability in the operation of processes appears to be a widely occurring norm.

Second, the operation of geomorphic processes in drylands has altered through time in response to climatic changes (Chapter 26).

Future rates of change may be significantly increased by anthropogenically enhanced global warming (Williams and Balling 1995). Notwithstanding debates about the precise causes, rates and magnitudes of anthropogenically induced global warming, Houghton *et al.* (1990) and others note that a doubling of atmospheric CO_2 levels over those of 1800 is a strong likelihood before the end of the twenty-first century, and perhaps sooner. Modelling of atmospheric and climatic responses to such changes is highly complex and a range of scenarios have been suggested and published (Houghton *et al.* 1990; Mitchell *et al.* 1990; Le Houérou 1992).

Though predictions of the impacts on specific regions are multifarious, several point to a significant decrease in annual rainfall amounts and rainfall reliability in low latitudes (Parry 1990; Le Houérou 1992; Muhs and Maat 1993), while all drylands are predicted to experience temperature increases. These may be up to $7°C$ in those in high-latitude areas of the northern hemisphere (Williams and Balling 1995). Such outcomes have great significance for the operation of geomorphological processes, and are summarised in Table 30.1 for major dryland areas. As Williams and Balling (1995) note though, interpretation of such outcomes should be cautious because the temperature-enhancing impacts of rising CO_2 levels may in part be counteracted by the possible effects of the increases in SO_2 levels that are simultaneously occurring in the atmosphere.

Conclusion

Not only is it impossible to ignore the fact that much remains to be achieved in the field of arid zone geomorphology, but it is probably still the case that less is known about drylands than about other major environments. One indicator of this is the frequency with which 'equifinality' is used in the explanation of arid landform development. It is likely that some very similar forms are genuinely the outcome of different, convergent, developmental routes; but for others this explanation is a consequence of the use of confusing terminology (for example, pans and playas, see Chapter 14), the application of a simplistic classification scheme and/or insufficient knowledge about the formative processes (as in the case of linear dunes, for example). As more data become available, forms and processes become more distinguishable.

Although a lot has been achieved in recent decades, and many developments have taken place in our understanding since the first edition of this book appeared in 1989, it remains that there is still much to be understood about the geomorphology of the arid zone. As human populations in drylands increase and concerns about the impacts of anthropogenically induced global warming grow, the relevance of striving for even greater understanding has never been greater.

References

Ash, J.E. and Wasson, R.J. 1983. Vegetation and sand mobility in the Australian desert dunefield. *Zeitschrift für Geomorphologie*, Supplementband, 45: 7–25.

Bagnold, R.A. 1953. The surface movement of blown sand in relation to meteorology. *Desert research*. Proceedings of an international symposium, Jerusalem, May 1952. Research Council of Israel, Special Publication, 2: 89–93.

Baker, V.R. 1985. Relief forms on plants. In A. Pitty (ed.), *Themes in geomorphology*. Croom Helm, London: 245–259.

Chorley, R.J., Schumm, S.A. and Sugden, D.E. 1984. *Geomorphology*. Methuen, London.

Cooke, R.U. and Warren, A. 1973. *Geomorphology in deserts*. Batsford, London.

Goudie, A.S. 1985. Themes in desert geomorphology. In A. Pitty (ed.), *Themes in geomorphology*. Croom Helm, London: 122–140.

Goudie, A.S. 1994. Deserts in a warmer world. In A.C. Millington and K. Pye (eds), *Environmental change in drylands: biogeomorphological and geomorphological perspectives*. Wiley, Chichester: 1–24.

Greeley, R. and Iversen, J.D. 1985. *Wind as a geological process on Earth, Mars, Venus and Titan*. Cambridge University Press, Cambridge.

Houghton, J.T., Jenkins, G.J. and Ephraums, J.J. (eds) 1990. *Climate change: the IPCC scientific assessment*. Cambridge University Press, Cambridge.

Le Houérou, H.N. 1992. Climate change and desertization. *Impact of Science on Society*, 166: 183–201.

Mitchell, J.F.B., Manabe, S., Meleshko, V. and Tokioka, T. 1990. Equilibrium climate change — and its implications for the future. In J.T. Houghton, G.J. Jenkins and J.J. Ephraums (eds), *Climate change: the IPCC scientific assessment*. Cambridge University Press, Cambridge: 131–172.

Muhs, D.R. and Maat, P.B. 1993. The potential response of eolian sands to greenhouse warming and precipitation reduction in the Great Plains of the USA. *Journal of Arid Environments*, 25: 351–361.

Parry, M. 1990. *Climate change and world agriculture*. Earthscan, London.

Peel, R.A. 1960. Some aspects of desert geomorphology. *Geography*, 45: 241–262.

Peel, R.A. 1966. The landscape of aridity. *Transactions of the Institute of British Geographers*, 38: 1–23.

Tarr, W.A. 1915. A study of some heating tests and the light they throw on the disintegration of granite. *Economic Geology*, 10: 348–367.

Tchakerian, V.P. 1995. The resurgence of aeolian geomorphology. In V.C. Tchakerian (ed.), *Desert aeolian processes*. Chapman & Hall, London: 1–10.

Thomas, D.S.G. 1988a. Arid geomorphology. *Progress in Physical Geography*, 12: 595–606.

Thomas, D.S.G. 1988b. The biogeomorphology of arid and semi-arid environments. In H.A. Viles (ed.), *Biogeomorphology*. Blackwell, Oxford: 193–221.

Thomas, D.S.G. and Middleton, N.J. 1994. *Desertification: exploding the myth*. Wiley, Chichester.

Thornes, J.B. 1985. The ecology of erosion. *Geography*, 70: 222–235.

Tsoar, H. 1978. *The dynamics of longitudinal dunes*. Final technical report, European Research Office, US Army.

Tsoar, H. 1983. Dynamic processes acting on a longitudinal (seif) dune. *Sedimentology*, 30: 567–578.

Warren, A. 1987. Arid land and aeolian geomorphology: introduction. In V. Gardiner (ed.), *International geomorphology 1986*. Part 2. Wiley, Chichester: 1209–1210.

Williams, M.A.J. and Balling, R.C. 1995. *Interactions of desertification and climate*. Edward Arnold, London.

Index

Index compiled by Lucy Heath